Read Ancient African scripts from any current African language

Kem Mesha Wour

Read Ancient African scripts from any current African language

Volume 1

LES ÉDITIONS DU NET
126, rue du Landy 93400 St Ouen

ISBN : 978-2-312-06281-5

Ntour Ngaŋ

A-su bə-bond bàmà

O-lughu nn à tiène nti Zamba ngul mə-sə.
Ma tari fwa və ù-nsoŋ'o nənə bə nana bə tara ùn-ntour ù-lughu à su yi nâ bə yn mə ngwal, bə lərə mà mə fəgh bi-kàngà. Ma və fə bo ngoul ù-lughu à su mi-'mfang mə-lebgha mábá. Ma və ngang à mu bə nga voili ma à-bôgh à-sə ù-til à-ndakh dinà.

Ma və fwa fə ùn-yəghlə nn wàmà, i-yəŋ mərə Nti **Anta Diop** ù-ntour ngaŋ. Un-ntour ù-lughu à bo yi nyè àmù mi-'mbu mi-'mbu à za koui ànà, à-təb'lə ma bə-bəbəla ya si ndón; à lərghə ma zn suswo à ***mAa.t***.

Un-mfang ù-lughu ù bô fwa fə ùn-yəghlə à-kəŋ bi-kanga wama tara **Obenga Mwené Ndzalé**. Ma və nyè à suswo ùn-nəm ngoul o-lughu à yi ù-laŋa à mou yi nâ à nga yâ bya mə zn ya fəgh mi-'ŋkɔbɔ. A man bi təble mə-kəŋ ya mi-'ŋkɔbɔ myàn mə dûghèn bə-dzîn. I ŋgoul i-si à nga boh ya yèn ngul o-lughu à mu i-si t ya yi bo nâ bi simzèn bə tara bə nana à kêghè ù-kobà à za koui à-mos y'ana.

Ma və fwa fə ù-nsamba bə-yəghlə wama u-sə ngaŋ à-su à-kəŋ i-si mor ù-sə.
Mə və nda bor dyàma i-sə ngaŋ à kêghè bə-bond bama bə-sə.
Be mvou bàmà ngaŋ i bo fwa fə boh.

Heartfelt Thanks

To my Ancestors

I would like first of all to express my thanks to his majesty **Hr** the primordial ancestor of the African peoples as well as my ancestors who were kind enough to assist me in drafting this document.

I pay a particular tribute to ***wsjr*** Anta Diop, my teacher who, through his work, has succeeded in forging my spirit and guiding me on the path of ***mAa.t***.

I want to express my deep gratitude to ***sS*** Thophile Obenga who has guided us in African linguistics, allowing current and future generations to take the true measure of our historical and cultural heritage..

I want to express my gratitude to all my different trainers who have worked directly or indirectly to my slow and progressive building by their knowledge.

My thanks go to all my family and my circle of friends and loved ones

Abbreviations

Acc.:	accomplished
Adj.:	adjective
Adv.:	adverbial
An.:	anaphora
aux.:	Auxiliary
C.:	consonant
Cmr	Cameroon
Con.:	connectives
Dm.:	dmonstrative
exp.:	Expansion
ina.:	unaccomplished
inj.:	injunctive
Mod.:	modality
N.:	noun
Ng.:	negative
PPA.:	adjectival prefix
Pl.:	plural
PN.:	nominal prefix
Poss.:	possessive
PP.:	personal pronoun, pronominal prefix
Rac.:	root
rad.:	radical
Sg.:	singular
Suf.:	suffix
V.:	vowel

Signs

+	Associate the different elements of the statement
/ /	Phonological slashes
| |	Morphological bars
[…]	Phonetic hooks
#...#	Limit of a phrase
–	Separation of morphemes in statements

Charts

Summary

Foreword

Remarking that, in the chapter of African antiquities, the history of humanity could not scarcely go beyond the mere stage of the interrogative register, in the context of Unesco's efforts to promote the mutual understanding of peoples and nations, his French General Director, Mr. René Maheu was authorized at the General Conference of the Organization in 1964 to take the necessary steps for the elaboration and the publication of a General History of Africa. Of course, this project was initiated by newly independent African countries and supported by UNESCO to wring havoc with the prejudiced and racially developed rhetoric about Africa to justify trafficking, colonization and neocolonialism.

The first step of this project (1965-1970) execution consisted mainly in works carried out in Africa itself and elsewhere (such as the Guide of the sources of the history of Africa) aiming at gathering elements of documentation, writings and oral. At the same time, international scientific consultations were organized to study the methods to be used.

They resulted in a number of recommendations made by experts during meetings held in Paris (1969) and Addis Ababa (1970), marking the launch of the second phase of the project, that means the preparation and writing of the General History of Africa, in eight volumes, under the sole intellectual and scientific responsibility of a scholarly body, the International Scientific Committee for the Writing of a General History of Africa. It is in this context that professors Anta Diop and Obenga were invited to give a scientific point of view on the writing of the second volume of the General History of Africa. A few years earlier, at the request of Unesco, Professor Ki-Zerbo had already directed the writing of the first volume[1] of the General History of Africa.

The assessment that emerges from the groundbreaking scientific work of professors Anta Diop and Obenga exposed in Cairo in 1974, points out that the general contribution of Africans to their classical humanities has remained rather timid until our days.
Fields of investigation remain numerous but fallow as the approach of a "global Africa" referring to a diopian perspective (Cheikh Anta Diop) exemplified by an African presence in pre-Columbian America. Another approach combines a broader reading of African history, integrating cosmologies, migratory movements, multidimensional contributions from Africa to the world, and so on.

The legitimate question arises regarding what do the self-proclaimed followers of Diop are doing? Intellectual curiosity dictates the apodictic evidence that the majority of African intellectuals, whether or not they are supposed to be, remains singularly encamped on the reseach work of Anta Diop. But, the further research work to which Diop invites Africans, no one really looks at it even in an unsatisfactory way.

In fact, for imperfect works, a later improvement is always possible. But without a sketch of previous work that is unsatisfactory, it is difficult or even impossible to make any improvement. We find ourselves in front of a status quo claiming all diopians, including me.

How can we explain the African billion people lack of reaction regarding a situation that is so critical at this stage?

1. Sous la direction de Joseph KI-ZERBO, Histoire de l'Afrique Noire: Méthodologie et préhistoire africaine, Unesco, Paris, 1980

Persuaded that this situation is dramatic in some aspects, but convinced that it is not irreversible as strange as it looks, this work has therefore prescribed on me as a duty, even as a necessity. The need to further the decolonization of the concepts, paradigms and terms used about Africa in social sciences and humanities. A perspective, which has been recalled since 1964 with the writing of the first volume of the General African History, even if the epistemological break was not limited only to this decolonization, invited to introduce new understandings of the world based on an African projection. It is enough to read the Universal History[2], recorded by Ibn Khaldûn in the 14th century, which structured the unconscious of many generations of men and women, to realize the extent of the damage.

> *"[...] In the south of the Nile River there are black people called Lemlem. They are pagans with stigmata on their faces and on their temples. The people of Ghana and Tekrour make incursions into the territory of this people to take prisoners. The merchants to whom they sell their captives lead them to the Maghreb, a country most of whose slaves belong to this Negro race. Beyond the country of Lemlem, in the direction of the south, one meets a not very considerable population;* ***the people who compose it resemble to wild animals rather than rational Human beings****. They live in wooded swamps and caves; their food consists of herbs and seeds which undergo no preparation; sometimes they devour each other:* ***they do not have to be counted among human beings.****"*[3]
>
> *(Ibn KHALDUN-1377)*

Since ancient times, Africans had made the observation of nature an art that led to the deduction of laws describing the functioning of nature and to formulate them in a cosmology way.
Following the footsteps of my ancestors, I was able to point out the apodictic evidence that an ocean is composed of a myriad of water droplet. Therefore, it is the sum of all these droplets that gives the ocean its extent as well as its volume.

The moral of this observation is that the individual contribution of each one to the collective effort of change is a requirement. If African people want to proceed the phase inversion of their current peculiar situation, each African must be instrumental by his contribution to changing and have to refuse the reach out, wait for help, adopt a wait-and-see attitude and comfort himself in a follow-the-leader attitude and intellectual weakness.

What defines a human being first and foremost is his identity as an anthropological and sociological entity. This notion of identity is often extended in social psychology as the recognition of what the individual is, by himself or by others.

The French sociologist Jean-Claude Kaufman notes in his book *L'invention de soi* that:

> *"Individuals did not show particular curiosity about what we now call identity, so powerful were their sense of belonging to the social structure in which they were embedded and their dependence related to it"*[4] .
>
> *(KAUFMANN, 2004)*

Freud, the first, had questioned in the 1920s on the identification mechanism by which a human subject is constituted. But his approach, which was almost solely centered on the individual as an autonomous whole in relation to his social environment, had remained at the stage of the theoretical investigation.

Widening the psychoanalytic approach by integration into a more open approach to the social dimension, the German psychologist Erik Erikson was then interested in this decisive moment of the personality genesis of puberty (Childhood and society, 1950), reiterating this later in 1972 to define identity as follows:

---

2. Al Muqqadima: Universal History
3. Ibn KHALDUN, Speech on Universal History Al Muqqadima or Prolegomena, Sindbab, Paris, 1377, p. 170-171.
4. Jean-Claude KAUFMANN, Invention of oneself A theory of identity?, Paris, A. Colin, coll. Individu et socit, 2004.

"The subjective and tonic feeling of personal unity and temporal continuity"[5]

(ERIKSON, 1972)

Except the African thought, which for some has remained undeveloped, the most convincing development of knowledge about human identity seems to be the result of studies conducted and published since 1970 by philosophers such as Levinas, Ricoeur and sociologists, including Kaufman, already cited, or Ehrenberg.

In the Freudian psychoanalytic tradition, identity is a construction characterized by discontinuities and conflicts between different instances (the ego, the id, the superego).
These two psychoanalytic conceptions speak of identity as a diachronic construction[6]. This approach indeed bases the personality on a temporal continuum, in a very interesting time scale. It is this particular approach concerning the African subject that lead to curiosity in this foreword.

For Weberian sociology, the four foundations of identity are material physical existence, biological species, cultural or community belonging, and individual personality[7].

Philosophy considered as the mother of all sciences, by the way, it is interesting to look a little bit at the philosophers' point of view. We find the notion of temporal continuum related to identity on the philosophical level. Indeed, for philosophy, a subject is an individual at once distinct from all others (uniqueness or synchronic identity) and remaining the same throughout time (diachronic identity).

This objective approach to personal identity responds to an objective and subjective perspective, the one of the subject who experiences his identity. Most often, it is considered that the lived experience of identity covers a temporal dimension by which the subject relates to his previous states on the mode of an autobiography continuity ("me" narrative) and a corporal dimension by which the subject experiences "First person" of his singularity ("me" corporal).

With these different approaches, however, scholars in sociology agree to characterize an individual or a human group from identity markers.
Because oral expression is part of the cultural register because it is a means of communication within a social entity, it is de facto one of these individual and societal identity markers.

Oral expression generally enrol individual or human beings given group in historical continuum of the language facts in an oral tradition. It may therefore be useful if it is not necessary to know one's language well in order to access its history by its own investigation through the reading offered by the wide source of diachronic data that we often call tradition.

Inarguably, knowledge of one's language provides access to both, written and oral records of knowledge. This approach is crucial because it allows the Human to be in possession of himself or to reclaim his identity through defective elements of his cultural heritage.

What is repossession of oneself at the end? Seneca, as a famous Stoic, suggests an answer ready for use in his letter to Lucilius, on "*The use of time*." To take possession of oneself, he tells us:

"[...] start by becoming time owner..."

(SENÈQUE, 1861)

5. Erik ERIKSON, Adolescence and crisis. The quest for identity, Flammarion, Paris, 1972.
6. Sous la direction de Gilles FERROL, Sociology Dictionary, Armand Colin diteur, dition 2010
7. Max WEBER, *Economy and Society*, Plon, Paris, 1922 (rimpr. 2003).

That means, taking back possession of oneself is becoming master of time. However it still to know how to master it. The idea could look as utopia, but only diachronic sources of cultural datas make it possible to overcome this apparent difficulty.

The fact of proceeding to a self-centered heuristic approach makes it possible to get rid of the projection erected in truth on the basis of voluntarist experimentation but unfortunately deforming that others can do. Indeed, located outside the paradigm of the subject of the study and although voluntarist, the approach of others is by the fact, biased. By his approach, the other gives facts investigated, including with the greatest objectivity, an interpretation according to their own introspection and subjectivity. By the way, it appears as their own phenomenal reading on their reality part that serves as material for his study.
Consequently there is,unfortunatly, a significant gap between the subject of study in itself and the subject as it is perceived for the one who studies it.

This voluntarist decoding without a proper grid reading can unfortunately lead to analyzes that are only stereotype if not mere conjectures. There is therefore a danger for a given individual or a group of individuals to be characterized solely through the subjectivist voluntarism of others. This great danger is called for Africans, intellectual guardianship. This intellectual tutelage breaks self-initiative and self-centered autonomy. It unconsciously places the one who suffers it in a perpetual situation of waiting. This is why this posture is criminal for Africa. This is why, despite the dramatic situation, a billion people are waiting and extorting money in front of foreign aid.
This singular situation is the one of Africa for which voluntarist African researchers have been locked in a practice singularly stamped with the official seal of ethnoscience when their peers themselves agree to qualify as science practitioners. This unusual situation somewhat has been noted by many African researchers. But it seems to me it was mostly highlighted by Binam Bikoi in his book: *Mpomo, le prince de la grande rivière. épopée nzimé du Cameroun.*

> *"[...] While, in general, research subjects are sufficient in their scientific and intellectual perspective, African studies, in America, Europe or elsewhere, have led us to believe that their perspective was always outside their subject. and could only find their balance at this level. Thus, the oral tradition was studied "for ...", the traditional literature "for ...", the African music "for ...", the African thought "for ..." And we suddenly had the impression that behind this recurring "for" was a concern to demonstrate, by comparison with a model established once and for all, the ways, forms and means of the accessibility of the studied effects to a rank, to a dignity, to a unic position, the only one able to carry or not the recognition of unidentified censors. Another impression suggested that of all that could be concerned with African studies, only the "functional" nature of things, events, data, African productions."*[8]
>
> *(BINAM BIKOI, 2007)*

Despite the efforts of African researchers regarding what has been renamed the ethnoscience despite a common sense reason, this singular situation can sometimes lead to some implausibility when the previously unidentified censors, the ones referred by Binam Bikoi, those who are depositaries of the norm and who are flocked with the label of supposed specialists in a field, strongly denounce the remarks of their peers under the pretext that they appear to them Afrocentric[9].

Thus, do they not hesitate to carry a sharp criticism on a work of research resulting from what they call ethnosciences under the pretext that, these works of research do not fit with their view of the science they are supposed to master more than those who appear to them with a little distance as just mere laymen.

---

8. Charles BINAM BIKOI, Mpomo, the prince of the great river. Epic Nzime of Cameroon, Khartala, Paris, 2007, p.7
9. Afrocentricity is a school of thought developed by Professor Molefi Kété Asanté and Ama Mazama which consists in exposing the worldview concerning Africa from endogenous sources. It is not to be confused with Afrocentrism a child of Africanism, which is a diametrically opposite trend, a part of the Africanist approach, ie exogenous with paternalistic hints.

The French researcher Luc Bouquiaux gave an exemplification during his book review of the African researcher Victor Bissengué on a sociological study related to Baka pygmies.
In 2006, the Africanist Luc Bouquiaux, researcher in ethnolinguistics at the CNRS proposed, under the title caricature and tendentious "*Les Pygmées Aka, victimes de l'afrocentrisme?*"[10],
a review report of the work done by Victor Bissengué that he titled "*Contribution à l'histoire ancienne des Pygmées: l'example des Aka*[11]".

A lot has been said and many are today the writings on Africa, by contacts between nations, between biological varieties, between continents, Africa and Africans have become object of laballed studies of foreign scientific without the fact that, Africans have not felt the need to study or label others even in the name of ethnoscience. The points of view are, in this respect, both diverse and evolving.

Since antiquity, they have taken various forms throughout history, the glances that crossed the continent and its inhabitants, not without translating, each time, a more or less marked ideological burden. We are therefore barely surprised by the position of censor adopted by Luc Bouquiaux about Victor Bissengué's research.

Reading Bouquiaux's review of Bissengué's book, we are somewhat surprised by the pitch of the man who, in a despotic and barely paternalistic way, lavishes "advice" on one of his peers on the followinf pitch:

> *"One does not improvised linguists and if you want to use linguistic arguments, you have to speak to specialists (sic). When "we devote a part of our activity to the defense of African cultures" (page 13), we first of all take care to note their languages with respect by correctly accounting for their particularities. Finally, there are no alleged guarantors who "bothered to read the manuscript and made the necessary criticisms" (p.9) even though they did not read at all, much less approve the content, but only given their deposit for the transcription of Sango (personal communication with Marcel Diki-Kidiri)." (P. 234)*
>
> *(Luc Bouquiaux)*

One can legitimately enquiry the reasons of this charge and the roughness of the purpose.

Without having to flip back to the deluge, and without stopping on each subject of polemic as those of Luc Bouquiaux led, it is necessary to observe that:

> *"[...] The African past has not always benefited the objective solicitude of men of opinion or scientists who have been interested in its study. Highlighting this phenomenon, Ahmadou Mahtar M'bow, a historian and former General Director of UNESCO, believes he can link his source to two sad events for the whole of human consciousness: the slave trade and colonization. In his opinion, by promoting the emergence of racial stereotypes generating contempt and misunderstanding, the slave trade and colonization swallowed the history of "African peoples to the rank of an ethnohistory where appreciation of historical and cultural realities could that be distorted" (...)"*
>
> *(BINAM BIKOI, 2007)*

Did Luc Bouquiaux had a paternalistic upset regarding a poor African? Was his approach a simple freak resulting from his title of specialist of African professor? Title that could only give him the legitimacy to espouse an acceptable official position on ethnolinguistic issues especially when they concern Africans?

Could his specialist label have led him believe that he knew more than independent African researchers who were not liable to the diktat of the university institution?

How, by a sporadic practice of secondary sciences, born of contacts more or less planned, founded by curiosity and driven by the need to know better and better the object of their efforts, a follower of sciences perceived as new on Can the classical epistemological chessboard (ethnolinguistics,

10. Luc BOUQUIAUX, L'Homme, Revue française d'anthropologie (N° 179), Juillet/Septembre 2006, pp. 227-235
11. Victor BISSENGUE, Contribution à l'histoire ancienne des Pygmées: l'example des Aka, L'Harmattan, Paris, 2004.

ethnobotany, ethnomathematics, ethnozoology, ethnophilosophy) assume a censor role with regard to Africans when its research is at most at the superficial level?

How could Luc Bouquiaux believe that he has more perspective than independent African researchers on their own reality? We could continue the enquiring and find out if Luc Bouquiaux was born among the Bakas or if he lived with the Bakas or even in their neighborhood to decide on a sociological topic that concerns them till to become their external defender and supervisor?

In my opinion, the questioning is allowed but one thing is certain for the outside observer that I am, all this leaves quite perplexed and interrogative.

But all this, although is very curious, is otherwise reminiscent of a comment that is made by Edouard Burton:

> *"The pure negro is placed in the human family below the two great races Arab and Aryan "(most of his contemporaries would have arranged the last two in the reverse order) and" the negro, taken in mass, will not improve. not beyond a certain point, which will not deserve respect; he remains mentally a child..."*[12]
>
> *(BURTON, 1864)*

The unconfessed motives of Luc Bouquiaux can better be understood by integrating all the mental schemas related to the raciological vision of Gobineau, Vacher de Lapouge, Hegel and so on. or reading Ruth Scheps in *La Science sauvage* especially on page 8:

> *"[...] The learning of oral language, from its comprehension to its translation, materializes the exemplary approach of a science of the man who fights against the separation. But it is difficult not to separate the knowledge that the West has made sciences more and more specialized. The savage sciences themselves do not escape the reductionist discourse which prefixes them "ethno -. ". Apprehended by science, the knowledge of others will thus be ethnosciences, diversifiable at will"*[13]
>
> *(SCHEPS, 1993)*

Without promoting an absolute indictment nor animate petty squabbling, which oppose Africanists to Afrocentrists, I just note by this simple example, the difficulty that can face Africans to be the leaders on their own history, following only their investigations, without seeking the approval or receiving the blessing of third parties. Some people do not seek the advice of others to decide on their own history . My goal is not to oppose the official science to ethnosciences but it is simply to take in consideration the merits of this dangerous segmentation of human knowledge which places on one side those who know and in another hand the one who must comply with doxa enactment.

If this statement is permitted, I do observe that it is inconceivable to others that Africans may give any point of view even erroneous or ethno-scientific opinion about their own reality or about their sciences. De facto, if we deny to a certain fraction of humanity science practice and if even by resolving to do endogenous ethnoscience, this fraction has no right to give a point of view with its errors on a subject without suffering the wrath and censorship of the "standard" fraction, what is left to the Africans people by the way? I just wanted to draw attention to this particular issue. By the way, that is the purpose of my legitimate questioning.
As Nelson Mandela says:

> *"Education is the most powerful weapon which you can use to change the world"*
>
> *(Nelson Mandela)*

Education is the key to eliminating gender inequality, to reducing poverty, to creating a sustainable planet, to preventing needless deaths and illness, and to fostering peace. And in a knowledge economy, education is the new currency by which nations maintain economic competitiveness and global prosperity.

---

12. Richard BURTON, Mission to Gelele, King of Dahome, Cambridge, Cambridge University Press, 1864.
13. Ruth SCHEPS, La Science sauvage, Des savoirs populaires aux ethnosciences, Seuil, Paris, 1993, p. 8.

Education is an investment, and one of the most critical investments we can make. This is true not only for the Africa, but also for countries all over the world.
Very often, the contact between two consciences, one of which is hostile, aggressive and the other assaulted, can lead to the attacked and traumatized conscience, invisible but profound sequels.

The apparent opacity surrounding African historiography after the violent slave raids (Arab-Berber [14] and western [15]) has led to an important identity loss of the African human being. One of the major consequences of this state of affairs has been the structuring of Africans in a historical context with significant time percolations of African subjects leading to a gap with their past and therefore a loss of a significant part of their identity. Despite the efforts of Unesco since 1964, regarding to the writing of the General History of Africa, the data collected on the eight volumes remain rather meager because incomplete. Moreover, it is not complacent to write an History. Yet it should be taught and popularized so that Africans could be aware of it.
Up until now, in a pragmatic and practical way, very few Africans can give the broad outlines of their history over five generations. African antiquity is still a mystery. The Africa of great empires seems to be a vast nebula. What really happened between Mpumalanga (-200,000 BC) and ***Km.t*** (-10,000 BC)? Even the greatest supposed specialists are not able to answer that question. Why African history remains trapped? Why did the Berlin-born onomastic in 1884-1885 not be abandoned? Etc. There are many questions, the answers nonexistent. It is simply held that the committee of scholars writes the General History of Africa. So let's find out for whom and why?!!!

Though, to know where we are going, we must already determine where we come from. Thus, if we do not know where we come from it can lead to significant personality disorders because we end up losing a significant part of its identity and we rebuild by adopting loan memories. So it's no surprise that John Henrique Clark highlights this drama for Africa:

> "... ***The cruelest thing slavery and colonialism did to the Africans was to destroy their memory of what they were before foreign contact.*** *Africans have not dealt forthrightly with invaders, slave traders and colonialists, who came among African people as guests and stayed as conquerors. The strongest thing about African people is their respect for the humanity of other people and the hospitality they have shown to strangers. In most cases, Europeans and Western Asians have come into African societies as guests and stayed as conquerors. Africans have never had a strong armed force. They assumed that they did not need one because they had no intention of conquering other people."* [16]
>
> *(J-H CLARK, 1990)*

Nowadays, this anachronistic situation is observed everywhere in Africa. In an excellent book of sociology, enlightened by brother Emmanuel Vangu Vangu, he takes a critical look at this drama that is being played out in Africa.

> *"[...] One element of another symbolic system can not be imported into another one without harming the importing system. Pushing a people to cultural asphyxiation is a crime against humanity! In fact, the pollution of the spirits is worse than those of the beaches, even those of the cities."* [17]
>
> *(VANGU, 2012)*

This critical situation is observed everywhere in Africa and sometimes reached peaks in Central Africa for example where a local fraction of the population has ended up building a borrowed identity based on the language of the English settler (yet arrived in 1918) as an identity marker, defining themselves strangely as anglophones. When an African is not defining himself by the borrowed languages, he often uses other exogenous marketing segmentation criteria such as imported religions. So, some are Muslims, Christians before being Africans etc. By the way there is a real problem. If populations define

14. Tidiane NDIAYE, Le génocide voilé enquête historique, Gallimard, Paris, 2008.
15. BWEMBA-BONG, Quand l'africain était l'or noir de l'Europe: L'Afrique, actrice ou victime de la "traite des Noirs"? Menaibuc, Paris, 2005.
16. John Henrik CLARK, *Can African People Save Themselves?*, 1990.
17. Emmanuel VANGU VANGU, *Sexualité, initiations et étapes du mariage en Afrique -Au cœur des rites et des symboles*, Editions Publibook, Paris, 2012, P.14

themselves in Africa through only exogenous and arbitrary criteria of segmentation, that means there is a severe loss of identity and a highly mental disorder and confusion. How to build a nation when individual and collective consciousness is severely affected and fragmented?

All those who are aware among Africans have understood this thorny problem. Anta Diop at the Niamey conference in 1984 said at the attention of his kind listeners:

> *"[...] I believe that the harm that the occupier has done to us is not yet cured, that is the essence of the problem. Cultural alienation ends up being an integral part of our substance, our soul and when we think we have got rid of it we have not done it completely yet. Often, the colonized looks a little, or the ex-colonized himself, to this slave of the nineteenth century who, released, goes to the door and then comes back home, because he does not know anymore where to go. He does not know where to go ... ... Since the time he has lost his freedom, since the time he learned subordination reflexes, since the time he has learned to think through his master (...)"*
>
> *(Anta DIOP-1984)*

This comment by Diop is very relevant because it sets with great lucidity the scene that leads to the current deadlock. Some people then naively believe that it will be enough to sell African wealth to a lower bidder to overcome the situation. By the way, we pour water with all our strength into a bottomless pit by refusing to consent the intellectual effort that pushes to ask the true analysis or to listen to those who require this true psychoanalysis of the situation.

According to the theorists of colonial positivism, the evangelizing and civilizing missions are often presented as having had positive contributions. Indeed, the colonizing mission in its "long-suffering" has undertaken among other things to train Africans. But in hindsight, it was meant to serve one's own purposes. The real reasons for these trainings could be questioned and issues may arise to a certain extents.

In a normative and linear approach of time, history being a knowledge of the past, based on orality or writings, it guides future choices based on the nature of past experiences (happy or unhappy).
On this account, the statement regarding Africans is brilliantly drawn by Anta Diop once again:

> *"[...] When a people ceases to be master of its educational system, it regresses. Egypt, from the moment when it was occuped by Cambyses II, it sufficed for seven continuous centuries of foreign occupation. Cambyse in -525, Alexander of Macedonia in -333 and then the Romans in -47. And then juvenal Roman satirical poet of the 2nd century AD, describes the Egyptians as a people without industry, a people who fell back into savagery, a barbaric people who practiced cannibalism. And it is enough that the same Egyptian people who had not even emigrated yet, the very one who created the civilization in the same valley, staying on the spot and losing only its educational system, remaining under the yoke of the stranger who prevented him from leading his destiny as he wished, that was enough for him to fall back into barbarism. This is how people regress."*
>
> *(ANTA DIOP, Conference of Niamey, 1984)*

To this day, Africa has become a simple consumer of ready-made ideas invented by others for its supposed benefit. Therefore inexorably, Africa continues its regression phenomenon with the active participation of its intellectual "elites" zombified by school programs designed by others. Any attempt by Africans to acquire their intellectual autonomy is severely controversial by those who are legitimized by the current regulations.

> *"[...] In Black Africa, the West Christian has imposed behavioral patterns on Negro-Africans, as if their morals were inferior. The cultural vampirism of the West is a slow but steady collective killing of African societies. In fact, the notion of "single thought" is to be taken seriously. It reports the loss of belief in our own ability to act effectively on social organization, and this, whatever the damage of this "mega culture" in the social field. Every society creates its customs, practices and some usages which are peculiar to it, where certain traditions follow."*
>
> *(VANGU, 2012)*

Although Hegel's direct influence on the African history development was weak, the opinion he represented, was accepted in the historical orthodoxy of the nineteenth century. This obsolete and unfounded opinion is not lacking in followers even today. Has not some thirty-five years ago, H.R. Trevor- Roper a Regius Professor of Modern History at the University of Oxford, in a moment of condescension that quickly became notorious declared to an audience on the BBC:

> *"It is fashionable to speak today as if European history were devalued: as if historians, in the past, have paid too much attention to it; and as if, nowadays, we should pay less. Undergraduates, seduced, as always, by the changing breath of journalistic fashion, demand that they should be taught the history of black Africa.* ***Perhaps, in the future, there will be some African history to teach. But at present there is none, or very little: there is only the history of the Europeans in Africa. The rest is largely darkness****, like the history of pre-European, pre-Columbian America. And darkness is not a subject for history. Please do not misunderstand me. I do not deny that men existed even in dark countries and dark centuries, nor that they had political life and culture, interesting to sociologists and anthropologists; but history, I believe, is essentially a form of movement, and purposive movement too. It is not a mere phantasmagoria of changing shapes and costumes, of battles and conquests, dynasties and usurpations, social forms and social disintegration. If all history is equal, as some now believe, there is no reason why we should study one section of it rather than another; for certainly we cannot study it all. Then indeed we may neglect our own history and amuse ourselves with the unrewarding gyrations of barbarous tribes in picturesque but irrelevant corners of the globe: tribes whose chief function in history, in my opinion, is to show to the present an image of the past from which, by history, it has escaped; or shall I seek to avoid the indignation of the medievalists by saying, from which it has changed? ..."*[18]
>
> *(Hugh Redwald Trevor-Roper, 1963)*

The Hegelian dialectic of master and slave has reduced a whole continent to the rank of sub-men. Also, when one analyzes the contents of the textbooks, Greece is by its scientists and philosophers considered as the well of the knowledge of the current world; but taking into account the historical facts one realizes that it was in reality only the pupil of an ancient Negro Egypt. It is well known today that Africans do witchcraft and magic when others do science. Some pseudo-African researchers take these black-and-white cants to promote them in Africa attracted by greed. I present the ultimate specialists in the intellectual scam. Find below the chapters in Kelman's book *Je suis Noir et je n'aime pas le manioc.*

> *"I am civilized (P27), I am assimilated (P40), I am Burgundian (P65), I am an executive (P84), I do not like Blacks (P120), I have a little one (hear a little penis) (P137), I'm Black and I'm not proud (P 153), I'm Black, and I'm caring (P176)."*[19]
>
> *(KELMAN, 2003)*

Let's give the floor to the great Voltaire who was a slave trade living-man:

> *"Men are everywhere equally crazy; they made laws to the extent, as one repairs breaches of walls. [...] Aristotle's partisans and enemies were excommunicated alternately, and those who wore long hair and those who wore them short. We have in the world of perfect law only to regulate a kind of madness, which is the game. The rules of the game are the only ones which admit no exception, relaxation, variety, or tyranny. A man who has been a lackey, if he plays lansquenet with kings, is easily paid when he wins; everywhere else the law is a sword, the strongest of which cuts the weakest in pieces."*[20]
>
> *(VOLTAIRE, 1817)*

Alain Mabanckou for those who do not know him is a French-Congolese writer born in Pointe-Noire on February 24, 1966. The others gave him a Renaudot literature prize in 2006 for his novel *Porcupine Memoiries.* He writes on porcupines when we must mobilize consciousness for the African Renaissance. He never wrote even a word in his own native language.

> *"[...] I am black, and obviously it visible. So the blacks I meet in Paris often call me "my brother". Are we really? What do an Antillean, a Senegalese, and a Black born in the Xth borough have in*

---

18. Joseph KI-ZERBO, *Histoire Générale de l'Afrique*, volume 1, Unesco, Paris, 1980, p.47
19. Gaston KELMAN, Je suis noir et je n'aime pas le manioc, Max Milo, Paris, 2004.
20. VOLTAIRE. Œuvres complètes de Voltaire: Dictionnaire philosophique Tome septième. Paris: Hachette Livre, 1817, P.669.

common, if not the color to which they complain of being constantly reduced? I forget obviously the genealogy they have forged, that of misery and humiliation - slave trade, colonization, living conditions of immigrants ... Because beyond the skin, what brings them together, they are their sobs. I do not dispute the suffering that the blacks have suffered and are still experiencing. I challenge the tendency to erect this suffering into signs of identity. I was born in Congo Brazzaville, I studied in France, I now teach in California. I am black, with a French passport and a green card. Who am I? I would have a hard time saying that. But I refuse to define myself by tears and bitterness."[21]

(A. MABANCKOU, 2006).

After having read Chamoiseau's *Texaco* and its creolities, Mabanckou and his porcupines, Kelman and his cassava, Beyala in short all the gang of the Francofoolish courtiers, one can thus rigorously measure the damage extent regarding the genocidal methodologies of the abbot Gregoire the author of *Rapport sur la nécessité et les moyens d'anéantir les patois et d'universaliser l'usage de la langue française (1794).*

If Kelman, Mabanckou and their henchmen do not know what they have in common with other Blacks except the shared far skin color, it is because from their memory was erased the triptych: community of origin, community of suffering and destiny community so important to Elder Marcus Mosiah Garvey. It is therefore not surprising that some people write books that please others by recommending a fallacious argument justified by Comtean's own positivism.

After this highly disturbing picture, I suggest that Make a stunning spot on the anathema that strikes the black scientists of prehistory:

"It is striking that almost no Egyptian scholar's name has survived. On the other hand, almost all of their Greek disciples have passed on to posterity by claiming the inventions and discoveries of their anonymous Egyptian masters. This is what emerges from the passages of Jamblicus which precede, and from the writings of Herodotus, alluding to Pythagoras who pretended to be the inventor of the ideas of his masters."

(ANTA DIOP, 1967)

De facto, locked in the intellectualist and intellectualizing matrix of the others, the pseudo African elites seem to have difficulty to take the real measure of the danger range that objectively emphasizes the alerts launchers as Diop. The cataclysmic situation of the continent is testifying and yet it does not interpellate anyone. We simply content ourselves with trompe-l'oeil solutions or various handicrafts, if we do not share distressing banalities. Unfortunately this observation can be made in all areas of the social fabric.

However at liability of these pseudo intellectuals, because the African education system is subject to a ruse that puts Africa under the compulsory and forced cut of a measurement standard, imposed by the West. So any opposition or giving up lead to exclude oneself from the system put in place by the diktat and the colonists cunning.

It is therefore understandable why African universities appear as Western universities branches. They inevitably become inoperative entities that consume the public good and appear to be simple machines good for making lunatics and mass unemployment. In truth, they have no wiggle room for the efficient training of Africans students.

In a country that I will not name, in order to target no one because the problem is so much widespread on the continent. The public expenditure for education has been estimated at nearly 17,500 billion CFA over 35 years for a distressing result, not to say neal, but that does not change the game or question the effectiveness of the process. On the other side, the colonial system is placidly perpetuated without an endogenous control logic devoid of self-criticism or self-control..

21. Alain MABANCKOU. Mémoires de porc-pic. Paris: Le Seuil, 2006.

Let us remember that in 1917 one of the theorists of the moral conquest of Africans, Georges Hardy asserted:

> *"To convert the primitive peoples of our colonies, to make them more devoted to our cause and useful to our enterprises, we have at our disposal only a very limited number of means and the surest way is to take the native from childhood, to get him to attend our schools assiduously and to undergo our intellectual and moral habits for several years in a row, in a word to open schools where his mind is formed in our intentions."*[22]
>
> *(G. HARDY, 1917)*

Today we have a better understanding why the concept of the chosen immigration is proposed to the African countries. A proposal welcomed with open arms because of a major critical deficit on the continent or for fear of the arsenal repressive serving as safeguards and using integrated mechanisms of subordination.

Young people with real skills are sent to the West with everyone blessing including the top of the state so that, they have to be proud to widen the huge gap that separates their host country from their country of origin.
So to believe that Africa is investing in the training of its children to send the best to modern sto washy. How by inner mediocrity concentration Africa can expect to challenge others in those circumstances ?

Those who engage in such acts in African academic bodies are benevolent subjects, out of the Western mold and coopted and then dubbed by Africanist academics. The so-called African specialists are charged with work by their Western counterparts to ensure that they have no wiggle room either to think or to carry out research that could revise the official views.

As the orientations of the African education system are subject to the diktat of the Africanist intelligentsia, textbooks are mere empty shells from the African point of view. All references in these manuals are exogenous to Africa. No African has ever made science, chemistry, philosophy of mathematics etc. It is a real problem when Africa is chained home to follow a path dictated by others that leads regrettably only to its death end and to its loss.

It follows several things: a loss of individual and collective identity and especially an inability to achieve failure to realize the prophecy of Leopold de Saussure French navy officer, sinologist and astronomer not to be confused with his brother Ferdinand de Saussure, the founder of modern linguistics, who laid in his time the foundations of semiology. Leopold de Saussure supported this in the last century by speaking of the colonies:

> *"[...] The politics of assimilation are based on very seductive reasoning, which will always have access to the Latin spirit, as long as the philosophical dogmas with which it has been impregnated have not been modified. If the natives, we say to ourselves, are refractory to the benefits of the civilization we bring them, it is because their prejudices have not yet enabled them to understand the advantages they may derive from them. These prejudices are maintained at home by the vestiges of their old state, by their beliefs, their institutions, their languages. Let's remove these remnants from an abolished past. If they are too inveterate in the present generation, let us address education to future generations. Let's teach our children our language, let us instill our ideas and France will soon have millions, if not new citizens, at least loyal and grateful subjects.*
> *This reasoning, apparently clever and generous, is in reality inapplicable, oppressive, and as harmful to our interests as to those of our subjects.*
> *The most skilled conquering races have understood the need not to offend the beliefs of their subjects and to respect their institutions. The French, on the other hand, try to transform native societies even before they have conquered. They profess that the institutions, the beliefs, even the languages,*

22. Georges HARDY, Une conquête morale: l'enseignement en A.O.F, Armand colin. Paris, 1917

> *maintain the hostility of the natives against the new state of things, and that to obtain their sympathy or their resignation, there is only one effective method: the assimilation.*" [23]
>
> *(L. De SAUSSURE, 1899)*

This sentence of Saussure echoes another emanating from Raphael Lemkin who coined the word "*genocide*", offering to this notion the approach of a "*cultural genocide*":

> *"The term "ethnocide" is also used. In a Declaration of San Jos of 11 December 1981, Unesco defines it:*
> *"Ethnocide means that an ethnic group is denied the right to enjoy, develop and transmit its own culture and its own language, whether collectively or individually. This involves an extreme form of massive violation of human rights and, in particular, the right of ethnic groups to respect for their cultural identity.""*

This statement therefore underlies a premeditated act based on a desire to harm and subjugate by using what critical sociology represented by thinkers like Bourdieu calls immaterial weapons. Claude Catalan who has studied the subject says this about immaterial realities following his investigation:

> *"[...] By translating the untranslatable ambiguously, they complicate the language, thus the exposition of opinions and projects. This multiplies the number, the importance and the influence of who knows how to use the words by possibly diverting their meaning. They can become by a non-naive use, keys to penetrate the consciences and to disturb them. Words can serve as weapons capable of influencing behavior and satisfying the power desire of those who master them.*
> *Vocabulary participates in the power of dialectics or rhetoric which makes us pass from the word to the sentence, to the syllogism, to the demonstration, and finally to the persuasion by the discourse."*[24]
>
> *(CATALAN, 2016)*

This act, which has been presented as a savior to Africans, reveals itself to critical analysis, surrounded by highly doubtful motivations.

Since the Cairo symposium in 1974, although a small handful of Africans seem to have taken the measure of Diop's work, surprisingly the implementation of his teachings as part of the training of African children seems to be absent. The senior cadres of the training of Africans since 1974 all seem to be struck by amnesia when some do not espouse DIOP's contrary theses which were swept in Cairo, people like Tidiane Ndiaye.

To the discharge of these teachers because they are the fruit of postindependence training whose content has always been thought and wanted by Africanists in order to harm Africa as much as possible.

> *"African youth leave somewhere; to drive them without worrying about where they come from, is not educating, but breeding."*[25]
>
> *(KI-ZERBO, 1990)*

Africa has become an open-air asylum where today's youth is trained to feed mass unemployment statistics. The best elements of this school being trained to become mere subjects in the pay of others under the chosen immigration. It is difficult to do otherwise because basically they are the fruit of a Machiavellian plan designed by others. The idealization of the other and of his model of society produces in the alienated subject a motivating signal linked to the mere idea of having to work for others, including against his own. This is one of the consequences of the alienation process o implemented through the educational system of others.

---

23. Léopold de SAUSSURE, Psychologie de la Colonisation Française dans ses Rapports avec les Sociétés Indigènes. Paris: Félix Alcan, 1899. P.28.
24. Claude CATALAN, Au-delà du matérialisme - Les réalités immatrielles, éditions Baudelaire, 2016, p.103.
25. Joseph KI-ZERBO, Eduquer ou Périr, l'Harmattan, Paris, 1990, p.79.

After years of independence, Joseph Ki-Zerbo already noted that "*poisoned dishes served on the school benches*"[26] *are making their effects*. Intellectuals have exhausted their intellectuality. Executives can no longer "lead" Africa. They have all fallen into the trap, unable to invent new paradigms, indulging in the status quo while waiting for exogenous solutions to endogenous problems. The African intelligentsia perpetuates these habits at school. We are satisfied with the iteration of what the settlers say or think.

> *"[...] I do remember that when I was passing the oral tests of the college patent in Cameroon, in history, the examiner asked me to talk about the success of nationalities in the Balkans and in Italy. In which school program in the West are we talking about the Mandingue epic of Mali? This historic battle of Kirina near Koulikoro, where Soundiata Keita, defeated Soumangourou Kanté and seized the kingdom of Sosso. If this African story were taught to the little Westerners, some misunderstandings would be dispelled, and many racists instead of always presenting bananas to blacks, would present them with pens."*[27]
>
> *(ELONO, 2014)*

If the educational system and its school content are already toxic, the religious books perfectly play their partition of the Africans general "dumbing down" by reinforcing the identity loss phenomenon by false beliefs.

It becomes more than urgent if is not necessary for a continent that will be nearly 4.5 billion human beings in 2100, in the near future, to reclaim its real history and its identity by participating in the challenges of his time.

It is not a matter of warning others that Africa is going to raise up again, but rather to shake Africans. Fanon had already sounded the alarm in these remarks:

> *"Each generation must in relative opacity, discover its mission, fill it or betray it"*[28].
>
> *(FANON, 1970)*

It may be time to discover each one's own mission and collectively find the keys to rebirth.

To a certain extent, if one complains in the 21st century that thousands of Africans are dying in the sea at Europe borders, some mechanisms are more pernicious and use the semantics of words and neuroscience techniques to proceed in silence to the memory erasure of the African and indirectly to the erasure of their collective memory.

As Karl Marx, Elie Wiesel or André Malraux assert with different words but by sharing the same key idea :

> *"Those who do not know their past are intended to experience it again"*
>
> *(André Malraux)*

Not only what these pernicious techniques do in silence but they also proceed to replace the source memory with a borrowed memory. This is the best illustration of alienation that is to say a death on the identity level.

> *"... You must know that the opponent is killing you morally, killing you intellectually, culturally before killing you physically. That's how whole groups were faded away ..."*
>
> *(ANTA DIOP, Niamey's conference, 1984).*

Often, the manipulation of the poisoned concepts used by Africans they do not realize unconsciously the immeasurable damage to which they expose themselves.

Simple methodological requirements in linguistics slide over time and become physical data of anthropology. The case of Bantu is quite illustrating. Following the "work" of classification of idioms in

26. Joseph KI-ZERBO, Histoire de l'Afrique Noire, Hatier, Paris, 1972, p.28.
27. R. ELONO, La cosmogonie des beti be nanga: Croyances ancestrales – Traditions. Paris, Editions Edilivre, 2014, p.53.
28. Frantz FANON, Les damnés de la terre, éditions Maspero, Paris, 1970, p.197.

subequatorial Africa by Bleek, Meinhof, Meeus, Kadima, Guthrie, Westermann, Baumann and others, there are nowadays in Africa Bantu populations. Better than that, a whole jargon has emerged around the concept of Bantuistik attributed to Wilhelm Heinrich Immanuel Bleek since 1859.

Rigorously when does a random criterion pertaining to the technicalities of linguistic classification give rise to an operating category in physical anthropology?
This is only possible if two ideas are skilfully joined. In this particular case, the first idea is classifying a group of languages cleverly flanked with the Bantu label and, by semantic shift, those who are enlisted in this classification despite themselves have become Bantu peoples. This is simply Machiavellian.

But in reality, it is this ability to associate ideas in a scholarly way but sophisticated technics that produces and seals the desired and induced thinking errors.

It is hardly worth mentioning that before the arrival of Bleek only in the 19th century with his concept of bantuistik, the populations concerned had a past, knew how to name themselves, name their territories, how to write etc. They did not wait for Bleek's arrival to become Bantu through performative trap. Homo sapiens sapiens until proven otherwise was born in Africa and studies estimate that it is since at least -300.000 years. So what is this mascarade on the Bantu taught in African universities? We come to the paradox of being named according to fanciful canons at the risk of not being understood any more. I sincerely believe that it will be necessary to stop these distressing banalities and to deconstruct all this system.

These human groups are created in ideology laboratories whose existence is inscribed in sociological truth in textbooks and then taught by the pursuers of the colonial model. These groups are then instrumentalised by forces outside Africa when political or geostrategic issues will be engaged in on the continent. We can therefore rigorously ask the question to those who promote the colonial style with the taxpayer's money.Where is then the cultural unity demonstrated by Diop in *Negro Nations and* CULTURE since 1954?

It is therefore urgent to deconstruct this system. This system with deep roots that draw their sap in the simple musicality of words with toxic esoteric content for the unsuspecting minds of Africans. These poor Africans credulous who are fooled by sleepy speeches based on a Western pseudo-universalism.

The current challenges are therefore for Africans to substitute the Western lexicon, an African technical glossary based on the similarity between "Africa" today and the "Egypto-Nubian", valley. Diop invites us at any price when he says:

> *"And African studies will emerge from the vicious spiral in which they move, to regain all their meaning and all their fertility, only by moving towards the valley of the Nile.*
> *Conversely, Egyptology will emerge from its age-old sclerosis, from the hermeticism of the texts, only from the day when it will have the courage to explode the valve which isolates it, doctrinally, from the vivifying source which constitutes, for it, the Negro world."*
>
> *(ANTA DIOP, 1967)* [29]

Better than that DIOP specifies what to do and how to deal with it:

> *"To the extent that the Egyptian-Nubian civilization marks the oldest stage of our culture, force-us is to reconnect with this one in all the fields if one wants to build a body of human sciences. Egypt is to the rest of Black Africa what Greece and Rome are to the West. The new African humanities will have to be built on the foundations of the ancient Pharaonic culture. Ancient Egyptian and Meroitic will have to replace Latin and Greek in the programs. Egyptian law will have to take the place of Roman law.*
> *Egyptian philosophy (the two systems of heliopolitan and hermopolitan cosmogonies) should be taught jointly with Greek philosophy especially (Democritus, Epicurus, Plato, Aristotle etc.) to*

29. Cheick ANTA DIOP, Antériorité des civilisations nègres – mythe ou vérité historique?, Présence africaine, Paris, 1967, p.12.

better highlight the contribution of Black Africa to Western thought. . Is it necessary to recall that under Thw.t-Mss.w III, under the XVIII dynasty around 1470 before our era, 800 before Homère, flourished already in Egypt that is to say in Black country, a classical period written towards. Similarly, African education must systematically integrate the study of the arts in Egypt: civil and military architecture, music, sculpture, painting, and so on."

(Anta Diop, 1976)[30]

Figure 1: Pharao Mn-Kaw-Re + wife Epouse Khamererneb ty. IVème dynastie.

It is thus necessary to close the tap which suspends the African youth of damaged knowledge. On this account, Africans must themselves look at their anthropology and historiography by following the advice of Anta Diop to get out of the intellectual tutelage that has become collectively suicidal.

Concretely, nowadays this leads to a serious interest in the question of the decipherment of the "Meroitic" by Africans and of course to interest Africans to the Nile Valley through reading of Egyptian texts according to Afrocentric point of view by making use of current spoken languages in an African mindset and paradigm.

It is therefore for Africans to proceed above all else to an epistemological break, to base their truth on their own investigation and to conceive the real according to their own paradigm without having to be content with the intellectual supervision of a third party.

This work is therefore part of this approach which is not exhaustive on a technical level but which should allow everyone to take the measure of the possibilities offered by the technical tools. The purpose of this work is therefore to give ourselves the means to think by ourselves, to question our past ourselves, to access it through our investigation, to allow on second thoughts the writing of our theories and the production of our textbooks with a educational harmless content. This is not a pastist approach, but a means for Africans to find the deep inside keys for the genius that has helped for the building of the great civilizations of the Nile Valley.

Development is a deceptive concept invented to mislead Africans. Africa is not underdeveloped, it is in regression. And, having made the right diagnosis, we can easily bring effective drugs and not ineffective solutions such as development policies based on completely mental traps and empty concepts.

To move forward Africa must be reconciled with its past as on the picture above.
To move forward Africa must rely on itself, on its children.
To move forward, Africa must do the asceticism of aids and intellectual guardianship Africa has not need others help to lead the world for more than 4000 years.

Africa must unite its children and put its ancestral values in the heart of its sociology.
To move forward, Africa must put the left foot located heart-front and get support on its back foot. This responds to the universal principle of perpetual motion ***xpr of*** its ancestors.

"In each African sleeps a builder of Nations whose good education should be able to awake"

(Anta Diop à Niamey, 1984)

30. Cheick ANTA DIOP, Parenté génétique de l'Egyptien pharaonique et des langues négro-africaines, IFAN, Dakar, 1977, p. XXV.

I. General Introduction

"Follow your heart as long as you live."

Ptah-Hotep

At a time when Africa is dealing with the supposed help of China to hang the last car of the "development [1]" flocked with the new label of Emergence at... and that some macroeconomic indicators are subject to concern for everyone, (4.5 billion inhabitants by 2100), this paradoxical situation that on one side we still agree that the future of the world is in Africa while on the other, we complain about the dwindling of the resources of the entire planet. What solutions does Africa propose to deal with the major structural problems of tomorrow? Is it waiting for the West to vote a Marshall Plan aids?

If we have learned how to train Africans according to the Western paradigm, the eminent African researchers (Anta Diop, Ki-Zerbo, etc.) have noted important training formal defects and have advocated a return to our roots. Yet, if their genius seems indisputable to the rationality over their researches subject, Africa postindependence continues to follow its way continuing the impetus of the colonial model of the 50s. It prohibits itself any questioning of colonial institutions.

According to Professor Greenberg's classification, there are about 2000 languages in Africa and almost as many ethnic groups.
Faced with the challenges of managing nearly 4 billion people in the near future, the question of culture and language inevitably arises because the future of Africa is subject to the realization of a pan-African state. The question seems inevitable as colonial languages act as iconoclasts of African culture and languages. The paradox that erects them in national languages in the majority of African countries has not so far led to the question of the need for self-determination.

While many issues related to the management of 4 billion people remain unresolved, the issue of the pan-African language in a common state remains unavoidable. According to me, it might be time to take care of it. Surprisingly, in his book the *Federal State of Black Africa the only way out*[2], the African specialist in linguistics Theophile Obenga addresses all aspects of the construction of this state and curiously avoids the question of the common language. Can African countries make a real difference by remaining in colonial languages? It seems to me that this question everyone must ask and better than all, everyone must really meditate on it.
In chapter II of his book *Les fondements économiques et culturels d'un Etat fédéral d'Afrique noire*[3], As well as in *Alerte sous les Tropiques*, Cheikh Anta Diop very happily approaches the issue of the choice of this language within the federal state and its elevation to the level of a modern language. It therefore seems appropriate for Africans whose past is related to Nubia and ancient Egypt, to begin to look at the decipherment of the "Meroitic" and the ***mDw nTr***. This must be a prerequisite for writing the future federal language. All African languages are genetically linked to Pharaonic and Nubian

---

1. In the context of international geopolitics, Truman in 1949 claimed to offer development to all peoples, "We must launch a new program that is bold and that brings the benefits of our scientific advance and our industrial progress to the service of the world. improvement and growth of underdeveloped regions. More than half of the world's people live in conditions close to misery. They do not have enough to eat. They are victims of diseases. Their poverty is a handicap and a threat, for them as well as for the most prosperous regions. One might think that he had a very commendable step. But his project was hardly philanthropic: it was part of a desire to force Europeans to share their colonial empire. Truman used what critical sociology calls an immaterial weapon, thus making his undisclosed purpose coincide with a need so as to remove it from sound criticism.
2. Thophile OBENGA, l'Etat fédéral d'Afrique noire: la seule issue, l'Harmattan, Paris, 2012.
3. Chekh ANTA DIOP, Les fondements économiques et culturels d'un Etat fédéral d'Afrique noire, Présence Africaine, Paris, 1974, p20-29

languages. They should therefore comparatively be read through these characters such as Chinese characters are read within the sinosphere[4].

The different works of African researchers on the genetic kinship of their mother tongue with Pharaonic is a given, it is still essential to give Africans access to their ancestral past through a methodology that allows everyone to read the ***mDw nTr*** or the Meroitic in his own language.

This document does not claim the completeness of the 2000[5] African linguistic varieties. It is thought as a step away from the colonial model. It is an example to serve as a guide for other approaches that will aim to access their past through the practice of local languages.
African ancient writings are being studied to provide Africa with helpful building materials of a body of science.
To carry out this exercise, it can only be done by confronting our languages spoken in Africa with the ancient texts of Africa because, a language is a crystallization of a people ontological vision of the world.
The language being stable through time and space, it allows by its morphogenetic stability to enlighten the science trough its prospective dimension.

The result of this endogenous research will allow Africans through a macrostructure to collect and to get at least 5000 years of data may be more on Africa's past. Access to these data can allow comparative studies in all current scientific fields to correct exogenous data and reconcile Africa with its past and history. Access to these data can of course allow Africa to be helpful for the entire humanity.

The prerequisites for the understanding of this approach are based on a technical jargon borrowed from previous works either in Egyptology and also in linguistics research. Thus, I will speak about the branches of the Negro-Egyptian languages, the *Kekwe* and *Kweke* languages brought to light by brother Jean-Claude Mboli and which have the characteristic of operating by phonemes inversion. In the context of the endogenous classification of African languages, I will also sometimes speak about the *beer* and *bere* languages which are characterized by the elision or not of a suffixal vowel.
We must start from the premise that Homo sapiens sapiens was born in Africa and learned how to speak and communicate. We have for long times identified Negro-egyptian as the basic language of Homo sapiens sapiens. The work of Anta DIOP and Obenga made this hypothesis technically possible. The research of Mboli allowed the reconstruction and setting up of the Negro-Egyptian grammar.
So we agree that Homo sapiens sapiens came out of Africa with a language. Semitic languages appear to us as slangs between ancient African languages spoken by Aryan populations and African old native languages. The present works will have in heart besides proposing the reading of the *mdw nTr* but to show that on the etymological level, the other languages can be understood from the current African ancestral strains.

We will therefore try to inscribe this research on the diachronic level since the ancient languages and to compare them with the current languages.
At first the document will expose the technical aspects of the working language Kóló or Ewondo which is an Ekang language sorry not Bantu. I will gladly have passed this part because it brings nothing in this heuristic but it seemed to me necessary since it comes under technical constraints.

Historically, after the invasions of the so-called Semitic peoples from the Middle East, we can observe on the maps of the world settlement the absence of blacks in these geographical areas. Nawadays, we find Blacks people from South Sahara to South Africa. Taking into account the work of Diop stipulating that the current Maghreb and therefore pharaonic Egypt were negroes, it became clear to question the

---

4. Xu DAN, Introduction to Chinese Syntax, l'Asiathèque, 2015.
5. Joseph KI-ZERBO, Histoire générale de l'Afrique, Méthodologie et préhistoire africaine, Unesco- Paris SP, 1980, p.321

past of the Ewondo populations and to use the language gateway that unites them to the Pharaonic Egypt in order to propose an endogenous alternative of ***mDw nTr*** reading.

If the choice is based on the author's mastery of the language, this choice should enable many African researchers who practice the so called Bantu languages (south Sahara to Cape Town) to be able to grasp the methodological requirements that make it possible to read the African holy characters according to their own language. It therefore requires for the Africans who will approach this work to be interested in their mother tongues but also to have knowledge on the African or Egyptian-Nubian paradigm.

In order not to induce confusion in people's minds, this work will try to explain that this method does not belong to Egyptology, that is to say, made by Westerners for the benefit of Africans. This method tries to rely on an endogenous logic of Africa by using the highly praiseworthy achievements of Egyptology to give an African point of view on the issue of reading and mainly understanding of the ***mDw nTr*** being understood since 1974 that Egypto-Nubian languages are African languages.
In a second step, this work will consist in presenting the method and will end with concrete explanatory examples.

The conclusion of this work should hopefully open up readers to other perspectives of research that are not undertaken as part of this heuristic. For the purpose I hope to complete this word by a complete comparative dictionary between the two languages.

II. Presentation and Criticism of Previous Documents

"Maliga ma nkɔyɔp njiis di li tɔb bɛ"

Basa'a's Wisdom

The international symposium on Egyptology organized by UNESCO from January 28 to February 3, 1974, in Cairo, Egypt, was a major turning point for the history of humanity at first and for African historiography in a second time.

Despite the new doubtful revisionist works of Africanists relayed by some Africans like Professor Tidiane Ndiaye, the conclusions of the Cairo symposium on the issue regarding the settlement of ancient Egypt and the decipherment of Meroitic writing are unequivocal.

Until this work, the fundamental problem for the organizer of the colloquium (Unesco) consisted in knowing which phenotype was the people who had built the most brilliant civilization of humanity; Africans as the professors Diop and Obenga supported it? The leucoderms represented by an aeropage of Western specialists or a civilization of Semitic essence?

The conclusion of this colloquium is, for any conscious African, an estimable datum. Here it is recorded below for the attention of honest people.

> *"The very detailed preparation of the papers by Professors Cheikh Anta Diop and Obenga did not, despite the details contained in the preparatory working paper sent by Unesco (see Annex 3), always have an equal counterpart. There followed a real imbalance in the discussions.*
> *These, however, have been very positive for a number of reasons: in many cases, they have shown the importance of the exchange of new scientific information; they highlighted, in the eyes of almost all the participants, the insufficiency of the methodological requirements used until then in the Egyptian research; they revealed examples of new methodology which would make it possible to advance, in a more scientific way, the study of the question proposed for the attention of the symposium; in any case, this first confrontation should be seen as the starting point for further international and interdisciplinary discussions, as the starting point for new research that appeared necessary. The very number of recommendations reflects the desire of the symposium to propose a future program of research; Finally, the conference allowed specialists who had never had the opportunity to confront their points of view to discover other approaches, other methods of information and other avenues of research that they were accustomed. From this point of view as well, the outcome of the conference was undeniably positive."*[1]

Any serious work concerning the study of the Pharaonic must be based on the research that has helped to highlight this language since the revolutionary work on Champollion's grammar as the immense research work done by the Western research centers. Names that stand out as safe values in the field as professors Leclant well known for his work on the so-called oriental languages and Egyptology, specialist in history and Pharaonic civilization, especially the XXVe dynasty, Jean Vercoutter, Gustave Lefebvre, Nicole Blanc, Pascal Vernus, Jean Yoyotte, the grammarian Serge Sauneron, Gaston

1. Sous la direction de Joseph KI-ZERBO, Histoire générale de l'Afrique, Etudes et documents, le peuplement de l'Egypte ancienne et le dchiffrement de l'écriture Meroitique, Paris, 1978, p.98.

Maspero, Georges Soukiassian, Joseph Francois Parrat, Amandine Marshall, Auguste Ferdinand Mariette etc. We can not be content to quote only the French school but we can talk about:

- The English Egyptology with Goodwin, Alfred Wallis Budge, Alan Gardiner, etc.,
- We can also mention the influence of the German school with: W. Kaiser, Gunter Dryer, Kurt Heinrich Sethe, Adolf Erman, Hermann Grapow, etc.
- The american Egyptology with, Herbert Eustis Winlock, James Henry Breasted, Gordon-Jaquet, Edwin Smith, Kent R. Weeks etc.
- Swedish Egyptology with Säve Söderberg, Johan David Åkerblad, etc.
- A Russian school with specialists like: Vladimir de Bock, Vladimir Golenichtchev, Alexandre Piankoff, Boris Piotrovski, Boris Touraïev
- Chinese Egyptology with Xia-Shang-Zhou
- Spanish like Ferran Iniesta, Victor Regalado-Frutos, Eduardo Toda y Güell etc.

We therefore see that, the research on pharaonic language remains very lively since the closing of the last Egyptian temples under Justinian in the sixth century of our era having resulted in the abandonment of the writings of Pharaonic Egyptian (hieroglyphic as well as hieratic or demotic) . Only language survived in the modern form of Coptic and its variants, but written sources gradually became dead letters.

From the Western point of view, it was not until 1822 and the discovery of Jean-François Champollion (1790-1832) that we once again had access to ancient documents written by the Egyptians themselves. Did he really translated or get help from a Coptic informant? According to us , we will simply say: no comments.

In the light of the work done by the various chairs of Egyptology around the world and the vitality of the research, one can not pick up the pen to dare any criticism just because we are Africans and due to the fact the symposium of Cairo paid in favor of Anta Diop and Obenga.

Can we rigorously criticize the proactive impulses of others when Africa is completely absent from History research or when its pseudo specialists, like Elikia M'bokolo, of Pap Ndiaye, are attached to diktats and to the servile exercise of iteration within the universities of others? It would be difficult to criticize without being concerned by the conclusions of the Cairo symposium and without taking the real measure or the international scope.

At the precise moment when Africa is trying to emerge from its black night after 500 years of slave trade and 100 years of colonization, Africans can be excused for their absence. Those africans who are more on the look-out of the quest for physiological needs in Maslow's undestanding, while other nations are looking for hegemonic positions on the world stage.

For African researchers who are dealing with the Pharaonic thesis, needless to tell them that it is requisite sharing the advances account of Western Egyptology without stirring up unnecessary and counter-productive chapel wars.

First of all, the goal is therefore and foremost to go along the road of advanced research and to be interested in the work of others but also the work of serious African researchers such as: Obenga, Mboli, Bilolo, Pfouma, Ngom, Diop, Moussa Lam, Oum Ndigi, Sarr, Soussoum Sougan etc

The obvious interest for Egypt and the Kush's kingdom allow to enquiry about what is nowadays the truth and no more longer just speculation or theory.

Access to the past of Africans is a complex issue for those who are questionning the topic, It involves to mobilize a lot of knowledge in linguistics, ethnonymy, toponymy, oral traditions, comparative mythology, history, linguistics comparative, in archeology, in Saharan prehistoric art, in literary criticism, in Egyptology, in sociology, in anthropology, etc. in order give a particular and an accurate perspective.

As the specialists point out, this coalition of sources is even more involved in dealing with parenting difficulties, for example in chronology. Carbon 14 dates are not always available. Although these are interpreted and compared with other data such as metallurgy or pottery (materials and styles). As in the north of Chad, masses of ceramic debris are still not available to construct a typology representing a six-level chronological scale.

Without minimizing the importance of chronology, the backbone of historical material, and without abandoning efforts to establish it on rigorous bases, must we succumb to the psychosis of precision at all costs, which risks then be a false precision? Why insist on writing 1086 for the fall of Koumbi Saleh instead of saying "at the end of the eleventh century"? Not all dates have the same importance. The degree of precision required for each of them is not the same, and each must not be erected as a statue.

Despite these precautions and difficulties, we are entitled to enquiry about a certain number of facts leaving us doubtful.

How comes that some studies on ***mDw nTr*** are based on comparative linguistics? If comparative linguistics is enlightening, was it used at the past to read the pharaonic characters?

Why is the general methodology of Egyptology not revised since Cairo, 1974 and especially with the publication of numerous works which clearly establish the genetical kinship between African languages and ancient Egyptian?

2.1. CHRONOLOGICAL PARADOXES

Among the questions that give rise to questioning, there are several types and orders, some of them induce a paradox compared to the official discourse.

2.1.1. Giant hieroglyphs of Nekheb

Figure 1 : Giant hieroglyphs of nekheb

Giant hieroglyphs [2] 5200 years old were discovered at ***nxb*** "Nekhab". According to the researchers, their use dates before the first pharaonic dynasties. Not to doubt, they are among the oldest known to date. Their large size, up to 70 cm, is exceptional

p

This is an incredible and unexpected discovery made by members of an archaeological mission conducted by Yale University and the Royal Museums of Art and History of Brussels. The region where these important discoveries were made is that of ***nxb*** or Elkab between the Nile and the desert 90 km south of ***wAs.t*** "Thebes", on the right bank of the river in front of ***nxn*** "Hierakonpolis". They came across proto-hieroglyphs engraved on a cliff in Egypt.

These proto-hieroglyphs were engraved on a rock wall about 5 250 years old and of a totally new size. We see a bust with a bull's head followed by two African jabiruses back to back and surmounted by a bald ibis.

2. https://news.yale.edu/2017/06/20/yale-archaeologists-discover-earliest-monumental-egyptian-hieroglyphs

Forms have been clearly distinguished, such as a bald ibis, a bird pointing to the "***akh***", a symbol of serene life after death. Archaeologists have explained that this arrangement of symbols is relatively common in later Egyptian representations of the solar cycle and the concept of luminosity. "These images could express the concept of royal authority over the ordered cosmos," noted Professor Darnell, who co-directed the study project at Yale University in the USA.
A very elaborate symbolism to believe Pascal Vernus, director of the School Practice of the Hautes Etudes (EPHE) and specialist for whom, it would be a "hieroglyph translated the Egyptian notion of ***bA***, the capacity for a human being to evolve itself in different aspects ".

According to the Egyptologist John Coleman Darnell, professor in the Department of Literature and Civilization of the Near East who co-directed the study project at Yale ("Elkab Desert Survey Project"):

> *"This newly discovered rock art site retains some of the oldest and most important signs of the formative stages of the hieroglyphic script and provides evidence of how the ancient Egyptians invented their unique writing system".*

According to authoritative sources, at the end of the fourth millennium BC in Upper Egypt, hieroglyphic writing was used until Roman times, for more than three thousand years. The knowledge of the hieroglyphs would be lost with the closing of the pagan places by the emperor Theodosius I about 380 after our era.

Günter Dreyer's authoritative work on the supposed discovery of ***mDw nTr*** in Abydos are dated between 3250-3200 BC. This 1200-year-old differential curiously reminds the backdrop of the official historical question breakdown already pointed out by Professor Anta Diop.

Indeed, at the time when these glyphs of ***nxb*** were drawn, according to legally sources, were Africans in History or in Prehistory? Would not we still be told that it's the writing that forces the way of humanity over prehistory to history? The characters of ***nxb*** are they ***mDw nTr*** or not? Why do official sources not realize the chronological data on the occurrence of ***mDw nTr*** and the documents dealing with historical boundaries?

2.1.2. GOSFORD'S CONTROVERSIAL GLYPHES

Curiously, we find in Australia also under the name of the glyphs of Gosford, characters similar to the Egyptian hieroglyphs. These characters were discovered in the 1900s in Australia, numbering 250. When examined, it is found that the characters of Gosford were carved into the rock. Presumably, they were an integral part of local culture and folklore for about a century and are dated to nearly 5000 years ago.

Gosford's Glyphes

Strangely, these characters that look like hieroglyphics are not yet considered as a writing and even less like the oldest in the world. The hieroglyphs would officially be 3400 years old and on the other hand, the same glyphs found in Gosford are dated to 5000 years BC. It ends up losing its Latin. Yet, some experts are unanimous on the question:

On the contrary, a recent deciphering (October 2014) of the glyphs, done by the experts Mohamed Ibrahim and Yousef Abdel Hakim Awyan, made it possible to conclude.

> *"The glyphs of Gosford are legitimate, because the scribes have correctly used several ancient hieroglyphs and grammatical variants which, capital fact, were not even referenced in Egyptian hieroglyphic texts before 2012, which makes it possible to immediately refute the theories of the" hoax ". "That have been around for a long time. The specific style of the hieroglyphs also makes it possible*

to position them in a linguistic chronology attesting to the Egyptian presence in Australia at least 4,500 years ago, whereas the translated text is so detailed that it even makes it possible to identify the former scribes by their name and profession."[3]

2.1.3. An Egyptian Calendar: Before Writing?

Another curiosity will draw our attention particularly with the dating of the Egyptian writing to -3400 years. Let's read R. A. Schwaller de Lubicz on the topic:

> *"The majority of Egyptologists therefore admitted the date of the founding of the Egyptian calendar as the year 4242 BC. our era, but for a quarter century the question has been discussed again, because it is obvious that the establishment of this calendar supposes long astronomical preliminary observations."* [4]
>
> *(Lubicz, 1963)*

It would have taken at least 6000 years of observations and calculations to arrive at this result, according to Plato, but they would have received everything from the gods who initiated them. These precisions prove that already 6254 years ago the ancient Egyptians possessed a significant astronomical science and perhaps even astronomical spectacles. They named the constellations of animal names.

> *"The most accurate calendar established in Egypt dates back to 4242 BC. NS, it is based on the heliacal rising of Sirius or* ***spd.t*** *(Sothis), that is to say when this star, invisible to the naked eye, appeared above the horizon. In these circumstances, the adoption of the calendar, which dates from 4242 BC. NE, at the end of the Neolithic, is a feat quite remarkable."*[5]
>
> *(M. WEISSEN-SZUMLANSKA).*

In the year 46 before NS, Julius Caesar, on the advice of the Egyptian astronomer Sosigene of Alexandria, decided that the year would be 365 days, as in Egypt. It is therefore curious that the Egyptian writing according to the discovery of Günter Dreyer in Abydos dates back to -3400 BC.

Diodorus of Sicily's Testimony:

> *"The local inhabitants of Thebes in Egypt say they are the oldest of men and claim that philosophy and astrology were invented at home ... The Egyptians also boast of having sent colonies all over the world. Belus, believed to be the son of Neptune and Libya, led one to Babylon. It is said that, having fixed his stay on the banks of the Euphrates, he instituted priests on the model of those of Egypt, whom he exempted from all taxes and public offices, and whom the Babylonians call them Chaldeans. They devoted themselves to the study of the stars, to the imitation of the Egyptian priests, naturalists and astrologers ... They also seem to know how to calculate the eclipses of sun and moon so that they can predict with certainty all the details."*
>
> *(Diodore de Sicile, I, L.)*
>
> *"There is no country where the positions and movements of the stars are more accurately observed than in Egypt. The Egyptians, taking advantage of favorable conditions, made astronomy their own science, which they studied first. They have been keeping records for an incredible number of years, where these observations are recorded. It contains information on the movements of the planets, their revolutions, and their stations; moreover, on the relationship of each planet with the birth of animals, and finally on the stars whose influence is good or bad. In predicting to men the future, these astrologers have often found just; they also frequently predict abundance and scarcity, epidemics and diseases of herds. The earthquakes, the floods, the appearance of comets, and many other phenomena that are impossible for the vulgar to know in advance, they foresee from observations made for a long time. It is even claimed that the Chaldeans of Babylon, so renowned in astrology, were emigrants who had learned their astronomical science in Egypt."*
>
> *(Diodore de Sicile, I, LVII, et LXXXI.)*

---

3. Graham HANCOCK, Magiciens des dieux: la sagesse oubliée de la civilisation terrestre perdue, Pygmalion, Roubaix, 2015.
4. R. A. SCHWALLER DE LUBICZ, The Egyptian Miracle: An Introduction to the Wisdom of the Temple, Inner Traditions/Bear, 1985.
5. WEISSEN-SZUMLANSKA Marcelle, Origines atlantiques des anciens Egyptiens, Les Editions des Champs-Elyses, 1965.

The Sirius star enjoyed an exceptional veneration among the Egyptians: the first month of the year began ritually at the appearance of this star: as she warned of the imminence of the floods of the Nile, they called it ***spd.t*** "Dog" (Sothis in Greek).

Its appearance in the morning announced the first heat at the same time as the great flood of the Nile. The year began under the sign of Aries (Imn-Re) but, with the precession law of the equinoxes, we have a lag of about three months. Their calendar of spd.t (Sirius), 6,254 years old (in 2017), proves it and allows to bring to more than 8000 years the beginning of the Egyptian civilization!

So how could the Egyptians without writing or counting have a written and accurate calendar that named the seasons and the months?

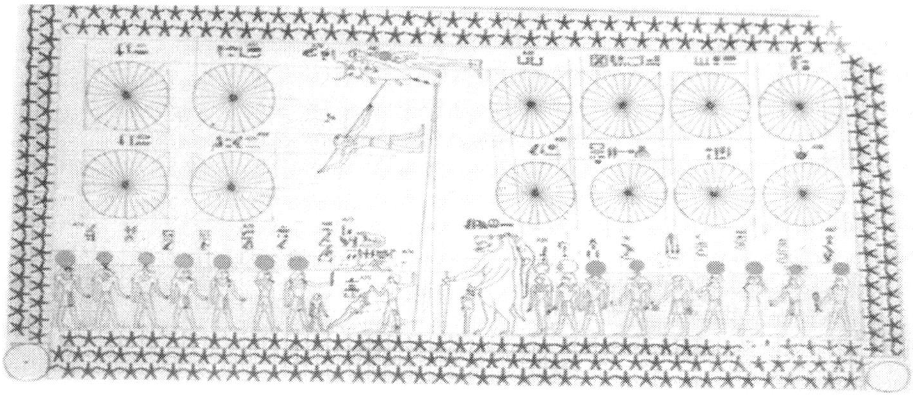

Image 4: Egyptian Calendar

2.1.4. Mpumalanga And His Petroglyphs

The general age of African glyphs could also be questioned, since that at the Mpumalanga site in South Africa, Michael Tellinger talks about the discovery of ruins around 200,000 years old. These ruins would include 3 large cities of some 100 square kilometers each, one of the most important places would be what is called the calendar of Adam, a stone monolithic calendar that could mark the time through the days.Curiously, the rock is engraved on the surface, symbols that remind us those who can be admired in Egypt:

Ankh cross in a sun *(Mpumalanga)*

Ankh cross *(Egypt)*

This site features glyphs that would be dated to -200,000 years ago. Here above, anyone can recognize engraved, the Ankh symbol in a sun. Among the ruins, a first pyramid can be seen, and the details carved into the stone as the Ankh Egyptian symbol.

"This is thousands of years before the Egyptians ..." reported Tellinger.

Two Tellinger books communicate greatly on the subject, "*Temples of the African Gods, 2010*"[6] and "*African Temples of the Anunnaki, 2013*"[7] without the slightest echo on the international level or even within African universities specialized in repetition.

If the ☥ ankh symbol is part of the ***mDw nTr,*** and is considered as a writing character, dated to -3400 years. How comes that the same symbol ☥ in South Africa is considered to be a petroglyph and all of these characters dated to -200,000 years are petroglyphs (not a writing) including the ☥ symbol?

2.1.5. Zinder's pyramids

On February 29, 2012, the Niger researcher in Egyptology, Maître Souley Garbare, discovered pyramids in Zinder, Niger.

If the settlement of Africa was from the Great Lakes region, as Professor Anta Diop has always maintained and demonstrated, are the Zinder pyramids in Niger, which are further south of Egypt are they not before those of Pharaonic Egypt?

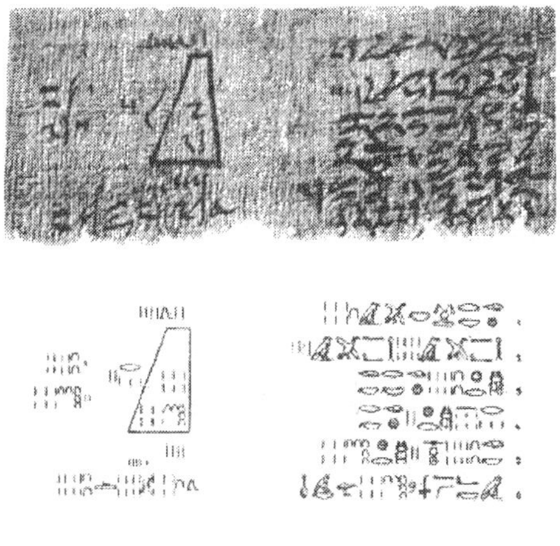

Papyrus says of Golenischev Calculations of truncated pyramids around 1850 BC.

What about the questioning to know if, for the construction of the Zinder's pyramids, the builders had ignored the mathematical sciences, paving techniques, large volume tugging, engineering, architecture, etc. used to build pyramidal buildings without writing but only with musing and speculation? The papyrus of Golenischev opposite suggests the opposite though.
Are Egyptian glyphs dating based solely on the work of Günter Dreyer to synchronize them with the cuneiforms of Mesopotamia relevant to ask a chronology and thus a division of the African humanities on the sole basis of the discoveries made in Egypt? It is time for Africans to take up these subjects lead on and do not wait for others to write and say their history.
Japanese people do not ask Chinese to write their story. French do not ask Germans to write their history, English people do not ask anyone to write their history. It is time for Africans to understand that teaching is a weapon by which the spirit of a people could be perverted.

Who is doing aerial archeology in Africa when supposed African specialists sit behind the desk or act as simple compilers?

2.2. CONVENTIONAL LIMITS

"Ngə i-dyɔm ya dzôbèn i-nə ùn-mfi ya yi woghbèn"

Tara Mvɔghɔ

2.2.1 War Paradigms

Africa having been the victim of numerous attacks from the barbarian peoples, it has lost the possibility of giving its opinion on its identity and its own history. If the question of decrypting *mDw nTr* was asked to Africans, the answer would depend heavily on the others truth. What Anta Diop named prosaically to get access to the truth through the channel of indirect knowledge. An acquaintance he said was

6. Michael TELLINGER and Johan HEINE, *Temples of the African Gods*, Zulu Planet Publishers; 2010, p.6.

7. Michael TELLINGER, *African Temples of the Anunnaki*, Zulu Planet Publishers; mai 2013.

dangerous because it can be confined and tinged with ideology. Acquaintance has several dimensions, including an exoteric one. Most often this one is handed over to feeble-minded.

In *Ecce Homo* Ferdinand Léopold Oyono an African intellectual through. *Hommage à un classique africain* of Gervais Mendo Zé brilliantly illustrates the subtle way in which ideology operates under the affable features of physical anthropology. It highlights the conceptual inconsistencies that abound between the mental representations that underlie the words, the facts and the gestures:

> *"[...] Handermayer (VN, 61): for Vandermayer. A common phenomenon in beti-boulou-fang, where the consonant [v] becomes [h], the consonant [f] becomes [p] and the consonants [p] and [b] alternate in certain contexts. This phenomenon is the one that would have led to the creation of the word pahouin from baboon. It is impossible, as some African and Africanist scholars have us to believe, that the word fang has become a pahuin in the mouth of a European, even a Frenchman, an Englishman, a German, a Spanish or a Portuguese. It is likely that the European called the populations of southern Cameroon and northern Gabon baboon and they themselves have turned baboon into pahuin."*
>
> *(G. Mendo Ze-2006)*

This illustration clearly raises the problem of the authenticity of the sources and the relevance of the given texts to be treated as a substratum of truths to be teached in Africa including in university. What does historical chronology says?

a- Ibn Wahshiyah and Arabic decryption (VIII$^{-ième}$S.)

Most historians believe that Islam originated in Mecca and Medina at the start of the 7th century.
If the decline of Rome coincides with the birth of the Arabo-Byzantin Empire, if it is unquestionable that the Arabs entered Egypt around 640 CE following the Roman occupation, they were inevitably faced with the translation of hieroglyphs.
Did they translate the hieroglyphs before Champollion in the early nineteenth century?

In his book titled "*Egyptology: The Missing Millennium. Ancient Egypt in Medieval Arabic Writings*" Dr. Okasha El Daly responds in the affirmative way to this question.
We learn that it is with the breakthrough in the research and analysis of the works of Abu Bakr Ahmad Ibn Wahshiyah, an alchemist of the ninth century that Dr. Okasha El Daly of University College London's Institute of Archeology in London did this discovery.

Indeed, the work of Ibn Wahshiyah on ancient systems of writing has demonstrated that he was able to correctly decipher many hieroglyphic signs. Alchemist and non-linguist, his main interest was the identification of the phonetic value and the significance of hieroglyphic signs in order to access ancient Egyptian scientific knowledge inscribed in the sacred African characters.
Ibn Wahshiyah is credited, among other works the writting of the *Kitab Shawq al-Mustaham*[8],
which offers a translation of Egyptian hieroglyphs, a book that discusses several other alphabets, which was later read by Athanasius Kircher in the seventeenth century and translated into English by Joseph Hammer in 1806, this text was also known by Silvestre de Sacy.
So, for Dr. Okasha El Daly of London's University College London's Institute of Archeology some hieroglyphics were decoded by Ibn Wahshiyah eight centuries before those of Champollion and Rosetta Stone.
To prove the validity of his theory Dr. Okasha El Daly declares:

> *"In particular I looked at the Egyptian Grammar of Sir Alan Gardiner which has a sign list at the end, it revealed that Ibn Wahshiyah understood perfectly well the nature of Egyptian hieroglyphs."*[9]

For the story of the fall of the Roman Empire Around the year 700 AD (the end of the first century of the Muslim era that began in 622 CE), an important cultural phenomenon took place in the empire byzantine era called the Movement of Translation.

8.Marie-Nicolas BOUILLET et Alexis CHASSANG (dir.), "Kutsami" dans Dictionnaire universel d'histoire et de géographie, 1878
9.Okasha EL DALY, Egyptology: The Missing Millennium: Ancient Egypt in Medieval Arabic Writings, UCL press, 2005.p.72

Carried away by this momentum of exchanges and communications, the Byzantines wanted to access the written knowledge external to their empire. Emissaries were sent all over the world to collect books, bring them back to Byzantium and have them translated into Arabic. Especially in the empires that had taken over from Egypt namely Macedonia and the Roman Empire. This approach was also part of a concern to preserve these writings and not only to translate them to be able to read them, because the disaster of the supposed burning of the Great Library of Alexandria with the ultimate destruction of the library of Alexandria in 639 when the Arab troops of the Caliph Omar besieged Alexandria remained alive in the minds.

ARAB DRAWING OF MDW NTR
hieroglyphs reproduced by ibn wahshiyyah, paris, national library, ms arabic 6805 folios 92b. ff

It was on this occasion that Arabic-speaking Muslim scholars gained access to texts written in Coptic. These texts have had for them the same use as the Rosetta Stone later for Champollion.
The Coptic is indeed a synchronism of the Pharaonic Egyptian and was written with hieroglyphs. With the help of the Coptic texts, they were able to determine that the hieroglyphics were also phonetic and have deciphered a good part and managed to access the knowledge recorded in many writings that escaped destruction.

b- **Jean-François Champollion and Western decoding** (XIXth CENTURY)

It is under the pen of Jean Lacouture that we will be guided on the biography of Champollion until the completion of his work on the decryption of the Egyptian from Coptic.
According to the author, one day in 1808, on the way to the College of France, one of his comrades announced to Champollion "Le Jeune" that the archaeologist Alexandre Lenoir has just published a complete decipherment of Egyptian hieroglyphs. Although this publication is whimsical, it decides Champollion to carry some of his studies on the deciphering of hieroglyphics[10].

It happens, thanks to the abbot of Tersan, a collector, to obtain a copy of the Rosetta stone, but prefers to study first papyrus in cursive scriptures. He discovered in 1808 the principle of ligatures (grouping) signs. He then postulates, on analogies with one of the Coptic dialects, the absence of vowels in Egyptian writing and obtains on August 15, 1808, as he explains in a letter to his brother, his first conclusions.[11]:

> *"I did a pretty big step in this study:*
> *1 ° I proved by approximations that all papyri belong to the same writing system*
> *2 ° that I have the value of all the letters by the inscription of Rosette that they are absolutely the same*
> *3 ° that I have deciphered the beginning of the papyrus engraved in Denon, plate 138 [...] which in Coptic means word for word: "Say: rest in peace, oh Egyptian, fill your last destination, escape the Darkness of the tomb and of death."*

But he dwells on Etruscan history. He writes: "The Etruscans are busy with me at this moment, tongue, medals, engraved stones, monuments, sarcophagi, everything is engraved in my head; and why? Because the Etruscans come from Egypt. "But his brother reminds him to order:" So study one thing instead of wandering all over the world and touching the matter. " He returns to his studies fearing that Etienne Quatremere deciphers the hieroglyphs before him in his Critical and Historical Research on the

10. Chantal GRELL, *L'Egypte imaginaire de la Renaissance à Champollion*, Presses de l'Université de Paris-Sorbonne, 2001, p. 132
11. Jean LACOUTURE, *Champollion, Une vie de lumières*, Grasset,1988, p. 121

Egyptian Language, published in June 1809. In the spring of 1809, he begins to write a Coptic grammar and studies the demotic text of the Rosetta Stone[12].
He makes "clean slate" of previous attempts to decipher hieroglyphs: he says indeed members of the Commission of Egypt that "their explanations are precisely that of sausage ...", and launches into great diatribes against the other researchers: "All that has been said about the obelisks Kircher, Jablonsky, Warburton, etc., serves only to prove that nothing can be heard and that nothing will ever be heard ". In the summer of 1809, it begins a grammar of "language thebain-sahidic",
that of the Coptic community of Saïd region. He writes to his brother in March or April 1809:

> *"I devote myself entirely to the kopte [...] I want to know Egyptian as my French because on this language will be based my great work on the Egyptian papyri. [...] I only do kopte, Egyptian. [...] I did:*
> *1) a saidic Theban grammar (the only one that exists).*
> *2) a memphitic.*
> *3) the concordance of the two dialects.*
> *4) I transcribed the Arabic grammar saïdique of a Coptic manuscript.*
> *5) I cofoot the texts.*
> *6) I made the letter "A" of a dictionary saïdique.*
> *7) I have completed seven letters from a root memphitic dictionary."*[13]

After studying the Bible for a long time, he criticized:

> *"The ignorant and superstitious Jews who want to maintain that the good father Abraham and the Hebrew priests have learned the four rules of arithmetic and all the arts to the Egyptians".*

He continues to work on trying to decipher the demotic text of the Rosetta Stone. But the task is not easy and it stumbles on many obstacles. His letter to his brother of June 17, 1809 is a good example:

> *"The attempt on the Egyptian text did not produce any results ... The proper names that I read as Äkerblad (although different on how to ungroup the simple letters) are not in exact agreement with the Greek text ... Thus the march that you have indicated to me it is not practicable, since it is based on the complete conformity of the Greek and Egyptian texts ..."*

From 1821, he deciphered the first royal cartouches, including that of Ptolemy V on the Rosetta stone, then that of Cleopatra on the basis of an obelisk and a bilingual papyrus. A friend, the architect Jean-Nicolas Huyot, having sent him reproductions of details from the temples of Abu Simbel which had just been discovered, Champollion spotted there in a cartouche the solar sign of Ra (Re), another sign that he knew to be M and two S: RâMSS, so Ramses, which at the same time means "Re gave birth to him". Ditto for ThôtMS, Thoutmôsis: September 14, 1822, he can also read the Egyptian names, exclaim "I hold my business" and then according to family legend (hagioscript son of Jacques-Joseph Champollion, Aime-Louis (1812- 1894)) fall into a coma a few days. This decipherment sign the birth of a new science, Egyptology. On September 27, 1822, he wrote the letter to Mr. Dacier on the alphabet of phonetic hieroglyphs in which he shared his discovery of a system of deciphering hieroglyphs:

> *"It is a complex system, a writing that is at once figurative, symbolic and phonetic, in the same text, the same phrase, I would say almost in one word."*

It will take another two years for Champollion to publish his Precis of the hieroglyphic system of the ancient Egyptians and open the doors of scientific Egyptology. His discoveries, however, provoke controversy and criticism from his contemporaries, including his former master Silvestre de Sacy, for whom Hieroglyphica Horapollon was the bible in the field and discourages him by communicating information to his competing colleague Thomas Young . He also gets many supporters, such as Wilhelm von Humboldt or relatives of the king (Duke of Blacas, viscount of La Rochefoucauld). In 1822, he wrote and published in his Letter to the editor of the Revue encyclopdique, relating to the zodiac of Dendrah, where without going forward to give a date, he contradicts that of Jean-Baptiste Biot, namely the year -716. But Biot is not the only one to advance such dates.

---

12. Jean LACOUTURE, *Champollion, Une vie de lumières*, Grasset, 1988, p. 126
13.Jean LACOUTURE, Champollion, Une vie de lumières, Grasset, 1988, p. 51.

So who deciphered the sacred characters first? This question will remain an enigma for many but for us Africans, we think that there was an avant-garde approach certainly but which is far from the hermeticism of these characters which are accessible only to those for whom they in are not an object of study but a way of life.
If the Coptic is a pharaonic synchrony read and written since by Africans how can we talk about the deciphering of hieroglyphs by ignoring Africans who have kept the practice on a daily basis?

END OF PLAYTIME?

Why do since 2005, the technical files communicated by the Egyptian Antiquities Ministry to Western specialists on dating made concerning the civilization of ***km.t*** are subject to the code of silence?

Was it therefore necessary that on February 21, 2018, Dr Khaled Saad, Director General of the Egyptian Antiquities Ministry, received by the Egyptian television media, indulged inincredible revelations about new dating concerning the Egyptian civilization?

The civilization of ***km.t*** is not only 7,000 years old, as he says:

> *"[...] We have in Egypt archaeological sites dating back to -150 000 years, and even other archaeological evidence that goes back to -500 000 years ago!".*
>
> *(Khaled SAAD, 2018)*

So it's a whole page in the history of Egypt that is being rewritten. Dr. Khaled Saad, who has represented Egypt in more than 9 international congresses and led many research teams in Egypt, reported dating and works not only on the discovery of sites of an ancient civilization dated to the very least. to -15 000 years, but also on traces of lives carrying them over -150.000 years to -500.000 years. 86 state-of-the-art circular houses were discovered in the Nawamis region of Ain Khundra in south Sinai: houses built 1500 years before the first known pyramid using the same techniques as the absence of cement or binding materials. In addition we find the same astronomical technology, coloring processes used in Egypt, the same technologies for the manufacture of materials and tools.

North of Lake Haroun, Dr. Saad reportedly discovered flint bow arrowheads. To the question of why did you make double-headed arrows if Egyptians did not have advanced medical knowledge to conclude that this type of arrow can kill faster than a single-headed arrow?

Concerning astronomy, Dr. Khaled Saad explains clearly that in the necropolis of **Mererouka** they discovered writings that explain the functioning of time over an annual period.
Dr. Saad compared to the archaeological site of **Göbekli Tepe** reported during his speech the oldest prehistoric clock dated -12000 years discovered in **Nabta Playa** and visible at the Historical Museum of **Aswan.**

Dr. Saad also reported on discoveries in prehistoric computer skills and digital literacy.

Surprisingly Dr. Saad asserts that to his knowledge the greatest architect in the world is an Egyptian: ***Hm iwnw*** architect under the reign of Fara *Khufu*. The only one so far able to build a pyramidal building of the size of Giza with a minor error of 1%. Which in itself is a real technological feat.

Why all those information are kept secret and out of knowledge of the wide public?

Aren't the dates advanced by Dr. Khaled Saad more consistent if one relies on the revision and the chronological division that he and his team propose? Does not this new chronological breakdown definitely lift the veil on the chronological paradoxes we have rigorously pointed out here?

Why do some people impose to others the study of erroneous information? Why are they deliberately corrupting human history? Who of Arabs or Westerners first discovered sacred scriptures?

2.3. AND AFRICANS IN ALL THIS?

Over nearly 14 centuries, the hieroglyphs have been more interesting to Arabs, Westerners, Americans, Italians, English, French, Czechs, Dutch, Russians, Chinese, etc. more than they seem to have interested Africans who should be the first to be concerned.
Why such a distance from a past that alone can allow Africa to get out of its silence?

The declines of the great African empires around the 11th and 14th century, the slave raids from the 7th to the beginning of the 19th century, colonization (19th century - 20th century) and neo-colonialism (since the independence of 1960) are for many in this distance put by Africans with respect to their own history.
Africans are fighting more for survival than they are spending energy reconnecting with their past.
You must have read Fernand Braudel [14]to take stock of the current situation in Africa. Notably on the evolution of Africa since the 10th century period of the great empires until 21s of so-called independent states. Braudel brilliantly describes the functioning of the economy against the backdrop of a three-tier economic triptych: the extreme floor, the subsistence level, the local market and the last of the world economy. If the slave raids had the effect of beheading the upper level of the African world economy, the colonization with his lot was to folklorize the middle floor. The survival of Africa is therefore due to one thing only, the hypertrophy of the primary stratum by increased solidarity.

The African renaissance, as Diop emphasized, can only be achieved by drawing resources from the lower stratum that has preserved African values. But this stratum is undermined by the local ferment folklorized by the colonial entrism and by the poison of Jules Ferry's school which is mandatory. Gradually Africa is dying and becomes an exuvia of civilization.

Cheick Anta Diop clearly states this in his book *Alerte sous les tropiques* when speaking about the deplorable mental confusion that strikes Africans from the stratum folklorized by colonization towards those of the lower stratum.

> *"The opinion considering that European languages are more spoken in Africa than native languages is false. What is happening is that European languages are spoken in large centers by a handful of intellectuals, which gives an appearance of extension, while the whole mass of the interior retains its maternal expression, and that is out of ease, laziness, lack of will, lack of decision-making, morbid taste of intellectual and moral subjugation, which we seek to satisfy ourselves with European languages and not by a practical calculation. It is often argued that the mass is uneducated and can not read: it is easier to teach him an alphabet of a foreign language."*[15]
>
> *(C. A. Diop,1990)*

Since the 1960 independence, the *africanoclast* pressure is less important and in fact many Africans have taken over the question of their past and the reading of hieroglyphs under the leadership of the first Panafricans among the most illustrious Professor Anta Diop and Obenga. The results up until now seem timid but encouraging.

According to African researchers, it turns out that they face incomprehensible blockades. In particular, as Professor Bilolo points out in his excellent book *Vers un dictionnaire Cikam-Copte-Luba.*[16] I quote him:

---

14. Fernand BRAUDEL, Civilisation matérielle, économie et capitalisme: XVe-XVIIIe siècle, Paris, Armand Colin, 1979
15.Cheikh ANTA DIOP, Alerte sous les tropiques, Présence africaine, Paris, 1990, p.36§37
16.Mubabinge BILOLO, *Vers un dictionnaire cikam-copte-luba: Bantuïté du vocabulaire égyptien-copte dans les essais de Homburger et d'Obenga*, 2011, Afrobook, Munich-Freising-Kinshasa, p.46.

"The ancient grammars of the Egyptian language had left a research project indexed or censored at this time. It is about the highlighting of the "African linguistic substratum" of ci-Kam. This perspective of research, evoked in the "introductions" to the grammar ci-Kam, has been stifled for more than 100 years. The Semitization thesis, without a statistically representative basis, stifled the thesis of ciKam's kinship with the languages of Central Africa, defended by Mr. N. Reich, at the end of the 19th century. We have been looking for this thesis for more than 30 years, but to no avail. We will come back to this when discussing the position of Abbée Amélineau (1916)."

If what Professor Bilolo said is true, we must admit that African attempts are not facing only with violent censorship but also fierce opposition on different levels. A constant that Professor Anta Diop had already noted well before Professor Bilolo, in a slightly different introductory statement:

"We close our eyes, we do not see the obvious. They rely on your complex, on your alienation, on the conditioning, reflexes of subordination and so many factors of this kind. And if we do not know how to emancipate ourselves from such a situation on our own, but there is no salvation ... we are leading and fighting the most violent struggle against us. More violent than the one that led to the disappearance of certain species. And it is precisely that your intellectual sagacity goes so far"

(Anta Diop confrence de Niamey, 1984).

Is it curiously about this war that Anta Diop was talking?

"My decision to destroy the authority of blacks people in Santo Domingo (Haiti) is not so much for considerations of trade and money, but rather because of the need to block forever the march of blacks in the world (...)"

(NAPOLEON BONAPARTE)

A text by Serge Carfantan on political cynicism inspired by the work of the American Aldous Huxley speaks for itself. I suggest you to explore it a little bit.

"To stifle any revolt in advance, do not do it in a violent way. Methods like those of Hitler are outdated. It is enough to create such a powerful collective conditioning that the very idea of revolt will not even come to the minds of men.

The ideal would be to format individuals from birth by limiting their innate biological abilities. Then, we would continue conditioning by drastically reducing education, to bring it back to a form of professional integration. An uneducated individual has only a limited horizon of thought, and the more his thought is limited to mediocre preoccupations, the less he can revolt. It must be ensured that access to knowledge becomes increasingly difficult and elitist. That the gap between the people and science is widening, that the information intended for the general public is anesthetized by any subversive content.

Above all, no philosophy. Here again, we must use persuasion and not direct violence: we will massively broadcast, via television, entertainment always flattering the emotional or the instinctive. We will occupy the minds with what is futile and playful. It is good, in a chat and incessant music, to prevent the mind from thinking. We will put sexuality at the forefront of human interests. As a social tranquillizer, there is nothing better.

In general, one will make sure to banish the seriousness of existence, to deride all that has a high value, to maintain a constant apology for lightness; so that the euphoria of advertising becomes the standard of human happiness and the model of freedom. Conditioning will thus produce of itself such an integration, that the only fear that will have to be maintained will be that of being excluded from the system and thus no longer being able to access the conditions necessary for happiness.

The inferior man, thus produced, must be treated as he is: a calf, and he must be watched as a flock must be. Anything that makes his lucidity sleep is good socially, which would threaten to awaken him must be ridiculed, smothered, fought. Any doctrine involving the system must first be designated as subversive and terrorist and those who support it must then be treated as such. It is observed, however, that it is very easy to bribe a subversive individual: it suffices to offer him money and power."[17]

---

17. This text was written by Serge Carfantan as part of a course on political cynicism, in which he is inspired by the works of Aldous Huxley, the Brave New World, and Gunther Anders, the Obsolescence of the man. Serge Carfantan is an associate professor of philosophy, a specialist in Indian philosophy and the pedagogy of philosophy. He taught in Bordeaux, Libourne, Parentis, Mont de Mar-san, Pau and the University of Bayonne.

(*Aldous HUXLEY, 1939*)

Figure 7: Stone *Statue of Jean-François Champollion*[18]

How could we understand the attitude of the human rights country where has been implanted a stone statue on the passaway of the great college de France. A statue of the supposed father of the Egyptology representing Champollion crushing the head of fara Râ-Mssw II?

Despite the heavy handicaps of Africa and research tools that can not be found on the part of their peers, some Africans and Afro-descendants such as: Aboubacry Moussa Lam, Adama Dramé, Babacar Sall, Bilolo Mubabige, Dr. Booker T., Coleman, Coovi Gomez, Pathé Diagne, Dou Kaya, JC Mboli, Gilbert Ngom, Harold Johnson, James Small, Kaba Kamene, Martin Delany, Mbombog Mbog Bassong, Mbokh D., Mouhamadou Nissire Sarr, Pierre Nillon, Omotundé K., Oum Ndigi, Dr. Gabriel Oyibo, Ousseynou Ndiaye, Rashidi Runoko, Sal, Oscar Pfouma , Yosef Ben-Jochannan etc. have made notable progress on the basis of the current rules of classical Egyptology.

Even though this is laudable in some respects, it stills insufficient from the African point of view because the holy characters are written in an African language. It would be a good idea to be able to read these characters in any African language since all the African linguistic variants are derived from the Negro-Egyptian as well as the Pharaonic. The majority of these African scholars are subject to the diktat of the university institution and Egyptology. It becomes very difficult to get out of this conceptual matrix. It is found that by the standard method, these characters are addressed in other languages: Chinese, Japanese, German, English, French, Russian, Swedish, etc. It is therefore an anomaly not to be able to read this writing directly in an African language but only by following the barbarism invented by Egyptology to express themselves in their own language.

Amadou Hampat Bâ raises a pertinent remark by this questioning:

> *"Is it not an element of cultural anthropology (the initiatory text of the Peul pastoralists) that allowed some prehistorians to correctly interpret the enigmas of the Tassili frescoes: animals without legs of the table called the Beef to*

18.Statue of Jean-François Champollion at Collège de france, 11 Place Marcelin Berthelot, Paris 75005

> *the hydra, the magic of Ouan Derbaouen, etc. Thus, beyond 10,000 years of parenthesis, the rites of today make it possible to identify the five mythical sisters of the seven sons of the ancestor Kikala, in the five marvelous dancers of the frescoes of Jabbaren."*[19]
>
> *(Hampaté BA and Germaine DIETERLEN, 1961.)*

It seems that the willingness of others is not enough to study and decipher African knowledge. It is necessary to be associated with voluntarism of the others, an endogenous approach even if it is ethno-scientific, in order to allow a more rigorous approach of the ancient African knowledge.

2.4. WAS IT POSSIBLE TO DO ANOTHER WAY?

If we can praise the work of research done on the decoding of the Egyptian pharaonic by Arabs or Westerners, it was enamelled of methodological choices and therefore many conventions that are from the African point of view a little questionable over time.

For lack of better, these choices were made by default to allow a reading and an understanding of a language considered dead because supposedly more spoken and little written.

In his book "*Towards a Cikam-Coptic luba Dictionary*" Professor Bilolo posits as a paradox a statement based on the fact that a language could no longer be written, does that prevent it from being spoken? If the Nile Valley still had more than 12 million inhabitants in Year 0, the language has not disappeared by magic. This remark makes it possible to have doubts about the mobiles and choices made by some to begin the linguistic study of Pharaonic Egyptian.

The new onomastic issue of the Berlin Congress of 1884/1885. The map of Africa and the evaluation of its area without counting the tendentious division concerning the history of peoples located below the Sahara to isolate them from the continuum historical with the Nile Valley. By inventing supposed technical data like cataract locks where the texts themselves speak of a country composed of three entities ***tA sty***, ***tA mry*** et ***tA nTr***. And of course to defend positions tinged with ideologies, the positioning of the country of ***Pwn.t*** (actual Ouganda) has been the object of a hardly believable crafts.

These doubts are confirmed by the technical analyzes of many African writing systems (cf illustration N°1) by specialists from both Africa and others who have demonstrated close links with ***mDw nTr*** and many African writings such as (Vai, Kpelle, Shu-mom, Mende, etc.). It was therefore possible to carry out real research in situ with the local populations and written documents to do a real job of decrypting the pharaonic Egyptian if the question of decryption was part of a process aimed at a universal goal. It was also possible to rely on the Arab, Greek and Roman heritage to do this job with the utmost seriousness.

If some documents can not be found, especially the papyri of the Egyptian libraries, if some works can not be found by the general public, especially those of Abbée Amélineau, others are brushed aside, particularly the works of MN Reich, at the end of the 19th century concluding that Pharaonic and so-called Bantu languages are genetically linked. So how and why do we come to these blockades?

One is entitled to ask questions about the real motives of this disappearance and of these oriented choices. One might wonder why Egyptology is pushing to promote what Professor Bilolo calls Egyptian Esperanto instead of promoting access to direct knowledge by Africans?

19.Hampaté BA and Germaine DIETERLEN, 1961.

Serious researchers like Simon Battestini or even Anne Zali and Annie Berthier after a remarkable job. The last two have established a sort of cartography of African writings and here is what they say:

> *"The earliest of these writings is the Vai script, which appears on the borders of Liberia and Sierra Leone in 1833. Its two hundred and twelve characters are revealed in a dream by a "white man" and transmitted in the form of a "book". to its inventor, Momulu Duwalu Bukele, who read the Latin alphabet. The inventor of Mende writing, Kisima Kamara, a Muslim tailor of Mandika origin, like the Arabic alphabet and vai. The hundred and ninety-five characters of the mende are inspired by him in 1921. At the same time in Liberia, Dr. Thomas Flo Lewis, of Bassa origin, also draws on the syllabic Vai to present a system of thirty-five signs , able to transcribe the tones. In Liberia in 193, Wido dreams that God is with him and hesitates to give him the writing, lest men move away from tradition and neglect the secret rites of initiation. God agrees to give the writing to Wido provided he does not reveal it to women. The one hundred and eighty-five signs of the loma alphabet appear to him the next day. In the same year, Gblili, a traditional hanoye chief, exposes the eighty-eight signs of Kpelle's writing to seven years of incurable illness and regains power.*
> *In Cameroon, King Njoya lives in contact with Hausa and Fulani literate in Arabic. From 1903, he invents an alphabet of eighty signs. The script practice is inspired by a premonitory dream. The Bagam dialect island (tsop) belonging to the Bamileke ensemble, which is inspired by this alphabet to establish, around 1915, an alphabet that is clean"*[20]
>
> *(Anne ZALI- Annie BERTHIER, 1997)*

After this advocacy of authorized specialists, among the few African languages spoken and written that have similarities with hieroglyphs we have to mention only a few that have alphabets and systems of various signs:

- **Shü-mom or Bamoun scripts** *(cf illustration n°2)*

The seventeenth ruler of the Bamoun kingdom between 1892 -1896 and 1933, has gone down in history for saying they created a writing ex nihilo: the famous bamun alphabet called *Shü-mom*.

The conception of a writing probably occurs at this period (1895-1902). The beginnings are laborious, and several tests necessary to the creation of Lewa, version of nearly 800 signs. Essentially ideographs and pictographs, borrowed from the traditional iconographs register and the motifs of precious textiles called "Ndop A second version of about 450 signs is developed a few months later. Five more will follow, in the direction of a growing reduction in the number of signs and a simplification of articulation rules.[21] Alexandra Loumpet-Galitzine (1917-1937).

This alphabet was initially pictograph. Officially, some people, close to the sovereign, had to draw a figure or an object and give it a name. The signs were drawn on boards with charcoal or juice from a liana. In the test of its implementation, it will be modified six times as it is used and will eventually reach 80 signs that will be further simplified later, the latest version dating from 1918, Cheikh Anta Diop has identified 23 signs[22] common between Egyptian scripts and Bamoun writing. The Bamoun being descendants of Egyptians, it is thus the writing of their ancestors which was perpetuated from the valley of the Nile to the banks of the river Noun with an adaptation by Ndjoya after many tests.
Dozens of texts and books will be written later with this alphabet and the royal administration will use it until the early 1930s.
The sultan wrote a book of traditional medicine as well as a History and custom of Bamouns. The artists, calliscripters who collaborated in the creation of this alphabet are the same ones who will excel in the field of drawing art. Examples include Nji Mama, Nji Fransawaya and Ibrahim Njoya etc.
Either strategically or in a completely stupid way, the Bamoun script is said to be borrowed from Arabic writing. All this can make ussmile but the amaizing thing in Africa is the fact that no one is longer surprised by the intellectual filth that some people produce and also propagate.

---

20. ZALIA andA. BERTHIER, L'aventure des écritures, naissance. Bibliothèque nationale de France, 1997.
21.Alexandra LOUMPET-GALITZINE, Njoya et le royaume bamoun: les archives de la Société des missions évangliques de Paris, 1917-1937, Karthala, paris, p.16
22.Anta Diop,Negro nations and culture, Paris, Présence Africaine, P.345

• Mendé's writing, Sierra Lone

Language spoken in Sierra Leone, Liberia and Guinea Conakry. It has a writing composed of 212 signs [23] according to Professor Westermann, p. 443. Here again, according to the official story, she would have revealed herself in a dream to a Manding tailor named Kisimi Kamara in 1921. Theophile Obenga and Cheikh Anta Diop have updated 15 common signs with the ancient Egyptian. Kisimi Kamara did not invent, the Mende writing. The strong similarities that this writing presents with the hieroglyphic characters necessarily make it descend from the Egyptian writing. In truth, it has been preserved in the lap of Sierra Leone following the migrations that followed the Arab invasions from the Nile Valley.

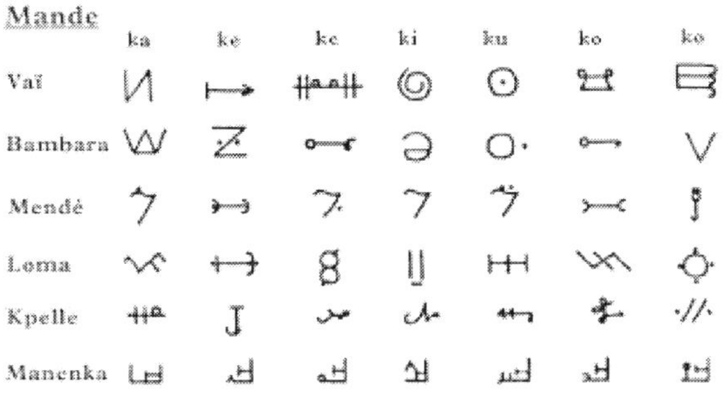

Figure 8: Common characters between mende and egyptian

• Bete's script, Ivory-coast

The Bete belong to the Krou cultural area which extends to the center-west, south-west of Ivory Coast and southern Liberia. The term krou refers to a linguistic division established on both sides of the Sassandra River. We distinguish oriental krou (Bete, Dida, Godie and Neyo) and western krou (bakw, kroumen, neyo, wan).

In 1958, Theodore Monod published for the first time an unknown Alphabet of the general public. A few years later, the script work of the same author will be revealed on the international art scene when André Magnin presents this time some of his drawings at the exhibition Magicians of the Earth designed by Jean-Hubert Martin at the Center Pompidou and at the Greate Halle de la Villette in 1989 at Paris.

You guessed it, it is about the thinker, philosopher, researcher, pedagogue, prophet, poet, artist, creator, encyclopedist Gbeuly Gboagbré [24].
He introduced Africans and to the entiry world, the bete writing. He will have in 1948 alone resurrected 401 pictograms for his language to the point of using them to write tales. A pictograph writing that relies on figures.

Further study can easily demonstrate the presence of common signs with other African writings. If Gbeuly Gboagbré has highlighted this writing, it's a safe bet that he may not have been the original inventor. He was certainly an insider of the myriad of secret societies of Africa. He has during his initiation teached the basics of the writing of his ancestors.

23.Gordon Innes. A Mende-English dictionary. Cambridge University Press, London, 1969, 154 p.

24. Frédéric Bruly Bouabré (deformation of: Gbeuly Gboagbré), born in 1919 in Zépréguh, circle of Daloa, Ivory Coast. Primary studies in Daloa, Dimbroko and Bingerville (1931-1939). Recruited for the service of the national navy in A.O.F., he landed in Senegal in 1941 and served, until the end of the war, as a sailor in Dakar. Railway-clerk at the Rufisque (1945-1946), then back in Dakar, attached to the judicial identity until the end of his stay in Senegal in 1957. Two important episodes mark this moment of his life: the "Divine vision" of March 11, 1948, following which he created a new religion: the Order of Persecuted; and the invention of African writing. Repatriated to Côte d'Ivoire in 1956, he was assigned to the Ministry of the Interior in Abidjan, then, from 1958 to 1973, to the direction of scientific research (he "underserved" on the Ivorian branch of the IFAN), from 1973 to 1982 when he retired to the Institute of Ethnocociology at the University of Abidjan. During these last assignments, he wrote an abundant scientific work. In 1989, finally, he is recognized as a writer (edition of his autobiography, We do not count the stars) and as an artist (participation in the great exhibition The magicians of the earth, in Paris).

Figure 9: Bete script

• **Ethiopians Amharic and Geez** [25] *(cf illustration n°3)*

These are probably the only writings in Africa that are still very much used and alive to the present day. The Geza or Geez, sometimes called ancient or classical Ethiopian, is an ancient language found in the classification of Greenberg in the family of Chamito-Semitic languages spoken up to the fourth century in the Horn of Africa and originally from regions of southern Eritrea and northern Ethiopia. Gueze was the official language of the kingdom of Aksum and then of the empire of Ethiopia. According to ancient analyzes, it would be for many languages of the region (including Amharic and Tigrigna) what Latin is to French, but it seems today that it is in fact sister languages that have evolved in parallel.

Today, the Geza is the main language used in the liturgy of the Ethiopian Orthodox Church and the Eritrean Orthodox Church. However, in Ethiopia, Amharic (the main lingua franca of modern Ethiopia) and Eritrea and the Ti gray region of Ethiopia, tigrinya and other languages can be used for sermons. The Geza[26] was existing before the beginning of the Christian era. Ethiopia owes the preservation of its heritage to Emperor Menelik II, a man who drove Italian settlers out of his territory. Today the Geza is absolutely everywhere in the country of Haile Selassie.

• **The Gicandi script, Kenya**[27] *(cf illustration n°26)*

The Gicandi is an ancient Gikuyu composition of enigmatic poetry presented to the public by two poets in a dialogue of back-and-forth battle of intelligence. Composed of more than 150 stanzas, the singer of the Gicandi had to submit a precise preparation and learn by heart the high number of stanzas containing so many enigmas: this explains why there were very few Gicandi singers even in the past and why they have practically disappeared. These initiated singers have appeared in different markets passing from one village to another.

The presentation of the dialogic challenge was made formally by "spreading" Kwara Gicandi in an arena in front of an audience. One of the candidates would first propose an enigma, and the other would explain it and propose the next one in turn. The competition would continue until one of the two failed

25.MUNRO-Hay STUART, Aksum: An African Civilization of Late Antiquity, Edinburgh, University Press, 1991 (ISBN 978-0-7486-0106-6), p. 57

26.GIRMA AWGICHEW DEMEKE, The Origin of Amharic, Addis Abeba, Centre français des études éthiopiennes, études éthiopiennes n° 5, 2009

27.KIMANI NJOGU,Cultural Production and Change in Kenya: Building Bridges, Twaweza Communications Kimani Njogu -G. Oluoch-Olunya, 2007, p.107

to give the performance and thus lost the game. The losing party handed the musical instrument to the winner.
It was first reported in 1921. It is the work of the Kikuyu people, who constitute the largest population in Kenya. Three common signs have been identified between the Gicandi and the ancient Egyptian.
The drawings on the Gicandi are as enigmatic as the words and constitute a mnemonic-pictorial system which is the only type existing among the Bantu tribes and reminiscent of the hieroglyphic writing system of ancient Egypt. This type of memory device uses a pictorial symbolism that proceeds through simplified images, tracing only a part of an object or a conventional image. A small number of images are enough to record a happening, to suggest to a medical man the formula of magical practices and to a singer the object and verses of his song. Kimani Njogu, a modern student of Gicandi, writes that "the interpreter considers the written text as an integral part of his performance and thus constantly refers to the pictograms of the poem. In most Gicandi the interpretations, the inscribed text and the Gicandi guard itself (with the seeds therein) and the composition of the poet merge indistinctly.

• **Nsidibi Script, Nigeria-Cameroon** *(ref. illustration N°41 et 41bis)*

About this writing, the researcher P.A. Talbot tells us:

> *"The nsidibi script is certainly of considerable antiquity, and is to a large extent, pictograph"*
> *(Maes DIOP, 172)*

Battestini's work on writing, among which nsidibi, semiology and scripts related to African characters, is undoubtedly the most successful to date. I quote him in his concern to replace Africa in its place far from the clichs:

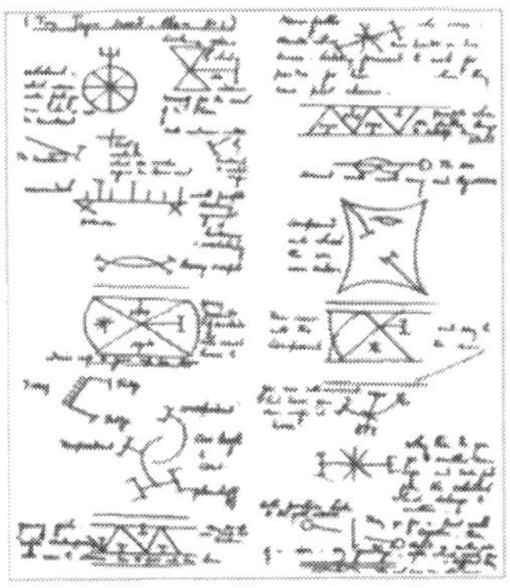

> *"The unquestionable importance of orality in Africa has masked the less obvious and rarer presence of the inscription of meaning. The tens of thousands of wall paintings and engravings, the first of which are 50,000 years old, the landscapes of various symbols, the millions of so-called "desert" manuscripts, endogenous writings, and modern postcolonial orthoscripts, bear witness to this. The imperialist ideology found its account. The old clichs of Africanist discourse such as "primitive" "without writing", "without history", "without literature", "of oral tradition" "tend to disappear". They are still widely used in journalistic texts and popular discourse, and with caution in dictionaries and general works."*[28]
> *(BATTESTINI, 2006)*

The dozen mutual signs between the Nsidibi and the ancient Egyptian also pushes to seek its origin in the valley of the Nile. The comparative works of Professors Talbot, Darell, McGregor etc. are pretty illustrative of the topic.

• **Metal writing**

"The metallic alphabet" or more precisely "the African metal pictograph in miniature" reads on the weights to weigh the gold of the Akan peoples.

It has been brought to light by the Ivorian ethnologist Georges Niangoran Bouah. According to the latter. One of the main functions of these weights was to be used to weigh the gold powder and to determine precisely, according to a convention known and accepted by all, the weight value of the gold sold or purchased. The use of the system requires two varieties of weight: weights-proverbs (already yèbouè) and weights geometric form (dja kpya). The proverbian weights are made of brass, they embody common themes of akan philosophical thought. Regardless of their weight, these weights are remnants of an ancient system of writing and thought transmission.

Figure 10:metal writing

28. Simon BATTESTINI, De l'écrit africain à l'oral: le phénomène graphique africain, L'Harmattan, Paris, 2006

We can not pretend that Africa has an oral and not written tradition whereas here Africans actually have a system of writing on metal. This writing is the equivalent of the hieroglyphs of ancient Egypt, with the difference that among the Akan, a single figurine represented a whole message[29].

Niangoran Bouah and Bouabré Bruly worked together to decipher the metal pictograms. It is in all legitimacy that Bouabré Bruly as initiate of the African spirituality that he has studied the question to determine the meaning of the glyphs engraved on the system of weighing of the gold.

Having proposed and implemented a methodology for reading the Bété script from so-called Bekora stones, He submited also the proverbial weights to the same type of analysis and concludes that the engraved or modeled figurines constitute a "talking pictograph". where is externalized a "teaching conceived, concretized, delivered by its wisdom to human curiosity ..." (a concept close to that of Georges Niangoran Bouah, who believed that "these weights constituted a complete system of education during the pre-colonial period").

As for the weight of geometric form, it confirms their proper weight function, and highlights a very specific index: the existence of an ancient accounting. In fact, the Solution-Bruly to the weight-of-gold or African-book that he wrote in 1972[30] is no more and no less, by his own admission, than an African accounting treatise.

Several alphabets have emerged in sub-Saharan Africa: for example, five syllabaries for the only Mande cultural domain. The oldest, that of Vai writing (Liberia) dates from 1833, the other four from about 1950 (after D. Dalby, presentation of the exhibition Africa and the letter, 1986).

- **Skin writing**

In addition to papyrus, metal, animal skins, Africa has also invented a writing system in scarification shaped on human skin.

Once again Bouabré Bruly has worked extensively on the subject. It is enough for Bouabré Bruly to look around him, to look at the humanity that surrounds him: Thus, he discovered the way or how all writing, and in particular the African writing could easily appear on the human body as multiple adornments, with all their mean currency meanings, and obeying both the circumstances and the fashions in time. It can thus be said that scarification is a writing printed on the human body for the need of the expression of the favorite marks and currencies, admitted to be worn in ornaments or in a means of recognition.

He highlighted in "The Museum of the African Face" (which was the subject of two handwritten editions .This 1965 contains 38 paintings which, according to the author, do not pretend to reproduce the reality: they are "the work of my own imagination, the work tinged with a traditional myth, a localized customary myth" (p.76).
The 1976 edition, more documentary orientation, more ethnoscript, includes 86 tables. Some of these paintings illustrate One does not count the stars.

29. Niangoran Bouah Georges, L'Univers Akan des Poids à peser l'or: Vol. II - Les poids figuratifs. p. 206.
30. Bouabré Bruly, Solution Bruly aux poids-d'or ou Livre africain, 1972, p. 112-113 et 115

Fig 11:Intrepid (the human or animal prey taunting the hunter, his traps and his weapons)

Fig12: The Mourning Nightlight in Crying: "the points put in six vertical lines on the breasts are crying: symbol of great mourning or long sadness"

- **Sacred writing of art objects**

While observing the artistic creation of his country, the Jesuit priest Engelbert Mveng of Cameroon concluded:

> *"This art is both, sign and meaning. As a meaning, it is essentially wisdom ... Bamoun art possesses a very sophisticated technique: it creates its signs and symbols according to secular laws ... Art thus appears to us like a great book written by the man for himself. to recount ... Art appears to us thus as a language, and not the simple language of every day, but a sacred language ..."*[31]
>
> *(Engelbert 1963)*

So we see that, not only Africa is at the origin of writing, but this practice remained alive after the decadence of the Nile Valley civilizations. The list of writing systems above is quite telling. There are probably many others that have disappeared during the destruction of Africa in the past 500 years. It is also very interesting to see that many of these writings have their roots in the Nile Valley, which reinforces the united identity of the Blacks people. Therefore Africa has not only an oral tradition, but also a long written tradition, no offense to ideologists of all kinds [32]

31. Engelbert MVENG, L'art africain d'hier et l'Afrique d'aujourd'hui, Présence africaine, p. 36-37.
32. Et l'Afrique offrit Dieu à l'humanité, par African History-Histoire Africaine

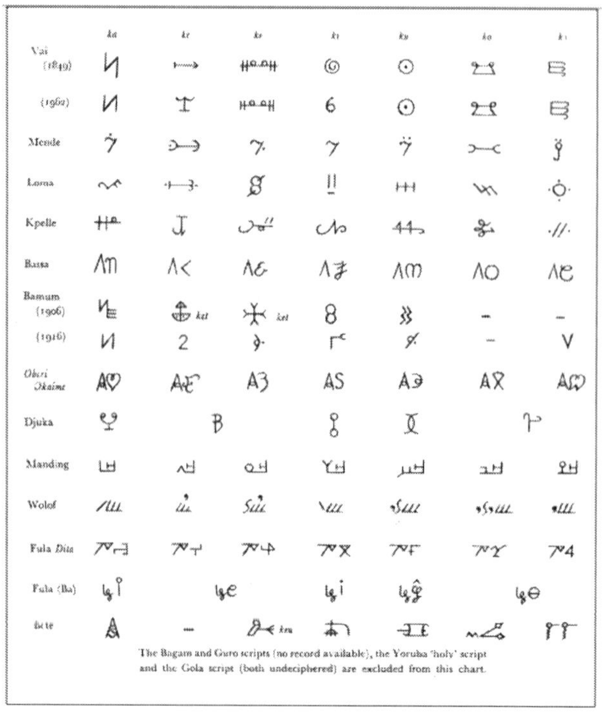

Figure 13- samples of several ancient African writings
(tir de: D. Dalby, 1970, pp.110 -111.)

Professor Pathé Diagne has also made it possible to highlight the similarities between these writings and the pharaonic Egyptian in Volume I of the general history of Africa. [33].

With all these writings which have similarities with the African sacred characters, technically it was therefore possible to approach the linguistic study of the Pharaonic Egyptian in a more rigorous way by relying on Africans and the operative studies and spoken languages with a writing that has maintained genetic links with its past in the Nile Valley.

> *"In practice, the vast majority of Black African people know The use of the ideogram, either through divination techniques, By the use of the ministers of religion, the engravers of works art, etc.."*[34]

The motives that led to dismissing these writings as relatively young in particular in the case of Shumom as highlighted by Cheikh Anta Diop in *Nations Negroes and Culture* on page 345 that these reasons are misleading and dates purely fanciful.
For Battestini, if the writing results from a faculty of the human species diversely exploited, it is multifaceted and heterogeneous, it rejects any ethnocentric, evolutionist and biblical attempt.

Battestini's speech while recognizing some relevant aspects of the Diop C.A, Obenga and Senghor refute any cultural diffusionism that makes Egypt the epicenter of the world. Would this explain the cautiousness of official researchers to take into account comparative studies on African scriptures and ***mDw nTr*** in order to enlighten the whole world?

By the way, Battestini[35] shows that African peoples invented systems of writing and others not; that is why he invents Africans to more modesty and vigilance so as not to fall into an ethnocentrism that does not benefit scientific knowledge. Africa is neither inferior nor superior, it is and exists and it must be taken with its achievements and its weaknesses.

33.Pathe Diagne, Histoire Gnrale de l'Afrique, volume 1, Chapitre 10, Paris, 1980, pages 280-286.
34.Pathe Diagne, Histoire Gnrale de l'Afrique, volume 1, Chapitre 10, Paris, 1980, pages 280
35. BATTESTINI S. Ecriture et texte, Contribution africaine. Paris: Présence africaine. Quebec: Les Presses de l'Universit Laval, 1997.

Moreover, it is found that the Coptic is written using Egyptian sacred characters, we can also see that these sacred characters were not lost by the Africans but remained in the hands of a handful of insiders.

So as an African, we can ask ourselves the question of knowing these characters that have always been mastered by true African initiates, like Boubr Bubry or Sultan Njoya etc. How can we talk about deciphering mDw nTr? This simple question is highlighting.

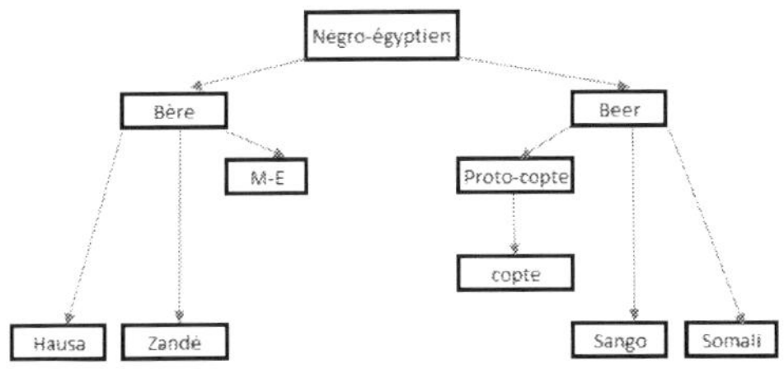

Figure5.3.7 Arbre généalogique du négro-égyptien

What may be a problem today is that Egyptology, with its conventions, has managed to convince even African researchers to follow in its path and its conventions. We will return to the incongruity of some.

2.5. COPTIC: A SKETCHY METHODOLOGICAL CHOICE.

After the work of Professor Obenga recorded in his cult book titled, *Origine commune de l'gyptien ancien, du copte et des langues ngro-africaines modernes*, the results obtained from a close comparative study made on a panel of nearly 16 languages by (JC Mboli, 2010) lead to the observation that the Middle Egyptian appears closer to current African languages such as Haussa or Zandé classified by Mboli in the Bere language group while the Coptic who was chosen as the standard for studies of the Pharaonic Egyptian belongs to another group of African languages that he names Beer.

If one were to approach the Middle-Egyptian in a linguistic study obviously, it would be more appropriate that it be the Hausa, the Zande or even the Proto-Coptic at most.

According to Mboli's ranking, the Coptic is closer to languages such as Sango, Ewondo or Somali because belonging to the language group he calls Beer.[36]

This question is needed because in: "Genetic relationship of the Egyptian pharaonic and negro-African languages" Anta Diop operates a comparative study between Walaf and Pharaonic without having especially needed the approval of the Coptic.

The African linguist Théophile Obenga in his book " *Origine commune de l'gyptien ancien du copte et des langues ngro-africaines modernes* " also agrees to bow to the use of Coptic to validate his statement.

What is the technical duty after the terms of the Cairo Symposium of 1974 which forces the study of the Coptic synchrony for any approach of the Pharaonic while serious studies have highlighted a link between the so-called "Bantu" languages and the pharaonic?

A simple methodological requirement or a desire to keep Africans away from their pharaonic past? In any case, it must be recognized that the Coptic study has rejected or even handicapped many African researchers.
This pitfall of the Coptic has led some African researchers, however serious, to achieve timid results in their research. No one will be quoted so as not to offend the egos.

But serious researchers who have tried comparative studies, particularly with regard to Douala or Basa'a languages, have at times glimpsed doubts about the validity of translations in the name of possible differences between their mother tongue and the sacristal Coptic.

36. Jean Claude Mboli, ORIGINE DES LANGUES AFRICAINES - Essai d'application de la méthode comparative aux langues africaines anciennes et modernes, L'Harmattan, Paris, p.305

If African languages are genetically linked to Pharaonic, what need do we have to continue to bend to the use of Coptic to conduct a serious comparative linguistic study between an African language and Pharaonic?

The Egyptological convention which imposes the knowledge of Coptic as the only known synchrony of the Egyptian and thus the unique way of access to the understanding of the Egyptian leave to this interrogative account on several levels because the Pharaonic Egyptian is an African language in the same way as the others. Are not these watertight barriers between sub-Saharan Africa and ancient Egypt part of a desire to isolate Egypt from the Black African heritage and make it an exogenous heritage for Africa? If one is entitled to question oneself, doubt is also allowed.

The fact that Africans can not access their past, through direct knowledge, leads here to all these doubts expressed here because they are the voluntarist trial of others, the ideology of the infantilization of Africans. It is therefore time if we want to make the African renaissance, to stop the intellectual tutelage and to start already out of the trap of the folklore tended by the colonization and to take back in hand our History and our past without waiting more than Non-appointed censors give us their approval of the accuracy or otherwise of our research and our remarks. If our contribution is not strong enough, we have to improve it gradually through the continuous excellence process . We will improve it with the contribution of future generations.

2.5 PROBLEM

Chinese characters (or sinograms) are the units of the logograph writing of Chinese languages. Their use extends over a wide geoscriptal area, called the "sinosphere"
The question is simple here. Since African languages have a common Negro-Egyptian ancestor or Paleo-African or African-common according to the authors, since Africans have a writing that is mutual to them because disseminated within different African scripts Why Africans can not have a common writing to all Africa and that everyone can read and vocalize in its linguistic variant just as the Chinese do with sinograms?

Why from the African point of view, is it necessary to be compelled to vocalize the Pharaonic according to the conventions of Egyptology? What is the urgent need for Africans to vocalize a language that has evolved to become the Coptic today? What is the purpose for Africans to indulge in Egyptology instead of looking into the possibility of reading the holy characters simply in the different current language synchronies? How to understand that Russians, Chinese, Germans, French, English etc. read the ***mDw nTr*** in their language and how does this becomes an incongruity when talking about an African approach?

The problematic of access to information recorded through sacred African scriptures is at several levels.
We think that there is a world between reading an information and understanding perfectly what one has read according to the levels of understanding and the languages used.

We believe that the African sacred characters are not neutral. They are part of a paradigm and involded an ontological vision of the world which alone is really understandable by a person who is impregnated with African philosophical thought.

We therefore believe that the proactive approach of others is highly commendable but biased to this effect.
Biased at several levels including those we have just presented but also on the level of conventions that founds science called Egyptology.

2.5.1 Methodological limits of Egyptology

The various scientific advances owe much to the research in Egyptology.

Thanks to Egyptology and its methods, the whole world has benefited from many discoveries related to research since the Chaldean ibn Wahshiyya. It is therefore difficult to criticize Egyptology and its methods. This for several reasons. The object of this work is not to try to vocalize Pharaonic in its synchronic version of the middle-empire. My approach is first of all for speakers of African languages and especially those who have a good command or want to strengthen their level of practice of African languages. I do not say that this method is not suitable for other languages. I do believe that worldwide we are usine a common lexicon but the deep understanding of those signs is more simple with an African perspective.

The purpose of this work is not to criticize the voluntarist approach of Egyptology. The purpose of this work is not to inscribe this work in a principled opposition regarding Egyptology on the fallacious pretext that Africa must give an opinion even though it is erroneous or inappropriate. I want it to be clear. The work of researchers in Egyptology is very important and, I learned a lot from their voluntarist approach. So I have no reason to biting the hand that feeds me.

When we see the pain that many Western researchers give themselves by searching the sand in the desert at the risk of their lives, it is completely absurd to want to criticize the pain they give themselves. The real research is done in the sites But Africans shine more by their absence.

It is certainly more comfortable a repeater position in an amphitheater. We are happy to relay what the others say even by dissolving any critical sense and any truth in relation to the source data that teem everywhere in Africa.

They are in the felted lounges to give lectures where they spread their impressions more than they are on the sites where the research is really taking place. If ancient Egypt is an African heritage, why Africans do not do the job?

Personally, considering the Cairo conclusions in 1974, Ido believe it is time for Africa to also look at this issue with its certainties and potentially, its mistakes but to give its position on the issue. As a general rule, I advise those who do not speak African languages to follow the teachings given in training centers of Egyptology. It is therefore proof that I have no reservations against this science but simply a great deal of admiration. I sometimes learn new words of my own language through Egyptologists researchers trough foreigners. It is a big shame for me.

However for someone who reads me and who does not know the Egyptian method, I will brush it quickly. It is based on a technique that consists of transliterating characters and then proposing a vocalization by adding vowels to skeletons of consonants obtained by transliteration. It has based on this set up grammar rules and experts have a good knowledge of pharaonic through ages and synchrony.

Operating logic

The reasoning logic used by the West as a rule, is a logic based on an axiomatic algorithm that posits an absolute in all data. We could therefore approach this thought on all-or-none principle.

African thought is relativistic. It is based on complexity, the mathematical fractal model. The reasoning logical binary appears therefore minimalist and reductive compare to African thought.

The ***mdw ntr*** being recorded in an African language, therefore are heavily imbued with the African complex thought. We therefore think that if the western method is usefull for reading, it is useless for understanding. The Egyptologist method is not adapted to understand or enlight the understanding of an esoteric content of a text.

In Africa, knowledge is a massive weapon. Therefore, ancient African society was based on an organization that took into account the division of labor, thus leading to castes. Thus true knowledge is transmitted in the secret societies of these castes. So everyone does not have access to knowledge including in a simple intended speech

It is therefore impossible for an uninitiated person to extract the quintessence of information in a seemingly harmless speech.

However, ***MDW ntr*** do not escape this endogenous logic. They codified in such a way that only a few true African initiates can really measure them.

Transliteration

Transliteration is the transition from a writing system to another system, an operation that involves writing letter by letter each scripteme of a system corresponding to a scripteme of the other system, without any concern. pronunciation.

In Egyptology we use alphabetical symbols, so that each of these symboles corresponds to a hieroglyph (or its hieratic or demotic equivalent), and vice versa.

It is important not to confuse the transliteration and the transcription (action to copy, to post a text).

Transcription is the operation which consists in substituting for each phoneme (we speak then of phonological transcription) or on each sound (phonetic transcription) of a language a scripteme or a group of scriptemes of a writing system. It therefore depends on the target language, a single phoneme that can correspond to different scriptemes according to the language considered.

One is entitled to ask questions about the arbitrary choice of these consonants on the panel of 6000 characters that counts ***mdw ntr***. What are the criteria that define the arbitrary phonetic value given to each character? Is there a protocol for defining these phonetic values?

In accepting this axiom, the ancient Egyptian possesses twenty-five consonants similar to those of other African languages. These include pharyngeal, emphatic, fricative and affricated consonants. It has three long and three short vowels, but their number has increased in late periods to nine. Some words from the Kikuki branch of the Negro-Egyptian who gave birth to the Semitic and Berber languages are shared between Egyptian and African languages. These words are easily recognized because they are transliterated as trilitera or biliteral of consonants and semi-consonants.

According to African thought a big tree doesn't get more than nine holes.

Suffixes are added to form the words. The conjugation of the verbs corresponds to the person. For example, the triconsonnant skeleton ***s-Ḏ-m*** is the semantic core of the word "hear"; his basic conjugation is ***sḏm =f*** ("he hears"). If the subject is a noun, the suffix is not added to the verbs ***ḏm ḥmt*** ("the woman hears").

Adjectives are derived from nouns through a process that Egyptologists call nisbation because of its supposed similarity to Arabic. The order of words is predicate-subject in verbal and adjective sentences, and subject-predicate in noun and adverbial sentences. The subject can be moved to the beginning of the sentence if it is long, and it is followed by a resurgent pronoun. For verbs and nouns negation is indicated by the particle n, but nn is used for adverbial and adjective sentences. The tonic accent is placed on the final syl-labe or the penultimate, which can be open (CV) or closed (CVC)

Egyptology: Reading direction

The ***mDw nTr*** or hieroglyphs are written from right to left, from left to right or from top to bottom, the usual direction being from right to left. According to Egyptology, the reader, in order to know the meaning of reading, must consider the direction in which the asymmetric hieroglyphs are turned. Transliteration tables (uniliterals, Biliterals, triliterals, etc.) must be used.•

Uniliteral: character with a single consonant serving as a support for vowel elements, speaking of a word. Examples of uniliterals: d: D; r: r; w: w; g: g

• Biliteral: A character with two consonants serving as a support for vocalic elements, speaking of a word. Examples of Biliters: ***mt***; ***šw***

• Triliterals: character with three consonants serving as support for vowel elements, speaking of a word. Examples of Triliteres: ***nTr***; ***nfr***

For example, when human figures and animals, easily spotted, look to the left, read from left to right, and vice versa.

Example 1: So this script below will become transliterated ***Imn***

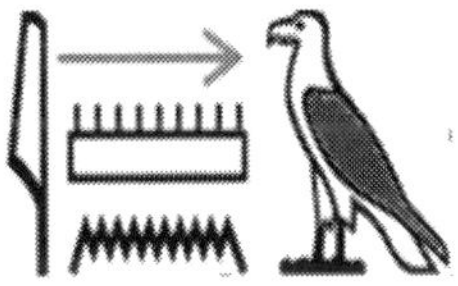

Example 2: This script will transliterate ***Min***

Typically, words are not separated by whites or punctuation. However, some characters appear mostly at the end of the word, so that it is sometimes possible to distinguish words by this means. It is obvious, however, that only a solid knowledge of the language and its syntax can cut a text into words.

Since the Cairo symposium in 1974, it is commonly accepted that ancient Egyptian is an African language.

Numerous comparative studies have been carried out with various languages, notably with Wolof by Professor Cheick Anta Diop, Mbochi with Professor Obenga, Duala with Professor Ngom, Basa'a with Professor Oum Ndigi, Ciluba with Professor Bilolo and of course with Zandé Jean-Claude Mboli etc. definitively accrediting this hypothesis.

Greenberg's comparative mass method has been heavily criticized by its fallacious side for making ancian Egyptian a semitic language and support the theses of Africanists. I will not come back to that in this writing.

The method of mass comparison, also called multilateral comparison, was developed by Joseph Greenberg in the 1960s. This method was intended to go beyond the limits of classical methods of historical linguistics, in order to reconstitute families of languages that diverged there. is more than 10,000 years old.

The rules used by the mass comparison method are therefore very different from conventional methods. First of all, it is not a question of comparing languages two by two (so-called bilateral comparison) but of comparing many languages simultaneously. For example, Greenberg searches for matches between language groups to show that they are related.

In addition, Greenberg generally has phonetic similarities that are less precise than those traditionally accepted in historical linguistics. This choice is justified by the fact that it works on longer durations, and therefore that the words could have evolved more than if one imposes a limit to 10 000 years for example.

It was this method of mass comparison that enabled Greenberg to classify all African languages into four families, then all American languages into three families.

Later on, Merritt Ruhlen will push Greenberg's method further, in particular by trying to reconstruct roots common to all families of current languages, and from a mother tongue of all current languages, which in his view would have appeared. about 50,000 years ago.

The mass comparison method is criticized by many linguists, for various reasons:

- use of infant words: see George Murdock and Roman Jakobson
- use of onomatopoeic sounds
- role of chance
- roots too short: see Jean Aitchison
- rules of resemblance too vague: see Hans Heinrich Hoc

Jean-Claude Mboli research effort, especially in his masterful book *Origine des Langues Africaines*, made it possible to draw up a new cartography of African languages spoken today and to depart definitively from the Greenberg classification.

To the thorny question of knowing since the Egyptian is an African language derived from the Negro-Egyptian as a number of current language synchrony, if there is any interest for the African 21s to vocalize the ancient Egyptian according to the Dogmas of the Egyptian Esperanto?

In my opinion there is no interest in this approach. Since all current synchronies are from the Negro-Egyptian just like the Pharaonic Egyptian is. On the other hand, since the ancient Egyptian is an African language, and the African languages are still spoken today, it is possible for any African to read the ancient Egyptian in his own language by mastering his mother tongue, some fundamentals of the linguistics and comparative linguistics in particular.
The challenge of this study is to build bridges between a random African language and demonstrate that it can be read using ***mDw nTr***. This approach, which seems to me to be appropriate, should encourage all Africans to read these characters by getting rid of the diktat of Egyptology in order to carry out an autonomous African approach.

The language taken as an example is not an exception, but it simply responds to the need for control and verification at every stage of this heuristic process.

III. Presentation of the Methodology

"MAat-maat is the correct method of investigation to know all that exists in the nature, each mystery, all the secrets."

*scribe **IaH-Mssw** who copied this copy*

For the great professor Mwené Ndzalé Obenga, *"the language has an oral tradition independent of writing*[1]"; that means, it consists of an autonomous system, having its own properties, regardless of whether it is written or not. These properties could therefore be studied for themselves, but also compared to those of other languages of the same space-time.
In general, languages evolve, even if they are unwritten; both in space and according to the migrations of their speakers, and in time because of the historicity of societal phenomena.

According to the African specialist, one should speak of a "genetic kinship" common to languages only if, on the one hand, they have inherited similarities between them; on the other hand, if the regular evolution in time and space of said similarities can be demonstrated by the method of historical linguistics. If it is commonly accepted that geographically, Nubia and Egypt are located in Africa and if we consider until proven otherwise that the sociology of Africa before any foreign entryism was sedentary type and if the Ekang population are also African descent, in this case we have a common ancestor language for Ewondo, pharaonic Egyptian and Meroitic.

I allow myself to assert without being a specialist in linguistics but only on the basis of the scientific discoveries that the anatomically modern Man was born in Africa and by the way had acquired the speech and developed varities of languages. Nowadays African languages therefore appear as ancestors of languages spoken throughout the world. We can thus find a genetic link between any African language and a language taken at random in the world. All the languages spoken in the world derive indisputably from Africa and particularly from the Negro-Egyptian. The phenomena of grammaticalization that made it possible to get current synchonies are perfectly identified by historical comparative linguistics.

It can be stated by the way that, if there is only one humanity, all the languages in the world are using the same lexicon. But the mindset of human beings depends of their sociology. The people who wrote meroitic and ancient Egyptian were imbued with African thought, daughter of sedentary type sociology.

It is therefore Darwinian theories imbibed in the raciological current that tend to imagine a polygenic source of humanity with consequently different and genetically different languages.

The aim of this work is not to demonstrate that the Ewondo language is genetically linked to the Pharaonic or to establish all points of convergence between the Ewondo and the Pharaonic. The goal is to start from this postulate since the demonstrations have already been done by the specialists on this main point.

1. Théophile OBENGA, Origine commune de l'égyptien ancien, du copte et des langues négro-africaines modernes, d. L'Harmattan, Paris, 1993, p. 14

> *"Professor Vercoutter said that for him, Egypt was African in its writing, in its culture and in its way of thinking."*[2]
> *"Professor Leclant recognized this same African character the temperament and way of thinking of the ancien Egyptians."*[3]

So knowing that Ewondo is an African language but not a martian one and therefore it is genetically linked to Pharaonic and knowing the technical elements of Ewondo, I postulate that we can go from one language to another but mainly, read the middle-egyptian from Ewondo.

I choose to rely on this language that I know well for auditing purposes. Moreover, the resulting hypothetico-deductive reasoning offers the precision that is not observed with globalizing analyzes on large quantities of languages. But the weakness of this method may be to believe that the comparison is between two languages que nenni.

In general, the comparative method appears strong when it mobilizes several languages in order to reconstruct pre-dialectal ancestors. On this account, the preliminary work done here and there on the continent on synchronies like Fang, Bulu[4], Fon[5] etc. and of course the comparative work done in the comparative studies on the languages Duala, Walaf, Basa'a, Mbochi, Fong etc.and especially the study which mobilized 16 African languages and highlighted by Jean-Claude Mboli[6] will serve as a guarantee to the methodological level to consolidate and establish the comparisons that will be proposed as part of this work.

So at first, I'm going to be content to present elements of the synchrony of the Ewondo language and secondly to mobilize the comparative tools of comparison added to the comparative linguistic to demonstrate that we can easily move from one language to the other and vice versa.

3.1. COMPARATIVE LINGUISTICS

Comparative linguistics (or comparative linguistics, historical linguistics or comparative grammar) is a field of linguistics that studies the history and languages evolution (taken individually) or families of languages.
"Historical linguistics" consists of a diachronic perspective of the study of languages in order to account for the dynamic nature of linguistic phenomena. For the purpose, this scientific discipline mobilizes the descriptive data made available by synchronic linguistic studies (lexicology, phonology, morphology, grammar, etc.), comparing them from one language to another (or from language to language). within a previously determined corpus through empirical observation of some similarities. This comparative approach aims, in his view, to test the similarities, in order to know whether they are "fortuitous", "borrowed", "convergent" or "inherited"

The relationship between two languages is qualified genetic type when the correspondences are numerous and are verified for complete systems; this is the case of the personal pronouns of the Romance languages for instance.[7]
It is a discipline that uses comparative techniques to enlighten the diachronic, especially when it comes to classifying languages.
But the synchronic aspect is also to be considered when it comes to comparing two or more languages, at a specific time, from a purely grammatical point of view. Comparative linguistics comes from

2.Le peuplement de l'Egypte ancienne et le déchiffrement de l'écriture Meroitique, acte du colloque du caire, du 28 janvier au 3 fevrier 1974, Paris, Unesco, 1978, p.87
3.Le peuplement de l'Egypte ancienne et le déchiffrement de l'écriture Meroitique, acte du colloque du caire, du 28 janvier au 3 fevrier 1974, Paris, Unesco, 1978, p.87
4. Georges L. BATES, The Handbook of Bulu, Halsey memorial presse, 1926.
5. Dominique FAÏDARO, Parlons Fon langue et culture du Bénin, l'Harmattan, Paris, 2001.
6. Jean Claude MBOLI, ORIGINE DES LANGUES AFRICAINES - Essai d'application de la méthode comparative aux langues africaines anciennes et modernes, L'Harmattan, Paris.
7. Y.PERROT, Linguistique générale, p.80.

philology, a term that sometimes has to be understood as a synonym even though the two disciplines are different.

The synchrony of the current Ewondo language and the synchrony of the Pharaonic (Middle Egyptian) having a common ancestor, the Negro-Egyptian, so the couple of language varieties have necessarily kept a genetic link and, by the way knowing one and having the elements of the ancestor, we can easily go from one to the other.

If comparative linguistics gives excellent results on a technical level. It allows to be precise on the grammatical filed for example. It will remain limited because it will only deal with devitalised associations of consonants where it is a question of extremely sharp operating concepts as we will see later in specific examples of scripts reading. According to specialist egyptian characters are ideographic that's why dealing only with devitalised consonant is a tremendus technical error.

On a lexical level, comparative linguistics presents another weakness that it tries to compensate by reconstructing predialectal ancestors between supposedly different languages. This weakness is linked to the lack of knowledge regarding the lexicon common to "languages". The other problem leading to error is the postulate that states that different ways of speaking are necessary different languages.

Two phenomena are observed when a comparative lexical study is carried out between two given languages. Homophony is observed between certain words. Comparative linguistics is therefore effective in explaining the variations observed in the unique framework of homophony. However, another loop exists between lexicons, the one of synonymy which is poorly understood and which often leads to reconstructions of fictitious predialectal ancestors. The use of comparative linguistics is less appropriate for this particular case.
The homophony loop and the synonymy loop are nested.

Two "languages" will be easily understandable one another when the homophony and synonymy relations are very little like, Spanish, Italian and French, .
Consequently two "languages" will be considered as different languages when the synonymy relations are much more important. In this account it is difficult to understand each other because the words used, associated with the tones are not those expected by the brain. Because the mechanisms solicited by the brain during communication phases are not the habitual ones.

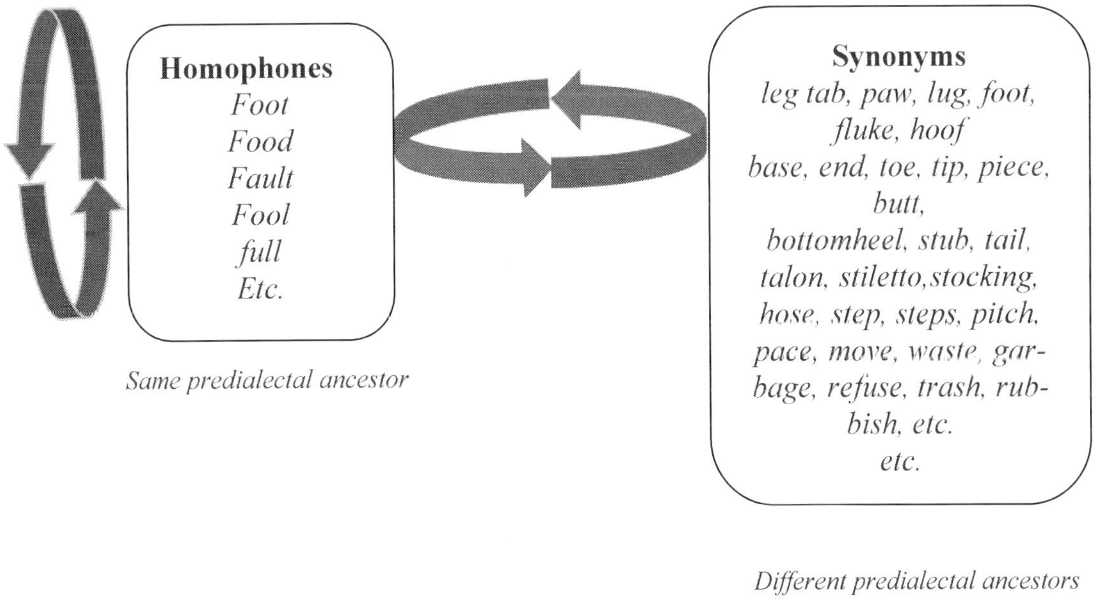

Example of semantic relationships between words within the human single lexicon worldwide

This comparative method presents a great interest by its comparative technical tools on the homophonical loop. This reconstruction is crucial insofar because it makes possible to get rid of the constraint posed as a norm by transliteration. If the reconstruction of a predialectal ancestor of a word by serving as a skeleton of consonants resulting from transliteration is in itself limiting, but the result obtained allows to identify all the possibilities of evolution related to the senses and the vocalization. In this sense, it is thus approaching an endogenous reading method built outside any normative, and exogenous framework. This method makes it possible to better understand the lexicological, morphological, syntactic, etymological, semantic and grammatical dimensions of any word.
A gross error implies that the reading of ***mDw nTr*** necessarily follows the Western languages reading logic by unique association of letters to form a word and the meaning of reading. All these errors are not detectable by the sophisticated techniques of comparative linguistics.

But this method can lead to a methodological tremendous error because a reconstruction of predialectal ancestors from the several languages of the study can be done with synonyms, antonyms, metonyms and therefore rigorously different meanings. A relatively common occurrence in variants of the African language that share a cannibalized common lexicon.

According to me, words of a same ancestors often share a radical phonetic proximity. It's a mistake to build an ancestor of two words synonymes. The question is then how comes that two words are synonymes? The answer is simple because all the words worldwide are synonyms. They are built from one an only one word. And semantic shift in meaning creating a new meaning. It is the school of thought that adapt to the shape of what is called in Africa the *mbokh* spirituality.

In my opinion, two words with phonetic proximity are semantically related according to the African prism. The ontological framework, semantic especially the complex African thought, greatly help to highlight the link of this phonetic proximity

Unquestionably, in order to write a grammar of a language or to make a comparison between the grammars of two genetically linked varieties of languages, this method is obviously interest.

Also the tools of the comparative linguistics will be mobilized to bring a light in case of need and to control the results obtained by a purely endogenous alternative method.

AFRICAN READING METHODOLOGY OF MDW NTR

In order to allow a better approach to the reading of ***mDw nTr***, I will therefore draw on the numerous studies carried out by African specialists and mainly on the results of Jean-Claude Mboli's work based on comparative linguistics to propose a close reading more adapted from the current African languages so as to allow any African eager to learn the ***mDw nTr*** in his current language to do so without any risk of vocalizing barbarism that does not represent any referent for him.

African Language Genealogy according to Mboli:

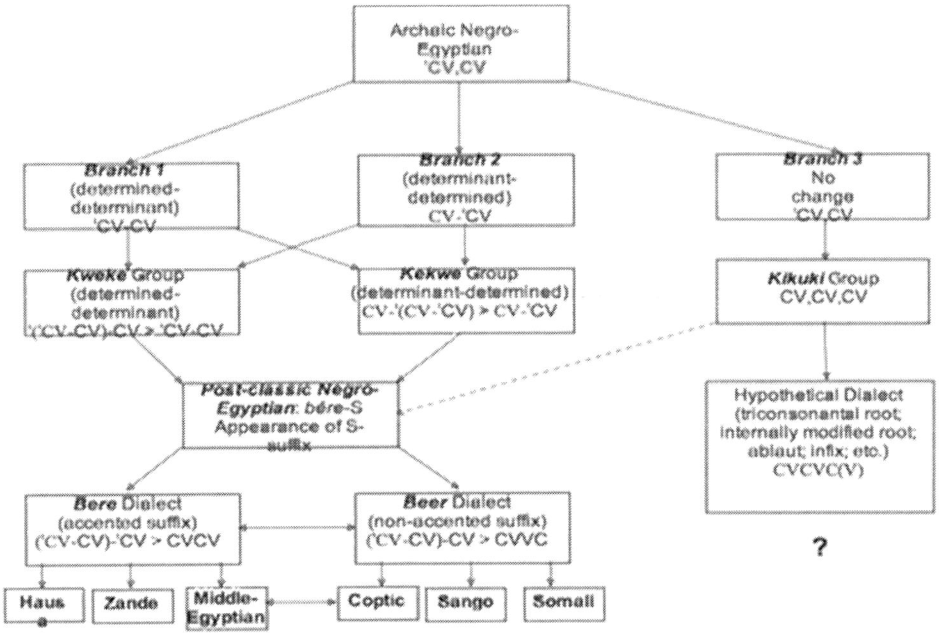

Genealogical tree of African languages [8]

We note that the Middle Egyptian is obtained mainly from a combination of two groups of current African languages (Kwéké and Kékwé) which have the particularity of operating by consonant metathesis (inversion of phonemes).

This simple remark is enough for Africans wanting to read the ancient Egyptian language to best adapt the reading methodology of the Middle Egyptian from the African point of view. We must therefore move away from Western Egyptology not by racism backward but by pure realism because the current postulate based on the issue of reading words seems a bit outdated. So for on transliteration, i can find in my language, Kekwe words, words from the Kweke branch even from the Kikuki branch. That means from negro-egyptian the same word could have 3 aspects depending of his associated branch.

I wish to make it clear that the present approach is not to judge the truthfulness and results achieved by Western Egyptology far from it. This approach is simply an alternative and endogenous path that's all. Through this new channel it allow Africans through their current languages to access directly to the information recorded by their ancestors in the sacred characters.

But first and foremost, every serious step requires understanding which links exist at the lexicological level, and so on. between the current dialects and the Middle Egyptian.

Indeed, lexicology is enlightening because it is the field of linguistics devoted to the study of words. It is interested in their nature, in their etymology but also in the systemic (especially semantic) relations that characterize them.

In this context, two sides appear. On the one hand the partly formalizable structuring of the units studied by lexical phonology, morphology and semantics. For the latter, the systematization work focuses on the syntagmatic and paradigmatic combinatorics of homonymy, polysemy and monosemy, superordination (hyperonymy and hyponymy), synonymy and antonymia..

8. MBOLI, Origine Des Langues Africaines-Essai d'application de la méthode comparative aux langues africaines anciennes et modernes, L'Harmattan, 2010, p.443.

The Egyptian method which imposes a one-way reading of the scripts is really unsuited to the current African varieties of the language because according to one belongs to the group of languages Kekwe or Kéwké one can not approach the reading of the same script in the same way. For the method imposed by classical Egyptology will end up in one language group with a correct reading and in the second group with inverted phonemes. Of course things will be complicated with the typology by reading entire sentences with grammatical structures, prefix or suffixal conjugations etc. To avoid unnecessary complications, we will only analyze a few scripts to start.

3.2. PRESENTATION OF THE STUDY LANGUAGES

3.2.1 The Kolo Language Or Ewondo

History And Men

The Kolo or Ewondo is defined as belonging to the group of African populations Ekang [9].
The Ekang people are Negro-African people that can be found today in Central Africa. They are founded in Cameroon, Congo, Gabon, Equatorial Guinea, Angola and even the Republic of Sao Tome where the 9, 5 % of a population estimated at 175,883 are Ekang and still speak this language despite the strong pressure of the Portuguese (the official language of this archipelago) and all the Creoles among which the Creole santomen or forro (used by 81, 7 % of people), the Creole angolare (spoken by 3,7 % of santomens) and monco creole (used by 2.9 % of people).

Many theses have been put forward by researchers to locate their origin. Lagneau, Cheikh Anta Diop, Theophile Obenga, to name but a few, located the origin of the Ekang in the Nile Valley and more precisely in Egypt. We will see that this is confirmed by the reading of certain scripts.

The area where the Ekang settled is from south-central Cameroon heading north-east, south-west in the upper part of Gabon; the Ogoou river, tracing in a bigs manner their southern limit. (See map N° 3).

Originally, these important migrations of Ekang would be the consequence of the push exerted by Arab Muslim conquerors, then Fulani in Cameroon, during the second half of the 13th century.

This Fulani thrust pushed back the savanna populations towards the forest zone. Most legends are located far away, in the North-East of Cameroon, the first country inhabited by Ekang.

The Language

The Ewondo or more exactly the Kɔlɔ is a language spoken in the central part and in the southern part of Cameroon. It is similar, with a few exceptions, to other languages of Cameroon such as ìtón (or Eton), nnanga, Bulu, fang (Ntumu, etc.). But in the fact it is similar to any african language. The differences observed with other languages could easily be technicaly explained.

It is the mother tongue of the Bəti bə Kɔlɔ (or Kɔlɔ Bəti) commonly called, by misuse of language, Ewondo. This last name comes from an alleged mistake made by the first settlers. More precisely, this error, which dates from 1895, is supposed to have been introduced by Georg August Zenker, then taken over by the Pallottines and later by the colonial administration[10].

9.
The term Ekang is used here is preferred to the term Bantu because it is endogenous and therefore constitutes an important key of historical research that can not be replaced by Bantu dating back to the 19th century. Peoples can lose the memory of a writing but they know how to name themselves beyond millennia.
10.M. NGESSIMO, Glimpses Of African Cultures- Echos des cultures africaines, L'Harmattan, Paris, 2011, p. 190-191

Among the Kɔlɔ Beti, Ewondo is just one of their lineages (and not a language), just like Edzoa, Bene, Emombo, Etudi, Tsinga, Enoa, Etenga, Yanda, and so on. These different lineages speak Kɔlɔ. Strictly speaking, the Ewondo language does not exist in the same way as Cameroonian, Belgian or Swiss.

Moreover, around the Ewondo, we find lineages such as: Mvog Atangana Mballa, Mvog Tsoungui Mballa, Mvog Fouda Mballa and mvog Ebanda Ndzongo. It should be noted that Mvog Atemengue, Mvog Betsi, Mvog Mbi, etc. are Mvog Tsoung Mbala families

The Bənə form 14 lineages (or families): Elnd, Mvog Manga Mombo Ndzie, Mvog Amugu Mombo Ndzi, Mvog Belinga Mombo ndzie, Mvog Ndi Mbi, Mvog Nnomo, Mvog Zambo, Mvog Man Ze, Mvog Ndzouli Fouda, Mvog Evoundou Baana, Ehondo , Otoloa'o, Eboa. They are distributed for the most part in the department of Mefou-Afamba in Mvinla in Nyong and So'o.

Mvog Abena, Mvog Ekala, Mvog Evuna and Mvog Ahanda make up the bulk of the Etudi lineage.
In the Yanda lineage, we have the Mvog Ngenda, Mvog Biako, Mvog Ebela Kunu lineages.
Similarly, among the Itón Beti, we find the lineages Essele, Mbokani, Mendoum, Menyagda, etc. The latter speak ìtón.

Kɔlɔ or Ewondo being the language that will serve as a support for this work, it is therefore useful to proceed to its synchronic study at first and afterwards only to pose technical elements of comparison with its cousins. Personally I will have left this study because it is useless to the extent that one allows oneself to consider the Kɔlɔ like a separate language distinct from the other synchronies of African language.

LOCATION (SEE MAP N ° 4)

3.2.2 The Middle- Egyptian

LOCATION (SEE MAP N ° 5)

Ancient Egypt is an ancient civilization of North-East Africa concentrated along the lower Nile River in what is now Egypt.

According to authorized sources, the civilization of ancient Egypt took shape around -3150 with the political unification of Upper Egypt to the south and Lower Egypt to the north under the reign of the first king and developed on more than three millennia. What about ***Smsw-Hr***?

Its history is peppered with a series of politically stable periods interspersed with several more troubled intermediate periods. Ancient Egypt reached its peak under the New Kingdom (1479-1425), particularly under the reign of reformist fara ***DHwty-Mss (w) nfr-xprw***.

It should be pointed out that under the New Kingdom the territory of Egypt extends outside the Delta, encompasses many current lands such as Syria / Lebanon, old Mitanni, Macedonia including Nubia etc. (see map N ° 6).

After the middle-empire the country enters a period of slow decline. The country suffered the repeated assaults of foreign powers in this late period and the reign of the pharaohs officially ended in -30 when the Roman Empire conquered Egypt to make it a province.

THE CIKAM OR PHARAONIC

The evolution of Pharaonic Egyptian over more than 4,500 years can be divided into six major periods that are identified by phonological, grammatical, lexical and orthoscript peculiarities. To date, the cikam is subdivided into six synchromes or states of language[11]:

The Egyptian says of "the first phase":

1- ***The archaic Egyptian*** (3200-2600 av. N.-E.), that means:
   a) The language of the thinite time (3200-2780 B.O.E.: 1st et II^{-nd}dynasties)

2- ***The old Egyptian*** (2600-2065 B.O.E.) that means:
   a) The language of the Old Kingdom (2780-2265: III^{-th}-VI^{-th} dynasties)
   b) The language of the First Intermediate Period (2260-2065 B.O.E: VIIth-X^{th} Dynasty)

3 – ***The Middle-egyptian or classical Egyptian*** (2065-1375 B.O.E) that means:
   a) The language of the Middle Kingdom (2065-1785 B.O.E: X^{th}-XIIth Dynasties)
   b) The language of the Second Intermediate Period (1785-1580 BC: XIII^{-th}-XVII^{-th} Dynasty)
   c) The official language of the New Kingdom until the end of the reign of ***Imn-Htp III-***rd (1580-1375 BC, 18th Dynasty)

4-***The neo-Egyptian*** (1375-750 B.O.E), that means:
   a) The popular language of the New Kingdom (1580-1085 BC NS: xv^{-th}-xxx^{-th} dynasty)
   b) The official language of the end of the New Kingdom since Imn-htp IV or Ax-n-Itn (1375 B.O.E: 9th-20th Dynasties)
   c) The language of the Third Intermediate Period (1085-750 B.O.E: XXI-XXIVth Dynasty)

5-***The demotic*** (750 B.O.E – 473 B.O.E date of Philae registration*)*, that is to say, the language of documents of everyday life, administrative acts and secular literature from the XXVth Dynasty to the Byzantine Roman period.

6-***The coptic*** (IInd century B.O.E. – XV^{-th} century B.O.E.) that is to say:
   The language of the scholars and priests of Egyptian temples closed by Theodosius I in 392 A.O.E.
   The language of the Christian monophysists, and stills today the liturgical language of this first Christianity

Of the Coptic synchrony, there are several dialects. The main ones are:

- Bohaïric
- Sahidic
- Fayoumic
- Oxyrhynchite (or middle-gyptian)
- Akhmimic
- Lycopolitan (or subakhmimic)

Pharaonic in its Coptic form was spoken in the sixteenth century in Upper Egypt. It was still spoken in the nineteenth century in monasteries.

Many works rely on Coptic to study other synchronies. However the purpose of this work is not to make a study of the Pharaonic but to read it using an African language here by the Ewondo language. Using a bridge through another synchrony did not seem necessary to me but irrelevant.

11.BARD, KATHRYN A.; Steven BLAKE SHUBERT (1999). Encyclopedia of the archaeology of ancient Egypt. Routledge. p. 274.

PHARAONIC SCRIPT

Jean Bottro affirms in his book this:

> *"Our ancestors the Mesopotamians invented the writing and, thanks to it, cast a new glance on the universe around them, developed a new way of thinking it, analyzing it, ordering it, as would have never allowed the simple oral tradition - the own lineaments of what, taken over, deepened and systematized by the Greeks, became our rationality, the true armature of our Science. In search of the last reasons to be of this universe and the ultimate sense of our existence of men, they have built a sumptuous and learned mythology, which already announces, on more than one point, what Israel, inventor of monotheism, will compose his "theology", which is still ours, even when we seek to get rid of it.*
> *After the birth of God, who was studying the origins of one of the most striking and singular features of this civilization, Jean Bottro wanted to go back up, in the same line, to the extreme horizon of history - which begins, indeed, in Sumer, since the writing and the document were born there - and, in the enormous treasure of the cuneiform tablets, hitherto inventoried by the only tradespeople like him, to discover other more archaic stammerings of our own philosophy"*[12]

Except Professor Jean Bottro and some illumined visionaries, the scientific community considers that ancient Africa is the cradle of writing. By this invention it has allowed humanity to move from prehistory to history.

In ancient Egypt we encountered four types of scripts, the sacred or 𓌃𓂧𓅱𓊹***mDw nTr***, the hieratic, the demotic cursive and the coptic.

The 𓌃𓂧𓅱𓊹***mDw nTr*** are the first signs of writing that appeared in Egypt based on Gunter Dryer's findings on the Abydos site. The brief inscriptions or some signs accompanying images go back to the years 3400 BC. BORN. This date is that of the oldest engraved objects found, but the writing must be much earlier, because several elements represented are archaic or prehistoric and no longer exist at the time mentioned above. Its origins are authentically African, because all the animals represented are African. According to the authorized literature, the writing does not take off and becomes rich only from 2700 BC. BORN. (Third Dynasty), later to the few inscriptions of the Thinite Period.

Hieroglyphic signs are 𓌃𓂧𓅱𓊹***mDw-nTr***, "Words of the gods", which hold in themselves the power of words. These contain the power of creation and to possess the key means to access the possibilities of the divine world: the word, written or read, can create the bread offering for the deceased, the resurrection or the rebirth for eternal life. To this end, writing is born and developed, since it is initially used only for commemorative inscriptions.

Hieroglyphic signs *(cf. illustration n°21)*

The close link between writing and image is demonstrated by the fact that in ancient Egyptian, one word means "writing" and "drawing". In a text of the Old Kingdom, the general name of "gods" is given to the hieroglyphs, because thanks to them, the intangible becomes script.

Initially, the hieroglyphic writing is almost exclusively ideographic (the drawing represents the object). It evolves then towards a mixed system with ideograms and phonetic signs, whose drawing transcribes one, two or three consonants. The signs that transcribe a single consonant are twenty-four and today called "alphabetical", although Egyptians do not use a pure alphabetic system. Also appear the determinatives, without which the Egyptian writing would be impossible to read, because it does not include vowels or separations between the words.

---

12. Jean BOTTERO, Mésopotamie. L'écriture, la raison et les dieux, Gallimard, Première parution en 1987

To clarify this concept, let's take the example of the word "char"; in a linguistic system such as that of Egyptian (but also Semitic languages, such as Arabic and Hebrew), it is written "chr". It is therefore difficult when one does not speak the language to understand if it means "char", or "choir", or "cher" or "chier" etc. The determinative then represents, respectively, a small chariot, a falling man, meat etc. Schematically, this is the system used by the ancient Egyptian.

However, we must not forget that the determinatives represent a set of objects or ideas: thus, the determinative of the sun can designate ***Ra***, the solar disk, or the day. Many old ideograms retain their original value. For example, the sky (a ceiling), god (maybe a flag on a mast, or an ax). According to the authorized literature, the signs used for the "gods" are true symbols: the hawk for ***Hr***, the throne for ***As.t***, etc.

Signs formed by analogy are also, to a certain extent, symbolic; thus the red color, but also the blood, are represented by the image of a beetle, and the green, which signifies prosperity, by the plant of the papyrus.

If the 24 alphabetic signs, we add about 80 determinatives and 640 hieroglyphs most used, we get the 750 most common signs that the scribe must know perfectly. To these signs are added others, more rare, especially used in the sacred texts.

Hieratic Writing *(cf. illustration n°20)*
Cursive development of hieroglyphic writing, this is a simplification of the hieroglyphic signs in order to accelerate the writing process. Appeared under the Third Dynasty, it is especially used to write on papyrus and on many documents extracted from legal, literary and sacred texts. It is used until the end of the Third Intermediate Period (about 747 BCE) and for religious texts until the end of the Pharaonic period.

Although originally written in vertical columns, hieratic writing tends to be written horizontally from right to left. Gradually, it acquires a particular style and moves away more and more from the original signs. A late form, the "abnormal hieratic" writing, is very far from its origins and extremely difficult to understand, just like the rock inscriptions.

Demotic writing *(cf. illustration n°20)*

Literally, the "popular" writing (from Greek demos). It appears under the XXVIth Dynasty (664-525 BCE) and derives from a further simplification of the hieratic. Thus whole groups of signs form, in the demotic, only a few isolated signs. It is the only form of writing widely used during the following millennium. It is a reflection of popular language and is widely used by leaders and in legal texts. The central band of the Rosetta Stone (Rashid) is demotic.

Although the origin of the demotic is probably to be found in administrative documents in hieratic, it is much more developed than the latter. The demotic has a grammatical structure, a new vocabulary, as well as many additional signs, abbreviations and related letters.

The Coptic *(cf. illustration n° 20)*
Language spoken for a time by the Coptic population and partially preserved in religious rituals. These are regional dialects of the ancient Pharaonic language, transcribed in an alphabet using Greek letters and to which are added some new letters to make the sounds that Greek does not have. This alphabet is adopted to overcome the lack of vowels of the Egyptian. Coptic is the language used from the Christian era to write thousands of texts on papyrus or other media. They allow to get a precise idea of post Pharaonic Egypt.

> *" Hieroglyphic writing is a complex system, a writing both figurative, symbolic and phonetic, in the same text, the same sentence, I would say almost the same word. "*
>
> *Jean-François Champollion*

What does the word hiroglyph mean?

Cette appellation vient des Grecs *hieros* veut dire sacr et *glyphein* graver. En gyptien, cette criture se dit ***mdw-ntr*** ce qui veut dire "paroles de dieu" selon l'acception officielle. Une traduction en Ewondo donnerait entre autres acceptions /mədzo mə ù-ntour / "les paroles de l'accompli"; /mədzo mə ù-ntouar / "les paroles de l'unique"; /mədzo mə ù-ntr / "les paroles de le handicap"; /mədzo mə ù-ntour / "les paroles de celui qui se suffit à lui-même" etc. On reviendra plus en deepness sur cette question.
This name comes from the Greek *hieros* meaning sacred and *glyphein* engraved. In Egyptian, this writing is called ***mdw-ntr*** which according to the official position means "words of god" according to the official meaning. A translation into Ewondo language would, among othe meanings give / *mə*-dzo mə *ùn*-ntour / "the words of the accomplished "; / *mə*-dzo mə *ùn*-ntouar / "the words of the unique "; / *mə*-dzo mə *ùn*-ntér / " the words of the handicapped "; / *mə*-dzo mə *ùn*-ntour / " the words of the self-sufficient one" etc. We will come back to this question more deeply later.

General principles of writing

Egyptian writing is constituted by figurative signs. It can be written in rows or columns, from right to left or left to right. The orientation of the signs indicates the reading direction.
The signs look towards the beginning of the text; thus, signs pointing to the right indicate a read from right to left and vice versa.

The signs are arranged in a precise way. Two horizontal signs are one above the other, even when the word is written horizontally. The Egyptian scribe still felt the need to group the signs harmoniously by shift in meaning them into imaginary squares.

Always read the signs from above before the one below and then go on to read the next square.

Transliteration

Each elementary sound conveyed by hieroglyphic writing is transposed by Egyptologists into a phonetic script called "transliteration ".

The vocalization of the Egyptian language does not correspond to what the ancient Egyptians were dealing with. It is an Egyptological convention to bind consonants whose phonetic value is known by "e" more or less accentuated. However, the transcription of Egyptian words in other contemporary languages, noting the vowels, and the knowledge of a more recent language state, the Coptic, can restore some vowels (a, i, or,) and tonic accents.

If this convention is useful by some aspect, it can be handicapping by its postulates tinged with ideology because instead of a convention, I postulate that the whole humanity uses the same lexicon. In fact, the word derived from the convention, exists in all variants of human languages we will see that later. Only intellectual weakness and bad postulates prevent us from finding the right word in our own language and from being content with conventions that sometimes lead to heresies. I therefore posit that transliteration is useful. It can become a danger for the neophyte who does not have the basics in morpho-syntax and lexicology.

Ideograms and phonograms

The Egyptian writing is composed of two kinds of signs:

- signs giving an indication to the category or idea to which the word can be approached: ideograms
- signs giving indications regarding the pronunciation of the word: the phonograms.

1.***Ideograms***

In my opinion, there are no signs that are in themselves phonograms and others ideograms. All signs are at the same time ideograms and phonograms. For the simple reason that they are related to a signified. When they are known, they therefore offer vocalization. But this vocalization is linked to an idea associated with them.

So the sign itself has an alphabetic value by modeling and simplification. The same sign has a phonographic value and an ideographic value. But there is no sealed barrier for signs with phonographic value and signs with ideographic value.

In Egyptian, a word can be written using an ideogram.

Example:

Middle-Egyptian: 𓋴𓂋𓃱𓀁 ***sr***
Ewondo: /*á*-tóŋ/ "to feed"

What is notable with the giraffe 𓃱 represented by the sign E27 of the gardiner's list is the length of his neck. The giraffe states the idea of length considering his neck. To qualify what is long and high like the giraffe's neck, the Beti Ewondo expresses it by the word / *un*-ntôŋ /. So we have the representation of the giraffe in the script, the best illustration of the idea of length by the scribe.

If the mouth is used to express something, it also serves for feeding. And the action of feeding is qualified in Ewondo language by :

Ewondo: / *à*-tóŋ / "to feed"

A banana leaf still wrapped serves as protection like the loincloth on the skin represented by the 𓋴 S29 of the Gardiner's list. This banana leaf still rolled is called in Ewondo language :

Ewondo: / *un*-ntoŋ /. " wrapped banana leaf "

By vocalizing the idea of length linked to giraffe's neck, we can access the idea of feeding oneself or even protecting oneself, one simultaneously vocalizes the notion recorded by the scribe through the determinative (the sign on the far right).

The position of this sign 𓀁 with the hand on the mouth and the presence of the mouth on the script on the left indicate that it is an action that falls within the verbal field, the action of talking in broad terms (talkative).

In fact the homophone of / *ùn*-ntôŋ / "length" is / *á*-tóŋ /"to tell, to narrate, to speak" . So this is how to read this script in Ewondo. The transliteration :

Middle-Egyptian : 𓋴𓂋𓃱𓀁 sr "predict, announce; to make known"
Ewondo : / *á*-kàr /; / *á*-tóŋ / "to tell, to narrate, to speak ; predict, announce; to make known "

The word orbtained by transliteration is a synonym of the word obtained by the endogenous method.

We can assume that the giraffe is also called / *un*-ntoŋ'o / "the long pole " . As i have emphasized the word exists in every human language. So you have :

Tshiluba: ntengu "giraffe"
Oluganda: ntugga, entugga "giraffe"
Kinyarwanda: Twiga "giraffe"
Etc.

2. ***The phonograms***
Phonograms represent consonants. There are three types of phonetic signs in ancient Egyptian:

- the uniliterals
- the biliterals
- the triliterals

This point is highly debatable. I suggest that one can model for simplification needs a sign by a sound but it is not univocal.

Sign	Convention	Ewondo Language
	**a** : *arm, hand; region, province; condition, state; article, piece; track, trace*	**wa** : *arm, hand;* **wa** ; **mbo-** : *one ; 1* **waé ; yui :** honey ***wai :*** *water in maori* ***mw :*** *water in Middle-Egyptian* **fɔkh ; foà :** *one ; 1* ***á*-fàê :** *to offer* **fôe :** felids *(cats, panthers, lions etc.)* **fwé :** *news* **á-wàê :** *steal, subtilize* **á-wàô :** *wash, purify* **á-wu :** *die* **mbogh ; mboa :** *one ; 1* **boé** : *bee* ***ùn*-mbèh :** *door* ***i*-yəkh :** *dam, closing* ***à*-yokh :** *Courageous, brutal, proud, violent, ...* ***i*-ŋguŋ :** *Pride, desire to appear, to rise, too much assurance, presumption, insolence.* **ngon** : *panther, woman, moon etc.* ***i*-nàm :** *member; piece ; part* ***ùn*-nàm :** *région, province; état; pays* ***ùn*-nəm :** *heart* **ndà :** *house,country, state* **doé :** *palm of the hand* ***i*-bəm :** *hut ; stubble ; cabin; shack* ***i*-mbgəm :** *lion* ***à*-dzô :** *speech; history; case* ***i*-dzo :** *member; piece ; part, fragment* **á-dzàè:** *ask ; beg ; cadge* ***i*-dzoé :** *authority, psychorigid* **kîŋ :** *track, voice, speech; neck ;* ***ùn*-nlàŋ** : *story* ***ùn*-nlan :** *track ; network* etc. **i-làŋ** : *which is used to read, buttocks; cynicism* **zén :** *piste, runway ; way, road etc.* **ndim :** *blind, water* **ndəm :** *standard of the cubit; mark, trace* **ndəm :** *sperm; Prostatic Specific Antigen, or PSA* (Prostate Specific Antigen), which is a substance naturally made by the prostate. (fountain woman)

mə-nyu : *fingers, hand*
mə-nyù : *drink; water*
à-nyu : *mouth*
i-yé : *cloth; lip; eyelid*
i-yi : *desire; conclusion*
à-yəm : *mark of recognition*
i-yəm : *knowledge; crippled, animal etc*
i-yàm : *cover, coitus; which is used to cook*
i-yámèn : *cooking*
i-yàmèn : *coitus* ☞ french : *hymèn ; mouillé*
i-yàl : *cover, coitus;*
byàl : *dugout*
dyal : *region, province; country; town*
dyaŋ : *maternal village*
i-yəl : helpless, *impotent*
i-yələ : *volatile*
i-lèh : *glass, three, tree*
ya : *that of*
i-yoé : *immature; green ; borrowing*
à-yol : *bitterness*
Toyin' à-wom : *ten thousand* (10.000)
i-yolé : *name*
I-yamèn Éq. French : mouillé ; hymèn

The unilitals

The Egyptologists created the unilitals phonograms from some ideograms for the phonetic interest of the words that they represented. Unilitone expresses one consonant at a time.

These signs are few. Egyptologists have classified them according to a fixed order adopted to store words in lexicons and dictionaries.

Uniliterals have about thirty signs corresponding to simple sounds. They compose a basic "alphabet" for Egyptologists. [13]

Hieroglyph sign		Transli-teration	Represented Object		Phonic Value	
		*A*	Vulture		a	kh
		*i*	Flowered reed		i	
		*y*	Double reed	2 features	y	
		a	Forearm		ââ	
		w	quail	Cord	w	
		b	Foot		b	
		p	Seat		p	
		f	horned Vipere		f	
		*m*	owl	rib	m	

13. P.Du BOURGUET, Grammaire égyptienne, Louvain, 1971.

	n	water	Crown	n
	r	Mouth		r
	h	house Plan		h
	H	Braided linen rope		h
	x	Placenta		kh
	X	Cow vulva		kh
	s	fabric	lock	z
				s
	S	Bassin d'eau		sh
	q	Sandy slope		q
	k	basket		k
	g	Jar holder		g
	t	Bread roll		t
	T	hindering	Pestle	tsh, ti
	d	Hand, snake		d
	D	Serpent		dj

Chart: Egyptian uniliterals

Alphabet or not alphabet, the question is raging between different specialists. Some people rightly argue that ***mDw nTr*** are not just alphabetic, they can be phonetic or ideographic.
In fact, we can not talk about alphabets.
For me, there is no debate here because, as Erman says:

> *" [...] There is no doubt that this alphabet (...) one of the most great creations of the Egyptian spirit (is) the prototype of the Phoenician alphabet which serves as a basis for all the writings of the modern world. "*

If the Phoenicians did not invent writing but inherited it from ancient Egypt *(cf illustrations n°4 & 5)*, it is not useful to debate on the subject.

We can use the word "alphabet" to be understood even if etymologically (beefhouse) means nothing. It may be remembered, however, that the source of the characters is African. We can if we want to go back to Negro-egyptian to demonstrate but it is not worthy here.

EWONDO
Lexicon
n-**yâ :** *cow*
ùm-**bè :** Door
i-**cè :** chase
i-**dî :** door
á-**hə :** to give
i-**fâ :** fork
ŋgama :partial
i-**hoé :** hole
i-**yi :** need
i-**kè :** help
i-**lè :** goad
maï : water *i*-ya**m :** water

And to say that Africans are taught the alphabet is an insult to the common sense. To say that we hold scholarly debates on the content of these characters without asking the Africans their opinion is an endless contempt.
The denomination *alphabet* denotes carelessness and craft but nothing more.

Since I argue that all languages could be used by the present method, i will take the Duala language to explain the word alphabet.
Duala: (Mu)-lópō "head"□Ewondo: Ún-loh "head" □ semitic: Aelph "beef" □ Greek: Alpha
Duala: Ndabo "house"□Ewondo: Ndah "house"□ semitic: Beth "house"□ Greek: Beta

If between (Mu)-lópō "head" and Aelph (beef for cow's head) there is an obvious phonetic proximity, in the second case the phonetic proximity does not seem to sink.

In effect, the two terms proceed by phonematic inversion. One comes from the Kekwé branch (ndabo) of the Negro Egyptian and the other from the Kweké branch of the Negro-egyptian (beth).
***Semitic**: Beth "house"*
***Ewondo**: i-tobo or nda "house"*□by phonematic inversion you get /boto/ *"people; inhabitants"*
***Greek**: Beta*

So here, I am closing this hollow debate which is not one properly. It looks like what is kwnown as a strategy of diversion and of distraction. So for me, I keep useful in this heuristic the idea of an Egyptian "alphabet".

Some attempts to distract by creating unnecessary debates that are, according to me, misguided, adventurous, and without a great interest.

According to the linguistic studies in Egyptology, the pharaonic Egyptian does not have any particular signs to vocalize the liquid consonant "l" (J. Vergote, 1945)[14]. He makes up by using the 𓂋 ***r***, more rarely by 𓈖 ***n*** or the 𓄿 ***A***. Generally the three phonemes 𓈖, 𓃭, 𓂋 so confused in Egyptian, the linguistic facts walafs confirm this[15]. (Anta Diop, 1976).

The respective value of the alphabetic hieroglyphic signs was surely determined by Coptic (the last stage of the Egyptian language), which Coptic is written with the Greek letters (free from accents and spirits) to which were added signs representing unknown sounds of Greek , and drawn from the demotic[16].

However, the hieroglyphic Egyptian also has signs representing two, three letters. These are the biliterals and triliterals. Overall we have been able to count nearly 6000 characters to date.
Egyptologists classify this /*mə*-dzo *mə*-Ntour/ ***mDw nTr*** « hieroglyphic » following this order of transliteration: A i y a w b p f m n r h H x X s S q k g t T d D. I want to note that I have nothing against this classification.

Points / Modes		Labiales		Apicales		Sifflantes		Dorsales		Laryngales	
		bi-labiales	labio-dentales	ordinaires	affriquées	ordinaires	chuintantes	palatales	vélaires	pharyngiales	glotales
Spirantes	*sourdes*		f			s	S	ç	x	H	h
	sonore									ε	
Occlusives	*sourdes*	p		t	č			k	q		'
	sonore	b		d	ğ			g			
Nasales		m		n							
semi-consonnes		w							j		
liquide				r	3						

Chart 2: *Consonantal system of the Middle-Egyptian*

It is agreed to recognize 24 consonants with phonological status, divided into 5 orders and 5 series.

Characters called hieroglyphs (from Greek hieros, "sacred, divine", and glyphein, "incise, engraved"): for monuments (tombs, temples). They can be used by priests, on papyrus, in a "cursive" way.

- hieratic writing (Greek hieratikos, "priestly"): widely used by all until the first millennium, for administration, stewardship.
- The demotic writing (from the Greek dêmotikos, "the people"): uses from the first millennium for administration, stewardship.
- The Coptic script (from the Greek Aiguptios, "Egyptian"> Coptic Kuptios): uses from the third century, mixture of ancient Egyptian and Greek

[14] Jozef VERGOTE, *phonétique historique de l'égyptien. Les consonnes.* Louvain, 1945, P.10 et suivantes.
[15] Anta DIOP, *Parenté génétique de l'Egyptien pharaonique et des langues négro-africaines*, Ifan, 1977, p.71.
[16] Gilbert NGOM, *La parenté génétique entre l'égyptien pharaonique et les langues négro- africaines modernes L'exemple du duala*, Ankh N°2, 1993, p.5.

The monumental writing (which is read from left to right or from right to left, "watching animated beings in the eyes", or "going to meet them", and always from top to bottom for the columns) serves to transcribe the language from the fourth millennium and remains in use for the sacred texts (the language of the "Middle Kingdom").

3.3. ELEMENTS OF WONDO SYNCHRONOUS

"Raise your words, not your voice.
It's the rain that makes the flowers grow, not the thunder."

Rumi

3.3.1. PHONOLOGICAL STUDY

Phonology is the branch of linguistics that studies sounds organization in different natural languages. It is complementary to phonetics, which is interested in the sounds themselves, regardless of their use. If phonetics is concerned with sounds as acoustic units produced by a physiological mechanism; phonology focuses on sounds as part of a system.

Phonology releases all phonic facts from the criterion of relevance. It ranks them according to their function in the language. Thus, the contingent is never imposed at the expense of the essential. The feature of relevance is therefore fundamental. It makes it possible to distinguish what is essential, because distinctive, in each language or each use. And what is incidental, that means determined by the context or certain circumstances.

It is therefore necessary to carry out a preliminary phonological study of Kɔlɔ or Ewondo as A. Martinet (2005, 64) understands it, namely to proceed to the analysis of the utterance in phonemes, the classification of these phonemes and the phoneme examination of their combinations to form the signifiers of the concerned language.

This inventory is made by the commutation test. The principle is to extract one element from the chain, then to substitute it by another, without modifying the context. This makes it possible to observe how an element is different from all the others that may appear in its place. On this account, the switching takes place on the paradigmatic axis. It must be distinguished from the permutation that takes place on the syntagmatic axis.

This is a prerequisite for the desired approach, which takes advantage of the genetic links between the different languages of our subject of study in order to establish bridges that make it possible, by comparison, to switch from one language to another.

What is important beforehand in this approach is to take advantage of phonology in order to determine free forms to which are respectively associated a non-decomposable unique sense. This approach brings us back to the very definition of "word" in the traditional sense.

The weakness of this methodology is to rely on an exogenous spelling analysis of the words used for the discriminant analysis. By the same token the analysis is biased. In fact, the words in a sentence fit into schemas of chords and by the same token are not written just as they are pronounced but as they are technically thought to justify these schemas of chords.
I therefore proceed with the analysis in the context of this heuristic by respect of academic standards and not by conviction. For my part we can not separate a word from its so-called nominal class and spell it only on the basis of simple phonetic impressions or exogenous grammar rules.

The word by definition being the smallest element that can be pronounced in isolation with a semantic or pragmatic content. This significant element can also:

- Take a plural index:

<u>Example:</u> Kɔlɔ→ English

	Singular	Plural
Kɔlɔ	/i-béh /	/**bì**-béh/
English	"pit"	"Pits"

Chart 1: The significant can take a plural index

- Being attached or linked to possessive pronoun

<u>Example</u>: Kɔlɔ → English

	Possessive Pronoun
Kɔlɔ	/í-béh dy-àma /
English	"My pit"

Chart 2: The significant can take a possessive pronoun

- Precede a verb at the imperative mod

<u>Example</u>: Kɔlɔ → English

	Imperative
Kɔlɔ	/Dzâ ù-dzoùé /
English	"Eat the banana"

Chart 3: The significant can precede an imperative verb

If one wants to grasp the functioning of a language at the level of its significant units, it is desirable, when approaching the phonological study, to take into account the variation of these significant units according to the context of appearance and of realization by a speaker.

General phonetic inventory

The abundant documentation related to what is so called the "Bantu" languages and the Ewondo in particular allows an inventory of 27 consonants according to the specialists.

In order to approach this phonological study a rigorous way, we will be guided by the methodological outline previously announced which distinguishes the following aspects in the presentation of each phoneme:

1. Determine the phonological status using minimal pairs that demonstrate the existence of oppositions.
2. Present the different achievements of the units in the respective packages.
3. Analyze positions of relevance.

The identification of the distinctive units will be done on the commutation test, which is carried out according to their oppositional possibilities in identical context. Recall that, two phonic sequences of different meanings but identical to a segment or suprasegmental close technically constitute a minimum pair.
It therefore appears useful to describe the phonological facts starting from these units to understand the operation of the segments that do not belong to the same experience facts. I will bring no criticism on the methodological plan but, i shall show the shortcomings of this approach later in the development of this heuristic.

This analysis is based first on the syntagm necessary. That means, the whole composed of a lexical element or grammatical and also an index, which refers to one of the classes of the language, to release the units of analysis.
The index or class prefix that can take different forms:

- Zero (Ø), (i.e. any class prefix)
- C-, V-, N- (Syllabic nasal), CV-.

While the radical can appear as:
–CV; CVC; -CVCV; -CV_1V_2; -V; VCV; -VC.

The prefix isa vowel (V-), the radical is a semi-consonant or structure:
–CV; -CVCV; -CVC.

When the prefix is C-, the radical takes the shape -V, -VC, -VCV. The prefix and the radical are in this case closely linked and articulated in a single voice broadcast, without pause, not even virtual. Only considerations of morphological order make it possible to distinguish them, take the word vocalised [dzɔm] for example. His spelling can be problematic if one limits oneself only to drawing parallels to the language of the colonist. We would be tempted to write dzom but it would be a big mistake.

Examples:

Singular		**Plural**	
Kɔlɔ	***English***	**Kɔlɔ**	***English***
/**dy***Òm*/	*"heap"*	/**m'**-*Òm*/	*"heaps"*
/ **dyé**h /	*"nail"*	/**bì**-iéh/	*"nails"*
/ **d**ɔp /	*"Belly button"*	/ **m'**-ɔp /	*"Belly buttons"*
/ **m**òt/	*"Personne"*	/**b'**-òt /	*"Personnes"*
/**by**al/	*"Canoe"*	/**m'**-al/	*"Canoes"*
/**dy**âl/	*"Village"*	/**m'**-âl/	*"Villages"*

Chart 4: comparative of singular and plural words

If the index or prefix is a morpheme zero signifier, to this account, the radical does not know other relations and it is either initial consonantal or initial vowel.

Example:

Meaning 1		Meaning 2	
Kɔlɔ	**English**	**Kɔlɔ**	**English**
/ nlÓŋ/	"rattan"	/nlɔŋ/	"Nyon"
/ ntÓŋ /	"stick"	/ ntɔŋ /	"guardian"
/ í-bɛ̏ /	"two"	/ i-bèh /	"Bamboo"

Chart 5: morpheme zero signifier

The prefix is a V-, CV-, N- form, the radical is exclusively consonantal initial. It is in this specific case, assignd a tone and assumes a syllabic function.

Examples:

Good Kɔlɔ	Wrong Kɔlɔ	English
/ *ùn*-ŋgoro/	/ngoro/	"Lezard"
/ *ùn*-nloh /	/ nloh /	"head"
/ *ùn*-nlɔh/	/ nlɔh/	"fly"
/*ùm*-ôn/	/môn/	"child"
/*ùm*-ingà/	/minga/	"woman ; wife"
/*ù*-wà/	/wa/	"hand ; arm"
/fɛ̃ /	/fɛ̃ /	"Machete"

Chart 6: A supposed initial phonetic consonantal.

In the wider of this document it is through the radical channel that the approximations will be made to ensure the identification of phonemes.
The phoneme is a sound form consisting of a bundle of distinctive features. It has a signifier and no signified according to the Saussurian point of view of the term. By switching, it helps to produce a change of meaning for higher rank units, the "words". It has a distinctive function. It is consider as the smallest linguistic unit.

As a general rule, the radical will be the necessary syntagma, which will make it possible to identify the phonemes, to define them and to classify them. In the course of this analysis, appearances of semi-consonants will be notified when the language testifies to them as to the determination of the phonological status of the first mentioned elements.

Consonantal Examining units will sweep away all positional relevance cases that follow:

1- Absolute initial
2- At the intervocalic position
3- And in absolute final position.

The same scheme will be used for the analysis of vowel units.
However, since Kɔlɔ is a so-called tonal language, it also obeys the technical constraints of tonal languages. So, in order to identify a vowel segment, you will have to know your tone. This study will devote a large part to the tones analysis because it is necessary to know the tones difference determining the same segment to get the semantics. Linguistically, tones are considered as prosodic units. It will therefore be useful to treat this point in order to understand their influences on the phonemes to which they are contiguous whether they are vocalic or consonantal. In truth, this is an uninteresting research as we will see later.

A CONSONANT SYSTEM

The phonemes

As a rule the consonantal and vocalic phonemes of the Ewondo are treated as second articulation units. They are an integral part of the spoken channel. On this account, they allow to define and distinguish the messages to each other.

Thus, phonemes are not an isolated representation. In fact, as the French linguist, grammarian and lexicograph Dubois points out:

> *"[...] They will be differentiated by their features according to the context, the conditions of emission, the personality of the speaker"*[17].
>
> *(Dubois, 2002-360)*

Thus, as pointed out above, the identification that follows, is going to be the result of the phonemes that emanate in most cases, from the initial of the radical.

Identification of consonant phonemes

In order to identify the phonemes of the Ewondo we will make use of two distinct but complementary operations:

1. The switching
2. The sound environment: initial position, median position and final position.

The structural linguistics initiated by Ferdinand de Saussure in 1916 conceives the language as a system, a structure, elements in opposition or in equivalence with each other.
This work is in the tradition of the structuralist school and also of the Prague Circle. Contemporary phonological theories are more complex to approach, they are not taken into account in this work which is not intended to make anaccurate phonology experience but to use it as a tool for a better approach of the Pharaonic Egyptian and Meroitic regarding an African language in this case, the Ewondo language.

On this account one has to respect a certain methodology from the moment when it carries out a structural analysis. We are thus required to define what are called minimum pairs. Indeed, in structural phonological analysis, a minimal pair is a couple of words that are distinguished (and thus opposed) one another by only one phonetic difference.

As a general rule, two sounds switch if the replacement of one by the other results **in the change of meaning** of the two words in which the two meanings are found. This must be valid in all three positions (context).

Example:

Positions	Minimum Paire	
Initiale position	Par /**p**aʁ/	bar /**b**aʁ/
Median position	Apport /a.**p**ɔʁ/	abord /a.**b**ɔʁ/
Final position	Crampe /kʁɑ̃**p**/	Crabe /kʁɑ̃**b**/

Chart 7: changing meaning by consonant's switching in a minimal pair

Let's proceed to the identification of the base pairs in Ewondo by taking inspiration from the approach used by D. Creissels and C. Grégoire for the elaboration of the phonological theories which, moreover, are based on the generative method. I still want to point out that this work. I do it by respect of the standards. It is useless in a global and general vision of a search for a unique African language. Every African knows how to speak his language and does not need a study like this one.

17.F. DUBOIS-CHARLIER Structures verbales, Aix, 2002, p.360.

Initial Position

phoneme /b/

Using the example above, this phonological identity is apparent from the following reconciliations:

b/t

Position	*English*	Ewondo minimum pairs		*English*
Initial	"to scold"	/bam /	/tam /	"to wish"
Median	"Vice"	/à-bíŋ/	/à-tíŋ/	"to plait"
Final	"to push"	/ á-lab/	/ á-lat/	"to sew"

b/k

Position	*English*	Ewondo minimum pairs		*English*
Initial	"Vice"	/biŋ /	/kiŋ /	"neck"
Mediane	"board"	/*mə*-bam/	/*mə*-kam/	"Pepper"
Final	"advisor"	/leb/	/lek/	"to dump"

b/kp

English	Ewondo minimum pairs		*English*
"climb"	/beʁ /	/kpeʁ/	"Arched"
"kola"	/à-bel/	/à-kpel/	"Scarify"
"to fish"	/lôb/	/à-kpwe/	"Viper"

b/m

English	Ewondo minimum pairs		*English*
"testicle"	/-bín/	/mín/	"to swallow"
"sting"	/à-bem/	/à-mem/	"to confess"
"sloping"	/nteb/	/ntem/	"a branch"

b/n

English	Ewondo minimum pairs		*English*
"Vagina"	/-bɔ-n/	/-nɔ-n/	"bird"
"refuse"	/à-bén/	/à-nén/	"to swell"
"Massage"	/màb/	/màn/	"to finish"

b/mb

English	Ewondo minimum pairs		*English*
"ugly"	/bep/	/mbep/	"peeler"
"Refuse"	/á-bén/	/ùn-mbol/	"meal"
"to knead"	/màb/	-	

b/f

English	Ewondo minimum pairs		*English*
"ugly"	/bep/	/fep/	"paper"
"gather"	/àbough/	/àfough/	"to mix"
"to knead"	/màb/	-	

b/w

English	Ewondo minimum pairs		English
"true"	/bele/	/iwele/	"biscuit"
"gather"	/àboro/	/àworé/	"to hit"
"to knead"	/màb/	-	

b/v

English	Ewondo minimum pairs		English
"true"	/bele/	/vele/	"to wake"
"rotten"	/ibôl/	/ivôl/	"mute"
"to knead"	/màb/	-	

b/d

English	Ewondo minimum pairs		English
"to cover"	/buri/	/duri/	"to drive"
"to break"	/ábugh/	/ádugh/	"to mislead"
"to share"	/ákab/	/ ákad/	"to say"

Recap on the phoneme /b/:

The phoneme / b / presents, at the initial position, three realizations considered as combinatorial variants defined in terms of contexts:

In front of the vowels / i /, / e /, / u /, / ɔ /, / a /, the phoneme / b / is realized phonetically like a occlusive, bilabial, oral, sonorous, exploded consonant [b].

In front of the former non-rounded variant / y / of the anterior vowel [i], it is realized as an affronated consonant, bilabial, oral, sound [by].

Example:

Kɔlɔ	*Realisa-tion*	*English*
/byàŋ /	[biiàŋ]	"drugs"
/ byàm /	[biiàm].	"mine "

2- The phoneme /t/

Like the previous phoneme /b/, the research for its phonological identity is based on the above-mentioned positional analysis, namely: initial, intervocalic, and final position. This analysis is obtained from the reconciliations below:

t/d

English	Ewondo nminimum pairs		English
"old"	/tukh/	/dukh/	"wc"
"to cut"	/àtigh/	/àdigh/	"to burn"
"say"	/àkat/	/ àkad/	"say"

At the final position, in the Kɔlɔ language and its variants cousin, at the final position, the / t / is often pronounced as a / d /or as a / r / because the Kɔlɔ language is situated in one of the african isoglossal. That is why arguing that it is enough to change the position of consonants to change the meaning of two

words could seam for us a semi-truth. For the example above i can have in Ewondo language /ka**r**/ = /ka**t**/ = /ka**d**/ "say". I can also have /lad/=/lat/= /lar/ "bind" etc. But this truth is not only for the last position it can occur even at the first position, envolving /v/; /w/; /h/. We can say:
/á-və/= /á-hə/ " to give".

t/k

	Ewondo		
English	minimum pairs		*English*
"spoon"	/tokh/	/kokh/	"crush"
"to boil"	/àtogh/	/àkogh/	"to crush"
"to say"	/àkat/	/ àkak/	"to load"

t/n

	Ewondo		
English	minimum pairs		*English*
"calculation"	/tang/	/nang/	"to grow"
"to feed"	/àtong/	/ànong/	"to take"
"to say"	/kat/	/ kan/	"swear"

t/s

	Ewondo		
English	minimum pairs		*English*
"weaving"	/tíng/	/síng/	"fight"
"palm nut"	/itóng/	/isÓng/	"take"
"jute bag"	/nkut/	/ nkus/	"widow"

t/l

	Ewondo		
English	minimum pairs		*English*
"calculating"	/tàng/	/làng/	"to read"
"scrub"	/àtugh/	/àlugh/	"to marry"
"skinny"	/nkot/	/ nkol/	"hill"

t/nt

	Ewondo		
English	minimum pairs		*English*
"write"	/tilí/	/ntilí/	"writer"
"scrub"	/àtugh/	//	-
"skinny"	/nkot/	/ /	-

t/ts

	Ewondo		
English	minimum pairs		*English*
"spoon"	/togh/	/tsogh/	"to think"
"to boast"	/àtam/	/ àtsam /	"to fight"
"skinny"	/nkot/	/ /	-

When it is followed by the unclosed anterior vowel, / i / in intervocalic position, the phoneme / t / in Kɔlɔ can be realized in [ts] all depending to the speaker. But this is not really a problem for us it can just be an issue for a foreigner.
As a general rule in Kɔlɔ the phoneme / t / is realized as an occlusive, oral, apico-alveolar, deaf consonant [t].

t/b

This case is not treated above again because the comparison has already been made with the analysis of the bilabial occlusive consonant voiced / b /.

3- The phoneme /f/

Like the previous phoneme, the search for its phonological identity is also based on the aforementioned positional analysis, namely: initial, intervocalic and final position. This analysis is obtained from the reconciliations presented below:

☞ **f/b**

	Ewondo		
English	minimum pairs		*English*
"wound"	/fəl/	/bəl/	"kola"
"scrub"	/àfígh/	/àbígh/	"to get riche"
	/ /	/ kab/	"to share"

☞ **f/k**

	Ewondo		
English	minimum pairs		*English*
"wound"	/fəl/	/kəl/	"abscess"
"to gather"	/àfar/	/àkar/	"to wean"
	/ /	/ kak/	"to load"

☞ **f/s**

	Ewondo		
English	minimum pairs		*English*
"to change"	/fuali/	/suali/	"to hide"
"goyava"	/àfélé/	/àsélé/	"to slice"
	/ /	/ ibàs/	"scale"

☞ **f/l**

	Ewondo		
English	minimum pairs		*English*
"to close"	/àfét/	/àlét/	"hard"
"to mix"	/mfugh/	/nlugh/	"to get married"
	/ /	/ kal/	"sister"

☞ **f/ts**

	Ewondo		
English	minimum pairs		*English*
"to mix"	/fit/	/tsit/	"animal"
"to scrub"	/àfigh/	/àtsigh/	"to cut"
	/ /	/ /	

☞ **f/m**

	Ewondo		
English	minimum pairs		*English*
"to grow"	/fê/	/mê/	"to balk"
"to inspect"	/àfias/	/àmias/	"to scatter"
	/ /	/ kam/	"wonder"

☞ **f/v**

	Ewondo		
English	minimum pairs		*English*
"wound"	/fól/	/vól/	"mute"
"mix"	/ifugh/	/ivugh/	"silliness"

☞ **f/k**

	Ewondo		
English	minimum pairs		*English*
"wound"	/fəl/	/kəl/	"abscess"
"gather"	/àfar/	/àkar/	"to tell"
	/ /	/ kakh/	"to load"

The phoneme / f / is realized as a fricative consonant, labiodental, oral, deaf [f]. There is no final position in Kɔlɔ.

The phoneme /l/

Like the previous phoneme, the research for its phonological identity is also based on the aforementioned positional analysis, namely: initial, intervocalic and final position. This analysis is obtained from the reconciliations presented below:

☞ **l/k**

	Ewondo		
English	minimum pairs		*English*
"to tell"	/lê/	/kê/	"to start"
"pass"	/àlor/	/àkor/	"slim"
"sore"	/fôl /	/ fôkh/	"one"

☞ **l/d**

	Ewondo		
English	minimum pairs		*English*
"to marry"	/lugh/	/dugh/	"to deceive"
"leave"	/àligh/	/àdigh/	"to burn"
"Sister"	/ kal/	/ kad/	"to tell"

☞ **l/n**

	Ewondo		
English	minimum pairs		*English*
"to build"	/long/	/noŋ/	"to take"
"palm tree"	/àlen/	/ànen/	"large"
Beard	/zəl /	/ zən/	"a path"

☞ **l/nt**

	Ewondo		
English	minimum pairs		*English*
“wedding”	/lugh/	/ntugh/	“loss”
“fright”	/melighben/	/mentighben/	“weaving”
	/minal /	//	

☞ **l/nts**

	Ewondo		
English	minimum pairs		*English*
“trap”	/lam/	/ntsam/	“spoiler”
“the reclaimed”	/ melibèn/	/ mentsingben/	“bouttons”
“sorting”	/ifial /		

☞ **l/t:** connection already done about /t/;
☞**l/f:** connection already done about /f/;

☞ **l/nl**

	Ewondo language		
English	minimum pairs		*English*
“instill”	/lɔ/	/nlo/	“head”
	/ /	/ /	
“tri”	/ifial /	//	

Phoneme /mb/

Like the previous phoneme, the search for its phonological identity is also based on the aforementioned positional analysis, namely: initial, intervocalic and final position. This analysis is obtained from the reconciliations presented below:

☞**mb/b**

	Ewondo language		
English	minimum pairs		*English*
“drum”	/mbè/	/bè/	“thigh”
“stayed planted”	/tembe/	/tebe/	“standing”
/bottle/	/lombi/	/ùnlob/	“fishing”

☞ **mb/k, b/k**

	Ewondo language		
English	minimum pairs		*English*
“drum”	/mbè/	/kè/	“help”
“to pick up”	/bor/	/kor/	“redeem”
“to be catched”	/abili/	/akili/	“to withdraw”
“to sting”	/àlàb/	/àlàk/	“excited”

☞ **mb/nl**

	Ewondo language		
English	minimum pairs		*English*
“race”	/mbíh/	/nlíh/	“cleaner”
“been baked”	/mimbùmèn/	/minlomèn/	“been sent”
“”	/xxxxmb/	/xxxxnl/	“”

☞ **mb/mv**

English	Ewondo language minimum pairs		English
"nut"	/mbaŋ/	/mvaŋ/	"vaccination"
"wrinkles"	/mimbor/	/mimvor/	"fur"
""	/xxxxmb/	/xxxxmv/	""

☞ **mb/nt**

English	Ewondo language minimum pairs		English
"wrinkle"	/mbor/	/ntor/	"unic"
"races"	/mimbil/	/mintil/	"writtings"
""	/xxxxmb/	/xxxxnt/	""

The phoneme / mb / presents two combinatorial variants in its realization:
A- before the vowels / i /, / u /, / ε /, / ɔ /, / ɔ / and / a /, the phoneme / mb / is realized as an occlusive, semi-nasal, pre-nasalized, bilabial consonant, sound [mb];

Phoneme /v/

Like the previous phoneme, the search for its phonological identity is also based on the aforementioned positional analysis, namely: initial, intervocalic and final position. This analysis is obtained from the reconciliations presented below:

☞**v/b**
This comparison has already been done during previous analyzes of the phonemes on the position b/v

☞**V/f**
This comparison has already been made during previous analyzes of the phonemes on the position /f/ dixit (f/v)

☞**v/d**

English	Ewondo language minimum pairs		English
"to give"	/ve/	/de/	"anyway"
"clearly"	/àvim/	/àdim/	"to arase"
""	-xxv	/àld/	"harde"

☞**v/w**

English	Ewondo language minimum pairs		English
"epilate"	/vooughè/	/wôughè/	"laugh"
"taciturn"	/ívolo/	/íwolo/	"brave"
""	xxv	xxw	""

The phoneme / v / is realized as a fricative consonant, oral, labiodental, sound [v].

The phoneme /w/
Like the previous phoneme, the research for its phonological identity is also based on the aforementioned positional analysis, namely: initial, intervocalic and final position. This analysis is obtained from the reconciliations presented below:

☞**w/b**

The link has been already done when processing the phoneme /b/ dixit (b/w)

☞**w/v**

The analysis has been already done when studying the phoneme /v/ dixit (v/w)

☞**w/y**

English	Ewondo language minimum pairs		*English*
“to crawl”	/wang ʔa/	/yang ʔa/	“fry”
“to scrape”	/àwam/	/àyam/	“to cook”

☞**w/kp**

English	Ewondo language minimum pairs		*English*
“to crawl”	/wang/	/kpang/	“cockchafer”
“swollen”	/ne-ŋgwim/	/àkpim/	“full”
“”	//	//	“”

☞**w/m**

English	Ewondo language minimum pairs		*English*
“crawl”	/wang/	/mang/	“the sea”
“to scrub”	/àwam/	/àmam/	“to massage”
“”	/xxw/	/xxm/	“”

☞**w/ngb**

English	Ewondo language minimum pairs		*English*
“”	/wxx/	/ngbxx/	“”
“”	/xxwxx/	/xxngbxx/	“”
“”	/xxw/	/xxngb/	“”

☞**w/ng**

English	Ewondo language minimum pairs		*English*
“to crawl”	/wang/	/ngang/	“witchery”
“to gratify”	/àwan/	/àngan/	“to refuse”

The phoneme / w / presents two realizations considered as combinatorial variants defined in terms of contexts:
In front of the vowels / i /, / ə /, / O /, / a /, it is realized as a consonant, oral, labiovelar, sound [w].
Example:
/ wə / "you" is realized [wə];

The phoneme /y/

Like the previous phoneme, the search for its phonological identity is also based on the aforementioned positional analysis, namely: initial, intervocalic and final position. This analysis is obtained from the reconciliations presented below:

☞**y/w**

Reconciliation already done during the study of the phoneme / w / dixit the case w / y

☞**y/k**

	Ewondo language		
English	minimum pairs		*English*
“to frizzle”	/yang/	/kang/	“osteoarthritis”
“to cook”	/àyam/	/àkam/	“to wonder”

☞**y/nk**

	Ewondo language		
English	minimum pairs		*English*
“cuticle”	/yóng/	/nkóng/	“rank”
“to cook”	/àyam/	/minkóm/	“captives”

☞**y/v**

	Ewondo language		
English	minimum pairs		*English*
“to cry”	/yon/	/von/	“ax”
“bitter”	/biyol/	/bivol/	“mutes”

☞**y/nl**

	Ewondo language		
English	minimum pairs		*English*
“baldness”	/yan/	/nlan/	“the path”
“”	biyu		“”

☞**y/dz**

	Ewondo language		
English	minimum pairs		*English*
“mamba”	/yang/	/dzang/	“to miscarry”
“to know”	/àyem/	/àdzəm/	“to dance”

☞**y/ny**

	Ewondo language		
English	minimum pairs		*English*
“mamba”	/yang/	/nyang/	“to suck”
“oignons”	/meyang/	/menyang/	“milk”

☞**y/l**

	Ewondo language		
English	minimum pairs		*English*
“mamba”	/yang/	/lang/	“anus”
“to fry”	/àyang/	/àlang/	“read”

☞**y/ɲ**

English	Ewondo language minimum pairs		English
"hot	/yong/	/ɲong/	"to swim"
"to fry"	/àyang/	/àɲang/	"to grow"

The realizations of the phoneme / y / are distinguished in front of the vowels / u /, / ɔ /, / ə /, / a /, where, /y/ phoneme is realized as an oral consonant, dorso-palatal, sound [y];
In front of the anterior vowel of first degree of non-rounded aperture / i / and in front of the rounded former variant [ü] of the central vowel of first aperture / µ /, it is realized like a fricative consonant, sibilant, oral, pre-palatal [j].

Phoneme /nk/

Like the previous phoneme, the search for its phonological identity is also based on the aforementioned positional analysis, namely: initial, intervocalic and final position. This analysis is obtained from the reconciliations presented below:

☞**nk/nl**

English	Ewondo language minimum pairs		English
"hill"	/nko^{h} /	/nlo^{h} /	"head"
"roots"	/minkang/	/minlang/	"stories"

☞**nk/nf**

English	Ewondo language minimum pairs		English
"mère in law"	/nkiè /	/nfiè /	"light"
"roots"	/minkang/	/minfang/	"importants"

☞**nk/ng**

English	Ewondo language minimum pairs		English
"sticky"	/nkwan /	/ngwan /	"girl"
	/xxnkxx/	/xxngxx/	

☞**nk/nts**

English	Ewondo language minimum pairs		English
"root"	/nkang /	/ntsang /	"sore"
"roots"	/minkang/	/mintsang/	"boutons"

nk/nt

English	Ewondo language minimum pairs		English
"chest"	/nkoug /	/ntoug /	"waste; mess"
"hills"	/minkol/	/mintol/	"elders"

The phoneme / nk / presents a realization in terms of contexts:
The phoneme / nk / present in front of the vowels / u /, / ɔ /, / ə /, / a /, / i /, is realized as an occlusive, semi-nasal, pre-nasalized, dorso-velar, deaf consonant [nk].

Examples: / mìnkol / "hills" is realized in [mìŋkol]
/ mìnkul / "tam-tams" is realized in [mìŋkul]

Phoneme /ng/

Like the previous phoneme, the search for its phonological identity is also based on the aforementioned positional analysis, namely: initial, intervocalic and final position. This analysis is obtained from the raccomodement presented below:

☞**ng/nk**
Reconciliation made during the study of the phoneme /nk/ dixit nk/ng.

☞**ng/w**
Reconciliation made during the study of the phoneme /w/ dixit w/ng

☞**ng/ngb**

English	Ewondo language minimum pairs		English
""	/ngxx /	/ngbxx /	""
""	/xxngol/	/xxngb/	""

<u>Note:</u> Whistling omissions.

In the Kɔlɔ language there is not as in its cousins variants including the Fang wisps omitted. Indeed in the variants fang, Ntumu etc. when b, d, t, g and k are followed by a / u /, the vocalization between the two letters is realized with a whistle not reproduced by the writing system, and that resembles according to the cases to a f or a v: bu = bvu (nine), du = dvu (mouse), tu = tfu (chair), gu = gvu, (pig) ku =: kfu (foot). Examples: edu (hammer); ŋgu (shield).
When g and k are followed by ü, there is always a whistle between the two letters: g ü = gyü, kü = kyü. Examples: ŋgü (wild boar), akü (upstream)[18].

At the end the characters used to write the Ewondo are the following ones:
In addition to the characteristics indicated for the vowels, the emission of the consonants causes a partial or total occlusion of a point of the phonatory apparatus (the point of articulation).

A sound will be considered as a consonant of the Ewondo language if it is attested as an autonomous phoneme. As a rule, the consonant will serve to delimit the syllables. Following the study of phonemes above, we can distinguish the following consonants:
b d f g, gb, gh h, j k, kp, kh 1 m n ɲ ny ny n̲y r s t v ṽ w y z ẑ.

B, d, f, k, l, m, n, s, t, v, w, y, z: articulation similar to the French or English ones.

Note: There is, a Yaoundé standardized spelling, since (1970) concerning what William Bleek has called Bantu languages and that we african people we do call Ekang languages.

Phoneme	**Example**	**Signification**
/b/	í-**bó**l	Decay
/f/	á-fàb	wing

Phoneme	**Example**	**Signification**
/d/	à-**dú**rú	To drive
/k/	**k**ómbó	Uterus

18.SAMUEL GALLEY, grammaire fang,édition Shenri messeiller, Neuchatel, p562

/l/	à-**l**àm	To trap
/n/	un-**n**em	heart
/s/	à-**s**úgh	To greet
/V/	á-**v**ín	stink
/y/	à-**y**ín	To pack

/m/	à-**m**ín	swallow
/t/	à-**t**élé	To rise
/w/	**w**aŋ	To crawl
/z/	**Z**à'a	come

Chart 8: recap of Ewondo phonemes

Kɔlɔ or Ewondo has a consonantal system typical of many Negro-African languages. He has roughly twenty [19] consonants divided into 4 orders and 5 series:

1. Bilabial order: Plosive sound / b / b; nasal / m / m;
2. Labiodental order: Fricatives deaf / f / f and sound / v / v;
3. Labiovascular order: Plosives deaf / kp / kp and sound / gb / gb; nasal / ŋʷ / nw; semi-consonant / w / w
4. Alveolar order: Plosive deaf / t / t and sound / d / d; fricative deaf; / s / s and sound / z / z; nasal / n / n; vibrating / r / r; lateral / l / l;
5. Palatal order: Nasal / ɲ / ny; semi-consonant / d / y;
6. Velar order: Plosives deaf / k / k and sound / g / g;
7. Glottal order: Fricative / h / h.

Speech

Consonants:

b, d, f, k, l, m, n, p, s, t: speech similar to English and French;
h: strongly aspirated
j: affricated (English *John)*
g: always prenasalized except in loan words
ŋ: nasal velar (English: *singing)*
ɲ: nasal palatal (french: *agneau)*
w, y: semi-vowels, sometimes noted between vowels without phonological value
l: intervocalic: allophone of / d /; ex.: judi / juli = "satiation"

The consonant system

Points / Modes	Bilabiales	labiodentales	Alveolaires	palato-alvéolaires	Palatales	Vélaires	Labio-vélaires
Plosives	b		t d			k g	kp gb
fricatives		f v	s z	tʃ dʒ			
semi-voyelles					j		w
Nasales	m		n				
Latérales			l		ɲ	ŋ	

Chart 9: The Ewondo consonant system

19. I do not want to get involved on experts useless debates regarding the real number of consonants.I tjust take about twenty consonants and i do deal with them for what i want demonstrate here. That's all…

We can agree to distinguish around 20 consonants with phonological status, divided into 4 orders (labial, dorsal, palatal, alveolar) and 5 series.

My goal here is not to demonstrate that I am a specialist in the definition and analysis of consonants.

B- Vowel System

The IPA defines a vowel as a sound which occurs at a syllable center.

The Ewondo vowel system is similar to those of most Negro-African languages. It consists of 7 oral vowels and 5 nasal vowels. However other studies like those of Angenot (1970), Abega (1972), Alexander (1967), Tabi-Mange (1992) estimate that the Ewondo has eight vowels. I leave the specialists to their specialties and I take for this work the bet to expose on 7 vowels.

Oral vowels have 4 degrees of aperture and are perfectly balanced. On the other hand, there is neutralization between the near-closed and the near-open nasal in favor of the latter. But in an earlier analysis of the Bulu language, Alexander writes: "The centralized vowel ə is not a phoneme, but remains more than a mere epenthesis[20]. It most often corresponds to a suffix I often an unstable prefix I that disappears in certain positions by causing a tonal sandhi by regressive or progressive assimilation. (1955: 23).

The debate of experts is rather sustained on the issue as well, other inventories report height vowels, recognizing (Essono, 2000) or misrecognizing (Ongun Essono, 2012) modulated tones. The solution lies, on one hand, in the origin and the nature and, to the other hand, in the status of the elements in question, that is to say according to others whether they are structural or lexical. More recently, in a constant reference to Essono's studies (1984), Tabi-Manga (1992) restricts to five the number of fundamental vowels of Ewondo: | a |, | i |, | e |, | o | and | u |. This point is fundamental for the grammar of this language[21].

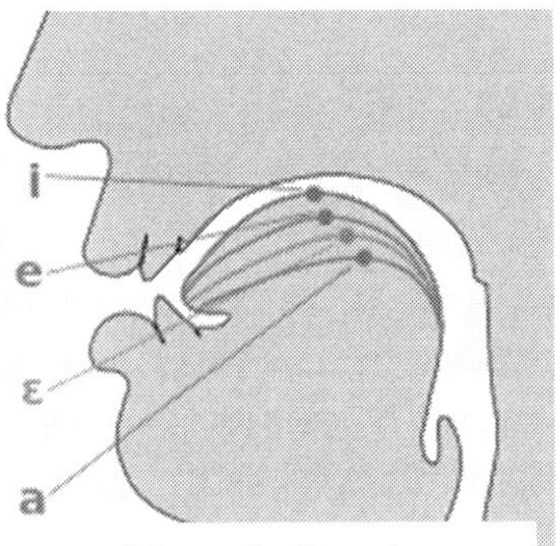

Figure 9: Vowels

Indeed, each of these vowels can serve as a topical vowel; which is not the case of | ə- |, | Ɛ- |, | ɔ- |. No doubt this is a proof that these vowels are modifications of some of the previous one.

The object of this work is not to discuss the vowel system of Ewondo. In the context of this study, the choice is to consider that Ewondo has a vowel system composed of 7 oral vowels and 5 nasal vowels as found in the literature of numbers of African languages.

	Oral Vowels		
	Front Vowels (*Palatal*)	**Central Vowel**	**Back Vowels** (*Velar*)
1st degree of aperture (*closed*)	/i/		/u/
2nd degree of aperture (*near-closed*)	/e/		/o/
3rd degree of aperture (*Near-opened*)	/ɛ/		/ɔ/
4th degree of aperture (*Opened*)		/a/	

Chart 10: The Ewondo oral vowel System

20. The epenthesis is the insertion in the speech of an additional sound (phoneme) which makes it possible to clarify, facilitate, or make more "natural" the speech. An example in French is the t of go or are there.

21.Jacques Evouna, *La morphologie nominale en Ewondo: La double préfixation*, Connaissances et Savoirs, P22.

	Nasal Vowels		
	front	Central	back
Closed	/ĩ/		/ũ/
Mid	/ɛ̃ /		/ɔ̃ /
Opened		/ã/	

Chart 11: The Ewondo nasal vowel System

Two successive homophone vowels can be heard in the hight run. They can also insert a near-vowel resulting in welarization (ua,ue etc.) or yodisation (ia, ie, iu etc.) phenomena.

Vowels Description

The phonetic tables present: 7 vowels for the first and 5 for the second.
As it has been said among, recently reference to Essono's (1984) studies, Tabi Manga (1992) restricts the number of fundamental vowels of Ewondo to five: | a |, | e |, | i |, | o | and | u |.

|a| nkándi: "underpant" |u| à-lùgh: "to marry"
|ə| à-ləb: "to advise" |o| à-lòd: "to skip"
|e| ú-títíe: "star" |ɔ| í-bɔgh: "haunch"
|i| í-dímí: "ignorance" |ɛ| à-lɛ̀: "draw"

However, we transcribe, with a near-closed line, the average vowels. These vowels can be classified according to three parameters:

1. The pivot point: as pronounced in front of the mouth or behind. Thus, the vowels produced in front are called front:

   [i] [à-hibi] ->"to suck"
   [i:] [à-kili] -> "sulk"
   [ə] [bebè] -> "frog"
   [e:] [à-kê] -> "to begin"

Beside the front vowels, we have those made behind and called, therefore, back vowels:

   [u] [à-fùlen] "to resemble"
   [u:] [*u*-mbûghi] "fracture"
   [ɔ] [*i*-lÓgh] "herb"
   [ɔ:] [ndÓô] "mango"

2. Another feature that better describes vowels is rounding effect. The vowels that respond to this characteristic are all posterior above presented. In phonetics, vowel roundedness refers to the amount of rounding in the lips during the articulation of a vowel. It is labialization of a vowel. When a rounded vowel is pronounced, the lips form a circular opening, and unrounded vowels are pronounced with the lips relaxed. In most languages, front vowels tend to be unrounded, and back vowels tend to be rounded. However, some languages, such as French and German, distinguish rounded and unrounded front vowels of the same height, and comparatively to Ewondo,Vietnamese distinguishes rounded and unrounded back vowels of the same height.

3. Vowels can also be described by the degree of opening of the mouth; thus, we will have features: closed, near-closed and open; a single vowel has a maximum opening:

[a] [zâ] ->"come"
[a:] [minkàmâ] -> "astonishment"

The vowels height

Vowel height is named for the vertical position of the tongue relative to either the roof of the mouth or the aperture of the jaw.
Differences in duration are perceptible and affect all vowels as well as certain substantives with dissyllabic and polysyllabic themes:

Vowels definition

The International Phonetic Alphabet defines seven degrees of vowel height, but no language is known to distinguish all of them without distinguishing another attribute:

- close (high)
- near-close (near-high)
- close-mid (high-mid)
- mid (true-mid)
- open-mid (low-mid)
- near-open (near-low)
- open (low)

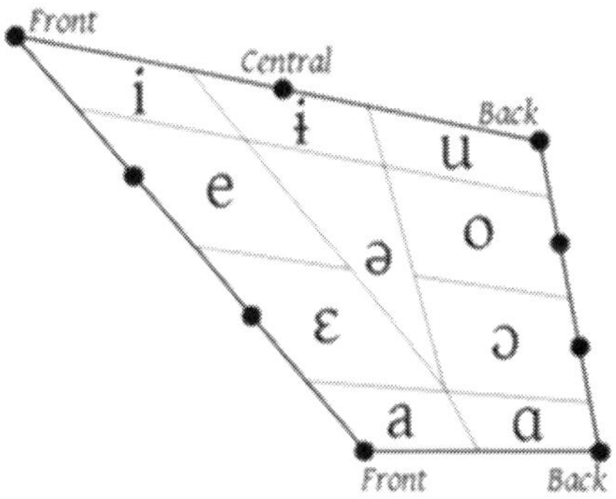

By the way, the ones define by Ewondo a those below:

[i] and[i:]: Front vowels, non rounded, orals, low and high respectively, closed;
[Ɛ] et [Ɛ:]: Front, non-rounded, oral, low and high vowels respectively, near-closed;
[u] et [u:]: back, rounded, oral, low and high vowels respectively, closed;
[ɔ] et [ɔ:]: back vowels, rounded, oral, low and high respectively, half-closed;
[a] et [a:]: central, non-rounded, oral, low and long vowels respectively, opened

The inventory of Ewondo vowels has two fundamental orders:

- The front stretched order
- The rounded back order.

The Ewondo language only have one vowel of maximun aperture.

Ewondo Phonological system

After studying the consonant and vowel system, at the end, i can approach the phonological Ewondo system as well:

	Points / Modes	Labiales		Apicales		Dorsales		
		bilabiales	labio-dentales	Alvéolaires	Palato-alvéolaires	palatales	vélaires	Labio-vélaires
Consonnes – Occlusives	sourdes			t			k	kp
Consonnes – Occlusives	sonore	b		d			g	gb
Consonnes – Constrictives	Fricatives sourdes		f	s	tʃ			
Consonnes – Constrictives	Fricatives sonores		v	z	dʒ			
Consonnes	semi-voyelles					j		w
Consonnes	Nasales	m		n				
Consonnes	Latérales			l		ɲ	ŋ	
Voyelles	1ère aperture					i	u	
Voyelles	2ème aperture					e	o	
Voyelles	3ième aperure					ɛ	ɔ	
Voyelles	4ième aperture					a		

Chart 12: Ewondo phonological system

C- Tonal and Syllabic System

We say that in the African languages and Ekang in this case, there are several elements of differentiation at the phonic level that distinguish them from each other: the accent, the tone, the deviation and the flexion. Although the analysis of the roots of words by the basic lexicon or genetic lexicon is more enlightening, it does not absolve us from requesting an analysis of the tonal system.

> *"Prosody is complementary to phonology in that it redefines itself as the set of phenomena that escape the division of the spoken string into phonemes, namely; rhythm, accent, intonation and syllabic quantity."*
>
> *Di Cristo, A. (2013)*

It is the different prosodies of the African language that give the uninformed people the impression that they are multi-lingual and make understanding difficult between the different varieties of the unique African language.

The Ewondo language as a certain number of African languages is considered as a tonal language.

The "tone" is defined in the authorized literature as a basic unit in tonology. It should be noted that as with his cousins, African tones are always punctual tones. In other words, there are no "melodic" tones like in Chinese or other Asian languages.

> *"[...] It should be known that absence of melodic tones does not mean absence of contours. A modulation on a syllable corresponds to the realization of several tones (two or three tones maximum) or the contextual realization of a given tone (for example, the downward realization of a low tone after a high tone)"*[22].
>
> *(A. RIALLAND, 2000)*

The tone organizes the melody and the transcription of the utterances. It is the manifestation of the semantic differences induced by the vowels.
Tones are discrete units in the same way as phonemes, but they are not segmentable: the tonal unit, called *toneme*, can not be perceived without the support of phonemes and does not exist without them. So they are considered as *suprasegmental* units.

22.Annie Rialland,Tonologie africaine et modélisation prosodique, paris 8, 2000

When one is a layman, one might think that the Ewondo as many of its sister languages are sung more than they are spoken. Indeed, the tones used are not used by the mere fact of chance, but they are in a very precise way. We could say more on the subject but it is not the object in the context of this work.

If we proceed to the semantic analysis of the suprasegmental units that compose the term: [kabʁa] which corresponds in English to the words (goat, etc.) we have:

- /à-kab/ →(to share) 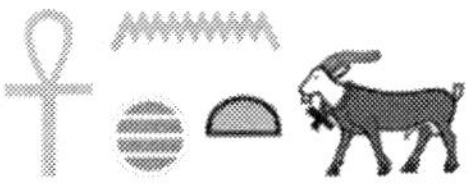
- /ʁa/ → (living energy) •

By the meaning of the word [kab: ʁa] rigorously, the purist Beti does not associate the goat but above all the sharing of vital energy. Which explains why no sacred ceremony can be done without the goat in African societies.

This example illustrates what is an Ekang language I mean not a simple juxtaposition of tones but a juxtaposition of elementary tonemes with non-invalid meaning. And it is this juxtaposition of suprasegmental units that organizes the lexicon. It is therefore important to master each basic unit to understand and know how is built an Ekang and an African language.

It is thus purposely that the scribe will represent the goat as the manifestation of the sharing of vital energy.

Suprasegmental units are therefore not devoid of meaning quite the contrary. Tonemes are bricks that structure the understanding of the African language and Ewondo in this case. The tremendous mistake according to me, was to study African languages using an inappropriate colonial technical matrix.

Some words are always spoken in a higher tone than the rest of the speech and others in a lower tone. On a lexical level this translates into a diatonia. This process often helps to distinguish what misnamed homonyms. In this regard, will see later how it all fits conventionally. It would obviously be very useful to indicate these differences in writing, but in reality it is not as simple as that because one does not write as one speaks. African languages have been written like those of specialists. Also the lack of perspective on the grammatical level leads to the deadlocks that are encountered today in the technical approach of these languages.

Also the grammatical requirements induce unsuspected realities on the technical level. Orally, there is: the high number of tones (5), border tones whose realization often comes out of the range of realizations of tones stricto-sensu, undergoing and inducing processes similar to those of other tones, so-called "floating" tones, ie tones without fixed association to a syllable or a more[23] and some of which are manifested by a lowering of the tones that follow (or "downstep[24]"). The "floating tones" that can be encountered in any point of a derivation, whether in the constitution of morphemes, words, phrases or even sentences.[25]etc.

It is only through the use of language that one can manage to put these tones in the ear. In this heuristic approach, it would be desirable to give each word its tone.

23.The **more** is an elementary sound emitted during phonation. Its name comes from the Latin mora meaning "delay" or "delay". Like many linguistic technical terms, its exact definition is the subject of discussion..

24. The Downstep consists of a lowering of the register of realization of the tones. The downstep can be a lowering triggered by a low tone carried by a carrier unit (syllable or more). We will then speak of " automatic downstep " or "downdrift".

25. Annie RIALLAND- P. SAUZET et A. ZRIBI-HETZ (eds.), Théories linguistiques et langues sub-sahariennes, L'Harmattan, Paris, 2004.

In practice this is not true because everything will depend on the prosody of the speaker. Also it will be the tone of the most important syllable of the word that will be the tone of the word. Many grammatical questions remain unclear on the question of the formalization of tones in Ewondo. The questions that are the subject of the debate concern prenasalization or semi-nasalization, the origin of "yodisation" or "wedisation", vowel neutralization, cumulative tones, augment, etc.

By neglecting to know the tones skills of the words, the unsophisticated (foreign) speakers who would make the effort to express themselves in Ewondo expose themselves to not being understood by the Ewondo allocutaries who would make of their subject a simple jargon / be- lob'lobo / or / ndzwabi Ewondo / when we guess mixtures of tones. I will be very curious to see Noam Chomsky speaking Ewondo.

The pitfalls in which unsuspecting people fall are numerous, related to homophones, processes such as apophony [26], tonal copy (copy of a tone on a syllable without tone), tonal propagation or tonal polarity (inversion of one tone with respect to another) etc. Indeed, certain suffixes may be in themselves devoid of tone and receive a tone-tonal tone, confusion in nominal-class chords, and so on. Abomo Morin has perfectly exemplified the purpose in his book in the following words:

> *"[...]The pitch usually indicates the relative height with which the vowels are pronounced. It is this tone that gives meaning to words and distinguishes them from each other.*
>
> Èbóŋ: the search for the sexual partner
> Èbɔ̄̆ŋ: shell mold
> Èbɔŋ: fruit of àdzàp tree
> Èbón: female sex
> Èbɔ̏ŋ: lover, concubine
> (…)"[27]

(Abomo-Maurin 2006)

These words seem different simply from the point of view of Western thought, but in reality taken in the African context, it is the projection of the same idea into different registers of perception of reality. It is not just a matter of tones. This is the way to express an idea according to the reality in which it is projected. That's why trading on tones here is boring misplaced or even a waste of time for me. This example alone is enough to challenge the supposed theory of the ideas world of Plato.

It is clear from this example of Abomo that a single reality is projected into the physical world while keeping the same vocalization by a simple tonal game.

All these words maintain each other a metonymic relationship. This point is important if you want to understand how to deal with ***mdw nTr***. That's why one draw in a sign will envolve another idea by homophony. This point highlights the limitations of the consonant analysis approach often used by comparative linguistics. The signified in this example is the same. This is why we do not understand the idea that initiates the notion of homonymy.

According to the legal standards, the Ewondo language does not escape the requirements of the languages with tones to know that, the melody is mainly determined by the realization of the tones and can realize any contour on any word. The contours are imposed by the tones and resist to the deformation. Thus, the more tones a language has, the less it will leave room for intonation-type changes.

The intonation fits with the degrees of freedom left by the tonal systems as well as in some points of the statements, in general, the statements of utterance, may be left to the expression of some kind of intonations. We must therefore be very careful to believe that tonal rules from other languages are fully

---

26. In phonetics, apophony is a phonetic modification consisting of a change of quality (or "timbre") of a vowel in a word during its history.

27. Marie-Rose ABOMO-MAURIN, Parlons boulou Langue bantou du Cameroun. Paris, L'Harmattan, 2006. P16.

transposable in Ewondo, it would be a serious mistake. / ùn-mbòkh /; / ùn-mbōkh /; / ùn-mbókh /; / ùn-mbôkh / ùn-mbǒkh / etc. Are all Kolo words with precise meanings and which are comprehensible by an exercised ear. The main issue here is not the tones but how do ancient Africans conceived their tonal sytem regarding their application fieds? Describing tones of a language is for my point of view a waste of time. The most important thing is the general rule that grounds the specific use of these tones.

Without a good practice of african languages and Ewondo in particular it is difficult to vocalize each sentence with all the tonal variations dictated by its good practice.

In the linguistics point of view, Ewondo being a so-called tonal language, in spite of everything and all proportion, it is possible to approach this language on a technical level like the other languages with tones. Thus we can proceed to a determination by tonal analysis in order to highlight their role in this language. This approach does not use a sonogram for tonal analysis nor the ToBi method[28] developed by J. Pierrehumbert et M. Beckman.

On this account, their identification according to the standard rules is carried out in Ewondo by the suprasegmental units in the sequences CV, CV: CVC, CVCV. According to some experts the Ewondo language identifies two types of tones; single tones and double tones:

- Double Tones → 2 identical spot tones
- Modulated tones → 2 different spot tones.

Depending on the variation levels, there are five tones: three punctual tones and two modulated tones. The average tone is strongly disputed by some specialists like Onguéné Essono.[29]

Whatever the effective height, it is sufficient to pronounce more or less acute, regardless of the starting point and no matter what the inflections of the voice.

Spot tones are known for their distinctive function in a set of homophonic signifier elements that appear in the language.

Examples:

- /Wa / "you"; /wâ/ "perish"; /wā / "hand";
- /zam/ "leprosy"; /zàm/ "appetite"; /zǎm/ "raphia palm tree";
- /ngwàn/ "cucumber"; /ngwān/ "moon"; /ngwan/ "young girl"; /ngwǎn/ "sparrowhawk".

When the same syllable carries several registers, there is at the same time modification of the musical height by passing from one register to another and modulation. In this specific case, which is relevant, it's as much height as modulation. For example, a high → low modulation is opposable to a low → high modulation, although the same heights are involved.

In Ewondo, there are, besides the punctual tones mentioned, modulations on the same syllable:

In the literature the tones are indicated using diacritics on the vowels:

---

28. ToBi: "Tones" and BI "Break indices" (pauses, glottal stops…). To & BI are the two sets of units used in the ToBI system. In the To BI system, two basic types of tones are distinguished: pitch accents and boundary tones. Let us add that we also distinguish "phrasal tones", which are of the level of prosodic groups. (Annie Rialland, Colloque "Langues subsahariennes et théories linguistiques", Paris 8. 2004).

29. Onguéné ESSONO, Grammaire de l'Ewondo. Essai de grammaire méthodique, Yaoundé, Cerdotola, 2012

- The high tone is indicated with the acute accent: ‹ á ə́ ɛ́ í ó ɔ́ ú ›;
- The average tone is indicated with the macron: ‹ ā ē ə̄ ɛ̄ ī ō ɔ̄ ū ›;
- The low tone, the most frequent tone, is indicated by the absence of diacritic: ‹ a e ə ɛ i o ɔ u ›;
- The tone is indicated with the antiflex: ‹ ă ĕ ə̆ ɛ̆ ĭ ŏ ɔ̆ ŭ ›;
- The falling tone is indicated with the circumflex accent: ‹ â ê ə̂ ɛ̂ î ô ɔ̂ û ›.

It is unthinkable to write the Ewondo language without the precisions offered by the tones. A copy of the exogenous rules as colonial languages would lead to not being understood for sure.

3.3.2. PHONETIC STUDY

A- PRESENTATION OF THE EWONDO PHONETICS

Rigorously on the basis of all the previous works (Diop, Obenga, Mboli, Ndigi, Ngom, Bilolo, etc.), which demonstrated a genetically related relationship between African languages and the Pharaonic language, this chapter seemed to me superfluous. I just submit to this exercise for form's sake. I will propose after this heuristic a complete comparative dictionary with all the words in the couple of languages.

.

If the two languages are derived from the Negro-Egyptian therefore all the Ewondo's words are obviouly in the Pharaonic language and vice versa. It's not just a matter of taking a few sets to make a hypothetical phonetic connection to show that there is a genetical link. To submit to this exercise is to accept the idea that Pharaonic is a non-human language that would have implanted itself at *km.t* by chance. This idea is totally absurd. To submit to this exercise is to accept the idea that the barely masked transliteration produced from the Oxyrhynchite Coptic is the only one that is valid for giving a vocalization of these characters. So what about other Coptic dialects? These demagogic and ideological positions are highly incapacitating for research and are useless for humanity.

That's why I'm not going to limit myself to the basic vocabulary, but I'm going to translate all the words of Pharaonic into Ewondo. As I said, I do not seek to vocalize the Pharaonic what interests me is to read Pharaonic from an African language and demonstrate that it is possible just as do the other dialects kopte, Swedish, American, Chinese, German and others. I therefore allow myself to record below a very small lexicon.

	Middle Egyptian	**Ewondo**
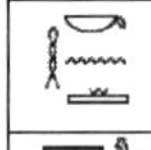	"hkn": to cheer, cheer	í-kaŋ: cheer; á-kaŋ: sing praises, to glorify
	"sndj": loin cloth	sánda: loin cloth
	"Haw.t": joy	mi-ntag: joy, pleasure
	"Dab.t": pile of burning charcoal	Mâgh/dagh: pile charcoal kam: smoky residue at the bottom of a cooking pot after cooking.
	"w'": one	wa, mb-wa, f-wa: one
	"w'b": purety, purify, to clean; purification, pure	á-woba, á-wâba: to bathe á-Wâren: purify
	"Dr": all, the universe (the hieroglyph represents a filled basket)	á-dzàli (verb): fill to the brim dyar: basket dyal: the village; country; city etc.
	"wD": order, to command	í-dzo: power, command. Á dzo (verbe): to command.

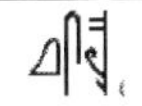	"ks"/ "qs": bone	í-vs: bone
	"mDw": words, speech	mə-dzó: words, speech, issue, things etc.
	"rn": name	ndán: sort of tambourine code name
	"kmA"/ "qmA": to create, to produce	á-kom: to create, to produce
	"bA": ram	ntomba: ram
	"bA": leopard	Zə: panthera ou leopard
	"Ab.w": Elephant	Zokh: Elephant
	"Ab.w": Ivoiry	u-mbang [Mbaŋ]: ivoiry á-sóŋ: tooth
	"bA": smashing; hoeing, weeding; destroy, devastate	á-bakh: to fell Í-bakh: hoe, shovel
	"Abt": family	Nda-bot: family (Human house)
	"yr": to supplant, to oppress	á-yir; á-yit: beat, correct with a chicotte
	"wS": to fall,	á-ku (verbe), fall., drop i-vus: offense; sin; peccancy; trespass
	"iA.t": place where the blessed live	á-yat: "the other side"
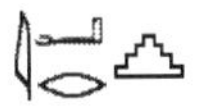	"iar": to climb	á-yələ: to fly, go up in the air. Á-yana: Climb above, reach the summit, put on it
	"nwy": flood, rising waters	á-yəmlə: flood, rising waters
	"Hwn": l'enfant, jeune homme	I-vôs, í-vÒm: nubile, jeune Mvóŋ: jeune pousse, germe, graine.
	"Hr": horus	á-su: visage, face, countenance ùn-nloh: head ùm-or : human being di-ís: eye Kírí: morning
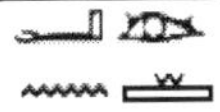	"an": être beau	Nyeb: beautiful; fine; nice; pretty
	"axm": extinguish, quench, destroy	á-dim: extinguish, quench, destroy
	"ibA": dance	á-bokh: dance yba/tchouba: thrust interjection while danc-ing
	it: father	í-ʒià: father
	"aA": great	í-wogh: big; tall ù-nya: mother
	"anx": mirror	í-yəna: mirror
	"anxy": eternal life	í-nyiŋ: life
	"anx.t": goat	í-kəlà: billy goat
	"war": to fly, to flee, to flee; fly against; to leap forward	á-yələ: fly away á-wulu: walk
	"wnm": to eat; wnmw: food	Ú-nám: puts, food, meal
	"wnn",: to be; to exist	nə: to be

	"wgi": chew	á-di: to eat
	"wA",: to be far, remote,distant	Ú-yap: far
	"wsm": butcher knife	Ú-kəŋ: knife
	"tni": eldest son	Ú-nto: eldest son
	"bn bn": sexual act, engender, have an erection; overflow	í-bóŋ: sexual act, layer á-bómbèn: sleeping with a woman Ú-nkón: the phallus á-kn: have a hard-on á-bin: purses, testicles
	"msi": give birth; to lay to be born; create; produce; make, manufacture, shape	á-soui: *lay, spawn; churn out* me-soui: *spawn* á-biálí: *birth* à-biálí: *be born, to spring*
	"bb.t": shore swirl	á-Kole: shore swirl
	"bin": bad, demoniac	à-bh: bad, ugly í-bira: bad, ugly à-nyàŋ: bad, ugly, desaster
	"bb.t": hole, cavity, niche	í-bh: *hole, cavity*
	"mr": to love, to bind, to link	á-díŋ: *to love* á-tíŋ: *a link, fastener*
	"m.t": virile, red-blooded; manly; potent	ndəm: *sperm*
	"kA.t": *vagina*	í-kokh: *coccyx* ù-ŋkat: *examination* à-kàŋ: *to adore* i-tûghà: *what is usually rubbed* ŋgout: *sexual area of a human body* à-kat:*to say* í-bón: *vagina; what makes barking*
	"kA.t": *work*	Ú-kwakh: *together work* i-tughà: *workman; boy* bí-sié: *work*
	"stp": *choose, cut, chop*	á-tob: to *choose* á-saŋ: *cut, chop*
	"DA": *forest, firelighter*	ndzàgh: *dead wood used to light the fire*
	"X.t": *breast of Isis, Isis's Blood, uterus*	Tou: *chest* Mə-ki: *blood*
	"sx": fraction, cut out	á-səkh: *cut, split, cut out*
	"isw": reeds	í-sóŋ: reeds
	"bni": A pleasant taste tuber	à-boura: sweet potato u-mbōŋ;[mbōŋ]: cassava
	"wDA": to be healthy	Mvouaè: to be healthy
	"qAA": *hill*	Ún-nko(l); [nko(l)]: hill
	"D": kind of bread	à-di: to eat i-di: food
	"Dadw": courtroom	Ndâ mə-dzəh: courtroom; tribunal
	"ky": *the other*	kala: *such; the other*

	"ky": *monkey*	Ko: *monkey*
	"khkh.t": *cough*	í-koukwè: *cough*
	"gAf": *long-tailed monkey, cercopithecus*	Nguî: *gorilla* nsakh: *monkey with yellow hairs*
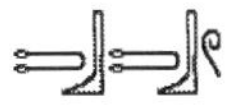	"tbtb": *to raise , increase*	á-təbə: *to raise* à-bét: *to increase, to climb* á-təbə-təbə: *get up; to rise (bulu)*
	"kA": *bull*	nyakh: *ox* Ú-nkáng, [nkáŋ]: *bull*
	"Dnd": to *be angry; submit; anger*	á-káng, [àkáŋ]: *to get upset*
	"nX.t": *strong, champion, a strong man*	ngul: *strength, power*
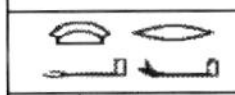	"xar": to *be angry; submit; anger*	Ú-lun, [oloun]: *anger*
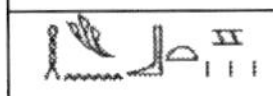	"Hnb.wt": limits	Mbgwakh: *borders* Bə-səŋ: *les limits*
	"km": to *finish*	á-kom: *finish repairing or manufacturing*
	"dwA.t": *prayer, worship*	Doumà: *prayer, worship* á-kaŋ: *to pray, to adore* í-kaŋ: *prayer, worship*
	"iAw": *prayer, worship*	í-kaŋ: *prayer, worship*
	"bnty": *pair of breasts*	Me-bèh: *pair of breasts* Me-nyaŋ: *pair of breasts; sucklers*
	"tAw": *to be attached, to capture, to bind*	á-tíŋ: *clip, link* Únkwal, [nkwal]: *clip, sting, rope*
	"nsw.t/sw/nsw/swtn": *ruler of Upper Egypt*	Ndzou, í-dzou: *authority* nkÓŋ: *authority; grade*
	"iw": *dog*	Mvu: *dog*
	"Sa.t": knife	Úkeŋ: *knife* à-saŋ, à-sàl: *cut out, split* à-tigh: *cut out, split*
	"ntr": God?	Ú-ntour, [ntour]: *full, holy* Tárá: *father* Ú-ntouar, [ntouar]: *unique* Ú-ntol, [nto(l)]: *Elder* Ndon: *all that is pure, without primer*
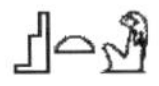	"As.t": Isis	Ú-ntour-etoa: *Holy see* Ú-ntour-ngwan: Holy (*gilr, woman, moon)* Ú-ntour-toŋ, [toŋ?o]: *holy-horns* Ú-ntour-í-dzo: *holy-authority* Ú-ntour-Ú-ndî: *holy-pillar*
	"wsjr": Osiris	Ún-nsuru: *The black* Ún ndimi: bedtime Ún nsoro: *phallus* Ún-nsor: *scratched* Ùn-nso dys: *single-eye* *Nso râ: the only one of râ*
	"kA": spirit, soul	Kon: ghost; phantom

	"Hr": horus	á-su: *face* d-is: *eye* Ú-nloh: *head* Elo/Eyo: *primordial ancestor* Kiri: *morning* Ú-nkor: *the savior, the one who dries* Ú-nkóŋ: *the top, the chef*
	"Râ/Rê": *sun, day*	Râ: *solar vital energy* Elo/eyo: *primordial ancestor, the vomited* Ú-nloh: *The head* Dzô /yop: *sky* Ùm-bokh: *channel*
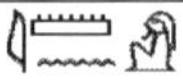	"Imn": *Amon*	Zámbà: *primordial ancestor* *I-mn: him-self* Tara-Zámbà*: primordial ancestor*
	"Imn-Râ": *Amon-Râ*	Zámbà-Elo:,*Dzô; primordial ancestor*
	"bes"*: Basa/Bès*	Zámbà: *Dzô; primordial ancestor*
	"zAb/sAb"*: jackal*	Zámbà: *primordial ancestor, Dzô*
	"inpw" *Anubis,sphynx*	Zámbà: *ancestor primordial, dzô* Ane-mvû: *similar to dog;*
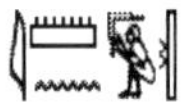	"Imn": *secret, hidden; hide, conceal*	à-sobo: *to hide* à-buri: *cover, shield* à-wôn: *badly hide* à-moŋ(lo): *hide, conceal*
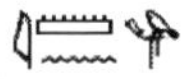	"Imn": *right, west, right side*	Ùn-Nómó: *right* í-moŋ: left; left-handed
	"stX": *Seth*	Tar'Ziŋ'i: *the jealous (ancestor)* Ngoul: *strength, power* á-saŋ: *to cut* á-souŋ: *to quarrel* á-sóŋ: *trap, question, to quiz*
	"Nb.t-hw.t": *Nephtys*	í-kari: crown, belt nab-naba: belt to climb trees ndáh: *house* ùn-mbon ù-nloh: *onction oil* à- fola Ntombà: the place of the lord I-lom, [elom]: *cobra* Anaba: *african name* Bâna; Naba: *african names; anointed*
	"Pr": House; household; palace; temple	á-fólà: *place, locus* ndáh: *house*
	"Pr-aA": *pharaoh*	Folà: *place* Nda-ze'e: *house of the panthera* Nda-íkôan: *unity palace*
	"pr-bity" Palace of the King of Lower Egypt	Nda-bot: the family Nda-fôŋ: *House of the bee* Nda bwity: Bwity's House
	"pr-nsw": palace of the king of high Egypt	Nda-ídzou: *house of authority* Nda-í-kaŋ: *House of Ekaŋ, Akang*
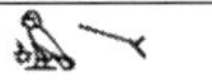	"bA": *soul*	Mbgwə(l): *metamorphic power*
	"nxx": to be old; old age, old age	Ún-ŋkoukh: *to be old; old age, old age*
	"db": hippopotamus	Ngub: *hippopotamus* Zoa me-ndib: *hippopotamus*
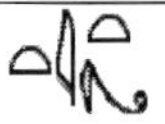	"tit": picture; form; figure; project; sign	á-tíl: *to write* ù-tíl: *writing*

	"kfa": to trap; to capture	á-kiabi: *catch, capture*
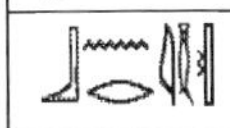	"bnr >bni": beautiful, nice, good	à-bəŋ: *beautiful, nice, good*
	"wa: bras": hand	Úà, [wa]: *hand*
	"km.t": Black country	Si-Kongo, si-Komo, La'a-Kam, Si-Úngola: Ngola's country
	"km.tyw": Egyptians	I-káŋ, mi-nkóm Be-Kongo, be-kam, be-komo,
	"nHsy": nubians	í-kaŋ, [kaŋ]: *kang people,* mi-nkóm: *boys*
	"nar": catfish	Ngo(al): *catfish* İ-yindi*: catfish*
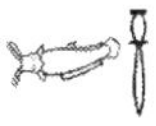	"Nar-mr": Narmer	Nyar-moro: *man-buffalo* Ngola: *royal fish* Mvás: *Clarias sp* Ú-ngwaŋ/ á-ngwan: *small finfish catfish*
	"km.t" carbon	Mâgh: *carbon*
	"Km-wr": the great black	Ùn-ŋkom-boro: *Dzô creator of Ekang*
	"Km-wr": flood, stagnant water	Ngoul ù-kos: *great storm*
	"Itn": Aton	à-twan: *circle, point, round* í-ton: *Eton* Nti: *lord* Be-'ti: *Beti, those of the lord*
	"niw.t": city, country, Village with intersecting roads	A-folà: *place* Mân: *roundabout* Un-nàm: *country, region* Dy-àl: *village*
	"ws.t": Thebes	A folà-bitoa: *the city of powers* A folà-ivs: *The city of Dzô*
	"Aby": panthera	í-bibi, [ebibi]: man eater zə*: panthera*
	"Aby": panthera	Ngu: brotherhood of the panther-men
	"f" [fy]: horned viper	Fì: *horned viper* À-kpə: *viper*
	"tS": sputum	Mə-tè: *sputum*
	"anx": life	i-ɲiŋ: *life*
	"sn": brother	Mà-dzáŋ: *my brother*
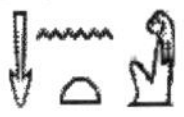	"sn.t": sister	Kál: *sister* Ngon: *girl*
	"Hm.t": wife	ùn-Mínga: *wife, woman*
	"Hm.t": servant	í tougà: *maid* Ún-nẐÔŋ: *maidservant*
	"Hm.t": copper, firmness of character, bronze	ŋgô: *Copper, bronze* ngut: *firmness of character; not easy to get on with.*
	"bA": skin	í-kob: *skin*
	"bA": smashing; hoeing, weeding; destroy, devastate	A-bakh: *smashing* i-bakh: hoe à-soum: *to hoe*

	"pth": the one who is the Cause of things	Tara: *the father* Úm-velghe: *the one who reveals*
	"Xnm.w": Khnoum	Ún-ŋkoum: *le pillar* Ún-ŋkóŋ: *l'authority* Zamba: *God*
	"imAx.w": venerable; reverend	Ún-ŋkùkùma: *Chief, Sovereign, Venerable*
	"nxx": flagellum	à-kpwàgha: *flapper* ùn-nkas: *swift* à-kpweñya: *flapper*
	"pth": create	á-kom: *create* á-vele: *to reveal, to awake*
	"tm": Atum God the creator	í-tam: *solitary, alone* ùn-NtÚm: *power, authority of the sovereign* í-tōbō: *the seat, the residence* Zámbà: *god*
	"tm": complete, entire universe	í-kóm: *a copy, a manufacture* à-mán: *to complete*
	"tm": close (mouth)	à-fət: *to close*
	"tm": die	à-toum: *to burn*
	"sA": worm; maggot	í-toum: *worm; maggot* Ún-nsɔŋ: *worm; maggot*
	"sA": hindrance for cattle	Mə-sɔŋ: *hindrance for cattle*
	"rA": stuffy; opening; speech, remarks, statements, speeches; language; intention	á-kàr: *to tell* Ún-ŋkàr: spe*ech examination, judgment*
	"bw": place, foot	à-bo: *foot, leg* vom: *place, locus, location* à-ko: *foot, leg*
	"D.t": body; image, body shape	Nyo(l): *body*
	"sA": son	í-sokh: *placenta* ùn-món: *son; child, kid; boy; lad* ùn-mɔngɔ: *young, nubile*
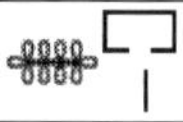	"sA": cowshed	Mə-sóŋ ou ndà-mə-sóŋ: *cowshed*
	"sA": back	Mvous: *spine, reverse, back etc.*
	"wHm": donkey	Ún-mbokh: *draft animal, donkey*
	"aA": door	Ún-mbèh: *door* Ún-mbokh: *prison*
	"aA": right here; there, there	A-lıà, á-và, và: *here,there*
	"SA.i": pig	Nsóŋ'ó: *snout, pork* Ngou: *pork, pig, swine, hog*
	"Styw": tortoise	Kùlù: *tortle; tortoise*
	"ib": heart seat of consciousness	Ún-nəm: *heart* i-vù: *the soul, resemblance* à-yi: *wish, desire*
	"Haty": heart	Úŋ-kumden: *the throbbing; heart* Ún-nəm: *heart* à-kout: *to hit, to flap*
	"Snkkw": darkness	à-kîghdi: *darkness* ùn-ŋkùkh: *darkness*
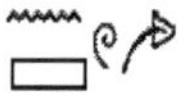	"nSw": flow	à-lène: *flow* Ún-nsəkh: *flow*

	“iwms”: inaccuracy, lie	Mə-seghga: *flattery; sweet talk* Mi-nswas: *inaccuracy, lie*
	“xrw”: *men*	Bor: *people, human* Bi-ŋgaŋa: *forbidden men of meat* bi-ŋkôna: *mature men*
	“bk.t”: Territory, constituency etc	Mvokh; mvogh: *country, village, city* Ngap si: *Territory, constituency etc*
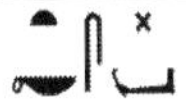	“tks”: torture, torment; rack; grill	à-doukh: *agonize with suffering*
	“swhn”: demolish	á-ndamn: *demolish; destroy etc.*
	“sbi”: *to rebel; rebel, revolted*	á-tsame: *to rebel; rebel, revolted*
	“sSnw”: ropes, ropes	mi-ŋkwal: *ropes, ropes* mi-ŋsinga: *locks*
	“ibH”: tooth; defense	Úmbaŋ, [mbáŋ]: *defense* á-sóŋ: *tooth*
	“dA”: to shake, to shiver; to quake	á-sugh: *to shake, to shiver; to quake* kām: *quake*
	“Dar.t”: scorpion	Olərе: *scorpio*
	“siwy”: announce; to complain	á-dzo: *to say* bi-dzódzô: *words without control*
	“mDH”: stonecutter; Carpenter	Mə-keŋ: *stonecutter; Carpenter*
	“gws”: squew; squint	á-yerbe: *squew; squint* Ún-nləŋ: *squew; squint; crooked; awry*
	“nmw.t”: net	á-vam: *mesh of the net* á-voar: *net* ù-mvînya: *net*
	“Ssp”: take; accept; to receive; to attribute to oneself to catch; buy	à-kap: *to catch; to snare*
	“Hnw.t”: mistress; paramour	Ùn-Mingā: *wife, woman*
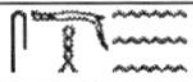	“smHy”: flood; to drown; to sink, to sink	á-sôle: *flood*
	“wr.t”: sacred cow	I-mvul: *buffalo*
	“mSa”: soldiers, army; infantry; team	Ún-wana: *soldier fighter, warrior* á-sap: *surround a village of soldiers* mə-sè: *feathers* á-tsàm: *fight* Ún-mfân: *arc; bow; longbow* á-koŋ: *lance, javelin* ngap: *fraction, share, part*
	“sqbbwy”: Bathroom	Ngab-mə-waba: Bathroom
	“st.t” soil; sol; ground	Tsia, si: *sol, soil, ground* À-Ko: *place, locus* Mvokh. *land country* I-to: *locus, sol; soil, ground; seat*
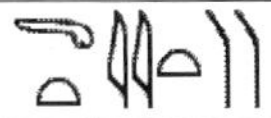	“mtyt”: straightness, accuracy	Bəbəlà: *straightness, accuracy*
	“xtw”: continued, subordinates	Bi-tùgha: *the servers*
	“qni”: to embrace	A-wubèn: *embrace; to kiss*
	“nqwt”: humidity; moisture	mi-‘nkout: *clowds*
	“fAk”: bald	Mfiaŋ-‘nlô: *bald* Fyor-‘nlô: *bald*

	“aAa”: negative influence	á-bh: *bad* í-kyeŋ, [kyeñ]: *harmful influence*
	“ssm.t”: mare	í-kabli: *horse*
	“qaH.t”: beef shoulder	un-ŋka(k)h: *upper end of an animal spine.* Un-ŋkakh: *the one who ties*
	“wsD”: to get tired of, to be tired, to get tired; die.	Á-təgh: *get tired, be weary* som: *tired* à-wou: *to die*
	“nmi”: travel; cross	A-daŋ: *travel; cross* A-daŋ mi-‘nam: *travel*
	“Hryw-say”: nomads	Ahu-Tsè: *those who swallow dust* Hausa: *dead of dust*
	“nms”: nmes	Ún-ntóm: *hat, headgear* Afakh: *hat, headgear* í-kóngó: *helmet*
	“sm”: walk	Mə-sima: *beginning of the child walking* Ndome: *walk with stilts, long legs* á-zəmlə: *to make someone walk*
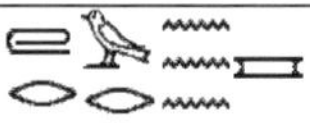	“pXr-wr”: Euphrate	à-wulú-‘nkwará: *the meanders*
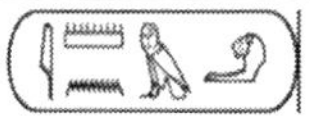	“Imn-m-HA.t”: Amenemhat	Zamba à-neu à ‘nem: *The first Path* Zamba à-neu à ‘nkukh: *In his own heart*
	“sA-Rê”fils de Rê	í-sōkh Elo-itam: *the unique son of Dzô*
	“Ax-n-Itn”: Akhenaton	Nti-okn: *Lord of the ring* Ú-kn-nti: *the lord’s ring* Nkn: *proper name of peoples Beti* á-twan okn: *the bud of the sun* á-twan-ikaŋ: *Ekaŋ excellence, the leader* of *the Ekaŋ* Nti-Ekaŋ: *Lord of belivers* á-kəŋ-à-twan: *The brillant* Ú-kéné-à-twan: *The point of the ring* Nkáŋ’a-à-twan: *the very valiant* nkáŋ-à-twan: *Whoever praises the brilliant* Etc.
	“itrw”: Nil, river	Ún-ŋkoŋlo: *along, roaring* Ún-nloŋ: *The nyong (cameroonian river)* Ún-ŋkoŋ’o: *kongo, du leopard’s place*
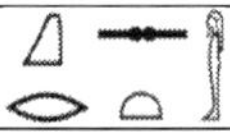	“krs.t”: mummy	Ún-ŋkoro: *the dried*
	“krs”: to bury, to inter	Soŋ: *tumb*
	“sqr”: hit, molest, beat	á-kut,[kwt]: *hit, molest, beat*
	“qrsw”: coffin; basket	Ùn-nkoun: *basket*
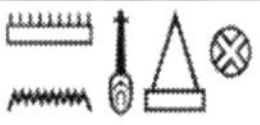	“mn-nfr”: Memphis	à-beŋ-‘nam: *the city of beauty* à-beŋ’ə: *The beauty*
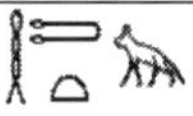	“HTt”: hyena	Zoé: *hyena*
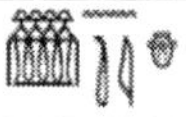	“xnti-ib”: happy heart	Ún-nəm-mintag: *happy heart*
	“ama”: anoint, rub, coat	á-wobo: *anoint*
	“sgm” grind, spray	á-kokh: *to crush*
	“pAD”: knee	à-bóŋ: *knee*

	"HkAw": magic, magic words	Mə-ŋgāŋ: *magic*
	"ib": thirst	i-kor: *thirst* i-vèh: *thirst*
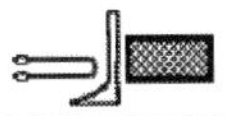	"Tb": cage	Ún-ŋkwab: *box*
	"mDw-nTr": words of god	Mə-dzo-me-ntour: *the holy words* Mə-dzo-me-ntouar: *the words of the unique* Mə-dzo-me-tara: *The words of the father*
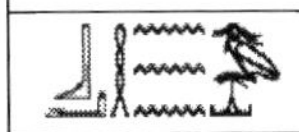	"baH": many, abundant	à-boui: *many, abundant*
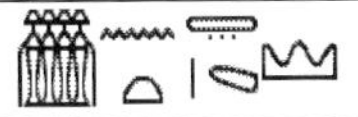	"xnt-tA": southern country, Nubia	Ùn-ŋkot ùn-nam: *dry country, desert*
	"mD.t": oil	Ùn-mbon: *oil*
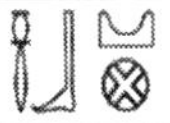	"AbD.w" Abydos	Nkolzokh: *elephant hill* À-fola zokh: *Dzô's place*
	"Geb": Earth	à-dzəb: *to bury* Si ndon: *Earth* Dzô: *primordial ancestor*
	"di": give (white bread)	à-twé: *to give* à-ke: *to give* dwé: *palm* í-di: *food*
	"niw.tyw": citizen	me-yoŋ: *lineage, family; household*
	"iwn.yw": The Anous	Anyû: *those of Dzô* Agni: *people of Ghana* bə-nti: *those of the Lord* etc.
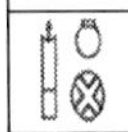	"Iwnw": Iounou, Hliopolis	À-folà i-won: *the place of Dzô*
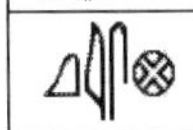	"Qis": Cusae	À-sw à kô: *the hiding place of Dzô*
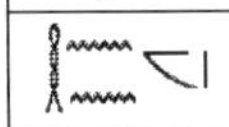	"Hnn": hoe	í-ngwàŋa: *hoe, big adze*
	"smA.w": darkness, dusk	Ne-kighdi: *darkness, dusk* á-dikh: *burn, coal*
	"mH": hold, grasp understand, capture	á-bih: *hold, grab, capture* á-vəm / à-həm: *intake (tabacco)* á-kab/à-kiabi: *hold, grab, capture*
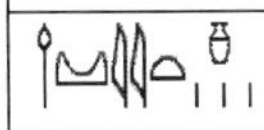	"HD.wyt": lamp, candlestick	mi-'ŋkoukoui: *rays of light*
	"nr": shepherd	Un-mbâla: *shepherd; herdsman*
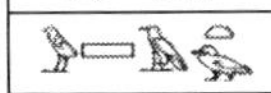	"wSA.t": whistling duck	Ún-nsōŋ'o: *the beak, the duck*
	"nm.w": dwarf, small man	bə-koué: *pygmies* man: *small*
	"nfr": zero	Nkor: *dry, empty (comes from the expres-sion wo unkor)*
	"Hr.yw-Say": nomads	Bə-Tsè-mə-sû: *Hausa*
	"dmw.t" operation, incision	à-toubi: *pierce, drill*
	"pA" Having done in the past	à-nga: *he has accomplished* à mbə: *he has accomplished*
	"pA": to fly	à-fap: *wing* à-bét: *climb, ascend, to grow up*
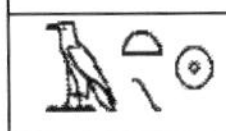	"A.t": moment, instant; time	í-tié: *moment, instant*

	"tA": burn	à-dikh: *burn*
	"iAk.t": garlic, onion, leek	à-yàŋ: *garlic, onion, leek*
	"HqA": sovereign, ruler, dynast	ùn-ŋkóŋ: *authority*
	"qr.t": door lock	ngaragh: *entrave, hamper, door lock*
	"antyw": dried myrrh	u-toui;[otui]: *myrrh; encense.*
	"xbA": destroy, devastate, ravage; get around	à-bakh: *fell, tear down* à-tsam: *devastate*
	"bw.t": abomination	à-béh: *mauvais, bad, abomination, evil* à-gnàmàh: *abomination*
	"nw": hunter	ùn-nsôm: *hunt; hunting*
	"m.t": thick	à-tigh, à-lighdi: *thicken* ù-ntip: *thick* A-fip: *thick*
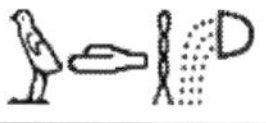	"wdH": pour, empty	à-souh: *pour, empty* à nekh; à nyukh: *pour, empty*
	"TA.ty": vizier, prime minister.	Un-ŋkukuma: *chief*
	"ir.t": vine	á-Yokh: *vine, wine* Kwara: *fermented corn alcohol*
	"iwrti" ou "irt": Yareth	Si-rÂ: *heath of The Black* Si-nkor: *land of the savior* Si-nti: *land of the lord*
	"srw": goose: duck	Soghe: *anactida, duck, goose* Tsir: *meat* Ù-nkor: *savior* i-di: *food* i-loli: *duck, goose* Kos: *parrot*

3.3.3. Comparative Morphology

3.3.3.1. Numeral and Ordinal Adjectives

A-1. The Ewondo and Pharaonic Notation Systems

The digital system is the system of writing numbers that we usually use in everyday life. The representation of numbers, limited to positive integers, such as 532 650 is a succession of digits.
Among the elements of the basic vocabulary, the names of numbers occupy a good place because of their frequent and ancient use in everyday conversations. It may therefore be useful to compare the notation systems between two languages that are supposed to be genetically linked to be sure that they are identical so that they can easily be switched from one language to another.

But the comparative question of notation systems in Africa is a complex issue that has been the topic of extensive research [30] from researchers like Dominique Gugan[31], Abdoulaye Elimane Kane, Henri Magdalena etc.

30. Gorges BALANDIER et Jacques MAQUET (dir.), Dictionnaire des civilisations africaines, Paris, F. Hazan, 1968, article "Numération", p. 299.
31. Dominique GUEGAN, Enseignement et mathématiques en langues africaines, Paris, ACCT.

In the context of this heuristic it is just to try to investigate if the two languages have the same notation system, to know if the terms used in one system are also used in the other and are semantically applicable. This approach inquires in the backdrop the issue of the notation systems that have been used in Africa: binary, quinary, octal [32], decimal, duodecimal, or vicesimal [33] since the Homo sapiens was using the Negro-Egyptian language etc.

The whole interest of this endogenous notation systems study could make it possible, from the mathematical documents still available. It can help to understand the mathematical vision of the world as stipulated by the scribe /Ngamessi/ ***IaH-mss***, "Iâhmessou" in the Rhind papyrus exhibited at the British Museum in London since 1865.

> *"Correct method of investigation in the nature to know all that exists, each mystery, all the secrets. By the Ahmès scribe who copied this copy".*
>
> *IaH-mss/Ngamessi*[34]

From the moment when one can know all that exists by numbers, it is possible then to operate for Africa and humanity of great upheavals.

There is therefore a double imperative here, the first is to trace the parallels between the Ewondo notation system with the one used at ***km.t*** certainly, but above all, to restore, where appropriate, the morphological and semantic links that could explain the apparent differences in vocalizations.
This therefore involves mobilizing a number of technical tools, including comparative linguistic tools at certain times, to allow for a broader and wider approach on the continent.

Syntax of Cardinals

In passing, it can be noted that, on the syntactic level as a rule the numbers are placed generally after the name they determine.

Middle-egyptian: || ***ngAw*** "2 bulls or 2 oxen with long horns"
Ewondo: /nyakh i-bèh/ "two (2) oxen"

But just note that in Ewondo the numeral can be ante or post-posed in relation to the substantive without fundamentally changing the meaning of the sentence.

Ewondo: /bor à-wôm/ "10 persons"
Ewondo: /à-wòm bor/ "ten people"

So let's simplify note this study that, the general construction is done according to the diagram:

Substantive + numeral

Concerning the agreements between the substantive and the numeral in Ewondo, let's take a simple example. Let's build it on the model above educated (substantive + numeral).

1- ***ù***m-or ***ù***m-bókh "one person"
2-**B**'-or **bə**-bèh "two persons"
7- B'-or zambgwa "seven persons"
8- B'-or mwôm "height persons"

32.COMPIEGNE, Louis DUPONT, marquis de, Afrique Equatoriale Gabonais Pahouins Gallois II, Okenda Bangouens Osyeba, Paris, 1875-1876

33.Abdoulaye KANE, Les Systèmes de numération parle des groupes ouest-atlantique et Mandé: contribution à la recherche sur les fondements et l'histoire de la pensée logique et mathématique en Afrique de l'Ouest, ANRT, Lille, 2001, P25.

34. This name is often worn by women. The prefix IaH means moon but also wife because every woman is associated with the moon or the panther. We have at the XVIIIth Dynasty King's wife "Thutmosis I" called iaH-mss. The famous scribe who recopies the mathematical papyrus certainly bears the name of his mother. This name Ngamessi is common in Ekang country.

3-**B**'-or **bə** -lèh "three persons"
4-**B**'-or **bə** -nyî "four persons"
5-**B**'-or **bə** -tàn "five persons"
6-**B**'-or **bə** -samèn "six persons"
9- B'-or i-bul "nine persons"
10- B'-or à-wôm "ten persons"

It is easy to observe that from 1 to 6, in Ewondo language, the numeral is read with the prefix of the substantive by a connective particle generally /**bə** / which makes the accordance and the link in gender and in number with the substantive. And from 7 to 10 in the example above, there is no mark of link. But in reality this rule could be applied to all numerals greater than 6. We can therefore arbitrarily define the first group as the one of the dependent numerals and the second (the right one) as that of the independent numerals. Is this the consequence of a change between notation systems? We can assume that.

It should be noted that this famous genitive connective particle  ***m*** or ***n(y)*** is attested in Pharaonic in different synchronies by specialists like Gardiner or Lefèbvre. They do believe that this genitive particle is observe only when the numerals are related to large numbers.

Examples:

Middle-Egyptian: 120 *n Abw* "120 elephants" (Urk.IV, 893,15)
Middle-Egyptian: *HHw nw rnpw.t* "millions of years" (Urk.IV, 358)

A-1.1 ***The cardinals adjectives***

In linguistics, the natural whole numbers: zero, one, two, three, etc. are called cardinal numeral adjectives. In Africa there are different notation systems such as binary, quinary, decimal etc. The current notation system of the Ewondo is decimal.

N°	Ewondo	Middle-egyptian		N-E	Sens
0	Ún-nkór/àloŋ/mɔ-mɔ	*nfr*		*$k^{j}un^{w}$	*Absence*
1	Fógh/fokh/f-wakh/foa'	*wˁ– (w / it)*	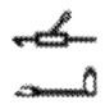	*$k^{w}ək^{h}$-	*alone*
2	Bãh/bèh	*sn– (wy / ty)*		*$k^{w}ək^{h}$-	*other*
3	Lãh/lèh	*ḫmt– (w / t)*	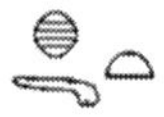	*$xiř^{w}ə$-$kiř^{w}a$	*tree*
4	Nyĩ(n)	*fd– (w / t)*		*$xiř^{w}ə$-$tiř^{w}$	*fingers*
5	Tán	*dỉ– (w / t)*		*-$k^{hj}un^{w}$	*Calculation*
6	Samâń	*sỉs– (w / t)*			*gather*
7	Zamgbwa(l)	*sfḫ– (w / t)*			*full*
8	mwÔm	*ḫmn– (w / t)*	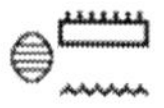		*Stack, pack*
9	I-bu(l)	*psḏ– (w / t)*			*The multitude*
10	A-wóm	*mḏ-(w / t)*			*Stack, pack*
20	Mewóm mebãh	*mD-te-(w / t)*			*2Stacks,packs*
100	*Ún*-ntét, [ntét]	*Snt-(w / t)*			The spiral
500	*Mi*-'ntét mitáń		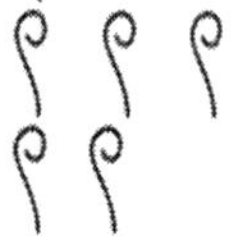		5 spirals

10^3	Tóyina= **(mH-tA)**	*xA-(w / t)*		1000 cubits of earth
6×10^3	Tóyina I-samań	*sìs-xAw*	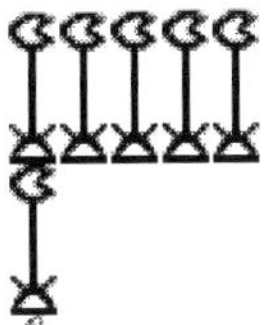	Six thousand 6000
10^4	Tóyina- Awóm	*DbA*		The finger
10^5	*ùn*-Ntét- Tóyina	*ḥfn-(w / t)*		tadpole
10^6	I-xùx;[è-kùkh]	*HH (w/t)*		the trunck
10^9	i-ndùndùm			the multitude

Chart 13: comparison between the Ewondo and Egyptian numeral systems

At first sight, apart from the decimal system of notation, one might think that the Ewondo and the Middle-Egyptian have nothing jointly. In fact, we must begin by a comparative study between the Ewondo terms and those of the Middle-Egyptian to see if the two systems are identical in all respects.

As a reminder, the Pharaonic Egyptian and the Ewondo language are all stemming from the Negro-Egyptian. So the reconstruction of the notation since the common ancestor can explain the forms observed in both languages, similarities and apparent differences.

The zero digit

It is related to the idea of an absence. And therefore, it can be translated as a negation. In this account, there are several Ewondo expressions to translate this idea of non-attendance:

Ewondo: /tèghè i-kpwap/ "without anything"
Ewondo: /tèghè-nyîx/ "without the slightest"
Ewondo: / tèghè dyôm/ "without any thing".
Etc.

It is especially the lemma Ewondo: /tèghè/ which plays the role of a negation tagger.
In addition to this type of negation, there is another who has taken the typical morphology of the African negative. This negation looks morphologically as a superlative construction and can be recognized by a repetition of syllables.

Ewondo: moh-moh □ "empty-hands"
Ewondo: si-si □ "empty, freedom, released, unattached"
Ewondo: zəzə □ "without; nothing, zero"
Ewondo: zeŋ-e-zeŋ□ "nothing, zero, empty".
Ewondo: əʔɳ- əʔɳ □ "no, nothing".

Apart from this morphological structure inherited from the Negro-Egyptian, the idea of zero can also be translated in Ewondo language by its scriptural shape: / à-lɔŋ / "the hole".
Certainly this expression of zero is no longer present in all the panel of African languages, it certainly can be found in many of them.

But one of the side that has persisted in Pharaonic Egyptian is the expression Ewondo / *ùn*-ŋkor / "dry", this idea has given rise to a semantax that has persisted until today in many languages such as English, namely: "To be dry" This expression refers to "to be broke, to flat broke etc".

Middle-Egyptian: ***nfr*** "zero"
Middle-Egyptian: ***sSrr*** "lessen, diminish, shrink"
Ewondo: /Ún-ŋkɔr/ □/wa ù-nkor /"dried hand"
Ewondo: /Ún-ŋkɔr/ □ /mi-ŋkor mi-moh/ "dried-hands"

Ewondo: /à-lɔŋ/ "zero"
Ewondo: /à-fɔkh/; /*ùn*-mfɔkh/ "vacuum; blank,; hollow, bare; zero"
Ewondo: /*ùn*-mfər/ "the closed; zero"

If we first of all focus on the expression ***nfr*** "beautiful; good", we could suggest an explanation using the sign F35 of the Gardiner list, but I will approached this issue from another angle. On this account, let us ask ourselves the question of what makes it possible to affirm that ***nfr*** in Pharaonic corresponds also to the word [nkɔʁ] "dry" in Ewondo language?

The answer to this question will kill two birds with one stone. It will explain why this expression is related to the beauty / good and at the same time, to explain why it also refers to zero.

From my point of view, the answer to this question is correlated with the african mythology character ***wsjr***. Indeed according to the abundant literature on the character ***Wsjr,*** he bears the qualifier of ***wnn-nfr*** that is to say, the perfect being. In short, the one who is good and beautiful.

Ewondo: /ù-nə/ "you are" equals the **Middle Egyptian**:*Wn* and **Middle-Egyptian**: *nfr* equals to **Ewondo**: /á-nyep/ "beautiful" in Ewondo.

By the way, let us observe the phonetic proximity between the term *Unnefer* (wn-nfr) and the word *Universe.*

After this, let's remind that feelings in general are in the realm of the heart.

-He is good person.

This sentence generally means: "he is kind-hearted".
The semantax related has therefore gone through the ages.

Ewondo: *ùn*-mbəbə *ùn*-nəm "to be kind-hearted"
Ewondo: *NE-FE-FE-FE*[35] *" Desire of the heart, thirst for something"*

Example 1: */Mezim ma bo me ùn-nlem* ***ne-fe-fe-fe/*** ***"****I wanna drink water".*
Example 2: */ Byôm bia bo me ùn-nlem* ***ne-fe-fe-fe/*** ***"****I have the desire to have wealth".*

Note: The tripling of the particle "fe" in the morpheme /nefefefe/ simply marks here a very great desire, and corresponds to the three slashes of the plural /// commonly used in Pharaonic scripts as plural tag.

According to the legend, ***wsjr*** was mummified by his son ***inpw*** *(cf illustration n°25)*. He is also called in the funeral register the mummy. The words for strips are strangely similar to the term ***wn*** associated with it. The purpose of mummification is to make the body rot-proof. The body is dehydrated. It becomes dry at the end of mummification operations. In the current Ewondo synchrony the word for dry is [nkor]. It is thus that we must understand how dry and beautiful/good are linked in the African

35.Samuel GALLEY, Fang-French dictionary. and French Fang followed by a fang grammar, éditions Henri Messeiller, Neuchâtel, 1964, p.394

ontology. The words beautiful / good and dry share the same character. In the african ontology ***Wsjr*** is the dried mummy and also the best meaning the beautifull and the good human being.

Middle-egyptian: ***wnxyt*** *"Mummification strips"*

What strikes the most when we compare the two written forms mobilized here is, the presence of the hare on both scripts.
Without any knowledge in an African language, we could observe this:

Figure 5: WSIR and his hare ears

Middle-egyptian: : the hare is associated with the heart
Middle-egyptian: : the hare is associate to strips or placenta.

What characterizes most the leporidae [36] in general is to possess long hind legs, very muscular which allow them to propel themselves by very fast leaps when necessary.
Let's question the lexicon of Ekang languages on the question of the jump or regarding hares.

Ekang: *NE-FAR*[37]*: (vb. Far, jump). "jumping idea"*

Example1: */Zire a far **ne-far/** **"**the classy jumps **ne-far".***
Example2: */Falge ŋgwène **ne-far/** **"**remove the iule with a stick".*

So the good reading of the triconsonant triliterate ***nfr*** is ne-far or nefe in the Ewondo language.
Nefar is basically the onomatopoeia translating the noise emitted by the friction of the grasshoppers wings during their movement through the ruts. By idea association, one can note a morphological similarity between the thighs of locusts and those of leporidaes. It is therefore this similarity in my opinion that leads us to retain this idea here.

The best interpretation to give to the consonants skeleton ***nfr*** is in the meaning / fara nənə / "the high priest".This interpretation is the best when we are talking about African spirituality.
Wsjr is the perfect ***nfr*** he is also the mummy ***nfr*** "dry". It is by this way that one must understand the semantax of the scribe which associates the script zero ***nfr*** with the trilitera ***nfr*** of the beauty. The Ewondo's expression / *mi*-'ŋkər *mi*-moh / "dried-hands" lights up here with a new day how to interpret the idea of zero.

We could also have remember that ***Wsjr*** is likened to the solar star from his purely cosmosgraphic point of view.

Middle-egyptian: ra "sun"
Ewondo: /ùn-ŋkor/ "the dryer"
Ewondo: /*mi*-'ŋkor *mi*-mwa/ "dried-hands; zero"

---

36. Leporidae are a family of lagomorphic mammals including hares and rabbits.
37.Samuel GALLEY, Fang-French dictionary. and French Fang followed by a fang grammar, éditions Henri Messeiller, Neuchâtel, 1964, p.394

In addition to these explanations, one could add the fact that ***wsjr*** was trapped in a sarcophagus during a repast organized by his brother ***stX*** "Seth". The rounded shape of the zero reminds for sure the idea of a curved line that closing in on itself. There is thus in Ewondo language, an homophonic relationship and a proximity of ideas between the shape of zero and the idea of confinement of ***wsjr*** in the sarcophagus.

Middle-Egyptian: ***wsir*** "Osiris"
English: *zero; wiser; wizard; wise; etc.*
Ewondo*:* /*ùn-m*fərèn/ "confined, close in"

Middle-Egyptian: ***nfr*** "zero"
Ewondo*:* /*ùn-m*fərèn/ "zero"

The causative[38] ***sSr-*** is also linked with ***wsir*** who is called the scratched /*ùn*-nsor/ in the African mythology.

Middle-egyptian: ***sSrw*** "linen cloth; linen bags; cereals; things, actions"
Middle-egyptian: ***sSr*** "pronounce, express; to dry, to dry; arrow"
Middle-egyptian: ***sSr.t / sASrt*** "cake or bread"

This is why, in my opinion, the paradigmatic dimension is an absolute necessity to understand these characters. We can not be satisfied with a mere alphabetical dimension or simple conjectures based on devitalized observations.

To end about zero, the Indians called the zero: *sunya* that means, emptiness, the word also exists in Ewondo is /*ùn*-nsɔh/"nude". Translated into Arabic the word gave *sifr*, which was repeated in Latin a few centuries later giving *zefiro*. We forgot the *fi* and we got *zero*. This *sifr* finally designated the entire collection of symbols for writing numbers, numbers: 0, 1, 2, 3, 4, 5, 6, 7, 8, 9.
On the diachronic plane we can suggest this evolution of the term used to designate the zero: **nkr > nfr* and zSr->zSr>*zfr > sfr > zr*

Etymology

The Arabic term ***sifr*** is etymologically linked to:

English:*suffer* "bear, brook, ache etc."
French: *soufre, frousse*; *subir*; *souffrir* "bear, brook, ache etc."

Indeed, before his death, ***stX*** inflict suffering to ***Wsjr***.

Historically and aside of the field of intellectual fraud, Zero is an African invention found in the papyrus Boulaq 18. Indeed, The Papyrus Boulaq 18 is an Ancient Egyptian document found in 1860 AD in the tomb of the scribe of the great enclosure Neferhotep.

---

38. In linguistics, the causative is a grammatical scheme that makes it possible to express a subject that is the cause of the action performed by a distinct agent and described by the verb. Technically, we observe that the construction increases by a causative actant the valence of a verb. In English, one typically employs doing (which also sometimes makes doing), for example:
Example: I had him repaint the wall. I gave him a ride.
Some linguists make a distinction between factitive and strict causative according to the transitivity of the verb affected: the factitive then applies to a transitive verb and indicates that the subject makes an action by an agent other than himself, while the causative is applied to an intransitive verb and indicates that the subject determines a transformation affecting the second actant. Others treat these two terms as synonyms.

The numeral 1

When the numeral follows the adjective numeral cardinal it is in accordance in gender and in number with the noun in Ewondo.

The type "fɔkh"; "fɔgh", "fwà" generally served for notation.

One thing and only one thing:

As a cardinal numeral adjective, there is not only one and only way of naming numeral 1. As a general rule, it enters into a scheme of accordance respecting the morphemes associated with it. In fact, its vocalization will be associated with the gender and the number of the name with which it is associated.

The different types of expressions for the numeral 1 that are used at the femine in Ewondo are: |d'–á|, |dý–á| and |dỳ–á|.We will see how to deal with each of them farther

Ewondo: di-dýá di-dýá □ "one after the other"
Ewondo: dy-óm di-dýá □ "only one thing"
Ewondo: dy-óm dýá □ "one thing"
Ewondo: dzōm d'-á □ "one heap"

Next to the previous forms of the numeral one, in its notation system, the Ewondo also uses the masculine singular form */ùm-bɔ*kh/ "one"

Example

Ewondo: /*ùm*-or *úm*-bɔkh/; [mbɔkh] □ "one person"

Middle-Egyptian:  ***wa*** "one" the scribe offers both characters:

Note

At the beginning of the words,  ***w*** hieroglyphic followed by occlusives or especially nasals n, m, regularly falls in African languages. It seems that this is a phonetic law.

If phonetically we can admit this law, at the grammatical and scriptural level is the tree that hides the forest because the ***w*** placed as initial morpheme position is a marker of gender in African languages. It indicates the masculine gender in our languages. This is why i will not fail to respect the good grammatical rules imposed by writing. I will come back to it later.

Phonology:

Since the Egyptian characters are phonological, let us try to give a good illustration with the Ewondo.
Middle-egyptian:  ***wa*** to name the numeral **one** (1) through the arm's appellation
Ewondo: /wâ / "used term to designate the numeral one and the arm"
Ewondo:  /ùm-bokh/ "term used to refer to the numeral One or the hoe"

After this aside, let's return to our subject on the number, we can try the following reconstructions to explain these variations.

Negro-gyptian: **kʷəkʰ- "one"*
Ewondo: *fɔkh<*bɔkh<*hʷəkʰ<*kʷəkʰ-□ form *okh- "one" common to many languages of the group Ekang (bantou group A / B).*

Ewondo: *wá< *wəh-< *hʷəkʰ-< *kʷəkʰ-*"one" □ *form * a (fully shared with pharaonic egyptian to say "one")*

In many African languages as well as in Ewondo, we find that on the morphological and semantic level, the numeral "One" is related to the significant lexemes:

Ewondo: "lone": ŋgob, ŋgo (k>ng>m)
Ewondo: "quarantine": ŋgo (k>ng>m)
Ewondo: "other", í-nyo *ùm*-bɔkh (the other one)

De facto, it becomes interesting to reconstruct the original form of the word. The Negro-Egyptian, gives the word:

Negro-gyptian: **kʷəkʰ- "one"*
Middle-Egyptian: **kʷəkʰi- > *waki > *waʔi>wa* "one"
Ewondo: **kʷəkʰu- > *hʷəkʰu- > *bokʰ(o)> *bokʰ* "one, alone"

Let us note concerning the numeral one that the emphatic expression "self" used in myself, etc. and the expression of the absolute superlative form in Egyptian by soliciting the numeral a ***wa or wa.t***.
The Ewondo language uses different expressions such as:

Ewondo: /i-tam /"alone, lone, single, only"
Middle-egyptian: ***tm*** "Atoum"

Note:
The egyptologists transliteration ***tm*** "Atoum" equals the items:

French: Matin "early; morning; chief"
English: tome; tame; team; mouth; time; thymus; adam; item etc.

The ***ntr*** ***Itm*** is considered in ***Iwnw's*** cosmogony like the one who was alone (cf illustration N°27).
The one who is alone is **one** in a binary notation system. So to believe that because the Ewondo use the expression / i-tam / and not / wa / because of this point of view it is discriminating and by the way totally wrong, that not true.

Ewondo: /-mén / "-self"; ☞**Ewondo**: /ma-mén/ "myself"
Middle-egyptian: Imn "Amon"

Whoever is **one** is therefore **himself** obviously. I close here the debate which could lead to search for the expression / wa / in Ewondo to translate the Egyptian emphasis ***wa***.
/Wa/; /-mén/; /i-tàm/ etc. are simply synonymous in the Ewondo language.

Note: This item exists also in several languages worldwide such as in French and in English for example:

English: immanent **French:** immanent

Let's illustrate this with the following expressions founded in the grammar of Lefebre [39]:

Middle-egyptian: *nsw.t pw aHA wa.w* "a king who fights alone"
Ewondo: /*í*-dzoué yà yirèn *i*-tàm/. " a king who fights alone"

39. G. LEFEBRE, Grammaire de l'égyptien classique, Ifao/Inst.Franc.Archeol, Le Caire, 1955, p.176.

We also find these other expression in the following sentence:

Middle-egyptian: *nnk **tm**, wnn-i **wa** kwi* "the universe was me in the state of One"
Ewondo: / dy-àma i-tam, mə mbə wâ à mə-kê/ "I was alone at the beginning"

<u>Comments</u>

Middle-egyptian: ***nnk*** "(this) belongs to me, (it is) to me; for my part, on my side"
Fulfuldé: ndiyam "water"
Ewondo: /*ndim/ "water" ☞**Ewondo**: / dy-àma / "for me, mine etc."

Note: the word transliterated here from ***nkk*** is the word Ewondo: /*ùn*-nkùkh/ "obscurity; blindness" this word is link with the word **Ewondo**: /ndim/ "blind"

Middle-egyptian: ***wnn*** "to be, to exist"
Ewondo: /i-bèh/; "thirst →water"; "two" ☞**Ewondo**: / mbə/ "to be"
Ewondo: /bèh/ "2 thighs"; "2 for long ears" ☞**Ewondo**: / mbə/ "to be"

Middle-egyptian: ***tm*** "Atoum"
Ewondo: /i-tàm/ "alone"

Ewondo: /á-siŋli/ "to Drag a heavy object on the ground (trunk, canoe, stone) and roll it"
Ewondo: /i-səŋlo/ "a sled"
French: solo "one"
English: single "one"

It is difficult to argue over these figures by evacuating the deep meaning they convey. I will not repeat here all the metaphysical and ontological dimension related to the number one. But the previous illustrations on the emphasis and the skeleton consonant ***nfr*** in my opinion are self-sufficient. In the Ewondo language, the numeral 1 "fokh" can be homophone and therefore synonymous with emptiness / ùn-mfɔkh / "emptiness, absence". In a logic of complexity the zero state can be identified to the first state. This word Ewondo: / *ùn*-mfokh / "emptiness, absence". Has to be linked with the item Arab: فقير, [faqīr] "poor"

Capitalization: Some readings

In order to vocalize numeral 1 in the languages below, there is no need to be compliant with the transliteration **wa**. We must be content to vocalize in the manner below without having brain being in knots.

Moyen Egyptien	*wˁ (w./it)*	«un, 1»
Akan	*Kün*	«un, 1»
tìsòlòngò	*kosì*	«un, 1»
Bambara	*kele*	«un, 1»
koniaké	*Kelc*	«un, 1»
En Sango	*Akó*	«un, 1»
Somali	*ków*	«un, 1»
Poular	*Gö'o*	«un, 1»
wolof	*Tus*	«un, 1»

Haoussa	*D'aya*	«un, 1»
Malinké	*deèm*	«un, 1»
Lumbu	*Imweghe*	«un, 1»
Basa'a	*Pòog*	«un, 1»
Kikongo	*Mosi*	«un, 1»
làrì	*mòsí*	«un, 1»
tìntàndú	*mòsí*	«un, 1»
Tupuri	*bɔ́ŋe*	«un, 1»
Mooré	*iémbō*	«un, 1»

tìyòmbé	*mosi*	«un, 1»
Mahoré	*Moja*	«un, 1»
Bamoun	*Mó*	«un, 1»
Soninké	*Bané*	«un, 1»
Béti	*Fóg/fwa*	«un, 1»
Fang	*Foghe*	«un, 1»
Songhaï	*Iffô*	«un, 1»
yemba	*ùɛ́-mɔ̀ʔ*	«un, 1»
kɔ́ɔnzime	*mpwár*	«un, 1»

<u>Note</u>

In linguistics, we usually define a morpheme as the smallest significant element, isolated by segmentation of a word, most often without linguistic autonomy.

For example, the word ***singers*** is composed of three morphemes: sing- "related to song", **-er**- "he who does" and ***-s*** (plural tag).

We could precise that, various denominations make it possible to designate the notion of autonomous unit constituting the lexicon of a language: *lexical item, lemma, lexie or lexical unit*. It is a chain of signs forming a semantic unit and being able to constitute a dictionary entry. The constituent signs can for example be made by characters, sounds, gestures.

In the current vocabulary, we will speak more often of "word", a notion which however is lack of clarity. Statements are constructed with lemmas; lemmas are made of morphemes.

prefix	lexeme	suffix	
pro-	*-consul-*	*ship*	*-s*

Generalley, the Ewondo language attaches to the radical of a lemma an affix which defines the number. More or less present, this particle could be associated with the radical and plays the equivalent of the definite or indefinite article.

Prefix		Lexeme	Suffix	
GP	*NP*	*Radical*	*GP*	*NP*

Example: feminine Particule

	Number particle	Gender particle	lexeme	lexicon
singular	*l*	*i*	**lé**	a tree
plural	*b*	*i*	**lé**	the trees

Chart 14: organisation matrix of a feminine noun

The apparent absence of the gender prefix is revealed by the schemes accordance, notably by the possessive sentence.

Possessive: /ilé dy-àmà/ "my tree"

So you can have an equivalence between the Ewondo affix of feminin ***di/da/dya*** "one" and **a**/***the*** "one" in English.

Example: Masculin Particule

	Gender particle	Number particle	lexeme	lexicon
singular	*ù*	*m*	**-on**	a child

	Number particle	Gender particle	lexeme	lexicon
plural	*b*	*ə*	**-on**	children

Chart 15: Matrix of a masculine Noun

The apparent absence of the gender prefix is revealed by the schemes accordance, notably by the possessive sentence.

Possessive: *ùm*-on ẁ-àmà "my kid"

So you can have an equivalence between the Ewondo affix of masculin ùn-/ùm "one" and:

English: *one* "single"
French: *un* "one, single"
Portuguese: *um* "one, single"
Ancient Greek: εἶς (*) (heîs) masculine, μία (*) (mía) feminine, ἕν (*) (hn) neuter
Kinyarwanda: *mwe* "one, single"
Anjouanese: -moja "one, single"
Picard: *eun* "one, single"
Italian: *uno* "one, single"
Berber languages: *yiwen* "one, single"
Welsh: *un* "one, single"
Dutch: *n* "one, single"
German: *ein* "one, single"
Corse: *unu* "one, single"
Etc.

By participating in the actualization of the kernel's name the indifinite article simply indicates that: the represented, the referent (the thing, the animal, the person in question), exists but remains unknown to the actants of the enunciation: it is the standard tool of the incomplete determination. He thus opposes to the definite article, which presupposes that the referent is known to actants of the enunciation.

The singular indefinite article is the same type as the cardinal numeral adjective ("one "). If this kind globally returns an unknown referent of the speaker, it is the article. If, on the other hand, this form refers only to a unit quantity, it is the numerical adjective. It is the linguistic or extra-linguistic context that alone will allow this deduction.

No matter how the indefinite article is vocalized in different human languages, we say that it is the African words for the mother that are mobilized.

Historically and aside of the field of intellectual fraud, one (1) is an African invention.
In the 1950s, the Belgian geologist Jean de Heinzelin de Braucourt discovered these bones in layers of volcanic ash on the shores of Lake Edward in the region of Ishango in the Belgian Congo (now the Democratic Republic of the Congo), near the border Ugandan.
Jean de Heinzelin de Braucourt followed the trail of fishermen who used harpoons (-100.000 years) from Katanga (southern DRC) to Egypt (-10.000 years). It is therefore by looking for these harpoons that the Belgian scientist will fall on the bone of Ishango whose mathematical significance is indisputable to this day.

It is therefore the shape of these harpoons that will inspire the writing of the number 1 that we know today.
Indeed, the harpoon is used to kill fish.
Ewondo: / ì-wé / "what is used to kill"
Ewondo: / wà / "one, 1"

prehistoric harpoons shaped one

The numeral 2

In the philosophy of ***ptAh Wr*** curiously called Pythagoras, the supposed Greek philosopher whose real name is Mnesarchos studied in Egypt around 547 BC, according to him, the dyad is the second thing created. So literally it's the other "ONE".

With regard to the number "two", the Ewondo equivalent word can also be explained from the root of the Negro-Egyptian: ***kʷəkʰ-** " ONE" as it is semantically means the other ONE.
This observation leads us to believe that this is an atavism of the binary numerical system. Indeed, in a binary system, "other" is semantically equivalent to that follows the first, that means, by the ordinal point of view the second. Thus, by semantic shift, we come to the cardinal notion of "two". We thus find different type to translate the idea of **two** in Pharaonic Egyptian as "one" and "other", "one ... the other" etc. as Mboli points out[40]:

Middle-egyptian: *wa...wa* "one...another"
Ewondo: wa wa "you and you" (you meaning other one)

Middle-egyptian: *wa...ky* "one...another"
Ewondo: wa ɲyo "you and another"

Middle-egyptian: *ky...ky* "one...another"
Ewondo: ɲyo ɲyo"this one and another"

Middle-egyptian: *wa.sn-nw.f* "one another" (literally: "one ... his second")
Ewondo: wa ɲyè-fe "you and another"

Hausa: *Gooyiwa* "twins"

The Hausa expression *Gooyiwa* to act out twinness is very interesting because it considers the twins to be ONE (ogh / go) and the other ONE (wa). Hausa language has kept perennial the old atavism idea of the African binary system.

In the first three expressions, it is clear that there is nothing to distinguish "one" from "another"; the fourth expression confirms the relationship between "two" and "other".

Negro-egyptian: **kʷəkʰ-*
Middle-egyptian: **kʷəkʰi- > * kʰikʷə- > *cĩkʷ- > *siŋ->sn-w(ii)* "two"
Ewondo: **kʷəkʰi- > * kʰikʷə- >hihʷə> *ibè - >* "two"

Quid?
Is there any link between the skeleton of consonants ***sn*** "two" and the word of Ewondo / bèh / "two"? On the surface, one might think that these two terms are different, but this is true only if one takes the Egyptian reference as an absolute norm. This is true only if we base our remarks on the sacrosanct phonetic proximity. How to deal with synonymy in that account?

Middle-egyptian: ***sn*** *(wt/ it) "two"*
Ewondo: bèh "two (2)"

The Ewondo offers a varied semantic field on this subject. Let's just take few words:

Ewondo: /ibè-ibè/, "Needless stinging action".

Example: /A bè ma yo i-bèbè/ "He stung me with his tip for no reason", "because there was no thorn"

Ewondo: /à bəm/, /à bəmbə/ "to sting"
Ewondo: /i-bèh/ "a spade, a bamboo with a point, lance, which stings"
Ewondo: /*ùn*-mbèh/ "the door"
Ewondo: /i-bèh/ "thirst"
Ewondo: i-bèh "two"

40.Jean Claude Mboli, ORIGINE DES LANGUES AFRICAINES - Essai d'application de la méthode comparative aux langues africaines anciennes et modernes, L'Harmattan, Paris, p.314

Note:

Ewondo: /i-bèh i-nə *ùn*-səŋ/"a pike is a a sting"

So for the scribe, it's the homophony between thirst, spades, stinging, linked to the spear which is for us the logical bridge between the transliterated numeral ***sn*** "à-koŋ" and the word bèh (2) of the Ewondo language which is here mobilized. If we remember that ***tA Sty*** "The country of the archers is the brother-country" of *km.t*. Then we understood this script and we read it without any effort. Without this data we are very annoyed.
The skeleton of consonants ***sn*** which is also closer to the Ewondo / *ùn*-nsoŋ /, the sting, the beak, the spear finds by homophony, the meaning of /i-bèh/ (the sting).

Moreover, the consonants skeleton ***sn*** can be approached by /á-sàŋ/ "to cut; to divide". The action of cutting in Ewondo justifying the meaning of two. Indeed, to pass from one to two, the scribe implies an action which consists in cutting the one. Which on the mathematical plane tends to approach by an algorithm based on what is more known by the initiates under the name conjecture of Syracuse. In short, any number can be divisible by two indefinitely. We find the consonants skeleton ***sn*** meaning to divide with the same skeleton of consonants ***sn.***

Middle-egyptian: *nis x xnt y* "dividing x by y"
Ewondo: /à-səŋ *x* ngap *y*/;/à-kap *x* bi-fas *y*/ "divide x by y"

As we can see, therefore, the two terms are linked without any doubt.
The sign T22 of the Gardiner list is mobilized in different scripts (see listing below) because they are related to the maternal figure:

sn: *brother*
sn: *smell, breathe*
sniH: *angler*
snaa: *smooth, polish*
snw: *two*
snw:*companion,similar; counterpart;*
snw: *offerings*
snw: *jar, container*
snwt: *brothers and sisters, siblings*
snwt: *sanctuary, chapel*
snwty: *companions*
snm: rot, *consume, provide*
snsn: *to be fraternal; fraternize; etc.*
sngr: *Babylon*
snt: *sister*
snt: *mast, shaft*
snt: *base (of statue)*
snt: *feast of the 6th day of the lunar month*
snTr: to incense; devotesnTr: *encens*
qbH-snw.f: *Kbehsnouf*

It is therefore necessary, in order to translate these words, to focus on the sign T22, the spear in our languages to apprehend different meanings not hidden but whose use has been lost on the philosophical level. I give an example in the analysis of the script "brother" ***sn.***

Capitalisation: Two (2) in african languages

In order to vocalize numeral || (2) in the languages below, there is no need to compliant with the ***sn*** - (wy / ty) transliteration. We must be content to vocalize in the manner below without having brain being in knots. And, if necessary, find the obvious link between the sign and the vocalization used. These vocalizations seem different from each other because they are synonyms. Those who are close to each other (Bambara, Koniake, Soninké) are homophonic. But we can explain rigorously why this or that vocalization is engaged and also corresponds to 2 in each variant of the unique african language.

Moyen-Egyptien	*wꜥ– (w / it)*	«two,2»
Akan	*gnon*	«two,2»
tìsòlòngò	*zole*	«two,2»
Bambara	*fila*	«two,2»
koniaké	*fila*	«two,2»
En Sango	*ousè*	«two,2»
Somali	*làba*	«two,2»
Poular	*dhidhi*	«two,2»
wolof	*bénn*	«two,2»

Haoussa	*buw*	«two,2»
Malinké	*pinya*	«two,2»
Lumbu	*wali*	«two,2»
Basa'a	*ibä*	«two,2»
Kikongo	*zole*	«two,2»
làrì	*zole*	«two,2»
tìntàndú	*zole*	«two,2»
Tupuri	*bɔ́kè*	«two,2»
Mooré	*mbù*	«two,2»

tìyòmbé	*wadi*	«two,2»
Mahoré	*mbili*	«two,2»
Bamoun	Mbà	«two,2»
Soninké	*Filo*	«two,2»
Béti	*bèh*	«two,2»
Fang	*bèh*	«two,2»
Songhaï	*hiNka*	«two,2»
yemba	*mi,puè*	«two,2»
kɔ́ɔnzime	*bā*	«two,2»

Etymology

Ewondo: /bèh/ "two"
English: two

These two words seem different but in fact they are synonymous and it only depends on the reading's grid. Let's understand how.

Power is symbolized in Africa by a seat ***As.t***. We call the seat in Ewondo language: / i-toa / "seat". So this word is here phonetically associable with the English word *two.*

Ewondo: /bèh/ "two"
French: deux "two"

Concerning the comparison above with the French term, we say that on the ornithological level, the ultimate african maternal concept is the vulture *A*. In Ewondo we call that vulture for the maternal figure / ndo /.

If in Appearance one could believe that these terms were different, using an African reading grid, one discovers in fact that they are simply synonymous.

Abaknon: duwa "two"
Aka-bo: wāsi "two"
Alabama: tòklo"two"
Albanais: dy "two"
Allemand: zwei "two"
Altaï du Sud: эки "two"
Aroumain: doi "two"
Azri: iki "two"
Bambara: fila "two"
Basque: bi "two"
Bavarian: zwoa "two"
Binukid: duwa "two"
Bosniac: dva "two"
Corse: dui "two"
Reunion Creole: d "two"
Croatian: dva "two"
Danish: to "two"
Spanish: dos "two"
Esperanto: du "two"
Finnish: kaksi "two"
Francoprovençal: dou"two"
Irish Gaelic: dhá "two"
Gagauz: iki "two"
Galician: dous "two"
Gallo: deûz "two"
Georgian: ორი "two"
Persian: دو (fa) "two"
Peul: ɗiɗi "two"
Pirahã: hoí "two"
Plautdietsch: twee "two"
Polish: dwa "two"
Portuguese: dois (pt), duas (pt)
Quechua: iskay "two"
Romanian: doi "two"
Russian: два (ru) "two"
Samoan: lua "two"
Sardinian: duos "two"
Sicilian: dui "two"
Silesian: dwa "two"
Slovak: dva, dve "two"
Slovenian: dva "two"
Slovio: dva "two"
Etc.

Note: Only someone speaking african languages and Beti particularly can understand the Subtility of this sentence:

Ewondo: /*ùn*-mfɔkh ù nə i-béh / "a vaccum is a hole"
by shift of meaning on homophony background we have:

Ewondo: / fɔkh i-nə i-bèh / "one equals two"

Meaning *one* and *two* are linked. By the way then we are not dealing with binary thought (all or nothing' tolling) but with complexity. How comes that, for african people, two and one are similar? Only by what is so called relativity or complexity thought. Could you agree with me that two is the other one?

The Numeral 3

According to Jean-Claude Mboli, the Negro-Egyptian root reconstructed has the following form: * *kiʀʷa*. It is identical to the word meaning "woman" - which also expresses the idea of collective. We hold therefore the etymology of this numeral from Negro-Egyptian since in pharaonic, the collective is expressed sometimes with the the feminine tag.*t*, sometimes with the three slashes of the plural. So we have:

Negro-egyptian: **xiʀ̃ʷə-kiʀ̃ʷa*
Middle-Egyptian: **xiʀ̃ʷə-kiʀ̃ʷa >*xm-Tw>xm-t(w)* "three"
Ewondo: **kiʀ̃ʷa-xiʀ̃ʷə > * hiʀ̃ʷa-hirʷə > * hila-hilə*>*ilalə>ilahə>*ilèh* "three"

Is there a link between the aspect obtained by convention, transliterated ***xm.t*** "three" and the vocalization / i-lèh / to designate the numeral three in Ewondo?

Middle-Egyptian:  *xm*.t "three, triple"
Middle-Egyptian:  *Hm.t* "woman, wife"
Ewondo: i-lèh; i-lah; etc. "three"

One might think that these two terms are different at first glance. This is the conclusion that many Africanist sociologists would draw because locked in the matrix of impressions towards Africa. But in reality it's not the purpose.

Middle-Egyptian:  *xm*.t "three, triple" ☞**Ewondo**: /i-kom/ "maker"
☞**Ewondo**: /i-kùm/ "parent tree"

Could you agree with me that the parent tree is also a tree?
As you can see above by phonetic shift , phallus terms in our languages often revolve around the semantic field of the tree:

Ewondo: /*ùn*-ŋkàŋ/ "the root"
Ewondo: /*ùn*-ŋkàŋ abîn/ "perineum of man; penis" litt. Scrotal root.
Ewondo: /I-sîŋgaŋ/ "Bubinga: wood rotproof three; penis"
Ewondo: /I-yô/ "the spine; penis" which is often represented by the obelisk.
Ewondo: /ùn-ŋkoum/ "the mast, the pillar"

The word /i-lèh/ "three, 3" of the Ewondo is homophone of / i-léh / designating a tree.

Middle-egyptian: *xm.t* "three, triple" **Ewondo**: /i-kom/; /i-yàm/ "phallus"
Middle-egyptian.: *iAm* "tree" **Ewondo**: /i-yàm/ " cooker; tree"
Middle-egyptian: *nhw.t* "trees" **Ewondo**: /tin/; /*ùn*-ŋkaŋ/ "root"
Middle-egyptian: *x.t*[41] "wood, beams, tree, stick, cane, stake, mast"

English: three (3) and tree "shaft" phonetic proximity maintained.

The term / i-lô /; / Elo / is used to refer to the primordial ancestor of the Ekang peoples. Elo is the third component of the trinity, the son. The word / i-lèh / is therefore close to the word / i-lô /. The scripts above clearly demonstrate the link between the two terms.

The sign ***Aa1*** of the Gardiner list corresponds to the homophone placenta of the African tree Garcinia Lucida.

Ewondo: I-sòkh ou i-sôgh "placenta"
Ewondo: I-sókh "peels of Garcina lucida tree"[42]

The placenta representation can be used by the scribe to name the tree Garcinia lucida in Ewondo. *(cf illustration n°30)*

TRINITY: son-father and mother

The determinative of the glyph ***nhw.t*** "trees"
indicates **three trees** as well as numeral **3** and thus the associated vocalisation. The meaning ***nhw.t*** retained by Egyptology corresponds to the word Ewondo / *ùn*-ŋkaŋ / "root".

In African thought, the three results from the union of the binary pair. This is why the scribe mobilizes organs of the semantic field of reproduction on the script and not the tree.

The best illustration of this idea is the trinity: **the mother, the father and the son** (sorry not the holy spirit, he is not yet involved. It could be only in another sociology but not the africain one).

I just note in passing that in a comparative analysis, the consonant ***x*** of Egyptology can also be likened to the bilabial nasal voiced ***m*** in the Ekang languages.

Philosophy: How does the event occur?

"Absolute activity is Unique, it is "Causal Cause", not "cosmic cause", it is Unity and not yet the Ternary Unit"[43]

(S. de Lubicz, 1977)

The ternary unity being inherent in the manifestation.
The Ternary Unit is the Cubic Perfection of the Egyptians and is found in the deviations of African thought that gave rise to religions.

$$1 \times 1 \times 1 = 1^3 = 1$$

41.FAULKNER p. 198, Wb. III p. 339, 10-14 sq.
42. Confère illustration N°29
43.Schwaller de LUBICZ, Propos sur l'Esotrisme et le Symbole, Paris: Dervy-Livres, 1977, p. 36-37

That is why what is called in those religions, the mystery of the Holy Trinity evacuating the maternal figure, God in three Equal Persons Father-Son-Holy Spirit, is represented geometrically by the equilateral triangle with a point turned up.

Among the Fulani, 111 is the ideal sacred number. It corresponds to the unity found on each of the three planes of the human being, the physical plane, the relational-emotional plane and the mental plane. Continuing the multiplicative series, the 3 is in the center of a mirror that reflects the 12 in 21 for the return to unity:

$$111 \times 111 = 12321$$

Thus, the consciousness of the Two (2) allows to reach the point Three from which is seen the two polarities of the binary manifestationnel. It allows to find the UN.

Capitalization: three in African languages

The various types of the word three (3) here can easily be read in Middle-egyptian highlighting the link with the signs of the spelling. One thing is clear the few words below are synonymous of ||| (3) three in the African language variant Ewondo.

M.E	*ḫmt– (w / t)*	«trois»	**wolof**	*ñar*	«trois»	**tìyòmbé**	*tatu*	«trois»
Akan	*nsän*	«trois»	**Haoussa**	*Ouk'kou*	«trois»	**Mahoré**	*Trarou*	«trois»
tìsòlòngò	*tatu*	«trois»	**Lingala**	*mísáto*	«trois»	**Bamoun**	*Ntèt*	«trois»
Bambara	*saba*	«trois»	**Lumbu**	*Tatu*	«trois»	**Soninké**	*Siko*	«trois»
koniaké	*saaba*	«trois»	**Basa'a**	*aa*	«trois»	**Béti**	*Lãh*	«trois»
En Sango	*Ota*	«trois»	**Kikongo**	*Tatu*	«trois»	**Fang**	*la*	«trois»
Somali	*sáddex*	«trois»	**làrì**	*tàtú*	«trois»	**Songhaï**	*l'hin'nza*	«trois»
Poular	*tati*	«trois»	**tìntàndú**	*tatu*	«trois»	**yemba**	*meu.tet*	«trois»
Mooré	*Tan*	«trois»	**Tupuri**	*swa*	«trois»	**kɔ́ɔnzime**	*lɛ̂l*	«trois»

Etymology:

Middle-egyptian: ***xm.t*** "three, triple" **English**: three (3)
Middle-egyptian: *nhw.t* "trees" **English**: tree (3)

The English language has maintained a phonetic link between the two terms.

Let's see what about the French language for example:

Middle-egyptian: ***xm.t*** "three, triple" **French**: trois (3)
Middle-egyptian: *nhw.t* "trees" **French**: arbre "tree"

Be honnest with yourself, you do really think that i'm traped. Don't you?
As i told you, over time the phonetic proximity may have been maintained with the African strain or else the relationship between two ideas has been kept by a synonymous link.
In the case of French it is therefore synonymic but more exactly the metonymy that has been mobilized. I will show you how.

Meroïtic: ***abr*** "Ancestor, elephant, family etc."
French: arbre "tree"
Ewondo:/*à*-bwàrà/ "the smallest"

In African sociology, the roots represent the father and the mother at the origin of the trunk represented by the son. As a basic principle the son is associated with numeral 1 and not 3 as can be conceived in Western logic. The son therefore is at the origin of all the branches following the fractal mathematic model**.** The word to designate the smallest /*à*-bwarà/ as the ancestor in Ewondo language that has been kept also in meroïtic language ***abr*** is then giving the right undestanding of the real etymology of the the french word *arbre* "tree". Therefore we have a better undestanding of the link with **tree** and **three**. Tree can be approached with the Ewondo word, /tara/ "parents"
By metonymy the French word and the English word are therefore identical because they relate to the same African signified.

Summary:
When french people often said ***arbre*** meaning the tree in english, to african point of view they just say ancestor or the smallest. That's all.

Position N°1: the mother
Position N°2: the other one = father
Position N°3: the son = one = ancestor = trunkh= tree.

Note: Only someone speaking african languages and Beti particularly can understand the Subtility of this sentence:

Ewondo: /i-bèh i nə i-léh/ "a bamboo is a vegetal"
by shift of meaning on homophony background we have:

Ewondo: /i-bèh i nə i-lèh/ " two equals three"

Meaning you cannot get three without two. Two and three are linked. By the way then, we are not dealing with binary thought (all-or-none principle) but with complexity. How comes that, for african people, three and two are similar? Only by what is so called relativity or complexity thought.

African socidological tree

All the words used among could then be explained by using only any african variant of language.

Numeral 4

The different types of the numeral "four" are derived from the Negro-Egyptian * xiʀʷə-tiʀʷ. This brings us to the evolutionary processes below:

> **Ngro-egyptian**: **xiʀ̃ʷə-tiʀ̃ʷ*
> **Middle-egyptian**: **xiʀ̃ʷə-tiʀ̃ʷ > *wu-tiw>f-dw* "four"
> **Ewondo:** **xiʀ̃ʷə-tiʀ̃ʷ> *hə-diw > *hə-ɲiw>i-ɲyi* "four"

Comparison between the middle-egyptian and Ewondo

Let's make a comparison on the semantic step between the term dedicated to numeral 4 in Ewondo / i-nyi / and the Egyptian consonants skeleton ***fD.t*** "four".

The phonetic proximity that is engaged here is the following:

Ewondo: nyô ou fì "snake" ⇨ **Middle-egyptian:** *D.t* "cobra"
Ewondo: nyô(l) "body" ⇨ **Middle-egyptian:** *D.t* "body"
Ewondo: ù-nyù "finger implied hand / foot" ⇨ **Middle-egyptian**: *Dba* "finger"

Ewondo: /i-nyin/ "which serves to strike"
Ewondo: /i-nyî/ = /*ùn*-nsɔŋ/"sharp, piercing; snake; worm"

Ewondo: /*nyî/ "four"

On the purely phonetic plane for the scribe, the snake, the body and the finger are homophones and the toes of the foot or the hand are named in the same way. I believe that this is how we must understand the link between the word 4 Ewondo / nyî (n) / and the spelling which records the 4 in Pharaonic.

Where transliteration is associated with **fedj** vocalization, it must be read in Bohairic Coptic: ftōw "four" reminding the words **foot** or **toe** of English. But this does not exclude other African vocalizations anyway:

M.E	fd– (w / t)	«quatre»
Akan	nnan	«quatre»
tìsòlòngò	ya	«quatre»
Bambara	nani	«quatre»
koniaké	nani	«quatre»
En Sango	Ochió	«quatre»
Somali	áfar	«quatre»
Poular	nayi	«quatre»
wolof	Ñétt	«quatre»

Haoussa	Houd'ou	«quatre»
Malinké	nani,	«quatre»
Lumbu	Yine	«quatre»
Basa'a	na	«quatre»
Kikongo	Ya	«quatre»
làrì	yá	«quatre»
tìntàndú	ya	«quatre»
Tupuri	hira	«quatre»
Mooré	Nancé	«quatre»

tìyòmbé	ya	«quatre»
Mahoré	Nné	«quatre»
Bamoun	Mpwà	«quatre»
Soninké	Narato	«quatre»
Béti	Ngnīnn	«quatre»
Fang	ni	«quatre»
Songhaï	l'Taatchi	«quatre»
yemba	Li.kwueu	«quatre»
kɔ́ɔnzime	nâ	«quatre»

Philosophy and mathematics

On the mathematical and arithmetic level especially the numeral 4 says the scribe is obtained by association of the hand (5) and his discard . We can therefore assume that the sign equals the numeric value -1.

There is indeed an Ewondo expression: / à-dzaŋ à fì / "to disappear completely"
Indeed, the viper is / fì / and the rejected in Ewondo is also the term / fì /.

Does not the French semantax expression *faire fi* "to flout; to turn up one's nose at etc." find here its true etymology?

To end on this subject, according to the Ekang thought, the number 4 is to be associated with the entry, femininity in the African traditions.

Numeral 5

Concerning the numeral "five", we must associate it with its primitive form of the Negro-Egyptian to the value * $k^{hj}un^{w}$ and the meaning of "all, complete". This one must, indeed, mean something like "quantity" or "number", and is thus morphologically and semantically related to * $k^{j}un^{w}$ "all" of which it is a simple weakened variant[44].

The Middle-Egytian has preserved the traces of this semantics since it presents the word qnw ~ Tnw "number, quantity", exact corresponding - as well on the phonological level as that of the morphology - of the sango wungɔ "number, quantity" and who, like him, is related to the word meaning "all" (Middle-Egyptian: km, Ewondo á-mâŋ). Hence the following phonetic evolutions:

Negro-egyptian: *-$k^{hj}un^{w}$
Middle-Egyptian: *-$k^{hj}un^{w}$ > *-*hĭwu*>*d-iw* "five"
Ewondo: -$k^{hj}un^{w}$ > *-*Thĭun*^{w} > *Tán* "five"

The etymology of the word "five" prompts us to suppose that the Negro-egyptian numbering system was a quinary time as it is the case in many Negro-African languages [45]. One example is the Foŋ[46] of Bénin or The Walaf language of Senegal.

Middle-Egyptian:: ***d-iw*** "five"
Ewondo: *dwé* "Palm"
Tupuri: dwi "five, 5"
Pular: jow- "five; 5"
Fulfuldé: jowi "five; 5"
Bohairic: tiw "five; 5"

This approach is quite interesting because it establishes in a clear and indisputable way the link with the old quinary notation system. This reading based on the sovereign transliteration does not offer a reading of 5 scilicet / tàn / as used in the actual Ewondo synchrony. But it names the palm of the hand: / dwé /. It is only by semantic shift in meaning that the palm of the hand by his fingers designates the numeral 5. Of course this implies 5 units like the 5 fingers of the hand. But especially in my opinion as the calculation tool / taŋ / "calculation".

Let's bend to the diktat of the sovereign transliteration ***diw.t*** equivalent of / dwé / "Palm" in Ewondo. It then remains the question to find the link betwwen /dwé/ and / tan / of Ewondo but also to clarify wether is was also used in Egyptian or not? To answer this tricky question, we must mobilize the homophones of 5 / tan /.
The script presenting the fishing net is quite illustrative. It presents the palm of the hand as being homophone of the fishing net. Indeed, the fishing net is called:
Ewondo: tan "fishing net"
Wolof: tan "vulture"

44. The aspiration of the velar is perfectly explained by the action of the agent prefix *niʀ̃ʷa > *hɨ: *kiʀ̃ʷanʷ "fat" >*hɨ-kiʀ̃ʷanʷ > *kʰiʀ̃ʷənʷ "quantity, number".
45. Abdoulaye KANE, Les Systèmes de numération parle des groupes ouest-atlantique et Mandé: contribution à la recherche sur les fondements et l'histoire de la pensée logique et mathématique en Afrique de l'Ouest, ANRT, Lille, 2001, P14.
46.Dominique FAÏDARO, Parlons Fon langue et culture du Bénin, l'Harmattan, Paris,2001, p.33

Ewondo: tan "five, 5"
Middle-egyptian: ***Had*** "fish-trap; bow; fish net"

The answer is also found in the following script: ***Sf-bd.t*** "month fifth"
Surprisingly, we do not find ***d-iw*** the word devoted to five.

The sign N37 "the garden pond"
The well or drinking water point is called in Ewondo: /i-tàm/ "well".
The snake is called /nyo/ but by reversing the reading direction the ɲ is transformed in ŋ.
So the snake can be used to designate the month: *ngo* as the snake of the well namely the catfish *ngon / ngol* "moon". The moon is the consecrated term for the month in Ewondo. The expression */tan/* to say five is thus to approach with wells / i-tam / in Ewondo. Visibly the bilabial nasal consonant voiced final **[m]** and voiced alveolar nasal consonant **[n]** could they be confused?

This approach is confirmed with the script below:
Middle-egyptian: ***rwdw*** "staircase, grave well"

What is interesting in this script is the fact that the term *well* is translated ***dw*** exactly like the expression ***diw*** devoted to numeral 5. According to the scribe, well and 5 are homophones. We also have:
Middle-egyptian: ***sdx*** "well, water hole".

We have here the well S21 /i-tàm/ because it's homophone of the lone /i-tám/ which is exactly the same as the numeral 5 /tàn/. I just precise for information that the two words are identical because they indicate the same meaning /ta-/ often vocalized /ta-râ/ "the father" .

Facetious, the scribe draws here the sign S29 given for the folded loincloth. This loincloth can be considered as a skin, / i-kob / in Ewondo which is an ananym of the word / *ìn*-mbòkh / which means among others, One but also hole which is validated by the determinative S21 .

So we have here the strong evidence that ancient egyptian scripts reading is possible both sides in our languages. In addition, transliterations induce a significant loss of information because of the dillution of synonymy related to semantic fields. In short if the scribe draws a skin to desingate a hole it is because these two words are for him and therefore for us one and the same word whose meanings vary according to the reading direction only.

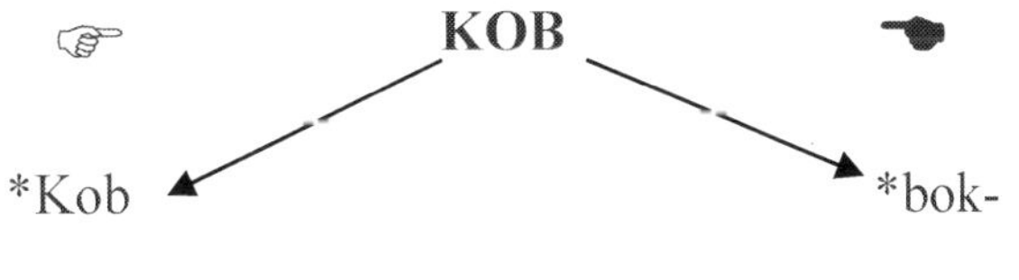

So this word ***sdx*** should be read in Ewondo / *ìn*-mbokh í-tàm / that is to say "the well's hole".
As already, it has been pointed out in the previous pages that, the sign Aa1 returns in our languages and to the Ewondo the phonetic correspondence of the letter **m** or **k** because this sign comes from the proto-version of Negro-egyptian ***kw***-.

The question of the comparative numbering systems is a little tricky because it is clear over time that Africa has integrated within it the different types of numeration systems, (binary, quinary, octal, decimal etc.). According to the zones of Africa these systems are still alive. It is therefore difficult to establish with certainty a direct link between the Negro-egyptian and all the forms of numeration that have existed or still exist. What is certain, we find fossil traces of these systems in the Pharaonic lexicon. The most plausible idea is that, these systems have been developed according to migrations and necessities. Is the decimal number of the Ewondo an atavism of the Egyptian system? Or is this system anterior to the Egyptian numeral system?

For example, the comparison of the terms / tan / (5) and / gifted / (palm of hand) may suggest an antecedence of the tan form. Any conclusion here would be hasty.

The authoritative literature claims that ***km.t*** invented the decimal system. But the observation offered by the analysis of the scripts and the taking into account of the anteriority of the synchronies of current languages make it possible to note that Africa already had system of counting including decimal before Egypt.

The East African ethnic groups: Luo of Kenya and the Yorubas of Nigeria as well as the Manjaque ethnic groups in Guinea-Bissau and Wolof in Senegal, the Foŋ in Benin still use the Quinary system. The decimal system prevailed in most African languages, and then the conversion took place either by borrowing number names from other languages (for example, borrowing from Swahili to Arabic) or by combining the decimal base with the quinary sub-base (example: chichewa).

To finish on this point, let us just note that in its esoteric dimension, this figure is associated with the point, the star, the calculation and therefore the masculine for the Ekang peoples.

The following vocalizations offered for the numeration of the number 5 by the spelling  are synonyms or homophones of each other. There is no need to focus on transliteration which is just one among many possibilities.

M.E	*dỉ– (w / t)*	«cinq»	Haoussa	*Biyar*	«cinq»	tìyòmbé	*tanu*	«cinq»
Akan	*nnou*	«cinq»	**Malinké**	*lulu,*	«cinq»	**Mahoré**	*Tsano*	«cinq»
tìsòlòngò	*atanu*	«cinq»	**Lumbu**	*Raanu*	«cinq»	**Bamoun**	*Ntièn*	«cinq»
Bambara	*dourou*	«cinq»	**Basa'a**	*tan*	«cinq»	**Soninké**	*Karako*	«cinq»
koniaké	*lolou*	«cinq»	**Kikongo**	*Tanu*	«cinq»	**Béti**	*Tán*	«cinq»
En Sango	*Onkou*	«cinq»	**làrì**	*tànú*	«cinq»	**Fang**	*tane*	«cinq»
Somali	*shán*	«cinq»	**tìntàndú**	*tanu*	«cinq»	**Songhaï**	*Ag'gou*	«cinq»
Poular	*dio'i*	«cinq»	**Tupuri**	*dwy*	«cinq»	**yemba**	*Meu.taa*	«cinq»
wolof	*Ñint*	«cinq»	**Mooré**	*Noun*	«cinq»	**kɔ́ɔnzime**	*tɛ̂n*	«cinq»

Etymology

Origin of the french word cinq.

Unique variants of the African language proceed by synonymy. The loss of bridges between different words often leads to believe that languages are different.

Compared to Ewondo, which serves as the reference language for the Comparative Study, the french word *Cinq* comes from Ewondo:

French: cinq "five; 5"

The french word cinq can be approached with the Ewondo word:

Ewondo: *siŋ* "fight; struggle"

Wrestler murals in tomb of Saqet III tombs of Beni Hassan Middle Egypt Egypt Africa North Africa

As we can see above here, during the middle empire (-2033 à -1786) ancient egyptians were wrestling. The theme of wrestlers had a unique development, at the point of occupying entire walls. Of the 39 funerary chambers excavated 4,000 years ago in the cliff overlooking the Nile at Beni-Hassan, 7 of them contain scenes of wrestling. The most famous are the tomb of Khety, which contains 122 pairs of wrestlers and that of Bakhti III which shows 219, drawn on the eastern wall of the burial alongside soldiers taking a fortress (which would tend to think they are fighting as part of a military training).

With the Greek and roman occupation of Egypt (-330 to 44), the practice will persist and become the Greek-roman struggle.

Greco-Roman wrestling is a form of wrestling in which wrestlers **can only use their arms** and can only attack the upper body of their opponents, unlike freestyle wrestling, where they can also use their legs and hold their opponents. below the belt.
This struggle still exists nowadays in Africa in Senegal or especially in East Africa.

Greco-Roman wrestling is particularly popular in Europe but is practiced around the world.

Origin of the word five

Middle-Egyptian:: [hieroglyphs] ***d-iw*** "five; 5"
English: five "5"
Ewondo: *tan* "five"

The etymological analysis must absolve impressions and must be based on the roots of words.
We say that numeral 5 is used to designate the vulture as a symbol of motherhood in Africa. The names for the vulture and those for the numeral 5 share an indisputable homophony.
The Vulture is still named by metonymy, the mother of the Wing even the feather/pen.
The word five can therefore be approached in Ewondo by the words:

Ewondo: *à-fàp* "wing"
Ewondo: *á-fəp* "to ventilate"

Just as /*doé/ designates the Vulture and palm in Ewondo, Tan the Vulture in Wolof; and five in Ewondo language, English uses here the term five"wing" which is a metonymy for the vulture.

Numeral 6

Let's try to propose a predialectal ancestor that explains how the two words were formed.

Middle-Egyptian: *$*s^{w}js^{w}$-t > *sĭsun > *sĭsuh> *sĭsʔ>* *sis* "six"
Ewondo: *$*s^{w}js^{w}$-t > $*s^{w}jn^{w}t$ > $*s^{w}$ĭbəd > saman* "six"

Idea of reuniting, of being together at the same place and at the same moment.

Ewondo: [à-saman] "reunite"→**Middle-Egyptian**: ***smA***
Ewondo: [i-saman] "the meeting; amphimixis"
Ewondo: [saman] "six"

The root of the word in Ewondo is **sam-.** We find the notion of ***smA*** used in the script **smA**-*tA.wy*. The same word is transliterated here ***sis*** That could be vocalized for me /zaza/[47] or /sasa/ in Ekang languages. This is because the Pharaonic of the Middle Kingdom is a Bere tongue that is to say a language where words are ending with a final vowel. The SIS consonant skeleton is actually a morphological type word (CVCV).

Middle.Egyptian: ***sis*** "six-thread linen fabric" *(cf illustration n°28)*
Middle.Egyptian: sis(w/t) "six"
Ewondo: i-samén "six; what has been woven"
Fang: sam "six"
Fang: be-səsam "six by six"

The Ngiemboon language spoken in the western region of Cameroon confirms the approch of egyptologist because the fabric is named like below indicated**.**

Ngiemboon: *Saza* "fabric"
Yemba: *asɛnza* "loin cloth; grass skirt"
Mboh mpahdong: *Sarah* "loin cloth; grass skirt"
Ewondo: sàndà "loin cloth; grass skirt"
Middle.Egyptian: ***SnDwt*** "loin cloth; grass skirt"

We think that for the idea of *fabric* the scribe has associated it with the word *feather*. Indeed, the feather gives the impression to have been woven. Thus morphologically the word six related to *fabric* is as a superlative of the word feather (*sA*) due proportionally to its size. Thus the fabric appears to the scribe and to us simply as a big feather. That's why for the consonantal structure of the feather, we thus obtain the words seen above related to the skeleton of consonants sis "fabric; six".

This piece of fabric is to be cut, by the way this bring the idea of the cut incarnated by the lion. A lion and a fabric can be associated in African thought as in Bbaledha language spoken en Rdc.

Middle-egyptian: ***rw*** "lion"
Middle-egyptian: ***ssrw*** "fabric"
Bbaledha: rw "cloth, loincloth"
Middle-egyptian: ***sis*** "six-thread linen fabric"

The loincloth enveloping the body, leads to the idea of wrapping, of assembling because it encloses the body.

47.ZAZA (bh) (lg) n.l, pl. **besasa.** Insecte **(mvane)** qui met ses oeufs dans les palmiers, ce qui produit le gros ver blanc fo (h) qui mange le coeur du palmier et est lui-même mangé par les indigènes. (cf illustration N°27)
Samuel Galley, Fang-French dictionary. and French Fang followed by a fang grammar,éditions Henri Messeiller , Neuchâtel, p.383

Hope that you have understood why egyptologist are talking about *sis(w/t)* and i'm saying that in Ewondo language is /sam-/ "six".

The sign M17 *i* of Gardiner list represents a reed in bloom. The word for flower in Ekang languages is /à-sàm/. By homophony this word can also mean: to assemble as we have seen
just before. In this account it is therefore the idea of putting together that is highlighted here.

We could take note also that the feather and the flower are the same person regarding phraonic Egyptian.

Middle-egyptian: ***sSA.t*** "seshat"
Middle-egyptian: ***MAa.t***
Middle-egyptian: ***As.t***

By a cross-fertilization of ideas, the thin copper wire to make wristbands is called in Ekang languages / *ùn*-nsa / "bracelet". The most important in this point of view is the string.
We have the confirmation that the two words are transposable to each other because the word bracelet is transliterated ***mss*** which is ananym of /*sam-/

If we remain a little bit on the idea of the string, in this account, the egyptologist term /sis/ appears to be as a projection of the idea of collegen strings of a sinew or a nerve. At this account a string can be associated with this idea and refers therefore to this other term Ewondo: / *ùn*-nsis / "nerve, tendon, vein, string; thread etc.".
That could mean the best idea to express the concept of a fabric according to egyptologist and regarding ancient egyptians is the hamstring.
Let's take note that for the purpose, we can vocalise /*nsi/ or /*nsis/ "tendon, nerve, thread etc.". The first version /*ùn*-nsi/ shares a phonetic proximity with the word six in French or six in English. But it also shares a phonetic proximity with /*ùn*-nso/ "nude". Explaining here why the scribe has used the sign S29 of Gardiner list. According to us this sign refers to a folded fabric that is worn over the shoulder to cover nakedness and symbolizes the true African nobility (cf. illustration N°10).

From this idea of nudity associated with sign S29, I would like to make an aside regarding the loincloth.

Tshiluba: ku-sanda "to fornicate"
Middle-Egyptian: *SnDwt* "loincloth"
Ewondo: /Sanda/ "loincloth"

The loincloth is named / Sanda / in Ewondo. Egyptologists give the consonant skeleton *SnDw.t* or *SnD.t* or *Sndy.t* to name the loincloth. The word SnDw.t vocalized sanda = loincloth gives by consonantal metathesis the words:
English: dance in
French: danse "dance"
For the purposer you have to Read the consonants skeleton *SnD.t* from right to left. Note two important things, most of african dances as Assiko; Baïkoko, Mutwashi etc. are danced with a loincloth around the waist and better than that the loincloth is a garment of the traditions of ancient Egypt.

In order to evacuate the nauseating ideology that teaches that others to help discovered niggers manking naked in the bush, nomadic people could not invent cotton weaving.

We can observe that the vocal dance = 1 in our languages ie "God". We thus discover here a highly spiritual and sacred dimension related to the dance. This is why sacred rituals are done in music with sacred dances. For those who speak Beti, we say:

Ewondo: *Zamba* "God"
Ewondo: *Samba* "feast, party".

Kɔ́ɔnzime: Nziem "God"
Ewondo: á-dzem "to dance"

After this aside once again, we can affirm to the initial question to know whether yes of no, we are talking about the same word or the same idea. The six reflects the idea of mixing, gathring. This idea can be understood to our point of view by a flower, a fabric, a tendon; a nerve as we have explained among.

We must therefore get rid of the absolute search of phonetic proximity with the standard vocalization of Egyptology and we should therefore consider that the vocalizations in different varities of the only african language are simple synonyms.
Sometimes they can be phonetically closer at other times they will not be any proximity because the couple of words are just synonymous. But sharing the same semantic field therefore they are linked by their signifed.

M.E	*sìs– (w / t)*	«six»
Akan	*nsian*	«six»
tìsòlòngò	*nsambanu*	«six»
Bambara	*woro*	«six»
koniaké	*wôlô*	«six»
En Sango	*Omenè*	«six»
Somali	*líx*	«six»
Poular	*diêgo*	«six»
wolof	*juroom bénn*	«six»

Haoussa	*shid'da*	«six»
Malinké	*woro,*	«six»
Lumbu	*Isamunu*	«six»
Basa'a	*sama*	«six»
Kikongo	sambanu	«six»
làrì	*sàmbànù*	«six»
tìntàndú	*sambanu*	«six»
Tupuri	*hira*	«six»
Mooré	*Yobé*	«six»

tìyòmbé	*sambanu*	«six»
Mahoré	*Sita*	«six»
Bamoun	*Ntoû*	«six»
Soninké	*Toumou*	«six»
Béti	*Samaṅ*	«six»
Fang	*sam*	«six»
Songhaï	*Iddouh*	«six»
yemba	*Ntog'o*	«six»
kɔ́ɔnzime	*saman*	«six»

Etymology:

Before concluding on this topic, i wanna make this observation:

English: tissue "fabric"
French: tissu "fabric, clothes etc."
Ewondo: sùt "cotton"
Ewondo: sandà "loin cloth"

The word *tissue* seems to be an ananym of the africain word */sut/* "cotton". We will not repeated enough that cotton yarn is a production of sedentary societies. Therefore the real etymology of the word tissu has to find its cradle in Africa. Before the arrival of Moorish populations in Europe the word was unknown.

English: Tissue (n.)
mid-14c., "band or belt of rich material," from Old French *tissu* "a ribbon, headband, belt of woven material" (c. 1200), noun use of tissu "woven, interlaced," past participle of tistre "to weave," from Latin *texere* "to weave, to make," from PIE root *teks- "to weave," also "to fabricate." The biological sense is first recorded 1831, from French, introduced c. 1800 by French anatomist Marie-François-Xavier Bichat (1771-1802). Meaning "piece of absorbent paper used as a handkerchief" is from 1929. Tissue-paper is from 1777, supposedly so called because it was made to be placed between tissues to protect them.

Numeral 7

Negro-egyptian: * $s^{w}p^{h}x^{w}$ *"seven"*
Middle-egyptian: * $s^{w}p^{h}x^{w}$> **zùphùxw >zfxw>sfxw "seven"*
ϯⲁⲥⲡⲓ ⲛⲣⲉⲙⲣⲏⲥ: *ϣⲁϣϥ "seven"*
Bohairic: šaš^{ə}f "seven"
Ewondo: **$s^{w}p^{h}x^{w}$- >*zəfgə- > *zà'mbà(l) > zambga(l) "seven"*

Let's try to understand how this script could be read in Ewondo:

Middle-egyptian:  sfT *"seven"*

Ewondo: / à-nẑomba / "liana used to make harp strings"
Ewondo:  /i-zombe / "what fell, what is grounded"
Ewondo:  /ùn-nzomlo / "the one who moves like a stream"
Ewondo:  /i-zombo / "the mother because she always has to support"
Ewondo:  /Zamba Elo / "quail chick symbol of ***Hr***"

What is very interesting with the scripts is that the scribe mobilizes a battery of techniques to express his ideas. Here, for the Ewondo that I am, it brings up the problem like a charade. The question is to determine who is the handicapped as a snake that knots its body in the placenta ?

In the Ekang cosmogony, Zamba is the eldest ancestor he is also called *ùn*-ŋkomboro (the one who makes humans). We call him in Ewondo: Zamba Elo, Eyo, Ntondobe.
Numeral 7 is called in Ewondo language Zambga (1). So here it is this notion that has been expertly put into perspective by the scribe for who understands or who knows African thought inferring that the quail chick designates the primordial ancestor.

On the esoteric level, this notion is also to be compared with the achievement because Zamba does everything according to his will:

Middle-egyptian: *sxpr* "to provoke, to accomplish, to arise, etc."
Ewondo: *à.twé* "to accomplish"
Fang: *à.tsȋie* "to accomplish"

Seven corresponds to the seven days of the week, to the seven planets, to the seven degrees of perfection, to the seven celestial spheres or degrees, to the seven petals of the rose, to the seven heads of the naja of Angkor, to the seven branches of the tree cosmic and sacrificial shamanism, seven musical notes, seven heart muscles, seven colors of the rainbow etc.

Seven was at ***km.t*** the symbol of eternal life. It symbolized a complete cycle, a dynamic perfection. Each lunar period lasts seven days and the four periods of the lunar cycle (7 x 4) close the cycle. Philo of Alexandria observes in this connection that the sum of the first seven numbers:
(1 + 2 + 3 + 4 + 5 + 6 + 7) reaches the same total: 28.

Seven indicates the meaning of a change after a completed cycle and a positive renewal.
It appears in innumerable Greek traditions and legends: the seven Hesperides, the seven gates of Thebes, the seven sons and seven daughters of Niobe; the seven strings of the lyre, the seven spheres, & c. There are seven emblems of the Buddha.
The circumambulations of Mecca inherited from the cult of black Sabeans [48] include seven rounds.

48. Anta DIOP, Negro nations and culture, Présence africaine, paris-Dakar, 1954 et 1979, P.195

The week includes six active days, plus one day of rest, figured by the center; the sky six planets (in the ancient comput), the sun being in the center: the hexagram six angles, six sides or six branches of stars, the center playing the role of a seventh; the six directions of space have a median or central point, which gives the number seven. It symbolizes the totality of space and the totality of time.
Associating the number four, which symbolizes the earth (with its four cardinal points) and the number three which symbolizes the sky, seven represents the totality of the universe in motion.

> *"The septenary also summarizes the totality of the moral life, adding up the three theological virtues, faith, hope and charity, and the four cardinal virtues, prudence, temperance, justice and strength."*[49].
>
> *(L.-C. de Saint Martin, 1946).*

The seven colors of the rainbow and the seven notes of the diatonic scale reveal the septenary as a regulator of vibrations, vibrations of which several primitive traditions make the essence of matter.
This sentence is attributed to Hippocrates: "The number seven, by its hidden virtues, maintains all things in being; it dispenses life and movement: it influences even celestial beings ".
Seven is the number, it was noted first, of cyclical completion and renewal. The world having been created in six days, God slept the seventh and made it a holy day: the Sabbath is not really a rest outside the creation, but its coronation, its completion in perfection. This is what the week evokes, the duration of a lunar district.
The sevenfold perfection of the senile rhythm is also familiar to Islam, and especially to Ismailism: the solid has seven sides (the six faces plus its totality - which corresponds to the Sabbath).

A Hindu tradition which is a continuation of km.t was the fruit of some priests who left Km.t to found the civilization of Mohenjo-Daro and Arrapat. This civilization, which has preserved an important part of the past of km, attributes to the sun seven rays: six correspond to the directions of space, the seventh being in the center.
The rainbow does not have seven colors but six: the seventh is the white, the synthesis of the six others.

The number 7 is universally the symbol of a totality, but of a totality in motion or a total dynamism. He is, as such, the key of the Apocalypse (7 stars, 7 spirits of the original entity, 7 seals, 7 trumpets, 7 churches, 7 thunders, 7 heads, 7 flails, 7 bowls, 7 kings, 7 colors , 7 musical notes, 7 days of the week etc ...)

Vocalization of the script in some African linguistic variants.

The vocalizations in the different African linguistic variants are either homophones or synonyms and nothing more. The fact that the Bohairic Coptic vocalizes šaš^{ə}f "seven" should never become a rule since in the other variants of Coptic the vocalization is not the same.

Akhmimic Coptic: sah^{ə}f "seven"
Bohairic Coptic: šaš^{ə}f "seven "
Sahidic Coptic: ϣⲁϣϥ "seven "

M.E	*sfḫ– (w / t)*	«sept»
Akan	*nsò*	«sept»
tìsòlòngò	*nsambaadi*	«sept»
Bambara	*wolowila*	«sept»
koniaké	*Wowiyan*	«sept»
En Sango	*Mbrambra*	«sept»
Somali	*toddobá*	«sept»
Poular	*diéddhi*	«sept»
wolof	*juroom Ñar*	«sept»

Haoussa	*Bakoy*	«sept»
Malinké	*worôula,*	«sept»
Lumbu	*Isambwali*	«sept»
Basa'a	*saambok*	«sept»
Kikongo	*Nsambwadi*	«sept»
làrì	*nsàmbwàdì*	«sept»
tìntàndú	*sambwadi*	«sept»
Tupuri	*harnam*	«sept»
Mooré	*Yopé*	«sept»

tìyòmbé	*tsambwadi*	«sept»
Mahoré	*Saba*	«sept»
Bamoun	*Sâmba*	«sept»
Soninké	*Niérou*	«sept»
Béti	*Zamgbwal*	«sept»
Fang	*nzangbal*	«sept»
Songhaï	*Iyyéh*	«sept»
yemba	*Sa'mbiè*	«sept»
kɔ́ɔnzime	*támbɛl*	«sept»

49.Louis-Claude de SAINT MARTIN, Charte naturelle des rapports qui existent entre Dieu, l'homme et l'Univers, Rochefort sur Mer, 1946.

Etymology:

The words are to express the idea of seven below:

> **English**: seven,
> **French:** sept,
> **German:** sieben,
> **Dutch:** zeven,
> **Portuguese**:sete,
> **Russian**: семь,
> **Romanian:** şapte
> etc.

therefore find their origin in the african commun language because of the antecedence of African languages. Consequence of what changing the dictionaries and encyclopedias should not be scary for anyone on the contrary.

Numeral 8

If we should propose a reconstruction from the Negro-egyptian, it could look like this:

> **Negro-egyptian**: *X^{w}uŋ^{w}> *-xùmnu > xmn "eight"
> **Ewondo**: *X^{w}uŋ^{w}> *g^{w}uŋw > mwôm "eight"

The term / mwom / "eight" denotes packets (here 2 packets of 4). It is an atavism of the quinary and octal sytems. We know that the numeral 10 will mobilize the same semantics. The 10 will be named /à-wôm / "the heap".

> **Middle-egyptian**:  "a type of boat"
> **Ewondo:** byal "canoe, boat, barque"
> **Ewondo**: *bi*byal "placenta" à-byali "spring" related to the uterus

> **Middle-egyptian:** 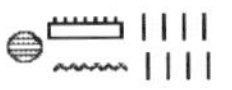 "eight" **Ewondo:** /mi-nsaŋ "bunches"
> **Ewondo:**/i-kóm/ "tas", /dy-ôm/ "heap" /kôme/ "placenta"
> **Ewondo:** mwôm "Huit",

> **Middle-egyptian:**  "city of the eight"
> **Ewondo**: à wom, "to peel, to molt"
> **Ewondo**:/A-folà bi-wôm/ "the place of the chicfs"

The term transliterated  ***xmnw "khemenou"*** in Pharaonic could offer, among other things, reading in Ewondo: / bi-kóm / "bunches, packets" and so by synonymy one has / mə-ôm / "heaps" homophone of / mwôm / "eight". We can also read / bi-wom /; / Bi-Wome /.

We have the phonetic laws that make the symbol x of the pharaonic correspond to the velar ***K*** in Ewondo or to the bilabial nasal consonant voiced ***m***. This is easily understood if we admit that the ancestor of the noted consonant ***x*** Middle Egyptian is ***K^{w}*** in Negro-egyptian.

To finish with the comparative step we have  and the M36 symbol:  the bunch. One should read here: /dyôm/ "bunch".

On another side, the eight is the figure of equilibrium and cosmic order, according to the Egyptians. He represents the infinite perfection.
In mathematics the symbol of infinity is represented by an 8 lying. This is the figure of immutable eternity or self-destruction. It is also the stopping point of the event. A figure expressing matter, it is also the symbol of incarnation in matter which itself becomes creative and autonomous, governing its own laws. He is the symbol of the New Life, in the worship of ***wsjr***, of the final resurrection. It represents the totality and the coherence of the creation in movement.
Represents the earth, not in its surface but in its volume, since 8 is the first cubic number $2^{\wedge 3}$.
The Pythagoreans made number 8 the symbol of love and friendship, prudence and reflection and they called it the Great Tetrachtys.

In Babylon, in Egypt, and in Arabia, it was the number of the duplication dedicated to the sun, whence the solar disk adorned with an eight-armed cross.
The number 8 means multiplicity, for the Japanese, in China, the 8 expresses the totality of the universe. Favorable number, associated with prosperity.
This is the number of rest, after the 7th day of creation

Vocalization of writing in some African languages

Since transliteration is not a dogma, being simply a small subterfuge inspired by the kopte, it simply translates a possibility among all the possibilities that exist. Speech is thus vocalized in each of the variants of the unique African language.

M.E	*ḫmn- (w / t)*	«Huit»
Akan	*môtchuê*	«Huit»
tìsòlòngò	*nana*	«Huit»
Bambara	*Ségui*	«Huit»
koniaké	*sé*	«Huit»
En Sango	*Miabè*	«Huit»
Somali	*siddéed*	«Huit»
Poular	*diêtati*	«Huit»
wolof	*juroom Ñétt*	«Huit»

Haoussa	*Takoss*	«Huit»
Malinké	*segi,*	«Huit»
Lumbu	*Inane*	«Huit»
Basa'a	*ndzoèm*	«Huit»
Kikongo	*Nana*	«Huit»
làrì	*nàná*	«Huit»
tìntàndú	*nana*	«Huit»
Tupuri	*ninma*	«Huit»
Mooré	*Nii*	«Huit»

tìyòmbé	*dinana*	«Huit»
Mahoré	*Nané*	«Huit»
Bamoun	*Fâmė*	«Huit»
Soninké	*Ségou*	«Huit»
Béti	*mwÔm*	«Huit»
Fang	*omôm*	«Huit»
Songhaï	*Yaahâ*	«Huit»
yemba	*Lefah*	«Huit»
kɔ́ɔnzime	*mwom*	«Huit»

Etymology:

In Africa, we find scattered everywhere remnants of the temporal cut including the week. It turns out that among many African people, this week lasts 8 days.

Curiously the Bohairic that served as a matrix to Egyptology says:

Middle-egyptian:  *xmn* "eight"
Bohairic Coptic: ϣмнн/*šmēn* "eight"
Sahidic Coptic: šmoun "eight"
Ewondo: /*ùn*-nsəmèn/ "sin"

Latin: septimana "group of seven mornings"
French: semaine "week"

As always, this proximity leads back to Africa, so I ask a simple question whether standardization of documents is not possible and necessary?

Numeral 9

Can one rigorously quibble about the nine numeral and defeat it from its esoteric dimension?

$$1 + 1 + 1 + 1 + 1 + 1 + 1 + 1 + 1 = 9$$
$$1^9 = 1$$
$$111\ 111\ 111 \times 111\ 111\ 111 = 12345678987654321$$

The symmetry is broken; however, Suite 123 reappears in the center

Middle-egyptian: *pwt^{wh}ĭ-> *pùsùdj> psD "nine"
Ewondo: *p^{W}t^{wh}ĭ-> *P^{W}S^{w}> *bÙsV´ > *bù(l)ù "nine"

Middle-egyptian:  psD "nne"
Ewondo: i-bù(l)ù "nine"

Some definitions:

ABÎ, ÔBÈ: Roof mat (butterfly), palm straw raffia,
WÔBE: Peel, moult, change skin (cockroach, snake, spider, crab). Ekô za wôbe, the skin peels. ***Nyo sa wôbe,*** the snake comes out of his skin, he is moulting. See wôp (b).

MBEÑY NDA: crossbows or horizontal bamboos fixed on the roof, to which are attached the straw in raffia palm tree.

CROSSBOWS: roof straw in raffia palm tree, roof mat, abî, ôbè. Make straws, twe ôbè, tw abî.
Middle-egyptian: ***psD*** "nine"
Ewondo: i-bu(l) "nine"

The word nine in Ewondo: / ebu (l) / is to approach phonetically to the straws: / i-bui / which can be placed on a roof but also can be worn around the waist during a dance.

Concerning the I10 sign  of Gardiner's list, is rather the word reserved for the exuvia that is here mobilized by the scribe. It is not therefore the snake itself that is concerned but its ability to moult, to change skin as confirmed by the lexicon of variants of the African single language:

Bassa: *ùβùl* "peel"
Setswana: *àpòlà* "undress"
Nilamba: *ulula* "remove the skin"
Bamanan: *wɔrɔ* "undress"
Silozi: *-yebula* "remove the skin"
Kiswahili: *-ambua* "remove the skin"
Buma: *obiɛl* "remove the skin"
Bobangi: *kuböita* "remove the skin"
Sakata: *ɛvula* "remove the skin"
Kisuundi: *kùvuula* "undress"
Bemba: *fúúl* "undress"
Soninké: *boora* "undress"
Silozi: *-bua* "flay"
isiNdebele:*-ebula* "remove the skin"
Lori: *obúl* "remove the skin"
Bukusu: *òòβùlà* "remove the skin"
Bukusu: *fùùla* "undress"
Kilaangi: *kovola* "remove the skin"

It was possible also to approach it by the protoword / i-mvul / (buffalo) used in Negro-egyptian to designate all dangerous animals for a human being.
And to finish on the comparative plane between the script and the Beti, phonetic what is wrongly called the 9 arches of Egypt mobilize the terms crossbows are related to the crossbows also.

Middle-egyptian:  psD "gleam, shine"
Kɔ́ɔnzime: ebyɛ́bla "gleam"

English: blue "color of the sky where shines the sun"
French: éblouir "dazzle, blind etc."
Duala: pɔ́limɛyɛ "wink, glance"
Ewondo: nyeb "beautiful, brilliant, bright, good" **French**: bien " beautiful, good"

On the semantic field the concept of the nine (9) is linked to the multitude because it is the association of the 4 feminine principle and the 5 masculine principle. The union of these two principles ipso facto engendering a new entity. Thus in humans the gestation of the child lasts nine months.

Ewondo: /à-bui/ "many"; /à-bulben/ "to multiply; breed", /á-biali/ "to be born, spring" etc.

The Egyptians called the number nine the *Mountain of the Sun*: the great novena was made of the evolution in the three worlds, divine, natural and intellectual, of the Trinitarian archetype ***Wsjr-As.t-Hr***, representing the Essence, the Substance and Life.

> *"[...] In the human order, the number nine is (indeed) that of the words necessary for the completion of the fetus, which is nevertheless completely formed in the seventh month. (We can also observe that the number 6 is that of the completion of creation, which culminates on the sixth day with the appearance of man)".*[50]
>
> *(J. BOUCHER, 1953)*

Three being the innovative number, its square represents universality. It is significant that so many stories, of all origins, express the infinite, overestimate, by the repetition of the new, such as the 999,999 Fravashis of the ancient Iranians: they kept the seed of Zoroaster, which were to be born all the prophets.

The Ouroboros, the snake biting its tail, image of the return of the multiple to the One, and therefore of the primordial and final Unity, is scriptally related to the reproduction of the number nine in multiple alphabets: Tibetan, Persian, hieratic, Armenian, Egyptian, etc.

Mysteriously this meaning of the new is apparent to the Hak of the Sufis, the supreme stage of the Way, bliss leading to the fana: the annihilation of the individual in the totality found; or, as Allendy says, the loss of personality in universal love. There are nine universal principles in the teachings of the oldest philosophical sect in India, the Vaïses-hica. It is known that Sidarta transferred them to the Indus Valley after having been rejected by Cambyse II to -525 BC.

The Orphic initiation would have likewise admitted three ternaries of principles, the first included the Night, the Sky, the Time; the second, the ether, the light, the stars; the third, the Sun, the Moon, Nature; these principles constituted the nine symbolic aspects of the Universe. The number nine, says Parmenides, concerns absolute things.

The nine museums represent, through the sciences and the arts, the sum of human knowledge.

The novena represents completion, the complete time. It existed in the Mazdean cult, it is found in the Zend-Avesta[51], where many purificatory rites are achieved through a triple ternary repetition: thus the clothes of a dead person must be washed nine times, three times with urine, three times with earth and three times with water. This triple ternary repetition is found in many rites.

The Indian tradition specifies this redemptive meaning of the symbol Nine, with the nine successive incarnations of Vishnu, who, each time, sacrifices himself to the salvation of men. Similarly, after the

50.Jules BOUCHER, La symbolique maçonnique, 2ème édition, p.227, Paris 1953.
51.The Avesta is readily seen as a collection of philosophical traits, religious maxims, and laws, intertwined with history and legends. Perhaps, originally, did he present something of this mixture found in almost all sacred books of antiquity.

plagiarism made to the texts of the spirituality ***wsjr*** to be called Jesus is crucified at the third hour, begins his agony at the sixth hour (twilight) and expires at the ninth.

Numeral 10

On the diachronic level one can suggest this evolution since Negro-egyptian.

Negro-Egyptian: *ŋ^{jw}k^{wh}ĭ->
Middle-egyptian: *ŋ^{jw}k^{wh}ĭ-> *ŋ^{w}t^{wh}ĭ-> mùtjù> mD “ten”
Ewondo: *ŋ^{jw}k^{wh} ĭ-> *ŋ^{w}t^{wh}ĭ-> *gwùmo> *gùm> *vom> *wôm “ten”

Moreover, the comparison between pharaonic and Ewondo is based on the following analysis:

Middle-egyptian: mD “ten”
Ewondo: dy-ôm “heap, package”
Ewondo: à-wôm “10, ten”

Middle-egyptian: ***mDwt*** “cowsheds”
Ewondo: nda bi-yəm “barns”

The signs V19; V20 are arches of ropes attached to a wooden bar tied the cattle.

So we have below here:

Ewondo: /à-yəm/ “to tighten, to clamp”
Ewondo: /i-yəm/ “beast, brute, animal, what serves for clamping”

We can observe that the root of the word here is ***om**.

Ten is the number that is the sum of the first four numbers (1 + 2 + 3 + 4 = 10).
It has the meaning of totality, of completion, of the return to unity, after the development of the cycle of the first nine numbers. The decade was, for the Pythagoreans the most sacred of numbers, the symbol of universal creation, on which they took the oath, evoking it.

Recapitulation of the number in some African languages

	Ewondo	Bandjoun	Mina	Malink	Kouyou	Foŋ	Kikongo	Basa’a
1	Fokh	*Ye’m*	*Eve*	*Kilĩn*	*Ohogo,*	*D*	*Mosi*	*Pok*
2	Bèh	*Ye’poue*	*Etou*	*Fula*	*Abaha,*	*W*	*Zole*	*I ba*
3	Lèh	*Ye’ta*	*Fnè*	*Saba*	*Asaro,*	*Atɔŋ*	*Tatu*	*Aa*
4	Nyî	*Ya’pfwe*	*Atõ*	*Nani*	*Ana,*	*Ẽnẽ*	*Ya*	*Ina*
5	Tan	*Ye’to’*	*Adein*	*Lulu*	*Atano,*	*Atɔɔŋ*	*Tanu*	*Itan*
6	Samèn	*To’oc*	*Adrein*	*Woro*	*Otoba,*	*Aïʒiŋ*	*Sambanu*	*Isamal*
7	Zambga	*Sombwe*	*Egni*	*Worõula*	*Tsambo,*	*Tiŋw*	*Nsambuari*	*Isambok*
8	Mwôm	*Ho’m*	*Gnid*	*Segi*	*Mwambi,*	*Tăŋtɔŋ*	*Nana*	*Juem*
9	I-Bul	*Ve’*	*Ewo*	*Konõto*	*Ibwa,*	*Tiŋŋiŋ*	*Wa*	*Bôô*
10	À-Wôm	*Ha’m*	*Eve*	*Ta*	*Dzomi...*	*Wɔ*	*Kumi*	*Jôm*

Chart 16: Some african Digital System vocalization

If we understood how to count from 1 to 10 in Ewondo and also in any african language compared to ***mDw nTr***. We can thus understand here that the enumeration made here appears to be nothing else except **Hr's** descriptors.

It suffices to realize that sometimes the same root is mobilized randomly according to the dialects. Then read each word and ask an analysis without knowing everything.

The 1 is related to the uniqueness of the ancestor. It then mobilizes the root *okh- or *koh-
The 2 will be associated with dimension of Love (*fila, fula, zol*) which characterizes him.
The 3 its dimension as ancestor, root, tree.
The 4 brings in the notion of greatness, life, entry.
The 5 the seat of power, calculus, the masculine, the star
The 6 the mixtion
The 7 is the Lord's number * Amba-; the completion
The 8 the figure that highlights the notion of old, rebirth
The 9 the number of the multitude or the unique
The 10 the unique or the return to the unit.

The Large Nombers

Numeral 30

Middle-egyptian:  ***mabA*** **"**thirty"
Ewondo: /mə-wom lèh/ "thirty"

Regarding Ewondo speakers, this script seems to reveal a notion for the word 30 which is rarely used and which would be related to the harpoon:

Middle-egyptian:  ***mabA*** **"**harpoon"
Ewondo: /A-venya/; /à-kɔŋ/ "harpoon"

All this seems very strange. But in reality it is not so because the mother is the handle of the family, / *ùn*-ntɔŋ / "sleeve" and / *ùn*-ntɔŋ'o / "nurturer". She is linked here to the harpoon. She is also / *ùn*-nya / "mother" homophone of / nyô / "the snake" (see script below):

Middle-egyptian:  ***mDwt*** "cowshades"

Middle-egyptian:  mD "ten"

The term for the exuvia of a snake in Ewondo is / i-wome / "moult" homophone of:
/ à-wôm / "ten".

The moult / i-wôm / appears as what emanates from the mother, what is given by the mother  /A-venya/ "What the mother gives; the awakening of motherhood; spiritual awakening"; /à-kɔŋ/; /à-venya/ "harpon".
The mother gives to the child  /i-vu/; /i-wou/ " metamorphic spirit, resemblance"

So I discover here, a name for the word thirty (30) which is not used at all but should be; because it refers to the term of African spirituality to notions that are beyond the scope of this statement.

> *"And Jesus himself began to be about **thirty** years of age, being (as was supposed) the son of Joseph, which was the son of **Heli**,,(...)"*
>
> *(Luc 3:23, King James Bible)*

The word *Heli* of this quote above is simply the word **Hr**; /i-lè/ "3"; /Elo/ or simple /i-hələ/ or /i-vɔlə/ or /i-wùlù/ the awakening; the sunrise.

Comparison between The Africain language variants

Middle-egyptian: ∩∩∩ *mabA* "thirty"
Lingala: Tuku Mi*sato* "thirty" (Tuku=dozen, misato =three)
Kikongo: MaKumi *Tatu* "thirty"
Mbènè: Móm a'a "thirty"
Mousseye: Dogo hindi "trente"
Poular: Thiapan-*tati* "thirty"
Bafanji: *tii* gwuŋ "thirty"
IsiZulu: Amashumiama*thathu* "thirty"
Wolof: Fan wrr "trente" (The number of days in a month)
Goun: *Gban "thirty"*
Podoko: *Djim-macra "thirty"*
Nzebi: Makumi ma *tata* "thirty"
Bambara: Bi-saba *"thirty"* (bi= multiple of ten and saba =3)
Ngiemboon: moomta "thirty"
Tamashek: Karadate timarwene *"thirty"* (three dozen)

From this approach through the riches of African thought, we withdraw several ideas: The digit thirty is not only 3 dozens. It is better then that. Some use it more to designate the month, it is the awakening and spiritual maturity in Man or also the awakening of motherhood.
The Bambara uses the word *Bi-saba* to designate the numeral 30. We find in filigree here various notions: the Greek word σοφία "Sophia, wisdom"; σοφός "Sophos, wise" at the origin of *philosophy*. We find the notions as sabbat, saba, we find the etymology of sábado "day of the saba", "the day of the dog".
If the age of 30 is the age of wisdom, the age at which one is master and one teaches, how could it be found in certain exegeses, whereas this has meant since the beginning of time the number 30 in African thought as here illustrated by the Bambara language?

Bréhier quotes MASSON OURSEL, a specialist in comparative philosophy, who already underlined this fact:

> *"La mentalité nègre [est] la toile de fond de la civilisation pharaonique"*[52]
>
> *(OURSEL, 1957)*
>
> *"The negro mentality [is] the backdrop of the pharaonic civilization"*[53]
>
> *(OURSEL, 1957)*

All these notions of the fragmentary current thought can enrich our different synchronies and sharpen the subtlety of spirit that has characterized our ancestors since the dawn of time.

Numeral 100

Middle-Egyptian.: ***S.t*** *ou* ***Sn.t*** "Coiled string or curved crown ornament"
Fang: Siŋ or *ùn*-ntér or *ùn*-ŋkama "hundred"
Ewondo: /ùn-ntér/ "hundred"

52. MASSON OURSEL: Histoire de la philosophie, 1er fascicule supplémentaire: "La philosophie en Orient", Paris, 3e d., 1957, p. 42.
53. MASSON OURSEL: Histoire de la philosophie, 1er fascicule supplémentaire: "La philosophie en Orient", Paris, 3e d., 1957, p. 42.

How do we establish the link between / Siŋ /or /*ùn*-nsàŋ/ and / *ùn*-ntér /? Besides the fact that Fang and Ewondo are very close and therefore we could be satisfied with the Fang's meanings to demonstrate the link with Pharaonic. But let's try to establish this link otherwise.

Middle-Egyptian: ***Sn.t*** "anger; wrath; ire"
Fang: *à*-Siŋ "Fight, fight with anger; struggle"
Fang: *ùn*-ŋkər "Anger in the eyes"
Ewondo: /*ùn*-nter/ "hundred"

It is thus the synonymy that exists between *ùn-ŋker* "anger" and ***Siŋ** "anger". The homophony between ***Siŋ** "anger" and ***Siŋ** " hundred " thus authorizes to establish clearly this link.
The spiral is here chosen by the scribe to translate the idea of deployment of the center of the spiral, also called the eye of the spiral.

Middle-Egyptian: ***SnT*** "to be the opposite"
Fang: à-Síŋ "to oppose, hate, jealousy"
Ewondo: /à-Síŋ/ "to oppose, to hate, jealousy"

So the term ***S.t / Sn.t*** can be vocalized / *ùn*-nter /, / *ùn*-ŋkama / or / Siŋ / in Ewondo as in Fang and return to the numeral hundred; without any misinterpretation or overinterpretation.

According to the African paradigm, a spiral is composed of a couple of forces [54]:
(***stX*** et ***Hr***)[55]. The two forces responsible of deploying the spiral are opposed in their actions. One is responsible for deploying the spiral up and the other down.
Snt / snt / SnT consonant skeletons refer to the register of violence, anger and destruction. All this is correlated with the characteristic of ***stX***. The african ontological approach associates this strength to the forces of "evil" since they do not elevate consciousness.

The other component of the spiral is ***Hr***. What characterizes this property off of having opposed ***stX*** according to African mythology, it is above all to be born as any child, weak of the lower limbs. The word dedicated to translate this reality into Ewondo is /*ùn*-mbokh/. According to the Ekang paradigm, a disabled person "*ùn*-mbokh" is also called, / *ùn*-nter / or / *ùn*-ntét / validated by the script ***.nitit*** "embarrass, slow down; delay, hesitate"

So we better understand why we have at least two words for the numeral 100. Therefore we find so far below the couple of forces that is responsible of the deployment of the spiral. These forces are still qualified in the tradition of the Nilotic peoples by the two lords.

M.E.: ***nbwy*** "the two lords"
Middle-Egyptian: ***nbwy*** "the two lords"
Ewondo: /mi-mbièh/ "the tweens"

You might think that i am amnesic about the term fang / *nkama / which is also found in many other varieties of the african language like Lingala and which is used to designate numeral 100. In reality, **Ewondo**: / *ùn*-ŋkàmà / also refers by Homophony to the one who defends, the one who avenges. In the case of the African myth of ***wsjr, Hr*** is the one who avenges. He is / *ùn*-ŋkàmà /, the avenger.

54.(cf illustration n°29)
55. Mbog BASSONG, Les fondements de l'état de droit en Afrique précoloniale, L'Harmattan, 2007, Page 361

So we have exhausted the terms used by Beti and close neighbors. It would be interesting to have other meanings to allow a better linguistic permeability between the different linguistic variants used in Africa.

Kɔ́ɔnzime: nter "100" to bring closer to the meaning of fang/Ewondo
Lingàla: *mo*-nkámá "100" bring closer to the meaning of Ewondo [nkama]

Basa'a: mbógô*l* "100" bring closer to mbógh "feet handicapped"
Hausa: ɗ-àri "100" bring closer to ***Hr*** (weak consonant H gives *à-)

Fulfuldé: *tem*erre "100" bring closer to ***Hr*** (tA-mry?)
Fulfuldé: ɓoggol "corde /spirale" bring closer to

Isizulu: ikhulu "100" *khulu the proto-version of ***Hr,*** the chief.
Yemba: ŋkhʉ "100" bring closer to the word foot "akhu" legs-disabled.
Etc.

Note that *hundred* can be linked with /*ùn*-nter/ when the French *cen*t can be linked to /*ùn*-nsaŋ/ "numerous; cutted; cutter"

Ngiemboon : nkio'o "cord; **"** and ☞**Ngiemboon :** nkio'o **"100"**

After this aside between the different linguistic varieties presented as ethnic groups by some specialists, let us end our remarks.

The scribe has mobilized here the constituents of the spiral to name the numeral 100. Let's note to finish that only digitally, 100 = 1 + 0 + 0 = 1. The word used refers to the eldest ancestor, the small, the first, the old, the handicapped, etc.Moreover ancient was also called tA-mry the real word exists in Hausa language and it's temerre "hundred" but according to what we have said it be assimilated to the spiral. The spiral is deep concept in African philosophy.

The number 1 is voluntarily associated with the ancestor in a cosmological level. The eldest of Ekang people is so far ***Hr.*** Therefore by naming one hundred (100) / *ùn*-nter / in Ewondo, we recall by homophony and semantic shift in meaning the father of the multitude. For this purpose, it's easy to observe that / *ùn*-nter / gives the skeleton of ***ntr***.

The voluntarist approach of defining a skeleton of consonants by simple trial seems very inappropriate in this event.

> *"Determination of stringing and activities related to this term, hence the phonetic value Sn, taken from Snw 'filet' Not to be confused with Z7"*[56]
>
> *(Allan GARDINER, 1957)*

Numeral 1000

Middle-egyptian: *xA* **"**Stem, leaf and rhizome (stem or root) of lotus"
Ewondo: *To "lotus"
Basa'a: ko'o "lotus"
Middle-egyptian: *x* "human placenta"
Ewondo: nya or kômbo, sôkh, kue, bibyal "placenta, uterus"

56.http://projetrosette.info/popup.php?Id=1015&Hiero=V1

Middle-egyptian: *xA* (1000)
Ewondo: Toyina "thousand"

The Ewondo expression toyina is therefore understood as / to yi na / literally the lotus and the mother. But if we speak well an Ekang language it refers to / nto ù-nyu / ie to the "major finger". So 1000 is the finger. Other dialectal variants will focus on the placenta or only on the lotus .

Other expressions exist to say 1000 as / à-kura / or / toasin /. Overall, the logic remains the same:

Ewondo: /à-kura/ = Ako + Ra. /à-ko/ "foot" and ra "mouth" or /à-nyu/ or /*ùn*-nya/
Ewondo: /tosin/ = to "foot, seat, place" and /sin/ or /*ùn*-soŋ/ "beak, mouth"

So we see that no matter the word, we are talking about the same reality.

In tupuri language we have simply :

Tupuri : Ndakré "Lotus" and by phonetic proximity **Tupuri** : Ndaré "1000; thousand".
Ngiemboon : *Tosse* "lotus" and by phonetic proximity in **Ngiemboon:** *Tosse* "lotus; 1000"

Numeral 10.000

Middle-egyptian: mD "ten"
Ewondo: dy-ôm "heap, package"
Ewondo: à-wôm "10, ten"

The root of the word being here ***om**, it would be more appropriate to note the package not by ***mD*** but by ***mw*** or ***wm***.

Middle-egyptian: ***DbA*** (10.000)
Ewondo: Toyina-awôm (ten thousand) or à-wom toyina "ten thousand"

Semantically, the cubit is a fold:

Ewondo: /i-nyîgha/ "a fold"
Ewondo: /à-nyigh/ "to fold"
Ewondo: /*ùn*-nyu/ "finger" or i-nyigha "that can be folded"

The scribe indicates folds by contortions of the snake applied to the hand or foot.
10,000 is therefore semantically the 28 folds of the hand or the foot, the fingers.

So for the scribe, the best idea to represent the 10 fingers and therefore the numeral 10.000 is therefore the finger.

If 1000 indicates the finger, then 10,000 represents the 10 fingers of the foot or the hand.

Ewondo: sign D58 /à-ko/ to vocalise /*to/ here because the foot also indicates a place and serves as a nominal class for places in our languages as a rule. / ù two vé? / literally, "Where do you live?"; "Where is your place? ". So you have to read not / ko / but / to / in Ewondo language.

Middle-egyptian: ***mnD*** "brest; bosom" **Ewondo:** /á-nyié/ "breast-feed"
Ewondo: /nyoa/ homophone of /*nya/ or /na/ "mum"

Ewondo: /to-yi-na/ “the place and the mother”

Ewondo: /wa/ read /wom/ for snake exuvia or for 10 fingers.

So we have so far /toyine à-wom/ or /à-wom toyina/ “10. 000”

Let’s explain it in another language :

Tupuri: *Do bong* “hand (finger)” **and we have in Tupuri:** *Do bong* “10000; ten thousand”

Numeral 100.000

Middle-egyptian: ***ḥfn*** (100.000)
Ewondo: *ùn*-ntét-toyina (hundred thousand)

May you please remember that the numeral 100 is linked to the spiral. It is represented here by the sign V28 of Gardiner's list or the sign Z27 .

Ewondo:/ùn-ntòŋ/ “stream”
Ewondo: /*nya/; /nyo/ “snake”

For the scribe, the tadpole can be considered as an animal that ripples like a snake in water.

Middle-egyptian: ***fnT*** “snake; worm; to be filled with worms, crooked”
Middle-egyptian: ***HfAw*** “snake”
Middle-egyptian: ***Hfnr*** “tadpole; fried, fry; pollard”
Middle-egyptian: ***Hfn*** “hundred thousand; multitude, infinity etc.”

Ewondo: *ùn*-nyəŋlə; nyama; myen “fry, fried, pollard, tadpole”
Ewondo: nyô(l) “snake”
Ewondo: *ùn*-nyəŋlə “flexible, that changes easily, that can be softened, tender, etc.”

So if the finger / *nyu / here represented by the snake / nyo /, the mouth /à-nyu /; a drink / à-nyú / equals 1000 so if we add 100 which is the spiral, we have 100,000.

So reading from left to right I have: / *ùn*-ntét toyina / "100,000".

Numral 1.000.000

Middle-egyptian: ***HH*** (1.000.000) million
Ewondo: *i*-kùkh (thousand, thousand times)

The term highlighted above is / *ùn*-ŋkùkh / "the trunk" that term is homophone by its root /*kukh/ of /i-kukh/ meaning millions.
One can also consult the script translated by ***HH*** associated here with the idea of praises.

Ewondo:/ *i*-kàŋ/ “praises, joy, party, celebration”,
Ewondo: /*à*-təgh *ùn*-ŋkùkh/ “to celebrate, to dance, etc.”

Because the praises are considered infinite then the transliteration ***HH*** returns the idea of infinity.

Morphology

Morphologically, the cardinal numbers are placed after the nouns they count and behave like adjectives.

Ewondo: nyakh *i*-kùkh "1.000.000 cows"

Their position in the sentence is therefore perfectly explained by their quality as determinants. This is also corrected by the fact that names are often changed to the singular (their basic form). This expression can be used with an indirect genitive to signify a lot.

Middle-egyptian: *n HH* "alot, mutiple, many, much etc."
Ewondo: bə nə i-kukh "they are very numerous"

OTHERS: special cases

Numeral 0.5

Ancien-Egyptien:  ***gsw*** "half"
Ewondo: /i-fas /; /i-toun/; /*ùn*-mfakh/ "half; halfback"
Ewondo: /*à*-ŋgos /"heap" or **Ewondo**: /*i*-sàngà /"a split"

The word often retained and dictated by classical Egyptology in transliteration corresponds to several words in Ewondo but all those words share the idea of division / i-tighâ / "a tip, a cutout"
The sign  W11 of the Gardiner list is a ring stand for a jar, a red ceramic pot.
Every pottery object is called in Fang / Beti language: / *ùn*-ŋkakh /. The term to indicate a place, a locaus in theses languages Fang / Beti is / *ùn*-mfakh / "side; part".

Example: Where do you live? / u two *ùn*-mfakh *ùn*-mbéh / literally where do you live? On which side is your place?
The sign S29 of Gardiner's list refers to the folded fabric. It is a loincloth of African notability often worn on one side: /*ùn*-mfakh/ (*conf. Illustration N°10*).

The sign Aa13i  of the Gardiner list refers to an animal's ribs. The term to indicate a fang / beti rib is / *mfakh /, / mviè / etc.

Ancien Egyptien:  ***gsw*** "half"
Ewondo: /*ùn*-mfakh/; /i-fas/; /ngap/ "half, halfback"

As always we can mobilize all the synonyms related to the term /*ùn*-mfakh/ as:

Ewondo: /*i*-fas/ "half, halfback"
Ewondo: /*i*-toun/ "half, part, portion; section etc."

Egyptian Notation: $= \frac{1}{2}$ "½; half"

Fraction: 1/5

In order to construct a fraction the Ewondo will use the word designating the part and indicating the denominator that served for the division to the unit. Thus, the Ewondo fraction appears as a sum of unit fractions, thoses fractions having numerators equal to one (1) and positive integer denominators, with these denominators all differents.

Ewondo: /i-fas/; /ngap/ “part”

So, the fraction1/5 below will be read / i-fas tana / or / ngap tana / since in Ewondo / tan / "five"

Middle-egyptian:  $= \frac{1}{5}$ “a fifth”

The egyptian fraction[57]

We observe that the Egyptian system works with the same logic as the Ewondo fraction, namely:

An Egyptian fraction is a sum of unit fractions, that is, fractions that have numerators equal to one and positive integer denominators, with these denominators all different. All positive rational numbers can be written in this way and in an infinite number of different other ways.

Indeed, it is trivial to express any fraction by a sum of unit fractions by allowing to repeat the terms as in the example above:

The sign D21 ***r*** was used to represente the numerator 1 because:

Middle-egyptian: r “mouth”
Ewondo: /*ùn*-mbokh/ “hole ”
Ewondo: /*ùn*-bokh/ “1; one”

Fractions were written with the sign D21 above and the denominator was indicated below. So ¼ for example was written:

Example:

Middle-egyptian:  $= \frac{1}{4}$
Ewondo: /i-fas nyînà/ / “a quater” litt. One part of a division by four

There were special symbols for the most common fractions as it was seen on it with fraction $^1/_2$ and for two non-unitary fractions $^2/_3$ and ¾.

Middle-egyptian:  2/3 “two thirds”
Ewondo: /bi-fàs bibèh yà ngap ilèh/; /i-bèh ya bilèh/ “two-thirds”

Middle-egyptian:  3/4 “three quarters”
Ewondo: /bi-fàs bilèh yà ngap i-nyîn/ “three quarters”

If the denominator becomes too wide, the sign D21 was placed just at the beginning of the denominator:

57. L. BEECKMANS, The splitting algorithm for Egyptian fractions, vol. 43. J. Number Theor., 1993, p. 173-185.

Middle-egyptian:  1/331 "a three hundred and thirty-first"
Ewondo: /i-fàs mi-nter mi-lèh à yi mə-wôm lèh é dyà / "1/331"

Fraction greater than unity
If we want to name a fraction greater than unity in Bəti languages like the fraction 7/4 for example, it will be: / bi-fas zembgwa bi tyène à yi ngap i-nyî / literally seven parts belonging to a division of the unit in 4 parts.
Or simply / i-fas nyîna, zambgwal / or / zambgwal, i-fas nyîna / literally a quarter seven times.

Special case of the cubit

The concept of cubit has been used extensively in surveying. So this notion is used for the number 10,000 but also for the cubit itself. Let's try to see how and why.

Middle-egyptian*:*  ***Dba*** "finger, thumb toe; 1/28 cubit; 10'000"
Ewondo: /ù-nyu/ "fingers even toes"

The Royal Cubit

It is defined as the distance from the middle finger to the outer fold of the elbow. It is worth 52.36cm. Since the Egyptian numeral system is decimal, it is based on the 10 fingers of the hand or foot.
The fractional system is indicated by the phalanges of each hand (14). 3 phalanges for each finger and two for the thumb.
In total two hands 28 phalanges with sharp divisions represented by folds.

The 1/28 cubit is actually the average length of a phalanx so ration 52.36cm / 28 is 1.87-1.88cm. From the XXVth Dynasty this value will be raised to 1.89cm.

Ewondo:/i-nyîgha/ "fold"

10.000

We must take into account the 3 signs because we have three syllables to say 10,000.

Ewondo:/toyine à-wom/ or /à-wom toyina/ "10'000 "

Understand the vocalization of each sign

Ewondo: sign D58  /à-ko/ to be vocalized /*to/ here because the foot also indicates a place, a place and serves as a nominal class for places in our languages as a rule. / ù to v? / Where do you live? literally where is your seat? We must read as in Ewondo / to/.
Ewondo:  /nyoa/ homophone de /nya/ ↔ /na/
Ewondo:  /to-yi-na/ "the place and the mother"
Ewondo:  /wa/ read /wom/ for snake exuvia.

We have so far /toyin' à-wom/ or /à-wom toyina/ "10'000"

Multiplication
Middle-egyptian*: wAH tp m x r sp y* "***x*** multiplied by ***y"***
Ewondo: / bi-ŋguma **x** *ùn*-ntîghèn bifas **y /** "Integers x cut into parts"

Division

Middle-egyptian: ***wAH tp 500 1/10.f m 50*** "take 1/10 de 500 = 50"
Ewondo: / i-fas à-wômo yi é bi-ŋgùma 500 bi nə 50 /

If I have contented myself in this translation just by pointing out that the numerals reading was perfectly possible outside any constituted framework, then it is notable that African thought as it transpires through this numeration seems to distance us from Western absolutist thought.

African complexity through the reading of the reality by numbers thus induces a new approach. This approach makes Western science a point in the infinite possibilities offered by African thought.

The African numeral system is different of the Western system. It would therefore be more sensible for Africans to take the measure to advance science and knowledge for the humanity benefit.
What appears to be a special case for independent researchers like brother **Kombo Diama** who revolutionized mathematics, especially by working on the thorny problem of prime numbers. I am convinced that any African-born having some knowledge can indisputably lead to discoveries that will surprise the world.
Diama Kombo simple literary just gave a masterful proof and a lesson to mathematicians around the world. Far from the comfort of international centers and flanks, he revolutionized the world and will serve as a model for young Africans.

***

A-1.2 ***The ordinals numbers***

Ordinals numbers are used to express numerical order, the order of beings or objects.
As in most languages of the world, the ordinal numbers are derived from those of cardinal numbers. Ordinal numbers are considered grammatically as adjectives obtained from the implementation of the derivation on the names of numbers.
The numeral connective theme is done by suffixing the final vowel of the cardinal numeral.
Unlike the cardinal numbers whose derivative or suffix carries a high tone, that of the ordinal numerals is at low tone.

N°	Cardinals	Ordinals	Comments
1-st	*okh	[Os]-Ús-**Ù**	*First*
2-nd	Bãh/bèh	bɛ̂–(è)	*Second*
		bâ–(à)	*only for Second Wedding*
3-rd	Lãh/lèh	lál–**à**	*Third*
4-th	Ngnĩ(n)	nyìn–i or nyìn– **à**	*Fourth*
5-th	Tán	tán–**à**	*fifth*
6-th	Samâñ	sàmàn–**à**	*Sixth*
7-th	Zamgbwa(l)	zàmgbàl–**à**	*Seventh*
8-th	mwÔm	mwòmò–(ò)	*Eighth*
9-th	I-bu(l)	èbùlû–(ù)	*ninth*
10-th	A-wóm	àwóòmò–(ò)	*Tenth*

Chart 17: Comparative of Ordinals numbers

These ordinal adjectives are formed by suffixation to the cardinals equivalent of the proto-version * ŋ^{jw}ə inherited from the Negro-egyptian. The closest form that has certainly been used in an archaic Ewondo is found in Fang with -ye[58]. This simplified form serves as connective to the numeral theme to form ordinals in the Fang language.

> **Negro-egyptian**: **ŋ^{jw}ə-*
> **Fang**: *nje>-ye*
> **Ewondo**: *-jà>-ià>-à (feminin)*
> **Ewondo**: *-jw>-iẁ>-Ù (masculin)*

As wrongly pointed out by Mintsa [59]in 2013, there would be another way to express order and numerical order, namely, by putting forward the connective of the numerical theme "ê" which is actually a simplified form of i-nyi; [Ê-nyi].

Example:

> -ê mot bè (the second personne) →ê-nyi bê (the second)
> - ê si à-wôm (the tenth country) →ê-nyi à-wôm (the tenth)

Syntactically this particle /**ê-**/ corresponds to what the specialists often call the *enhancer* or in French *augment*. Since the *enhancer* is a refresher of the noun / pronoun, we must not look here for any intrinsic link with the enunciation of rank.

Special case of the ordinal adjective first.

The Ewondo root of the cardinal numeral "one" is * okh. Regarding its morphology, it comes from the Negro-egyptian Kwéke branch.

The ordinal numeral adjective formed for the ordinal /ù-sù/ "one" which is also vocalized [o-sù] is realized as a result of a loss of an amuizement of the consonant k: (k> g> s) .
On the diachronic plane, we have:

> **Ewondo:** *okh->ù-kù>ù-gù> ù-jù> ù-zù >ù-sù.

In terms of African ontology, we think that this term refers to the primorial ancestor Hr which is associated with the face, the head.

> **Old-gyptian:**  "Horus"
> **Ewondo**:/à-sù/ "the face"
> **Ewondo**:/ù-sù/ "in front; ahead, before"

May you try to observe that /Mə-**kê** mə nə *ù*-**sù**/ " the beginning is the first time". So saying the beginning or the first time what is the difference according to you?
There is therefore an honorific anteposition which associates always the idea of the first with the eldest ancestor.

We can observe that this cardinal numeral is used in many African languages as:

- **Kikongo**: m-osi "one",
- **Lari:** m-ósi "one"
- **Tiyombé:** m-ósi "one"
- **Titandu**: m-ósi "one"

58.Joe MINTSA, Le Complet de La Langue Fang, Lulu Press, 2013, p.143
59.Joe MINTSA, Le Complet de La Langue Fang, Lulu Press, 2013, p.143

The Ewondo thus uses the evolved style [o-sù] shared with its sibling languages for an ordinal logic and the archaic form * okh for the cardinal numeration. This may suggest according to linguists skills that the Ewondo has evolved by itself or uses an evolved form borrowed from its sister languages according to their possible contacts. But in reality it is simply a synonyms link for characterizing the same signified.

I therefore recall that the variant [o-sù] recalls and shares the same root with the word face [o-su], it is to be put in correspondence with the Egyptian operative concept ***Hr*** "visage; face" of the Middle-egyptian who associates to the face, the idea of head ***(tp)*** equivalent of (tp.y) in Middle-egyptian for first, chief ***(Hr),*** of first (ancestor) etc.

The variant * ***oko*** still exists in languages like Basa'a in wich they use the item *pɔ́k* for "first", *ɔ́kɔ* in Sango for "first", kowaad in Somali for "first"etc.

On the tonal point of view, the ordinal theme at low tone is superfluous for the base | -fɔg | in the same way of the cardinal at high tone is for the ordinals.

Construction of the ordinal in Pharaonic Egyptian:

Middle-egyptian: ordinal = cardinal + *nw* (masculin)
Middle-egyptian: ordinal = cardinal + *n(w).t* (feminin)

The ordinal theme is obtained by suffixing the numeral theme of the proto-version the agent affix * *ŋʲʷə* "being", at the end we associate to the numeral a verb of state. Third is to be the cardinal numeral three.

***

3.3.3.2. The Gender

B-1 ***The feminine comparison of common names***

The Ewondo as some African languages including Ekang languages is said to be devoid of grammatical gender. In reality does it deprived of gender or not. We want to indulge ourselves in the assumption that if the Egyptian presents a morpheme for the feminine it is possible that he inherited it from someplace. What about Ewondo language used here as a support? It has it or it does not? To realize it, it is technically necessary to carry out reconstructions from the Negro-egyptian and to make a comparison with, the Middle-egyptian which according to specialists makes the feminine by suffixing to the radical the particle: ***.t***
Indeed, in pharaonic ***.t*** feminine is used to highlight the notion of collective: the word ***rxy.t*** is an example of collective "people"

Middle-egyptian*:* feminine = radical + ***.t***

The feminine tag in Pharaonic Egyptian is obtained by grammaticalization and, the Ewondo being more archaic than the middle-egyptian, this could lead to a problem to say whether or not it is possible to find a genetic link between a language that makes its feminine with a ***.t*** and a language which is supposed deprived of any feminine affix.

I suggest first of all to list a Middle-egyptian series of feminine words and by the way to find their equivalent semantically in Ewondo. And consequently to see if there is or not a feminine remaining tag in Ewondo or not.

1-**Middle-Egyptian:** *ib.t* "thirst" →**Ewondo**: *i.vè* "thirst"
2-**Middle-Egyptian:** *Dr.t* "hand" →**Ewondo**: i.*dzÓ*"member"
3-**Middle-Egyptian:** *pr.t* "graine" →**Ewondo**: *fər* "graine"
4-**Middle-Egyptian:** *xAw.t*"skin, leather" →**Ewondo**: *i.kob*"skin"
5-**Middle-Egyptian:** *wD.t* "decree" →**Ewondo**: *à-dzÓ* "decree"
6-**Middle-Egyptian:** *gr.t* "leg" →**Ewondo**: à-*kÓl*"leg"
7-**Middle-Egyptian:** *x.t* "offrandes" →**Ewondo**: bi.*kâ* "offerings"
8-**Middle-Egyptian:** *s.t* "place" →**Ewondo**: sí "place, seat, hearth"
9-**Middle-Egyptian:** *n(w)-.t* "water" →**Ewondo**: à.*nyú* "drink"
10-**Middle-Egyptian:** *mr.wt*"love" →**Ewondo**: i.*dîŋ* "love"
11-**Middle-Egyptian:** *ir.t* "eye" →**Ewondo**: *i.iɛ́r*"glaces"
12-**Middle-Egyptian:** *X.t* "belly" →**Ewondo**: *ŋgùt* "lower abdomen"
13-**Middle-Egyptian:** *Hm.t* "woman" →**Ewondo**: *i.hôm* "marriageable"
14-**Middle-Egyptian:** nb.t "mistress" →**Ewondo**: i-buàn "paramour"
15-**Middle-Egyptian:** *sfA.t* "hate" →**Ewondo**: i.sapákh "unpopularity"
16-**Middle-Egyptian:** *sH.wy* "meeting" →**Ewondo**: i.sə́kh "*meeting*"
17-**Middle-Egyptian:** *Dw.t* "bad" →**Ewondo**: i.dzô " *problem, bad, naughty, dirty etc.*"
18-**Middle-Egyptian:** *niw.t* "place, city" →**Ewondo**:*nyá "*motherland*"
19-**Middle-Egyptian:** *sip.t* "portion" →**Ewondo:** i.fás "*portion; part*"
20-**Middle-Egyptian:** *mis.t* "liver" →**Ewondo**: i.səkh "*liver*"
21-**Middle-Egyptian:** *nay.t* "mooring pile" →**Ewondo**: i.ny "*keeper*"
22-**Middle-Egyptian**: *A.t "moment, instant"* →**Ewondo**: i.yóŋ; i.tôa*"moment, instant"*
23-**Middle-Egyptian:** *bb.t "hole, cavity etc."*→**Ewondo**: i.béh; i.bəbə "*hole, cavity*"
24-**Middle-Egyptian:** *Dab.t "heap of coal"*→**Ewondo**: dibi "*obscurity*"
25-**Middle-Egyptian:** *bnr.t "date -palm"* →**Ewondo**: i.bérèn "*climb-palm*"
26-**Middle-Egyptian:** *db.t "brique":* →**Ewondo**: i.béd "*builder*"
27-**Middle-Egyptian:** *dwA.t"stellar matrix"*→**Ewondo**: ndá "*house*"
28-**Middle-Egyptian:** *Xst "weakness"* →**Ewondo**: á-sùgh "*weak; feeble*"
29-**Middle-Egyptian:** *nx.t "victory"*→**Ewondo**: á-kiná" *victory party* "
30- **Middle-Egyptian:** *mDA.t "papyrus"* →**Ewondo**: mə-ndakh "*words*"
31-**Middle-Egyptian:** *Dab.t "heap of coal"*→**Ewondo**: à-dùmbà "*nest, heap*"
32-**Middle-Egyptian:** *mis.t* "liver" →**Ewondo**: m'is "*eyes*"
33-**Middle-Egyptian:** *sfA.t* "hate" →**Ewondo**: i.fâs "*rejection*"

This series concerning words having the same meaning in the couple of languages clearly tells us that the Middle-gyptian ***t-*** tag of the feminine systematically corresponds a high tone on the last syllable of the corresponding word in Ewondo. Better than that, we have before the noun a form in / i- / or in / a- /. On this account, the rules of phonetics inform us that: the fall of an occlusive in final position can provoke a tonal elevation after having become a glottal occlusive. We are therefore led to conclude here that the suffix of the feminine 𓏏**.t** of the Middle-egyptian is an atavism of the Negro-egyptian and that its culmination in Ewondo is an occlusive that we can identify as a 𓏏**.t** in Middle-Egyptian.

On the other hand we see that what appears as a suffixation (by the choice of the reading sense) of the Middle-Egyptian, obtained by pure convention, is an [í-] or [à] anteposed in Ewondo. In short, if the choice of reading had been made from right to left the **-.*t*** the suffix would be in anteposition as the feminine tags [í-] or [à-] of the Ewondo language.

This information is true only on the basis of the words we have carefully chosen. The current Ewondo speakers having lived before the migrations north of the Sahara it is therefore possible that we find Ewondo words that have kept the traces of atavism **-.*t*** which has been identified as a feminine tag. The example below shows that some feminine words in Ewondo have kept their final tag **-t**.

12-**Middle-Egyptian:** *X.t* "belly" →Ewondo: *ŋgùt* "lower abdomen"
--**Middle-Egyptian:** *sw.t* "cotton" →Ewondo: *sùt* "cotton"
--**Middle-Egyptian:** *swH.t* "shroud" →Ewondo: *ùn-ŋkùt* "jute bag"
--**Middle-Egyptian:** *wdn.t* "heavy" →Ewondo: *à-dit* "heavy"
Etc.

In order to validate this assumption we will rely on the observation that, the Ewondo's high tones correspond to the consonants ***k*** and ***A*** of the Middle-egyptian. The phoneme **-A** being derived from Negro-egyptian * k^h. Thus this phoneme can be realized in * k^h *or* * g^h in Ewondo in final position.
Example:

M.E: ***i.bA*** (dance) vocalised /à-bókh/ in Ewondo. Then we have so far (b-=b- and -A=-kh)
M.E *inbA* (mute) vocalised /mvukh/ in Ewondo. Then we have so far (nb-=mv- and -A=-kh)

It is therefore logical to identify the consonant of the feminine affix in proto-Ewondo to the ***k*** occlusive velar of Middle-Egyptians at the final position. Consequently we could imagine that the Negro-Egyptian affix of the feminine ipso facto has the format * ***-kia***.

On this account, we can suggest the following evolutions to explain the issue due by the presence of ***.t*** as suffix marking the feminine in Middle-egyptian whereas a priori the Ewondo does not have an affix acting as feminin tag.

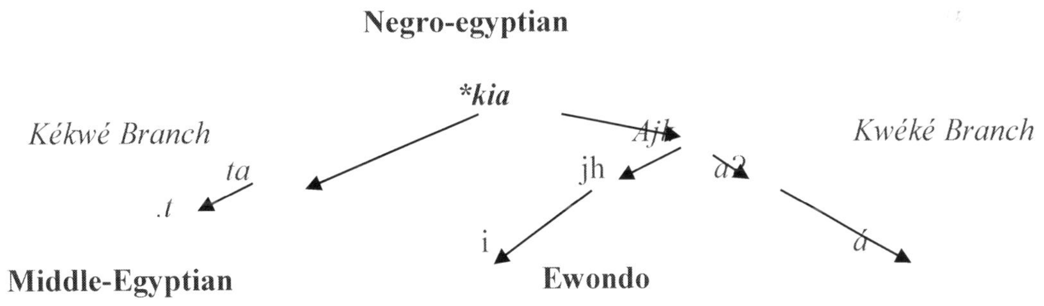

Chart 18: Evolution of Ewondo feminine affixes from the negro-egyptian branches

As a rule the current Ewondo synchrony is making the apocope generally on the final vowel. We can therefore identify the feminine only by the gender prefix stemming from the Negro-egyptian. By the way, this atavism of the proto-form * ***kia*** is found mainly at the beginning of the words in the gender prefix ***à-*** or in the variant ***í-***

Before any evolution of the synchrony, the affix of the gender was also suffixed.

Ewondo: à-kombà "temptation"
Ewondo: à-korà "Variety of raphia palm tree"
Ewondo: à-babgà "Variety of frog"
Ewondo: i-bêghà "Handle; handful"

Ewondo: à-banà "Pearl necklace"
Ewondo: à-bà "jealousy"
Ewondo: à-bebenyà "Big hardwood tree, quite rare"
Ewondo: à-bənga "boat pitching"
Etc.

The evolution of the language that led to the loss of the suffixal vowel led specialists to conclude that this language is not a "Bantu" language since it is devoid of a suffixal vowel. But the atavisms of the presence of this vowel are still present in the suffix form, in the morphology of many words and better than that in the accordance schemes.

Technical Reservation

I want to emphasize a fundamental difference regarding the suffixation of the 𓏏-*.t*. If we admit that the bread cake can indicate a feminine we also think this tag can or can not be vocalized and even undergo a weakening in a final position. let's take the example of two anthroponyms:

Middle Egyptian: 𓇋𓏠𓈖𓏏𓁐 ***imn.t*** "amonet"

We therefore say that this transliteration refers to two potential nouns on the basis of phonetic proximity and not semantic study.

Fulfulde: Amina
Fulfulde Amin**ta**.

These two anthroponyms are still used today and are worn by women. So the 𓏏-*.t*. is not only a feminine tag it can be a syllable in the final word.

I wan to take another example:

Middle-Egyptian : 𓂝𓏏𓄹 ***a.t*** « member »
Middle-Egyptian : 𓂝 **a** « hand »

As it can be seen from this couple of scripts above, one could not here confine oneself to wandering about the supposed gender of a hand or a feminine of the hand. In both scripts, it is a question of the hand. I will therefore offer a healthy reading in Ewondo of these two scripts to fit the requirements of the above transliteration.

Middle-Egyptian : 𓂝𓏏𓄹 ***a.t*** « member » ☞**Ewondo** : /*i*-twàè/ « what allows to take »
Middle-Egyptian : 𓂝 **a** « hand » ☞**Ewondo** : /wà/ « hand »

As you can see in these examples also, 𓏏-*.t* is not only a feminine tag from the African point of view. So according to me, we do not have to neglect the suffixal 𓏏-*.t* during the reconstitution of the final word.

B-1.1. *Creation of Feminine in Ewondo*

The crossing from the old Ewondo language to the current synchrony of the Ewondo leads to the observation that the prefix of number (PN) is often erased (partially or totally) on the surface in the singular giving the impression to observers that each stuation is a particular type of class prefix (CP).

→ **Old Ewondo**: lə-‘búm “belly”
→ **Ewondo**: Øà-búm “belly”

In addition to this phenomenon, the Ewondo is realized by a partial or total loss (number prefix and gender) and the total loss of the suffix vowel to the neighboring languages of group A of Bantuists such as Douala, Bassa'a, Eton, Lingala, Yemba or Kikongo etc.:

Standard	Duala	Basa’a	Ewondo
Guthrie	A.24, group C.20	A.43a and A.43b.	A72a
Bas-tin/Coup/Mann	A.24	A.43a et A.43b.	A72a

Chart 19: Classification of Ewondo and few sibling languages

Let's go back to our previous example to illustrate this point:

Proto-grassfield: *búm, *là

Old Ewondo: *lə*-búm “belly”

Basa’a: *li*-búm “belly”

Yemba (West-Cameroon language): *lə*-vem “belly”,

Lingala: *Li*-búm**u** “belly”,

Bangala: *Li*-búm**u** “belly”,

Duala: *di*-búm “belly”,

Ewondo: *à*-búm “belly”,

If we had to rebuild the proto-word, it would take the style: *prefix* + lexem + *suffix* with:

Prefix: (gender prefix + number prefix)
Suffix: gender prefix

Depending on the evolution of language synchronies, we have lost more or less some affixes. In order to facilitate the understanding of the purpose, we will consider that the current Ewondo synchrony is apocoping on the final vowel like its sibling languages of the *Beer* group.

Negro-egyptian: **ku-hu* “belly”

Proto-bantu: **ku-hu* >**ɔ̃ b^{w}u-ebu* *“place of viscera”*

Proto-grassfield: **là; *búm-,*

Oluganda: *olubuto “belly”*

Proto-Bantu: **búm-,“belly”*

Vieil Ewondo: *lə-búm "belly"*

Ewondo: *à-búm "belly"*

*"[...]The Proto-Bantu prefix of the form * ǒ bʷu-, in the light of the Negro-egyptian facts and the data of the Bantu languages themselves, serves both to form place names (see tswana bo-tswana "land of Tswana, luganda ubu-ziba "middle of the lake") and abstract (see above)."*

(Mboli 2010)

Clarification

The semantaxe related to the belly refers to the idea of proportion (à-bím), related to the middle, to the heart, inside in our languages here the proportion of the body. The consecrated expression is /à-bím dili/ or /à-bím-di/, /à-bím-té/ "this proportion".

The word equivalent to / à-bum / in Middle-egyptian is ***m-ab*** (in the middle), ***hr-ib*** (face of the interior), ***ab*** (meal) which corresponds to the word / à-bum / in Ewondo, the expression for gut: ***qAb*** "belly cord, inside, middle"

The word * vu> * hu returns the idea of the air and the lungs and by semantic shifting the idea of viscera and therefore of the heart.

The term *ihundi / ivundi* "ventilation / air", in Ewondo will be at the origin of the English term window and not the opposite as one might think. Without constructing a building and having a nomadic life style, it is difficult to invent a word of which one does not have the use and the necessity.

Gender and accordances schemes

Thus, since the current synchrony of Ewondo undergoes a total or partial elision of the prefixal part in which the gender prefix and the suffixal part are situated, this suggests to unsuspecting observers the absence of grammatical gender.

However, since there are schemas of accordances in this language, to identify the gender of a word, it is just need to follow it with a presentative or a possessive and check the type of accordance resulting. This exercise will determine if the word is feminine or masculine. Because technically, for a langue which dealing with the gender like French language, the presentative or the possessive makes accordance in gender and in number with the name. It seems more expressive in French:

Example in french: mango

Singular Possessive: **ma** mangue "my mango"; ☞pluriel: **mes** mangue**s** "my mango**es**";
Singular Presentative : **cette** mangue "this mango";; ☞pluriel: **ces** mangue**s** "these mango**es**";

Example in french: donut
Possessive: **Mon** beignet"my donut"; ☞pluriel: **mes** beignet**s** "my donuts";
Presentative: **Ce** beignet "this donut"; ☞pluriel: **ces** beignet**s** "these donuts";

If we do apply the same principle to Ewondo we will have:

Example 1: feminin gender Prefix *í*

Ewondo possessive singular: **í**-toa ***dí*** –àmà: "*my chair*";.
Ewondo possessive plural: ***bí*** -toa ***bí*** –àmà "*my chairs*"

Ewondo Presentative singular: **í**-toa ***dí***: "*this chair";*
Ewondo Presentative plural: *bí*-toa *bí*: "*these chairs*"

On the diachronic level it would be more appropriate to postulate this evolution only for the prefix:

kaí-toa > **(d)í**-toa >**(l)í**-toa>**(h)í**-toa>**í**-toa: "*chair*"

We can observe here that the prefix of number in the singular **(d)** has disappeared in Ewondo. The agreement synchrony of Ewondo suggests that the word chair was in the past: ***di***-toa.
We argue that for two main reasons, the first one is linked to the accordance as we have seen above and the second reason is linked to the fact that we can find this prefix in Ekang's neighbouring areas where sibling languages are spoken.

Example 2: feminin gender Prefix í

Singular: **í**–lé: *tree;* **Plural** : b**ì**–lé "*tree(s)*"
possessive☞ **Ewondo**: **í**–lé **dí** –àmà "*my tree";* → (d)**í**– lé>(l)**í**–lé>**í**-lé: "*tree*"
Presentative ☞**Ewondo**: **í**–lé **dí** "*this tree";* → (d)– **í**lé >**í**–lé: "*tree*"

If phonetic laws are taken into account, then this possessive sentence should be written:

Ewondo possessive: **í**–lé dy–àmà "*my tree" because* **i** + a = ya *there is then "yodisation".*

Example 3: feminin gender Prefix à

Ewondo: à–lɔŋ "*hole";* mə **à**–lɔŋ: ☞ mə–lɔŋ "*hole(s)*"
Ewondo possessive: **à**–lɔŋ **də** –àmà ☞ **à**–lɔŋ **d'**–àmà "*my hole";*
Ewondo Presentative: **à**–lɔŋ **dí**"*this hole";* → (d)**à**–lɔŋ > **à**–lɔŋ: "*hole*"

If phonetic laws are taken into account, then this possessive sentence should be written:

possessive **Ewondo**: **à**–lɔŋ d'–àmà "*my hole*"

Example 4: feminin gender Prefix Ø

Singular☞**Ewondo:** kùb: "*hen";* Plural ☞**Ewondo:** bə–kùb "*hen(s)*"
possessive☞ **Ewondo:** kùb **dý** –àmà reconstruction ☞ **Ewondo: dí**–kùb **dý**–àmà "*my hen";*
Presentative☞ **Ewondo** Presentative: **í**-kùb ny**í** "*this hen";*

B-1.2. ***Creation of feminine proper nouns***

1) **M-E:** *Niw.t* "city" → **Ewondo:** *à-yóŋ* "county", -> name, Ayóŋho>*Ayóŋ'ó*
2) **M-E:** *Hm.t* "wife" → **Ewondo:** *i.kom* "sterile woman" -> name, Ikómó
3) **M-E:** *D.t* "body" → **Ewondo:** *-niol/nio* "body" -> proper noun, Nióló
4) **M-E:** *x.t* "sting" → **Ewondo:** *nkol/nko* "string" -> proper noun, Nkóló
5) **M-E:** *Hr* "head, visage" → **Ewondo:** *à-su* "visage" -> proper noun, Asú'ú
6) **M-E:** *ban.t* "neck" → **Ewondo:** *kiŋ* "neck" -> nom propre, Kíŋ'i
7) **M-E:** *sA.t* "girl" → **Ewondo:** *ngon* "girl" → proper noun, Ngónō
8) **M-E:** *xy* "placenta" → **Ewondo:** *ive* "viscera" → proper noun, Ivú'ú
9) **M-E:** *Dab.t* "charcoal" → **Ewondo:** *dâghe* "charcol" → proper noun, Dâghà
10) **M-E:** *pHar.t* "race" → **Ewondo:** *mbîl* "race" -> → proper noun, Mbílí
11) **M-E:** *Sdt.t* "paste" → **Ewondo:** *ikôh* "paste" -> proper noun, Ik*ó'ó*
12) **M-E:** *niw.t* "country" → **Ewondo:** *mvogh* "place" -> proper noun, Mvógó
13) **M-E:** *s.t* "place" → **Ewondo:** i*tÓ* "place, seat" -> proper noun, It*ó'ó*/Eto'ó

14) **M-E:** *n(w).t* "water" → **Ewondo:** me-*nyú* "drink" -> proper noun, Menyú'ú
15) **M-E:** *iA.t* "middle" →**Ewondo:** zaŋ "middle" -> proper noun, zaŋ'á

To achieve a proper noun from a common noun, we suggest the following equation:

Proper noun = commun name + copy of last vowel.

By the way, a proper noun is réalized from a common name always ends with a vowel in final position. So the points of view arguing that Ewondo language is not a "Bantou" language and a Bere language because it doesn't allow a final vowel, that point of view is disputable. According to me Ewondo out of proper nouns, Ewondo is using many words with a final vowel.

B-2 *Creation of Masculine in Ewondo and in Middle-egyptian*

If the Ewondo like many African languages seems to be devoid of grammatical gender according to specialists. We see that the middle-egyptian makes the masculine by adding a suffix particle: ***w*** or ø (*the morpheme to mean zero*).

According to Egyptologists, there are a number of words that, even if they end with a **.t** are not feminine, they are therefore masculin. They are few but they do exist:

Middle-Egytian: ***xt*** *"wood, tree"*
Middle-Egytian: ***twt*** *"statue"*
Middle-Egytian: ***it*** *"Barley"*
Middle-Egytian: ***mwt*** *"death"*
Middle-Egytian: ***it(i)*** *"father"*

If I take the first example ***xt*** *"wood, tree"* we do say: / i-lé / "tree" or / ù-kyé / "leaf" these two words can respond to this transliteration and designating a tree. That's why gender theories are highly debatable here. But I present them just for fun. All these words presented as masculine by specialist can find a masculine or feminine equivalent feeting to this transliteration. The reason is that we are not in an all-or-none principle but in complexity when dealing with African thought. A tree can be consider as leafs, wood even roots etc. depeding of circumstances.

The word written ***xt*** *"wood, tree"* is a tree according to me just by metonymy but this word is for leaf. In african ontology a tree looks like a foot therefore by mtonoymy the word written here is foot also /à-ko/. As we have seen in the previous chapter /à-ko/ is feminin but /ù-kyé/ is masculine. Both are dealing with the tree here by metonymy relationship. The tree is a deep concept in African philosophy and sociology.Then some people may pay attention with their certitudes.

On this account, is there a genetic link between a language that is masculine as above and a language that is supposed not to have a characteristic affix of the masculine?
It is worth noting that, in Middle Egyptian feminine and masculine are the result of grammaticalization phenomena by prefixation or suffixation of affixes to the lexeme according to the specialists.

As for the previous case, let's take a series of words that are validated as masculine in Middle-Egyptian and find their equivalents in Ewondo language then observe their behavior at a purely morphosyntactic level.

1-M.E.: kA *"bull;* taurus" → Ewondo *un-n*káŋ "taurus"
2-M.E.:: *Ab. W* "elephant" → Ewondo: *u*-mbàŋ **"**Tusk"
3-M.E.: *Apd* "duck, fowl" → Ewondo: *u*-pkwal"partridge"

4-M.E.: *nH*"guinea fowl" → Ewondo *un*-nkáŋ "guinea hen"
5-M.E.: *ntr*"god" → Ewondo *un-n*tour"holy"
6-M.E.: *qAA "hill"*→ Ewondo *un-n*kō"hill, mountain"
7-M.E.:*tni* "eldest son" → Ewondo *un-n*tó"eldest son"
8-M.E.:*wnmw* "food" → Ewondo *un-n*am"meal, food, dish"
9-M.E.: *bA* "ram" → Ewondo *un-n*tómbá"ram, tup, aries"
10-M.E.: *Hr* "head" → Ewondo *un-n*lóh"head"
11-M.E.: *kA "taureau"* → Ewondo *un-n*káŋ"taureau ; bull"
12-M.E.: *xar* "to be furious" → Ewondo *u*-loun"anger, furious"
13-M.E.: *HqA* "souverain, dirigeant" → Ewondo *un*-ŋkóŋ "ruler"
14-M.E.: *antyw* "dried myrrh" → Ewondo *u-toui* "incense, myrrh"
15-M.E.: *nw* "hunter" → Ewondo *un*-ŋsôm"hunt"
16-M.E.: *wSA.t* "whistling duck" → Ewondo *un*-ŋsóŋ'o "the beak"
17-M.E.: *nfr "zero"* → Ewondo *un*-ŋkór "dry"
18-M.E.: *Tb "cage"* → Ewondo *un*-ŋkwab "box"
19-M.E.: *itrw* "Nil, river" → Ewondo *un-n*tour "the holy"
20-M.E.: *Sn* "ring" → Ewondo *u*-kéné "ring"

We notice that all the words that are grammatically identified as words of the masculine gender in pharaonic (**w** or ø), these words all start with a type of prefix particle: ù-lexeme in Ewondo.
If by convention the terms of the masculin gender of pharaonic are recognizable by their suffixes (w or ø), undoubtedly Ewondo we find the same prefixal particle that is attached to the lexeme.

Cheick Anta Diop explains that:

> *"[...] When ( ) =wa "one" precedes the substantive, we connect it to this one by the adjective of the genitive n(y) n = w'n*
> *which corresponds to the French indefinite article and whose usage is common in Coptic. (...)"*[60]
> *(Anta Diop-1977)*

Example:

Ewondo: á-**lób** "to fish"
Ewondo: *ún*-n**lób** **"**a fishing rod; a fishing party**"**
Ewondo: *ún*-n**lób** *um*-b**wà** **"**one fishing rod**"**
Ewondo: *ún*-n**lób** *um*-bé? **"**Which fishing rod?**"**
Ewondo: *ún*-n**lób** *ù*-àmà **"**my fishing rod**"** ú-**ámà** → **wámà** because *(ú+ámà = wámà)*

A masculine term in Ewondo is: *ù-substantive.*

This prefix is highlighted by accordances schemes as above demonstrated and by the way, gives a better understanding of the accordances we observe for sentences expressing the possessive or the demonstrative or even the interrogative.

Example 1: masculin gender Prefix: ù vocalised [y]

Ewondo:Ù–wá: "*hand*"; mə **ù**–wá; m'–wá [mwa]"*hands*"
Possessive ☞**Ewondo**: **Ù**–wá **ù**–àmà"my *hand*" better write (ù-wà wàmà)
Presentative ☞**Ewondo** : **Ù–wá ù**-í "*this hand*" better write *(ù-wà wi)*

60. Cheick ANTA DIOP, Parenté génétique de l'Egyptien pharaonique et des langues négro-africaines, IFAN, 1977, Dakar, P.8.

If phonetic laws are taken into account, it is necessary to write this phrase:

Ewondo Presentative: ù-**wá** **wí** "*this hand*"(because we have u +a = wa and u +i =wi)

this phenomenon is called "welarization".

We finally see that:

- The number prefix (PN) is morphologically consonantal but has partially disappeared into the singular Ewondo whereas it is maintained in the plural.
- When the word is masculine, it can take this form: ù-*substantive.*

We can then suggest that in the couple of languages, Ewondo **ù** = 𓅱 **w** hieroglyphic followed by occlusives or especially nasals ***n, m,*** the sound [y-] also falls in Ewondo as in the other African languages. If this sound falls on the phonetic plane, I reiterate that it is a grammatical and technical necessity because of the accordances phenomena related to it.

- When a word introduces the masculine gender initially, its accordance is made with /U-/ particle in Ewondo language.
- If the gender prefix is feminine /i/, /a/, then accordances are done with /i/ or/a/.
- For names that have lost the prefix of genus and number initially, the developers of these prefixes remain the numeral adjective cardinal, the presentative, the interrogative or the demonstrative.

Ewondo: ø-zokh "elephant"
Possessive: zokh ***di***-àmà "my elephant" written /zokh dy-àmà /(di+a= dya)

To understand the phenomena of rection [61] concerning the nominal zokh for "elephant" we must go back to its proto-version:

Negro-egyptian: kia-*ŋʲʷəkʷɨʀ̃(u)>di-ndjoku>zokh.*

One might think that the suffixal ***-u*** of the proto-version is a gender affix marking the masculine of the word, but that is not true. It is simply a tag signifying "the one who embodies" a tag dealing like the one for a proper noun (here it marks: force, authority, fear). Elephant in african ontology is not only an animal it's also the one who embodied the chief's function.

if: **Negro-egyptian**: *-kʷɨʀ̃ equals force*
Negro-egyptian: *-kʷɨʀ̃ (u) is the one who embodies strength.*

This principle is used to create the proper nouns starting from a common name as we have seen above.

If the word Ewondo / zokh / have lost its genitive phonation "***n***" it has still preserved it in the accordance schemes over time.

By the way, in Ewondo presentative we will have: /í-zokh **ny**i/ "this elephant". We thus find the atavism of Negro-egyptian ****ŋ***ʲʷə-.

The word Ewondo for elephant, which is still used to make the actual accordances, was therefore certainly ***di-ndzokh***, a non-used word that is a proto-version. The difficulty of the technical study of

---

61. The rection is the property that a morpheme has to cause the presence of this or that grammatical category. So it is a unidirectional relationship between two constituents of a phrase, called the rector constituent and the governed constituent

Ewondo lies in this stiffness of understanding the accordances of the language only from a synchronic point of view.

If we consider the gender theory, could you tell me what is the gender of this word minga or mininga:

Ewondo : ùm-inga "wife, woman" plural: **Ewondo** : b'-inga "wifes, women"
Possessive : **ù**m-inga **ù**-àmà "my wife, woman"

If you are carefull you will answer that this word is masculine according to the rules above. Answering this way you will fall in the trap. How comes that to point a woman we do use a masculine gender? That's a tricky question.
The root of the word is *inga. As you can see here this word is feminine, it is the famous African panther. The grammaticalization resulting by the mixing with the genitive particle *ùm-* allows masculine e schemes of accordance giving the impression that the word is masculine.
From the western point of view, [minga] is the wife and the woman. But the facts allow here to think that this word means those of the panther. So this ***ùm-*** particle has to be correlated with the genitive particle 𓅓 ***m*** of the middle-Egyptian.
We can be sure of that because we have :

Middle-Egyptian: mabA« harpoon »

I have explain how the concept of harpoon is linked with the morther when dealing with numeral 30. But I will explain it again further. The root ***Ab/bA*** "panther; leopard". Let's summarize a bit: the root of the word is feminine "inga" and the prefix particle "ùm-" is not a gender prefix but a genitive particle allowing to think that the word is masculine by grammaticalization. It's a nonsense to point the feminine gender by a masculine word in a feminine sociology.

So I grant you that the question was a little bit tricky. Congratulations if you have found the answer.

A-3.1 Creation of a proper noun in Ewondo

1) M-E *wxA* "night" →Ewondo *un-ŋkugh* **"**darkness" -> proper noun, *Nkughu>Nkughû*
2) M-E *Ab.w* "elephant" →Ewondo *ùn-mbogh* **"**elephant" -> proper noun, *Mvogho*
3) M-E *xA.w* "herbs" → Ewondo bi-lokh "herbs" -> proper noun, *Biloha> Bilao*
4) M-E *nfr* "beautiful" →Ewondo *A-beŋ* "beauty" -> nom propre, *Abeŋ'he*
5) M-E iw "dog" →Ewondo *mvu* "dog" -> proper noun, *Mvu'hu>Mvû*
6) M-E *bA* "panther" →Ewondo *Ze* "panther" -> proper noun, *Zəhə>Zəə*
7) M-E *bA* "skin" →Ewondo *ikob* "skin" -> proper noun, *Ikobho>Ikobo*
8) M-E *Apd* "fowl" →Ewondo *ù-kpwal* "partridge" -> proper noun, *ùkpwala*

These proper names are formed by the suffixation of the group ***-hi*** of the Negro-egyptian to the lexeme. This group will therefore generate the presence of the final vowel that will operate a copying of the penultimate vowel giving a new final vowel e, a, u, o, i by elision of h.

***

3.3.3.3. The Numbers: Plurial, Dual

C-1. *The Plural*

For the ancient Egyptian, the plural is expressed grammatically in two ways:
First, there is the classical plural, starting from three units. The termination [62] ("variable element that adds to the radical". Le Petit Robert) is the the 𓅱 or 𓏲 ***w*** for masculine substantives and ***.wt*** for the

62.Termination is the variable element that adds to the radical. - Le Petit Robert feminine ones. To this termination of plural, will be added the plural determinative ׀׀׀, the "three vertical bars or three slashes"

Why do the plural is identified by 3 in african philosophy?

In a binary system, 2 is the other 1.
Actually coming out of the binary system for the higher system (quinary, octal, decimal etc.), the three is therefore the moment when we begin to speak of the plural or else of ***Hr*** the father of the multitude, the third person of the trinity.

The pharaonic synchrony of the time of the pyramids indicated the plural by repeating the determinative 3 times. At the time of the classical language, that of the Middle Egyptian, some texts have kept this repetition, especially in the case of words with a singthe sign.

There are also words that, because they express the idea of plurality (a crowd, a troop, a crew) or things that can not be counted or measured (beer, wine, gold , the oil), are written with the determinative of the plural, while being singular on the grammatical plane.

Ewondo: /mə-yokh/ "the wine" litt. The drinks
Ewondo: /mə-nyùèn / "the drink" litt. The drinkables
Middle-egyptian: ***nnw*** "primitive water Noun"

Note

The three vertical bars that are only considered as a marker of the plural can appear from my point of view as an indicator of the superlative. To this count the word is vocalized by a tripling of the first syllable. It is an atavism of the Negro-Egyptian. We still meet it in some Ewondo words like:

Ewondo: *NE-FE-FE-FE*[63] *"huge Desire of the heart, thirst for something"*
Ewondo: à-*TA-TA–TA-TA-TA TA* [64] *"non-stop, talkative"*

C-2. *The dual*

The duel or Nisbe is an ending (\\ = y) which allows to create nouns or prepositions epithets adjectives marking a relationship or dependence. The real name of that sign is not a nisbe but /ùn-mbiàs/ "twins". For en endogenous point of view i'll prefer the word /ùn-mbiàs/. That's why i'll use it at the following steps of this document.

On the grammatical plane they are real nouns.
The duel is used when it comes to 2 (two) objects / people; this is the case mainly for body parts or for what usually comes in pairs.
Think at expressions like: the two legs, the two eyes, the two entities (Horus and Seth), the Two Lands (Upper and Lower Egypt), and so on. The dual's tag is ***wy*** (*for the masculine*) and ***ty*** (*for the feminine*).

63.Samuel GALLEY, Fang-French dictionary. and French Fang followed by a fang grammar, ditions Henri Messeiller, Neuchâtel, 1964, p.394
64.Samuel GALLEY, Fang-French dictionary. and French Fang followed by a fang grammar, éditions Henri Messeiller, Neuchâtel, 1964, p.394

In Middle-Egyptian, there are two kinds, the masculine, the feminine, and three numbers - the singular, the dual and the plural. Names and adjectives make accordance in gender and number.

Here is a table of the endings of the plural and the dual, masculine and feminine.

Sens	Ewondo	Midd.-egyptian	Sens	Suff.	Syntax
Father	*Í-sià*	***Itf***	a father	-	Masc.sing
Glass/ mirror	*i-yer*	***ir.t***	an eye	*.t*	Fm. sing
glasses	*Bə*-i.yer→B'iyer	***ir.wt***	Eyes	*.wt*	Fm. pluriel
fathers	*Bə*-í.siá→b'isia	***it.w***	Fathers	*.w*	masc. Plur.
2 fathers	*Bə*-ísiá *bə*-bè	***It.wy***	2 fathers	*.wy*	Duel masc.
2 glasses	*Bə i*-yer bə i.bè	***ir.ty***	2 glasses	*.ty*	Duel fm.

Chart 20: Comparative of plurials Ewondo and ancient Egyptian

I just wanna remind that the Nisbe item imposed by the Esperanto-Egyptian is derived from the Semitic languages. It corresponds semantically to the word Ewondo *ùn-mbiás* (meaning twins). The term Ewondo / *ùn-mbiás* / "twins" is therefore more appropriate for the African than I am.
We will see how this term is used in both languages later.

Ewondo: *um*-mbiás (one of the twin pair)
Ewondo: *mi*-'mbiás (twins)

Compared to Pharaonic, the notion of duality in Ewondo does not induce the forming of particular types of nouns. One agrees to use a classical plural (*me, mi, be, ba, bi*) whereas when this notion is used as a superlative, it can induce a repetition of syllables and thus finds its full expression in Ewondo.

Middle-Egyptian: ***mAa.ty***
Ewondo: maâ-maâ "the fairest"

Middle-Egyptian: ***DHw.ty***
Ewondo: *úŋ*-kəkəŋ "the cleverest"

***

3.3.3.4. Connective Particles

D-1. *Coordinating Conjunction "And"*

In grammar, *inflection* or *inflexion* – sometimes called *accidence* – is the modification of a word to express different grammatical categories such as tense, case, voice, aspect, person, number, gender, and mood. The inflection of verbs is also called *conjugation*, and one can refer to the *inflection* of nouns, adjectives, adverbs, pronouns, determiners, participles, prepositions, postpositions, numerals, articles etc., as *declension*.

An inflection expresses one or more grammatical categories with a prefix, suffix or infix, or another internal modification such as a vowel change. For example, the Latin verb ducam, meaning "I will lead", includes the suffix -am, expressing person (first), number (singular), and tense (future). The use of this suffix is an inflection. In contrast, in the English clause "I will lead", the word lead is not inflected for any of person, number, or tense; it is simply the bare form of a verb.

The Ewondo language is classified as inflectional by specialists. It gives the posibility to use the coordination conjunction y/ or báà/ban as "and" in order to unite two sentences, two sub-sentences or to unite within an independent sentence, two elements of the same syntactic function and generally also of the same grammatical nature.

	above	with		Lexicon
Ewondo	Yi/ye	báà/ban	və	AND
Middle-egyptian				
	Hr/on	***m***/with	***Hn***/with	

Chart 21: Equivalent of And

The idea of conjunction is found in the Ekang languages. It can be translated by: *ye, ba, ve*.
These three words, according to me are interchangeable with the Egyptian signs.

It is admitted that the Egyptian does not know the coordination structure like the one of the French language, but only a *juxtaposition* style, when it comes to expressing **copulative coordination**, for lack of having a morpheme equivalent to that known in French ***and*** "addition".
The notion of addition is expressed semantically by replacing the last syntagm of the juxtaposition structure with a prepositional phrase introduced by the preposition ***Hr*** "on" and also the preposition ***Hn*** "with":

Example:

Middel-Egyptian: ms-w-i ***Hn*** ' sn.w-i "my children and my brothers"[65]
Ewondo 1: /*bw*-an *bə*-ámà **yé** *bə*-ànyáŋ *bə*-ámá/ ☞ Eyo "eldest ancestor"
Ewondo 2: /*bw*-àn *bə*-ámà **báà** *bə*-ànyáŋ *bə*-ámá/ ☞ bA "spirit"
Ewondo 3: /*bw*-àn *bə*-ámà **ve** *bə*-ànyáŋ *bə*-ámá/ ☞ à-və "to give"
English: "my children and my brothers"

D-2. The Conjunction "Or"

There exists in Egyptian as in Ewondo the coordination structure, to express a disjunction, then using the qualified coordinator in Egyptian Esperanto by ***r-pw*** "or" this conjunction is in Ewondo / ŋguə kigh /; / ŋguə fə/; /ŋguə fwà/ "or; else" followed by the conjunction of subordination / náà / "that":

<u>Example</u>:

Middle-Egyptian: m nb, m sn, m ẖnms **r-pw**
English: "as a ruler, as a brother, **or** as a friend" [66]

Ewondo: /à mou *úm*-myè**,** à mou màdyáŋ**, ŋguə kigh náà** à mou mvoué/

	or	lexicon
Ewondo	*ŋguə / ŋguə náà/ ŋguə kigh náà*	*Or, But, If*
Midd.Egyp.	**r-pw**	*or*

Chart 22: Equivalent of or

65.= "enfant-Plur-je" "avec" "frère-Plur-je" (d'après Gardiner, 1994,69)
66.D'après Gardiner, 1994,69

D-3. The other Conjunctions are:

Ewondo	*Middle-egyptian*	*Lexicon*
Yí?	***In***	*Is?*
Ane, ùn-nti à ne	***kA***	*so, then, for, since*
Ngə, vədà	***Swt***	*but, if, or,*
Í-vôm, í-vôkh, um-bol,	***HA***	*Since, as*
À mou, *à-sou*	***M***	*because*
Nn, nyin	***Imi***	*Since, for*
Fana, fana ho na, vè bo na,	***M***	*Although, whatever*
E bo na, e yi na,	***N***	*So that*
A su dàbà	***n sn***	*For them*
à mbar' nâ,	***Nri***	*for fear that,*
íyóŋ	***HA; ir***	*... if*
ndə	***kA***	*...Then.*
Iyóŋ	***M***	*when*

Chart 23: other conjunctions

***

D-4. GRAMMATICAL'S FUNCTIONS OF EGYPTIAN NOUN

In Pharaonic nouns are generally used as:

- Subject
- Direct object complement
- Indirect object complement
- Circumstantial complement

Its admitted that in ancient Egyptian, the noun in a sentence can take six functions:

- apposition
- genitive
- coordination
- disjonction,
- adverbial function
- Adjective's complement

D-4.1. Apposition.

The pharaonic Egyptian often put in apposition[67] two (2) nouns, the second serving to qualify the first:

Example:

Middle-Egyptian: ***sS*** ***Taty*** "a scribe vizir" (here vizir is in apposition to scribe)
Ewondo: / *ùn*-ntén bi-kangà/ "a sign designer; a scribe"

67. Apposition is a name or group of words that are read by a name or a pronoun. Unlike the addition of the name apposition is above all independent. It can be placed before as after the name itself. It is a process that can be used to inform more accurately than the completion of the name or that can serve as a qualification of the name or pronoun.
Example:
My son, that nice little rascal, is sitting on this table.
My son is sitting on this table, that nice little rascal.

As a general rule, the synchrony of Middle-Egyptian uses the apposition for:

- Measurement (quantity, volume)
- Or indication of a place

So, a wooden table will be translated:

Middle-egyptian: ***wDHw x.t*** "table + wood" (an apposition to material).
Ewondo: /téb'lé à-bàm/ "wooden table"

D-4.2 Genitive

Commonly the genitive takes on the meaning of "which belongs to". By juxtaposition with the second owner.
In the case that the possessor is an ancestor, Pharaoh, or a person whom one wants to honor, then we carry out according to the convention an honorific anteposition, which consists in placing the owner in first position, as in:

Middle-Egyptian: ***mDw nTr*** "the holy words"

<u>Example</u>

Ewondo: /Ndah mvoué/ "his friend's house"
Ewondo: /Ndəkh bit/ "honey pot"

According to the authorized literature, the Egyptian genitive presents two genitive forms, one direct, the other indirect.

The direct genitive

As in the case of apposition, the two substantives are placed side by side, the second playing the role of the first complement.

In the following example, the expression Ewondo /*ùn*-ŋkôm Tara/ ***Hm ntr*** "Prophet" consists of two substantives, servant and "God", the last being the complement of the first. It is interesting to note that the direct genitive is mainly used to realize compound words, common expressions.
Grammatically, compound words using the direct genitive are considered as entire entities by themselves. Thus, they can be accompanied by a determinative and followed by an adjective that agrees with the whole, as in:

Ewondo /à-bəŋ *ùm*.inga-ndàh/ "the beautiful housewife"
Middle-egyptian ***nb.t*** ***pr*** ***nfr.t*** "the beautiful housewife"

Always speaking of "compound words" here is an example that allows me to introduce a small interesting element about the order of signs in writing. This is what Egyptologists call "honorary anteposition" But in reality this is a myth from the Ewondo point of view.

Let's take our example first, which is

Ewondo: ***Hwt-nTr*** translated (castle) of the "god", that is to say temple.

Now, seeing how the expression is written according to the stardards ***nTr-Hwt***, supposedly out of respect for the divine. It will be the same when it will be question of the other deities in the names and other expressions. In reality, if we read from right to left, we will say in Ewondo [nda ntour] or [nda Tara] "The father's house" without any need for an honorific anteposition since this morphological structure is expressing the idea of a place or a seat The principle is simple, we call the syntagm which designates a place and we attach its direct genitive.

Ewondo: /ndàh bi-yem/ "the house of animals" →the stable
Ewondo: /ndàh ú-dòb'rò/ "the medical home" → hospital
Ewondo: /ndàh Zibi/ "Zibi's house"
Ewondo: /ndàh mə-kùt/ "madhouse"

Hope that you will get your own opinion on the purpose now

Indirect genitive

It is the most used and is constructed from the adjective ***n (y),*** called /ùn-mbias/ (nisbé). Asi have advised, this term of nisbé comes from the Arab language and defines a type of adjectives generated from substantives or prepositions. Its general meaning is "who belongs to". Like any adjective, n (y;) / nyo / has its own specificities, depending on whether it is masculine, feminine, singular or plural. The Chart that follows shows you the declension.

		Signs	Translit.
Singular	Masculine	𓈖	N(y) = n
	feminine	𓈖𓏏	N(y).t= n.t
Plural	masculine	𓏌𓏤	N(y)w = nw
	feminine	𓏌𓏏	N(y)wt = nw.t

Chart 24: Indirect genitive summary table

In reality, in Middle-egyptian, the endings tend to disappear and to be reduced to 𓈖 ***n*** for the masculine (singular and plural) and 𓈖𓏏 ***n.t*** for the feminine (singular and plural), as in the example below:

Middle-egyptian ***HqAw*** 𓈖 ***n*** ***xAswt*** "the chiefs of foreign countries"
Ewondo / Bi- dzoé ya *ùn*-nam *mi*-'nkol/ (an expression traduced Hyksôs in French).

The coordination is rendered in French using the conjunction and. Now, in Middle-Egyptian, this ***and*** is not usually expressed. The words are simply placed after each other. On the other hand, this idea of coordination, when expressed, uses two different propositions.

Middle-egyptian ***Hna*** "with", or **Middle-egyptian** ***Hr*** "on".

The same goes for the disjunction. The expression "or" embodies the shape of:

Middle-egyptian ***r-pw*** "or"
Ewondo: / ngue kigh/ "or"

This word ***r-pw*** is often placed at the end of the enumeration as we have seen in a previous example.

Middle-egyptian ***sn*** ***sn.t*** ***Hnms*** ***r-pw*** (a brother, a sister or a friend).
Once again, if the reading is from right to left we would have exactly the same structure as in Ewondo.

Ewondo: /ŋgue kigh mvoué, kal, *ùm*-anyaŋ / "a brother, a sister or a friend".

The noun can also be adverbially used to indicate time. In this case, it is not preceded by a preposition. For example :

Middle-egyptian: ***D.t*** "eternally" ***n*** and ***ra*** ***nb*** "each day".
Ewondo /*ùm*-myene à-mos *ú*-sə *à*-kwam/ "lord of the day forever"

If the complement of the adjective is not very common in French, it often comes back in Ewondo as in Middle-Egyptian. It is used as beautiful type of shape, clever of his fingers, etc., whose construction is: adjective + substantive.

3.3.3.5. Comparative Morphology: The Demonstratives

E-1 *The Demonstratives in Ewondo Language*

In grammar, the demonstrative pronoun is a subcategory of pronoun expressing an idea of monstration. As a representative, it replaces a nominal group preceded by a demonstrative determinant.

As for its deictic value, it makes it possible to indicate that the object is found, either in the text, or in space or time, defined by the situation of enunciation. The demonstratives establish with the depth, the relations of types deictic and anaphoric[68] in Ewondo we have the same type couple of demonstratives.

In the category of determinants, the corresponding of the demonstrative pronoun is the demonstrative adjective. The demonstrative pronoun is therefore equivalent to a noun preceded by a demonstrative adjective.

Example

Ewondo*: /***Un***-ntóm* ***uí*** *mə nga kom* ***uó*** *I-kurbəsanda; ínyuo-lí me nga kom* ***uó*** *à Ngomdzap. /*

"This hat, I did it at Ikurbəsanda; that one, I made it in Ngomdzap".

The syntagm "This hat" is actualized by the demonstrative adjective [ui] → wi "this", while the demonstrative pronoun [í-nyuoli] "that one" equals "that hat".

On a morphological step, some demonstrative pronouns Ewondo vary according to the gender and the number, while others remain neutral. We will observe this in Pharaonic Egyptian too. Moreover, as with the demonstrative adjective, the pronoun forms can be reinforced by means of the adverbial particles "-here" or "-there" (for " that is here" and "that here")). We can therefore distinguish simple forms or compound forms.

They can also be distinguished according to the degree of remoteness (near, intermediate and distant)

> *"[...] The Egyptian demonstratives are the equivalents of French demonstrative adjectives (this, these) but do not have any morphological kinship in the French language. Negro-African languages must be sought to establish family ties"*[69]
>
> *(Cheikh Anta Diop, 1977)*

68.Denis APOTHLOZ, Rôle et fonctionnement de l'anaphore dans la dynamique textuelle, Genève, Librairie Droz, 1995
69.Anta DIOP, Parenté Génétique de l'Egyptien pharaonique et des langues négro-africaines, Ifan, 1977

	Feminine			Masculine	
	"i" Type	"A" Type	"Ø" Type	"Ú" Type	"Ø" Type
Ssing	í-lúm->**dí**	á-kúd->**dí**	Kúb-> ***nyi***	Ú-fumbi->**wi**	kombo-> ***nyi***
Pplural	Bə í-loum->**bí**	Mə-kúd->**ma**	Bə-kúb->**ba**	à-fumbi ->**dí**	Bə-kombo->ba
Ssing	də-is→**dí**	á-lo->**dí**	**D**yam->di	ú-lùn→**wí**	Mor->***ny***-o
Pplural	mə-is→**ma**	Mə-lo->**ma**	**Mə**-am->**ma**	mə-lùn→**ma**	Bor->ba
Ssing	I-díŋ→**dí**	á-lùgh→**dí**	Zén->***nyi***	ú-kəŋ→**wí**	
Pplural	bə i-díŋ→**bí**	mə-lùgh→**ma**	Mə-zén->**ma**	à-kəŋ →**di**	

Chart 25:The demonstratives

We have diachronically the substantival prefix of the feminine singular which evolves in:

*kja->*dia->*da->*də->**à-**
Or
*kja >*dia>*di>**í**

Depending on the evolution of the language one of these variants will be preserved, nouns may have lost all or part of their substantival prefix but the agreements will be made with one of these archaic variants. It is those variants that often enter into the schemes of accordance.

<u>Example</u>1:

Ewondo: **í**-lum "a fork"
Possessive: í-lum ***di***-àma "my fork"
Presentative: í-lum ***dí*** "this fork"
Ewondo Interrogative: í-lum ***i***-fé? "which fork?"
Ewondo Numeration: í-lum ***dia*** "one fork"

í-

<u>Example </u>2:

Ewondo: kub "fowl"
Ewondo Possessive: kub ***di***-àma "my hen"
Ewondo Demonstrative: í-kub ***nyí*** "this hen"
Ewondo Interrogative: kub ***í***-fé? "which hen?"
Ewondo Numeration: kub ***di***a "one hen"

<u>Example </u>3:

Ewondo: zén "the path"
Ewondo Possessive: zén ***dí*** àma "my road"
Ewondo Demonstrative: í-zén ***nyí*** "this road"
Ewondo Interrogative: zén ***í***-fé? "which road?"
Ewondo Numeration: zén ***día*** "one road"

<u>Example </u>4:

Ewondo: *ùm*-òd "man"; ☞plural **Ewondo** : *bə*-òd→[bòd]"men"
☞Possessive: *ùm*-òd ***w***-àmà "my man"
☞Demonstrative: í-mòd ***nyùo***"this man"
☞Interrogative: mòd ***úm***-fé? "which man?"
☞Numeration: *ùm*-òd ***ùm-bw***â "one man"

Example 5:

Ewondo: ú-fumbi "l'orange"; à-fumbi " oranges"
☞Possessive: *ú*-fumbi ***w***-àmà "my orange"; *à*-fumbi də-àmà →[àfumbi dàmà]
☞Demonstrative: *ú*-fumbi ***wí*** "this orange"
☞Interrogative: ***ú***-fumbi ***úm***-bé? "which orange?"
☞Numeration: ú-fumbi ***ùm***-bwa "one orange"

In this last example the possessive looks like a singular simply because in the plural we have two types in Ewondo language as in phoraonic egygptian, the plural in /à-/ is related to a collective and the canonical plural which must be here /mə-fumbi/ "the oranges".

Let's note that the demonstrative seems not to follow the rules of accordance on examples 2,3,4. In reality it is because the kind of the demonstrative returns to neutral. It will be seen deeply later during the comparative part with the Middle-Egyptian.

E-1.1. *The deictic demonstratives*

As a rule, we consider that the demonstrative deictic refers to an enunciative reference, including beings. These cumulative demonstratives of non-linguistic indices (a glance, a gesture) to justify the identification of the referent. It is about the construction with the discontinuous morpheme, elements either in the nominals or in the pronouns according to a position in the statement. It can be observed in a report in three dimensions: near, mid-close or distant.

In some literatures specialists introduce a pre-prefix index called "enhancer" or "augment" in French | í- |. This prefix is characterized by a high tone in the phrases.
It is useful because it helps to strengthen the context to determine the spatial location. This notion is technically questionable but I will not discuss it in this heuristic.

The demonstrative is discontinuous with suffixale vowel | -í | in some environments and allows a substitution with the nominal. According to J. Dubois (1965, 191) "the demonstrative pronouns play a role parallel to that of the personnel, but they bring a greater information and intervene in the removal of the ambiguities" They can be anteposed or postponed in Ewondo.

Example: general case

<u>*Close context*</u>

|*ùm*-ingá+à-díŋ+ nyí| "the woman likes this one"

<u>*Medium proximity*</u>

|*ùm*-ingá+à-yì+lug+ **í-nyí**| "the woman is going to marry **that one**"

<u>*Background of accented distance*</u>

|*ùm*-ingá+à-yì+lug+ **í-nyí-lí**| "the woman wants to marry the one who is **over there**"

The introduction of a morpheme of location in the latter case makes it possible to insist on the determination of the space suggested by the speaker.

E-1.2 *Postposition of the Deictic*
The post position of the deictic is characterized by the addition of Adjective morphemes of Conson-Vowel structure (| -ná |, | vá |, | vɛ́ | and | lí |) on the suffixed types.

These kinds allow the determination in space and insist on the precision to be made on what we talk about. The gesture that often accompanies them may be near, more or less distant, or distant from what is mentioned in the enunciation. Thus, the different types are combined in the following way:

To name someone or something close:

|í. ùmingà+**nyì**+à-díŋ +**í.nyíná**| "**This** woman who is here, loves **this one**"
|za'a tobo **vá-**| "come and sit **here**".
|vághe **íbí**+ bi nə í-nenekh +bí- -**vá**| "remove those **(here)** that are green **here**".

To designate what is more or less distant:

| -àfumbí+à nə **vá-** -káb+ -dwo | "Those oranges that are **there**, share them".
|íbà+ be nə vá- vâ mə bo - | "Those who are there, give them to me".
|twagí+íbà-be ne **vá-**| "pick up those who are there".

To designate someone or something that is distant:

|wá-bə dzàm lor à-lí| "you could go over there".
|mb+ù-nga diŋ á-lí| "which one did you like there?"
| kələ kùs+í-dzí-lí| "go buy that over there!"
|utÓng+uí+wà-mán sibi-àlí| "this stream is drying up there."

E-1.3 *Anteposition of the deictic demonstratives*

When in Ewondo they are predestined, deictics can replace the nominal noun and assume the functions of the noun. On this account, they are said in the discontinuous authorized literature and are manifested around the noun at the beginning of the utterance..

Near

|á-vá+| "here"
|í-binga+bá+ sɔ à-lɔg| "these women were fishing"
|í-bá+b-á-sə+à-mə-ndim| " these one are coming back from the well "
|í-bídi+ bíná+bi-man yi boué| "these foods are already rotten"

Medium

|Záà vá| or | Záà há| "Come here"
|ì-bílí - +*ùn*-ndamèn| "Those one a rotten"
|í-bílí+by-á+dzáŋ| "those ones are disappearing"

Far

|à-válí | | ou | á-hálí+| "Over there"
| í-lúm+d-í+ +lí í-nə+mvít| "That fork over there is dirty"
|í-b-á+ +lá - +be nɛ mí-'mfúfub| "those who are there are saints."

Summary table of deictics in Ewondo

	Here		**There**	**Over there**
Things/humans	*Inyí*	*ínyíná*		*ínyílí*
	Ibí	*ibíná*		*íbílí*
Things	*Iyí*	*íyíná*		*íyílí*
	Imí	*ímíná*		*ímílí*

	Idí	*ídíná*		*ídílí*
	Imí	*íminá*		*ímílí*
	Idzí	*ídzíná*		*ídzílí*
	Ibyí	*Ibyí-ná*		*Ibyí-lí*
Place	*á-vá /vá*	*á-vá ná/ vá ná*	*Avá-lá/ vá-lá*	*Aválí/ vá-lí*
Place				*W/àw*

Chart 26: summary of Ewondo deictics

E-2 Anaphoric demonstratives

The anaphora[70] consists in repeating one or more identical word (s) at the beginning or at the end of verse or sentence. The language Ewondo uses certain words as demonstrative. They therefore have an anaphoric function. The literature of few specialists qualifies them as dependent on reworking or demonstrations of recall (Tabi Manga, 1992: 66). The role in use
determine the resumption of a previously mentioned name. It is therefore more appropriate for the particle "-Te" from this point of view, more appropriate to approach by the word "said" rather than "this".

Unlike the demonstratives deictic, they serve to indicate the demonstrative without introducing a notion of distance. In reality they are more presentatives than demonstratives. As a general rule, they are post-positioned in relation to the substantive and accorde in gender but also in number.

<u>Example:</u>

| ndùm-*Ø-té*+ í-təbə | "May That fight ceases"
| mə-ndùm *mə-té*+ mə tàm təbə| "May that fights cease"

In Ewondo anaphoras are always attested in definitive syntagmas and not in pronominal use.

<u>Example</u>:

| -*Ún*-nám+ ẃí- | "this meal"

<u>Others</u>: Assimilated demonstratives

The literature presents another serie as demonstrative adjectives. These are pseudo-demonstratives:

Examples of Presentatives

Feminine Ewondo Word: PN+A-NOUN (Gender prefix = A)
Feminine Ewondo Word: PN+I-NOUN (Gender prefix = I)
Masculine Ewondo Word: U+PN+NOM (Gender prefix = U)

With the evolution of the language the syntagmatic prefixes (PG, PN) may have disappeared partially or totally. It is therefore important to reconstruct them because the demonstrative is composed of PG + PN.
It is crucial for the singular words to reconstruct the word form attached to the numeral adjective cardinal to grasp the nature of its accordance scheme. Since the words were written on the phonetic basis only. And that a phonetic law makes it possible not to vocalize.

> *"[...] A initial (at the beginning) words,). 𓅱=w hieroglyphic followed by occlusives or especially desnasales n, m, falls regularly in our languages. This is a phonetic law."*[71]
>
> *(G. NGOM, 1993)*

70.Michel Charolles, L'anaphore: problèmes de définition et de classification, Verbum, 1991, p. 203-215
71.Gilbert Ngom, Revue Ankh N°2, Paris, 04-1993, P.46.

Example 1:

Ewondo: /ùm-òd **um-**bwa/ "one person" (we determine the gender and the number).
Ewondo: /b'-òd/ "persons" (we go to the plural to determine the root)
Ewondo: /um–òd → ny-**o/** "here is the man "(we reconstruct the syntagm)
Ewondo: /Bə–òd/ →/í Bə-òd bə–**á/**→/í Bòd b**á/** "these men here"

Example 2

Ewondo: /àkali **da/** "one donut" (we determine the gender and the number).
Ewondo: /à–kàlì **da**–í /→/à–kàlì dí /-->"the donut here" we have dai>di
Ewondo: /mə–kàlì/ /mə–kàlì mə–á/→/mə–kàlì má/ "these donuts"

Example 3

Ewondo: /í–bén **díà**/→ "a instep"
Ewondo: /í–bén də-í/→/í–bén dí/ "this instep" we have **də-í** >di
Ewondo: /bə ì–bén bə–í/→/bì–bén bí/ "those insteps "

Ewondo: /dí–ióm **díà**/→ "one thing" (do not write dzom but dyóm)
Ewondo: /í-dí–ióm/ → /í-dýóm dý–í/ "the thing here" (di +i = dy)
Ewondo: /í-bí–ióm/ → /í-býóm bý–í / "the things here" (bi + i = by)

Ewondo: /sí/ –/sí nyì/ "this earth here"
Ewondo: /síí –síí ɲý–í/ou /mə sí mà/ "these earts here"

Ewondo: ndóm "brother"→/ndóm nỳ–ì/"here is his brother"
Ewondo: ndóóm "brothers"→/ndóóm ný–í/or /bə ndóóm bà/ "here are his brothers"

Example 4:

Ewondo: *ùn*–mfəg "the bag"→/*ùn*–mfəg ẁ–í/"this bag"
Ewondo: mì–úmfəg "the bags"→/mì–'mfəg mí/"these bags"

Ewondo: *ùn*–nkàd " The exam"→/*ùn*–nkàd ẃ–í/"This exam"
Ewondo: mì–únkàd "these exams"→/mì–'nkàd mí /"Here are the exams"

Example 5

Ewondo: ù–kəŋ "the knife"→/ù–kəŋ ẃ–í/"Here is the knife"
Ewondo: à–kəŋ "knives"→/à–kəŋ dí/"here are the knives"

Ewondo: ù–kàlà "equal"→ /ù–kàlà ẃ–í/"our/your equal"
Ewondo: à–kàlà "those equals"→ /à–kàlà dí/"here are the equals"

According to some African linguists, the hypothesis of a surface erasure concerning the prefix accordance with the nominal class is more likely. The búlù, the neighboring language of the Ewondo, provides arguments in favor of this vision. The demonstrative is preceded, even in the singular, by the prefix of agreement in class. The advantage of such an approach makes it possible to formulate a unique explanatory principle for the agreement of the determinants paradigm. But in reality, depending on the type of study, synchronic or diachronic one can have the impression of an erasure on the surface. It is a simple evolution of language which on a diachronic level preserves elements of the past and adopts new finery in its current synchronic state.

It is not, strictly speaking, determiners or even adjectives. The fact is that these elements considered as demonstrative adjectives can be combined with other determinants, possessive or numerals for example:

Ewondo: /Mvóé dỳ–àmà tè ný–ì/. "Here is my concerned friend".
Ewondo: /*ùm*-ádzáŋ *úm*-bɔg nẃ–à/. "Here is one of my brothers".

The status of these elements is actually the one of presentatives:

Ewondo: /Mvóé dỳ–àmà tè ný–ì lì/ná. "Here / There is my concerned friend".
Ewondo: *ùm*-ádzáŋ úm-bɔg nẃâ–là. "There is one of my brothers".

Moreover, as shown in the examples above, they are generally used in correlation with the exophoric particles (Essono, 2000) that is to say presentative.

E-3. Pharaonic Egyptian Demonstatives

The demonstrations of Pharaonic Egyptian are either feminine, masculine or neutral.

Masculine			Feminine			Neutral singular		
pw - pwy	*this*		*Tw-twy*	*that*		*nw*	*This, that*	
pn	*this*		*tn*	*That one*		*nn*	*This one, that one this*	
Pf- pfA	*That one*		*tf*	*this one*		*nf*	*This one That one This one*	
pfy	*This*		*tfA*	*That one*		*nfA*	*This one That one This one*	
pA	*This*		*tA*	*This*		*nA*	*this, that*	

Chart 27: Table of Ancient Egyptian demonstratives

Demonstratives consist of two easily identifiable elements:

A first element, which indicates the gender:
p → for the singular masculine,
t → for the feminine singular
n → for the neutral

Then there is a second element which indicates nuances of proximity:

-w et *-n* → for what is near
-f(A), *-A* → for what is far

According to the Mboli-2011 comparative study, the Negro-egyptian possessed demonstrative pronouns based on a root ** k^w*, to which are added vowel suffixes indicating the degree of proximity to the speaker. A Negro-egyptian word structure like ** k^hi^wr* meaning roughly "place" or something close. This nominal base then received vowel suffixes to mean "here" and "there".

Deictics:

Negro-egyptian: **$k^hi^w\tilde{r}$* "here" and "there".
Middle-Egyptian: *aA*" here "
Ewondo: *à-ha, à-va* " here "

Demonstratives

	Masculine	**Feminine**	**Neutral**
Negro-egyptian	*c^{hw}ŭ- k^{h}ɨ^{w}r̃	*c^{hw}a- k^{h}ɨ^{w}r̃	*ŋ^{jw}ə-k^{h}ɨ^{w}r
Middle-egyptian	*p-A*	*t-A*	*n-A*
Ewondo	*wi*	*Di*	*ɲỳ-o / ɲỳ–ì*
Meroïtic	*Qo-le*	*Qe-le/Qa-le*	*wɔɔ -li/wɔɔ -le*

Chart 28: Demonstratives from Negro-gyptian language

3.3.3.6. The interrogative

The interrogative[72] is a particle of a pronominal or adjectival nature that grammatically serves to interrogate. This interrogation proceeds with various processes such as intonation, the order of words, and so on. and is intended to ask a question or imply doubt. In English these particles are identified as interrogative markers: who, what, where, when, how, why.

It is therefore interesting to see if the Ewondo language also has these interrogative tags and so comparatively they are found in Pharaonic Egyptian or Meroitic and in which shape.

If the interrogative questions the place, the demonstrative answers the interrogative question and to this account the two entities are thus linked.

Example: interrogative- demonstrative

Ewondo: /***à***-kòl ***à***-fé? / "Which foot?"; ☞Plural **Ewondo** :/***mə***-kol ***mə***-fé? / "which feet"

Ewondo: /***à***-kòl **di**/ "this foot"; ☞Plural **Ewondo** /***mə***-kol ***ma*** / "these feet"

F-1. Interrogative in Ewondo

In the classical literature interrogative particles are associated with the nominal classes. So, if the nominal classes are poorly defined the remaining part is simply phantasmagoria.

The myth of the nominal class of words beginning with M

Example:

Ewondo: *ùm*-ôd "man"; ☞Plural **Ewondo** :*b'*-ôd "men"
Possessif: *ùm*-ôd **w**-ama "my man"; ☞Plural **Ewondo** : b'-ôd b'ama "my men"

The root of the word being ***-ôd,*** it admits in the singular a prefix of number ***m*** and a prefix of gender ***ù*** de facto the word must be written, not /môd/ as found in classical literature and authorized but rather: /*úm*-ôd/. This word exists in the archaic form in **isi-zulu**: um-untu "man". Unlike Ewondo, the apico-dental "t" of Zulu has evolved from "d" to Ewondo. And the "n" epenthetic zulu suffered a syncope and replaced by the circumflex accent on the "o" Ewondo. The Ewondo being a Beer language as classified by Mboli makes the apocope on the final vowel. If in appearance, môd and umuntu do not sound the same, semantically, morphologically and diachronically it is exactly the same word.

Interrogatif: ***um***-ôd ***úm***-bé**?** ***um***-ôd **úm**-fé**?** **"**which man?"

☞**Plural Ewondo : /*Bə*-**ôd ***bə***-fé?/→/b'-ôd bə-fé/ "Which men"

Presentative: i ***um***-ôd nyi-**o** "This man"; i bə-ôd ba→/bôd ba/ "These men"

72.FOERSTER, p. 77: l'**interrogatif**; p. 191: point **interrogatif**

Structure of nominal syntagm for mascline singular: (G PNP)-NOUN-suffix

Few examples:

Um-mfəgh	*Um*-ôd	*Um*-mbokh	*Un*-nom	*Un*-ŋkukh	*U*-kpwal
Um-mbəghə	*Um*-bondé	*Um*-mbôm	*Un*-ntòm	*Un*-ŋkandi	*U*-kəŋ
Um-mból	*Um*-mfughu	*Um*-mbobé	*Un*-ndzugh	*Un*-ŋkol	*U*-kpəŋ
Um-bè	*Um*-mfôl	*Um*-mbərbə	*Un*-nwan	*Un*-ŋkulu	*Un*-nam

Chart29: Few examples of masculine Ewondo Syntagms

As a result, other writings that are authentic to this day are to be forgotten because they do not meet the grammatical requirements of the Ewondo. The same reasoning can be used for words of the feminine gender and demonstrate the same deficiencies.

Why all this?

It has been decreed by those who know more than Africans on their own language that in the initial words of our languages, ***w*** hieroglyphic noted by the scribe followed by occlusives or especially noses ***n, m,*** falls regularly in our languages. This would be a phonetic law. If we add to this the natural evolution of language that leads to the total or partial loss of substantive prefixes, we understand better the difficulty that grammarians may encounter.

The fact of having deleted or lost this ***w*** initially for reasons that escape the most basic common sense leads to this grammatical impasse. This especially since this ***w*** is the masculine tag as recorded by the scribe. This sign ***w*** pronouncing really [wou] "flyer" in Ewondo designating here the chick like an immature bird synonymous of iyə (1).

	Singular		**Masculine plural**		**Feminine Singular**	**Feminine plural**
Which		**Um**-bé?	Which	**Bi**-fé?	í-fé	**Bə**-fé?
		fé?	Which	**mi**-fé?	**á**-fé?	**Mə**-fé?
	Which	**U**-fé?	Which	**á**-fé*[73]?		
		Um-fé?				
				Bám?		**Bám?**
How much?				**Mám?**		**Mám?**
Who?	Who?	Zâ ?		**bə-zâ?**		
Where?		Vé?/àvé?		Vé?/àvé?	Vé?/àvé?	Vé?/àvé?
What?		Na?		Na?	Na?	Na?

Chart 30: Ewondo interrogatives tags

73. * This plural is to be matched with a a group.

F-2. How are the interrogative pronouns used in Ewondo?

The characteristic of interrogative statements is to fit into a dialogic relation. De facto, the question asked can concern all the statement, in this case it is about a total interrogation which does not require justification.

Or, the question asked is simply a part of the statement, on this account it is a partial question that leads to a justification on the part of the interlocutor. It is important before going into a comparative logic to identify the statements that make it possible to highlight these sequences.

F-2.1 *The simple statement: of one element*

/zà?/ "who?", "which/what?"	/dzé?/ "what?"
/ná? / "how?"	/à-vé? / "where?"
/ ngà/ "is?"	/yə ...? / "is? How?"

Chart 31: Ewondo simple statements

These monosyllabic statements are sufficient in principle to formulate a question in a speech between two people. The statements in response are often rather simplistic, confined to an affirmative / owé/ "yes" or a negation statement / tèghè / "no".

F-2.2 *Other Simple Interrogative Types.*

Other variants that can be encountered in interrogative construction are:

- /Taŋ yá/ "How much?" When it comes to giving a price.
- /Bə nə bám fə? / "how many are they?" When it comes to counting.
- /à-kià à fé? / "what (kind, nature, species, style ...)?"
- /nə dzé/ "what ... which?"

The analysis of these different constructions is done as and when they are presented.

F-2.3 Total interrogation

The total interrogation leads to the units identification of on which the question is concentrated. The latter allow its introduction into the discourse and guide the questions where the interlocutor must answer with "yes" or "no".

The interrogative type / ngà / is preposited or postponed in the statement:

Ewondo: /ngà? / "Is?"

/yə ù- yəm fwà nyè ngà? /"Did you recognized him?"

/ngà ù- yəm fwà nyè? / "Did you recognized him?"

#PV-recognize +substitutive+interro.#

Ewondo: /yə/ "is?"

/yə bià yì dí/ "Are we going to eat?"

#Is + bi-à+yì+di#

#interro+ PP-present+the+eat#

/yə wa lugh/

/yə-wá à-lugh/ "Are you getting married?"

#interrogative-present + verbal modality + marry#

F-2.4 Partial interrogation

As a general rule, it is considered that, unlike the total interrogation, the partial interrogation deals only with one of the elements of the statement.

Ewondo: /zà/ "who? which?..."

This substantive form is always placed in the first position in the utterance.
Example:

/zá á-tam so líh sofià? /"who comes clearing?"

#WHO+PV-the- reclaim +PN#

If we consider the tonological aspect, the word /zà? / "who?", when isolated, takes a low tone.

The types /v? / "where" and /yà? / appear in different positions of the utterance, they participate in the identification of substantive or syntactic groups. They justify a development in the description of the situation that is mentioned in the speech.

Example1: /vé/ "where?" is a morpheme that we find in the final position of the statement.

/bià tòbò **vé**? / "Where are we sitting?"

#PP–aux-seat+interro#

Example2: /yá/ is often attested at the absolute final.

/wàà-dzo **yà?** / "What are you saying?" "how are you saying?"

#PP-say +interro. #

A-2.5 Morphology: Comparative table of interrogative statements tags

The statements tags identified in Pharaonic Egyptian are:

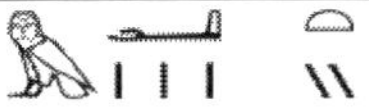	m--ty	Be… za?	Who, what, which one?
	P(w)--ty	Za …*ùm*bé?	Who, what, which one?
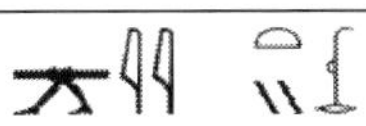	sy--ty	Zá.. tə?	Who, what, which one?
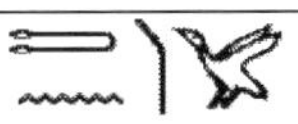	Tn	à-vé?	Where, from where?

-So here we have  ***sii*** or **zii** which corresponds semantically to the interrogative particle Ewondo / zá / "who, which"

We also have the tag □ p (w) that will correspond to the interrogative particle Ewondo / *ùm*-bé / "who, which".

Note: On the script above we have to link sign Q3 □ of Gardiner list with the door / *ùn*-mbè /.

- One also has the aspect  ***Tn*** which corresponds to the interrogative particle Ewondo / vé / or / á-vé / "where".

Note

On the script above, this notion is introduced by the landing of the duck which indicates a place. Water recalls thirst / i-vè / and the hindrance for horns bones on the script is called / i-vé(s) /

If I am consistent with what I have supported so far then the transliterated consonants skeleton ***Tn*** is hiding a synonym for the word Ewondo: / i-vé /.

On this account we have:

Ewondo: / à-tôlé / "fringale" which homophone of / i-tələ / "place; landing" synonymous of / vom / "place"

The query can be translated as:

Ewondo: / vom ùm-bé? / "which place?, where?"

3.3.3.7. The Adjectives

G-1. *Types and functions of the adjective*

As in western languages like French, we find the adjective as the qualifier. Moreover, it serves, as we have seen, as a predicate attribute in one of the non-verbal sentence statements. Finally, the ancient egyptian offers some particular uses that we will discuss in this section.

G-1.1 *The qualifying Adjective*

In the nominal group, the qualifying adjectives are placed directly before or after the noun, according to the usage. They are called epithets or attributes of the subject. The epithet adjective becomes gendered and numbered with the noun. The adjective can itself have an expansion: the complement of the adjective.

Examples:

Ewondo: ***mòr*** nda *ún*-ŋkukuma "**The spacious** chef's house"
Ewondo: ***à-yap*** byál "**The long** boat",
Ewondo: ***mán*** *u*-m-òngò "**The little** child"

Here are some examples of adjectives used to qualify a noun. In Egyptian, the adjective follows the noun which it qualifies and accords in gender and in number with it.
Note that the endings (endings: masc., Fem., Sing., Plur.) Are the same as for nouns:

Middle-egyptian	Ewondo	Transliteration
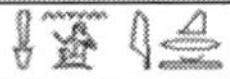	i-komba madzaŋ	***sn ikr*** "excellent brother"
	i-komba kal	***sn.t ikr.t*** "excellent sister"
	Bi-komba bobanyaŋ	***snw ikrw*** "excellent brothers"
	Bi-komba bə-kal	***snw.t ikrw.t*** "excellent sisters"

Chart 32:examples of adjectives

Note: the Egyptologists ***ikr-*** can be translated in Ewondo by /*i*-kul/ with the same meaning.
The endings writing of the adjective are often defective. It will therefore be necessary to get used to this phenomenon which has nothing to do with the grammar respect or not! The context of the sentence is usually enough to grasp the reality of the adjective's accordances.

- o The plural masculine ***w*** and the singular feminine ***t*** are often missing.
- o The same is true for the plural determinative (the three small bars) which is often absent.

G-1.2 *The Adjectives Ùn-mbias Called Nisbe*

Many adjectives in Middle Egyptian shall lie in accordance with nouns and prepositions. The termination (termination) in the singular masculine is ***y***, with the meaning of: relative to. Nisbe in Arabic is inspired in its grammar by this type of adjective which is still called relational adjectives.

To better understand, let's take a simple example: from a noun:

Middle-Egyptian.: 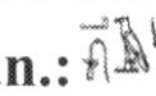***Awai*** "robbery, stealing"
Ewondo: /wub/ "robbery, stealing"
Ewondo: /á-wai/ "to steal"

We will have ***Away (t)*** the adjective relating to robbery, the adjective thief.

Middle-egyptian: 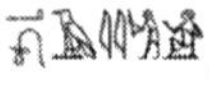***Away*** "thief"
Ewondo: /*ùn*-way/ "thief"
Ewondo: /*ùn*-wuwub/ "thief"

The couple of signs M17 corresponds as we have noticed to the realisation of the adjective by first syllable redoubling.

Thus, the sign M17A *ùn*-mbias "nisbe" for twins, is used to express a kind of relationship, of belonging, and they are quite numerous in pharaonic-egyptian. Some prepositions areused as contributors with the sense of who, is, in, on, with, above, in front, etc., while being able to govern a complement, keeping in their preposition role. Here are the most common by comparison with Ewondo:

M-E	Transliteration	Adjectives	Ewondo	Meaning
	m	imy	À-nəm	The one among to
	r	iry	à-kələ	The one related to
	Hr	Hry	À-yop	The one who is higher than
	xft	xfty	À-siŋ	Opposite, Opponent
	tp	tpy	ùn-təbə ù-sù	The first, the one who is on
	Xr	Xry	à-kol	The one under, carrying
	xnt	xnty	Ùn-ŋkɔŋ	The one who presides, who is in front

Chart 33: comparative adjectives realised from a noun or preposition

There are many expressions like:

Middle-Egyptian: ***HH*** "million" meaning a lot. **HH** is often associated with the indirect genitive ***n***
Ewondo: /i-kukh/ "numerous"

Middle-Egyptian: ***Dr.f*** "whole" until the end literally.
Ewondo: /ùn-ndzalèn/, /à-tùkh/ "full"

Middle-Egyptian: ***n-hwy*** "a few, a few" note the presence of the indirect genitive particle ***n***.

Ewondo:/i-nyungui/; /ù-bə/ "few; small"

Note: The adjective as Predicate could be consider as the non-verbal sentence already addressed.

G-1.3. *Exclamative Role of Adjective Attribute.*

How famous is the surveyor! It illustrates well the emphatic character of the language and the mentality of the ancestors. The idea of magnitude found in buildings and architecture in general was closely linked to everyday language. Everything was expressed with the notion of greatness.

To make the exclamation, ancient Egyptian were using the particle transliterated ***wy*** (var.) placed immediately after the last hieroglyphic sign of the adjective attribute.

Then we have for example:

Middle-egyptian: ***nfrwy pr n TAty***, "How beautiful is the chief's house!"**.**
Ewondo: / *ùn*-mfəfə à-folà be Tarà / "New beautiful chief's house!"**.**

This sentence can be rendered by: "How beautiful is the vizier's house! This is a so-called a non-verbal phrase with predicate attribute. This mindset is very common in African dialects. Just as we find them here in this Egyptian example.

Rigorously the presence of the // nisbe "un-mbias" here induces a repetition of the first syllable to mark a superlative. In fact we could also read the sentence above:

> **Middle egyptian**: ***nfrwy pr n TAty*** "The beautiful chief's house!".
> **Ewondo**: /úmbəmbə ndàh ùn-ŋkùkùma / "The beautiful chief's house!".
> **Ewondo**: /úmbəmbə ndàh i-nə é ùn-ŋkùkùma / "The beautiful chief's house!".

G-2 *The Comparison.*

To express the idea of comparison, such as in the following example: "The man is smaller than a tree", Pharaonic often use the preposition **r**, which means **<u>more than</u>**.
The Ewondo uses the expression /* lor/. The feet are used to walk, to go over / à -lor / or speech refers to / ùn-ŋkobo /. In African language the speech is the mouth / à-nyu /. So the mouth is related to feet. If we consider the two syllabes of the word / ùn-ŋkobo /, each syllabe means foot. /à-kô / "foot" or / à-bo / "foot". That's why the mouth is designed by the scribe to express the idea of comparison. Feet a used to walk, to flow, to go over etc. and those words mean also more than in Ewondo language. This word is used in English through the item ***oral*** referring to the mouth area.
Ewondo: /*ùn*-ndzwàli/ "talkative" is homophone of **Ewondo**: /*i*-dzali/ "full" and **Ewondo**: /*i*-lor/ "full; overtaking; exceeding" are synonyms.

Example:

Ewondo: /mə bələ *fəgh* mə **<u>loro</u>** wa/ "i am **more** *intelligent* **than** you"

This construction for example does not feet the French one:

> **Middle-egyptian:**  nDs s ***r*** Xt , "A Human is smaller than a tree".
> **Ewondo:** /mor à-nə inyunghi à **loro** i-léh "A Human is smaller than a tree".

Note: the transliteration ***nDs*** corresponds to the **Ewondo**: /*ùn*-nsəssàndà/ "smallest; little bird" or
Ewondo: /sàŋdà/ "smallest; african forest little red ant"
Ewondo: /i-tsetsa/ "brown little bird far of villages; dust" **Ewondo**: /tsè/ "dust"
Ewondo: / i-tsetso/ "little black forest bird";

The transliteration *nDs* can be be split in two parts. The ***n*** particle in front can then be consider as a genitive particle meaning the one belonging to Ds.

This kind of sentence is a non-verbal sentence with predicate attribute according to scholarly literature.

G-3 *The Superlative*

Superlative is an adverb or adjective degree of comparison expressing quality or modality to a very high degree, higher or lower than others (relative superlative), or regardless of any reference (absolute superlative). Term of meaning too strong, exaggerated: Abuse superlatives.

The superlative is involved in two types in Pharaonic as in Ewondo:

> The first expresses the absolute idea: **"He is very strong"**.
> The second, relative, corresponds to **"The most X of..."**

G-3.1 *The absolute Type.*

In Pharaonic, the absolute type is mainly used to read the funerary / bioscriptal inscriptions, even those which, through the dynasties, tell the exploits of the pharaohs. In these texts, the absolute superlative therefore plays a very important role, being presented in several kinds.
By using one of the following three adverbs: respectively **aA**; **wr** and **wr.t**, adjectives used as adverb and meaning in Ewondo / nén /; /ngoul/ ; /i-tour/"greatly; sorely; boss, mogul".

Example:

> **Middle-Egyptian:** , ***s.t nfr.t aA***,
> **Ewondo**: /Akia à-bəŋ mingà/ "*A so beautiful woman!*"

Some formulas use the noun ***wa***, **the unique, the only**, as in the following example:

> **Middle-Egyptian**: ***sS wa iqr*** "an excellent scribe at the highest level"
> **Ewondo**: /*ùn*-nsenzə i-kóm *ùn*-ŋkô/ "an excellent scribe at the highest level"

Note: Incidentally, that ***wa*** stands before the adjective and the one in the example above, it is a substantive in apposition to ***sS***.

G-3.2. *The absolute Superlative form by repetition*

The third form we encounter in Pharaonic is the repetition of the word (adjective, substantive, adverve or verb), usually indicated by the expression ***spy*** which literally means 2 times (times, 2). So:

> **Middle-Egyptian:** ***aSAspy*** "many times, i.e many, many." numerous
> **Ewondo**: */mi-'ŋkun, mi-'ŋkun/* or */à-bui mi-'ŋkun/* "numerous baskets"

It is a special case of the plural which is noted by the three vertical lines /// (indicating a triple repetition) and which are reduced to 2 for ùn-mbias "the nisbe", the dual.

We have already seen this case in the plural case analysis (dual and plural)

G-3.3. *The Relative Type.*

The relative form of the superlative, in the sense of the greatest of (Ahanda, Bikie, Mballa). Once again, the Egyptian shines through its multiple forms of expression. In this case, we have two:

> **1.** Using one of these two adjectives "nisb" to ùn-mbias (we will return to this type of adjective in the next section): respectively ***imy*** and ***nty-m***, "among". So:

Example:

> **Middle-Egyptian;** ***sS iqr imy srw***, "A scribe, the most reliable of the functionaries" or literally a scribe, reliable who is among the officials.
> **Ewondo:** /*ùn*-nsenzə *ùn*-ŋkô y'à-zàŋ mi-'ntómbà/ "a scribe, the most reliable of the noteworthies"

> 2. Two substantives can also be used, one complementing the other (direct genitive or indirect genitive). The example below comes from the Tale of Oasian (B1, 53)

Example:

> **Middle-Egyptian:** ***wr n wrw*** "The greatest of all" literally, the great of the great. We can note the use of the indirect genitive with ***n***.

Ewondo: /á-nə ìn-ntol mi-'ntol/ "The oldest of all" literally he is the former elder.

Other examples exist, in *The Adventures of Sinuhe*, among others, where the link between the two nouns is direct.

3.3.3.8. The noun classes

I have thought long and hard about this purpose because it is grammatically important. The correct definition of the nominal classes makes it possible to better understand and write the accords related to the noun.

In linguistics, a noun class is considered as a particular category of nouns. A noun may belong to a given class because of the characteristic features of its referent, such as sex, animacy, shape, but counting a given noun among nouns of such or another class is often clearly conventional. Some authors use the term "grammatical gender" as a synonym of "noun class", but others use different definitions for each. Noun classes should not be confused with noun classifiers

Jesko von Puttkamer said this in 1894 concerning the Ewé linguistic variant:

> *"The Ewe language is and remains a wild Negro dialect, extremely primitive, that it is useful to learn [...] in order to be understood with its workers. However, the language of a civilized people is far superior to the beginnings of all these Negroes, halfway between the language of men and that of apes."*[74]

The question of nominal classes can not be addressed by decontextualizing it from the spatio-temporal register which gave it birth. And Von Puttkamer's statement illustrates on a philological level the mindset that will preside over the birth of the idea of nominal classes.

The arrival of european settlers in the fifteenth, in charge of biblical writings in African languages will be strongly impregnated with the thought of Gobineau. We come to civilize the African savages. It was therefore the same mindset that in 1851 gave rise to the concept of "nominal class", invented by William Bleek as part of his classification of languages that he called Bantu.

Linguistic specialists have therefore approached the question in good faith or with an ulterior motive. But still, they validated certain truths and African specialists who hold their recognition of these specialists stick to these truths without ever daring or questioning. Is there a set of universal criteria applied to all languages of the world to define nominal classes? This simple question wipes out any logical reasoning.

The literature on the nominal classes of Bantu languages abounds with various and varied classifications. We can therefore deduce that the subject seems debatable for everyone.
The variation of the number of classes according to the languages, the criteria of classifications, the global logic, the linguistic method solicited etc. are all subjective factors.

Considered as a "Bantu" language by its characteristics, the Ewondo is an Ekang language that the others classify in the group of languages with "nominal classes" whose minimal syntactic unit can not appear alone, but only accompanied from an obligatory modality to "discontinuous signifier" determining in the utterance a specific accordance schema: the class accord
So I will be curious to know where the French language for example title articles that are predeposed relative to the determinant for example. Yes I know it's an advanced language. If the man was born in Africa then why, when we treat African languages, we change the rules of the game?

74. LANGE, Marie-France, Le choix des langues enseignées à l'école au Togo: quels enjeux politiques?, dans Politique africaine, Paris, no 27, 1987, p. 483-484

Martinet (2005, 209) declares in this respect that the establishment of moneme classes affects the syntax as defined here:

> *"From syntax, we will say today that it is, in the broad sense of the term, the examination of how the listener can reconstruct the unity of the message from the succession of significant units that is offered to him. It operates not with individual monemes, but with nominal classes. These classes are established by grouping the monemes of identical compatibilities, the class of nouns in French, for example, containing the monemes that can be determined by articles, the plural, adjectives, the class of verbs grouping what receives as determinants, the monemes of time, modes, adverbs".*

In linguistics, the term nominal class, or class of the noun, refers to a system of categorization of nouns. A noun may belong to a given class because of its referent's characteristic traits, such as gender, animated / unmanaged distinction, or form, but naming a particular class is often purely conventional.

Some authors use the term gender as a synonym for nominal class, while others associate a different definition with each of these terms. In this heuristic we will stick to this approach.

Classes are determined from three criteria according to (Kadima: 1969, 82)[75]:

- Accord,
- The aspect of the substantival prefix
- And the matching.

The distinction of two classes is made by a difference in their chords, and if, in the case of chord identities, their substantive prefixes and their matches are simultaneously distinct. In other words, two classes form only one if their chords are identical except when their substantive prefixes and their matches are different.

The question of the nominal classes in the "Bantu" languages and therefore the Ewondo language is found floundering between the conventions of the different classifications, a classification which is based on the work of the specialists (Guthrie, Kadima, Meinhof etc.). From which we count about ten classes.

According to Carl Meinhof, the Bantu languages have a total of 22 nominal classes. If we do not know any language that expresses them all, most of them have at least 10. For example, according to Meinhof's count, Swahili would have 15, Mesotho 18, and Luganda, 19.

Critics of Meinhof's approach note that his system of counting nominal classes divides the singular and the plural of the same name into separate classes. This seems to them to contradict the way other languages are are traditionally viewed, with the orthogonal number of the gender: according to these critics, an analysis in Meinhof's way would result in 9 genres for ancient Greek. If we stick to the more general linguistic tradition and consider singulars and plurals as belonging to the same class, then Swahili would have 8 or 9 classes, mesotho 11 and luganda 10.

The category of nominal classes replaces not only the category of the gender, but also that of the number and the case. The case being in the broad sense a grammatical feature mainly associated with the noun, the pronoun, the adjective and the determinant, and expressing the syntactic function of these in the proposition, or their semantic role in relation to the process expressed by the verb.

According to specialists, the Ewondo, like other Ekang languages, could be distinguished by its affix types, the functioning of its agreement, the distribution of its classes, the semantic value of its prefixes, the pairing of its classes, except the rental classes, without failing to determine the types listed in each

75.KADIMA, M., Le système des classes en bantu, Leuven, Vander, 1969, p.82

criterion. In appearance only some classes generally express the singular and others the plural. But in the distribution, we sometimes find classes, labeled in the singular, which serve as plural to others.

If instead of yielding to the authoritarianism of Bantuistik and providence of the resulting synchronic classification, we give ourselves time for a close study from the Negro-egyptian, as Mboli proposes, the question of the classes nominal appears to us in a new light. To understand the state of the synchrony of a language and Ewondo in the present case, a diachronic study is more indicated since the Negro-egyptian by reconstructing the proto-forms.

Unfortunately, in the plethora of accords proposed so far since Guthrie's sacrosanct method, the official nominal classes of the so-called Bantu languages are based on an arbitrary phonetic approach based on impressions as will be demonstrated.

The fact of having admitted that there is no gender (male or female) in Ewondo has led to the determination of nominal classes in a phantasmagorical way.
From the moment that it is demonstrated that there is a gender in Ewondo (masculine / feminine) and thus a gender prefix, we understand why the nominal classes defined in the authorized textbooks are completely obsolete. It is certain that no scheme of accords can be justified solely on the basis of phonetic impressions and the singular versus plural distinction.

The substantival prefix consists of a consonantal number prefix (PN) and a vowel-type genre prefix (PG) that indicates to which kind the substantive is related.

I also point out that the difficulty of studying the synchrony of the Ewondo is its evolution compared to its sibling languages. As a rule the Ewondo makes an elision of its substantive prefixes and the realization of the agreements escapes all logic for the least informed. Also to learn the Ewondo listening only gives terrible results to the use as the mastery of accord schemes seems surreal.

I precise again that, if the prefixes that have disappeared around the substantive have existed in earlier synchronies and are still used in other sibling languages of the Ewondo language.

It is therefore not enough to make a singular →plural agreement to determine the nominal class, it is also necessary to take into account the schemes of accordance including the presentative and the possessive or even the interrogative which have the characteristic of highlighting the PG and the PN. It is only that way we can determine the true nominal class.

> *"[...] All African classes are demonstrative forms corresponding to the different answers to the same question posed, according to the initial consonant to the term to which the interrogation refers; also the African demonstrative while serving to define and locate the object can not be confused with the definite article of Romance languages."*[76]
>
> *(Anta diop-1976)*

I therefore suggest to start from a classification model with 18 classes admitted as just. To this model I put forward to apply accordance schemes that reveal what the grammarian Jaques Evouna calls in a synchronic study of the Ewondo the pairing and the numeral disconnection[77].

But in fact it is an evolution of Ewondo's synchrony. Languages are dynamic, living, they evolve in contact with other languages or in time and space depending on the speakers or in contact with other languages. Thus Ewondo makes accordances with previous states of this same language. The Ewondo language has the characteristic to get rid in all or part of its substantival prefixes (PG + PN) including

---

76.Cheick ANTA DIOP, Parenté Génétique de l'Egyptien pharaonique et des langues négro-africaines, IFAN, 1977, Dakar, P.XXVI.
77.Jacques EVOUNA, La morphologie nominale en Ewondo: La double préfixation, Connaissances et Savoirs, P48

the suffixal vowel. But in fact these are simplifications of its actual synchrony which is found between midwater. It is found on one side with simplified types of names and having to use syntactic elements of another synchrony. One can easily find the archaic variants from which it draws its accords schemes by interrogating its sibling languages which still have their substantive prefixes such as Zulu, Duala, Lingala, Ci-luba, Basa'a etc..

N°	PG	Example	Possessive	Presentative	recons-tructed Form	PC cor-rected	Glossary
1	**ø**	nyâ	nya **w**-ama	I-nyâ n**w**-o	*ùn*-nyâ	W	A mother
2	**bə**	bə-nyâ	bə-nyâ b'-ama	bə-nyâ ba	ba-nyâ	Ba	mothers
3	**m-**	m-inga	m-inga ***w***-ama	I m-inga **nw-o**	***um***-inga	W	A woman
4	**b-**	b-inga	***bə***-inga ***ba***-ama	I.***ba***-inga ***ba***	***ba***-inga	Ba	women
5	**ø**	ḿbəghə	***ḿ***-bəghə ***w***-ama	***ḿ***-bəghə ***w***-i	***uḿ***-bəghə	W	A load
6	**mi**	***mì***-ḿbəghə	***mì***-ḿbəghə ***mi***-ama	***mí***-uḿbəghə ***mí***	***mì***-uḿbəghə	mi	loads
7	**à-**	à-mâŋ	à-mâŋ ***də***-ama	à-mâŋ ***d***i	dà-mâŋ	Dia>da/di	A jowel
8	**mə-**	mə- mâŋ	mə-mâŋ mə-ama	mə- mâŋ ***ma***	ma- mâŋ	ma	jowels
9	**d-; dz-**	dsis; dz-oé	dzis də-ama; dzoé də-ama	I.di-is ***di;*** I.dí-o ***di***	***di***-is→dys ***di***-o→dyo	Di	An eye; A nose
110	**m-**	m-is , m- oé	mə-is mə-ama; mə-o mə-ama	i.mis ***ma***; i moé ***ma***	***ma***-is; →mis ***ma***-oé→mwé	ma	eyes noses
111	**- dz-**	-j; dz-at	I j ***di***-ama; dí-at ***d***-ama	I j ***di***; Idi-at ***di***	***di***-j; ***di***-at	di	A garment; A basket
112 bi-	**bi-**	***bi***-jé; ***bi***-at	***bi***-jé bi-ama; ***bi***-at bi-ama	I.***bi***-jé ***bi***; I ***bi***-at ***bi***	***bi***-jé; ***bi***-at	bi	garments; baskets
113	**o-**	o-nyü	o-nyü ***w***-ama	o-nyü ***w***-i	***w***-nyü	w	A finger
114	**à-**	à-nyu	à-nyü ***d***-ama	à-nyu ***d-i***	***da***-nyü	Dia>di/da	fingers
115	**ø**	kûb	kûb ***di***-ama	i-kûb ***ɲi***	***di***-kûb	Dia>di	A fowl
116	**bə-**	kûb	***be*** kûb ba-ama	i-***be*** kûb ***-ba***	***ba***-kûb	ba	fowls
117	**ø**	ndâ	ndâ ***di***-ama	I-ndâ ***nyi***	***di***-ndâ	di	A house
18	**mô-**	mô-ndâ	***mə***-ndä ***ma***-ama	***mə*** -ndâ ***ma***	***mu***-ndâ	ma	houses

Chart 34: Ewondo standard nominal classes revised

The conclusion that emerges here is simple: classes 1, 2, 3, 4 once corrected are in fact one and the same class despite the impressions of the supposed specialists.

Classes 7, 8, 9, 10, 17, 18 are exactly the same classes.

We also note that the plural supposed to be **bə** is an evolution of the language of ba→ bə. We simply have a modification of the degree of aperture of the initial vowel **i** or **a**. Here is an illustration opposite.

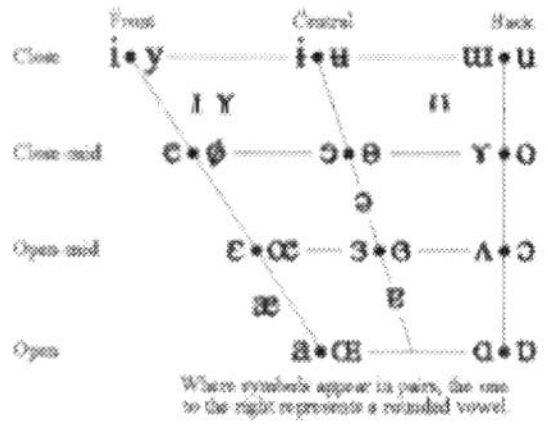

In some cases we have two accordance shapes in the same synchrony:

Ewondo demonstrative: I-ndâ ***nyi*** "this house"
Ewondo possessive: ndâ ***di***-àmà "my house"

In this example above normally we should have observed a similar accordance sheme but it's note the case. We don't say /ndâ nyâma/ but / ndâ ***di***-àmà/ "my house". Let us try to understand why.

In this case, I took the most archaic form to define the nominal class. We also have the 13/14 classes that i also maintained. But in truth it is a particularism, it is usually said, / *ún*-nyu / "finger" and / mə-nyu / "the fingers". This corresponds to a case already treated.

But there is this particular case where we say / *ún*-nyu / "finger" and / à-nyu / "fingers" but this case is not related to a phenomenon of contamination but simply to a **plural designating a collective, a mass, a common**. It is simply the fact that the current usage forsakes the canonical form of the plural for the collective plural.

Like the one of the class | a- |, the PG realized between the PN | d- | and a vowel in the initial position of the radical is at the origin of the semi vocalization of the PG: during this modification, it takes the form of | y |, the lexical base being a radical. Option is taken, despite this change, to keep his tone at this PG. It must again be observed that the tone of the semi-vocalized PC is similar to that of the nominal radical if it is vocalic:

Ewondo: /d–**è**–ióm/; /d–**ì**–ióm/→ d**ý**óm: *"thing"*
Ewondo: /d–**è**–ád/; /d–**ì**–ád/→ d**ỳ**–ád: *"basket"*

The modifications to this requirement should have led to putting **y** and **w** everywhere in the rewrite of the summary table of nominal classes. But for the sake of clarity I left **U** or **I**.
Syntactically, other phenomena make it difficult to highlight the PG, particularly in contact with what the specialists describe as augment.

There are vowel edicts related to the prohibition of the hiatus of two grammatical forms.

(d–**à**–àm); (d–**è**–àm); (d–**ỳ**–àm); d**ỳ**–àm: *matter*[78]

The analysis of all the particularities on the technical level could distance us from the subject of this study.

Since curiously we define two different nominal classes for the singular and the plural, I suggest this simplified schema.

	PLURAL				
	Mi	**Ma → Mə**	**Ba**	**Bi**	**Collective Da**
Feminine		di	di	di	
Feminine		da			da
Masculine	W		w-	w	

Chart 35: nominal classes rivisited

	PLURIEL				
	Mi	**Ma ¶ Mə**	**Ba**	**Bi**	**pluriel** Da
féminin		*di*	*di*	*di*	
féminin		*da*			
masculin	*w*		*w-*		*w*

Chart 36: nominal classes rivisited and simplified

78.Jacques EVOUNA, La morphologie nominale en Ewondo: La double préfixation, Connaissances et Savoirs, P64

Note:

If we consider that the feminine is realized since the proto-version of the Negro-egyptian syntagmatic prefix * ***kja-*** that can be simplified into **I** or **A** of our current languages.We could further reduce this table to its simplest expression namely, plural forms in B and forms in M.

And according to the nature of the prefix of the feminine one could deduce the plural: I (mi / bi) and A (ma / ba) with the particular case of the collective as in Egyptian-pharaonic also.

Note:

If the nominal classes existed, they would be clearly identified as pharaonic.
I do not know if the symbols:

Middle-egyptian: 𓊖 niwt "city"

Middle-egyptian: 𓉐 pr "house"

Middle-egyptian: 𓃀 "foot, place"

So it would have been enough for the scribe to draw the 𓃀 foot to indicate a place and any word translittered with the foot would necessarily be a place isn't it? Why do we have then to get confused with the other two signs as ***niwt*** and ***pr etc?***

3.3.3.9. VERBAL FORMS

I-1. *The temporal system*

The temporal Ewondo system is composed of about nine times of conjugation completely represented in the so-called indicative mode.
It is arbitrarily organized around three great moments defined as:

- The action is accomplishing
- The action is accomplished.
- The action will be completed

Thus it is possible to distinguish on the diachronic plane: the present moment, the past moment and the future moment. With the exception of the present which is one, the past and the future are divided in time symmetrically grouped together. So there is a very recent past that corresponds to a near future, a recent past that corresponds to a simple future, and a distant past that corresponds to a distant or uncertain future. We will return thereafter to bridge with the Pharaonic.

I-2 *Modal system comparison*

The question of the modal system is in Pharaonic the subject of debate among the grammarians experts of pharaonic Egyptian.
One observation is nevertheless admitted that on the inflectional point of view, Pharaonic Egyptian does not express a graduation of time (past, present, future). What matters according to Jules Farina professor of Egyptology at the Royal University of Roma:

> *"[...] it is above all the quality of the action from the point of view of its development, its continuity, its character of completion or incompleteness, the categorical, indicative, and so on. of the action. So in a mode of Egyptian correspond several times"*[79].

79. J. FARINA, Grammaire de l'ancien égyptien(hiéroglyphes), Payot, Paris, 1927, p.105, § 211.

(FARINA, 1927)

Some grammarians, however, very serious define the modes completed, unfinished, pseudo-participates and the prospective. For Professor Obenga,

> *"[...] It is rather curious to see the Egyptian grammar encumbered with notions of Semitic grammars: "perfective", "imperfective", "pseudo participle" ("old perfective"), "weak" verbs, "verbs secondae germinatae", "verbes tertiae infirmae" "etc."*[80].

(OBENGA, 1993)

Maspero seems to share Professor Obenga's point of view when he says:

> *"Two facts especially characterize the Egyptian conjugation:*
> *1. An extreme shortage of time and fashions, since time and mods reduce to two, which express in a general way, the first the idea of present action, and the second the idea of past action.*
> *2 A tendency to specify the verbal value, attributed to the conjugate root, by various artifices of language, addition of auxiliary verbs, intercalation of particles, accumulation and repetition of subjects. It follows from this tendency that each verb can conjugate the two tenses it possesses in three different ways:*
> - *By joining the theme of time, the subject whatever it is.*
> - *By adding one or more verbal roots to the verb that play the role of auxiliary.*
> - *By interposing between the subject and the verb, a preposition which marks the direction of the action accomplished or undergone by the subject."*[81].

(MASPRO, 1871)

To finish about this subject Professor Anta Diop recognizes only three modes, the completed, the unfinished and the pseudo-participle.[82]

The expert literature in Ewondo presents a modal system Ewondo comprising several modes that correspond to the various verbal forms obtained according to the way in which the action expressed is presented. According to the literatures of specialists there are about 9.

Indicative	*Conditionnal*	*Debitive*
Imperative	*Subjonctive*	*Intentionnal*
Potential	*Infinitive*	*Optative*

chart 37: hypothetical modal system of Ekang languages

This division in my opinion is not very rigorous and deserves to be modeled on the system highlighted by the simplest Egyptian grammar and certainly closer to our reality. The desire to produce modes in order to resemble the other grammars of the settlers therefore seems less indicated here.

I-3. *The verb and morphology*

The Infinitive

In Middle Egyptian, if the verb ends with a weak consonant [83], then we suffix the sign X1 ⏢ ***-t*** of Gardiner's list to get the infinitive. Otherwise, the infinitive remains identical to the radical of the verb.

Middle-Egyptian: 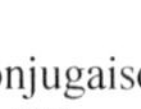 ***DAi*** " to transship, to cross"
Ewondo: /á-dáŋ/; /á-daè/ "to transship, to cross"

80. Th. OBENGA, Origine commune de l'Egyptien ancien, du copte et des langues négro-africaines modernes, l'Harmattan, Paris, 1993, P142.
81. G. MASPRO, Des formes de la conjugaison en Egyptien antique, en démotique et en copte. Hachette, Paris, 1871, 1, § 2.)
82. C. A. DIOP, Parenté Génétique de l'Egyptien pharaonique et des langues négro-africaines, Ifan, Dakar, 1977, P.42.
83. High consonants are P /p/, T /t/, K /k/, S /s/, /h/ and TH /tʰ/, weak consonants are B /b/, D /d/, G /g/, Z /z/, L /l/, R /ɾ/, M /m/, N /n/, F /ʋ/ and NJ /ɲ/.

Middle-Egyptian: 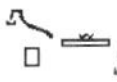 ***stp*** "to choose"
Ewondo: /á-top/ "to choose"

For the suffixes X1 ***-t, -i, -w*** to realize the infinitive, every verb in the infinitive in Ewondo is accompanied by the prefixal particle / á- /.

The carry out of the verb in Ewondo does not seem to totally obey to the traditional system of realization of the verb in the Ekang languages.

This is a dual process which is defined as follows:

a- Radical + derivative (grammatical suffix) = simple verbal basis.
b- Aspect[84] + simple verbal basis = verbal constituent.

By associating the a- and b- joints of the process described above, we finally obtain a shape that looks like this:

Verbal Constituent: Aspect + Radical + Derivative (grammatical suffix)

The derivation of the verb Ewondo can thus be illustrated on the following way:

Aspect	+ radical +	Derivative	=	Verbal Constituent	Glossary
à-	-wogh-	-lo	→	*A-woghlo*	Hear, listen
à-	-kor-	-o	→	*A-koro*	leave
à-	-Sil-	-i	→	*A-sili*	Request

Chart 38: derivation of the verb in Ewondo Language

Another articulation is conceivable in addition to the verbal constituent described above. It is the one that introduces the notion of verbal extension. Indeed, we then attach to the simple verbal part a morpheme called *extension* and we obtain an *extended verbal basis*. For simple bases with derivative ø, the extension is directly attached to the radical. We will have therefore:

Aspect + verbal base + *extension* = verbal constituent with simple extended base

Aspect	+ radical +	Extension	=	verbal constituent extended	Glossary
à-	-kore-	-bèn	→	*A-korebèn*	Redeem himself
à-	-kil-	-bèn	→	*A-kilbèn*	Turn around
à	-Sil-	-bèn	→	*A-silbèn*	To Rub

Chart 39:verbal constituent with simple extended base

The verb to be

Seven verbal themes expressing the idea of being and used as auxiliary in the pharaonic Egyptian conjugation.

84.DESCES, J-P, Quelques concepts relatifs au temps et l'aspect pour l'analyse des textes en Etudes Cognitives: sémantique des catégories de l'aspect et du temps. Warszawa, Académie des Sciences de Pologne- Institut de slavistique, 1994, p60.

: Iw/à-və
: tw/à-tə
: ***wn/à mbə***
: ***aha/à-hə***
: ir/i-nə
: pw/ à-mbə
: ***xpr/ à-kulèn***

Iw, tw, pw, wn, or ***nw*** make a special group, each term has its analog in the group of the suffixed personal pronoun of the first person singular ***i*** "i" and the articles ***pA*** "the", ***tA*** "the", ***nA*** "the/them".

Putting aside the ending ***w*** common to all the auxiliaries and the ending ***A*** common to all the articles, one finds at each element of the series an identity root between the auxiliary verb and the pronoun or the corresponding article.[85]

Iw = I+w	=*w*		= *i*
	pw		pA
	tw		tA
	nw		nA

Chart 40: Variant of auxiliary in the pharaonic Egyptian conjugation

The Aorist

The aorist affirms the attribution or the accomplishment of the action, therefore in the verbs of quality, of state, in French it corresponds to the present of general truth. In action verbs, the aorist indicates that the action is accomplished. It can be compared to the simple English present (I eat, you eat etc.) which describes a notion of repetition or generality.
The aorist is realized like this in both languages:

> **Middle-egyptian**: radical + personal pronoun suffix corresponding to subject
> **Ewondo**: personal pronoun prefix corresponding to the subject + radical.

Midd.Egyp.	*Pronoun*	*Transliteration*	Ewondo	English
		wnm=i	Mà á-min	I eat
		wnm=k	wà á-min	You eat (masc.)
		wnm=ṯ	Wà á-min	You eat (fm.)
		wnm=f	À á-min	he eats
		wnm=s	À á-min	She eats
		wnm=n	*Bỳá* á-min	we eat
		wnm=ṯn	*Mỳá* á-min	You eat
		wnm=sn	Bá à-min	They eat

Chart 41 : Aorist in middle-Egyptian

A little curiosity that we note here about the Ewondo is the use of an auxiliary, "à-". This feature is also found in Egyptian Coptic synchrony or Swahili.
For Professor Obenga and other specialists, it is possible to adopt another approach in writing that merges the pronoun with the auxiliary. To this account we write:

85. Gaston MASPRO, Des formes de la conjugaison en Egyptien antique, en demotic et en copte, Hachette Livre BNF, Paris, 1871, p.26

Ewondo: / mà di / "I eat" while the other possible writing gives / Mà á-di / "I eat" literally "I eat". Obenga (1993: 146)

At first glance one might think that these two languages are not matching and de facto, stop here any idea of comparing or switch from one language to another and vice versa.

1. The first thing that quickly enquiried is that the Ewondo proceeds by prefixal conjugation when Pharaonic uses a suffixal conjugation.
2. Another argument for this is that personal pronouns are not the same from one language to another.
3. There are Pharaonic forms that the Ewondo does not have (feminine in you etc.).

One could certainly find superficial discriminating technical elements justifying the difference between the two languages. According to me, I do not maintain that the two languages are identical in every aspect. Far from it. But as everyone knows since, I argue that those languages are genetically linked. And on this account, one can read in Ewondo what is written in Middle-egyptian just as one reads in French what is nevertheless written in Egyptian without the two languages being identical.
In fact, let us take up the different questions that lead to hesitation, to discrimination and try to provide technical elements to remove the doubt.

Pharaonic Verbal Suffixation versus Verbal Prefixation in Ewondo

Concerning the inclination on the suffixal conjugation in Pharaonic, this is a consequence of the reading convention put by Egyptology regarding the reading direction of ***mDw nTr***. If the Arabs had posited their convention potentially in the present case the conjugation would be prefixal and no longer suffixal. While remaining on the bases of the operative norm, one also finds in pharaonic a conjugation of prefixal type when the personal pronoun plays the role of subject and that it is said absolute[86].

Middle-egyptian:  ***ink rx*** "i know"
Ewondo: măà-yəm "me know" → "i know"
Ewondo: măà-kar "i say" whoever says is the one who knows

Note: One could also propose the reading **rx** by [tsokh] in Ewondo. This meaning is brought by homophony with the placenta which is called in Ewondo: / i-sôkh / and to think, / á-tsokh /.
In Pharaonic we see that the two types of conjugation, suffixal and prefixal are possible in the same sentence [87]:

The missing personal pronouns in Ewondo

Middle-egyptian: 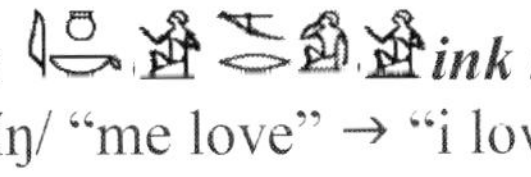 ***ink mr=i*** "i like"
Ewondo: /mă à-díŋ/ "me love" → "i love"

Why do we have to vocalize the consonants skeleton ***ink***, [mă] in Ewondo? Simply here because the subject is devoid of the use of its lower limbs. Which is equivalent to the expression [úmbokh] in Ewondo. This word [úmbokh] also refers to the gourd vase [ùmbokh] but also the numeral "a" [mbokh]. De facto the scribe indicates that the subject doing the action concerned is unique, that is also to say

86. Gaston MASPRO, Des formes de la conjugaison en égyptien antique, en demotique et en copte, Hachette Livre BNF, Paris, 1871, p.12
87. Gaston MASPRO, Des formes de la conjugaison en égyptien antique, en demotique et en copte, Hachette Livre BNF, Paris, 1871, p.13

himself. In a binary system, which has prevailed in Africa, the "one" indicates the ego. The second or the other being by the way the other one.
Let's note also that the numeral one [úmbokh] can be vocalized [wa] in Ewondo. We find the justification of the two vocalizations on the transliterated script "alone". The water is called in lingala *mái*, in Ewondo, the sea is designated by the word / máŋ /, the water by / mə-ndim /

Negro-egyptian: *mw "water"
Middle-Egyptian: mw "water"
Lingala: *maï* "water"
Tumbuka: *maji* "water"
Vili: *masi* "water"
Luganda: *amazzi* "water"
Fang: *məzim* "water"
Ewondo: *mə-ndim* "water"

The symbol w24 of Gardiner's List, transliterated ***nw*** is used to designate the water or primordial ocean . One can thus, without falling into the exaggeration say that the transliterated writing ink reads [mă] en Ewondo.

Since we have to submit to the sacrosanct transliteration we are saying that, the script transliteration ***ink*** carried by Egyptologists leads to the Ewondo item /i-kon/ "clay". Hope that you will agree with me that sign W24 of Gardiner list is of course clay but moreover shaped clay. Until now, the action of shaping the clay to produce an object is called in Ewondo language:
Ewondo: /á-me/ "to shape" by the way then, a pot clay shaped can be called /i-me/ "shaped"

So only the pot was enough to read the script /me/ or /ma/. That's why even the transliteration gives **ink** we call read if [mă] "me; I" because those two words are linked because they are sharing the same semantic field.

A complete analysis of all the personal pronouns is carried out in section **3.4.3.32**.

Rigorously here , one should read in Ewondo: [mə diŋ] "the One who loves". The links of the hoe bringing the word "to love" á-díŋ <á-tíŋ "knot, link" (á-tindi à-lugh) "to bind, to tie the marriage". Let's close here this little parenthesis.
Since the second person of the singular is the other ONE, then of course the Ewondo will say wa.

Middle-Egyptian: ***mr=k*** *(second-person singular masculine)*
Ewondo: /wă diŋ/ "you like"

Due to the fact that the other is the other One in African thought, and the One is vocalized as wa "alone".
The symbol V31 appears as a synonym for w24 in the **ink** transliterated script.

It can be vocalized / wă / without any problem because of the handle. There is a iron stewpan pot with similar shaped handles called in Ewondo: / i-wanga /. The decomposition of the word * wa-nga lets appear, / wa / and the / nga / another way to say **me** in Lingala. Let us allow ourselves the curiosity to compare the independent pronoun ***ink*** with the personal pronoun / ***ngai*** / of Lingala by shifting the K to G and suffixing the i. We would have ***nki*** vs **ngai**. The Curiosity can be allowed in this present case isn't it?

Middle-Egyptian: ***mr=T*** *(second-person singular feminine)*
Ewondo: /wă diŋ/ "you like"

I know, now is the tricky question, how am I going to get out of this issue and demonstrate the relation between the two languages on this particular point?

It is relatively simple here the sign V13 is considered as a rope to tie animals. The term for chord is /*ùn*-ŋkô (l)/ and the term hindrance is called [mbokh] too. The rope serves as an obstacle to beasts of burden / mi-'mbokh /. So we return to the first two cases. One can therefore be satisfied in Ewondo to say /wa/ "you" in Ewondo without distinguishing the feminine from the masculine at the second person of the singular isn't it? Let me juste remind you that in Ewondo/*ùm*-bokh/ "one; 1" and /wa/"one, 1; you"That's normal because two(2) is the other one in a binary system.

Middle-Egyptien: ***mr=f*** *(third-person singular)*
Ewondo: /ă diŋ/ "he likes"

The symbol I9 in Gardiner's list is vocalized in Ewondo / nyua (l) / "snake". Note that the scribe decomposes the characters ***n*** and ***t*** the symbol N35 ***n*** can be vocalized nyu when it represents a drink. The symbol X1 ***t*** represents the feminine tag in Ewondo namely / a /. So we get the reading /nyua/ meaning also: *the one who.*

This form / nyua / is derived from the Negro-egyptian * ŋʲʷ- and is often found attached as a prefix on the radicals related to animated beings [88].

When creating our languages from the Negro-egyptian, depending on whether the noun is masculine, it suffixed with / u / or with / ja /. Jean-Claude Mboli explains clearly how :

> *"The masculine and the feminine being the same ((*ə̆-u-, *ə̆-ja-, *ə̆bʷu-,*ə̆bʷa-) > (*ə̆-, *ə̆bʷa-))"*
> *(Mboli 2010)*

We also know that in Egyptian, the feminine serves to express the collective and the notion of the neutral.
The Ewondo would have kept this Negro-egyptian atavism here to express the collective without distinguishing the gender. Over time, the class prefix of the animated * ***ŋʲʷ-*** has given way to the particle of the collective in Ewondo. /à/

Middle-Egyptian: ***mr=s*** *(third-person singular feminine)*
Ewondo: /ă diŋ/ "she likes"

Here the S29 replaces the sign I9 of Gardiner's list. The folded loincloth is intended to be put over the body. It can be assimilated to the body by semantic shift in meaning. On this account we return to the previous reading namely: / nyua /. And we also use the following explanation to know that * ***ŋʲʷ-a*** ended up simply by vocalizing / a / since expressing the idea of a collective without distinction of gender.

Middle-egyptian: ***mr=n*** *(first person plural)*
Ewondo: /byă diŋ/ "we like"

The scribe recalls the three lines of the plural and puts the N35 signs that can indicate a drink (bia = beer). In short, the scribe says that it is the plural that is done with the first person but that is vocalized as the drink.

Middle-egyptian: ***mr=Tn*** *(second person plural)*
Ewondo: /myă diŋ/ "you like"

88. Jean-Claude MBOLI, Origine Des Langues Africaines - Essai d'application de la méthode comparative aux langues africaines anciennes et modernes, l'Harmattan, Paris, 2010, p.484.

The scribe recalls the three lines of the plural and puts again the sign V31 which indicates the other "one". The plural and the other lead us indisputably to you. So we're going to vocalize that mya literally / ba nya mbokh / "the others One".

We can observe several things here: the imposed transliteration ***Tn*** can lead to the word / ù-toŋ / "brook" in Ewondo / á-toŋ / "to follow" which admits the synonym / myè / "to join"
It could indeed be seen that the sign N35 of the Gardiner list refers to water and therefore to the word vocalized in lingàlà as seen above *mayi* "water" in Koonzimé: mîhé "eyes"; Koonzimé: emíha "urinating" the word /mə-hié/ "action of urinating or urine" this word remains little used.

Middle-egyptian: ***mr=sn*** *(third person plural)*
Ewondo: /bă diŋ/ "they like"

The scribe recalls the three lines of the plural and puts again the sign V31 which indicates nyu - "collective". The plural and the collective lead us indisputably to theirs.
So we make the plural here in / bă /

We can observe several things here: the imposed transliteration ***sn*** can lead to the word / ùn-nsoŋ / "sharp" in Ewondo / i-bèh / "sting" which admits the homophones: /bə/; /bà / "they"
It could have be seen also that the sign N35 of the Gardiner list refers to water and therefore reminds the feeling of thirst /i-bèh/ homophone of /bə/; /bà / "they"

Once again, the transliteration, if sometimes it could be necessary to set a starting frame, it can easily limit access to anyone in his language from the moment he is envolved in the standardization of the transliterated data. The characteristic of these characters is to have an ideography side which calls upon a logic peculiar to African philosophical thought. The transliteration is a standard not a goal.

Concerning the prefixation versus suffixation debate

The purpose here is not to study Pharaonic Egyptian according to standard rules but to see how it can be read knowing the Ewondo. Admittedly, some African languages use a suffixal conjugation, notably Wolof, Seereer, and so on. inherited the phenomena of grammaticalization of Negro-egyptian.

If we then accept for a moment that the current languages are derived from consonant metathesis, then the reading can be done in both directions, on this account, what is prefix in one sense becomes suffix in the other and vice versa.

Let me clarify the purpose a little bit. To say house French people use the following items:
French : *maison* "house"
French: *sonmai* "house"

So where is the good direction to read this word?

We have for the first person singular example:

Conventionnal direction

Middle-Egyptian: sDm i "hear i" → (i hear) according to the convention
Ewondo: voghlo mă "Follow-i" → ((*you*) follow-me) by suffixing the personal pronoun. This structure is used in Ewondo to traduce an imperative.

Ewondo: mă voghlo "i Follow" → (me follow (*you*)) by prefixing this time the personal pronoun. This structure is used in Ewondo to express a situation qualified by specialist as an aorist.

Note: The consonants skeleton ***sDm*** is the equivalent of the item Ewondo:

Ewondo: /á-simda/ " toThink about, imagine, remember, understand; Think to one another".

Non-conventionnal direction

Middle-Egyptian: i mDs "me hear" →(i hear)
Ewondo: [Máà-voghlo] "me hear" →(i hear)

"The structure  ***ni mi****: " to me "also exists in Pharaonic ".*

(Anta Diop 1976)[89].

In the light of these facts, this conjugation in Pharaonic appears therefore suffixal only by pure convention.

On the technical level, we have the Negro-egyptian which has given by grammaticalization two groups of African languages that proceed by ananyms, Kweké and Kékwé according to the work of Jean-Claude Mboli (2010). In fact, by grammaticalization, there has been a reversal of phonemes and consequently positions of affixes (verbal or nominal). This is why conventional one-way reading for some languages can be problematic. It is therefore necessary to make a reading in the sense that supports comprehension in one's own language. In Ewondo the imperative verb forms give the impression of a suffixal conjugation.

Example:

Ewondo: [vâ Máà] "Give me"

The historical facts all specify that the Ekang, large human group to which belong the concretes which are Ewondo speakers who in the recent past are from Egypt. De facto in the past the reading of the sacred characters ***(***  ***mDw nTr)*** was also possible to them as well as the conjugation although apparently the conjugation of the Ewondo is not suffixal. This hypothesis is only true if the reading of signs ***(***  ***mDw nTr***) is possible in both directions.

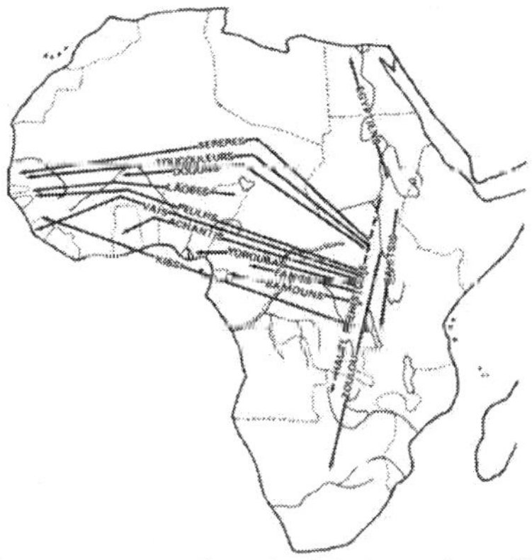

Map of Migrations after Arab's invasions 7th century BOE

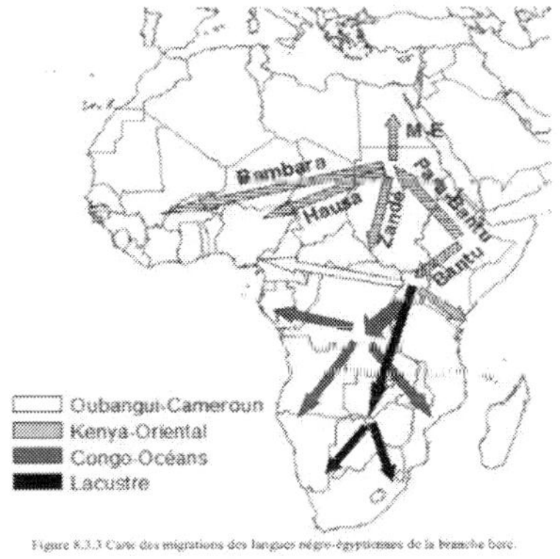

Map of Migrations of Negro-African Populations from the Upper Nile and Great Lakes Region

Map: *Populations of Africa - (according to C.A. DIOP)*[90]

89.Cheikh ANTA DIOP, Parenté Génétique de l'Egyptien pharaonique et des langues négro-africaine, Ifan, Dakar, 1976, P.25.
90. Cheikh ANTA DIOP, Negro nations and culture, Présence africaine, Paris, 1954, P.363.

Concerning the apparent differences between the personal pronouns, again, it must be remembered that the two languages Ewondo and Pharaonic are derived from the Negro-egyptian and consequently the pronouns recorded are in fact only evolved types of the archaic variants due to the grammaticalization of pronouns from the Negro-egyptian as JC Mboli points out[91].

The word Pronoun, in Latin, carries with it its definition, since it means that which takes the place of a name, which represents the name, almost as well as proconsul means who represents the consul, who takes the place of the consul. Régnier-Desmarais (1706: 126)[92]

Personnes	Type	Glossary
1ère sing.	*hi	"me"
2e sing.	*hi	"you"
3e sing.	*ha	"that", "he", "she"
1e pl.	*hi-k^{j}un^{w}	"we"
2e pl.	*hi-k^{j}un^{w}	"you"
3e pl.	*ha-k^{j}un^{w}	"they"

Chart 42: Reconstructed variants of the archaic Negro-egyptian pronominal

On the syntagmatic level, and especially the enunciative referentiality, the French uses personal pronouns subjects whereas the Ewondo uses personal pronouns complements or objects resulting from his filiation with Negro-egyptian.
From this table we can therefore explain the different variants of personal Pharaonic pronouns (dependent, independent suffix) as well as the personal pronouns Ewondo.

Person	**Negro egyptian**	**Old-egyptian**	**Pronouns Middle-egyptian**			**Ewondo**	**Meaning**
			Independ.	Suffix.	Depend.		
1st sing.	*hɨ	ink	ink	.i	wi	***M***–à	"me"
2nd sing. Mas.	*hi	Masc.:Twt	ntk	.k ~ .T	Tw	***Ẁ***–á	"you"
2nd sing. Fm.		fem.:Tnt	ntT	.T	Tn		
3rd sing.Mas.	*ha	Swt	ntf	.f	sw	à-á	"that", "he", "shc"
3rd sing.fm.			nts	.s	sii ~ st		
1st pl.	*hɨ-k^{j}un^{w}	stt	inn	.n	n	***Bỳ***–á	"we"
2nd pl.	*hi-k^{j}un^{w}	inn	ntTn	.Tn	Tn	***Mỳ***–á	"you"
3rd pl.	*ha-k^{j}un^{w}	ntTn	ntsn	.sn	.sn	***Bə***- á [Bá]	"they", "them"

Chart 43: evolution of personal pronouns since Negro-Egyptian

Regarding the non-existent forms of personal pronouns in Ewondo, this is not a problem to use the unique form Ewondo that does not distinguish the masculine from the feminine to the second and third person singular.

91. Jean-Claude MBOLI, ORIGIN OF AFRICAN LANGUAGES Test of application of the comparative method to ancient and modern African languages, Harmattan,Paris, 2010, P.360.
92.REGNIER-DESMARAIS, Trait de la grammaire française, Jean-Baptiste Coignard, Paris, p. 216.

All of this is covered in detail in Chapter 4, section 3, example N °32 to explain how to read the different pronouns of the Egyptian in Ewondo.

Example:

Middle-Egyptian: sDm=k "you comprehend" (masculine) et sDm=T "you comprehend" (feminine)
Ewondo: *Wá* à-simda "you comprehend"

Incidentally note here that the verbal translation in Ewondo is realized by verbal suffixes which group together according to the four phonatic structures which are:

Ø; v̏; -vc, -(v)c v̏

- Ø: RAD + Ø (monosyllabic verb)

Ewondo: -kom + Ø "arrange"

-v̏ (low tone vowel)

Ewondo: -ləmb + ə "to spy"
Ewondo: -bím + i "to strike"
Ewondo: -bór + ó "to gather"
Ewondo: -namb + à "to touch"

-VC (low tone vowel + consonne)
RAD + VC
Ewondo: -tind + en "attach"
Ewondo: -bím + en "to hit"
Ewondo: -bəl + en "to have"
Ewondo: -təbel + en "to order"

This Ewondo verbal suffix **-en** corresponds to the suffixed form 𓅱-w of the pharaonic Egyptian.
The quail chick is a small quail. The word little in Ekang languages is said / ùm- an /; mwene; mon, man etc.
The root of these different forms is vc type. In the plural we have / bwan /, for example.
To the question why this root is related to the passive form? The answer is relatively simple It is antonymia of 𓅃***Hr***. Certainly, he is the smallest but paradoxically being the primordial ancestor of Africans he is also the eldest. He represents the past.
According to the variants of the synchrony Ewondo one can have the forms: **-an; in; ən**

The quail chick 𓅱docs not vocalize **w** as in the so-called Niger Congo languages that introduce the **w** between the radical and the final vowel **a**.

Example: *Ku piga* "to strike" giving the passive form *ku-pigwa* "to be struck"

The equivalent Basa'a being done by adding **à**, **-i** suffixal. I suppose by amussement of voiced alveolar nasal consonant [n] of the variant **-an**.
One can also observe the variant **-bà** that I connect to the foot /à-bo/ is the other ending of 𓅃**Hr**

-(v)c v̏: -(V)CV ((vowel) + consonne + low tone vowel)

Ewondo: -ɲoum+ ulu "to smell"

Ewondo: -tib + ili "to maltreat"
Ewondo: -bom + olo "to nail, to tack; to spike, to pin"

The past Tense

The progressive Form

This variant describes an action being developed in the past or in the present. It can be compared to present or past progressive in English (You are working, he was reading). In French, it would translate into a present or an imperfect one, but more generally by a "he is in the process of ...". It is constructed in this way in Egyptian:

 ***m*** + infinitive

M will be used in Egyptian more readily with a movement verb, but in this case, he can also express an action about to begin or just started. It can be approached from the location:

Ewondo: / ndəm / "come from ..."
The owl can be called / ù-nəmə / in Ewondo when it is related to the world at night. He can also designate / ùn-yəmə / "the scholar" who offers by synonymy / ndəm / "mark; came from… "

 ***hr*** + infinitif

This expression is to match with the Ewondo (hə-/və-).

Middle-egyptian: 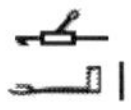 wa "lone, one, unic"

The sign z1 of Gardiner's list is common between ***Hr*** and ***wa.*** If the arm is called in Ewondo /wa/ and confirms the transliteration wa of Egyptologists, this arm can also designate the action of giving / à-hə /; / à-və / "to give". This is confirmed by the script  ***itm*** we read / i-tàm / "single, unique, the only"

The past conventionally breaks out into three distinct periods defined according to the distance between the moment of completion of an action and that of the utterance which expresses it.

We thus distinguish:

- Le past Tense1: it serves to express an action that has just been accomplished.

The recent past Pl is marked by the morpheme **Ewondo**: /***ndəm*** / "just"

Ewondo: /ù-ndamba ***á ndəm*** + kê/ "the match has ***just*** begun"

- The past tense 2:

The past P2 is used to express the actions that took place quite recently. A little less recent than the very recent past P1. It is marked by the morpheme ***hə/və.***

Ewondo: *ùm*-wan *à-**hə*** + yón "the child has cried"
Ewondo: *ùm*-wan *á-**və*** + yón "the child has cried"

Note: The words /**hə**/ and /**və**/ are simply variations of the same word. They are pronounced indifferently according to the geography areas of the speakers in the Ekang country in which they are living.

- The past 3:

The P3 is used to express the actions accomplished from the day before the day of the statement, even in the more distant past. It is marked by the morpheme ***ŋga.***

Ewondo: *ùm*-ingà *á*-***ŋga*** + sóh "the woman is or had come"

This last variant is that one may approach of the old egyptian. Let's try to illustrate all this calmly.

Ending *n* variant

Radical	Midd.-Egyp.	*Transliteration*	Ewondo	English
	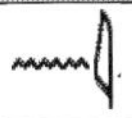	*ka.**n**=i*	Mə ***ngá*** ákwè	I threw up
		*ka.**n**=k*	U ***ngá*** ákwè	You threw up (masc.)
		*ka.**n**=T*	U ***ngá*** ákwè	You threw up (fm.)
		*ka.**n**=f*	À ***ngá*** ákwè	He threw up
	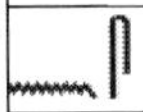	*ka.**n**=s*	À ***ngá*** ákwè	She threw up
	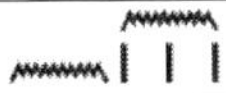	*ka.**n**=n*	Bi ***ngá*** ákwè	we threw up
	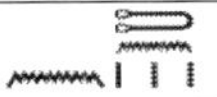	*ka.**n**=ṯn*	Mi ***ngá*** ákwè	You threw up
	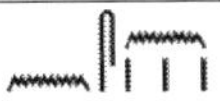	*ka.**n**=sn*	Bə ***ngá*** ákwè	They threw up

Chart 44: comparison of the past between the Egyptian and Ewondo

Middle-egyptian:  ***qa*** "to spew"
Ewondo: /à-yo/ "to spew"
Ewondo: /à-kẁè/ "to expectorate"

The reading / yo / is brought by the sign Δ N29 which represents the top / yo(p) /.
The reading / kùè / is brought by the sign Δ N29 which represents the top / ùn-nko /.

We notice above that the time markers are infixed in the verbal construction between the initial and the verbal radical.

The other endings used as past markers are:

- *.tw* = (passive form)

Example:

Middle-Egyptian: sDm.*tw* =i "I am listened"
Ewondo: /ma voghlo.*bèn*/ "I am listened"

Note: The Egyptologists *.tw* is equivalent of:
Ewondo :/tə/ "negation" synonym of **Ewondo** :/bén/ "negation"
As I have emphasized the transliteration is either identical or a synonym of the canonic item we often use.

- n-tw =
- w =
- w-n =
- in=
- in-tw =
- t=
- t-n=

The Futur

The future, is for a system, the set of possible configurations that do not exist (-not "still" - occurred). In the timeline, the future comes after the present and the past.
According to African thought, the future does not exist. The future simply appears as a mental representation that helps to represent ideas in the present moment. Curiously we find this worldview among the Piraha in the Amazon region.
The members of the human species, from a certain age, are distinguished by high cognitive abilities, conceptualization and imagination that allow them to consider the near future and to project themselves, but also to imagine possible futures farther away where will live what is now called future generations. Human beings can thus develop "anticipatory" strategies, individual and sometimes collective, called short, medium and long term, which are one of the governance stakes.
The future presents a hypothetical course of events to come. According to G. Guillaume (1968, 54) quoted by Teresa Tomaszkiewicz [93] (1988, 102):

Prospective is the multidisciplinary approach of estimating probabilities and scenarios of "possible futures" and probable or desirable; rather in the short and medium term, in the service of decision support. In Egyptian conjugaison, the **prospective** is marked by the **y** (). It expresses a nuance of future in ancient Egyptian.

Example:

Middle-egyptian	***mr*** "to love"	***mr-y*** "the beloved one"
Ewondo	*à diŋ* "to love"	i-nyo à díŋbèn

Chart 45: Ewondo-Middle-egyptian prospective comparison

Note: the transliteration ***mr*** is the equivalent of the French word *aimer, amour, mêler* "to mix" and the English item *to melt; amour etc.*
The equivalent kékwé word in **Koonzimé** is *lémé* "heart" and the kwéké word  in Ewondo language is :

Ewondo: /məli/ "grass cutting, work with a machete or a hoe for little herbs cutting"
Bulu: *ùn*-nləm "heart"
Ewondo: *ùn*-nəm "heart"
Ewondo: *i*-ngwanghà "adze"
Ewondo: *à*-nyàmi "amalgamate"

☞ **Ewondo**: *à diŋ* "to love"
☞ **Ewondo**: *à-ŋgoŋ* "bell"

The words to designate the adze and the bell are homophones. The heart is the seat of love, it is beating as a bell. That's how we do understand the link between the adze and love. Let's go back to our subject.

Like the past, the future is divided in Ewondo in three levels which indicate that the action envisaged is short, medium or long term. These are like the past: the future1, the future 2 and the future 3.

Future 1 or near future

The near future or F1 is used to express actions that will be accomplished in the very near future. I1 is marked by the morpheme ***nga yi ya*** "already want" that we can be approached by **soon**. It expresses the idea of "so that" / "for that".

93.Teresa TOMASZKIEWICZ, étude comparative de quelques indices du concept d'énonciation en français et en polonais, Poznań: UAM, 1988, p.102.

As a general rule said Obenga (150,1993), there is a type of sentence, not rare in Old Egyptian and Middle Egyptian, whose first element is either a name preceded by the particle in, or an independent pronoun, constituting the subject, and the second element is a verb with suffixed which summarizes or recalls the subject: this construction expresses the future.

Example1: independante particle **in**

Middle-egyptian: ***in*** *nsw ir.f st* "The king will do it", "the king will have to do it"
Ewondo: /*i*-nə *ùn*-ndzwé nyi à-yi boh / "The king will do it", "it's the king who will do it"

Old-Egyptian: ***in*** *Dr.t T wTs sw* "T's beam will support it"
Ewondo: / wa T wa yi nyi sukh / "T's beam will will support it"

Example 2: Either independant pronoun.

Middle-egyptian: ***ink*** *ir.i st* "i will do it", "i will have to do it"
Ewondo: /ma mə à-yi boh/ "i will do it"; "i will have to do it"; "i'm gonna do it"

Example 3: verb with suffix

Old-Egyptien: *ntsn rdi.**sn** n.i wy.sn* "They will give me their hands"
Ewondo: /ba yi me təgh mwà m'àbà/ "They will give me their hands"

	Middle-egyptian	*Transliteration*	Ewondo	English
 ***Pr.i***		*pr.**r**=i*	Ma áyi koro	I will go
		*pr.**r**=k*	Wa áyi koro	You will go (masc.)
		*pr.**r**=T*	Wa áyi koro	You will go (fem.)
	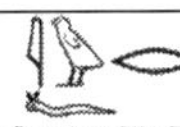	*pr.**r**=f*	À áyi koro	He will go
		*pr.**r**=s*	À áyi koro	She will go
	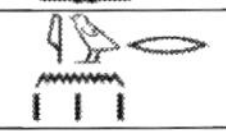	*pr.**r**=n*	*Byá* áyi koro	We will go
	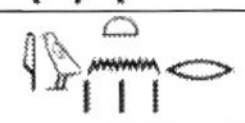	*pr.**r** =Tn*	*Myà* áyi koro	You will go
	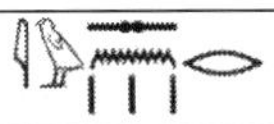	*pr.**r** =sn*	*B'**á yi*** koro	They will go

Chart 46: to leave at the future

Note:

> *"More usually still, and in the absence of the preposition*  *the auxiliary verb*  *takes the tags of the person at the same time as the attributive verb which follows the auxiliary: this Combination also constitutes a future of the indicative mode."*[94]
>
> *(C. Le JEUNE, 1836)*

The future is used to express the wishes, expectations or requests, or the purpose ("so that" / "for that"). It translates as a subjunctive present of the verb power (Ex: "may it") or by a simple future. Verbal roots are:

94. Champollion Le JEUNE, Grammaire Egyptienne, Principes généraux de L'écriture sacrée Egyptienne appliquée à la représentation de la langue parlée. Paris: FIRMIN DIDOT FRERES, 1836.

Verb Type	Example	Traduction
Strong verbs	sdm=f	May he understand
Gemini verbs (no repetition)	m3=f	May he see
Weak verbs	mr=f ou mry=f	May he like
Irregular verbs	di=f	May he give
Special cases	iwt=f	May he come
	int=f	May he bring

Chart47: verb types and expectative future

Notes:
Weak verbs often show y ending *y* in the first person.
The verbs "come" and "bring" take an extra ***-t*** in the future.
The verb ***mAA*** is thus presented:

Middle-Egyptian: *m3n=f.*

The future is often difficult to distinguish from the aorist out of context, however we do not use an auxiliary of enunciation (***iw***) with the future.
The verb ***rdi*** makes it possible to express causality "cause", "allow someone (to do) / let someone do something " when followed by a verb in the future.

The Pseudo participe.

Also called in the authorized literature: "Stative", "old perfective", "old perfect", or "accomplished transitive verbs", the "pseudo participle" is a verbal form characterized by specific endings.
As a rule, it essentially marks the state, or the result of an accomplished action.

The Endings

Personne	Construction	Examples
1-masc.sing.	Racine + , , **.kwi**	aHa.kw, "i'm up"
2-sing.	Racine + , , .ti	.
3-masc.sing.	Racine + , .w. (w(y))	.
3-fem.sing.	Racine + , ,	***Hry.ti***, "she is far"
1-plur.	Racine + .wyn	.
2-plur.	Racine + ,.**ti**wny	.
3-plur.	Racine + , .w (.w(y))	.

According to specialists, what characterizes this form is that it has no temporal value. It expresses the result of an action rather than the action itself.

Use in independent sentence

Middle-Egyptian 
Transliteration ***it.i*** n.i Ha.ti n ma nfrw.i
English Translation you have come (stative) to me and you are happy (stative) to see my perfection
Ewondo *A bə ma wa so ndô wa tag mə-yn à-bəŋ dàmà*

Chart 48: Example of independent sentence use

Use in subordinate sentences

Middle-Egyptian 
Transliteration di.i mA; sn Hm.k apr(.w) m Xkr.k
English Translation I make them see your majesty equipped (stative) of its adornment
Ewondo *Ma bɔ nâ bə yn ù-mfiê wo ù-bwarèn i-nyəbgha dy-o*

Chart 49: Example of subordinate sentences use

Explanations and comments

To facilitate the understanding of this translation, I suggest to analyze only the last example related to the use of pseudo participates in subordinate sentences. This analysis is used as a school case and shows as script by script one can read a whole sentence.

Middle-egyptian:  di.i "i do"

: Here we have the figure of  ***Wsir*** who is vocalized / ma / "I" in Ewondo because we have an honorific anteposition related to the mother. Being the brother of the mother and the depositary of the patrimony of the family, he is called in our languages and therefore in Ewondo: mother. And until nowadays the mother is called /ma/; /mama/

: this sign has been transliterated ***di*** "do" because it represents a conical bread. The word ***/di/*** corresponds to the food in Ewondo in general. In short, what is destined for the mouth / à-nyù /. But the mouth is related to the word / ùn-ŋkɔbɔ /. ùn-ŋkɔbɔ decomposed into toneme gives / à-ko / "foot" and / à-bɔ / "foot" so this term can be summed up in / à-bɔ / "foot". But / à-bɔ / offer the homophone / á-bɔh / "to do". So coupling the two signs we can then read:

Ewondo:  di.i "ma bɔh"
English: i do

Let's move on:

Middle-egyptian: 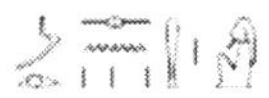 ***iri.sn Hm.k*** "they do see your majesty"

: we have here two signs: the sign  representing the mother / maâ/ or /ùn-nya /.

This sign is accompanied by the sign which serves to see /á-yén/. The couple of signs proceed by inverting phonemes **nya** vs **yen**. It is therefore necessary to read here / * yen / "to see.

: the three vertical bars of the plural are sufficient to vocalize / bə / "two" because the plural begins with /bè/ "two". The sign N35 could be read /bèh/ il represents /*nyu/ "drink; mouth". As a mouth, it's a hole /i-béh/. The Sign O34 is used to link /*bèh/.

: The sign is a pestle /*ùn*-ntoum/ likened to the club, it is called /*ùn*-mbəŋ/ it offers the homophony with the word /*ùn*-mbəŋ/ "beauty, goodness, greatness" etc.
We no longer comment on the sign representing the self or the you because in a binary system you is the other me.

Ewondo: /bə á-yén à-bəŋ dwé/ "they see your majesty"

And finally, let's take the last phrase:

Middle-egyptian: ***apr(w) m Xkr.k***
English: *"equipped (stative) of your adornment"*

: The sign would designate a counterweight, a pendant necklace or a tissue bag. In fact, it can therefore be called in Ewondo: / *mə*-'ŋgômda / or / *bi*-nyebgha / "jewels".
: often transliterated **bA** for spirit.
: The sign is used in ornamental architecture as frieze horses. So it will recall the words here to dress, embellish, equip, adorn in our languages.

So here is one of the overall way to read this sentence and understand it. Everyone can find the springs to read in his own language without major complications.

3.3.3.10. *Morphemes Of Negation*

There are different morphemes for negation in both languages: in Middle-egyptian and Ewondo, as we have already seen, this notion in some cases like the zero or the affix ***w*** as well as the markers of negation found during conjugations.

Since we have already pruned the subject, I will not logically resume all this in the details. The different morphemes serving as negation in Middle-egyptian are derived from the reconstructed form of the Negro-egyptian: *k/un^{w}* "refusal, negation".

Midd.-Egyp.	Ewondo	Use
nn	ʔən-ʔən; kɔkɔ; mɔmɔ; zeŋ-zeŋ	Not to be
W/Y	bə; kə; sara; sa-kigh; sa	
m95	bə; sara; sa-kigh; sa	Auxil. optative Proposition; imperative
Tm/s	À-təb; tèghè; tə;	
nfr	*Ùn*-ŋkər	
mkf 'w[96]	kə	

Chart 50: Comparative negative tag s

95. In the so-called Nubian languages one encounters the type M negation in the forms: m, m a, mam, mun, mi n, men. Westerman states that in the so-called Western languages of Sudan this form is prefixed "me" or suffixed m; imi; my
96. L'adverbe mkf 'w renforçait la ngation en nouvel gyptien ."

Small comparison between African variants:

Bambara.: *Ahi* "no"
Ngiemboon: *Ngang* "no"
Yemba: *Ngan* "no"
Ngeombale: *Ngah* "no"
Kikongo: *Ngo; kanig* "no"
Crole Martiniquais: *Awa.; an-an.;en-en* "no"
Ciluba: *bwaala; naanyi* "no"
Nzebi: *A-ah!; Agha!* "no"
Lingala: *Te* "no"
Obamba: *Kali* "no"
Moor: *Ayo*; *ayê; onh onh* "no"
Songhai: *na* "no"
Mɔnɔ: *ne(ne); ni(ni)* "no"
Mbanza: *nënë; në* "no"
Baluba: náni "nullement"**;** nanshá "no; without, not";
Nkundó: nyŏnyŏ "no".
Bakongo: *nábi "nullement"; kani* "ni; not yet"
Bayɔmbe: *nana* "no; none";

Note:
Beside these morphemes of negation the authorized literature is very timid on the markers of affirmations, yet our language knows how to mark the affirmation too. We find in both languages quite similar things:

Middle-egyptian: 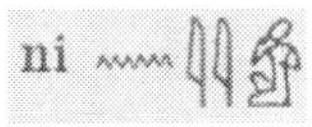

"Ebers Pap. 97, 13" (part.averment, yes)[97]
Ewondo: eŋ'hə "yes" adverbial particle of assertion.

Middle-egyptian:  I "Anastasi 1, 13, r" (surely, in truth, certainly)[98]
Ewondo: à-hâ "yes" interjection

Middle-egyptian: 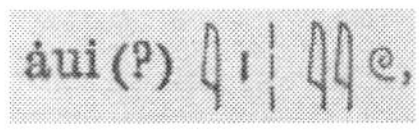

I "yes, surely; in truth, certainly")[99]
Ewondo: owé "yes" adverbial particle of assertion.

We also find words in Ekang languages like: /sâ/; /kə/ or /ka/; /yâ/ etc.
As we have seen above, / sa / is the pen, / ya / is brought by the double pen (y) so we can easily read it aside of the official framework.

Small comparison between African variants

Bambara.: *Õwo* "yes"
Ngiemboon: *Oon* "yes"
Peuhl: *Èeye* "yes"
Ngeombale: *Eng* "yes"
Kikongo: *he* "yes"
Crole Haïtien: *wi* "yes"
Lingala: *Èee* "yes"
Nzebi: *Èeeh* "yes"

97.BUDGE, Sir Ernest WALLIS. Egyptian Hieroglyphic dictionary, Harrisson and sons, London, 1920, p.503
98.BUDGE, Alfred WALLIS. Egyptian Hieroglyphic dictionary , Harrisson and sons London, 1920, p.189
99.BUDGE, Alfred WALLIS. Egyptian Hieroglyphic dictionary, Harrisson and sons, London, 1920, p.185

If we make a small comparison between the morphemes of negation and those used to make an affirmation, curiously in Ciluba (rdc), Eyo is "yes"; In Moor Ayo or ayê it is "no". Fortunately this term is known in Ewondo / Eyo /, it designates the primordial ancestor of the Ekang peoples.

If we sift through all the words that are used for both the negation and the affirmation, they all revolve around the nominatives of the mother. In short, negation or affirmation is associated in the collective unconscious with the mother. It's the mother who says NO or says YES in african thought.

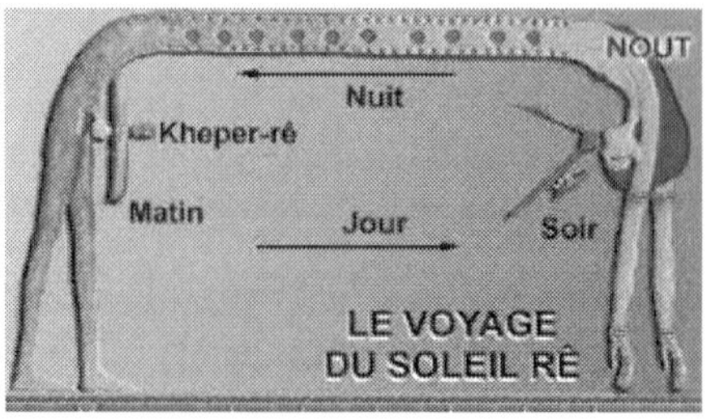

Figure 7: Eyo "the one who is vomit"

Etymological level:

Middle-egyptian: *nn* "no"
English: none "negation"
French: non "negation"

Middle-egyptian: *áui* "affirmation"
English: yes, yeah, yeap "affirmation"
French: oui "affirmation"

It is useless to quibble any longer about the question, things are pretty obvious here.

3.3.4. ELEMENTS OF LEXICOLOGY

A- TIME

A-1. TIME'S DIVISION

Ewondo	Middle-egyptian	English
Kikiríghi/kiri	***Hrw***	Morning, day
Zang-ômôs/u-nloh dzô	***Wp.t-râ***	midday
Ún-ŋgoghè/ngongog-lè	***xAwy***	evening
à-lú	***wx.t***	Night
à-mòs	***mn.t***	day
Swanda	***mD***	week
Ngwan	***abd***	Month
U-mbú	***Rnp.t***	year
bi.sep	***smw***	

búkh	***A.t***	Minute
à-wola	***wnw.t***	Hour
À-bôkh	***rk***	Period, age, time
À tigh à-wola	***mtt***	Moment
A bogh di	***m tAy A.t***	At the moment
angoghe /ongôghe	***sf***	yesterday
Ana	***aSA***	today
o-kirí, à-kiti	***dwA.w***	tomorrow

Chart 51: Comparative Table Of Time Division

The SIGN 𓄿 A in Ewondo

If we wanted to ask a diachronic analysis of the exact phonetic nature of transliterated signs A and often discussed by Egyptologists. Some scholars consider that they are vowels because such are their achievements in Coptic (Obenga: 1993). Others believe that we are dealing with a glottal occlusive (/ ʔ /) and a pharyngeal fricative (/ ʕ /) respectively. Fortunately I am not trying to read the ancient Egyptian so this problem does not concern me in fact. What I know is what is called the vulture in an African language and what it represents.
In the vicinity of the phoneme note 𓄿 A for the sign (Egyptian vulture) the deaf * **p**, * **t**, * **k**, * **kw** preserved in hieroglyphic are effectively sounded in Ewondo.

In reality the phoneme note comes from the Negro-egyptian kh-. In fact, it can adopt in Ewondo different vocalisations:

<u>**At initial**</u>: it can be the velar ***K*** or all the forms amuïes

Example 1:

Middle-Egyptian: ***Apd*** **"**fowl"
Ewondo: /ó-kpwal/ "partridge"

Example 2:

Middle-Egyptian: As "bald; hairness etc."
Ewondo: /á-yan/ "bald; hairness etc."
Ewondo: /Ún-nso-(unloh)/ "bald headed"

<u>**At the intervocalic**</u>: it can be a high vowel ***a, e, i,*** etc.

Example 1:
Middle-Egyptian: *AxAx* "stars"
Ewondo: /á-títi/ "stars"; /mi-nkükui/ "rays";

Example 2:

Middle-Egyptian: ***DAr.t*** "scorpio"
Fang: /á-dan/ "sting"; /o-kôr / "dart, sting"; /oler(e)/ "scorpio"
French: dard "dart"

In final position:

A can be **-kh** like:
Middle-Egyptian: ***ibA*** "dance" ⇨Ewondo: /à-bokh/ "dance"
It can be, **-ŋ; t; f** fun shapes of the velar **k.**

Example 1**:**

Middle-Egyptian: AA "pile of rubble, ruins"
Ewondo: /í-tough/; /ŋgoun/ "pile of rubble, ruins"

Example 2:

Middle-Egyptian: dwA.t "lower world"
Ewondo: /í-doukh/ "restroom, toilet"

Example3:

Middle-Egyptian: kA "cow"
Ewondo: /un-ŋkáŋ/ "taurus"

It can be the weak consonant -h as well:

Middle-Egyptian: ibA "dance"
Ewondo: /í-bà/; /yebà/ "shouts of joy pushed by dancing"

But we know that in Old Egyptian **A** sometimes alternates with ***r*** as evidenced by the words "Main": ***Dr.t, wDA.t***. (Gardiner 1999: 455). This clearly shows that at the beginning (beginning of the Old Kingdom) the two signs were phonetically close. But **A** also alternates with ***n*** in some words like ***HfA.w*** and ***Hfn.w*** which both mean "snake" or even "see" which is sometimes written ***mAn*** (Gardiner, p.299). So we are in the presence of a complex sound that has nothing to do with a glottal consonant.

Fortunately, the present approach is not based on the fate of this or that symbole transliterated by Egyptology. The method is based on endogenous character analysis. The need to get equivalences between characters and letters is certainly indicated for some but it does not really make sense. The laws we can find are true for some cases never for all.

B- THE DAYS OF THE WEEK

To establish a comparison on time division between Egyptian and what is currently observed in our latitudes, we must first of all look at the link between Egyptian astronomy and what has been inherited directly from the Greco-Romans and that we still use nowadays. With colonization, the slave trades suffered by Africa in the 15th century, it is clear that the division of time was strongly influenced by colonization. The present course is therefore to go back to Egyptian astronomy and to establish the link from the scripts and see what extent the words mobilized to state the days of the week conform in every respect.

This approach confronts various technical problems including the divisions of weeks in Egypt (7days, 8days, 10days) as well as the coherence observed through multiple endogenous calendars (Bamileke, Dogon, Bamoun, Kongo, Basa'a , Akan etc.).

Fortunately, Africa still today has people in charge of monitoring and maintaining endogenous calendars. Also, despite the colonization we can always refer to what is still done nowadays and with the endogenous data, to carry out a work of serious research. In the past, this question has been the subject of much research, and it is the unscrupulous Champollion Figeac who records an interesting passage in his book on ancient Egypt.:

> *"[...] It is to Egypt that we will ask for the solution of this singular problem; and we shall learn that in our time, as in those of all antiquity, the first day of the week was that of the moon, Monday, the second was that of Mars, the third of Mercury, the fourth of Jupiter, the fifth of Venus, the sixth of Saturn, and the seventh of the sun, or the day of God; while the astronomical order of the planets was quite different: the moon, Mercury, Venus, the sun, Mars, Jupiter and Saturn, that is to say, for the denominations of the days of the week, if they were analogous, Monday, Wednesday, Friday, Sunday (day of the sun or of God), Tuesday, Thursday and Saturday. An ancient writer, Dion Cassius, gave us the key to this enigma, and learned that the Egyptians had divided the day into four parts; that each of them was under the protection of one of these pla-netes, and that every day took the name of the planet which protected the first part. Thus the first day was that of the moon, because the four parts of that day were consecrated to the four planets, the moon, Mercury, Venus, and the sun; the next day was dedicated to the four planets Mars, Jupiter, Saturn, and the moon continuing to follow the series; the third day was necessarily that of Mercury, since the planet of Mercury was the first of the four which, in the order of these stars, belonged to this day, and so on, until the end of the week. The seven days of this period just exhausted the picture of the seven planets after four consecutive rolls; and it is to be observed that one would arrive at the same order in the denominations of the days of the week, and in the same complete exhaustion of the picture of the planets, 24 times repeated, by affecting a planet every hour (...)"*[100]
>
> *(CHAMPOLLION-FIGEAC, 1839)*

Nowadays, the current Ewondo week has 7 days like that of the lunar cycle of the Egyptian week. But some sources mention in the past a week in (4d – 5d)[101] , 8 days may be even 9 days. It is therefore extremely difficult to find reliable information on the subject.

> *"[...]*
> *Philippe Laburthe-Tolra gives an account of the importance of the number 9 among the Beti people (Fang) of South-Cameroon: "In fact the number 9 plays this role among the Beti as with their neighbors; the saying elen enen, mebog ebul: with big tree, nine notches, to signify that the more one is big the more one is attacked, does not make sense, as Tsala (Dictionary, p.186) points out that if nine means an infinity. The big T'sò lasts nine days, the guinea pepper is taken in drugs by 9 grains, the staircase which goes down at Esama Ndzigi in hell counts 999 steps, etc. Nine intervenes as soon as there is a will to reach fullness, to aim for the absolute, in a magical or religious perspective; he evokes the whole. The Beti have forgotten it, but among their neighbors Duala and Bulu, this exceptional character of the number 9 was explicitly linked to the week-old tradition of eight days came perhaps to the addition of two weeks of four days as among the Tikar and the Bafut (Ritzenthaler, 1966 p.123) and linked at least among the Duala, to an accountancy based on eight (Ardener, 1956 p.48, Meillassoux, 1971 p.343), in relation to which the 9 is the transcendent number, the return to the origin. For the Bulu, according to von Hagen (1914, p.111), who claims that his informant did not have immediate Christian influence, he was stopped working every ninth day for fear of irritating Zobe nya. Mebe'e God (von Hagen 1914 pp. 39 and 88). In Minlaaba, Benedict Bikw (25.12.1967) described to me the raffia checklist reproduced by Tessmann (1913, p.308), stating that it was used to count the days and to cheat the seventh in honor of Zamba's daughter. Nine is said to be in the language Beti ebûl, put to ebulü (Abega, thesis p.144), which allows perhaps to bring it closer to the root bulu "to multiply, to be in abundance", ebûl would be "the number in large quantity The old Bulu, according to von Hagen (1914, pp. 39 and 88) claimed that the ninth day was their day and*

100. J.-J. CHAMPOLLION-FIGEAC, Egypte ancienne. Firmin Didot Frères, Editeurs, Paris, 1839, P.96.
101. Philippe LABURTHE-TOLRA, P. Initiations et sociétés secrètes au Cameroun: les mystères de la nuit. Paris: Karthala ditions, (1992), p.332. .

> *that ebulu came from their name "(Philippe Laburthe-Tolra, Initiations and Secret Societies in Cameroon pp.42-43)."*[102]

The week's progress in Ekang country depends on whether one is part of the elite folklorized by colonization or the lower floor that has kept the memory of its past. With the immeasurable loss of data I will therefore stick to the week to 7 days revisited.

The division of time to the week mesh offers the following vocalisations with colonization:

Monday: mə-tari me sòndo

Tuesday: sòndo atoä melu metàn

Wednesday: sòndo atoa melu menyi

Thursday: sòndo atoa melu melè

Friday: sòndo atoä melu bè

saturday: sòndo atoä alu dä

Sunday: sòndo

The statement to the enunciation of the days strongly remembers the influence of the colonization because the days are stated according to the term / sòndo / "Sunday".

It is therefore more appropriate to be very cautious and to set aside this system in order to use less corrupted data.

Days	Ewondo	M-E	Lexicon	Kikongo
Monday:	*Swanda alu dâ*		Moon	*ngonda*
Tuesday:	*Swanda-melú-me-bèh*	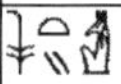	March	*kavita*
Wednesday:	*Ewûm*		Mercure	*hangia*
Thursday:	*Beletô*		Jupiter	*kizaka*
Friday:	*Nkúl wongha*		venus	*kazayu*
Saturday:	*wonghâ*	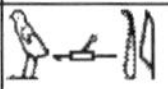	Saturn	*kimbenza*
Sunday	*Swanda*		Sun	*ntangu*

Chart 52: seven days of a week

A comparison with other time cuts allows the reading below:

For Bamiléké of Cameroon

Regardin the number of days of a week in Bamileke country it is explained that the Bamileke calendar is lunar. In the lunar cycle, the Moon being closer to the Earth, then come in order; Mercury, Venus, Sun, Mars, Jupiter and Saturn. Starting from the Moon and descending to the right of the figure this ancient order is respect. A line starts from the Moon towards Mars, then from Mars towards Mercury, from Mercury to Jupiter and so on until Sunday for the Sun. It is therefore this particular projection that has given the order of the days we still use today.

102. D. MBOCK, Kongo, Lulu, 2013, P.149

1-Nzinyam	2- Ntânbù'	3-Ntântə'	4-Ntânla'
5-Nsigha	6-Nsəmntə'	7-Nga	8-Nkotū

This division of the week in 8 days curiously reminds the octal base characterizing the cosmogony of *xmnw*. We will therefore be thinking of a logic that one day attributes to each entity of the ogdoade opposite

The Kikongo with its week of 4days (*Nkenge*, *nsona*, *nkandu* and *konzo)* presents the same logic regarding the entities of *xmnw* dealing by pairs. So we would have one day a week for each couple of entities of the creation.

This idea of 4 week-days thus joins this vestigial approach which survived in Kongo country.

	Mand[103]	**Bamana**	**Basa'a**	**Ngiemboon**	**Agni**
					Morofoe
Monday	*Kô don*	*Ntènè*	*ngwa 'njaňngumba*	*Madumze*	*Kissi*
Tuesday	*Kôba don*	*Taraata*	*ngwa um*	*Metuaze*	*Djrò*
Wednes-day	*Kunun don*	*Araaba*	*ngwa ng*	*Chatchu'nza'a*	*MMlan*
Thursday	*Bidon*	*Alamisa*	*ngwa 'mbok*	*Kuo'chua*	*Ouwouo*
Friday	*Sinin don*	*(ge)juma*	*ngwa koo'*	*Ngangna*	*Y aa*
Saturday	*SininKènè don*	*Sibiri*	*ŋgwà jôn*	*Cha'ngangna*	*Fouo*
Sunday	*Kari don*	*ka*	*ŋgwà nɔ̀y*	*Zemezeme*	*M'MIon*

Chart 53: Comparative weeks in some African lineages

Basa'a Calendar

The Basa'a language accounts for an amplitude of 4 days around today (d-day). This seems to be an atavism of the old Egyptian decimal week.

Symmetrical days with the same name.

Since 4day	*lɛn mána*
Since 3days	*mà u maa*
Other day, etc.	*nɔ̀mâ*
Yesterday	*yààni*
Today	*ŋgwà lɛ̂n*
Tomorrow	*yààni*
After tomorrow	*nɔ̀mâ*
In 3 days	*mà u maa*
In 4 days	*lɛn mána*

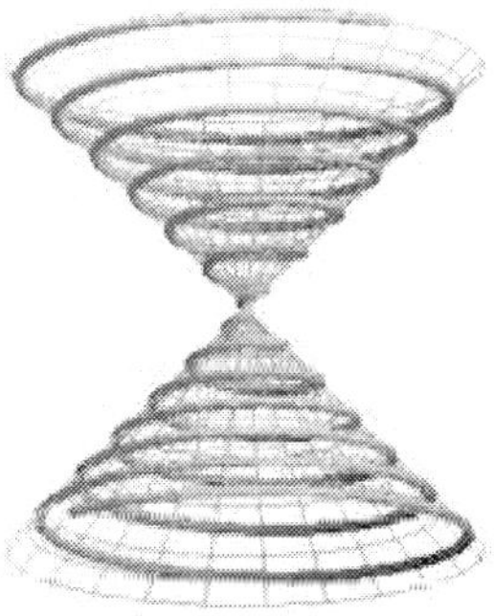

Chart 54: 9days of basa'a's week / Spirale double

The global logic that founds the old week basa'a makes it possible to understand why, at a time we evoke a 5 days week. In short, today we have more than ± 4 days.

103. Don=jour; Bi = aujourd'hui; Sinin=demain; Sininkènè= après demain

With the colonization set up an adaptation to the Western week. This step is to precede the name of each day of ŋgwà and, thus, from Monday to Sunday we have: *ŋjàŋgumbà, ùm, ŋgê, mbɔk, kɔɔ, jòn* and *nɔi.*

<u>In Mahoré</u>[104]

1- Monday = Fou mo rarou
2- Tuesday = Fou moi in
3- Wednesday= Fou mo tsano
4- Thursday= Yahowa
5- Friday = i djoumoi
6- Sunday = Fou mo tsi
7- Sunday= Fou mo vili

<u>The Akang</u>[105]

If one manages to find elements of its culture scattered everywhere, it is the Akan approach that brings the last touch to search for this time division.

Indeed, the Akan calendar is 7 days old.

The week of 7 days corresponds as everywhere on the continent to the 7 planets of the solar system as already underlined Champollion Figeac and the Agni for which they exist in their cosmogonic universe and which reign during these days.

From a technical point of view, the question comes down to understanding how to admit that seven (7) days can result in a week of eight (8) days?

Well it's simple. The Agni have two explanations for this issue. According to Akang they have a mathematical and astronomical approach related to the time division.

Mathematical approch
The Agni (and very singularly the Abron, Nzima, Baoulé, etc ...) name the week of seven days "èlèmotwê" or "elotchuê", which literally means eight days.The use of this term indicates that the number zero is used in In other words, **it means that the first day does not count in the enumeration of the days of the week.**

So, if on Sunday an Agni says "mlon ka bô ô balô, ti yi èlèmotchuè kun", that simply means in one week, from tomorrow on Monday.

To understand the symbolism of the eight-day week counting system, one must understand the system of time evaluation among the Agni (but more generally among the Akan). Indeed, in the Agni, time refers to a continuous phenomenon, it can only be measured over an interval. From this point of view the seven (7) days of the week designate spaces of time that these days cover. So we can only talk about Monday because we start counting from Sunday. Similarly we can talk about Saturday because we know that the time that covers Saturday goes to the di-sleeve. This is why the current Beti system takes Sunday / Sondò / as a reference and all other days of the week are indexed on the basis of this zero point.

It follows that the week at the Agni is composed of eight days (Sunday to Sunday) in time value. But in absolute terms, there are only seven (7) days, Sunday being counted only once in the interval concerned.
The astronomical approach

Agni time as explained above also leads to a cyclical phenomenon which assumes that everything starts at one place and usually finishes at the same place. It is this conception of time that preceded the

104. *Fou mo = jour de * Tsi = terre
105. http://www.cameroonvoice.com/news/article-news-31340.html#0jtEwfWk9WtvfrKx.99

conception of the lunar calendar by our ancestors. This is so true that in the Agni cosmogony, life is seen as a continuous and endless cycle that revolves around death and solar rebirth. This conception is formally based on very advanced astronomical knowledge. The Agni have indeed astronomical knowledge based on their observations of the sky that date back a very long time.

The meaning of figures 3, 4, 7 and 12 among the Agni

The Agni used a digital duodecimal system that is to say they counted by twelve.

They did it on the knuckles of their fingers. Try it yourself and you will realize that you get the number 12. Mathematically, it was more convenient and easier to do calculations (division, multiplication, etc.) because the number twelve is divisible by 3, 4, 6 and 12. It should also be noted that in the twelve-based system, numbers such as 12, 60 and 360 are "round numbers".
Here's how to count by slice of 12.

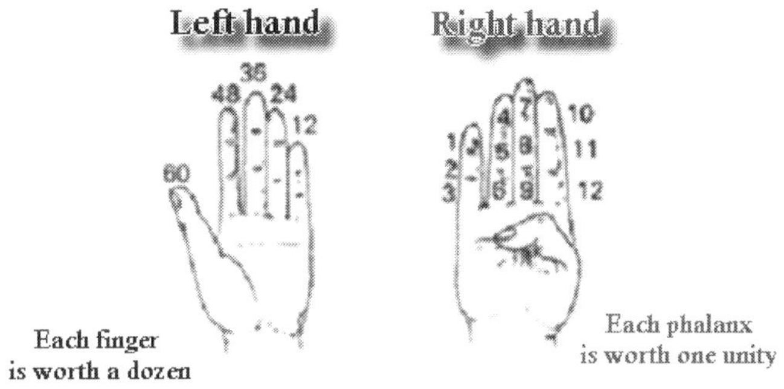

Here's how to count by slice of 12.

The numbers 3 and 4 (perfect divisors of the number 12) have a particular importance in the Agni cosmogonic universe. Indeed the number 3 is considered as the number of the man, while the number 4 is admitted as being the figure of the woman. Put together, these two figures give the number 7 which is the number of cycles renewal. You will notice that by multiplying 3 or 4 by 7 you have always got the woman's normal menstruation cycle. We also know that mating between woman (4) and man (3) leads to the renewal of humanity by the birth of a new being. We then understand by the way why the Agni do not admit the homosexuality phenomenon.

The importance of the number 7 also appears in the compulsory rituals step leading from one stage to another. For the Agni, the first step of initiation from "child-baby" to the child is at the age of 7 years. The second stage of transition from childhood to adolescence usually takes place at the age of 14, i.e 7 years after the first. Elders will also inform you that if you honestly take stock of your life, you will realize that big changes ("good" or "bad") that took place in your life took place at a rate of 7 years.
The Agni consider figures 3 and 4 as harmony figures. Thus, when a woman gives birth to three successive children of the same sex, the third of this series of children is called "N'Guessan" (or "N'San in Baoulé)." If the fourth child is the same sex then the child "N'Guessan", it is called "N'Dri." For the two children (N'Guessan and N'Dri) are attributed important mystical powers and special ceremonies are made to consecrate these powers. The "èfèwa" system, according to which the first two children of a dying mother do not receive large funerals, (we do not cry them either) also follows from the importance given to the number 3 in the number of Agni.

The week according to astronomical position

Based on their astronomical observations, the Agni discovered that it takes approximately 360 days before the cycle of the seasons and the constellations cycle. They also observed that this cycle process is punctuated by the cycle of the appearance of the moon called the SRAH. The Agni noted that it takes approximately 30 days between two moons. They have therefore logically concluded that a total of 12 moons will pass before the cycle of seasons and the constellations cycle again. You will notice that the numbers 12 and 30 are themselves perfect divisors of 360.

The cyclic year among the Agni is thus assimilated to a circle of 360 ° corresponding to the 360 days of the year.

Divided into four parts, the circle represents the four seasons mentioned above. Each of the parts in question corresponding to a 90 ° angle.

It should be noted, however, that among the Agni, there are 3 major seasons based on harvests in relation with UAE which is the essential element for life:

The flood waters, around mid-July to mid-November.

Reflux corresponding to the period of sowing. This period is around mid-November to mid-March

The low waters that correspond to the harvest period, and are around mid-March to mid-July.

But they admit however that the lunar year can be divided into 4 seasons.

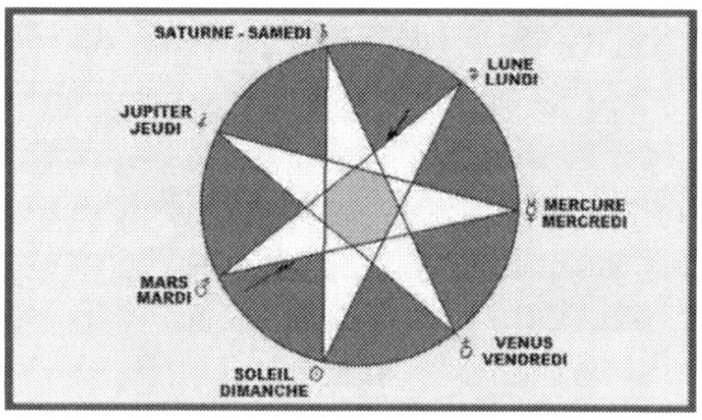

If we divide each of the four parts into three (3) subparts, we obtain a circle divided into 12 parts or houses with an angle of 30 ° each. Note that each of these twelve (12) parts represents each of the 12 months of the year, and the value of the angles represents the number of days per month, ie 30°.

Let's go further by dividing each of the twelve (12) parts of the circle into four (4) subparts. These four (4) subparts represent the four (4) weeks that each month and whose angles measure 7.5 °. As for the number of months, the reasoning is the same here and induces that a week is composed of 7.5 days is about 8 days.

In absolute numbers, we can say that the division of time in weeks, is to divide the circle of 360 ° in 48 weeks to 7.5 days each. The number of days with the decimal simply indicates that the day is measured in time space traveled from one day to another.

All this information is rich and can better understand what appears a bit like a cacophony over the days of a week.

Overall we have a week of 9 days that offers reading variants according to the reading grids and the terms of use.

C- THE MONTHS OF THE YEAR

From the moment we understand how the week is built, cutting the annual mesh becomes obvious.

Current system		New Ekang system		Km.t	
September	9th	Seble	Ngwan i-bûlu	***Hwt-Hr*** (17 sept. au 16 oct.)	3rd
October	10th	Otômle	Ngwan à-wômo	***kA Hr kA*** (17 oct. au 15 nov.).	4th
November	11th	Nomle	Ngwan à-wôme--dya	***tA aAbt*** (16 nov. au 15 dc.).	5th
December	12th	Nmfôme	Ngwan à-wôme í-bè	***pA n mxrw*** (16 dc.au 14 janv.).	6th
January	1st	Yâma	ngwan ù-su	***pA n ImnHtp*** (15 janv au 13 fv.).	7th
February	2nd	Fele	Ngwan bèh	***pA n Rnnwtt*** (14 fv. au 15 mars).	8th
March	3rd	Moas	Ngwan lâlah	***pA n xnsw*** (16 mars au 14 avril).	9th
April	4th	Akôn	Ngwan nyina	***pA n int*** (15 avril au 14 mai).	10th
May	5th	Bi-maha	Ngwan tanà	***ip ipi*** (15 mai au 13 juin).	11th
June	6th	Bi-nung	Ngwan samane	***Msw.t Ra*** (14 juin au 13 juillet).	12th
July	7th	Bi-yele	Ngwan zambgwala	***DHwty*** (19 juil.au 17 août).	1st
August	8th	Bi-kûle	Ngwan mwômo	***Pa n Ipt*** (18 août au 16 sept.)	2nd

Chart 55: Comparative table of the months

Note: The adaptation to the colonial system led to count the months of the year according to an ordinal system. In addition to this system, there is an unusual system that has its roots in borrowing from the colonial system. In the Ekang, as in the Agni (Akang), it is more accurate to approach the monthly system of weather conditions, field work and seasonal events since these were the basis of the elaboration of the months names. So we can have:

Bi-fola: the dews
Bi-wâa: the rains
Mə-bûghe: The torrents
Bi-fûa: The herons
Mə-dzîghi: ignition
Mə-tsam: Entropy

I-bèfer: seed
Be-kôn: Invocation
MAa: conformity
Bi-ɲung: malaises
Bi-lène: runoff
Bi-vouê: drying up

D THE HUMAN BODY

Artery
Ewondo: *un*-nsís
Middle-egyptian:rwD.w

*R̃^{w}t^{wh}-
**rwD.w* (M.-E).
ùn-nsis
mu-sisá (Duala)
sih (kɔ́ɔnzime)
mo-nsisá (lingala)

beard
Ewondo: ngouab-ze "goatee"
Middle-egyptian: xbsw.t

*x^{w}p^{wh}s^{w}-
xbzw.t (M.-E).
ù-ngouab-ze
nʒɛdu (Duala)
nzyel(kɔ́ɔnzime)
ndefu (lingala)

Mouth
Ewondo: à-nyú
Middle-egyptian: rA

*r^{w}̃ k^{w}-
rA (M.E)
ro (Copte)
Ra ; ri ; ro ; re
mu-dumbu(Duala)
num(kɔ́ɔnzime)
monɔkɔ(lingala)

Body part
Ewondo: i-nàm, bi-nàm
Middle-egyptian: HA.t , mn.t

*R̃^{w}n^{w}-
mn.t (M.E)
i-nàm
enama(Duala)
dipangu (tshiluba)
lemaane(Yemba)

Hair
Ewondo: í-sil,
Middle-egyptian: srt

*s^{w}R̃^{w}-
sr.t(M.E)
i-sil
ntsɔ'(Yemba)
hòŋ(ɓàsàa)
mpuo (kɔ́ɔnzime)

Ankle
Ewondo: mvongle, ves yà-ako
Middle-egyptian: xAA

*k^{wh}R̃^{w}-
xAA (M.E)
mvongle
lezēkɛ́khu(Yemba)
lìtós(ɓàsàa)
ekǎ koo(kɔ́ɔnzime)

Eyelash
Ewondo: ngoŋ d-ís
Middle-egyptian: inh

*nx-
inh (M.E)
ŋgoŋ di-ís
ŋgwuŋgwɛ́t(Yemba)
bìlèlèp(ɓàsàa)
kóho (kɔ́ɔnzime)

Heart
Ewondo: ùn-nəm
Middle-egyptian: íb, HAty

*k^{h}n^{w}-
ib (M.E)
un-nəm
á-'tə́(Yemba)
muléma(Duala)
lémé(kɔ́ɔnzime)
tʊma(proto-ekang)

Back bone
Ewondo: nkagh
Middle-egyptian:Ts, bqsw

$*k^h n^w$-

bqsw (M.E)

um-ŋkàkh ; *bi*-sûghà
ùŋka'ti (Yemba)
mo-kesa(Lingala)
i'e ɓaawa(Fulfulde)
cana(proto-ekang)

Body
Ewondo: nyô, nyol
Middle-egyptian:D.t

$*t^{wh}$-

D.t (M.E)

nyol
nnɛ̄t (Yemba)
nzóto (Lingala)
ɓandu (Fulfulde)
nyʊtʊ(proto-ekang)

Rib
Ewondo: (ibè)-mviè
Middle-egyptian:im

$*hk^w$-

im (M.E)

(ibè)-mviè
lekyɛ́mbɛ (Yemba)
lonkásá(Lingala)
wakkere(Fulfulde)
banja(proto-ekang)

Neck
Ewondo: kîŋ
Middle-egyptian: Tz

$*k^w s^w$ -

Ts/Tz (M.E)

kîŋ
ǹtɔʜŋ (Yemba)
nkíngó(Lingala)
daande(Fulfulde)
kiingo(proto-ekang)

Skull
Ewondo: í-kukü (nlô)
Middle-egyptian: Dnn.t

$*k^w k^w$ -

Dnn.t (M.E)

í-kukü-nlô
athū.(Yemba)
mokonzi(Lingala)
gumbal-hoore(Fulfulde)
*tú`(proto-ekang)

Brain
Ewondo: boo
Demotic:pi.t
Coptic: à-pè
Middle-egyptian: amm

$*hŋ^w ŋ^w$-

pi.t
(demotic)

boo
epɔŋ-ŋ̄thu.(Yemba)
bɔngɔ́(Lingala)
ŋgaandi (Fulfulde)
*bóŋ`(proto-ekang)

Thigh
Ewondo: à-bèh
Middle-egyptian: mn.t

$*k^w n^w h$-

mn.t
(M.E.)

à-bèh
eʙfəʜ (Yemba)
ebɛlɔ(Lingala)
daŋgo(Fulfulde)
*tám (proto-ekang)

Tooth
Ewondo: à-sòŋ
Middle-egyptian: Ts.t

*$k^w s^w h$-
- Ts.t (M.E.)
- à-sòŋ
- lə́-sɔŋ(Yemba)
- lĭno(Lingala)
- nyiindere(Fulfulde)
- *sòŋ(proto-ekang)

Finger
Ewondo: ù-nyú
Middle-egyptian:an.t
(Faulkner p. 43, Wb. I p. 188, 1-7)

*$t^{wh} k^w$-
- an.t (M.E.)
- ù-nyú
- ntsɔ̄'(Yemba)
- mo-sai(Lingala)
- honndu(Fulfulde)
- *mòk´(proto-ekang)

Back
Ewondo: mvús
Middle-egyptian: sA

*S^wʀ̃h-
- sA (M.E.)
- mvús
- nʙdzəʙm (Yemba)
- mbúsa (Duala)
- ɓaawo(Fulfulde)
- *buca´(proto-ekang)

Shoulder
Ewondo: tul, abáne
Middle-egyptian: wdi

*$K^w th$-
- **wdi** (M.E.)
- i-tul
- lekwɛ̄ (Yemba)
- dikata (Duala)
- balawal(Fulfulde)
- *tʊgo´(proto-ekang)

Stomach
Ewondo: o-sàŋ
Middle-egyptian: xm

*$X^w n^w h$-
- **xm** (M.E.)
- ù-sàŋ
- ajhʉ(Yemba)
- ebungá(Duala)
- reedu mawndu(Fulfulde)
- *gundu´(proto-ekang)

Buttocks
Ewondo: mə-kane
Duala: di-kandi/ ma-kandÌ
Kikongo: kandà
Middle-egyptian: xpd.w

*$k^w n^w th$-
- **xpd.w** (M.E.)
- mə-kane
- lemwɛt(Yemba)
- dikandí(Duala)
- pooɓe (Fulfulde)
- *kùn´(proto-ekang)

Liver
Ewondo: í-sekh
Middle-egyptian: mis.t/ms.t

*$S^w k^w h$-
- **mis.t** (M.E.)
- í-sekh
- lepi(Yemba)
- dibadi(Duala)
- heŋre (Fulfulde)
- *tʊkʊ ´(proto-ekang)

Knee
Ewondo: á-bóŋ
Middle-egyptian: pAd; pd (Wb I, 501, 8)

*$k^w t^w h$-
- pAd (M.E.)
- á-bóŋ
- akukwí'tí(Yemba)
- dibɔ́ngɔ(Duala)
- hofru(Fulfulde)
- *kʊmba´(proto-ekang)

Throat
Ewondo: í-ŋgōŋ
Middle-egyptian: bby.t
(Faulkner p. 82, Wb. I p. 455, 4)

***k^{w}k^{w}h-**
xx (M.E.)
í-ŋgōŋ
ǹtɔнŋ(Yemba)
mungɔngi(Duala)
kondondol(Fulfulde)
*pogo´(proto-ekang)

Hip
Ewondo: í.bokh
Middle-egyptian:msd.t
(Faulkner p. 118, Wb. II p. 153, 1-3)

***k^{w}s^{w}t^{w}h-**
msd.t (M.E.)
í-bokh
tóngo(Duala)
asaangal(Fulfulde)
*kʊndʊ´(proto-ekang)
aʙzoʔʜ(Yemba)

Intestin
Ewondo: *ùn*-yiâ
Middle-egyptian:qAb
Ewondo: "ùn-ŋkol aboum" → belly rope
(Faulkner p. 275, Wb. V p. 9, 17-20 sq)

***k^{w}r̃ ^{w}p^{w}h-**
qAb (M.E.)
ùn-ŋkol à-boum
nja (kɔ́ɔnzime)
mweya(Duala)
teteki(Fulfulde)
hanji(Hausa)

Leg
Ewondo: à-bo / à-koh
Middle-egyptian: bw
(Faulkner p. 81, Wb. I p. 450, 8-13 sq)

***k^{w}ʀ̃^{w}h-**
bw (M.E.)
à-kol
kʉ̌ (kɔ́ɔnzime)
ki-taːfaː-u(Basari)
kosŋgal(Fulfulde)
ƙafa(Hausa)

Cheek
Ewondo: à-máŋ/ mə-máŋ
Middle-egyptian: mnD.t/mnD.wt
(Faulkner p. 110, Wb. II p. 93, 10)

***k^{w}t^{w}h-**
mnD.t (M.E.)
à-máŋ
-mú (kɔ́ɔnzime)
ki-taːfaː-u (Basari)
wolo(Fulfulde)
kunci (Hausa)

Tongue
Ewondo: oyem
Middle-egyptian: ns
(Faulkner p. 271, Wb. IV p. 545, 1)

***n^{w}S^{w}h-**
ns (M.E.)
ù-yəm
jîm (kɔ́ɔnzime)
di-limbi-l (Basari)
ulimi(isiZulu)
harshe(Hausa)

Lips
Ewondo: biy –(bi-àɲou)
Middle-egyptian: sp.ty
(Faulkner p. 222, Wb. IV p. 99, 13-14 sq)

***k^{w}p^{w}h**
sp.ty (M.E.)
bi-kop
ekăn (kɔ́ɔnzime)
epopo (Duala)
udebe(isiZulu)
lēɓḕ(Hausa)

Jaws
Ewondo: mí-ʻŋkekh
Middle-egyptian: wgy.t
(Faulkner p. 71, Wb. I p. 376, 3-5 (wgj.t wgw.t) , Meeks, AL 77.1078)

***k^{w}k^{w}h-**
wgy.t (M.E.)
mí-ʻŋkekh
bɔ́ŋ(kɔ́ɔnzime)
mpe(Duala)
lombángá(Lingala)
haɓa(Hausa)

Hand
Ewondo: w-a car (ù + a)
Middle-egyptian: a
(Faulkner p. 36, Wb. I p. 156, 1-24 sq)

***n^{w}k^{wh}-**
a (M.E.)
wa
mbô(kɔ́ɔnzime)
di-á(Duala)
lobóko(Lingala)
hannu(Hausa)

Muscles
Ewondo: mi-nsón
Middle-egyptian: m.t
(Faulkner p. 120, Wb. II p. 167, 9-14)

***t^{w}k^{w}h-**
m.t (M.E.)
mi-nsón
muson(kɔ́ɔnzime)
su:x (kàllaama wolof)
mo-súni(Lingala)
tsōkā̀(Hausa)

Nose
Ewondo: mbôkh dyẃ
Middle-egyptian: Sr.t
(Faulkner p. 270, Wb. IV p. 523, 1-19 sq)

***k^{w}r̃ h-**
Sr.t (M.E.)
Dy-ẃ
muson(kɔ́ɔnzime)
bakkan(kàllaama wolof)
zolo (Lingala)
hanci(Hausa)

Navel
Ewondo: dóp
Middle-egyptian: XpAw
(Faulkner p. 201, Wb. III p. 365, 14-16)

***k^{w}t^{w}ʀ̃-**
XpA.w (M.E.)
D-óp
ɗɔ̄ɔ̀ŋ(kɔ́ɔnzime)
ʒumbax(kàllaama wolof)
kótwa(Baoulé)
cībìyā(Hausa)

Neck
Ewondo: dwa(l), á-kóp
Middle-egyptian: mHA; mkHA; HA
(Faulkner p. 119, Wb. II p. 163, 6)

***k^{w}t^{w}h-**
mkHA (M.E.)
á-kóp
ɗɔ̄lɔ̄ (kɔ́ɔnzime)
sòl(Mundang)
kɔmin(Baoulé)
sú:li(Tupuri)

Eye
Ewondo: d-ís
Middle-egyptian: ir
(Faulkner p. 25, Wb. I p. 106, 6 sq)

***t^{w}h-**
ir (M.E.)
d-ís
dî-h(kɔ́ɔnzime)
nìsí(Fali)
ɲĩ́-mà(Baoulé)
yiitere - yite(Fulfuldé)

Nail
Ewondo: dy (d-i +é)
Middle-egyptian: an.t
(Faulkner p. 43, Wb. I p. 188, 1-7)

***t^{w}kh-**
an.t (M.E.)
Dyé ; *i*-nyit
nyâ (kɔ́ɔnzime)
wu-kòsu(Guiberoua Bt)
kadyĩge (Sénoufo)
feɗeeŋgo(Fulfuldé)

Ears
Ewondo: mə-ló
Middle-egyptian: msDrwy;anxwy; bAbAw
(Faulkner p. 119, Wb. II p. 154, 13-16)

***t^{w}k^{h}-**
msDrwy (M.E.)
mə-ló; *bi*-simda
egwǎl(kɔ́ɔnzime)
toro (Soninké)
suk(Tupuri)
nofru(Fulfuldé)

Bone
Ewondo: í-vés
Middle-egyptian: qs, rDw.t
(Faulkner p. 281, Wb. V p. 68, 2-12)

$*k^w sh$-
qs (M.E.)
í-vés ; i-sûkh
jweh(kɔ́ɔnzime)
ewes(Duala)
hum(Tupuri)
*kúi(n)(proto-kang)

Toe
Ewondo: o-ɲú à-bo or o-ɲú à-ko (l) → "toe "
Middle-egyptian: Dba→(D= u-ɲú and ba =à-bo)
We find in the morpheme above the same semantaxe namely: the toe = foot finger .

Eyelids
Ewondo: biyé-bi mys
Middle-egyptian: sp.t
(Faulkner p. 119, Wb. II p. 154, 13-16)

$*k^w p^w h$-
sp.t (M.E.)
í-kop *dys*
ncémé (kɔ́ɔnzime)
akupɔ*lēzēk* (yemba)
dabool (Somali)
fatar *ido* (Hausa)

Skin
Ewondo: í-kob
Middle-egyptian: bA
(Faulkner p. 77, Wb. I p. 415, 7-11)

$*p^{wh} kh$-
bA (M.E.)
í-kop ; i-bəghə ; i-bâ
konó(kɔ́ɔnzime)
ŋ́'gúb(yemba)
kowa(Tshiluba)
isikhumba (Zulu)

Foot
Ewondo: á-bò ou à-ko
Middle-egyptian: bw
(Faulkner p. 81, Wb. I p. 450, 8-13 sq)

$*k^w \tilde{R}^w h$-
bw (M.E.)
à-ko(l)
kǔ(Kɔ́ɔnzime)
akhu(Yemba)
mguu(Swahili)
unyawo(Zulu)

Fist
Ewondo: í-kura
Middle-egyptian: fxa, xfa
(Faulkner p. 190, Wb. III p. 272, 16-19 sq)

$*k^w th$-
xfa (M.E.)
í-kura
kúra(Kɔ́ɔnzime)
bu(Yemba)
ngumi(Swahili)
juNg~o (Fulfuldé)

Chest
Ewondo: *ùn*-ŋkwkh
Middle-egyptian: gwA.t
(Faulkner p. 288, Wb. V p. 160, 7)

$*k^w k^w h$-
gwA.t (M.E.)
ùn-ŋkoukh
bɨr(Kɔ́ɔnzime)
kïùkhïù(Yemba)
wɔ̃ni(Mundang)
wiɓɓere (Fulfuldé)

Thumb
Ewondo: *un*-nôm-onyú "thumb"; metele-itam: "i'm alone"
Middle-egyptian: an.t

Lungs
Ewondo: bi-sekh bi-'nem "the viceroys of the heart" ou ùŋ-həbə ou ùŋ-vəbə" (the breath)
Middle-egyptian: wfA
(wfA Faulkner p. 60, Wb. I p. 306, 3)

***p^{wh}ʀ̃^{w}h-**

wfA (M.E.)

ùn-həbə
pupul(Kɔ́ɔnzime)
azāŋ(Yemba)
yak(Mundang)
poofɗe(Fulfuldé)

Kidneys
Ewondo: u-tám; *un*-mbáŋ u-tám
Middle-egyptian:dp.t (*m>p et t>d*)
(Faulkner p. 312, Wb. V p. 445, 13-16)

***t^{w}p^{w}h-**

dp.t (M.E.)

ùn-mbáŋ ù-tám
pupul(Kɔ́ɔnzime)
mbaŋ(Yemba)
t3sale(Mundang)
ɓoyre(Fulfuldé)

Wrinkles
Ewondo: mi-'mbwar
Middle-egyptian: HtAw
(Faulkner p. 179, Wb. III p. 182, 15)

***k^{wh}t^{w}r̃ h-**

HtA.w (M.E.)

mi-'mbwàr ; *bi*-təgh
nzóhó(Kɔ́ɔnzime)
mbhʉɛ́ŋgɔ̄ŋ(Yemba)
mstari(Swahili)
jema(Fulfuldé)

Spittle
Ewondo: mə-ndn
Middle-egyptian: nX
(Faulkner p. 139, Wb. II p. 318, 14-15)

***k^{w}n^{w}h-**

nX (M.E.)

mə-tèh
mecyel(Kɔ́ɔnzime)
məʙkiʜ (Yemba)
ma-te(Swahili)
tuuɗe(Fulfuldé)

Blood
Ewondo: mə-ki
Middle-egyptian: itH

***k^{w}jr**

***itH-**

mə-Kih
mə-Tchi
mə-tsih
mə-tih

Breast
Ewondo: mə-bèh
Middle-egyptian: bn.ty

***n^{w}ə-n^{w}-**

***bn-** M-E

mə-bèh
ébél(kɔ́ɔnzime)

Head
Ewondo: *Ún*-nloh; i-təbə
Duala: mu-lopó
Middle-egyptian: tp

***t^{w}p^{w}**

***tp** M-E

ùn-nlo
*lopo(Duala)
lû̂(kɔ́ɔnzime)

vein / artery / nerve:
Ewondo: un-nsis
Middle-egyptian: rwD.w

$*\tilde{R}^{w}t^{wh}$

***rwD.w**	*ùn*-nsis
(M-E)	mu-sisá(Duala)
	sih(kɔ́ɔnzime)
	mo-nsisá(lingala)

Belly
Ewondo: à-búm
Middle-egyptian: Ab

$*K^{w}n^{w}$-

Ab	à-bum
(M-E)	ebungá (Duala)
	mo(kɔ́ɔnzime)
	li-bumu (lingala)
	lìɓùm(ɓàsàa)

visage
Ewondo: à-sù
Middle-egyptian: Hr

$*x^{w}\tilde{r}$ -

Hr	à-sù
(M-E)	bo-só (Duala)
	mpǒm(kɔ́ɔnzime)
	elongi (lingala)
	sú(ɓàsàa)

THE ANIMALS

Animal

Ewondo:tsít or tsir "meat"
Middle-egyptian: Asr "roasted meat"

***Kir̃**

	tsit
	tsir
	tchit
Asr (M-E)	tit
	tsid
	tsir
	etc.

Bee

Ewondo: bw, mvufoŋ, fóŋ, mvem,
Middle-egyptian: bi.t

$*p^{wh}$-

bi.t	Bit; Abo, abé
	fóŋ
	bwé (bo +é)
bi.ty	bombo
	fofóŋ
	mvufóŋ

Lamb

Ewondo:*mwan ntomba*
Middle-egyptian: *sri.t*

Eagle

Ewondo:*dzü, í-fôŋ, nkuŋ, nyama-ŋkone*
Middle-egyptian: *tyw, A*

Donckey

Ewondo:*í-dzakásse, Ún-mbôkh, i-bôm, Ún-ŋkôm*
Middle-egyptian: wHm, aA

Spider

Ewondo:*á-bôbo*
Middle-egyptian: ibww, ibwy

Duck *soghe, lola, sua, lole*

Ewondo:*sokh, í-loli*
Middle-egyptian: sA, s.t, wSA.t.

Cat

Ewondo:*I-sínga, fôe*
Middle-egyptian: *sm,mi.w*

Horse

Ewondo:*I- kâb'li; un-nsisim*
Middle-egyptian: *ssm.t*

Goat

Ewondo:*kabat, kabrâ,* kaba
Middle-egyptian: *bA, anx.t, wa.ty*

Dog

Ewondo:mvú
Middle-egyptian: iw

Pig

Ewondo:ngwé
Middle-egyptian: SA, ipH

Fowl

Ewondo:*(Únôm)-kub*
Middle-egyptian: *Apd*

Crocodile

Ewondo:*nkôm; nzindi; ngán*
Middle-egyptian: *kAp.w; msH; mry.t.*

Elephant
Ewondo:*zokh*
Middle-egyptian: *Abw*

Giraffe
Ewondo:ùn-*nkókh*
Middle-egyptian: *mmy*

Frog
Ewondo:*nkóŋ*
Middle-egyptian: *abxn, pgg.t*

Owl
Ewondo:*a-kûŋ; i-kùmù*
Middle-egyptian: *km.t*
Copte: **m**

hippopotamus
Ewondo:: *ŋgÚb; zokh mendzim "elephant d'eau"*
Middle-egyptian: *xAb; db/db.t; dib*

Swallow
Ewondo:*kuleyebe, fulyebe, ndélé*
Middle-egyptian: *wr, mn.t*

Lezard:
Ewondo:*un*-nswé
Middle-egyptian: aSA

Hare
Ewondo:*o-kpweŋ*
Middle-egyptian: *sXa.t*

Lion
Ewondo:*í mbgwem*
Middle-egyptian: *mA.t*

Fly
Ewondo: *nlò, fifyè*
Middle-egyptian: aff :"àfôm: wasp hornet"
f-/f- et -f/-m

Mosquito
Ewondo: *í-nyunyoŋ* (nu +o)
Middle-egyptian: *xnws*

Sheep
Ewondo: Ú-ntómbá
Middle-egyptian: bA, waty

Bird
Ewondo: Ú-nuàn
Middle-egyptian: *mnw.t*

Panther
Ewondo: *zəe*
Middle-egyptian: *bA, Aby*

Perrot
Ewondo: *kós*
Middle-egyptian: *aSA*

guinea fowl
Ewondo:*o-kpwal*
Middle-egyptian: *Apd*

Fish
Ewondo: *kos, kwas* (ku+a)
Middle-egyptian: rm

Chip
Ewondo: *gnegnè; fifiè*
Middle-egyptian: *py* (The y explains the repetition of syllables)

Snake
Ewondo: *nyo*
Middle-egyptian: *D.t; niw.t, naw*

Monckey
Ewondo: *kwé*
Middle-egyptian: *ky*

Turtle
Ewondo: *kulu*
Middle-egyptian: *St.yw*

E- EARTH AND THE WORLD: NATURE

Air
Ewondo: í-vundou (etym. English window)
Middle-egyptian: tAw

Tree

Ewondo: í-l
Middle-egyptian: x.t

Silver
Ewondo: miáŋ, bifura
Middle-egyptian: qmA, rmn

Back of the house

Ewondo: í-su (ndá)
Middle-egyptian: hA

Branches
Ewondo: Ú-ntəm / mi-'ntəm
Middle-egyptian: smA.w

Fog
Ewondo: ú-nkúr / mi-'nkúr
Middle-egyptian: qri

Heat
Ewondo: á-yôŋ, ndúgrou, chî
Middle-egyptian: srf; tA.w

Sky
Ewondo: dzôp, yôp
Middle-egyptian: tp, p.t; pi.t

Smoke
Ewondo: ù-tsirá
Middle-egyptian: Hty

Water
Ewondo: mə-ndip, mə-nyu
Middle-egyptian: Sp; nnw, mi

Star
Ewondo: otiti
Middle-egyptian: sbA

Iron
Ewondo: í-ky / bí-ky
Middle-egyptian: aA.t

Fire
Ewondo: nduane
Middle-egyptian:DA, wdi

Leaf
Ewondo: ù-ky / á-ky
Middle-egyptian: stf.w, DrD

Flower
Ewondo: à-sàm; ù-sessàm
Middle-egyptian: sSn

River
Ewondo: nsiñ-nden; ù-nlóŋ; ù-sw
Middle-egyptian: xr.w; itr.w

Forest
Ewondo: á-fan'
Middle-egyptian: StA

Cold
Ewondo: á-vep; á-heb
Middle-egyptian: Hs; qb

Herb
Ewondo: í-logh, ù-but
Middle-egyptian: smw; Sw.t

Humanity
Ewondo: bôt; bôr; bôd
Middle-egyptian: pa.t

Island
Ewondo: í-bul, í-lóne
Middle-egyptian: iw;

Flood
Ewondo: ndne
Middle-egyptian: Ax.t; wDn.w

Lake
Ewondo: í-liba, í-tám
Middle-egyptian: mr; Hn.t

light
Ewondo: ù-mfiè; lám
Middle-egyptian: HD

Moon
Ewondo: ngwane
Middle-egyptian: iaH

Sea
Ewondo: máŋ, Ú-nkoma
Middle-egyptian: wAD-wr

World
Ewondo: sí, ndón
Middle-egyptian: s, tA

Mountain
Ewondo: u-nkôl, u-nkô
Middle-egyptian: D.w

Cloud
Ewondo: ú-nkúr / mi-ʻnkúr
Middle-egyptian: qri
Shadow
Ewondo: Ú-nsíssım
Middle-egyptian: syw.t

Gold
Ewondo: kông, ngôss, mvwat
Middle-egyptian: nb.w, rmn

Stone
Ewondo: i-nyor; à-kókh, ngokh
Middle-egyptian: iky, inr (having given the word of titulus INR supposed to be jesus of nazareth king of jews)

Seeds:
Ewondo: mvóŋ, fè
Middle-egyptian: pr.t

Rain
Ewondo: mveŋ
Middle-egyptian: mw;

Dust
Ewondo: tsiá; u-mbÚlùgh
Middle-egyptian: Sa; xmw

Root
Ewondo: ù-nkáŋ
Middle-egyptian: wAb; mnyt

Sand ù-nsəlegh
Ewondo: ù-nsəlegh
Middle-egyptian:Say

Sun
Ewondo: nlô dzôbo, viane, viène
Middle-egyptian:râ, Sw

Hearth
Ewondo: Tsia; me-tekhe
Middle-egyptian:tA

F- COLOURS

White
Ewondo: ù-mfum(lu); í-fum(lu), fəm
Middle-egyptian: HD, bAq

Black
Ewondo: nsur; í-vindi; kam
Middle-egyptian: km; Dab

Blue
Ewondo: nsine máŋ, nfîna
Middle-egyptian: xsbD; irtyw

Red
Ewondo: í-vele; à-vi
Middle-egyptian: dsr.t; dSr

Yellow
Ewondo: nfôô
Middle-egyptian: xnty; qni.t; qn.t

Green
Ewondo: zoghobon; mendzâ; í-nenegh
Middle-egyptian: wAD.t

All other colors are realized by association of ideas. In fact, translating them leads to idiomatic expressions.

Example: "persimmon: gnulif (viper color)"

The variants of green, are associated with the most expression idea

Dark green: "gnôul meki (cassava leaves)"

G- CARDINAL POINTS

Nord

Ewondo: Ú-sukü
Middle-egyptian: *mHt.t*

South

Ewondo: à-si'nkiè
Middle-egyptian: *xnty*

East

Ewondo: meküi me nlô dzop; nòmò,
Middle-egyptian: *iAb.t, Hr*

West

Ewondo: ùn-mbo-ngala; Ú-dzimi; í-mɔŋ; dibi
Middle-egyptian: *imnty; wsjr*

G- SEASONS

Short dry season
Ewondo: ó-yôn
Middle-egyptian: pr.t →(ndón = crue= wdn.w)

Long dry season
Ewondo: í-sep
Middle-egyptian: Sm.w

Short rainy season
Ewondo: Úmbu
Middle-egyptian:?

Long rainy season
Ewondo: sughu, o-khos, Su'u
Middle-egyptian: Ax.t

The temporal division of the seasons and their characteristics are summarized as follows:

Long dry season (/ i-sep / in Beti dialects) characterized by the drying of rivers and wells, drying and conservation, fishing at the dam and sowing of forest fields (I-səp), fires and bush fires, the proliferation of contagious diseases, the disappearance of certain herbs (king of herbs), the scarcity of insects. This period is from December to February.

Short rainy season (ù-mbu or ù-ŋkóm in Beti language) characterized by regular rains but their duration becomes variable, the work of preparation of fields, a climate favorable to the rebirth of life, l 'ban on hunting, greenery of nature and abundance pregnancies. This period runs from March to May.

Short dry season (ù-yôn in Beti language) characterized by cold weather, the appearance of dew on herbs, the drop in the quantity of water, limited problems for crops and harvest, inter-season crop, moisture, butterflies caterpillars, dragonflies and birds abon-dent. This is the period from June to August.

Long rainy season (Mvəŋ / Sugh'u in Beti language) characterized by sowing, high humidity, reappearance of insects and reviving vegetation, crop ripening (grows at harvest), floods and abundance rains, frequent winds and falling trees. This period runs from September to November.

The km.t calendar was based on the flood of the Nile responsible for life. The seasons have some constants like the heat or the heavy rains for the intermediate seasons peculiar to the equatorial zone it took in my opinion to be formalized to the climatic reality of the environment.

3.4. ENDOGENOUS READING METHODOLOGY

3.4.1. Scripts Reading Methodology And Practical Cases

From the African point of view, the issue of ***mDw nTr*** is based on two important pillars:
- Character reading
- And mutual understanding through the current synchrony

The exercise of reading mobilizes a certain number of tools on the technical level. The question of understanding is a paradigmatic question. So here are the levers that will be mobilized in the heuristic below.

Before reading entire sentences, it seems better to read each character well and to associate it with the next character. It is the methodological choice proconized.

The inevitable question is to always wonder what the scribe wishes to express as an idea. On this account, one must turn one's gaze to the determinative first. The question seems simple but it is not so much that because the African thought is complex, teleological and teleonomic that is to say that a produced character is by necessity and for several uses. Precision in the matrix construction of this language is a masters affair.

To determine the reading of the determinative (sign on the right here in red for example) 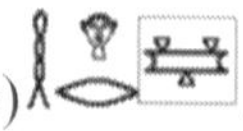.

Overall we can propose this functional diagram:

1- Recognize the different characters and name them in their language
2- Two possible cases, we recognize the right sign we call it
3- Or we do not know it. In both cases we will use the left signs to determine the phonetics in his language.
4- Put a character analysis on the left to find the link with the sign on the right.
5- Take into account syntactic and morphological constraints. if we have a ù-mbias // y (nisbe), we must take the first syllable of the word and associate it with the word read by a system of rebus (or + angel = orange)
6- Check the correct reading taking into account morpho-syntactic constraints, global logic.

Let's take a script referenced by *Faulkner p. 2, Wb. I p. 7, 15*, and try to read it from our languages. This first assumes that each one tries to read the spelling in his language and in the second time that he makes the effort to understand what an African offer says when he names the same signified. Reading in one's own language achieves the first objective, and making the effort to understand what is said by one's African brother also makes it possible to wait for second objectif.

In this case, I will use the Ewondo language demonstratively as the language of the exercise.This method is using symbols as phonograms.

3.4.2 ILLUSTRATION OF FEW CASES

Phonographic method or rebus

The African holy signs suggest three levels of approaches: phonographic, ideographic and alphabetic. We say that for reading needs, the alphabetic and phonographic approach are easy and indicated, the ideographic approach helps to understand the levels of languages and especially the esoteric information coded. Let's try to understand how the phonographic approach works. I allow myself to think that it can be approached by what is known as a rebus method.

When I have presented the topic of tonal study, I argued that suprasegmental units (tones) were not meaningless from the African point of view. On this account, we can therefore assume that each sign represents a primary tone. That is to say that if I want to read a script, I must attach to each sign of the script a monosyllabic sound that reflects the idea represented by the scribe.

Situation1: 

I recognize the right sign, here a pachyderm even a Chinese, can recognize it:

Traditional Chinese: 象 *Xiàng* "elephant"
Japanese : 象 *Zō* "elephant"
Javanese : *gajah* "elephant"
Hindi : हाथी *haathee* "elephant"

I think it is safe to say in the language devoted to the exercise the reading below:

Ewòndo: /zokh/ "elephant".

Note

However, I have a fifty-fifty chance of getting wrong about the good reading. If it's written left side elephant with the left signs then that's great for me. But this is riddle and luck. If it's not written elephant left side then it will be unlucky, i was wrong. Let's say according to the analytical abilities and the insight into the knowledge of the language I could propose a minimalist reading or a thorough reading. Did the scribe wanted to represent the pachyderm? The elephant-men caste? The defense function related to the elephant? The chief as elephant? The sacred representations of the eldest ancestor knowing that for Africans is the elephant? An aspect of the starry vault? A particularity in biology which best representation is the elephant? I think that we must be wary and humble in front of these characters. The evidence can be misleading

Situation 2: 

In order to remove doubt on the previous case, I must take into account the signs on the left. This is the most interesting case for we Africans. The signs on the left are not mere signs, they are signs loaded with an ontological dimension that only those who are used to call insiders understand.

-Sign U23 of the Gardiner's list "Chisel" in Ewondo: /ùn-mbokh/ "Chisel"
-Sign D58 "foot" in Ewondo: /à-boh/ or /à-koh/ "foot"
-Sign G43 "quail chick" in Ewondo: /kub/; /wu/ "fowl, hen"

Let's do an analysis right now.

According to the postulates of Egyptology, the sign G43 is used to tell whether the word is masculine. Here it is masculine. We will leave it aside to focus on the two other signs that constitute the root of the wordsert.

Sign U23 of the Gardiner's list "Chisel" in Ewondo: / ùn-mbokh / "Chisel"
This sign can be broken down into tonemes with non-zero signifiers in Ewondo in: *bo + *okh

Sign D58 of the Gardiner's list "foot" in Ewondo: /à-boh/ or /à-koh/ "foot"
Surprisingly we find the syllables that constitute the previous word.***Boh** and ***koh**.
It seems that the scribe has provided for both meanings.

So rigorously the good reading is done by associating to the root the prefix of gender and number:

Ewondo: /*ùn*-mbókh/ "beast of burden"

I note that by homophony, synonymy, metonymy and antonymy I can suggest other terms related to this term.
Curiously, a rigorous analysis already dismisses us from the certainties of the first case.

Situation 3 

Let's try another approach.

-Sign U23 of the Gardiner's list "Chisel" in Ewondo: /*ùn*-mbok/ "Chisel"
-Sign D58 "foot" in Ewondo: /à-boh/ in Ewondo: /i-dzô/ "member"
-Sign G43 "quail chick" in Ewondo: kub "fowl, hen"

Let's do an analysis right now.

According to the sovereign method, the sign G43 serves to say if the word is masculine. We will leave it aside to focus on the other two signs that constitute the root of the word. It is often transliterated **w** and associated with phonetics [w] we will take it for granted but I will demonstrate that a little later.

Sign U23 of the Gardiner's list "Chisel" in Ewondo: /*ùn*-mbok/ "Chisel"
Sign D58 "foot" in Ewondo: /à-boh/ in Ewondo: /i-dzô/ "member"

So rigorously the good reading is: / i-dzoué / "authority". According to the African paradigm, the elephant is associated with royalty (cf illustration N ° 7).
It is one of the representations of ***Km.t*** and therefore of the primordial ancestor of the African peoples. Hindustani assert that elephant is haathee. Could you please Remove doubts about the meaning of ***hat*** in English?

Hindi : हाथी *haathee* "elephant"

Middle-Egyptian: ***HAty*** " heart; thought "
Middle-Egyptian: ***HDt*** " Hedjet, the White Crown"

Is the crown a type of hat? Think about it.

Associating the root of the word with the gender prefix W, we have a blocking since in the specific case the gender prefix is W.
The word found / i-dzoué / begins with [**i-**] whereas the reading should be done with a word beginning with [w].

So we can stop there or else transform / i-dzoué / in / *ùn*-ndzou / nominative keeping the same semantics.

Ewondo: /*ùn*-ndzou/ "the authority" (the one who exercises the authority, the depositary)
Ewondo: /i-dzoué/ "the authority" (the owner of the authority)

These two political notions in Africa are not identical. The power is feminine essence but the exercise can be experienced by the depositary who is a man normally the brother of the power owner.

Phonetic proximity: The term / i-dzoué / allows the following reconciliations.

Ewondo: /dy-ò/ "The nose" (implied the trunk)
Ewondo: /dy-o/ "the name" "the speech" chief's symbol.
Ewondo: da-oè →/dwé/ "the palm of the hand"
Middle-Egyptian: *wr mdw Smaw* "the utmost"
Ewondo: /nda/ "house" because au of the sign D58
Etc.

Note

I would like to point out here that where many, including Egyptology, saw an elephant, the African understands that his ancestors may have recorded the idea of authority or the country for whom the African paradigm is known.
That is why in front of these signs and certainties we advocate modesty and humility.

Middle-egyptian: ***wr mdw Smaw*** "The 10 greats of upper-Egypt"

If we understood that power is symbolized by the hand / dwé /, we would not have translated hand by 10 here. We must therefore translate this: "the highest authority of Upper Egypt". Does that makes more sense? Doesn't it?

Situation 4 

Let's try another approach: In this case I will associate the signs with a rebus method. This method is the one that is closest to what Egyptology uses.

- Sign U23 of the Gardiner's list "Chisel" in Ewondo: /ùn-mbok/ "Chisel"
- Sign D58 "foot" in Ewondo: /à-boh/ in Ewondo: /i-dzô/ "member"
-Sign G43 "quail chick" in Ewondo: kub "fowl, hen"

Let's do an analysis right now.

The sign G43 transliterated **w** serves as a substantive discriminator. He informs us that the gender prefix is masuline here. We will leave it aside to focus on the other two signs that constitute the root of the word.

Sign U23 of the Gardiner's list "Chisel" in Ewondo: /*ùn*-mbok/ "Chisel"
Sign D58 "foot" in Ewondo: /à-boh/ in Ewondo: /i-dzô/ "member"

According to the fact that the two unit syllables are different in this count, the good reading is obtained by association of the two syllables *dzou + *koh.

This association would therefore lead to the creation of this kind of word:
ndzouko "elephant"

A word that in Ewondo, from the original beer language gives by apocoping the final vowel.

Ewondo: *Zokh.* **Basa'a:** *dʒɔ̀k* **lingala**:*nzɔku*
Duala:nʒɔu **kikongo**: nzau**yemba**: евʃʉнон

Ewondo: ù-*ndzoukh ou* ù-*ndzoughù* "effort"

Note: the item elephant also existe in Ewondo as /i-lɔbàn/ "the biter".
This word for *biter* is a metonymical word to name the chief. Let's remind that the real chief is a panther/leopard. This point enlights why the chief is a biter. As we can see below the morpheme for panther and elephant are sharing the same root ***Ab-***.

Middle-Egyptian:  ***Aby*** "panther; leopard"
Chinese: 豹 Bào "leopard"
Middle-Egyptian:  ***Ab.w*** "elephant"

All these words are close to the signified ***elephant***.
But when associating the root the gender prefix W, we observe an incompatibility which constitutes a blocking due to the fact that, the gender prefix is W. The script could have read */zokh/* but the constraints put in place by the scribe impose to take a meaning beginning with sound [u] or [w] and not sound [i] or [a].

The term / zokh / being devoid of a prefix of gender and number, the technical trick is to enter a scheme of accordances including using the possessive to determine its gender. It appears by the way that the term / zokh / is of a feminine gender because one says / zokh dy-àmà / "my elephant" and not / zokh w-àmà /.

So here the possible good readings are / *ùm*-bokh / or / *ùn*-ndzoué / or / *ùn*-ndzughu / "the pain" recalling the idea of a beast of burden. Recalling above all the role assigned to the leader as Man of service ***Hm ntr***. These words can accept antonyms, metonyms, synonyms and homophones.

Situation 5 

Let's try another reading: In this case I will associate the signs with a rebus method once more.

- Sign U23 of the Gardiner's list "Chisel" in Ewondo: /ùn-mbok/ "Chisel"
- Sign D58 "foot" in Ewondo: /à-boh/ in Ewondo: /i-dzô/ "member"
- Sign G43 "quail chick" in Ewondo: kub "fowl, hen"

Sign U23 of the Gardiner's list "Chisel" in Ewondo: /i-yô /"a sting"
Sign D58 of the Gardiner's list "foot" in Ewondo: /à-koh/ "foot, leg"

Because the two unit syllables are different then in this count, the good reading is obtained by association of the two syllables *yô + *koh.

This association would therefore lead to the word:

Ewondo: /i-yokh/ "Curse" hidden meaning related to sex.
Ewondo: /i-yokh/ "Who takes strength"
Ewondo: /à-yokh / "Authority, severity, brave, brutal, violent"

Situation 6 

Let's try another approach: In this case I will associate the signs with a rebus method once again.

- Sign U23 of the Gardiner's list "Chisel" in Ewondo: /ù-mbok/ "Chisel"
- Sign D58 "foot" in Ewondo: /à-boh/ in Ewondo: /i-dzô/ "member"
- Sign G43 "quail chick" in Ewondo: kub "fowl, hen"

Sign U23 "Ciseau" in Ewondo: /i-vokho /"enlarger"
Sign D58 of the Gardiner's list "foot" in Ewondo: /à-koh/ "foot, leg"

Due to the fact that the two unit syllables are different, the good reading is obtained by association of the two syllables *vokh + *koh.

This association would therefore lead to a type word as:

Ewondo: /i-vogho/ "big, mastodon, imposing, massive"
Ewondo: /i-vyəkh/ "very old; old-world"
Ewondo: /vyoukh/ "very black"

If we change the meaning of the sign U23 just by a new synonym we are able to generate a new word again of our basic vocabulary.

Sign U23 "Chisel; scissors" in Ewondo: /i-fokh /"expander; under-reamer"
Sign D58 of the Gardiner's list "foot" in Ewondo: /à-koh/ "foot; leg"

According to the fact that the two unit syllables are different to this count, the good reading is obtained by association of the two syllables *fokh + *koh.

This association would therefore lead to the realisation of a new type of word:

Ewondo: /*ùn*-mfokh/ "edentate; toothless" refering to ***Hr***
Ewondo: /fokh'o/ "The first" refering to ***Hr***

Conclusion.

We could continue to analyze this script and demonstrate all the wealth of information it offers to who understands the meaning of these small drawings from the endogenous point of view.
We have seen how words are created from the signs and meanings associated with them. What seemed to be an elephant at first sight is not so much when you push a little bit the analysis.

Principe:
Let's try to better understand all this beyond impressions and assumptions. Let's try to understand how this spreads around the continent.

The item Ewondo: / ùn-dzoué / "authority" obtained case N°3 can rigorously offer by synonymy the term Ewondo: / i-mbârà / "The Scary"; Ewondo: / á-mbàrà / "to be afraid".
We can also have Ewondo: /wâŋ/ "the scary" leading to:

Chinese:王 wáng "king"
Chinese: 彎刀 Wān dāo "machete" **Ewondo**: wàŋ ndà "house's fear"
Middle-Egyptian: ***mAa.t***
Toupouri:wang "chief, supra-lineage authority"
Ewondo: / wàŋ'à / "scary, warrior; king; chief etc."

Note: I want to undeline that:

Ewondo: / á-wàŋ / "to crawl" that's why the perfect king is a snake or a worm.
Middle-Egyptian: ***wsir***
Middle-Egyptian: ***wsir***

Note: if for others worm and warm do not share anything together we can underline here that the signified is the same, the sun for warm and the worm as a creeping.

Aside the preceding note above, we can therefore read these scripts above:

Chinese:王 wáng "king"
Ewondo: / wàŋ / "scary, authority; king; chief etc."
Ewondo: / kîŋ / "neck; ; neck of canines; voice"
English: king

Sign U23 "scissors; chisel" in Ewondo: /i-mbar/"ax; axe, hachet"; /i-mar/ "flat" chief's traditionnal tools
Sign D58 "foot" in Ewondo: /râ/ "the eldest ancestor"

But the word /* mbarã / is used for example in Zandé language to designate an elephant:

Zandé: mbarã "elephant"
Ewondo: /i-mbâra/ "the scary; the chief"

Zandé: gúnd "fear, scare"
Ewondo: /à-ngóndò/ "fear, scare, horror"

Note : Among African lineages we have some call *Bambara*. But the real meaning of those people is ☞*ba*-Mbarã "those of the elephant" as you can see it above.

Curiously, when reading the script of the first pharaoh *nr-mr*, Egyptology mobilizes the vocalization ***mr*** associated to the sign U23. Proof is therefore made that transliteration is useful but limiting and oriented vocalisations.

This example tells us at least about two things:

- The script was not about the elephant but as we saw it with the different examples.
- But moreover, the descriptors of elephant reality are mobilized everywhere in Africa in the shape of different signifiers. We thus conclude simply that, the different words to designate the elephant are in fact only synonyms whose use has been lost due to the slave raids of the XVth century and the Arab-Berber migration thrust of the 7th century but also poor vocabulary of the languages speakers.

Only, in the case above, the root / * mbara / does not serve only to designate the fear and therefore the authority in my language for example. It also serves to designate something else by homophony then by synonymy or even by antonymy and metonymy.

Meaning	Homophones	Synonyms
What is flat	/*mbar-/	*bame; bôme; zən;bôkh; yin; tsel etc.
What is wrinkled	/*mbwar-/	*ntômdo
Native ax	/*mbar-/	ù-vôn
The wet	/*mbar-/	bwé; bole; asep; loghlokh; wogh; logh
clothing	/*mbwar-/	fer tô; yem tô; fer ôkan
Cover	/*mburi/	vîr; burbe; bur-bur; vîrbe; vîr; yam etc.
To be toothless	/*mbur-/	byam.; mfokh; i-ŋgoleŋgole

Chart 56: Example of Homophones and Synonyms related with mbar

And yet again, each word in the series of homophones will admit in its semantic field a new lexicon based on synonymy, metonymy and homophony. All these words are likely to be mobilized to name the signified elephant.

We therefore observe that the reconstruction of predialectal ancestors previously experienced is biased, since two apparently phonetically different languages operate in realities by lexicographic choices based on synonymy. By the way, the difficulty of understanding each other is then exacerbated by also a typology inherited from the Negro-Egyptian.

One could know a little bit of African thought to say that kulu "foot" the primordial ancestor is also called ***Ra*** in its cosmosgraphic acceptation. Also know that the chisel allows to cut / à-bakh /. So we would have understood the meaning of the word zandé: mbarã "elephant".

Small Illustration about /**mbar-*/ and /ù-vôn/ "ax; axe"

> *"**MVÔÑE**: (h) n.l, pl. The mvôñe. Odoriferous plant with pleasant odor (elokhmesè). See another similar plant: amvôñ. Formerly they were worn on the neck to perfume. Le nyôkh (fô) this plant is eatable. Syn.: asam."[1]*
>
> *(GALLEY, 1964)*

So the word / *ùn*-mbar / "wrinkle", "ax"; "Authority" is associated with the back of the neck as well, because a plant / mvôñe / homophone of / *ù*-vôn / "ax".

Note: The English word ax recall the Ewondo word /à-sà/ "splitter" /i-sià/ "father, chief" and in spanish *hacha* "ax". Does this enlight more why *axes* are symbols of some peoples?
The famous tomahawk of Indians (Native American) or the Vickings axes. Think about it…

Vicking axes

Indians tomahawks

French coat of arms

Reminder: 

As we have seen above the ideas of **ax** and **foot** are linked. May we validate that with another language :

Chinese : 鏚 qi "battle ax" ☞**Ewondo**: /*à*-ko/ "foot; " ☞**Ewondo**: /i-kié/ "iron"

Chinese : 斧 Fǔ "ax; " ☞**Ewondo**: /fè/ "machet" ☞**Ewondo**: /i-fè/ "splitter"

Verification by example: Synonymy and metonymy

Ewondo: /*ùn*-mbwar/ "wrinkled"
Ewondo: /dol/ "neck"

This meaning can also be used to designate the elephant as the place that is wrinkled like the nape of the neck /dol/. This word /dol/ is an ananym of /lod/ in Ewondo meaning "to pass" justified by the sign D58.

1. S. GALLEY, Fang-French dictionary. and French Fang followed by a fang grammar. Neuchâtel: ditions Henri Messeiller, 1964, p.216.

So, without knowing the word above, one can, due to the context, knowing the mechanics of African linguistic variants, understand the purpose here. Sango, the national language of RCA, confirms this statement.

Sango: *dòlì* "elephant"
Ewondo: /*dôl*bo/ "the sommit" by metonymy, the chief with /bo/ "foot"

Was this word used in a former Ewondo synchrony to designate an elephant? At first glance no. But in reality it's the opposite. To understand all of this just take our basic method.

-Sign U23 of the Gardiner's list "sting" in Ewondo: /? /"chisel"
-Sign D58 of the Gardiner's list "foot" in Ewondo: /à-kol/ "foot"
-Sign G43 of the Gardiner's list "quail chick" in Ewondo: /kub/ "fowl"

Sign U23 of the Gardiner's list "chisel" in Ewondo: /? /"chisel"
Sign D58 of the Gardiner's list "foot" in Ewondo: /à-kol/ "foot, leg"

The word proposed by the Sango is:

Sango: dòlì "elephant"

Let's try a reconstruction

It is thus obtained by the association of the following tonemes: *dokh + *koli.
The Ewondo being a beer language and the Sango a bere tongue, the Ewondo makes the apocope of the final vowel. In fact the association of Ewondo is a type: *dokh + *kol.

The word Sango allowed two things: to regenerate and understand which word was mobilized by the Sango to get the word *doli* and especially the meaning and the use of the word which we lost the use.
Ewondo: /-dokh/ "hollow"[2]
The sign makes it easy to hollowness. I therefore postulate that we could call the elephant / dɔlɔ / because this word has been used in the past to name the elephant and has certainly fallen into disuse.

Curiously some animals of the African sacred bestiary are named in Ewondo / doro / pork, baboon etc.

What's going on?
The sign as has already been pointed out is used to name the foot /*kol/ therefore since everyone has understood the principle, it is mobilized in the script as a hidden sense of the determinative .

What are these hidden meanings of ?

There are full of hidden meanings for the word foot. But one of the hidden meanings is where mushrooms are plentiful. It's called / i-kol /> / e-kol /. So we have the idea: many, abundant, numerous etc. If we take many, abundant, which are the synonyms offered by Ewondo language?

Synonyms of Many	Synonyms of Many
i-nẑônẑôkh; ikol, aboui	*bul; beghbe; byerbe;*

2. S. GALLEY, Fang-French dictionary. and French Fang followed by a fang grammar. Neuchâtel: ditions Henri Messeiller, 1964, P.446.

kume; narbe; nare	*menbe; men; ŋgunbe*
i sam, i-ŋgaŋgam; burbe,	*ŋgambe; kum; be*
ŋgune; mbyerbe etc.	*Etc.*

Chart 57: Synonyms of many in Ewondo language

Middle-egyptian: *Abw* "elephant"
Coptic language: ⲉⲃⲟⲩ *ebow* "elephant"
Ewondo: /à-bui/ "many, numerous"; the chief represents the multitude.
Ewondo: /à-boh/; /à-koh/ "foot"
Ewondo: /à-koh/ "foot" ☞**English** : ax

Ewondo: /*mbar-/ ; /ù-vôn/ « Native ax »

- The sign D58 of the foot that we can vocalize /à-boh/ is used to translate the hidden meaning / à-boui / "many" by homophony.
- The sign D58 of the foot /à-koh/ is used to translate the hidden meaning of /-kol/ "many; manifold" by homophony

By surreptitiously relying on the transliterated coptic version, Egyptologists' unique meaning excludes the richness of the other possibilities offered by synonymy and metonymy in the context of the practice of other African linguistic variants.

A demerit point ought to be pointed however.

Middle-egyptian: ***Abw*** "elephant"
Copte: ⲉⲃⲟⲩ *ebow* "elephant"
Ewondo: /à-boh/ "foot"

The reading /à-boh/ here is an obvious meaning and therefore false the analysis a little bit. Since the sign D58 is supposed to be a hidden sense related to the determinative .

Thus, if during the reading we understood that the sign of the foot must be associated with the determinative , on this account, it offers the word Ewondo: / à-tín /. So we do not just deal with a foot here. We must understand that, we are referring to the legs of the elephant.

Homophony

Ewondo: /*tin/ "elephant foot"
Ewondo: /tin/ "parent tree, ancestor"

Applying this principle here we realize that the script is alluding to the eldest ancestor of the African people.

When a person has elephantiasis[3], this disease gives rise to hypertrophy of the genitals (bursa) and lower limbs .

---

3. Elephantiasis refers to a significant increase of the volume (several times the normal size) of a limb or part of the body. It is a hard, chronic lymphoedema, which is an effusion of the lymph outside the lymphatic system, into the subcutaneous tissues.
In tropical countries, this is a chronic and serious complication of lymphatic filariasis, a parasitic disease that in 2007 concerned more than 120 million human beings, a third of them on the African continent. The disease is caused by a nematode worm, whose larvae are transmitted by mosquito bite.

Middle-egyptian: *Aby* "panther"
Ewondo: / à-bya/ "testicular elephantiasis"
Ewondo: / mə-tin/ "feet elephantiasis"

The transliteration for ***Aby*** panther of Egyptology is associated in Ewondo with the elephantiasis of testes / à-bya /. We also see that the feet can be associated with the legs of the elephant following a pathology.
Is it a coincidence if this pathology bears the name of the elephant in european medecine as in African languages? I think that i have my own idea about it but I will keep it for myself.

<u>Elephant in the African language variants.</u>

Since we understand the mechanics of moving from one linguistic variant to another, at least I hope so, let's see if you are insightful:

Hausa: *gīwā* "elephant"; **Pulaar**: *Ñiiwa* "elephant"
Ewondo: zokh "elephant"

Take the case of the words above. At the reading of these three words, the right-thinking people come and inform you that these three terms are not identical and therefore your languages are different and that's why you belong to different ethnic groups.

They offer you a scholarly classification of what they call dialects, patois or vernacular languages, idioms, etc. and that you negroes are required to teach your children happily because they know more than you do with their complicated methods.

Question?

How comes that, speaking Ewondo language, I do say /wa/ for hand but the name of the elephant **Pulaar,** speakers said *Ñiiwa,* the Hausa said *gīwā* "elephant"?

<u>Reply from Greenberg</u>:

- First of all, the hand is not the elephant ok?
- Secondarily, you do not speak the same language. A fortuitous phonetic proximity without any doubt.

I call for help the scribe who recorded 6,000 years ago that in the African language, the word hand and the word proboscis (elephant) are one and the same word:

Middle-egyptian: ***Dr.t*** "hand, trunk, handle"[4]
Middle-egyptian: ***Dr.t*** "hand, trunk, horn; handle"
Middle-egyptian: ***Dr.t*** "hand, trunk, handle"
Middle-egyptian: ***Dr.t*** "hand; trunk, probosis, horn; handle"
Middle-egyptian: ***Dr.t*** "hand; trunk, probosis, horn; handle"
Ewondo: /i-dùrù/; /dwé/; /wa/ "hand"

Since the trunk is associated by metonymy with the elephant, this situation explains why here the word hand and the word elephant are interchangeable in the unique African language, we mean phonetically and semantically. This further confirms the idea that Africa

4. (FAULKNER p. 323, Wb. V p. 580, 3-15 sq.)

has a unique language that has lost its mastery and good use. The smart guys will tell me that the trunk and the hand are not the same word in their language. So I'm saying that you're just using synonymous forms.

I already hear doubts so far. Yes I know, it was a coincidence if I could explain that. So let's try something else:

Middle-egyptian: ***wsir***
Hausa: *gīwā* "elephant"

If I take the arbitrary transliteration ***wsir***, it offers to reading a word of the Kweke branch. In fact. If I invert the phonemes to find the word of the Kekwe branch and I have the consonants skeleton ***siwr***. Does this shock you? If so, I suggest this evolution diachronically:

siwr > *siwl* > *siwh* > *siwa* > *gīwā* >*hiwa* > *iwa* > *wa*

Middle-egyptian: ***wsir***
Hausa: *gīwā* "elephant"
Ewondo: /wa/ "hand "
Ewondo: /ùn-ŋgwal/ "trunk"

Is it also luckily if I get that?
The horn being the elephant's nose also serves as his hand.

Ewondo: /dy-wé/ "nose" (**d** nomber prefix and **i** gender prefix)
Ewondo: /wa/; /wo/ "hand" example /wo i-sié/ "work quality"

If I say finally that ***wsir*** is the elephant and I say he is / wa/ "the unique" homophone of /wa/ "hand" It's okay or not?

If it's shocking it's not a problem but the scribe saw things like i do.

Honestly, I do not think it will shock. All roads lead to Ngomdzap although Ngomdzap is not everywhere. I-dili fas…

Let's say I was lucky. So let's take another variant of the unique African language, Somali which according to the SAVANTS has nothing to do with Ewondo taking regarding their official classifications:

Somali: maroodi "elephant"
Ewondo: zokh "elephant"

Phonetically I already know, they are two different languages, do not heap up even more. So let's take a sibling language from Africa which has the particularity to maintain a close relationship with the African mother-language through its Sankrit but especially Greco-Latin roots.

French: *maraude; marauder; maraud* "Theft of food, especially fruit, vegetables, poultry, in the fields or on a farm."
English: maraud; marauding

French: *remède "drug"*
English: *remedy* "drug"

So if I understood everything correctly, the word *maraude* (r) of French and the word *maroodi* of Somali share a kind of phonetic proximity. So these two languages are not as far apart as this is presented in the books of scholars. Surprisingly French does not appear alongside Somali in the classification of Greenberg[5].
Let's try to see if the phonetic link is genetic between *marauder* and *elephant* is found in the triangulation Ewondo, Somali, French.

***THEFT: nzu** (h), wup (h).*"[6]

(GALLEY, 1964)

Surprisingly, I find here again the confirmation of my words. It seems that Men were born in Africa, isn't it? It reassures me so …

Question: does the word *maroodi* of Somali have an equivalent in Ewondo?

As we have said before, it's the same language so the answer is automatically yes.

Fang: /à-mara/; /mare/ "To fear oneself, to flee each other, to fear, to flee, etc.".
Ewondo: di-mbara >i-mbara "scary"

Note: The presence of an epenthetic **b** is found between fang, Somali and Ewondo. As a general rule, the appearance within a word of a non-etymological b is said to be epenthetic. We often find that it occurs every time we have: **m** and **r** (or **m** and **l**).

The epenthesis of **b** also occurs in Egyptian, Coptic and African languages under the same conditions: mr -> m**b**r. We must link the origin of this phenomenon b said epenthetic in the evolution of variants of the unique african language the **Negro-egyptian**: k^w>mb> m

If the Ewondo use the affix ***di-*** in prefixal position (prefix of gender and number), the same affix by grammaticalization is suffixed in Somali. Just remind Kékwé and Kweké branches of the negro-egyptian.
So that explains in Somalia the ***-di*** in suffixal position of the root ***mar-**.

Once again the word somali *maroodi* is very well known in Ewondo it's the same word used in the **Zand:** mbarã "elephant" that we have already peeled back in previous layers.

What about remedy?
It is simple *remedy* as *remède* in French are the kekwé branch equivalent words of the marauder of the kweke branch of negro-Egytian.
Those words are anagrams. They are linked by the fact that the eldest ancestor is the chief as the keeper /mbâlà/.
The word for remedy in Ewondo is:

Ewondo: /*mə*-bàlà/ "remedy" and **Ewondo**: /*ùn*-mbâlà/ "keeper, shepherd"
Other verification:

And what do the Middle-Egyptian scripts say? Do they make a connection between hand, trunk and steal? So you just have to notice a drawing of the hand in the three meanings.

Middle-egyptian:  ***dAr*** "to strip; to dispossess"
Middle-egyptian:  ***dr.t*** "trunk; proboscis"
Middle-egyptian:  ***siwH*** "to loot, to steal"

5. GREENBERG J. H., Languages of Africa, La Hague, The Netherlands, Mouton & Co, 1966.
6. S. GALLEY, Fang-French dictionary. and French Fang followed by a fang grammar. Neuchâtel: ditions Henri Messeiller, 1964, P.556.

I can not be more royalist than kings or more catholic than the Pope as you prefer. I therefore step aside from this science which is based on impressions and not on absolute truth.
I suggest that we continue our momentum to try to understand vocalizations that seem different to us but that are associated with the same meaning in our example.

In the Kwa languages spoken in West Africa we have:

Kwa: sro "elephant"
Ewondo: zokh "elephant"

The term *sro* of the linguistic variant kwa offers the phonetic proximity with the Ewondo: / * nsor- / "virile member" or / * "sor / "scratched" these two meanings refer us to the one who calls himself by the scholars ***wsir*** i.e the leader in our languages.

I noted in this script that its according to the endogenous model referred to the idea of leader, of authority in Case No. 3. From the point of view of the Ewondo, the Kwa speakers thus identify the elephant by its phallus [7]. By metonymy and antonymy they identify ***wsir,*** the one whose phallus is said to have been eaten by the catfish with the elephant.

We find out for sure with the words soninke or malinke languages spoken in Mali:

Malink: fɔrɔ "phallus" ☞**Ewondo:** /fara/ "priest, panties, string; picker"
Wolof: awra "phallus" ☞**Ewondo:**/awura/ "drop of sexual energy"
Soninké: xonto "phallus" ☞**Ewondo:** /*ùn*-nsot/ "phallus"
Bamana: wulu"phallus" ☞**Ewondo:** /à-wulu/ "to walk".

Other meanings that sound phonetically different from the Ewondo:

Kimongo: Nnamba "elephant"
Ewondo: /á-nàmbà/ "to touch" we find again the link with the hand .
Ewondo: Zamba "eldest ancestor"
Duala: /nyambè/ "eldest ancestor"

We find in this vocalization offered by the Kimongo the root ***amba*** /Amma, included in (Nyambé, Zamba, Nzambi, Zama, Nzambé etc.) related to the eldest ancestor of african people ***wsir***.

Ngiemboon: Chwoon "elephant"
Ewondo: sɔŋ/tumb/; /*sóŋ/ "hindering" synonym of /ùn-mbokh/ a drudge dixit (case N°2).
Basa'a: sɔ'ɔŋ "the father" ***wsir***.
Agni: È-swin "elephant"

Wolof: Ñay /ñey "elephant"
Ewondo: /Na/ "mother"
!Xóõ: ǂ'àn "phallus"

isiXhosa: indlovu "elephant" ***wsir***. (v=w; l=r; d=s)
Ewondo: /i-nda lo vyoukh/ "The ancient's house"

7. The phallus of an elephant can reach on average 1.95 m, ie the size of a man who ate a lot of soup.

The arbitrary transliteration ***wsir*** simply gives in Isizulu the consonants skeleton ***idlv.*** The word **isiXhosa**: indlovu is said in Ewondo /i-nda lo vyoukh/. Confirmed by ***wsir.*** Which can be interpreted visually as the seat of the ancestor.

Protect me from the wrath of scholars for whom some languages like **Xhosa** would be clicks languages[8]. They therefore have nothing to deal with the so-called Ekang languages, which are falsely described as Bantu. So if I could understand this word it's just because of the contamination. Some "Bantu" languages nearby have developed clicks; the most famous are those of the Nguni language group, including Xhosa, Zulu and Sotho.

Let's take the example above:

> **!Xóõ:** ǂ'àn "phallus"
> **Wolof:** Ñay /ñey "elephant"

The! **Xóõ** language is one of the Khoisan languages called clicks languages. Wolof suggests here by consonantal metathesis the same word by inversion of phonemes. We said that phallus is associated with the elephant.
How, then, can the Xóõ "Martian click language" by excellence shares the same lexical and semantic field with other African languages? Why is there a semantax that associates sex[9] to the elephant, the woman and the man? Are not we talking about elephant-woman among other things?

Answer:

Middle-egyptian:

At the time when the scribes wrote this, were the Martians already in southern Africa? Male and female sex clearly bear the same name in our language. If I name the elephant by the male sex, I name the female sex automatically.

As.t and ***wsjr*** are the male and female figures of the throne that may be occuped by the holder of the power or by the depositary (the mother of the country).

We then read this script /na/; /nya/; /maâ/ etc. for the mum.
The word /Na/; /maâ/ is associated to the Panther ***Aby***
as power holder in the royal charge. Should I have to remind that ***Ab.w*** "elephant" design and is the depository?

African supralignatory political authority: "Uncle of the country"

> **Wolof:** *nday* ji reew "Uncle of the country" litt. The house of Rê wu.
> **Seereer:** *Maad* xo:x "witness of the social contract; Uncle of the country"

---

8. The "click" languages in southern Africa: the example of xhosa, Khoisan languages such as! Xóõ,! Xũ (! Kung), zhu|'hõasi (Ju | 'hoan) and nama. .
In phonetics, a click is a sound produced with the tongue or the lips without the help of the lungs (it belongs to the non-pulmonary consonants, as well as the glottalized, injective and ejective consonants).
9. In the Kamasutra, man is divided into three classes, namely: the hare-man, the bull-man, and the **horse-man**, according to the greatness of his Lingam "phallus."
The woman too, according to the depth of her yoni "vagina", is a doe, a mare, or a **female elephant**.
The yoni (in Sanskrit योनि / yoni ("place")), in Hinduism, refers to the female genital organ (matrix or vulva); he is the symbol of feminine energy called shakti.
If for Hindu the Item YONI is the female genital, it refers to the word Ewondo /i-yon/ "what makes crying" for Ekang people the female genital is /i-bon/ "what makes barking"

Kongo: *Ma*m nsi "Uncle of the country, depository of political power"
Kinyarwanda: *Mua*mi "Uncle of the country, supreme political authority"
Yombé: Fumu à maka*nda* "chief, supralineage authority"
Mooré: Moro Naaba "chief, supralineage authority"
Maba: Kamkalak "chief, supralineage authority"
Guemba: Natche*ma* "chief, supralineage authority"
Pulaar: L*aam*do; Laamido; Ardo "chief, supralineage authority"
Toupouri:wang "chief, supralineage authority"
Kera: Koum*na* "chief, supralineage authority"
Beti: /*ùn* -ŋKukuma/ "chief, supralineage authority"
Malink: Fa*ama*; *Ma*nsa "chief, supralineage authority"

The words related to royalty are often related to weaving, spinning etc. The only animal that weaves a spiral is the tarantula. That's why this animal is sacred in African societies.

In Ciluba: "chief",

Language	**"weave, style"**	**Chief**
Bamana	Digi	Tigi
Nzebi	oufughu	Mfumu
Ciluba	Ku. Fuma	Mfumu
Ciluba	Ku Kuma	Kuma
Ciluba	Ku Tuma	Tuma
Ciluba	Ku Kuta	Kuta

Chart 58: chief and waving

The weaver / garment maker's staff is used in hieroglyphs to mean craftsman and chief.
A man who exercises power bears the name that is often misleadingly translated by the ethno-sociologists of the country's uncle because he is the depository of power. The most appropriate word to translate this subtlety of Ewondo is / *ùn*-ndzoué / "depositary of authority"

It is the sister (the panther) who holds the power / i-dzoué /. We inherit from our maternal uncle. This is why the word uncle is called mother / nya-ndome / "mother-brother" and we have both possibilities when we talk about the throne. It is not enough here to talk about Asset and Wosiris. These barbarisms are to be avoided from the mouths of Africans because they participate in the erasure of memories and their replacement by hollow dogmas

This word / Na /; / Ma / by honorific anteposition serves to designate the first person of the singular: "I, me " in our languages (see example N°32 - 3.4.3.32). The word has not changed in time since the 15th century and it has been preserved in the West Indies in Creole speech.
Crole Guadeloupen: *An* "me" **Ewondo:** /na / "mum"
Crole Guadeloupen: *Mwen* "i" **Ewondo:** /mwan-na / "mum's son"
Crole Guadeloupen: *Man* "me" **Ewondo:** /maâna / "mum's kid"

Let's flip back to our former comparative in African languages:

Songhoï: Tarkoumda "elephant"
Ewondo:/Tara/ "father, ancestor"
Ewondo: /ôkè-kü/ "extremely old"
Ewondo: /nda/ "house"

Bambara: Sama "elephant"
Ewondo:/I-sam/ "concession, paddock, flower"

Kikongo: Nzau "elephant"
Ewondo: zə "panther"; ngon "panther, earth";
Middle-egyptian: ***Aby*** "panther" ☞**Ewondo**: zə "panther"
Middle-egyptian: ***Abw*** "elephant" ☞**Ewondo**: zoa "elephant"

Bamougoum: Nseuh "elephant"
Ewondo: zə "panthère"; ngon "panther, earth"
Middle-egyptian: ***Aby*** "panther"
Middle-egyptian: ***Abw*** "elephant"

Sara: keur "elephant"; **Mbaye:** kédè "elephant"
Bùlù: nkot "savior"
Ewondo: /*ùn*-nkoré/ "saviour"
Middle-egyptian: ***wsir*** "eldest ancestor"

Lingala: nzoko "elephant"; **Kouyou:** ndzoku "elephant"
Ewondo: /à-ko zə/ "the place of the panther"
Ewondo: /zokh/ "elephant"
Middle-egyptian: ***Abw*** "elephant"

Tamashek: Elou "elephant"
Middle-egyptian: ***Abw*** "elephant" because ***Hr*** is the foot.
Ewondo: /Elo/ "Zamba"
Ancien Egyptien: ***Hr*** "Hor"

Dazaga: Kouwoun "elephant"
Ewondo: /à-koh ù-wun/ "place of the panther" here by metonymy /wun/ [10]
Ewondo: /à-koh/; /ndah/; /à-boh/ "place"
Ewondo: /à-wun/ "to become famous"
Ewondo: /ù-wun/ "the crane for its ability to carry heavy loads"

Ewondo: /*ùn*-nlod/ "scale" ☞**English**: load; loud
English: crane; cranium ☞**French**: crâne; craner; cran; crin
Etc.

The show off is related to a slow walk like the one of the elephant.

So we see that with a little effort and aside the constraints of prosody, we can understand what other people do say. It takes a little brain training, but it is possible.
In short then, it is the context that will guide the choice of the word to use. We could have just as well turned to other readings to name the right sign or accompany it with precise morphological elements in order not to name the elephant but to characterize it by a precise descriptor.

10. Woun: rumbling of the pantherid type big cats.

3.4.2. Examples Of Scripts Reading

The general logic underlying the good understanding of ***mDw nTr*** is a logic based on the complexity in the philosophical sense of some advanced researchers [11] of nowadays. It is also necessary to search in the polymorphology of the senses of symbolic thought, the elements of access to the ontological, scientific and philosophical truth which is recorded in these characters.
This supposes to make the effort to leave the binary logic of all-or-none principle and to integrate this kind of thought involving the relativity and that is correlatively complex, even destabilizing.

The tree is in the seed and the seed is in the tree. This is a concrete example from this way of thinking. If we transposed this idea in the philosophy field, we would say that there is a consubstantiality between the father and the son. The father or the mother is the son and the son is the father or the mother which is rather unusual and is a mystery in the religions of the book.

An unscrupulous reading of the African sacred texts will lead to introduce the notion of Greek Theos (From ancient Greek θεός, theós "god") in African texts denaturing by the same fact. We will invent a divine birth of some Fara as in the tale of Papyrus Westcar, describing the birth of the first three months of the 5th Dynasty.

We find this dynamic phenomenon in the portrayal concerning the Fara ***Ḫnmt Jmn Ḥȝ.t-Špsw.t*** "Hatchepsout" which would be born by a theogamous process in order to justify its legitimacy to the throne. She will be the first ruler who will solemnly represent in her great temple of Deir el-Bahari [12] scenes of his birth described as being abusively and dishonestly Divine. One of two things is either a will to deceive or a misinterpretation of the texts. In any case, this is not what the so-called theogamic birth gives to understanding from the African point of view.

I will therefore gradually illustrate my purpose with examples and after that, i'll densify the levels of understanding with examples.

I will therefore explain this method from the basic elements that allow understanding and show the links with metonymy, homophony, synonymy and antonymy etc.

A single character has different aspects. On the other hand, the perfect use of ***mDw nTr*** allows to use the right springs for the understanding according to the castes, the languages levels as well as lineages and levels of african initiations.

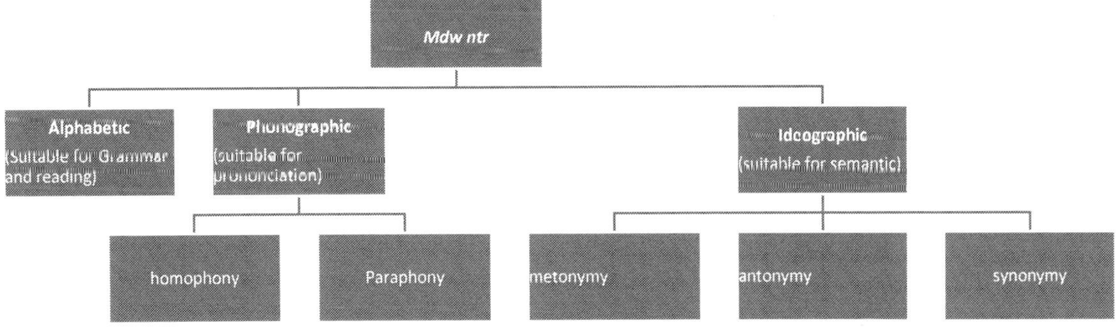

Basic principle of mdw ntr

11. The thinkers who have taken the measure of complex thinking nowadays are Edgar Morin, Ilya Prigogine Ren Thom, J. Ziegler, Jean-Louis Le Moigne etc.
12. Dayr Al-Baḥrī in Arab: "The convent of the Sea". He recalls the existence of the Coptic convent raised in this place.

Homophony and paraphony

In linguistics, paraphony is a variety of homonymy designating the relationship between two different words with the same pronunciation. In this respect, homophony is distinguished from homoscript, which refers to the relationship between two words of different meanings with the same spelling.

Sometimes we also speak of homophony between two or more words (or expressions) when these words (or expressions) have pronunciations normally different phonetically but sufficiently close (paronyms) to be sometimes confused phonologically according to the regional accents, or when they are pronounced in another language which does not make a clear difference between several phonemes, or when the pronunciation of the spelling is uncertain in this language. But we should rather speak in this case of paraphony.

Example:

Middle-egyptian: ***Abw*** "elephant"
Ewondo: /à-buî/ "multitude"
Copte: ⲉⲃⲟⲩ *ebow* "elephant"
Ewondo: /i-bɔh/ "rotten"
Yemba: / mbhū/ "dog"

Woldwide, regardless of country or language we say that if two words share a phonetic proximity, from the African point of view, they necessarily have the same signified This is why we do not retain the notion of homonymy because it implies different meanings. This true at the surface only. So if two words share some phonetic proximity they are keeping an archaic vivid connection of their distant African past.

Antonymy

Two lexical items are in antonymic relation if we can exhibit a symmetry of their semantic features with respect to an axis. Symmetry can be broken down in different ways, depending on the nature of its support. There are several media that are as many types of antonymy:

Complementary Antonymy concerns the application or non-application of a property ('applicable' / 'non-applicable', 'presence' / 'absence'): for example, 'inform' is antonym of everything that has a form as well as 'tasteless', 'colorless', 'odorless', etc. of all that could have flavor, color, smell,

Scalar Antonymia concerns a property that affects a calibrated value (high value, low value): for example, 'hot', 'cold' are symmetrical temperature values

Antonymy dual the existence of a property or element considered as symmetrical by use (for example 'sun' 'moon', or by natural or physical properties of the considered objects (for example 'male' 'female' , 'head' 'foot', ...).

This way of thinking is an African signature of conceiving that everything above is like everything below. In fact, this induces the notion of opposition. To speak from above or from below is only a question of degree, but the nature is identical in every respect. Everything has a top and a bottom. The thing is more important than the point of view.

Example:

Middle-egyptian: **w** "small" an aspiring fowl
Middle-egyptian: **w** "great, chief"

Numeral 1 also symbolizes what is integer, which is complete compared to proportion. So the 1 embodies the totality it is is big. It can be small while compared with other numerals.

Synonymy

Synonymy is a report of semantic similarity between words or expressions of the same language. Semantic similarity indicates that they have very similar meanings. Synonymically related terms are synonyms.

This is the key issue of misunderstanding between human tongues

Example:

Middle-egyptian: ***Abw*** "elephant"
Ewondo: /zokh/ "elephant"
Copte: ⲉⲃⲟⲩ *ebow* "elephant"
Zandé: mbara **"**elephant**"**
Sango: Doli "elephant"
Seereer: ɲig "elephant"
Bahasa Indonesia: gajah "elephant"
Nāhuatlahtōlli: elepantli "elephant"
Qusqu qhichwa simi: ilijanti "elephant"

Metonymy

A metonymy is a figure of speech that replaces a concept by another with which it is related by a logical link implied: the cause for the effect, the container for the content, the artist for the the city for its inhabitants, the location for the institution installed there ...

Middle-egyptian: ***Abw*** "elephant"
Ewondo: /à-bya / "elephanthiasis"
Ewondo: /à-bui/ "numerous"
Ewondo: /à-bo / "foot";
Ewondo: /à-bé / "dangerous"
Etc.

Phonogram

A phonogram is a written item which, contrary to morphograms, ideographs and logograms, is the arbitrary transcription of a sound. In linguistics, for example, the letter **a** is a phonogram because it notes the sound [a]. On the other hand, x is not always a phonogram: in animals, it does not note a sound but the morpheme of the plural.

Example:

Middle-egyptian: "they, the; their, theirs"
Ewondo: /be/; /mi/, /ba/ "they, the; their, theirs"
Ewondo: wou "fly, air"

Alphabetic

An alphabet (alpha and beta, the first two letters of the Greek alphabet) is a writing system consisting of a set of symbols, each of which represents, for example, one of the phonemes of a language.

Example:

Middle-egyptian: 𓅱 w "they, the; their, theirs"

If this is a pure convention, this methodological choice makes cringe but it is useful for writing a grammar. It can help for reading also.
This transliteration can become disabling when it encumbers the roots of words with the grammatical requirements of the morpho-syntax standards. At this point, the root of words are encumber with sounds that do not exist in African languages. ***imn.tyw*** does not exist. But we can approach the root ***imn-*** of known words.

If I do not exclude the use of transliteration unlike other African brothers who study the holy characters, for me, I want to distinguish the reading of a character, its vocalization, its understanding in its paradigm, and there are rules of morpho-syntax that are related to its use. We can not confuse the use of a character records on pain of stagning.

Since there is only one humanity, there is therefore a single lexicon worldwide. As a result, the roots of transliterated words are automatically present in all humen languages.

I do specify if necessary that, to be comfortable with this approach, one must imperatively have a good level of fluency to speak an African language for the simple reason that the ancient Egyptian language was written at least 6000 years ago and by scholars who used a way of thinking and an excellent command of language and writing. To allow non-practitioners of the language to study the accuracy and consistency of the process[13] Ewondo-English.

I'm not saying that this method does not allow you to access reading and a certain level of comprehension in any language spoken worldwide. I am simply saying that the subtlety of esoteric information is not accessible at the level of exoteric mastery.

For Africans I recommend a dictionary in their own linguistic variant.

3.4.3.1. <u>Example 1</u>: How To Deal with This Symbol 𓅱 *?*

The good vocalization of the transliterated sign 𓅱G43 w, representing the quail chick, is important because this sign is used in many scripts. Moreover, this sign is used in morpho-syntax as a gender prefix, as a plural marker, as a pronoun, but also as a negation tag in grammar. In fact, according to its position and use, it will mobilize precise vocalizations according to african variants languages. The objective in this example is to go through the different cases and to explain how by an endogenous method with a singthe sign, we can mobilize different vocalizations. I'm just saying that I'm exposing a principle not a law.
Meaning that, perhaps, the links established in Ewondo are certainly not the same in another African variant language. I articulate a very important principle that guides to find these links by any african language speaker.

<u>Reminder</u>: classic method

Middle-egyptian: 𓅱 "quail chick"
Transliteration : ***w*** "quail chick"
Vocalisation: [wou]

13. www.ddl.ish-lyon.cnrs.fr/projets/clhass/PageWeb/ressources/fang.pdf

Endogenous method:

We will try to describe the different steps and explore all the main possibilities related to the symbol 𓅱G43.

Vocalization [wou] and associated place in a syntagm

Step 1: *Name all the signs of the script*

Her we have just one signe a quail chick

> **Middle-egyptian**: 𓅱 *w* Sign G43 of the Gardiner list "quail chick"
> **Ewondo:** /*ùm*-wan kub/ "chick"

Step 2: pick up all the hidden meanings.

As we have emphasized so far, either the vocalization proposed by Egyptology is coherent in the spoken language or it does not hold. In the second case we have a synonym. In fact, between what we see and what we should vocalize, there is a metonymic relationship. On this account, one must resort to the hidden meanings related to what one sees. Either in an African language, quail chick is called / wou / so it is different.
The analysis of the sign G43𓅱 "quail chick" will therefore prevail
In quail chick, there is chick and there is quail. To find the hidden meanings is to ask the question of what, according to the African ontology, is a quail chick?
To answer this question we can mobilize descriptors. And, depending to the knowledge of African thought we will have simple descriptors or descriptors that escape the general average knowledge. Each of these descriptors is therefore a key for the level of understanding and for the good intellection of a text or the script taken in isolation.

Example: a quail chick is small, it is a bird, it has feathers, it is young, immature, it pecks, it is fragile, it is a quail etc.

Note : I want to emphasize that even though the scribe has drawn a quail chick the resultant level of information is not only linked to the fact of recognizing and naming the drawn quail chick .

1. **We will start by dealing with the chick meaning.**

This step will be based on reasoning centered on the African paradigm. The principle is for the scribe to note signs that are homophones of the determinative. At a certain level, what I see and name, represents a hidden meaning that I have to vocalize and then that the scribe wishes to express as the key idea.

What does the scribe expresses as an idea with 𓅱 ?

So I can not start thinking about something that I do not see. IF I act like this it looks like inventions. I am therefore obliged to name what I see before projecting myself into various reasonings.
This point therefore presents a major divergence from the official method because we will always first notice what appears to the senses and then we will highlight the secondary ideas that are attached to them.
A chick is an emerging bird, it's a bird immature, a young bird. Here, I see a bird.

> **Ewondo**: /i-yələ/ "bird"; **Ewondo**: /á-yələ/ "to fly",

In this case, the word "*fowl; bird*" thus offers a homophone which is one of the ideas that the scribe wants to express in this particular case.

But, the **Ewondo** Root: /I-yəl/ also serves to designate an impotent, that is to say, who has chronic weaknesses of the male member. One can associate the idea of impotent with a *death* of the phallus.

In this account we say in Ekang language: /à-wu byəm bi fàm/ literally "the death of men's things".

So the term /I-yəl/ ⇔ "death" (male genitals)

So we have the equation: *chick = bird = impotent = death.* It is one of the logical conditions that makes it possible to understand how one arrives in Ewondo to the vocalization [wou] of the standard method. Just because the rules of the standard method allow the association of the quail chick to the vocalization [wou]. Perhaps another logical sequence would have to be taken into account in another African language synchrony, but this one allows for example to be understood in the Ekang languages.

Step 3: Lecture and verification

Verification is quite simple here. It is enough to prove that phonetically, one has in both languages (Pharaonic and Ewondo):

Death = impotent = chick.

Death

Middle-Egyptian: ***Aw*** "death" (Z7 hieroglyphic adaptation of the abbrev. hieratic of G43)
Ewondo: /à-wu/ "death"

Impotent

Middle-Egyptian: ***iwy*** "Impotent"
Ewondo: /à-wu/ "death of mal member"
Ewondo: i-ye(l); pl.: bi yel "impotent(s)"

Chick

Middle-Egyptian: ***w*** "chick"
Ewondo: /i-yələ/; pl.: bi-yələ "bird" ☞ I see a bird not death.

This script can thus be read / wu / by synonymy of / * yel / in Ewondo. The purpose of this approach was to comply with the dictates of transliteration associated with the quail chick, the vocalization / wou /. In any case, I did not need the diktat of Egyptologists.

Note: We note that for the "impotent" acceptance, the scribe represents a boat simply because the couple of words are homophones in our languages.

So, we have:

Middle-egyptian: ***iwy*** "impotent"
Ewondo: /b-yal/ "boat"
Ewondo: /i-yələ/; pl.: bi-yələ "bird" ☞ I see a bird not death.

Both words are therefore homophones.

In this logic of complexity, we see that a chick corresponds to a boat, to death, to immaturity, to impotence etc.

Little cultural point

On a cultural approach, in the Beti thought, it is often discouraged for nubile men to approach the young birds to their reproductive organs so as not to end up impotent. For Ekang people, poultry is therefore death, immaturity or a boat etc. This clearly explains why **w** is vocalized [w] or [u] in my language and certainly in African languages in general in the name of African monoculture.

Morpho-syntax

In certain variants of the African single language, this 𓅱 *w* / wou / can serve as a affix and position itself as a prefix or suffix related to the radical. In the Ekang languages, which are abusively called Bantu as Ewondo, it is prefixed.
So we will have in Ewondo a structure like

> **Middle-egyptian**: radical-w
> **Ewondo**: (ù+pN)-radical

In reality the gender prefix / ù-/ is associated with the number prefix (PN) is therefore this pair that is associated with the radical. Note that the suffixation of the Egyptian is only by pure convention. But that convention is not a problem for us at all.

It must be admitted that this morpheme **w** can be vocalized differently in a prefix position. We will see it a little further during the treatment of particular cases because the current synchrony of the Ewondo is not either the old Ewondo or then the ancestor of the old Ewondo. But Pharaonic is a language less archaic than Ewondo so it defeats us from the constraint of looking for vocalizations. We can therefore be content with that of the current synchrony.

<u>Example</u>:

> **Middle-egyptian**: 𓂋𓅱𓃭 *rw* "lion"
> **Ewondo**: ***ùm-mbokh*** "chief, cavity, priest" (𓂋 *bokh + 𓅱 *bokh)
> **Ewondo**: ***ùn***-ndzoé "authority" (𓂋 *dzô + 𓅱 *dzô)

Note:

Although Gilbert Ngom rightly points out that:

> *"At the beginning of the words, 𓅱 =* ***w*** *hieroglyphic followed by occlusives or especially nasals n, m, falls regularly in our languages. This is a phonetic law"*[14].
>
> *(Gilbert NGOM, 1993)*

I have personally noted this **w** vocalized / wou / which disturbs those who impose the rules as I have above stipulated in order to keep some alignment with the Pharaonic. And especially to make more comprehensible the accordance schemes that are related.

Serious African grammarians who have done deep studies of languages such as Isizulu also adopt this approach without going against the phonetics sacrosanct law.

14. Gilbert NGOM, The genetic kinship between the Pharaonic Egyptian and the modern Negro-African languages, the example of the duala. Paris, Revue Ankh N°2, 1993, P.18.

Isizulu: umfazi "woman"
Ewondo: ùm-inga "woman" (thank you for not writing */minga/* anymore)

Should we note here [w] or [u] as i do it arbitrarily, this is an expert debate between phoneticians and grammarians. This issue can be debate during harmonization. Therefore, i close here this parenthesis.

A little Diachrony

If we had taken the word chick in the Pharaonic Sahidic synchrony we would have had the word: ϩⲁⲗⲏⲧ that would be closer to the wondo / à-sè / "feather". The quail chick is a bird whose body is covered with feathers.
We observed earlier that the quail chick was a poultry / kub / but the transliteration brought a sound / wou /. Why do we observe this difference phonetically? This is related to the fact that the two terms are simply synonymous. Those who were informed by the Copts copied what the informant was saying and nothing more.

Conventionally, on a technical level what's happening?

To understand what happens, we could reconstruct the predialectal ancestor word offering the transliteration **w** and vocalisation [wou] for poultry, chicken, and fowl.

Negro-egyptian: $*k^wk^w$-
Proto-kang: cucʊ; kʊmbʊ; cʊmbɩ; *gÚb` etc.
Sahidique: ϩⲁⲗⲏⲧ "little bird"
Ewondo: kʊmbʊ > kubu > kub "hen"

If the variant **kub** is used to designate a bird in Ewondo, its sibling variant * **wub** is associated with the semantic field of the fly. Starting from the proto-version, it reminds us well the Grimm's laws [15] with kw >w.

Ewondo: /à-wâ/ "To throw in the air"; /i-wâ/ "a flyer";
Ewondo: /á-**vu**ŋlu/ "ventilate"; /à-**vu**ŋ/ "fley"; /i-***vu***ndu/ "air";

The loss of the velar consonant **Kw> w** will bring the letter **w** to the initial position of the word. Thus the letter **w** at the beginning of the word is no longer pronounced / wu / w will tend to pronounce [v] as a v; or like [ʊ] with a slight voice. That's what's happening here. With the different African synchrony, there is no truth we can find the different cases: / w /; / v / and / ʊ /.

Ewondo: /à-wâ/ "To throw in the air"
Ewondo: /à-vuŋ/ "to fly"
Ewondo: /mvu*foŋ*/; /fu*foŋ* / "bee"
IsiZulu: Imbonga "bee"
Yemba: lepipi' "bee"
Yemba: mbumbuŋ "albino"

The variant resulting from the root * **wub** of the Negro-egyptian is therefore common to Ewondo and Pharaonic. It is thus the one that is transliterated by the Egyptology **w** and vocalizing [wou] with a complete amussement of the final ***-b*** *(k^w>wb>w).*

15. When two syllables containing aspirated occlusives succeed one another, there is a loss of aspiration of the occlusive of the first syllable. Thus phonemic is observed with occlusive labio-velar appendages of the so-called archaic languages including in the Indo-European primitive.

The variant * **wub** also admits an interesting homophone, meaning in Ewondo / á-wâè / "to steal, to snatch, to rob, to purloin etc. ". This explains why for example in French the two words appear homonymous whereas they are actually homophones even synonyms since backed by the same signified. It's just the depth of the reading grid that will make homonyms while a rigorous African reading grid makes them homophones and synonyms in complexity.

- If this hypothesis is true then we should find two things on a technical level in Middle-egyptian: an equivalence between flying in the air and robing, stealing in Pharaonic.
- Also a skeleton of consonant associated with bird, beginning with a velar and which would be the pendant of the variant * **kub** observed in Ewondo. This is to the extent that, Egyptology has not transliterated and retains only the variant **w**. But in reality, there exist all the weakened variantss between the labio-velar and the simplified variant **w**. I'm not gonna treat all these cases. So I'm gonna deal here only with few cases.

Verification 1: Equivalence fly in the air and steal, robe

Middle-egyptian:  ***wAr*** "fly against, fugitive, runaway"
Middle-egyptian:  ***wr*** "swallow; martin"

Middle-egyptian: 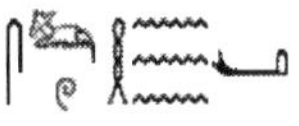 ***siwH*** " steal, rob, take, pick"
Middle-egyptian: 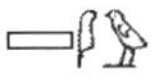 ***Sw*** "get up; rise"

Middle-egyptian:  ***tAw*** "steal, rob, take, pick"
Middle-egyptian:  ***tAw*** "wind; air; breath, respiration"

Verification 2: presence of a bird with a velar at the initial.

Middle-egyptian:  ***A*** "vultur, bird" A=kh^{w}
Ewondo: /i-kuba/ "vultur" (P.102; p.553)

Middle-egyptian:  ***kfA*** "rump, posterior"[16]
Ewondo: /fufukh/ "feather"; /i-bokh/ "haunch"

Other spellings can be added to support this point:

Middle-egyptian:  ***xw*** "breadth"
Middle-egyptian:  ***Hw*** "food, eating"

We therefore remember that the vocalisation [wou] of Egyptology related to the quail chick  transliterated **w** appears as a hidden meaning in Ewondo related to stealing (robbing) homophone of to fly. It is therefore through this channel that you should seek the vocalization / wou / associated with the quail chick but also to all the birds in general. We also find in the word flying in French for example, the root **vo-** associated to the suffixale ***-ler.*** In English we observe also a phonetic proximity between to *harrow and harrow*, *wing* and *wind*; *flay* and *fly; swallow and swallow(martin) etc.*

16. Faulk. 285, Wb. V p. 120, 6-8

So it is this Ewondo **vu-** that we find in the root **vo-** in French corresponding to the 𓅱**wou** of Egyptologists.

If according to egyptologists this z7 sign 𓏲 is similiar to the G43 sign 𓅱, according to Ewondo language, the spiral represented by the z7 sign is called:

> **Ewondo**: /i-vul/ "spiral; 𓏲"

As you can note, Ewondo: /vul/ and the French: /vol/ "fly; rob, pilfer, steal" are the same word. This demonstration enlights the reason why *vol* is used by French peoples as homophones to express the idea of flying or stealing.

It is therefore this ***vo-*** also that we often find as the third person of the plural **you** and serving as a sign of respect in the formal **vous** of French people. For the simple reason that **wou** is 𓅱***Hr***. And 𓅱***Hr*** is the leader. The **you** is bound to 𓅱 ***Hr*** as a symbol of the multitude.

So it is not enough to copy the oxyrhynchite Coptic language and sign words without a deep understanding of their true meaning and their real etymology.

After analyzing the first aspect of the quail chick, let's move on to the second.

2. **We will take the quail meaning**.

The word used for the designation of the quail in Ewondo is:

> **Ewondo**: /i-woré/ "quail"
> **French:** vol (we have a better understanding the origin of this word)
> **English:** swallow "martin"

A person who has a decrease in vital energy that can be manifested by the loss of capacity of the virile member. This important decline is called:

Ewondo: /à-wurâ/ "the death of râ" literally the death of vital energy.
We still find the idea of death that is associated with the quail.

> **Middle-Egyptian**: *Aw* "death"
> **Ewondo**: /à-wu /"death"

The word /i-woré/ "quail" offers:

The homophone:
> **Ewondo**: /à-woré/ "scratch; nibble"**;**
> **Ewondo**: /à-Wôlé/ "scratch"

The synonym:
> **Ewondo**: /wué/ "to skin";
> **Ewondo**: /*ùn*-wuà/ "that can be skinned"

It was also be possible to find the vocalization [wu] imposed by the transliteration taken in the oxyrhynchite Coptic using the quail descriptor associated with the quail chick. The simplest way here is to consider the quail as a bird and a bird can be skinned. The word to skin in the Ewondo language is /á-wué/.

Special cases

1- Plural in Ekang languages

The quail chick is associated to ***Hr*** and with the numeral (3) Three in terms of numerology. We have for the purpose in Ewondo: / i-le / "tree"; / i-la /; /i-lè/ "3, three".
The word / á-bui / "multitude" shares the same root with the word / ù-bə / "small".

> **Middle-egyptian**: or ***w*** "They, the, them, their etc."
> **Ewondo**: /à-bui/ "numerous" **Middle-egyptian:** *Ab.w*; *bA*; *Aby*
> **Ewondo**: /ù-bə/ "small"
> **Ewondo**:/i-vul/ "spiral" ☞ **Ewondo**:/i-bul/ "multitude; nine; 9"
> **Ewondo**: /bə/; /ba/ /bi/ "They, the, them, their, etc"→ plural

Note: The spiral /i-vul/ is also a hole /i-bé/ in African thought. So the spiral or the quail chick can be also vocalized [bə]. Such a vocalization is used for plural in Ewondo as we have seen it with class nouns.

Example 1:

> **Middle-egyptian**: ***kAmwtt*** "cob; ear"
> **Ewondo**: /***kyé** mə-**tét**/* "Vernonia leaves"

Other w plural

> **Middle-egyptian**: ou ***w*** "They, the, them, their, etc."
> **Ewondo**: /mi/ "whoever is not dwarf, the leader" → antonymy
> **Ewondo**: /mi/; /mə / "They, the, them, their, etc."

I just point out that the vocalizations for the plural of Ewondo noted above do exist in Pharaonic and are associated with plurals with different meanings.

> **Middle-egyptian**: ***mA*** "stern" cut into several pieces
> **Middle-egyptian**: ***mw*** "water; rain; seed" meaning multiple
> **Middle-egyptian**: ***mi*** "water"

Example 2**:**

> **Middle-egyptian**: ***qrf**.w* "wrinkles"
> **Ewondo**: *mi-**mbwar*** "wrinkles"
> **Ewondo**: *bi-**fəŋlo*** "what wanks"

It is the homophony with the word small that gives here the good vocalization of this spelling and thus its use for the plural.

2- Vocalisation of the negation associated to *W*

- Negation variant n°1: vocalisation /bə/

The quail chick associated with ***Hr*** is related to the idea of small / ù-bə /. This idea can be expressed by the synonymy / ù-tokh / "small" related to the small size of the quail chick compared to an adult fowl.

Middle-egyptian: *w* "do ...not (part. encl.)"
Ewondo: /ù-bə/ "small" → **Middle-Egyptian**: *bw* "place"
Ewondo: /bə/ "the recent negative past, not"

Example: / à **bə** laŋ/ "he has not read yet."
Note : we can also read /wo'o/ "not" Example: / à **wo'o** laŋ/ "he has not read yet."

- Negation Variant n°2: vocalisation /təkh/

Middle-egyptian: *w* "do not (encl.part.)"
Ewondo: /ù-tokh/ "small"
Ewondo: /à-təkh dẁùé/ "extend hand"
☞**Negation Ewondo**: /təkh/; /tèghè/ "do not " → **M-E**: *tkAw* "torch"

Example: / *təkh* di / "do not eat; without eating"

- Negation Variant n°3: vocalisation /kə/

The sign D35 induces the idea of asking, / kə mà / "give me". The G43 sign being associated with him as the foot / á-koh /, offers in his semantic field /á-kəh/ "to go, to leave".

Middle-egyptian: *w* "do not (part. encl.)"
Ewondo: /à-kə/ "give, help"
Ewondo: /kə/ "Negative imperative, or simple negation"

Example: / kə di / "do not eat"

- Negation Variant n°4: vocalisation /koko/

The sign D35 associate to the sign G43 /kô/ "foot" can induce repetition of syllables /ko/; /kə/ because we have 2 hands that are asking.

Middle-egyptian: *w* "do not (part. encl.)"
Ewondo: /koko/ "adverb of negation: NO".

Note: we can observe that the quail chick is a bird Ewondo: /ù-none/. This item /ù-none/ maks us better understand the meaning of the negation **none** « nobody, neither, no one" in English or **non** in French.

- Negation Variant n°5: vocalisation /bərâ/

The G43 sign is associated with Hr, the leader, the great / berâ / or the small / à-bwara /.
Even though the transliteration is done only with ***w*** and not with ***r*** added by egyptologist**,** here indeed, no consonant **r** is apparent and therefore transliterated, we say that from the African point of view this reading is necessary. However, we find this variant "great" ***wr*** which here is not mobilized and yet should be because associated to Hr. There is therefore only a reconstruction of a predialectal ancestor that highlight this potential ***r.*** The other way of looking at things is to say that all the occurrences that engage the **w** are mobilizabled.

Middle-egyptian: *w* "do not (encl.part.)"
Fang; Bulu: /berâ/ "Negation aux.: not yet" contraction of /bar kigh?/

Example: /Ma bera kə wé/ "I do not go there anymore"

- Negation Variant n°6: vocalisation /sâ/; /sâ kigh/

The sign G43 is here associated to the feather: /à-sa/. The vocalization /kigh/ is brought by the sign D35 introducing a repetition which is reminiscent of two (2) in two crevices /bi-kigh/. The association of these two signs can offer the negations /sâ/ or /sâ kigh/.

Why does kigh vocalization is here involved?

Hr being the potter offers the word:
Ewondo:/ à-kîgh/ "shape into a round, ball"

Hr being thirsty and at the breast of his mother, offers:

Ewondo: /à-kîghdi / "To drink a lot of water",

We have also :

Ewondo: / kîgh / "Interval between two nodes "
Ewondo: /à-kîkh/ "to cut"

Regarding the idea of the foot associated to ***Hr*** we have:

Ewondo: / *ùn*-ŋkîgh/ "when we walked a lot"
Ewondo: / à-kîgh / "Give a part of...",

Negation will mobilize here either one of the two variants or both simultaneously.

Middle-egyptian: ***w*** "do not (encl.part.)"
Ewondo: /sa/ "no"
Ewondo: /sa kigh / "no; not"

Example: /sâ kigh ma/ "not me"

<u>Little Wink:</u> Why LE and not WOU in French

The quail chick is associated to ***Hr*** related to the idea of death by synonymy as we have already shown. But besides death, ***Hr*** is associated by homophony to /i-yələ/ "bird" but also to:

- Three on numerological level: /i-lè/; /i-la/ "3, tree; three"
- the trunk of the tree: /i-lé/

"ELE: pl. bile. Dance to remove the mourning. Syn.: fôbla.. The Fañ believe that the dead stays first in the house, without leaving. If we danced for him, then he can go out. B a bo le."

<u>Note</u>: Ewondo: /À-fəblə/ "blow, ventilate"

We still have in the same semantic field:

*"ELE: Small bow for children, to kill **birds**. see nyigha."*

(GALLEY, 1964)

This also justifies the use in our languages of the affix **le-** before the radical often associated with the transliterated variant **w**. This **le** was used in an old synchrony of Ewondo but no more now. It is still found in the current synchrony of different African linguistic variants like: Iton, Basa'a or Yemba.

This is confirmed including by the Latin demonstrative **ille**. Shape that gave, il, elle, le, la, in French. In the trinity (mother, father, son), ***Hr*** is the third person he is associated with number three (3). Surprisingly ***il, elle,*** are also mobolized for the third person singular.

Does this explain why in some skies, some people have used malice intent when talking about themselves at the third person of singular?

So the quail chick can phonetically offer vocalisation ***le***, ***la*** without any doubt. Let's check it otherwise.

> **Middle-egyptian:** ***rê*** "Rê, sun, day" (this is an equivalent of ***w, wr, rA***)
> **Middle-egyptian:** ***rA*** "stuffy; opening; word, purpose, etc."
> **Ewondo:** /à-lê/ "tell, talk"

Phonologically, all this is explained by the deformity of voiced alveolar roll consonant, [r] to the liquid consonant [l]. This is a common phenomenon in phonetics. The place where some peoples like in Japan, one does not even make the difference between the [r] and the [l].

If we want to make phonetic diachrony from the Negro-Egyptian, the vocalization [w] associated with the quail chick can be explained by the proto-version ***ŋ*^{jw}*ə-* corresponding to** /i-nyo à -/ "The one who" in Ewondo:

<u>A masculine word and evolution of the gender</u>

The reconstruction around the masculine prefix **w** from the Negro-egyptian might look like this:

> **Negro-egyptian:** *hi-**ŋ***^{jw}*ə- Radical-hu* *(hu suffix)*
> **French:** *hi-**ŋ***^{jw}*ə->ille- > le -*

So that's the reason allowing to read the quail chick in French **le** and not [wou]. By questionning the Coptic and inventing a language that is a set of barbarisms here we have te prouve that Champollion has been informed. The Ekang people call ***Hr***, ***Elo***. Some call him **El** otherss **Allah/Eloï** isn't it?

This **Negro-egyptian** proto-version: *hi-**ŋ***^{jw}*ə-* gives in pharaonic the variants like:

> **Middle-egyptian:** ***rê*** "Rê, sun, day"
> **Middle-egyptian:** ***inr*** "stone; boulder; block, slab"
> **Moore:** Neere "crusher; masher; smasher"
> **Bwamu:** Nïlo "crusher; masher; smasher"
> **Ewondo:** /i-nyor/ "stone; boulder; block, slab"

Ewondo: /I-nyor/ "crusher; masher; smasher"

Note: To crush, to smash, to mash is said in Ewondo language:

> **Ewondo:** /á-nyor/ "To crush, to smash, to mash"
> **Ewondo:** /i-nyor/ "stone; boulder; block, slab; smasher; crusher etc"

So there is no **INRI** meaning Jesus of Nazareth King of the Jews. This acronym INRI represents the Latin inscription IESVS NAZARENVS REX IVDÆORVM (Iesus Nazarenus, Rex Iudaeorum), in English read as "Jesus the Nazarene, King of the Jews" and its supposed definition are just intellectual swindle.

If we stop ideology we could easily go forward. There is not a Latin humanity and an African humanity. There is only an Homo *sapiens sapiens* who has acquired the speech in Africa. So if you want to study world's languages rigorously, you have to study African languages and also associate Africans to this research. One can study a language, but perceiving its subtleties is a little complicated. Regarding the great pope of linguistics Noam Chumky, I am not sure that he can deliver a sentence in Ewondo or understand a sentence in a sustained register of an African language.

The aside I have done here on the masculine french morpheme can be done from the African point of view with any prefix of gender used to designate the masculine in the world. The couple of masculine French articles **un** and **le** as you have noticed are the same in Ewondo language. We do not use **le** anymore in the present synchrony but you do really find it in the old Ewondo language and through the sibling languages around Ekang area.

Important achievements

What has been imposed as 𓅱 ***w*** with a sound / wou / is from the African point of view only the visible part of the iceberg. If this script seems to correspond to **w** for Egyptologists. According to our point of view, it is not vocalized only [w]. As a result, the transliteration **w** is a special case and nothing more. It can indeed take the forms bə, ba, mi, kigh, kô, sâ, bərâ etc. Let's not give him a learned name of biliteral or x-litera.

When this sign 𓅱 is replaced by its synonym the sign D21 𓂋 because the chief is the mouth, it can take the value the prefix value: le-, lw-, la-, li- in our languages and replace the sound [w].

> **Vieil Ewondo**: *lə*-bum "belly"
> **Basa'a**: *li*bum "belly"
> **Yemba**: *le*-vəm "belly"
> **Bafoussam**. *l*"ɛʔʙ "cloud"

It is therefore this particle **le-** that has been considered as a nominal class to describe the parts of the human body in the books of scholars. And yet I name parts of the body in Ewondo and I do not find this supposed affix of the nominal classes.

> **Ewondo**: *à*-bum "belly" **Ewondo**: *wa* "hand"
> **Ewondo**: zut "buttocks" **Ewondo**: *à*-ko "foot"
> **Ewondo**: *ùn*-nsɔt "phallus" **Ewondo**: *di*-oé "nose"
> **Ewondo**: *di*-is "eye"

Here are some parts of the human body not presenting this particle relative to the nominal classes of the so-called Bantu languages. So far, I prefer to read what is written and content myself with saying what is written and not going into classifications biased or tinged with ideology. The defining criterias for nominal classes are completely arbitrary, are not they? Let's emphasize it towards those who are not used to this scholarly literature.

Quid?

Do we have to vocalize this sign 𓅱 [w] as in **ouate** or in [u] as in the word **doux**?

For me, this question is not important. Because there are more than 2000 variants of the unique African language and 7000 worldwide. Only specialists in linguistics and grammarians can give an accurate point of view on this question according to their expertise and the objectives to be achieved.

For the remaining part of this document I will use / ù- / attached to the radical as a masculine tag in Ewondo although the [w] is more indicated to have an alignment with all the African linguistic variants. I vocalize it as I usually do it every day that I express myself in an African language. For Africans and those who understand how all this is articulated, it is necessary to leave the diktat of 𓅱 *w* to be vocalised /wou/ and is an unilitera.

Note: The verbal suffixation of 𓅱 *w* to give the passive form in Egyptian and in our languages is not discussed here. There is no witch it takes the suffix **-èn** in Ewondo as *hen* in English. I'm talking about the quail chick.

In terms of the overall study of ***mDw nTr,*** this character is very complex. We will discover it through examples further.

Chick in some variants of the African language.

Ewondo: /ù-mwan kup/ "chick"
Godi: *koko yu* "chick"
Podoko: *Choueke* "chick"
Lingala: *Mwana nsoso* "chick"
Nzebi: *Mwana Tsoussou* "chick"
Shicomori: *Mwana wa kouhou* "chick"
Moor: *Nonbila* "chick"
Bambara: *Chèdein* "chick"
Baoulé: *Akô-wa* "chick"
Kikongo: *Muana nsusu* "chick"
Wolof: *Thioudje* "chick"
Bamoun: *Moh ngap* "chick"
Gouro: *Maan nin* "chick"
Tshiluba: *lukùmbukumbu* "chick"
Toubou: *Kogoyay addi* "chick"
Agni: *Acô ba* "chick"
Tamachek: *Akarte* "chick"
Bamileke: *mu-gop* "chick"
Mina: *Cloclovi* "chick"
Etc.

Capitalisation

This approach solves the problem reported in the *Faulkner dictionary*, namely: the incomprehension of the **Phonetic [w] associated with the quail chick judged for unknown reasons[17].**

We can thus observe that all the words invented by the Egyptologists supposed to be Egyptian words, the words associate to the quail chick with the letter **w** taking the vocalisation [wou] some of them are therefore false. The quail chick does not necessarily induce a sound [wou] in ancient egyptian languages. I am not against the idea of a modeling that associates with the quail chick the letter **w** for linguistic needs. I do understand that very well. I do accept that only at the condition that, this modeling does not become a phonetic true rule.

17.http://www.hierogl.ch/hiero/Sign:G43

Philosophical Step

I wanted to point out that, the weakness of the phallus / i-yə (l) / is the characteristic of men. So that's why the quail chick 𓅱 / i-yələ / "bird" / i-yəl /; / i-youe / "green, young, nubile, immature" here is a male marker.

We can therefore better understand why the original power is embodied by women in Africa because men are immature figuratively as well as literally.

3.4.3.2. <u>EXAMPLE 2</u>: HOW TO READ THE SYMBOL BELOW IN THE EKANG LANGUAGES?

Like the sign G43 𓅱 the good vocalization of the sign 𓏏 representing a bread is important because this sign is mobilized for various reasons on the technical level: in morphology as a feminine tag (you, your etc.), as a tag of the collective (adjective, noun), and in various grammatical cases. In fact, according to its position and use, it will mobilize precise vocalizations. The objective in this example is to go through the different cases and explain how regarding endogenous method with a singthe sign we can mobilize different vocalizations.

<u>Classical method</u>

Middle-egyptian: 𓏏 "Bread roll" Faulkner p. 292, Wb. V p. 337, 4
Transliteration : *.t* "Bread roll"
Vocalisation: [te]

<u>Endogenous method</u>

<u>First Part</u>

<u>Step 1</u>: Name all the signs of the script.

- Sign X1: /i-fən *ùn*-mfəma / "round loaf, Bread roll"

<u>Step 2</u>: sorting of signs and character analysis.

Here we have only one sign the round loaf. The analysis therefore pushes us to investigate its semantic field including the usefulness of this cake. It appears for many people since the invention of bread making in Egypt[18] that bread roll is a food. So if it is food, it must necessarily bear this meaning.

Ewondo: i-di "food"
Ewondo: à-di "to eat"
Duala: da "eating"
kɔ́ɔnzime: ɗẽɔ̀ "food"
Middle-Egyptian: DAi "devour"(contiguous presence of vowels /a/ and/i/)
Middle-Egyptian: m DAy "supply food"

18.http://www.sitefeb.com/la-federation/histoire-du-pain/

A little bit diachrony

In Negro-egyptian this variant derives from the proto-version ****kja-*** related to the mother. Here we mobilize the nourishing function of this latter. It will evolve from the proto-version to the current Ekang languages (proto-kang) into **cɩa-/*dɩa-.***

The negro-egyptian word ***dia** will give the root words * **da-** and * **di-** of actual synchronies. These variants will lose the consonant particle in the current Ewondo synchrony. The consonant serves on the morphosyntactic plane as a number prefix. By the way it just ramains a simple prefixal vowel / to / or / i /.
We can therefore propose the following evolutionary scheme:

> ***kja->*cɩa->*dɩa->*dɩ-> *hi->*ʔi->*ɩ-***
> Or then
> ***kja->*cɩa->*dɩa->*da->*ʔa->*a-***
> or
> ***kja->*cɩa>ci-** in the languages of the Kongo pool.
> Etc.

Those morphemes will serve as tags for feminine nouns by prefixing and/or suffixing [19] on the radical. Those morphemes are persisting in different stages in the current synchronies of Ekang languages. Due to the fact that these languages evolving at their rythms in space and time.

Tshiluba: Ci-
Ewondo: /i- /"with loss of the consonant particle which serves as a prefix of the number"
Ewondo:: /à-/ "with loss of the consonant particle which serves as a prefix of the number"
Kɔ́ɔnzime: ē-"with loss of the consonant part which serves as a prefix of the number"
Lingala: Di-.
Etc.

The different accordance schemes has prouved it easily as already explained.

What is the signification of the bread roll?

Middle-egyptian: ***.t*** "bread"
Ewondo: /bi-di/ literally "food"
Ewondo:/ù-mfəma/ "bread"
Ewondo:/ù-mfum/ "White"
Tshiluba: / mfùmù/ "chief"

Middle-egyptian: ***A*** "vulture"
Ewondo: /ndo/ same root as /*di/
Ewondo: /nda/ "House"

White crown of Egypt

Ewondo: /i-dzou/ "authority"
Ewondo: /*ndî/ "the pillar"

19. The Ewondo being a Beer language, loses its suffixal vowel except for the formation of proper nouns or adjectives by derivation from the noun and addition of the final vowel. The term ù-nkukuma designates the leader and ends with a final vowel. The pillar pillar, the very large pillar, the incarnation of the pillar. The word is a superlative. We can also find words such as ù-ntuba, ù-nkandi, i-burgha, i-yole, etc.

Bread is the representation of *wsjr* found in the Christian Eucharist. The idea recorded by the scribe through the egyptian bread is the chief's body .

The vulture embodies the maternal function in African thought. The mother is /*nyâ / and this word is also the counterpart of the mouth / * nyu / often used to nurse, to drink and to eat. The presence of the vulture in a script in my opinion may indicate a feminine principle or maternity or the action of eating. In fact the bread cake indicates the hidden meaning of the female characteristics of the throne I mean the earth. The cake of bread can thus introduce the reading food, ground, queen, panther etc.

A little wink

Middle-egyptian: *rA* "mouth; opening; speech, words,etc."
French : la

Phonologically, all this is explained by the change of voiced alveolar roll consonant, [r] to the liquid consonant [l]. Phenomenon common in phonetics to the point where in some peoples as in Japan, we do not differentiate between the [r] and the [l].
We can therefore explain why this script can be read in French and not only [.t] as is customary with Egyptology.

Limits of gender theory:

Middle-egyptian: *mAt* "female lion" **Ewondo:** /ùn-ŋgal i-singa/ "felid female"
Middle-egyptian: *mAi* "lion" **Ewondo:** / i-singa/ "felidae"

This example is very important from the pedagogical point of view because the word / i-singa / by itself indicates a feminine (.or). But in principle, there can be a male or a female gender.
If one sticks to the rules of classical nominal morphology, one could have trouble distinguishing these two words which appear both as indicating a feminine. And yet from our point of view, the indication of sex is in ewondo with / *ùn*-om / "male" and / *ù*ŋ-gal / "female"

The general rule is to state a generic class of animals and the precision of the sex is done according to the needs.
If in the feline class of the example above we wanted to focus on the lion, basically the term / i-mbgwem / "lion" is neutral. In case of need, only a male or female lion will be specified by prefixing it / ùn-om / "male" or / ùŋ-gal / "female". The position of the cake of bread on the feline thus leaves us thinking that the spelling will mobilize the word / ùŋ-gal / "female" in anteposition. So here we read from right to left.

Why are we reading / i-singa / here?

The sickle or billhook ***mA*** called / i-bâgha / in Ewondo. It is used for cutting. The term / i-sang'ha / is therefore the most appropriate for designating an object that serves this purpose. This term offers phonetic proximity with / i-singa / "cat, felidae". Etymologically this animal is presented as the one able to cut and to tear.

By consigning the D4 sign ***ir*** the scribe probably puts the accent on the sight of the felids. To See is said in Ewondo: / á-yən / it is the ananym homophone of / á-nyé / "to breast, lick" of / nya / "mother".

Middle-egyptian: ***irtt*** "milk" **Ewondo:** /*ù*-nyaŋ/ "glaucoma"
Ewondo: /á-nyaŋ/ "to suckle"

Ewondo: /á-nyé/ "to breast-feed"
Ewondo: /nya/ "mum"
Ewondo: /á-nyù/ "to drink"

So that's why we read this script in Ewondo: / ùŋ-gal i-singa / "felid female"
I translate this term by feline and non-cat as many would have liked because for a civet example, one would have in doubt, simply specified the term "wild" to distinguish it from the domestic cat: / i-singa à-fan /. And in the same way, sex would have been predated. The word / i-singa / therefore indicates a class of animals namely the felidae. I therefore close this parenthesis here.

Note:
We think that we must not confuse the genre of the word with the kind of thing. It is possible that both can be different.

I took note of this curiosity on a technical level. What is the root of the word female in Ewondo? / * gal / or / * al /?

Ewondo: /ùŋ-gal/ "female" →singular
Ewondo: /bə-yal/ "female"→ Plural

We can observe that the root of the word "female" is different in both cases (singular vs plural). This phenomenon is related to what is called consonant mutation at the initial.
But it can also be a metonymy between sperm and male in our languages.

Part II

The M17 symbol above from Gardiner's list is given as a reed in bloom.

Ewondo: /mə-sam/ "flowers" **Ewondo:** /mə-sèh/ "feathers"
Ewondo: /i-sóŋ/ "reed" **Ewondo**: /ùn-nsòŋ/ "top, pointed, beak"

Why do we say that the feather is commutable with the flower?

Middle-egyptian:  *sSAt* "The one of Seven petals"
Middle-egyptian:  ***mAa.t***
Middle-egyptian  ***Sw.t***

The flower and the pen refer to the same entity. The part that bears the flower is usually located at the top of the branch.

Ewondo: /*ùn*-ntòŋ/ "the top" **Ewondo**: /*ùn*-ntòŋ'o/ "feeder"

The synonymy between these different words establishes and explains the semantic porosity and the resulting transfers.

Ewondo: *ùn*-ntòŋ "feeder" dixit feeder root
Ewondo: i-ntòngha "food, to feed, to nurture etc."
Ewondo: tòŋ'o "horns"

Ewondo: i-di "food"
Ewondo: à-di "to eat"

Obama the feather /à-sèh/

Mum Ndo'o breastfeeding the feather

It is good observing the image above to take note of the origin of the phonetic proximity between feather, father and feeder of the English language.

We can thus vocalize this symbol as the precedent for the reasons mentioned here. We could have noticed that the feather is the own of the birds and thus of the vulture too.
The vulture is the ultimate maternal symbol in Africa. It's the feather and the feeder.
However, we must be careful to seek a vocalization that fits absolutely to the normalized alphabet of the others. These symbols are inscribed in a precise paradigmatic and ontological dimension which here, fortunately, is African one. It will be noticed just below that according to our point of view, the word feather = pen equals flower and related to the scribe.

Middle-egyptian: ***Hrrt*** "flower"
Middle-egyptian: ***Swt*** "feather"

Ewondo: /à-sùh/ "visage"
Ewondo: /à-sèh/ "feather"
Ewondo: /à-sàm/ "flowers"

The words: face, flower, feather etc. are identical in the field of complexity because they have one and only one meaning.

Morphosyntax: other uses of the bread cake

Vocalisation as YOUR

Middle-egyptian: ***.T*** or ***.t*** "you, your, yours"
Ewondo: /i-dĩ/ "sting"
Ewondo: /i-di/ "food"
Ewondo: /i-dzòé/ "authority"
Ewondo: /ı.ndi-òé/ "your" litt. /i-di wa/ "what belongs to you"

Vocalisation as You, your

Middle-egyptian: or "you; your, yours"
Ewondo: /ùn-mbokh/ "hindering" /ù-mbokh/↔ /wa/ "1"
Ewondo: /i-dzòé/ "authority" /ù-mbokh/ ↔ /wa/ "1"
Ewondo: /wa/ "you"

Note: this hindering is used on horns. Horns can be consider as stings.

Ewondo : /i-di/ « sting » and wa have that : /dià/ "1" and /wa/ "one, 1; you".

Example**:**

> **Middle-egyptian**: *sDm=T*"you hear (fem)"
> **Ewondo:** /wa wogh / "you hear" (/i-wough/ "the one who hoots")
> **Ewondo:** /wa simda / "you think"
> **Ewondo:** /wa simza / "you think"

Note: The sign ***t*** /wa/ "you" can be replaced by the sign (ù-mbokh/kob) also bringing back the word /wa/ "you" in Ewondo.

When this signal is engaged in a plural feminine, there is no specificity. This is the classical plural in ***be, ba, me, mi*** who takes the lead.

Example**:**

> **Middle-egyptian**: *SnTw* "conflicts, dispute; enemy"
> **Ewondo:** /*bi*-twam / "problems; conflicts"
> **Ewondo:** /*mə*-siŋ / "problems; conflicts"

Vocalisation .t ***as a collective tag***

In some specific cases, the sign ***.t*** is used for a collective. In this case it looks like a feminine singular although referring to a plural.

Example: /à-lwat/ "ducks" (indeterminate nomber, numerous)
We find this feature in Pharaonic with words as ***km.t*** which also designates a collective

Special case: male words ending in T

According to the legal literature, some egyptian nouns ending in "t" in the singular would be masculine gender:

To impressions, we say that all these words are feminine. They are feminine in the meaning that specialists give to this item.

Lexicon	Egyptian	Ewondo	Possessive
Barley, the grain	*it*	*mys*	**D**ys **d'**-àmà
The father, seat	*it*	*I-toa*	**I**-toa **dy**-àmà
The thickness, the niche	*wmt*	*À-fip*	**À**-fip **d'**-àmà
The death,	*m(w)t*	*À-wu*	**À**-wu **d'**-àmà
The camp	*xnrt*	*ndâ*	Ndâ **dy**-àmà
The breeze,	*swt*	*i-vuvu*	**i**-vuvu **dy**-àmà
The strip,	*wt*	*i-yalgha*	**i**-yalgha **dy**-àmà
The conduit, mother	*mt*	*ŋgut*	ŋgut **dy**-àmà
Wood,	*xt*	*i-l*	**i**-l **dy**-àmà
The face,	*xnt*	*À-sù*	**À**-sù **d'**-àmà
The image	*twt*	*i-vəghlə*	**i**- vəghlə **dy**-àmà

Chart 59: Verification array of supposedly masculine egyptian words

From the point of view of the Ewondo language, all these words are feminine. We will treat them as well. For us the rules that make it male words ending by ***.t*** is not justified.

Conclusion:

The bread cake represents the body of ***wsjr*** cut by ***stX***.
Bread roll that was consumed in memory of ***wsjr*** and symbolizing African vitalism. As food, the bread cake is associated with the maternity, the feminine, the throne, the authority, the action of eating.

On the sociological level, the feeding function in Africa is the responsibility of the mother because, it is the mother who feeds from the maternal breast. Man can provide food but this function is linked first to maternity, to the feminine.

For the remaining part of this document I will use /i-/ or /à-/ according to the circumstances to mark the feminine in Ewondo, in light of what has been explained.

3.4.3.3. Example 3: familiarization with the method

Classical method: What is done with the conventional method.

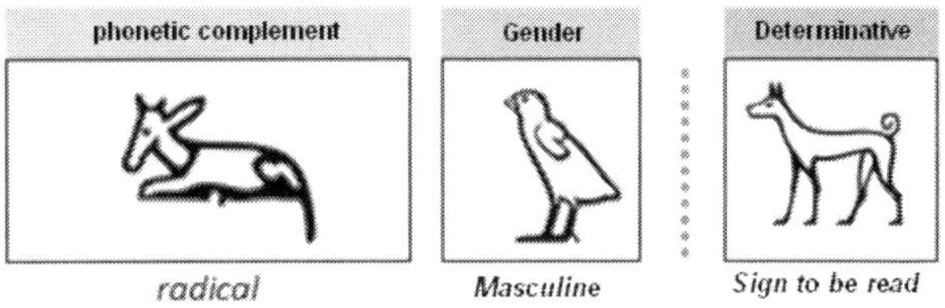

Transliteration :

Sign E9 ***iw*** of the Gardiner's list "Fawn of hartebeest"
Sign G43 ***w*** of the Gardiner's list. "quail chick"
Sign E14 ***iw***. "dog"

Vocalisation: the Sign E9 give the vocalisation to the Sign E14 [iwou]

Middle-egyptian: ***iw*** "dog" (*Faulkner p. 12, Wb. I p. 48, 3*)

Endogenous method: What we suggest and what changes.

- Sign E9 ***iw*** of the Gardiner's list

This is the sign on which the vocalization of the radical part of the determinative is based.

Ewondo: /i-mvu(l)/ "hartebeest"

- Sign 𓅱 G43 ***w*** of the Gardiner's list.

It does not just serve as a genre here, it is vocalized / i-wou / as we have seen in the previous parts. It gives a distinctive characteristic of the right sign 𓃡 E14 .

Ewondo: Sign 𓅱 G43 ***w***/ i-woré/ "quail chick"
Ewondo: /wou/ "barking of the sign 𓃡 E14"
Negro-egyptian: ***xu** "noise emitted by the voice"
Negro-egyptian: ***xu-** "dog"

We point out the fact that, if the transliteration had been correctly done sign to sign, it would have given: **iw.w.** Thus this transliteration would have given to this word according to the the standard rules a masculine gender. Surprisingly, we do not see the sign / i / in this transliteration explaining the ***i.w***. This pirouette is avoided in our languages and therefore in Ewondo because the hartebeest is called / i-mvoul / that's all. So / i-mvoul /, followed by 𓅱 G43 ***w,*** this last sign would be a marker of the masculine gender. This is Egyptology and its mysteries. Of course the sign 𓅱 G43 ***w*** is supposed to specify the vocalization of ***iw***. It will be argued. let it be so then!!!

The question for us here is to find in our languages, the homophone of hartebeest[20],
and which characteristic is feature related to 𓅱 wou "the foot, the flyer".

Reply:

Middel-Eyptian: 𓃡 ***i.w*** "dog"

Ewondo: /mvu/ "canid, dog"
Duala: mbɔ́ "dog"
Fulfuldé: rawaandu "dog"
Bété: ì-fè "dog"
Kɔ́ɔnzime: Mpyh "dog"
Wolof: xaj "dog"
Zulu: inja "dog"
Proto-Ekang "bʊa"
Chinese: 狗 Gǒu "dog"
kinyarwanda: imbwa "dog"
Basa'a: ŋgwɔ́ "dog"
Hausa: kàṛé "dog"
Baoulé: alwá "dog"
Téké: mvá "dog"
Yemba: mbhū "dog"
Senoufo: pónmúnò "dog"
Mandinka: wùlu "dog"
Japanese: 犬 Inu "dog"

Note : Chinese: 爱屋及乌 ài wū jí wū "Who loves me loves my dog"

One could easily demonstrate how one arrives to the pharaonic form ***i.w*** from the proto-versions. But to get familiar with the method it is not accurate on a pedagogical step.
But, from this series, it is clear that drawing parallels between the variant ***i.w*** and certain words above is impossible. Also, for the sake of simplicity, we have buried the subject in our ranks by contenting ourselves to relaying what is said by the others about a supposed Afro-Asian origin of the Pharaonic language.

Morphosyntaxic Verification

Middle-Eyptian: 𓃡 E14 ***i.w*** "dog"

Note that this word has an anteposed prefix, a remnant of the Negro-egyptian proto-version ****kja- >*hi-*** **.** At this point, the root word is well ***w***.

20. The hartebeest is the vernacular name commonly given to several distinct species of African antelopes. They are all bovines belonging to the subfamily Alcelaphinae.

This morphology is obviously confirmed with the feminine word Ewondo: / mvu /. Indeed, used in a possessive sentence, it induces accordance schemes of the feminine.
We thus say / mvu di-àmà / "my dog". The gender prefix is / i- / feminine tag and number prefix / d /.

However, note, that the current Ewondo synchrony has lost the prefixes of gender / i / and number / d / and only keep the radical / * mvu /.
It is only in the accordance schemes that those morphosyntactic elements are revealed. We can assume that the elements absent in the current synchrony was necessarily in the proto-version of Ewondo language.
When the root dresses up then the proto-version might look like:

> **Proto-Ewondo:** /***di***-mvou/ "dog"
> **Lingala:** ***i***-mbwa "dog"
> **Punu:** Mon-***di***"dog"
> **Lumbu:** Mon-***di***"dog"

Consequence of this, if the word had been masculine (ended by 𓅱 ***w*** in M-E) we would have had in a possessive sentence in Ewondo / mvu wama / "my dog" but, this is not the case. This variant belongs to / ndzouabi iwondo /, the rubbish Ewondo.

So the presence of this sign 𓅱 ***w*** here really indicates the root of the word and not its gender. The endogenous vision thus illuminates here the lack of clarity and the weird twirl of some people.

Summary of achievements:

This synthesis without going deep will focus only on the series of words
Caught randomly in African languages to designate the dog.

> **Baoulé:** alwá "dog"
> **Ewondo:** /à-loué/ "to call";/à-loho/ "ear"

> **Mandinka**: wùlu "dog"
> **Ewondo:** /à-wùlu/ "to walk"

Note: The Mandinka word for dog and the Baoulé word for dog are dealing by consonantal metathesis from the **Ewondo:** /à-wùlu/ "to walk"

> **Kɔ́ɔnzime**: Mpyéh "dog"
> **French**: pied "foot"
> **English:** paw

As surprising as it may seem, regarding the few examples taken at random above that, what characterizes the dog in the African unconscious is its particular penchant for riding. Obviously this penchant for the mop was part of the collective memory. Referring to the Ewondo language, we see that the Mandinka language has kept this characteristic of the dog associated to the mop, the gaby.

> **Mandinka:** /wùlu/ "dog";
> **Ewondo:** /à-wùlu/ "to walk".

But as surprising as it may seem, we can also note the phonetic resemblance between the word *Mpyéh* of Koonzimé and the French word pied "*foot*" even the English paw. Simple chance?!! I'm not so sure.

It seems to a naive observer that, the semantaxe that binds the foot to the dog is still alive nowadays. During canine training operations, there is often an iteration expression "at the foot" in French "au pied". The atavism of Ekang languages is therefore alive.
A truism notes that, on the other hand, the term pied "foot" is not associated with other animals.

The link between the foot and the dog is therefore archaic. It is a link inscribed in the collective unconscious. The reconstruction of the predialectal ancestor can offer us a better understanding because the reconstruction of the root ***xu** of Negro-egyptian can itself be derived from the protolexeme ***K**^{w}**-** "foot"

Negro-egyptian: *k^{w}i-xu-ŋ^{jw}ə > *hɨ-xu-n "dog"

Middle-egyptian: ***i.w***"dog"
Middle-egyptian: ***x.t*** "foot plant"
Middle-egyptian: ***i.d***"taurus"

Proto-kang: *bʊa "dog"

Ewondo: /mvou/ "dog"
Ewondo: /à-koh/ "foot"
Ewondo: /i-kéla/ "male"

If I can forward with the risk of offending the well-thinking people, the word pied *"foot"* of French obviously shares a phonetic proximity with the item *mpyé* of Kɔɔnzime, a language spoken in eastern Cameroon. Astonishing?!!! Not so much since the theory of evolution agrees on an African origin of humanity.

African Moorish people from the 10th century participated in the Western civilization of the world. That one finds traces of Ekang languages in the sibling western languages seems to be quite normal. And it is even more since the symbols of the first French Merovingian royalty are African symbols. I want to prove it through this testimony of Count Bernard Germain Etienne de Laville-sur-Illon:

> *"[...] The medical schools established in Salerno and Monte Cassino were in his kingdom of Naples. These schools, led by learned and pious Benedictines, worthy of all the gratitude of the friends of the sciences and of humanity, had been carefully enriched by all the knowledge which the Moors or Arabs preserved, cultivated, and taught themselves in the world. Muslim Spain, and especially in Cordoba.(...)"*[21]
>
> *(LACPÈDE, 1826)*

After this interlude, let's go back to our ***mDw nTr*** and on a rather delicate issue related to the good vocalization of a spelling. To highlight this problem, I will choose scripts that are similar after transliteration. It can be seen from the series below that all the words are transliterated by Egyptologists. On this account, it will be practically impossible to find an African linguistic variant that uses a single word to rigorously name all these spellings..

Middle-egyptian: ***w*** "quail chick" — **Ewondo**: /kuh/ "fowl"
Middle-egyptian: ***iw*** "is , are" — **Ewondo**:/*nə/"he is/they are"
Middle-egyptian: ***iw*** "hunchback" — **Ewondo**: /à-kul/ "hunchback"
Middle-egyptian: ***iw*** "come, come back vb. aux." — **Ewondo**: /à-wùlù/ "to walk"
Middle-egyptian: ***iw*** "island" — **Ewondo**:/i-bul/; /i-kor/"island"
Middle-egyptian: ***iw*** "to scream" — **Ewondo**: /à-bón/ "to bark"
Middle-egyptian: ***iw*** "dog" — **Ewondo**: /mvou/ "dog"
Middle-egyptian: ***iw*** "mischief, injustice" — **Ewondo**: /à-ku/ "sin"
Middle-egyptian: ***iwA*** "horned cattle" — **Ewondo**:/i-mvoul/ "horned"

21.Bernard-Germain-Etienne de LA VILLE LACEPEDE, General, physical and civil history of Europe, Mame et Delaunay-Valle éditeurs, Paris, 1826, p.56

It can be seen that if transliteration seems to offer a kind of comfort in the approach of these characters, it highlights by this serie an important methodological deficiency by the official vocalization [iwou].

Indeed, all these scripts are transliterated ***i.w*** but are they all vocalised identically in the current synchronies African linguistic variants? I'm not so sure. It will be difficult to justify at a technical level the morphosyntaxic links between the ***iw*** of the Middle-egyptian and the vocalizations of the equivalent variants that derive from them in Ewondo.

Several reasons can explain this fact technically. Languages are dynamic and evolve in space and time according to speakers and geoscriptal areas. These languages inherit specific grammatical aspects from Negro-egyptian. The choice of transliteration ***iw*** is also a simple methodological trick. The vocalizations used in the current synchronies of the African linguistic variants could also just be simple synonyms. For these reasons alone, it is almost impossible to vocalize all these words in the same way in a single linguistic variant, even if it were a sibling language of the middle-egyptian synchrony (Oxyrhynchite Coptic for example).

Thus, the only way in the traditional methods of studying these characters is to overcome this deficiency by resorting to diachronic linguistic techniques.
And it is these techniques that can help to understand why from the same supposed skeleton consonants ***iw*** in Middle-egyptian we can have all the vocalisations taken here as an example in Ewondo.

Quite malignant who will be able to demonstrate that in everyday life, to read these characters the ancient Egyptians were resorted to techniques of reconstruction of pre-dialectal ancestors. All the more so because the current Egyptian synchronies offer many dialects (*Bohairic, Sahidic, Faimumic, Oxyrhynchite or Middle-egyptian, Akhmimic, Lycopolitan or Subakhmimic*). It is more certain that everyone read these characters in their linguistic variant.
So it becomes useful to approach all this by another way; namely an endogenous and somewhat different path.

One of ***Hr*** characteristics according to African mythology, is to be born as any child without the direct use of its lower limbs. He was therefore qualified by metonymy in African languages as "The foot".

In many African regions the generic word for ***H***r and thus "foot" takes the shape of the root: * ***kul-*** "the paw"

From my point of view, representing a quail chick, we actually represent ***Hr***
There are several reasons to explain this:

- ***Hr*** *'s* main symbol is the feather. Feathers are covering also the body of the chick.
- The child in the trinity is the small "i-yəl" homophone of / i-yələ / "poultry".
- The monemes /ùn-mbokh/ "cripple" and /kub/ "poultry" are ananyms.
- The best representation symbol of the foot is the chick ***war.t*** "foot"
- In Ekang languages the moneme /kub/ is composed with /à-ko/ and /à-bo/ "foot"
- In Middle-egyptian ***wr*** represents a bird. The most emblematic being the hawk.
- One of ***Hr*** attributs is ***wr*** as eldest, ancient; great, important, etc.

Ewondo: ***w*** / i-woré/ "quail" **Ewondo**: ***Hr*** /ùn-ŋkoré/ "the savior"
Ewondo: **iwaw** /à-kul/ "tab"; /à-bèh/ "haunch; hind quarters"

So for all these reasons we conclude that the quail chick here is not a simple phonetic complement or potentially a mere gender prefix of the concealed masculine but, it indicates ***Hr*** himself, the radical on which one has to rely when reading these scripts. I make it clear here that we are talking about an operative concept and not about a person.

In the list below, the chick ***Hr*** gives by metonymy the idea of the foot that the scribe indicate also by the sign A2 of the Gardiner's list. Of course, he indicates indique a mouth but also a foot on the sign A2.

One could read this script while leaving the diktat of the transliteration, using other words. We note in the series below that the diktat imposed by Egyptology represents a single column. For example, we say in English that a bump is an outgrowth, a protuberance, a swelling, a tumor, a convexity, etc. is it fundamentally wrong? We think that, not at all since we reason more in a logic that names an idea rather than a logic based on an orthoscript corset. I could have requested other synonyms.

Quail chick		*w*	Kub	ùn-ŋkàŋ	i-wole	tsir	ù-bəŋ
Dog		*iw*	à-ko	á-kan	mvou	tsir	à-bón
To scream		*iw*	à-kur à-kar	à-kàŋ	Vouŋ à-vous	à-tar	à-bón
hunchback		*iw*	à- koùl	á-kan	i-voul	i-tur	à-bône
To come, to come back		*iw*	à-koul	à-kén	mvous wùlù	twara	bén
île		*iw*	i-kor ù-ŋkol	ù-ŋkàŋ	i-vele	i-tur si	i-bul
michief, injustice		*iw*	i-kor i-kot	à-koŋ ù-nson	i-vous	Aton i-tur	à-bh i-bén
Horned stock		*iwA*	à-koul	ù-kpweŋ	i-mvoul	i-tur	i-bèh

Let us note here that the series of words used in the above case by phonetic proximity with the transliterated form ***iw*** are not the only ones that offer a semantically correct reading. We can therefore rigorously mobilize all the synonyms that are related to the root ***iw*** on the semantic plane and the argument will be exactly the same.

Let's analyze the method of vocalization of these scripts and let us take at random one script:

Middle-egyptian: ***iw*** "misdemeanor, injustice"

The sign G37 given by Egyptology as a "Sparrow" (*Passer domesticus niloticus*), we name it in the beti languages / *ùn*-mvelkuma /. Literally the one who awakes kings or the wealthies.

Although the resemblance between the two passerines is very strong (*see illustration n ° 36*), we think that the sign noted G37 representes more in this script (*Pyenonotus gabonensis*) a small brown passerine singing morning and evening in our latitudes.

Ewondo: G37 /kolwore/ "*Pyenonotus gabonensis*"

As I said in this document, our African languages are built on the basis of association of basic tonemes representing operative concepts. Thus the expression / kolwore / can be decomposed in **Ewondo** : / kol / "foot" and / wore / "quail".

So reading this script is to answer the question of knowing, in what could be summerized the leg of a quail. So we apply that the leg of a quail is summed to a bone: / i-vés /.

The word consecrated to name a bone in Ewondo is /i-vés/. However the signs E9  and G43  being related to the sound [vou] will therefore lead us to read not / i-vés / but rather / i-vûs /. What characterizes African beef is the presence of a hump on the back / mvus /.

Middle-egyptian:  ***iw*** "misdeed, injustice"

Ewondo:  /i-vûs/ "mischief, injustice, error, sin etc."

The sign G37  /ù-tokh/ "Little green bird, family of the nightingale" is a little passerine. He offers by homophony the word / ù-tokh / "small". He embodies the idea of small.

In the Ekang languages, is called / va /, one of the small objects put together in a tight pineapple bag. So we find the phonetic proximity with / ivu- /

<u>Small History point</u>: Who is Haroëris for Egyptology?

Let us just remember that:

Middle-egyptian:  ***Hr-wr*** "Haroëris"

Ewondo:  /kolwore/ "*Pyenonotus gabonensis*"

We go through this account a lot of information in the so-called authorized literature, so I allow myself to record below some extracts of this literature:

> *"A superb falcon, Horus the Elder is one of the first gods worshiped north of Egypt. His name means "the Far One" because he lives far from gods and men. Sometimes he lowers his gaze towards the deities who never dare to look up at him. God of Air Spaces, his eyes shine with an unparalleled brilliance because they are the Sun and the Moon. When the two stars disappear, the darkness covers the Earth, Horus the Elder becomes formidable and fights with incredible fury. In the earliest times of Egyptian history, Horus the Elder gains importance because the kings of his city, Hierakonpolis, unify Upper and Lower Egypt.*
> *He becomes the patron of the pharaohs and the whole kingdom. Two proofs attest to his power. On the one hand, the Egyptians call the period preceding the unification of the kingdom "the Time of the Servants of Horus"; on the other hand, on the stelae of this distant epoch, they engrave the name of the king in the design of a palace on which perches a falcon."*[22]
>
> *(Viviane KOENIG, 2004)*
>
> *"Horus the Elder (or Horus the Great, or Horus the Elder) = Ḥr-wr or Ḥr smsw [Haroeris in Greek]. This hawk god is a very ancient celestial deity whose right eye represents the sun and the left the moon. He loses his sight more or less depending on the lunar cycle and even changes his name according to whether he is seer (Khenty-irty) or blind (Khenty-en-irty). He is the lord of Khem (Letopolis) in the Delta, where he is celebrated on the last day of the month of Epiphi and where he takes care of the "divine relic" the shoulder blade of Osiris. He was also called Ḫnty Gsj (the one who presides at Qous, in Upper Egypt) <2>. The Book of the Dead (chapter 112) states that Horus the Elder, who is a funerary form of the creator, had with Isis 4 sons: Hapy (Ḥpy) with a monkey's head, Douamoutef (Dw3-mw.t = f) with a head of dog, Qebehsenouf (Qbḥ-sn.w = f) with falcon head, and Amset (Jmstj) with man's head. Their heads are represented on the caps of 4 canopic jars <53>. The Heliopolitan theologians will belatedly be Horus the Elder one of the children of Geb and Nut. In the temple of Kom Ombo he is presented as Chou, the god of the air. He is also lord of Ombos and Qous in Upper Egypt <19>"*[23]
>
> *(Richard CHABY, 2014)*

22.Viviane KOENIG - Marie-Aude WAYMEL, Gods and Genii of Ancient Egypt, Livres de Poche, 2004

23.Richard CHABY - Karen GULDEN, Words and Names of Ancient Egypt: Volume 2: French - Egyptian, Reli, 2014, P178

"[...]Nut will then give birth to five children who are: Osiris-Isis, Seth-Nephtys, Horus the Elder. Horus the Elder, called by the Haroeris Greeks, is an aspect of the solar god Re and does not have an important role in this myth. On the other hand, the tribulations of the two couples: Isis-Osiris and Seth-Nephtys constitute what one could call the pre-humanity. They are considered the direct predecessors of men. Osiris, assassinated by his brother Seth, is the first god to know death, because before him, death did not exist."[24]

(LARRE, 1998)

As it can be seen, the content of these literatures is somewhat questionable if not surprising. In reality, there has never been a ***Hr-Elder***, it is simply ***Hr*** whose description in the word ***Hr-Wr*** is a focus on the foot and nothing more as we have shown calmly but surely. The bird that embodies ***Hr-Wr*** in ewondo is named / kolwore /. His performances are found alongside those of ***sbk*** at Kom-ombo temple. We call ***sbk*** "ùn-nkomboro" literally the humans creator.
We therefore see that, a different approach allows to better understand the links that unite these scripts and the most accurate way of vocalizing them while shedding new light on what could be akin to an intellectual swindle.

Was it possible to read this script another way?

It was possible without great intellectual gymnastics to read this script. Let's quickly see how:

Middle-egyptian:  ***iw*** "mischief, injustice"

It was enough to know or remember that the hartebeest is a bovid and that the bovid / nyakh / in the African paradigm, boils down to: / á-kúl / "the hunchback"; / to-kùl / "the paw";

We have proof on this subject that the word for bull is among others the leg, the foot:

Middle-egyptian: 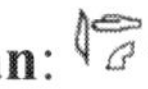 ***i.d*** "taurus, bull, cow"

So the enigma related to reading would have been summed up to find what is the paw of the chick quail / wore /. The following you know it better than me.

Few quick reads examples.

- How could we summerize the speech  of the foot  ***Hr***? (mouth = speech = foot)

  Middle-egyptian:  "lamenting, shouting"
  Ewondo:  /à-bî /;/à-yî/ "groan"; /à-kɔbɔ/ "to talk"

- Reconstruct the word hartebeest /i-mvul/ homophone of the upper limb /wa/.

  Middle-egyptian:  ***iwa*** "inherit (from); take over from; act as heir"

  Ewondo:  /à-yia/; /à-bia/ "compensate, succeed"; /à-bigh/ "get rich"

- Soft part or piece of flesh , of the small foot /obəh/ or the upper limb /wa/ refering to the numeral 1 so to the smallest again.

  Middle-egyptian:  ***iwaw*** "thigh, leg, haunch"

24.Christian LARRE, The spiritual heritage of ancient Egypt, Diffusion rosicrucienne, 1998.

Ewondo: /à-bèh/ “haunch”; /i-bah/ “cut, slicing”

- An action using a knife and vocalised as two or as a door in wondo.

  Middle-egyptian: *wpi* “to cut, to slice”
  Ewondo: /á-bèh/ “to cut, to slice” etc. when we mobilize the synonyms.

- Let’s find the word related to bovid or to hand and which designate a ring.

  Middle-egyptian: *iwaw* “ring”

  Ewondo: / ùm-mbokh /; /i-béh/, /à-lɔŋ/ “hole”; /i-vu/ “empty, vacuum” etc.

- Let’s find the word use to name the la protuberance of a mother.

  Middle-egyptian: *iwr* “conceive (a child); to become pregnant”
  Ewondo: / à-toa à-boum /; /Kîkîle/; /ndue/ “pregnant” etc.

- Find the word that is used to indicate a place and/or the mouth of the disabled /ùn-mbokh/.

  Middle-egyptian: *hrw* “day”
  Ewondo: / à-mos /; /à-bokh i-yén/ “day” etc.

- Find the word used to name the darkness, synonym of authority.

  Middle-egyptian: *Axw* “night”
  Ewondo:/ à-lou / (due to the ear lobe); /dibi/ “night”; /mə-kw/ “night”; etc.

- Find the homonym of placenta also used for darkness or smoked cassava.

  Middle-egyptian: *axxw* “dusk”
  Ewondo: / à-lou /; / á-ngôghè / “dusk, falling night” etc.

Logic of Complexity

Let's take a look at the dog script below.

Middle-egyptian: *iw* “dog”

We notice that within the same language, the same linguistic variant, the three signs share a certain phonetic proximity.

	Sign E9	**Sign G43**	**Sign E14**
Ewondo	imvul “horned beast”	Imvu(l)	mvu
Ewondo	kabat “chèvre”	Kub “volaille”	***sAb***: *zamba*

Chart 60:the three signs share a phonetic proximity

The sign G43 representes ***Hr*** one of the representations is the bull. That's why he is / i-mvul /. When he is associated to ostrich he becomes / i-mvul ù-nwan / "buffalo bird"

In Toupouri language we have: Way “dog”, Twy\twi\touwi\touwy “bird”, Fy\fi “goat”

In many languages, the dog is known sometimes just by one name. But by setting the triple equality above, makes it possible to realize that there are synonyms for the word dog in our languages. So we know the two associated meanings for example:

Middle-egyptian: ***sAb*** "jackal"
Middle-egyptian: ***sAb*** "dignitary"

If we apply the same method here with the new signs engaged for ***sAb***

	Sign G1	Sign D58	Sign O34
Ewondo	i-mvul ù-non *"bull bird"*	I-mvul *"synonym of Hr"*	I-vul *"roll, tie"*
Ewondo	**kuba** "vulture"	**ŋgob** *"shoe"*	***Ùn-ŋkab**la:* *"which serves to attach"*

Chart 61:the three signs share a phonetic proximity also

We notice here that the foot is associated with the shoe and in fact also admits all the associated homophones. One of the homophones associated in Ewondo is the ketartiodactyle below:

Middle-egyptian: *xAb* "hippo" **Ewondo:** ngup **"**hippo"

What I want to say is that in the complexity all the words involved here are synonyms and homophones. So I will take the last case and focus on the Ewondo that I do understand.

Ewondo: mvu "dog"
Middle-egyptian: ***sAb***

xapaitíiso: giopai "dog"
Ewondo: ngup **"**hippopotame"

The word dog can therefore be vocalized as the word hippopotamus in two variants of the same language. How can Indians in the Amazonian forest call the dog by the word hippopotamus of African languages without knowing this link that only the hieroglyphs bring to light? This fact is true in other languages.

Ewondo: mvu "dog" **Arabe**: الكلب *al*kalb "dog"
Middle-egyptian: ***sAb*** **Ewondo:** /Ùn-ŋkabla/ "impeded"
Ewondo: à-kol mvu "dog's foot"
Middle-egyptian: ***ssmt*** **Ewondo:** /i-kab'li/ "horse"

Ewondo: chèvre **"**kab'ra" **Ngiemboon:** Memvo "goat"
Middle-egyptian: ***iw*** **Ewondo:** mvu "dog"

Ewondo: ngup **"**hippo" **Ngiemboon:** *Noon Tse* "hippo"
Middle-egyptian: ***xAb*** **Ewondo:** *ù-nwan tsè* **"**terrestrial bird"
Ewondo: ngup **"**hippo" **Kotokoli**: Boudou "hippo"
Ewondo: i-bubur **"**soft, soft to the touch"

Ewondo: bod **"**people" ***Hr's*** qualifier
Ewondo: dubu **"**prisoner" ***Hr's*** qualifier

Ewondo: ngup **"**hippo" — **Mooré:** yémdé "hippo"
Ewondo: *ùn*-yemde **"**impeded"
Ewondo: nda i-yəm **"**the disabled's House "

Ewondo: ngup **"**hippo" — **Ewondo:** ùŋ-yinga mor **"**revolted, angry"
Soninké: Yingamer "hippo"

Ewondo: ngup **"**hippo" — **Wolof:** Leubberr "hippo"
Vieil-Ewondo: le-bor (o)"people" Hr's qualifier

Ewondo: ngup **"**hippopotamus" — **Maasai:** olmakau "hippopotamus"
Ewondo: mə-kô mə Hor "***Hr's*** *feet"*

Ewondo: /mvu/ **"**dog" — **Fulfuldé:** rawaandu "dog"
Ewondo: /i-warèn nda/ "purification of the house"

When there have been a violent death (blood) in a family, there is a ritual among Beti people consisting in throwing over the roof of the mourning house a young female canid to ward off bad luck.

The conclusion I draw therefore is that from an African language, I can understand words in a language that I have never spoken before just by using ***mDw nTr*** and simply by phonetic proximity with my maternal language.

I find that what we call *names* in reality are just descriptors of a reality from an endogenous language. Just take the measure of the words used here and there. These words are only synonyms.

If I take the example of the French word chien "dog", it will correspond by phonetic proximity to the word:

Ewondo: /à-sɔŋ/ "tooth" because to the dog, could be associated /à-sɔŋ mvu/ "dog's tooth"
French: chien
Note
So, if we want to reconstruct a predialectal ancestor between the word *chien* "dog" and the word Ewondo : /mvu/ "dog", most of the time, we are going to collect the two words borrowed to the two languages and sharing obviously the same meaning. Then we are going to apply linguistic tools to stick them. Here is then the trap.

We postulate the unconscious premise that Ewondo and French are two languages of different strains that have nothing in common, and that is precisely what we are mistaken. As I pointed out in this work, all languages in the world use the same lexicon.
This is according to me, the tremendous error of comparative linguistic. Is there absolutely any ancestor's word between the French word and the Ewondo word for dog? I'm not so sure. The fact is that the expression /à-sɔŋ mvu/ "dog's tooth" gave according to me the word *chien* in French.

It is thus indisputably the word Ewondo / à-sɔŋ / that is at the origin of the word *chien*. One can indeed have the morpheme / à-chəŋ / very close to the Fang vocalization. Let's prove it by the spelling:

Middle-egyptian:  ***ibH*** — **Ewondo:** à-sɔŋ "tooth"
Middle-egyptian:  ***sAb*** — **French**: chien

Middle-egyptian: ***ibH***

So we find here the link between the tooth and the dog but also with the goat. In short, the goat is similar to the dog and is also to the tooth. The French word *dent* is reminiscent of the word Ewondo / à-daŋ / "stepping over, jumping etc. ". All these words are related to the foot and the dog as we can see on the script.

If I take the example of the word Bambara *dosso* meaning dog for example, it will correspond by phonetic proximity to the word:

> **Ewondo**: dysi "eye"
>
> **Middle-egyptian**: ***gsw*** "to squint"
>
> **Middle-egyptian**: ***i.w*** "dog"

So we find that the sign G43 that links the two scripts clearly showing that the word *dosso* of bambara corresponds to the word / dysi / "the eye" of the Ewondo. Conclusion of what, I discover by the way a term to designate the dog that has fallen into disuse in the Ewondo language.

If I take another example, the word Toupouri *way* "dog" for example, it will correspond by phonetic proximity to the word Ewondo / mvu / "dog" or /à-woué/ "to kill" or / * wou / "death, poultry" brought by the quail chick G43 .

Note:
In western Cameroon, a female dog is called Maâ'vou (the mother of death in Bandjoun "Maâ" mother and "vou" dead/dog).
When a traditional chief dies in western Cameroon, the local custom is to beat a dog to death.

The words **Wolof**: *Xathie*; *koutie* "dog, puppy" that we find almost identical in
Tinigua: *xáḍà* "dog" remind us of the words of the female genitals in **Ewondo**:"ngout"; "zout". The only way to validate this is to find a script that certifies it.

> **Middle-egyptian**: ***s.t*** "woman"
>
> **Middle-egyptian**: ***kA.t*** "vagina"
>
> **Middle-egyptian**: ***sAb*** "jackal, canid"

The vagina is the woman, but also the dog. So I can understand this with ***mDw nTr*** and using only phonetic proximity.

Few examples out of Africa.

Language	Dog	Ewondo	Scripts	
Tinigua	xáḍà	*Ngout;zout*		
Witoto	huko	*À-koh (foot)*		
Zaparo	ariawko	*À-nyu A-koh*		
Quichua	aSku	*Àsè ko*		
Arawak	peero	*i-fər "lock"*		
Tupi	jaguar	*Ngən; i-dîgha*		
Wôpanâak	anum	*À-nyu "mouth"* *i-nyum "smell"*		
Tucano	diayi	*Dywé (nose)*		
Maku	Eme*ko7	*mə-koh (feet)*		
Siouane	rų·pa	*Fara (priest)*		

Chart 62: the word dog out of africa compare to Ewondo

Semantax

When a woman is considered to be morally corrupted, it is said in French in popular language that she is : "*une chienne*" litt. a female dog "a bitch". The phonetic proximity offered by the Wolof word for dog in Ewondo thus makes it possible to enlighten us on the secular link which remains in the collective unconscious between the frivolity and canines. I should almost have said hugs but that's something else isn't it?

Complexity:

If the mother is associated with canids for their frivolous nature, but she is not only frivolous. When studying the numeral 7 it was noted that the placenta related to the mother / *nya / is associated with the number 7 highlights her natural ability to be supporting.

One could highlight this resilience of the mother in the homophony that exists between the word Ewondo word /dibi/ "*night*" and the word /*à*-dibi/ *to support*. We have seen that the word naming darkness are also synonymous of maternal authority .

Middle-egyptian: ***Axw*** "night"
Ewondo: /dibi/ "night"; etc.
Ewondo: /à-dibi/ "to support"

Ewondo: /i-zombo / "the mother because she is the support"
Ewondo: /à-zombo / "to support"
Middle-egyptian: *w**Hm*** "jackass; donckey"
Middle-egyptian: ***Hm.****t* "woman"

We thus discover that women share donkey's attributes . The words to designate the donkey in African languages are automatically linked to the mother.

Bandjoun: *na'a* **"**donkey"	**Ewondo**: *na* "mum"
Seerer: *fãm* **"**donkey**"**	**Ewondo**: *fam* "man"
Kɔ́ɔnzime: *jakâh* **"**donkey**"**	**Ewondo**: *à-koh dza(l)* "country"
Fulfulde: *wamnde* **"**donkey**"**	**Ewondo**: wa *nda* "N°1 of the house"
Hausa: ʒà:ki "donkey"	**Ewondo**: à-koh dza(l) "country"
Wolof: *Mbaam* **"**donkey**"**	**Ewondo:** *à-bam* "board, scold**"**

In example N°7 EXAMPLE 7 (3.4.3.7.) We saw that a table is what is abused in africain languages. The word table draws in this respect its etymology to the Ewondo word / i-tib'li / "what is abused". So let's close this parenthesis that leads us from canines to perissodactyls. I can gear on the homophony between / zombo / "baboon" and / à-zombo / "to support" and we would find ourselves in the middle of great apes with ***DHw.ty*** also transliterated ***ana***. This last transliteration is confirmed by the word **Ewondo**: *na* "mum" seen above.

Etymology:

Middle-egyptian: *iw* "dog"
Mbaye (Chad): **dog** "buffalo"
English: dog

How is it possible that 6000 years at the time when the scribe draws what we can see above, the dog for someone can carry the name as the buffalo for others?
This is true only if they share the same reality. Is the buffalo a temperate area animal?

In any case we have a better understanding of the origin of the word dog. But you can observe that what we often call *god* is an ananym of *dog*.

If, as I have just proved, the french word *chien* for dog finds its etymology in its association with the word *tooth* in an African language, it would be desirable that the Greco-Latin humanity of dictionaries change their etymology and put an African origin of the word *dog*. In my opinion it is more logical if we consider that, the Man anatomically modern was born in Africa and has acquired the language there.

When British dictionaries note the word *asshole*, the word is rigorously understandable from the African prism with the spelling of dog . Indeed, the symbol G1 representes over the vulture but also the mother's figure. Its simplified variant is the feather or /à-sè/ "feather". The symbol D58 is vocalised /à-kol/. What is the link with *asshole?* It is established with the script ***kA.t*** for soft parts and therefore the female genital area.

The insult in the language **Ewondo**: / mə-bih ya zut / literally the crap of the genital area comes to reinforce this idea. Although we have lost the etymology, I just note that the anthroponym **Assako** has survived until today in Ekang country.

The word ***asku*** "dog" in Quichua is the equivalent of the word English ***asshole*** "asshole "by loss of K velariness.
As we have just seen it above, the words related to dogs are those of the anus, the smell, the stench etc., I can then emphasized that, the English word dog, is related to the word Beti: /i-dugh/ "toilets; facelift; washstand etc.".

Ewondo: / i-dugh / "where we often crap or shit"
English: dog "canid"

The french word *anus "asshole"* can also find here its etymology because the dog is assimilated to the mouth /à-nyu/ associated to the feather /à-sè/. The script ***sAb***.

If the semantax associates the foot with the dog by metonymy, we name the foot in Ewondo / à-koh / but the mouth being associated with the speech by the word / * kobo / . The two tonemes which constitute the word / * kobo / all of them are designating the foot / * ko / or / * bo /. We can therefore consider that the *dog* is a *mouth.* This is why, those who have the right to speak, i mean the dignitaries are associated with the dog. For the purpose we still have the semantax to bark as a dog. "- Keep barking".
Let's us note to finish that by metonoymy relationship, anus, female sex and mouth are smelling as dogs.

For South America peoples who speak **Wôpanâak**, the dog is summarized compared to Ewondo to the smells like the mouth or while its anal pendant. The African languages thus light up the word *anum* of Wôpanâak language which serves them just to designate the dog. In Ewondo, /*i*-nyum/ "odor". If speaking about a dog I use the word /*i*-nyumu/ "odor" I'll be understood.

One might think that the word *peero* of the **Arawak** to designate the dog is a loan made to Hispanics. That would arrange a lot of ideologists, but that would imply also that all the peoples of the earth could have known the dog except the Arawak and they knew the dogs just because the Conquistadors visited them and gave them the Spanish word: *perro* "dog". To this account, it would be necessary that one also explains me why the same word is vocalized like a lock in Ewondo and figure also on this script .
Allow me to name this sign in Ewondo: /i-fér/ "lock"

What about **Fataluku**, a Papuan language spoken in East Timor using the same root to designate the dog with the word *iparu*?

Why do in **Tupi** language of South America, dog are called jaguar? Is it an error or language misnomer? Hallucinations because of Amazon rainforest drugs?
To incantations, once again the ***mDw nTr*** twist the blow to the ideology and deliver the right information, for the benevolent who could be led to believe that it is a mistake.

Middle-egyptian:  ***mA-HsA*** "lion"
Ewondo: /ùn-saŋ'a/ " the one who tears " association dog and teeth /mə-sɔŋ/
Tupi: jaguar "dog"

Middle-egyptian:  ***nDA.w*** "splinter of wood"
Ewondo: /ndzâgh/ "wood/
Tupi: jaguar "dog"

Middle-egyptian:  ***mAaw*** " variety of wood "
Ewondo: /máàgh/ "charcoal/
Ewondo: /ndzâgh/ "wood/ ↔**Ewondo:** /ndzûgh/ "pain"
Tupi: jaguar "dog"

The idea of sex being embodied by the mother and therefore by metonymy associated with the dog, or the mother being a panther so, a felid can be associated with felines (leopard, lion, cat, etc.) as seen on the scripts above.

I then believe that, the linguistics logic that has founded supposed borrowing of certain languages from others should be reviewed on the basis of the *mdw nTr* evidences. I think that linguistics had a lot of fun basing its argument on declarative about the supposed borrowings made to languages as if the languages were coming from different humanities but also were fixative data.

If for many peoples of the earth, the dog can be a feline, an equine, an odor, an elephantid, a ketartiodactyl, a goat, the sex of a woman, the soft anal part, a tooth etc.

Basically are the standard classifications still as relevant as that? Which criteria has humanity used to structure the collective consciousness and therefore the art of naming beigns and things?

The hypothesis put forward by some serious researchers who have stipulated that Homo sapiens sapiens identifies wild animals by their dangerousness to humans, in this case the dog may as well designate a sex as an odor or even a penchant for frivolity or even a mother. This hypothesis becomes what?.

There is therefore a paradox between the information delivered by our mental schemes and the classifications offered by research today.

If the perspicacity of the ancient Egyptians is no longer questioned in view of their scientific progress, how could such clairvoyant people have allowed themselves to be abused by mere impressions? How could they correlate animals in languages around the world on the basis of simple impressions?

Perhaps then we only perceive shadows as was supposed to have said Plato in *The myth of the cave*.

From these illusions we have built models based on impressions of languages and classifications that are easily undermined here.
Why does Homo sapiens sapiens name dog entities that appear totally distant each other?
When Saussure evokes the notions of signified and signifiers, what is the signified here?

If all these realities have the same meaning then what are we talking about?

When studying embryogenesis, it is found that a totipotent cell initially gives completely different tissues in terms of appearance or functionality.
In a complex pattern the tree is in the seed and the seed is in the tree.
Does this diffraction of the real in a fractal logic could better explain what is essential as an undeniable fact of language?

3.4.3.4. **Example 4**: Familiarization With The Method

Classical method: What is done according to the conventional method

Middle-egyptian: ***niw*** "ostrich" (*Faulkner p. 125, Wb. II p. 202, 8-11*)

Endogenous method:

Step 1: Inventory of characters

Sign N35 ***n*** of the Gardiner's list. "water undulations; water corrugation"

Sign M17 ***i.*** of the Gardiner's list. "Reed in bloom"

Sign G43 ***w*** of the Gardiner's list. "quail chick"

Sign G34 of the Gardiner's list "ostrich[25]"

characters Analysis

Sign M17 ***i.*** of the Gardiner's list. "Reed in bloom" is called in Ewondo: /mə-sam/ or /mə-sèh/ "feathers, flowers" this sign informs us that the sign G34 is characterized by its feathering, it is a feathered animal. The word reed is / *i*-sòŋ / which shares the same root with / *ùn*-ŋsòŋ / "the beak". The plant shoot is usually spirally wound: / i-vuvul /. Morphosyntactically, the term / i-vuvul / can be likened to a superlative.

Ewondo: / i-mvul / "buffalo"; **Ewondo**: / *i*-mvuvul / "big buffalo"

The sign G43 ***w*** of the Gardiner's list "quail chick" representes /i-ŋgurukh/ a little bird or an emerging bird, a chick. Compared to the ostrich, the difference is related to the size of the two birds.

25.(Struthio camelus)

Thus the term that highlights the comparative size between the whole and the small is /i-ŋgur/[26] to designate the small size. Therefore the sign G34 will be vocalized by the homophone /i-ŋgura/ [27] designating the whole, what is total; full etc.

In addition to this detail, the sign G43 ***w*** informs us that the sign G34 is a volatile. We have the immature term / i-yel / which characterizes the chick and is homophone of / i-yələ / "poultry". This is why it is useful to have a good knowledge of African languages in order to avoid *belly dancing*.

Sign N35 ***n*** of the Gardiner's list. "water ripples" will inform us by:
Its phonetic proximity to the connection between the meaning / ù-soé / "river" and / à-sèh / "feathers"
By his / nyu / "drink, water" will give us / à-nyùh / "fingers, mouth"; the bird in question would have a special mouth and special fingers.
The mouth of a bird being a beak one will thus have in hidden sense / ù-ŋsoŋ / "beak", animal having a beak.

We also have the pleating fact or waving is called /bôm/ as a packet formed by the spiral symbolizing the sign M17 .

When the water is wrought because of the wind we say /mə-ndim ma bômla à y'i-vuvul/. The owrd /vuvul/ can also help to designate the bird concerned. By the root ***iw*** it recalls the buffalo / i-mvul / synonym of big and strong.

Overall the Sign N35 indicates that we have a volatile / ù-nwan /. We have as evidence of this:

Middle-egyptian: ***nwn*** determinative is a **bird**.
Ewondo: /*i*-nyièn/ "the entrance"

Ewondo: /*i*-nyùèn/ "drinkable; drinking"
Ewondo: /*i*-nyù/ "drink" **Middle-egyptian**: ***niw*** «water »

Ewondo: /*ù*-nwàn/ "bird"

This script is read from the bottom to the top:
Middle-egyptian: ***niw*** "ostrich" **Middle-egyptian**: ***niw*** «water »
wondo: /i-ngour' ù-nwan/ "the whole bird, total, accomplished"
wondo: /i-mvuvul/ "the buffalo bird"

Let's read another script to make sure the reading still holds.
Middle-egyptian: ***niw*** "ostrich"

Ewondo: /wa/ "arm" homophone of /wa/ "1, one, total, full"
Ewondo: /i-ŋgur/ "disabled" homophone of / i-ŋgura/ "total, full"

Ewondo: /ùn-mbokh/; / mvughla/ "spirale" homophone of /ùm-bokh / "one,1"

26. ÑGUR (b) n.5, pl. biñgur (vbñgurga b). Incomplet, mutil. ñgur wo, arm stump, one-handed. ñgur ônyu, doigt coup. ñgur ôzar, pantalon très court de jambes. Pour un arbre coup, on dit kum (souche). P.115
27. ENTIER: total, complet, ñgura (h).whole half, ñgur'tun (hb). Whole cassava, ñgura mbôe ne-vyokh (b).whole day, zô ñgura ne-vyokh. P.461

Ewondo: /ùm-bokh / "un,1" synonym of / i-ŋgura/ "total, full "
Ewondo: /i-nyièn/ "the entrance" because it's a hole"

Middle-egyptian: ***zA.t*** "Northern Pintail; duck"
Ewondo: / soghe./ "duck; placenta"
Ewondo: / nya/ ; /i-sôgh/"placenta; mum" ☞**Ewondo**: / *nyî/ "entrance"

Final reading

Ewondo: / i-ŋgur' ùn-nwan/ "ostrich" literally the whole bird.

Note: each sign could be read and is self-sufficient in order to identify the determinative. Therefore, i do not see the need of transliteration or the supposed reading direction.

Other variants

Hausa: jimina "ostrich"
Fulfuldé: ndaw "ostrich"
Gourane: Soughoun "ostrich"
Zulu: umqambi "ostrich"
Soninké: firige "ostrich"
Wolof: bànjóoli "ostrich"
Arabe: naam "ostrich"
Tupuri: jî "ostrich"
Somali: gorayo "ostrich"
Bambara: Kono sogoti "ostrich"
Massa: Reketta "ostrich"
Qhichwa: suri "rhea"
Ngambay: ɓū "ostrich"
Dazaga: sougoune "ostrich
Ntumu: Kondón kúm mi-nkənə "ostrich" *litt. perched poultry paws*

We can see from these examples that the transliteration ***niw*** is not sufficient to name this bird in the different African linguistic varieties. ***Niw*** is the equivalent of the word /*ùn*-nwan/ "bird" .We can also note that the approach of rebuilding the predialectal ancestor shows its limits here. African linguistic variants operate random choices of signifiers[28] because they do not name the signified as could be understood the Saussurian meaning of this term. They name characteristics of the signified[29]. When one says he rebuilds a pre-dialectal ancestor on what is it based on? Is he based his rebuilding on the : signifier or signified?

The case of Ntumu language is quite illustrative because by diachronic techniques, one would simply rebuild an ostrich word. The Ntumu characterizes this bird by its perched paws more than it could name it as we could imagine. The word ostrich can be approached by the ewondo word /ù-tsiri/ "trampling or a lot of meat". We have noticed above all that, this bird has special fingers. It is then quite possible that some people have made a focus on its trampling.

28. The signifier designates the acoustic image of a word. In classical logic we speak of significant voice by convention. What matters in a word is not its sound in itself, but the phonic differences that distinguish it from others. Its value derives from these differentiations. Each language constructs its lexicon from a limited number of phonemes, characterized as signified, not by their own and positive quality, but by what opposes them: rolling an "r" in French is of no consequence for comprehension; not doing so in Arabic leads to confusion, since this language has both a vibrating apical [r] ("r" rolled) and a sonic velar fricative [ġ] (close to the French "r" grassey). The words rasoul (messenger) and ġasīl (detergent) are distinguished only by the opposition r / ġ.
29. Ferdinand DE SAUSSURE, General Linguistics Course, Paris, Payot, 1972, 520 p.

3.4.3.5. Example 5: How to read this script ?

Middle-egyptian: *it (Faulkner p. 32, Wb. I p. 141, 10-19 it (itf))*

Many words can evolve in space and time but the words related to the family are in principle quite stable. They are part of the basic lexicon. For this script, I suggest to interrogate African languages first on this subject:

Isizulu: *ubaba* "dad"
Duala: *sango* "dad"
Basa'a: *sáŋ* "dad"
kɔ́ɔnzime: *sɔ́ŋ*"dad"
Mambwé: *atata* "dad"
Bassari: *-sàs* "dad"
Mòoré: *tã̀:bō* "dad"
Mandinka: *baabaa* "dad"
Krw: *isatu* "dad"
Denya: *ntɛ*"dad"

Ewondo: *i-sià*"dad"
Oluganda: *kita* "dad"
Hausa: *uba* "dad"
Wolof: *baay* "dad"
Ciluba: *-satu* "dad"
Kinyarwanda: *so* "dad"
Punu: *bitatu*"dad"
Tomba: *háto*"dad"
Ekégusii: *tata*"dad"
Fali: *dǎ*"dad"

Soninké:*fa:ba* "dad"
Baoulé:*sĩ*"dad"
Bété:*ì-kí* "dad"
Kotoko:*gákǔru* "dad"
Yemba:*mɔ'ɔ̄*"dad"
Dogon: *m̀ bá*"dad"
Vili:*ńkulu* "dad"
Téké:*tàará* "dad"
Rikpa:*Bɛ̀ll* "dad"
Dinga:*ta*"dad"

We notice several groups in these names, those in s/t/d; ba; son etc. Let's try to understand what happened. For that, let's take the transliterated script ***itf*** "father"

Step 1: name the signs:

Sign M17*i* of the Gardiner's list "Reed in bloom"
Ewondo: /i-soŋ/ "reed"; /me-sà/ "feathers" ☞ /i-sia/ "father"

Sign X1 *t* of the Gardiner's list. "roll bread"
Ewondo: / ùn-ntuba/; /i-nsia/ "banana's cake" ☞ /i-sia/ "father"

Sign I9*f*. "horned vipers"
Ewondo: /ùn-nsóŋ/ "viper; worm"; /tsia/ "earth" ☞ /i-sia/ "father"

Just by naming the signs we can notice to which descriptor of the script each linguistic variety is linked.

Step 2: Let's try to understand why we have different terms in the current synchronic.

The tree is in the seed and the seed in the tree, roots are in charge of the nourishing function according to the African sociological scheme, namely, the father and the mother. Let's try to rebuild the language ancestor word and see how the word has evolved over time.

By taking again the different words randomly listed one can thus propose the following reconciliations:
Negro-egyptian: *kja- "food, mother, father";☞ **Bété**: *ì-kí* "père"

Middle-egyptian: ***iti*** "dad";
Ewondo: /i-toà/ "chief, seat, place";
Mambwé: *atata* "dad";

Middle-egyptian: ***it*** "barley, cereals"
Ewondo: /mə-tè/ "saliva; spit"
Ekégusii: *tata* "dad"

Middle-egyptian: ***swt (zwt)*** "wheat";
Ewondo:/i-sià/ "father";
Ewondo:/zié/ "hunger"
Ewondo:/zàm/ "appetite; relish; zest"

Middle-egyptian: ***wSa*** "eat".
Ewondo: /i-nsià/ "rip banana plantain"

Middle-egyptian: ***dAdA***"head; figurehead; leading light; leading figure"
Middle-egyptian: ***rdi***"plant, place, implant, put, name,etc,"
Ewondo: /à-di/ "to eat"
Fali: *dǎ* "father"

Middle-egyptian: ***wSm*** "Wheat ear beard".
Middle-egyptian: ***spd r wSm*** "pointed like a spiked beard"
Ewondo: /ù-nsoŋ/ "sting, sharp, spike, worm, grave etc."
Kɔ́ɔnzime: *sɔ́ŋ* "father"
Basa'a: *sáŋ* "dad"

The sign I9 is read /ù-n*sɔ́ŋ/* "as worm" and /fì/ as "snake" no supposed debate is allowed here to know if the word should be transliterated ***it*** or ***itf.***
According to me, it is just a degree of precision on the criterion put forward by the scribe.

Middle-egyptian: ***bit*** "bread";
Ewondo: /à-bèh/ "to plant";
Isizulu: *ubaba* "dad";

Middle-egyptian: ***bit*** "lower Egypt's crown"
Ewondo: /ùn-mbaŋla/ "bread"
Mandinka: *baabaa* "dad" etc.

When one analyzes things from the unique angle of consonants and grammatical laws, one is saying that ***it*** is a masculine term ending with ***.t.*** An exception made by the word father, a male figure. This denotes inaccuracies related to arguments a bit misleading.

As we can see, ***it*** is ***Hr*** as the father's figure and as the leader who is assimilated to the seat / i-toà / by metonymy. As a parent, it refers primarily to African languages, the nurturing function that is linked to the parental root.

The main difference between ***it*** and ***itf*** is simple. *Itf* has to be used for the word *father* as a priest **fara/fata** and *it* is a normal father **tara**.

I say that this function is related to cereals (wheat / barley) which are ***wsir.*** According to the legend, sheaves of cereals would have sprouted where the pieces of his body after being killed and cut were scattered by his brother ***stX***. It must be considered that everything in the unitary form gives the impression of a division and this account is assimilated to ***wsir*** whose body was cut.

The upright manner of reading this scripture therefore imposes for us Africans, a good understanding of the operative concept presented by the Osirian myth in order to make the link between the current residual vocalisation in each of the languages with the elements of the spelling.

It can thus be seen that any attempt to vocalize in our languages on the basis of ***it / itf*** variant will lead to two alternatives: a happy coincidence or a bitter failure. The forced orthograph reading is not adapted but by an analysis of the operational concepts which are related to the paternal function preferred way of the African culture namely ***wsir.***

<u>Note</u>: Each sign once read is sufficient in itself to offer the vocalization of the determinative. I do not see the need for transliteration or the supposed arbitrary meaning of reading.

3.4.3.6. <u>EXAMPLE 6</u>  *(Faulkner p. 32, Wb. I p. 143, 3-14)*

<u>Step 1</u>: Name the signs

Middle-egyptian: Sign M17  ***i*** of the Gardiner's list "flowers, feathers"
Ewondo: /me-sà/ "flowers, feathers"

Middle-egyptian: Sign U33 ***ti*** of the Gardiner's list "Pestle"
Ewondo: /ùn-ntoum / "Pestle"

Middle-egyptian: Sign G7  ***hr*** "***Hr*** on the bulwark; pillar"
Ewondo: /*ùn*-ŋkoum/ "***Hr*** on the bulwark; pillar"

Middle-egyptian: Sign A23  ***ity***. "King with stick and mass"
Ewondo: /*ùn*-ŋkukuma / "King with stick and mass"

<u>Step 2:</u> sorting of signs and character analysis

Although some only consider it as a semanteme [30], the sign  G7 transliterated ***Hr*** offers in Ewondo the vocalization / *ùn*-nti / "lord, majesty".
This vocalization is not linked to a "god" or an argument of authority, but from the point of view of endogenous cultures, it is a formula of eulogy used to mark the respect due to the primordial ancestor. And so, this formula is used between descendants of the primordial ancestor and especially to formulate the respect related to the birthright that belongs to the sovereigns who are his followers.

The sign M17A  transliterated Y, designate for us / *ùn*-mbias / "twin; duel".
In Ekang thought, we say:

- /***yor***/ the smaller of the two twins. So we have good understanding of this sign.
- The word /yor/ offers an homophony with the root /****yar***/ and to the word /*nyar*/ "The buffalo" an ultimate symbol of ***Hr.***
- The Men's dance for the disappearance of the chief among the Ekang is called /*nyar*/.

Used in this script, this sign alone is enough to name the leader because it identifies him by the twin but also by the buffalo. Thus by synonymy and the game of homophony he can offer the full lexical pallet reserved for the word chief.

Used here, it can also constitute an invitation of the scribe to consider that the term will be built on the morphological plane like a superlative in our languages. Just by the fact that y= twin then leads to repetition. This repetition is done on the first syllable of the radical.

---

30. In linguistics, semanteme is a term used by some linguists to represent: according to Charles Bally, a sign expressing an idea "purely lexical", excluding "grammatical signs", and likely to take various forms. Ex: the radical lup-, the word simple wolf, the word compound werewolf or the expression hungry wolf all refer to the idea of wolf;
according to Bernard Pottier, one of the component elements (groups of semes) inside the sememe. Pottier distinguishes three types of possible groupings: the classeme (generic semes), the virtueme (occasional semes) and the semanteme (specific semes of the considered unit, making it possible to distinguish in all cases the unit considered from the other terms belonging to the same paradigm).

The sign U33 ***ti*** of the Gardiner's list is often considered in some minimalist analyzes as the radical sign, it would be the one on which the analysis is based which allows the good vocalization of the script. We will even consider this hypothesis in the present case.

Ewondo: /*ùn*-ntum/ "Pestle, cane"
Ewondo: /*ùn*-ntutum/ "Rainbow, great chief"
Ewondo: /*ùn*-ntum/ "chief"
Ewondo: /i-myar/ "long and thin bone"[31] bone is the other name of ***Hr***.
Ewondo: /nyar/ "chief, buffalo"

In general, it is considered that the African political structure is based on three pillars:

- The chief, king, regent (depositary or holder of political power)
- The administrative manager
- The person in charge of the ancestral ritesdes rites ancestraux

The sovereign is therefore the pillar of the pillars. One of the words designating a pillar is / *ùn*-ŋkoum / so we understand why the term / *ùn*-ŋkúkúmà / serves to designate the sovereign by the Bəti.

Among the Bəti, since the sovereign delegates the function of the public administration, the title of / ù-ŋkúkúma / can therefore be carried by the person in charge of the public administration, namely who manages the wealth of the country. He is therefore the rich of the rich. He is therefore / ù-ŋkúkúma /. In literatures dealing with Beti sociology, we retain this meaning, but in reality it is a delegated function of the leader.

Ewondo:/à-kümà/ "wealth"

The sociological structure of the Ekang peoples makes it a headless group[32]. The ekang simply use a spiritual guide and no leader. The acceptation / ùn-ŋkukuma / tends to designate a rich person rather than actually designate the leader in the African sense of the term.

Ewondo: /zomloa/ "spiritual guide"
Middle-egyptian: *Sms(w)-hr* "follower(s) d'Hr = just person"

Step 3: verification and validation of reading

We also find this script below transliterated ***ity*** and designating as the previous a sovereign or a monarch.

Middle-egyptian: ***ity*** "sovereign or monarch"
Ewondo:/*ùn*-ŋkukuma/ "sovereign or monarch, Prime Minister"

The Sign I3 of the Gardiner's list designates a crocodile.
Ewondo: /*ùn*-ŋkɔm/ "crocodile"

The fact that the scribe draw the sign I3 twice indicates a vocalization doubled of the item crocodile . We have the perfect illustration of the twin function transliterated Y.

31. Cubitus and fibula, ie the two thin, long bones of the arm and leg.
32. A headless human group is a group without a leader, which is the case of the Bəti in general. The heads of the current villages are administrative representatives. They have a secondary role as guarantor of tradition.

We get the word: / *ùn*-nkom-nkom /. This radical word will become a superlative by adding the final vowel and give / ù-nkom-nkom / + a. It takes into account the meaning of the person who embodies the ideal pillar, the pillar of the pillars. In the current synchrony this word becomes :
Ewondo :/ *ùn*-nkukuma /. This term being derived from the kékwé branch, it also admits its counterpart in the kwéké branch.

We can also vocalize this script / *ùn*-mbombokh / "handicaped" the crocodile is considered as weak of lower limbs, it can be seen as an handicapped. This term is reserved for the ritual function of the sovereign or then for the delegated function of the officiants.

Before closing the reading of this script, check that the good reaging is :
Ewondo :/ *ùn*-nkukuma /. and also designates the function of the prime minister.

Middle-egyptian:  ***TAty*** "vizier, ship of the vizier"
Ewondo:/*ùn*-ŋkukuma/ "sovereign, monarch, prime minister"

The sign G47  transliterated ***TA*** serves to name a duckling.
The equivalent word for ducking in Ewondo is:

Ewondo:/mwan i-loli/ litt. "the son of the female duck[33]"

Things are very subtly brought here by the scribe. Why does he involved the duck's figure?
According to the African ontology, life comes from an egg of anatidae. A further explanation is reserved later for this operative concept. But just remember this :

Middle-egyptian:  ***zA.t*** "Northern Pintail; duck"

Ewondo: / soghe./ "duck; placenta"
Ewondo: / *nya/ ; /i-sôgh/"placenta; mum" ☞**Ewondo**: / *nyî/ "entrance"

The word placenta and the word duck are the same phonetically. Among word to designate the placenta you have one meaning mother. By the way for the scribe, mum and northern pintail or duck are synonyms. The scribe can then draw a duck meaning mum.

In African thought, it is considered that the family unit is composed of the father, the mother that can be assimilated to the roots of a tree. It will be noted below that the phallus representing the parents is also a root. Indeed, the terms for the phallus are associated to the root in African languages.

Middle-egyptian:  ***m.t*** "mother"
Middle-egyptian:  ***hi*** "husband, lover"
Middle-egyptian:  ***hnn*** "phallus"
Middle-egyptian:  ***mny.t*** "root"

And therefore, using the image of the tree, the child is the trunk. So that's how in Ewondo we find also the term /*ùn*-ŋkum/ "trunk; pillar".

Ewondo:/*ùn*-nkum/ "trunk; pillar, mast"

33. Cane: bird of the family Anatidae, female duck

It is therefore useful here to remind Africans people who are reading these characters that the vocalization *djaty* or *tchati* should not be a psychological blockage as is often subtly suggested. We must stick to our realities and vocalize as it is today. This is how the right vocalization of the above script is in Ewondo:

Ewondo:  / nti *ùn*-ŋkukuma/ "His Majesty the King"

Moreover, it is considered that the vizier's boat is called ***TAty*** too. The mast of a boat is called / *ùn*-ŋkum /. The mainmast or the mast of the masts, the mast of the chief is therefore obviously: / *ùn*-ŋkukuma /. And so by semantic shift one arrives at the notion / *ùn*-ŋkukuma / as prime minister's boat.

Let's sum:

Middle-egyptian: 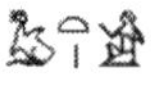 ***TAty*** "vizier, vizier's boat"
Ewondo:/*ùn*-ŋkukuma/"sovereign, monarch, prime minister"
Ciluba: *Ntita* / *Twite* "sovereign, monarch, prime minister"
Hausa: *sarki* "sovereign, chief, monarch"

I first allow myself this comparison here with the Ciluba language to clearly demonstrate that it is the same language despite the fact that / *ùn*-ŋkukuma / and *ntita* or *twite* do not sound the same way to the ear.

The sign  G47, shows a duckling who is limping. Whoever limps is qualified:

Ewondo: /*ùn*-ntet/ "the one who is limping, who has trouble getting around"
Middle-egyptian: 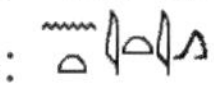 ***nitit*** "embarrass, slow down, delay, hesitate"**.**

It is therefore the acceptance / ùn-ntét / that has persisted in the Luba language. But at the end, we are talking about the same reality. With space and time remoteness the word Ewondo / *ùn*-ntét / took the aspect Ciluba: *Ntita* / *Twite*. Moreover, we can also approach it to the word /*ùn*-ntout / "the saint, the full" or even / *ùn*-nti / "lord, majesty"

Note that it is the same term / *ùn*-ntet / which is used to designate the numeral 100 in Ewondo: / *ùn*-ntét / "hundred; multitude; throng".

Beyond any debate, the term *Ntita* is still relevant in the kingdom's Luba or BuKam. It designates the head of government, who is, at the head of bi-lolo or ba-bilolo [34].
Let us just remember if it is still necessary that the word **Ewondo**: / i-loli / is for duck.

After comparison with the ciluba language, how will I establish the link between the Ewondo language (classified with the reference A72a by Guthrie, and the code 99-AUC-ca by Linguasphere) and the term Hausa *sarki* given for chief.
The question is legitimate because the censors class Hausa in the family of Afro-Asian languages (also known as Hamito-Semitic) and it belongs to the group of Chadic languages (sub-group of Western Chadic languages). How can I explain that this is the same language? Many are expecting me to fall in a trap. I do know that. But this is relatively simple you will see…

I sugget for this purpose, to start from this script below because it is a great use on the educational level.
Middle-egyptian:  ***TAty*** "vizier, vizier's boat"

34.The Bilolo or Ba Bilolo represent the legislative power the equivalent of (deputies and senators) are the guardians and interpreters of the laws next to Ntita. A third caste is that of Muadi'a Mvita (Ministers of War).

Ornithologically, the main symbol for designating ***wsjr,*** that is to say, the father of the duckling it is the duck itself. (*conf. Illustration N°32*). Let's take some scripts showing that ***wsir*** is the father of the duckling.

1. **Middle-egyptian**: ***wsir*** "osiris"
2. **Middle-egyptian**: ***wsir*** "osiris"

What do we notice in these two scripts? ***wsir*** designate the throne in the first and in the second, it is an anatidae ***wsir*** (duck).

The throne is the sovereign and the regent is the throne, the chief is the duck as in the transliterated script ***TAty***. So we could also name the throne for the healthy reading of this script. That is:

Ewondo: /*ùn*-nsoŋ/ "top, duck , beak"
Ewondo: /i-toa/ "seat, throne"

According to the mythology described as Egyptian, ***wsir*** was cut in pieces by his jealous brother, ***stX***. After being cut, the pieces were then thrown into the Nile. Subsequently, they were found and assembled by ***As.t*** his sister and wife. ***As.t*** ends up resurrecting ***wsir***...

After this little reminder on the paradigmatic level, I suggest that we summarize the purpose:

Hausa: sarki "sovereign, monarch, prime minister"

We can therefore disassemble it into ***sar*** + ***ki***. We said that the expression ***wsir*** designates the leader and thus vocalizes certainly as ***sar*** for leader in Hausa. So it remains the second part ***ki***.

Hausa: kai "tête"
Middle-egyptian: ***kAi*** "think about; develop, design"
Middle-egyptian: ***mkhA*** "back of the head, occiput"

So the ancestor of Hausa had to offer the word ***sarkai*** meaning the head chef, the chief of chiefs.
The proximity of the two vowels / a / and / i / ended up leading to the elision of one of the two in Hausa giving at the end the actual *sarki* variant.

The term Hausa *kai* "head" finds its etymology in Negro-egyptian with the word **** kaj*** / **** kja*** related to the mother or the father and here precisely to their nurturing function. We find this term in Middle-egyptian with ***kAw*** "Food, breads, food, provisions". We will better understand all of this now.

To celebrate the Bəti, / i-kul kone / "the novena" literally the exit of the ghost, what Egyptologists call "the return to the kA", there is a practice among the African people that consists in organizing and sharing a meal and drinks with the deceased, family, neighbors of the deceased etc. Surprisingly, this practice of sharing meals is called:

Ewondo: /sarka/ "sharing party with the deceased"

Let's set things back in their right place. The Bəti's / sarka / appears as the authentic name Ekang of what has become the Christian *Eucharist*. This Eucharist is the sharing of bread and wine in the memory of the body of ***wsir*** cut by his brother ***stX***.

If it is certain that this festival was practiced in ancient Egypt, I say here for us Africans that, this festival is called / sarka / in the Ewondo language and is practiced during the ritual of / ikul kôn /.

We read a lot of things that are very approximate, approximate or not to say false about it by unidentified censors.
African sources help better understand its meaning. ***wsir*** having revealed to men the benefits of agriculture and the making of bread, it also symbolizes that grain (grain, wheat) that is thrown into the ground dies and is reborn as another stalk of grain after 40 days. This vision represents the death and resurrection of ***wsir*** (There are also frescoes where you can see the wheat growing directly on the body of ***wsir***). Bread holds his spirit and breaking it releases his energy. His blood comes only to complete this religious ritual. This feast of the I-kul kon / translates what is called African vitalism.

The term *sarki* of the Hausa designating the king thus finds his connection with the Ewondo through the word / *sarka* / . This Sarka is nothing else than the authentic name of what took the folklorized form of the Christian *Eucharist*. For Haussa people, the main idea regarding the chief is the memory of the one whose body has been divided and shared.

There is therefore an ambivalence around the word sharing. In a sense the sharing recalls the carved body so shared ***wsir*** and sharing recalls also the sharing of the meal with the deceased (s).

There are many terms for king in the variants of the African language. Have fun finding the link with the variant you master, for sure, you'll see it's funny.

Etymology

We say that the equivalence of the word *sarki* is *grace* in English. The two words are obtained by phonemes inversion.
This word *grace* is homophone of *grass* meaning *herb*. The words *herb* and *grass* are indeed synonymous because *herb* corresponds to the word beti / i-berghà / "what we put above".

Middle-Egyptian: 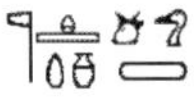 ***Htp*** "offrings"
Ewondo: bi-berghà "what we put above the table"

This term induces a double meaning the first one joins the idea of food explained above and the second joins the idea of the pyramidion namely the top of the pyramid and therefore by metonymy, the chief.

Finally, it should be noted that the green color is associated with *wsir* in african thought and the sheaves of wheat would have sprouted wherever the pieces of his body had been scattered by his brother *stX*.

We can observe that the name for grass in many african languages is link to the ancestor's item. All that words are sharing the same semantic field:

Duàla: Bekòkó "herbs, grass"	☞**Ewondo**: /ùn-ŋkukh/ "very old"
Bambara: Ngèsan "herbs, grass"	☞**Ewondo**: /ù-sàŋghà/ "citronnella; divided"
Tupuri: fy"herbs, grass"	☞**Ewondo**: /fì/ "snake"
Ewondo: /i-lokh/ "herb; grass"	☞**Ewondo**: /kùlù/ "ancestor"
Fulfuldé: kuɗol-huɗo "herb"	☞**Kinyarwanda**: urukundo "love"
Lingàla: litíti "herb; grass"	☞**Ewondo**: /ùn-ntét/ "ancestor; old, disabled"

Summary of achievements

We can see that the classifications studies of African languages made by some people that I will not mention (Greenberg, Bleek, DeLafosse, Meinhof C., Guthrie, H. Bauman and D. Westermann etc.), with some very questionable methods that I will not name (typological classification, mass classification, geographical classification, glottochronology etc.). Beyond the very pompous and very learned names, these methods prove very little rigorous because it missed them the keys of reading and the sufficient thickness to approach with serenity the study of the African languages. With all the respect we have for the trouble that have been given to each other, they appear to us with indulgence seen from Africa as simple gropings that we still greet the proactive character.

We have just proved that by naming the king in different African linguistic varieties we speak of the same reality with different descriptors. Either we talk about the king of kings ***wsir*** or we mobilize the lexical field of matriarchy.

This clearly demonstrates that consonant skeletons can not be posited as a condition and impose forced vocalization from this skeleton of consonants when we know that the characters are polysemic, ontological and paradigmatic. The Egyptological technique of transliterating is therefore good because it sets a frame but it is limiting because it keeps African away from the field of possibilities offered by the script or the associations of the different characters.

So it does not matter the word found for king in any area of Africa even woldwide, I think it's easy to find also its counterpart in the lexicon of our own language, of every African language if we have a good practice. I took at random and simply by phonetic and semantic proximity:

Shikamari (comorian): Mfaoumee "king"
Ciluba: Mfumu "king"
Ewondo:/ùn-mfum/ "white, unearthed, much, whole"
all qualifiers of ***wsir***

nsw.t, king wearing the white crown of Upper Egypt

This is why some people say *mfumu* associates by beti with white.
sn wsjr 1st
We can thus find the ***km.t*** origin of populations who use mfumu to say king.

Does this shed new light on the wearing of the white color for mourning in some parts of Africa and the world in memory of ***wsir?***
The word ***wsir*** is equivalent of the word ***Tsar*** is Russian. Do I say tsar for king in Ewondo language? No. I do say Ewondo: /ùn-ŋkukuma/ "king". Is there an equivalent of the word ***wsir*** in Ewondo? The answer is yes. **Ewondo**: /i-dzwari/ "the one in front of whom one bends, the simplicity".

Ewe(Ghana, Togo): fia "king" ☞**English**: fear
Bamougoum (Cameroon): Nfeû "king"
Ewondo:/*ùn*-mfiè/ "light, sun" one of the cosmological and cosmographic counterpat of ***wsir.*** We have ***sSp*** "to be sparkling white, to be brilliant; light etc."

Miènè (Gabon): Ogha "the king" litt. "The one" with *okh- >*ogh- ***nb wa***"only sovereign"
Lingala (kongo): Mokonzi: "king, leader, lord" litt. ***Moko*** "one" and ***nzi*** "leopard"
Ewondo:/ze fokh/ "the first panther" (conf. Illustration N°13)

Nzebi (Gabon, Congo, RSA etc.): Mu-koutou "the owl, the eldest, the boss or the chief"
Ewondo:/ùŋ-kut / "beaten, molested, burlap, shroud etc. "

Ewondo:/à-kut; / "hit, molest, beat" qualificatives of ***wsir*** *(Conf. illustration N°33)* ***krs.***

Moundang (Mayo-kebbi Ouest/Tchad; Cmr): Gong "majesty, respectable".
Ewondo:/à-ŋgong/ "padlock" the cadenace, one of the qualifiers ***wsir*** because he was locked in a sarcophagus then padlocked ***krs***.

Twi (Ghana): nana "king"
Ga (Ghana): Nii "king"
Ewondo:/na/, /nana/ "mum" an atavism of African matrial power.

The mother of the country translates literally even if it is about a man (the uncle of the country).

Fulfuldé: laamiiɗo "king "
Ewondo:/à-lam/ "trap, snare" litt. the trapped, one of the qualificatives of ***wsir***. Indeed, it is said that he was trapped by ***stX*** at a banquet then padlocked in a sarcophagus ***krs***. We find with the transliterated script ***aany*** the sign A24
(man striking with a stick) who makes the connection with ***wsir*** who was copiously molested according to african mythology.
This little exercise of style makes explode the official standard classifications, demonstrating if it was necessary the lack of seriousness in the approach and especially the inconsistency on the subject treatment.
After being unknowingly chosen by his father to succeed as chief, in western Cameroon, some of the chief's notables are sent to catch the new leader. Whether he agrees or not, he is led to take the succession according to the will of the ancestors. On the historical continuum, we find the approach put forward, namely that the leader is the one who is trapped.

So besides the fact that one can understand the words used in the different African linguistic variants, it was also possible to establish the link with sarki "king" in Hausa even more simply.

It was therefore necessary to remember that our languages are not a juxtaposition of onomatopoeia but a juxtaposition of operative concepts. Thus / sarka / or / sarki / consists of two operative concepts.

> **Ewondo:**/à-sar/ >/à-sal/ "to divide, to share"
> **Ewondo:** /kaï/ "gifts, foods, present"

Between / á-sar / and /á-sal / the difference is made on the final consonant the Rhotic consonant Uvular R which on the diachronic plane will evolve from the old synchrony to a lateral liquid consonant L in a new synchrony. This term translates well the phenomenon of mixtures of synchronies that one finds in the isoglosse of the synchrony Ewondo.
To say Man we use indifferently the words: mot > mod> mor

As I have already emphasized many times, Africans use operative concepts, so how to capitalize on this reading on the onomastic's level?

Onomastic

We will use phonetic proximity as a search marker.

Anthroponymy

wsir in his sense of sharing / sar / or / sal / carries the anthroponym: Sala, Tsala, Be-sala etc. among Ekang peoples.

wsir in his mummy sense /*ùn*-nkor/ takes the anthroponym: *Nkoro, Koro, Kodo* etc. among Ekang peoples.

wsir in his chief sense carries the anthroponym: *Sore, Sira, Sila, Sida* etc. among Ekang peoples.

wsir in his dimension of the trapped, he will take the anthroponyms: *Olama, lamido, Mala, Lembe* etc.

wsir in his pillar dimension he will carry the anthroponyms: Nkoumou, Ngoumou; Ndzié; Ilée; (Cameroun, Gabon, Guine quatoriale etc) etc.

Ethnonymy

wsir in his dimension of the trapped, he will give the ethnonyms: Lama (togo), Tulama (Ethiopie);
wsir in his dimension of pillar he will give the names: Ngumu (Mozambique); Ngumi (Angola) Etc.

Toponymy

wsir in his dimension of the trapped, it will give the place names:Olama (cameroon)

- ***wsir*** in his dimension of pillar it will give the place names: Ngoumou; Me-Ngomo (Cameroon), Etc.

3.4.3.7. **EXAMPLE 7**  (*Faulkner p. 302, Wb. V p. 339*)

Step 1: Name the script's signs

Sign R3P of the Gardiner's list "table"

Sign V13 of the Gardiner's list
Ewondo: /*ùn*-mbɔkh/ "animal hindrance"

Sign X1 of the Gardiner's list
Ewondo: /i-di/; /i-komba/ "bread roll"
Ewondo: /*ùn*-mbokh / "solid, soft, elastic"

Sign Z1 of the Gardiner's list
Ewondo: /*ùm*-bokh / "one"

Step 2:

The sign V13 serves to designate the device for hindering the beasts of burden. Signs Z1 and X1 come to clarify this reading by homophony.

Ewondo: /*ùn*-mbɔkh/ "animal hindrance"
Ewondo: /à-tib'li/ "to maltreat; batter; ill-treat etc."

As a general rule, an abused person suffers of hindering of his freedom. A beast of burden is a pet or carrier animal, it is a domestic animal used by man as a working animal to carry loads. A draft animal is used by humans for: transportation; agricultural work, including plowing; the war; industrial work; sport and recreation with competitions and team races. These animals have been deprived of their freedom and they are subjected to forced labor.

Ewondo: /i-tib'li/ "hindering; mistreatment; maltreatment; cruelty "

For the scribe, the  *table* is wood, that has been hindered somehow. Therefore, the words used to describe the table will thus marry the lexical field of restraint in our languages. Let's examplifly the:

Ewondo: /*ùn*-ntsomle/ "nail, tip"
Ewondo: /à-bàm/ "board" and its semantic counterpart ☞**Ewondo**: /à-bám/ "to scold"
Ewondo:/*ùn*-ntsimi/ "board" and its semantic counterpart ☞**Ewondo**:/ à-timi/ "discriminate"

We could extend the list but it is not the purpose. The most important think according to me is to understand the principle.

We therefore draw the following conclusions:
The good reading of this script is not necessarily / téb'lé / as it is customary in the current synchrony. One can suspect a colonial influence with the contamination with *table*.

Many linguistics specialists in the Ewondo language sometimes consider that this term is linked to a colonial loan, which is not accurate in light of this explanation.

There is no anteriority of colonial languages over African languages. It is an anachronism to consider that nomadic peoples would have invented tables running behind camels in the desert.

It seems to me more appropriate to read / *i*-tib'li /. This reading is more designated because it is associated with an endogenous logic.

The word table used in Western languages therefore finds here its true etymology, namely, an obstruction linked to beasts of sums or traits. If I can explain it so easily, I then challenge anyone who believes that this word is a loan made to the colonial languages to give me the real etymology of this word in those languages. Here is below what can be read concerning the word table.

The Origine of the word table

Table (n.)
late 12c., "board, slab, plate," from Old French table "board, square panel, plank; writing table; picture; food, fare" (11c.), and late Old English tabele "writing tablet, gaming table," from Germanic *tabal (source also of Dutch tafel, Danish tavle, Old High German zabel "board, plank," German Tafel). Both the French and Germanic words are from Latin tabula "a board, plank; writing table; list, schedule; picture, painted panel," originally "small flat slab or piece" usually for inscriptions or for games (source also of Spanish tabla, Italian tavola), of uncertain origin, related to Umbrian tafle "on the board."

That means before the arrival of Maurish people no one knows anything about the word **table**.

To make tables, you have to master the work of wood. Is the table furniture from a nomadic or sedentary sociology? How do you carry tables on the backs of camels? What do we do about organizing African caste work? Is it not more appropriate to find the table within the caste of African blacksmiths or carpenters?Was Jesus in a nomadic sociology a carpenter or is this another Western sleight of hand?

But in truth ***Hr*** being the chief /*ùn*-nloh/

admits the homophone /i-léh/ "tree" assimilated to him. By the way the punishment that ***wsjr*** seems to have undergone according to African mythology, the lexicological and semantic field of the tree will thus be organized around the parallel with ***Hr***.

I just remind that ***wsjr* = *Hr***. As an illustration here how quickly can we break down ***wsjr*** although it does not make sense.

Middle-egyptian: *Ws*

Ewondo: /ivés/ "bone"; /dys/ "eye"; /à-sè/ "feather"; /à-su/ "visage"; /si/ "earth" etc.

Middle-egyptian: Jr = ir= Hr =ra = rê = rA (eye, mouth, ear, head, speech, visage, sun etc)

A little history reminder of the concept *i-tib'li.*

Although beyond the common sense, Cinderella is portrayed as having a Western life, the authors of the deception (Perrault, Walt Disney etc.) would find it difficult to explain the story of cinderella facing the African tale of Dorishta.
Indeed, the story of Dorishta was found on a papyrus dating back to 3000 BOE. This tale highlights the abuse / i-tib'li / through storytelling. A wise way for African scholars to preserve the perennial myth of ***wsjr***.
The Egyptian Dorishta, is the equivalent of Cinderella except that in the wester fake, the glass slipper is nothing else than a leather sandal and the prince is a pharaoh.
In Egyptian Dorishta means, "the one who repels the secret".

The Ewondo name ***Dorothia*** transliterate "Dorishta" became "Dorisha" out of Egypt, the Greeks turned the name into "Rhodopis", "the one that is pink" ... (Rhodos = the rose). And by extension a "rhodope" became a girl to seduce, then, still by extension, a prostitute with pink cheeks ...

Middle-Egyptian : drSTA «Dorishta »

Ewondo : ***Dorothia***
French: Dorothée
English: Dorothy; Dorothee

Those who falsify history including at the level of the collective imagination should not they be ashamed?

Spencer Lewis was discreetly expressing by this famous Latin formula: "*Magna est veritas et provalebit*" (the magnificence is truth and prevails ...)?

3.4.3.8. **Example 8** *(Faulkner p. 141, Wb. II p. 349, 1-5)*

I took this script for two reasons mainly. I will reveal them after its reading. In fact, this spelling is not a problem. But it has an educational interest.

What matters is that we name certain realities without grasping the real meaning of the words we use. If I ask a Ekang language specialist the etymology of the item associated with beef, I am 99.99% convinced that few will give the exact meaning.

In the previous illustration, we had a hindrance for draft animals. In this script we have the animal but not the hindering. But keep in mind that both ideas are correlated.

Step 1: Name the signs

Sign W11 of the Gardiner's list.
Ewondo: /à-ko/ "ring holder for jar"

Sign G43 of the Gardiner's list
Ewondo: /wou/ "quail chick"; /à-koh/ "Hr"

Step 2:

The sign W11 used for reading here will recall on the semantic plane either the support or the vase drawed for this support, or elements of the semantic field of pottery. The vase stand is simply the place / foot of the beverage vase. /Akô(l) mə nyu/. This place offers the singular imperative, /nyâkh/ "drink!".

The Sign G43 represents ***Hr*** the chief / nyar / "buffalo" thus offers phonetic proximity with "nyakh".

Little comparative study

Ewondo: nyakh "beef, zebu; buffalo, taurus"
Lingala: ngombé "beef, zebu"

By reading these two words one could think that these two words come from two different languages!!! That is Greenberg and his friends the point of view.

For the scribe, this animal has a hump, a protuberance, it is vaulted.

Ewondo: /á-koúl/ "hunchback"; with his counterpart ☞**Ewondo**: /à-koul/ "animal's paw"
Ewondo: /à-ŋgoŋgbe/ "lie on the edge, curl, vaulted etc.."
Fang: *ùn*-ngôbe "lie on the edge for a board"

So if the board is on the field, it is considered vaulted. Being the perfect representation of the obstacle, one understands therefore how one appeals here to the board and to the synonyms of the obstacle that it represents in the different languages

The term / nyakh / of the Ekang languages is equivalent to / i nyo á-nə á-kúl / "the one who is hunchbacked".
Ngombe "beef" of Lingala is equivalent to the one with a hump, a vault. The Ewondo will say about it / *ùn*-ŋgoŋgbe / or / / i-nyo à á-ŋgoŋgbo /. Although this term is not used in Ewondo to designate the cattle in the light of this explanation it is clear that we are in the same semantic field. We have different signifiers but the same signified.

From the moment when we understand that the board is a hindrance in our languages, that the zebu is a beast of burden hindered, at this point we can rigorously understand each other in the different languages spoken because it actually acts as synonyms or homophones that are mobilized.
To finish on this subject let's check that this hunchback acceptation has been maintained in Pharaonic.

Negro-egyptian: *kʷiʀ̃- "foot, strong, fighter, dangerous for man"

This Negro-egyptian term will give several variants in the Ekang languages, * **kiʀ**-; variant * **niʀ-.**and the form * **wiʀ-.** The final voiced uvular fricative consonant R with time will evolve from the Negro-egyptian to a lateral liquid consonant L in the current synchronies.

Middle-egyptian: *iw*"hunchback"
Ewondo:/á-kúl/; /i-vul/ "hunchback"

Middle-egyptian: *iwaw* "thigh; leg, haunch"
Ewondo: /à-kul/ "animal's paw"; /à-bèh/ "thigh"; /i-dzô/ "piece"

Middle-egyptian: *nri* "scare, frighten"
Ewondo: /nyar/ "buffalo"; /nyar-moro/ "authoritarian" litt. Man-buffalo.

Middle-egyptian: *iw* "fawn of hartebeest" or ox.
Ewondo: i-mvul "hartebeest"

So we see that the term to designate an hunchback person is exactly the one used to designate a bovine. The two items are therefore to be read as explained and not otherwise.

3.4.3.9. <u>EXAMPLE 9</u> (*Faulkner P. 260, Wb. Iv P. 377, 8-12*)

<u>Step 1</u>: Name the signs

- Sign S29 of the Gardiner's list
  Ewondo: i-kob "skin"
- Sign D46 of the Gardiner's list
  Ewondo: /ùm-bokh dou/ "palm of the hand"
- Sign X1 of the Gardiner's list "roll bread"
  Ewondo: /*ùm*-mbokh / "solid, thick, dense, elastic"
- Sign Z4 of the Gardiner's list: /*ùn*-mbias / "jumeaux"
  Ewondo: /*mi*-mbokh / "the related" twins
- Sign F33 "queue; tail; stem"
  Ewondo: /*ùn*-mbokh/ "eye, hole"

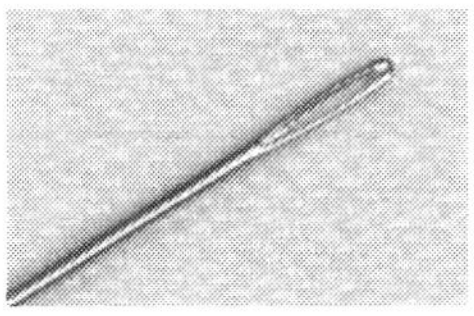
eye of the needle.

<u>Step 2</u>:

Signs S29 and D46 operate by phonemes inversion in Ewondo.

- Sign S29 of the Gardiner's list: *i*-kob "skin"
- Sign D46 of the Gardiner's list: /*ùm*-mbokh / "hole"

<u>Step 3</u>: determinative's reconstitution

We can propose two cases here taking into account the sign // y:

Ewondo: /*ùn*-mbombokh/ "the bedridden"
Ewondo: /*i*-bobokh/ "baby; kid, infant"

<u>Step 4</u>: Choosing the right word

The determinative gender of the sign X1 indicating a feminine, the word to choose here is the one starting by ***i***.
Taking into account the standard rules, the good reading here is therefore:

Ewondo: /*i*-bobokh/ "baby"

But in reality both terms are semantically justifiable in our languages depending on the context. A baby is a bedridden. So by metonymy we can use one word for the other.

Note: If this script had been approached in a mesh between Ewondo and Koonzimé, it could have been read more easily because the hand is called in koonzimé: /mbô/ oe /ebô/. The Ewondo has kept another word for hand but use /*ùm*-boh/ "the one who does". The root /*mbo-/ "one; 1/ is a synonym of /wa/ "one,1; hand/

According to the meaning of /*ùm*-boh/ "the one who does" it is supposed that doing is correlated with hands. In Ewondo language, the word for the palm of the hand is /dwé/. Is it therefore the real etymology for the English word ***do*** *"make; build; handle"*? I'll not be surprised anyway.

*3.4.3.10. <u>**Example 10:**</u> **Variation Of Sign And Sense***

Middle-egyptian:  (*Faulkner p. 260, Wb. IV p. 377, 8-12*)

<u>Step 1</u>: Name all the signs.

- Sign S29 of the Gardiner's list "Folded linen"
  Ewondo: /i-kob/ "skin"
- Sign I10 of the Gardiner's list "Cobra at rest"
  Ewondo: /nyô (l) / "serpent"; /*ùn*-mbombokh/ "handicap"
- Sign X1 of the Gardiner's list "roll bread"
  Ewondo: /*ùm*-mbokh/ "solid, thick, elastic"
- Sign A17 of the Gardiner's list "kid"
  Ewondo: /*ùm*-wan / "child"

Step 2: take care of the root of the word.

- Sign S29 of the Gardiner's list: /*i*-kob/ "skin"
- Sign I10 of the Gardiner's list: /nyô (l) / "snake; serpent"

The sign I10 represents a snake and can be read / *ùn*-mbombokh / "Disabled" because the snake does not have any use of its legs. He is therefore comparable to a paralytic. Indeed, just like a snake, a new-born does not have the use of his feet, he is also a sort of paralytic. He begins by crawling like a snake before walking. This kind of relationship is metonymy.

Step 3: Determinative Reconstitution

We fall back again in the case of the example N°2:
By the way we can have two cases here:

Ewondo: /*ùn*-mbombokh/ "the disabled; the paralytic"
Ewondo: /i-bobokh/ "baby" taking into account the right sign .

Step 4: Choice of the right word

The determinative gender of the sign X1 indicating a feminine, the word to choose here is the one starting by ***i***.

So the reading here is then:

Ewondo: /i-bobokh/ "baby"

Don't we have here the origin of the word baby with the amuissent of the final velar consonant **k**?

3.4.3.11. Example 11: DUAL MANAGEMENT

Middle-egyptian: ***mmy*** (*Faulkner p. 106, Wb. II p. 58, 14*)

Step 1: Name all the signs

- Sign G17 of the Gardiner's list "owl"
  Ewondo: /à-koung/ "owl"

- Sign Z4 of the Gardiner's list
  Ewondo: /ùn-mbias /; /mi-mbokh /; /mə-làrà/ "the related" twins

As a reminder, in the ontology of Ekang people, the owl is a nocturnal bird. It is considered in the world of the night as a human being. At this point, it bears the word:

Ewondo: /*ùn*-ŋkukh / "trunk, darkness"

This script is therefore read: /ùn-ŋkokh/ by phonetic proximity.

However the words for owls and intelligence being homophones, it is therefore considered that the owl is the symbol of intelligence in the African paradigm, it incarnates de facto intelligence.

Ewondo: /à-kəŋ/ "intelligence" or / *ùn* -yeme/ "connoisseur; expert"
Ewondo: /à-kouŋ/ "owl"

Therefore the sign E27 can be read [*i* -yam-i-yam]
who by homophony refers to the idea of great intelligence /*ùn*-yəmyəm/ or the great dumb / i-yəmyəm / or by synonymy /ù- mvuvugh/. The adult giraffe can be characterized by its placidity and silent character.

The presence of Sign Z4 (y) /*ùn*-mbias/ "nisbe" gives the inclination of reading by the repetition of the first syllable. As we can see, the scribe can name the sign E27 by purely endogenous descriptors. It is not a name but descriptors that characterize the determinative. What is considered as a noun is a qualifier and by metonymy or cross comparison becomes a noun.

I just note that the word *Nisbe* being subtilized to Arabic appears very exotic for Egyptology. But it is more indicated in Ewondo to speak of / *ún*-mbiás /; term reserved for the twinning. In the overall logic of this document it will therefore systematically replace the exotic barbarism *nisbe*.

Little sum.

We can agree that the sign Z4 (y) /*ùn*-mbias/ "nisbe" refers in phonetic to a repetition of the first syllable when It translates a superlative idea.
But when it establishes a correlation "what belongs to" then we have to use /nyo yà/. The reason is simple here. The serpent returns the idea of contempt in African thought. The word for scorn are sharing the same root with the word /**nyô/** "snake". We can have: /*á*-**nyi***be*/ ; /*á*-b**ya**/; /á-b**yan**e/, etc.
We have according to the sign Z4 above mentionned :

Ewondo : /*i*-biàs/ "contempt; disdain; scorn; lousiness etc."
Ewondo: /*ùn*-mbiàs/ / one of the twins/

3.4.3.12. ***<u>Example 12</u>: Management Of An Ostensibly Script.***

Middle-egyptian:  ***sn*** *(Faulkner p. 230, Wb. IV p. 150, 8-13 sq).*

<u>Step 1</u>: Name the signs

- Sign T22 of the Gardiner's list
  Ewondo: *à*-kɔŋ "spear"
  Ewondo: /*ú-n*tóŋ/ "spear stick"
  Ewondo: /*ú-n*sóŋ/ "spear peak"

- Sign N35 of the Gardiner's list
  Ewondo: /*úm*-aŋ/ "sea"
  Ewondo:/ *úm*-ón/ "infant; kid"

The reading invites here to consider the signs: T22 and N35

The sign T22 is composed of a handle and a point. The neck is called / *ún*-ntóŋ / in Ewondo. But / *ún*-ntóŋ'o / designate also by homophony the nurturer (male of female). Reading from the right to the left we have: / *úm*-ón *ùn*-ntóŋ / that we can approach by "the child of the nurturer". In the triangulation (father, mother, child) of African thought, the father and the mother are assimilated to the nourishing roots of the tree. These vocals are associated with the root or the phallus .

Middle-egyptian: *mw.t* "mother"
Middle-egyptian: *hi* "husband"
Middle-egyptian: *Hnn* "phallus"
Middle-egyptian: *wAb* "root, orbit"

However, in a matriarchal regime, the child is identified to his mother, the nurturer is therefore the mother, ie, /*ùn*-nyà /. We have to confirm this statement by the spelling ***mAw.t*** designating in Pharaonic the stick of the spear used with the sign T22. But, the bald-headed vulture refering to the mother is called in Pharaonic ***mw.t.***

So as the scribe rightly points out here, in Pharaonic the handle and the mother are semantically interchangeable in our languages.

Ewondo: /*úm*-ón *ún*-ntóŋ'o/ therefore becomes in our languages and in Ewondo.
Ewondo: /*úm*-ón *ún*-nya / "the mother's kid" or "the handle's kid".

We often say in the great African Ekang wisdom that:
Ewondo: /*úm*-ón *à sikigh ùm-biéh úm*-ón *à nə hə ùn-ntòŋ* /

Literally thats means: "A child is not the one to whom we give birth but is the one who we rear."

The expression / *úm*-ón *ún*-nyà / is used to refer to children of same-sex siblings. It is also used in the broad sense of the African family to designate children of the same family, of the same lineage. So, two brothers or two sisters will designate themselves by the contracted form / *úm*-anyaŋ/[35].

M.E.: sn →**Ewondo:** /*úm*-ányaŋ/ "brother-brother"
M.E.: sn.t→**Ewondo:** /kal/ "sister" because the spear handle is also called /*ún*-ŋkəl/
M.E.: sn.w →**Ewondo:** /*b'*-ànyaŋ/ "brother and brother or sister and sister"
M.E.: sn.wt→**Ewondo :** /*be*-kal/ "designation of the sisters of a brother"

The spear represented by the sign T22 is named / à-kóŋ / in Ewondo it indicates by homophony here a bloodline, a lineage:

Ewondo:/ *ún*-ŋkwáŋ / or / *ún*-ŋkóŋ / and by synonymy
Ewondo:/ dyàŋ/ designate the lineage, the maternal village.

The word written by egyptologist is:
Middle-Egyptian: sn "brother"
Ewondo: /*ùn*-nsəŋ/ "compound" ☞ brothers
Phonetically is sounds like /*ùn*-nsəŋ/ "sting". We can admit that a spear is a kind of sting.

35.S. GALLEY, French-fang and fang-french dictionary followed by the fang grammar.ditions Henri Messeiller, Neuchatel, p207

This transliteration is suitable phonetically to the English word: *son* "child; kid". But semantically this script above has to be read mother's son. Son for the spear and mother for the water.

In a matriarchal diet the mother is the handle of the family, of the lineage. The sign T22 thus designates the handle and its bloodline. The determinative simply specifies the element of the lineage of the handle. These levels of information can only be accessed with a good knowledge of the language. The simple analysis of the consonants does not allow to explore the potentiality of these details. You have to put these characters in a paradigm to pronounce them with accuracy.

*3.4.3.13. **Example 13**: Quick Analysis.*

Middle-egyptian: *SsA* (*Faulkner p. 271, Wb. IV p. 543, 7-16 sq.*)

Step 1: Name the signs

To determine the reading of the determinative (sign on the right) it is necessary to:

1- Recognize the different characters and name them in your language

Ewondo: Sign N37: /à-ŋgóngà/, "box, Pool" /mə-kəŋe/ "the edge of a hole"
Ewondo: Sign O34: /à-ŋgóŋ/, /*à*-sóŋ/ "lock; bolt" /*à*-keŋy/ "to assemble"
Ewondo: Sign aA18: /à-ŋgóŋ/, /*à*-sóŋ/, /*à*-kəŋ/ "lock for box with lid"
Ewondo: Sign G1: /í-ɱbgáŋ/, /ùn-nsóŋ/, "raven"; /*à*-kouŋ/ "owl"
Ewondo: Sign Y1: /á-ndakh, *ú*-kóŋ/; /à-fəb/ "papyrus, sheet"
Ewondo: Sign F5: /*ún-n*loh úŋ-kókh/ "hartebeest's head"

The scribe informs by the sign Y1 that it is an idea that characterizes the mother. But the support in question plus the head of the hartebeest indicates that it is an idea related to the characteristics of the scribe

The scribe is a head, the one who is intelligent / à-kəŋ /, the one who is sharp (*ún*-nsóŋ). It is the one who draws / à-kaŋ /. The head is used to think. The word sought is vocalized as hartebeest or giraffe and all other signs.

Finally

2- Find the link between the different signs

what I see	What the scribe wants to express
Ewondo: Sign N37 "itam"	/I-yəm/ "disabled"; /itam/ "alone"
Ewondo: Sign O34 /à-sóŋ/ "lock"	/I-yəm/ "lock; bolt";
Ewondo: Sign aA18: /à-ŋgóŋ/ "lock"	/I-yəm/ "lock"; /á-yəm/ " to know"

Ewondo: Sign G1 : /í-ŋbgáŋ/ "raven"	/I-yəm/ "animal";
Ewondo: Sign Y1 : /á-ndakh	/ùn-yəmə/ "intelligent "; /à-fəb/ "sheet"
Ewondo: Sign F5 : /ú-*n*loh úŋ-kókh/	/*un*-yemyem/"savant"; /*ùn*-nloh/ "head"

Chart 63: revelation Array of a script hidden meanings

Note: It is not obvious that the characters represented by the scribe are automatically associated with the idea that the scribe conveys in all African languages. We must therefore find the logical link that unites all the signs and very often this link is of a metonymic type. Therefore it is important to have in memory this mental set.

In our languages, camel, hartebeest and giraffe bear the same name, / *ún*-ŋkókh /. The giraffe is the great intelligence /*uŋ*-yemyem/[36]. In short, the sign of the papyrus and the head of hartebeest allowed us to read the spelling without using phonetic complements. So here we are talking about the great intelligence.

This script is therefore read*: /à-kəŋ/ou /ún-yemen/ by synonymy.*

3- Check the correct reading taking into account morpho-syntactic constraints.

Both terms perfectly meet the requirements on the technical level.

Ewondo: /à yem/ "to know" synonym of /í-yəm/ "animal" here the hartebeest.
Ewondo: /à-kəŋ/ "intelligent, clever, wise"

3.4.3.14. Example 14: Management Of An Ostensibly Script

Middle-egyptian: ***mri*** (*Faulkner p. 111, Wb. II p. 98, 12 sq.*)

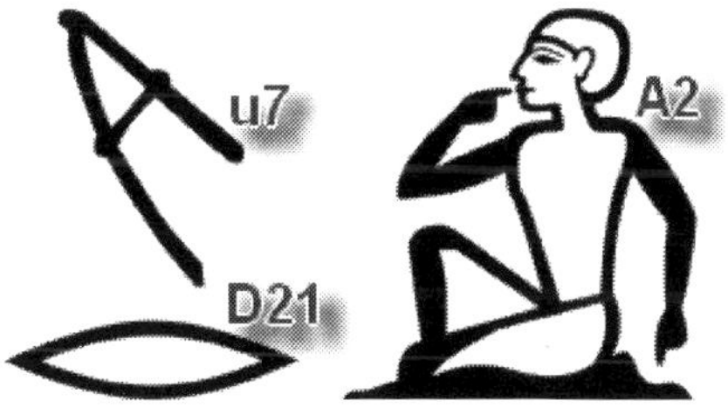

Step 1: Recognize the different characters and name them in your language.
Ewondo: Sign U7: /í-bakh/ "hoe", zíŋ "cutting object", à- *ŋ*gong, "adze"
Ewondo: Sign D21: /à-nyu/, "mouth" ùn-*ŋ* kóbó "speech"

Here I am not going to expose an analysis on the modalities of research and reading.
The Egyptologists who transliterated as a rule give this script the skeleton of consonants ***mri***.

Egyptology: mri "to love"

36.See example N°3

Wolof: mar "to lick"[37]
Ewondo: /à-diŋ/ "to live, to like"

Seeing a hoe and a mouth difficult to reach the meaning of love in my language at first sight. Moreover, the word "to love" in my own language has nothing to do with either the hoe or the mouth.

It's the good understanding of an African approach that gave me the keys to this reading.

The word lick in Wolof is *mar*, in Ewondo it is / á-nyié /. Until then no click. But Búlú, which is a sibling language of Ewondo does not say / á-díŋ / to love but / á-nyeʔe /. The Fang opts for / á-nyeghe / "to love".

Everything is clear now.

Wolof: mar "to lick→ to love"
Ewondo: /à-nyié/ "to lick"
Búlú: à-nyeʔe "to love, to like, to lick"
Fang: à-nyeghe "to love"

On the diachronic level we can suggest the following evolution of the word:

Ewondo: Nyeghe> nyeʔe>nyié "to lick"

The word *mouth* /á-nyù/ in Ewondo represented by the sign 𓂋 D21 of the script has therefore led to this series of words. And leads us to understand that for African the words of love are related to the semantic field of the mouth. This is why we say that "God" is Love since "God" is represented by the sign 𓂋 D21 vocalised ***rA/rê***.

Let's go back to our spelling for a better understanding of the links with the other sign on 𓌻 U7.

The Kɔɔnzimé language is an Ekang language but luckily it is also, my mother's language. It uses the following words:

Kɔ́ɔnzim: e-kpƐl "to love"
Kɔ́ɔnzim: e-bâ "marriage"
Kɔ́ɔnzim: e-bî, egbah "pummeling, weeding, hoeing"
Kɔ́ɔnzim: bɔ' "hoe, adze"
Kɔ́ɔnzim: é-bél "breast"
Ewondo: à-bal "hoeing, hoeing, weeding"

When we know that the pharaonic replaces the voiced alveolar lateral spiral consonant ***l*** by the voiced alveolar roll consonant ***R*** and that *m* and *b* are bilabial (nasal *m* and occlusive *b*). If, in addition, the words have the same semantics, we say that we are on the right track,.

If the word for love is / á-diŋ / in Ewondo, is there an expression Ewondo relating to love using the semantic field of the hoe? The answer to this question is yes.

Ewondo: / **á-bal** minga man í-səp à ŋgón/ which literally means hoeing a woman's field. But it's a prosaic way of talking about physical love.

37. Anta DIOP, Genetic kinship of Egyptian and Negro-African languages, Ifan, Dakar, 1977, p.33

The term *weeding, hoeing*, is used in this context to refer to love in the second instance in Ewondo. But the use has been reserved for romantic prose or prank depending of each one language registers.

In Ewondo, where does the term /à-díŋ/ "to love" comes from?

Ancient Egyptiand Wooden hoe realised from a wooden handle-shaft

In order to answer this question, we must focus on the three signs of the script and analyze them rigorously.

what I see	What the scribe wants to express
Ewondo: Sign U7: /í-bakh/ "hoe",	**Ewondo:** /*ùn*-ndíŋ/ "rattan"; **Ewondo:** /í-ŋgoŋghà/ "adze"; **Ewondo:** /ŋgwakh/ "adze"
Ewondo: Sign D21: /à-nyu/ "mouth"	**Ewondo:** /I-di/ "mouth"; **Ewondo:** /I-dià/ "a beloved"; **Ewondo:** /I-ŋgoŋ/ "throat";
Ewondo: Sign A2 /mor/ "a man"	***Basa'a:*** *murã* "woman"; **Ewondo:** /Mor/ " a man; human being"

Chart 64: revelation Array of the script hidden meanings

<u>How to read this script in Ewondo</u>

Signe U7:

Middle-Egyptian: Sign U7 ***mr*** "hoe"
Ewondo: /ù-ndíŋ/ "rattan";

By homophony: ☞**Ewondo:** /á-díŋ/ "to love, to like";

Note: Although the scribe has represented a hoe, it is certain that the word involved in Ewondo is not the hoe but a descriptor of the hoe also relates to the mouth .
The handle of the traditional African hoe could be made of rattan. So this word is here put in evidence. The word for rattan is homophone of love in Ewondo.

Signe D21:

Middle-Egyptian: Sign 21 ***r*** "mouth"
Ewondo: /i-dí/ "mouth";

By homophony: ☞**Ewondo:** /i-dyà/ "the beloved, darling";

Note: Although the scribe has represented a mouth, it is certain that the word involved in Ewondo is not the mouth but of a mouth's descriptor.

We generally use the mouth for eating. The word here highlighted is related to the descriptor of the mouth as a masticatory device. The word for masticatory device is homophone of love in Ewondo

If the purpose of this script was to highlight the love signified then we have two synonyms. But this script is more than two synonyms. Let's see how.

Other meanings

Ewondo: zíŋ "cutting object", /à- ŋgong/ "adze"

This word is to be compared to the links  of the hoe that say: / á-tíŋ /. In fact, for a marriage among the kang the expression used could be translated as "to attach the marriage" / á-tindi à-lúgh /.

Middel-Egyptian:  "weavers"
Ewondo: /à-tíŋ/ "to weave, to twill, to loom"

The object represented by the scribe  is called /à-zíŋ/[38] in Fang / Ewondo. We have on the diachronic plane the following morphological evolution:

Ewondo: *tíŋ> *díŋ>*zíŋ

*Tíŋ "wave"; *díŋ "love"; *zíŋ "hoe"
It may be noted in passing that the Fang uses the word / á-zíŋ / only to characterize love including conjugal love.

Comparison between African variants

Moyen-Egyptie: 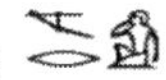 "to love; to like"
Ewondo: /à-díŋ/ "to love; to like"
Creole of Martinique: Inmin "to love; to like"
Haitian Creole: Renmen "to love; to like"
Creole of Guadeloupe: iném "to love; to like"
Nzebi: Ou tôdrô "to love; to like"
Peul: yidhougol "to love; to like"
Wolof: Bëgg "to love; to like"
Kikongo: Ku zola "to love; to like"
Mbènè (Basa'a): Ngwha "to love; to like"
Lingala: Ko linga "to love; to like"
Adjoukrou: Erurrr "to love; to like"
Bambara: Feè "to love; to like"
Tamacheq: Tarha "to love; to like"
Diola: Ka mamang "to love; to like"
Bamoun: ŋkɔ̃ "to love; to like"
Ngiemboon: mkwuŋ "to love; to like"
Bandjoun: Ne kouŋ "to love; to like"
Godi: wa "to love; to like"
Toupouri.: ndalgué "to love; to like"
Dschang: Ncheu laah "to love; to like"

38.Samuel Galley, Fang-French dictionary. and French Fang followed by a grammar fang,ditions Henri Messeiller , Neuchâtel, p.128

All these words come from a link with the mouth or with the hoe.

Etymology:

Wedding vs weeding

Before going further just observe the phonetic proximity in English between:

English: *weeding* "hoeing, scraping"
English: *wedding* "marriage"
Ewondo:/i-diŋ/ "love"

Lingui vs langue

Middle-egyptian: "to love; to like"
Lingala: Na lingui yó "i love you"

The link between the script ***mr*** "to love" vocalized *lingui* in Lingala and the French word *langue* "tongue" is established in the following way. The word *langue* designates two things, an organ for degustration and an evolving system of linguistic, vocal, script or gestural signs, which allows communication between individuals. Indeed, the tongue organ is in the mouth . The word *lingui* is therefore etymologically linked to the word *langue*. The language makes it possible to taste, to appreciate and therefore by metonymy to love.

Speech is the articulated human language intended to communicate thought, and is to be distinguished from various oral communications, such as cries, warnings, or moans. "Articulate speech" consists in creating audible the signs, the syllables, creating the words that constitute symbols. Metaphorically, speech has become any communication addressed to the mind (to speak with the gaze, the gestures, the silence) and all this comes from the mouth .

The Lingala word *lingui* is linked with the french word *langue* that is used to designate the tasting organ but also the expression.

The word *langue* is also to be linked with the French word *laguer* "to prune; to lop". We prune with this . We can also have the French word *gueule* related this time with the *sign D21* "mouth"

Let's deal with adze

When dealing with the signe U7, we noticed that :

Ewondo: /í-ŋgoŋgà/ "adze";

This is quite surprising. Not so much because the mouth and the *hoe* are homophones or synonyms. Indeed we have that:

Middle-Egyptian: signe U7 ***mr*** "adze"
Ewondo: /i-dzô/ "speech"; /á-dzô/ "to speak, to talk, to tell, say etc."
English: adze like /à-dzô/ "speech";

We can notice clearly the phonetic proximity between the english word *adze* and the Ekang word /à-dzô/ or /á-dzô/. Some people might think that there is no link between both then explain the link established by the scribe 6254 years ago by this way?
To get a better understanding, of the link between the adze and the mouth, let us remind just an old African ritual that's still practiced in Africa an precisely in West Cameroon nowadays : The Opening of the Mouth Ritual by Marie Parsons

"When an ancient Egyptian died, he was not buried into the ground, mourned and then forgotten. Nor was his grave simply visited at certain times and some token words spoken over it, so that once again he is forgotten until next visit. The ancient Egyptians believed that ritual existed which would bring sensory life back to the deceaseds form, enabling it to see, smell, breathe, hear, and eat, and thus partake of the offering foods and drinks brought to the tomb each day.

Priests would recite hymns such as this one, for Pa-nefer:
"Awake!..May you be alert as a living one, rejuvenated every day, healthy in millions of occasions of god sleep, while the gods protect you, protection being around you every day."
Once the deceased was rejuvenated back with all his senses, he could also interact and watch over the family members, affecting their lives. Letters have been found attesting to the continued contact, or at least, belief in the continued contact, between deceased and living. Letters such as this one, from the scribe Butehamun to his deceased wife Ikhtay, where he asks her to intercede with the Lords of Eternity on his behalf. "If you can hear me in the place where you areit is you who will speak with a good speech in the necropolis. Indeed I did not commit an abomination against you while you were on earth, and I hold to my behavior."
***The ritual that would re-animate the deceased was called** The Opening of the Mouth ceremony. It was an important ritual in both funerary and in temple practice. The Opening of the Mouth originated as a ritual to endow statues with the capacity to support the living ka, and to receive offerings. It was performed on cult statues of gods, kings, and private individuals, as well as on the mummies of both humans and Apis bulls. It was even performed on the individual rooms of temples and on the entire temple structure.*
The Opening of the Mouth Ritual
*The effect of the ritual was to animate the recipient (or, in the case of a deceased individual, to re-animate it). **The ritual allowed the mummy**, statue, or temple, **to eat**, breathe, see, hear and enjoy the offerings and provisions performed by the priests and officiants, thus to sustain the ka. .*
The Egyptian terms for the ceremony are wpt-r and wn-r, both translating literally to "opening of the mouth." The verb wpi denotes an opening that entails splitting, dividing or separating, and is used to describe the separation of two combatants, the dividing of time or even a determination of the truth. The verb wn emphasizes accessibility and exposure, used in contexts such as wn-hr, literally "open the face", but more correctly meaning "see" or "be seen".
The earliest Old Kingdom textual references to the ceremony date to the early 4th dynasty, to the Palermo stone and the decoration of the tomb of the royal official Metjen. At this time, the ritual seems to have been performed solely to animate statues, rather than to re-animate the deceased. The Palermo stone states that the ritual takes place in the hwt nbw, in the goldsmiths quarter, sometimes translated as "Castle of Gold, (or perhaps referring to the quarry of Hatnub). The textual formula for the ritual reference is written as "the fashioning and opening of the mouth of (a statue of god X) in the goldsmiths quarter/Hatnub."
The captions of the scenes in Metjens tomb mention that the ritual is performed four times, in conjunction with censing and transforming the deceased into an akh. In the Pyramid Texts and later funerary texts and captions, the rites are also said to be performed four times. The spells are repeated four times, for Horus, Set, Thoth, and Dwn-anwy.
It was probably not until the sixth dynasty that the statue ritual was incorporated into an Opening of the Mouth ceremony already developed independently as part of the funerary ritual. This ritual itself may have been a symbolic re-enactment of the clearing of a babys mouth at birth. The earliest implements used were probably the priests fingers, later replaced by finger-shaped iron blades. In many texts, reference is made to the fingers of Horus
Earliest references to the ritual comes from the Pyramid Texts, inscribed on the burial chamber of the pyramid of Unas, dating to the end of the 5th dynasty. One set of Pyramid Texts referring to the use of the fingers to open the mouth are PT 1329-1330, translated by Faulkner as "your mouth is

split open by Horus with this little finger of his with which he split open the mouth of his father Osiris."
The Opening of the Mouth Ritual
*Other implements besides fingers were added, as indicated by Spells 11-15 of the Pyramid Texts. They describe the Opening of the Mouth ceremony using the foreleg of a bull and an iron wood-working **adze.** Other inscriptions give an offering ritual in which two blades of meteoric iron, called the ntjrwy, are said to open the mouth. Faulkner translates this spell as "O Osiris the King, I split open your mouth for you gods iron of Upper Egypt, 1 ingot; gods iron of Lower Egypt, 1 ingot." The ritual was thus be performed with various implements, most commonly a wood-carving **adze**, which were touched to the lips by the officiating priest. An adze was an arched metal blade fasted across the top of a wooden handle with leather thongs, used in woodworking. The ceremonial **adze** was made from the metal of heaven, bi3 n pt, meteoric iron. The **adze** mimicked carving and sculpture, logical if the Funerary ceremony evolved from the ritual performed on a statue..."*[39]

Marie Parsons

As it can be read in this excellent text by Marie Parsons, the opening of the mouth ceremony once again puts the mouth in connection with the little **adze**. We thus find the indissoluble genetic link between the word Ekang /á-dzô/ and the English word used to name the ***adze***. Surprisingly, the opening of the mouth ceremony is a ritual of sedentary peoples and not nomadic peoples just as the adze is a tool of sedentary peoples. A tool that the metal part is shaped by the blacksmith caste. Should we have to imagine that the Moors brought in their luggage, adzes and the ceremony of the opening of the mouth?

Where is the origin of the transliteration mr?

Africans must be very careful with the certainties of the civilizing scholars of Africa.
As we have already said, this script deals with love indisputably. To the question of who we like the answer is quite clear to the scribe. He speaks of the Love of the human race.

Middle-egyptian: ***mri*** "to love; to like"

On this account we say that this transliteration from the Ekang point of view responds to the following approach:
A man can love a woman an vice versa. For the scribe, love is a triptych mouth, hoe, Man.

Ewondo: /mor/ "man; human being"
Basa'a: *murã*; *muta; muda* "woman";
Middle-Egyptian*:* *mw.t*
English: *mouth* "stuffy; bazoo etc."
French*:* meuf "woman"
French*:* dame "woman" ☞ananym

The person who has been informed by Coptics has not thus recorded notonly the word *Love* but also the word *Man* as the beloved We have the signification love here just by simple metonymy.
If Western philosophy points out that Man is reducible to his thought by his *cogito ergo sum,* here we have a different approach knowing that, man is above all Love.

Rigorously the consonants skeleton ***mr*** not corresponds only with the word *Love* but it is related to the word *Man* also in the meaning of the Human race according to our point of view. Here is how it works.

Middle-Egyptian : ***mr*** « hoe » ☞**Ewondo** : i-bèh « hoe ; what is used to plant »
But we have by homophony that ☞**Ewondo** : i-bèh « sharp ; splinter ; peaked »

39. http://www.touregypt.net/featurestories/open.htm

A sharp bamboo shows normally a shrinkage. We have then in Ewondo language:

Ewondo: / á-myar / "to shrink"
Ewondo: / i-myar / "what has been shrunk"

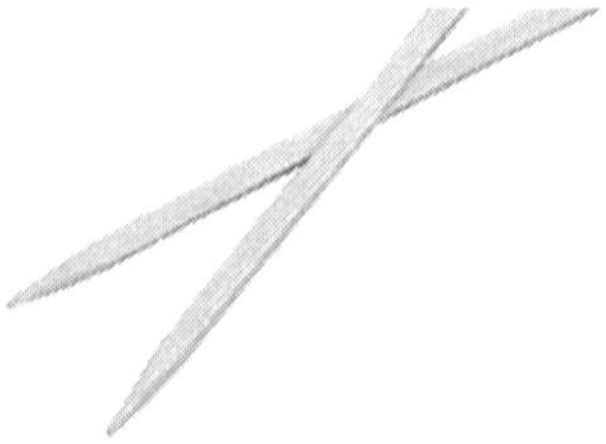

i-bèh

However the word / *i*-bè / "hoe" admits by homophony an item that can be absolutely associated with the idea of marriage and therefore with the idea of love by metonymy in our languages.

Ewondo : /Bè/ "Two" ☞Example : Bor *bə*-**bè**, two men. ↔ to get two = marriage.
Ewondo : /Bèa/ "To join by marriage between two families or villages.
Ewondo : / à-bè / "wedding; Family in law"

Driver: /*i*-bèh *I*-nə *i*-myar/ and / *i*-bèh i-nə *i*-bâ(kh) /

Generalisation
Some of African sacred animals are Horned "bull, lamb; ram etc.) because they are representing the eldest ancestor.

Sociologically, there were fishermen, shepherds and farmers at ***km.t*** .This script representing a hoe, expresses an idea associated with the socio-professional category of farmers. Also, among the shepherds this same sign will be represented by horned sacred animals.
As I have emphasized the idea of pointy can easily be associated to a horn and by metonymy to the animal draw by the scribe.

Middle-Egyptian : *mrw* ☞ **ram** — **French :** *Amour*, **Italian :** *Amore*, **Spanish:** *amor*

Ewondo: /i-ləb/ "sharp" ☞ **bull** — **Geman :** *liebe*

Ewondo: /*vul/ ☞ **bull** — **English:** *love*, **French**: *lover*, **Russian** : любовь *lyubov'*

Ewondo: /*n-yâ*/ ☞ **bull** — **Chinese:** 爱 *Ài* "love"
Ewondo: /*n-yâ*/ ☞ **bull** — **Japanese:** 愛 *Ai* "love"

Note: As I have previously underlined several times, the human being lexicon can be devided in two main groups of words: words from the kekwé group and words from the kweké group. Both are dealing by phonemes inversion.
So from any name of the african holy horned, to get the word linked to **love** in any language just invert syllabes: read from right to left the name of the horned animal.

The Japanese or the Chinese word for **love** recalls the French onomatopoeia for the pain ***aïe***; Ouch.

Love represented by a heart pierced by an arrow.

Maori : aroha "love"
English : Arrow

Reminder

Middle-egyptian:  ***mri*** "to love; to like"

In Egyptian conjugaison, the **prospective** is marked by the **y** (). It expresses a nuance of future in ancient Egyptian.

	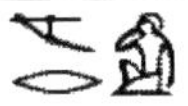	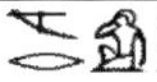
Middle-egyptian	***mr*** "to love"	***mr-y*** "the beloved one"
Ewondo	*à diŋ* "to love"	i-nyo à díŋbèn

Chart 65: Ewondo-Middle-egyptian prospective comparison

Note: the transliteration ***mr*** is the equivalent of the French word *aimer, mère, mari, amour, mirer; amitié, marie, mêler* etc. and the English item *to melt; amour etc.*

The equivalent kékwé word in **Koonzimé** is *lémé* "heart" and the kwéké word  in Ewondo language is :

Ewondo: /məli/ "grass cutting, work with a machete or a hoe for little herbs cutting"
Bulu: *ùn*-nləm "heart"
Ewondo: *ùn*-nəm "heart"
Ewondo: *ì*-ngwanghà "adze"
Ewondo: *à*-nyàmi "amalgamate"

☞ **Ewondo**: *à diŋ* "to love"
☞ **Ewondo**: *à-ŋgoŋ* "bell"

The words to designate the adze and the bell are homophones. The heart is the seat of love, it is beating as a bell. That's how we do understand the link between the adze and love.

Which conclusions can we draw from all this?

Conclusions are numerous.

Sociological level

The reading of this script above illuminates another day some African rituals like the sharing of the cola. Indeed, the cola is called / à-bəl / in Ewondo. It symbolizes love, friendship and sharing. Let's remind if necessary that the homophone / á-bal / "weeding, hoeing etc."

- *Bafia: ataras "kola"*
- *Bafoé: bobe, dibe "kola"*
- *Bakossi: ebeu "kola"*
- *Bakwéri: mbanga "kola"*
- *Bamiléké: tala "kola"*
- *Bassa: goro "kola"*
- *Ewondo: abeu goro "kola"*
- *Foulfouldé: worore "kola"*
- *Haoussa: goro "kola"*
- *Kɔ́ɔnzimé: bil "kola"*
- *Fang: abel, eya-abel, "kola"*
- *Ibo: oji "kola"*

Chart 66: kola in varieties of the unique african language

The homophony between kola and the words of the semantic field of marriage, mouth or hoe is not a mere coincidence. Its use in African rituals ans finds its origin there.

> *"Kola is also used in Cameroon in ritual ceremonies"*[40]
>
> *(MBOLO-1998).*

Kola is not simple fruit, it can be used in divination arts among Basa'a for example.

> *"The cotyledons on the ground are used among the Bassa to know a hidden truth"*[41]
>
> *(O. NDOYE-2006)*

In the traditions of the Nilotic peoples, the kola is specific to certain rituals like the one of marriage.

> *"Kola nuts are offered by the husband in some ethnic groups of Cameroon, to the beautiful family during wedding ceremonies."*.[42]
>
> *(E. MATIG-2006)*

Not understanding the metaphysics surrounding Kola and its ontological dimension, the Arab historian Shihab al-Umari (1300-1349) declared about the kola:

> *"The kolas are bitter, unpleasant to taste and the blacks only eat them"*[43]
>
> *(Shihab al-Umari-1928)*

The Kola nut is composed of two cotyledons strongly glued and which give a welded aspect like the sign U7. There is therefore no better representation of the link symbolizing the marriage bond than this fruit. That is why it is offered among others in the context of marriage by the African peoples.

Let's take our script again to have a better understanding of all this.

Middle-egyptian: ***mri*** "to love; to like"

The sign U7 is called in Ekang languages:

Ewondo: /í-ŋgoŋgà/ "adze"; offring the following homophony:
Ewondo: /í-ŋgoŋ/ "throat"; ☞ extension of the mouth
Ewondo: /à-ŋgoŋ/ "padlock; lock"; etc."

If we stick to what the scribe's writings say, marriage is a hindrance, a solid bond as a stone. The strength of the marriage is ensured by a padlock type security. This relation looks as a prison because the adze offers an homophony between digging, drilling and the word prison in african languages:

Middle-egyptian: ***mr*** "hoe; drill";
Ewondo: / i-ŋgwakh/ "small adze".; /*ùn*-mbôkh/ "drill; prison"

By homophony ☞ **Ewondo:** /ŋgwakh/ "stone, rock; Pebble".

Returning to Shihab al U-Mari, we totally agree with him that there is only a culture of kola nut consumption in Black Africa or among Afro-descendants. Vivien and Faure give a good illustration:

> *"The cotyledons of Cola acuminata are chewed as a stimulant"*[44]
>
> *(Vivien & Faure, 1995).*

40. MBOLO, Data collection and analysis for sustainable forest management: joining efforts and International Programs, EC-FAO Partnership Program (1998-2001), Forest Budget Line tropical B7-6201/97-15/VIII/FOR, PROJET GCP/INT/679/EC, Statistical product data non-timber forest in Cameroon.
41. IPGRI, The edible forest fruit trees of Cameroon,Bioversity International, p.171.
42. IPGRI, The edible forest fruit trees of Cameroon,Bioversity International, p.171.
43. Renaud H.P.J., The first mention of the kola nut in the medical field of the Arabs, Hespris, vol. 8, 1928, p. 43-57
44. VIVIEN, J. et FAURE, J. J. 1995. African Wild Fruit - Species of Cameroon. Editions NGUILA-KEROU, France, 416p.

This observation leads us to conclude that, where the kola tree does not grow, those people can not have built ***km.t***. On the other hand, those for whom this nut is pungent and unpleasant to taste, can not claim the paternity of ***km.t*** as well. Only the negro can understand what the Kola is sociologically because it is a negro cultural marker indisputably.

To return to the script, its reading allows a better understanding of the term foster breastfeeding which is related to the breast / á-bèh / in Ewondo, / é-bél / in Kɔɔnzim those words are homophones of /i-bal/ "hoe". Note that in a sedentary and agricultural society the hoe is used to make the fields and thus to feed. Love is thus nourishing one another, and the African dowel here lights up a new day. To love a woman and to bring the dowry is to bring food and drink to her family. It's feeding your family in law.

One could add the term ***ibA*** of the Pharaonic refering to the dance, / *á*-bokh / in Ewondo, *e*-bwal in Kɔɔnzimé.
Love is synonymous of joy and festivities. / *bí*-dí ; *mə*-nyù / "foods and beverages"

Anthroponymy

Middle-egyptian: ***mry.t*** "the other name of As.t, commonly called Isis"
Ekang Languages: Inyeghe, Bela, Bala, Mbala, Bêlè, Ebèlè, Idiŋhi etc.

All these words are attested in Ekang countries.

Toponymy

Middle-egyptian: tA mry "the beloved country"
Ewondo: Ta Bela, Ta Bala, mballa I &II (Yaoundé's quarter) etc.

The technical reservation which appears in the Wörterbuch (Wb. P.98,11) stating that the Pharaonic expression ***tA mri*** does not necessarily mean "the beloved land" rather should be likened to "the plowed soil" therefore finds here all its limit on the technical level. In African languages: loving, licking, plowing, hoeing, weeding, are interchangeable because they share the same semantic field. In Fulfuldé language for example: *temerre* means, hundred, multitude.

Reminder : I want to emphazied this below on a point already seen.

Somali: maroodi "elephant"
Ewondo: zokh "elephant"
Zandé : mbarã "elephant"

Phonetically I already know, they are two different languages, do not heap up even more. So let's take a sibling language from Africa which has the particularity to maintain a close relationship with the African mother-language through its Sankrit but especially Greco-Latin roots.

French: *maraude; marauder; maraud* "Theft of food, especially fruit, vegetables, poultry, in the fields or on a farm."

English: maraud; marauding

French: *remède "drug"*
English: *remedy* "drug"

So if I understood everything correctly, the word *maraude* (r) of French and the word *maroodi* of Somali share a kind of phonetic proximity. So these two languages are not as far apart as this is presented in

the books of scholars. Surprisingly French does not appear alongside Somali in the classification of Greenberg[45].
Let's try to see if the phonetic link is genetic between *marauder* and *elephant* is found in the triangulation Ewondo, Somali, French.

***THEFT:* nzu** *(h), wup (h).*[46]

(GALLEY, 1964)

Surprisingly, I find here again the confirmation of my words. It seems that Men were born in Africa, isn't it? It reassures me so …

Question: does the word *maroodi* of Somali have an equivalent in Ewondo?

As we have said before, it's the same language so the answer is automatically yes.

Fang: /à-mara/; /mare/ "To fear oneself, to flee each other, to fear, to flee, etc.".
Ewondo: di-mbara >i-mbara "scary"

Note: The presence of an epenthetic **b** is found between fang, Somali and Ewondo. As a general rule, the appearance within a word of a non-etymological b is said to be epenthetic. We often find that it occurs every time we have: **m** and **r** (or **m** and **l**).

The epenthesis of **b** also occurs in Egyptian, Coptic and African languages under the same conditions: mr -> m**br**. We must link the origin of this phenomenon b said epenthetic in the evolution of variants of the unique african language the **Negro-egyptian**: k^{w}>mb> m

If the Ewondo use the affix ***di-*** in prefixal position (prefix of gender and number), the same affix by grammaticalization is suffixed in Somali. Just remind Kékwé and Kweké branches of the negro-egyptian.
So that explains in Somalia the ***-di*** in suffixal position of the root ***mar-**.

Once again the word somali *maroodi* is very well known in Ewondo it's the same word used in the **Zandé:** mbarã "elephant" that we have already peeled back in previous layers.

The technical reservation appearing in the Wörterbuch (Wb. P.98,11) stating that the Pharaonic expression ***tA mri*** does not necessarily mean "the beloved land" rather should be likened to "the plowed soil" therefore finds here once again all its limit on a technical level. I therefore send you on illustration N°19 hope that you will see the elephant's head. If you do not see well, I'll confused you by this reading in Ewondo:

Ewondo : /*ùn*-nloh *i*-kama/ "elephant's head"
Ewondo : /si *i*-mbàrà/ "dread land"
Ewondo : /Tsè *i*-mbàrà/ "the chief's land"
Ewondo : /Tsè *i*-mbàr/ "the land of the ax"
Ewondo : /Tsè *i*-lob/ "god's land"

We have in the northern side of Cameroon the city of **Tokombéré** a toponym recalling to the transliteration ***tA mri.***
Tokombere covers an area of 552 km^2 in Cameroon's. It is one of the most beautiful region of the Far-North. Located in the northern part The municipality is head of department Mora in Mayo Sava.

---

45. GREENBERG J. H., Languages of Africa, La Hague, The Netherlands, Mouton & Co, 1966.
46. S. GALLEY, Fang-French dictionary. and French Fang followed by a fang grammar. Neuchâtel: ditions Henri Messeiller, 1964, P.556.

We must not blame our friends Egyptologists who are doing what they can to study the history of Africans while those africans shine by their absence in their own research.

Other approach

Middle-Egyptian.: 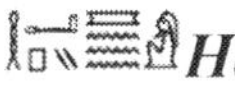***Hapy*** "Nile"
Ewondo: /mə-bèh/ "breasts" or ***itrw:*** / twé/ "breast"
The beloved river of ***km.t***

Ontology

In the land of Canaan, former suzerainty of ***km.t,*** was worshiped the "God" Baal. Rigorously this country was the one of the Natufians descendants in the meaning of Miss Dorothy Annie Elizabeth Garrod, the British archaeologist and prehistorian who excavated Es Skhul and El Wad, in order to enlighten humanity on the settlement of the Present-day Palestine during the Paleolithic period.

We therefore better understand what the term Baal, which appears in historical documents, means about ancient African cultural traditions.

Baal, a character who is omnipresent in the Torah and the Old Testament. This eponymous name was revered by the Jews and Phoenicians who were in Asir (Saudi Arabia next to Yemen) around 1000 years B.-C. with as opponents the Philistines.

The word Baal is also found in the Semitic language, notably in Akkadian "Belu" written in Ougarite (Lebanon) in the alphabet of the time without a vowel "B'L which means" Lord, Master ". to approach the idea of master with the term Ewondo nyamoro in the language beti (the accomplished Man) that one will find in Fulfulde language like nyaamougo (food, to eat).

With a little effort, the term adonaï finds its counterpart in wondo / á-díŋ / "to love". The word Yahweh finds its equivalent in the action of hoe hoeing / á-bal / with the forms (a-b, i-bal, a-bəl). The Jews worshiped Baal as much as Yaweh, who is nothing but Adonai (Lord) in Aramaic. The semantics of Adonai refer to Yahweh or Jehovah meaning Lord.

Baal as "God" was identified by an agrarian worship (as represented by the sign  U7, the hoe) which celebrates the death and burial of grains and their resurrection in the spring after 40 days underground. The supposed death and resurrection of Jesus must not be a pure coincidence, but certainly a bad plagiarism. Which is here brought to light. We understand better here the meaning of Lent. Esoterically, 4 is the number of the matter and 3 is the number of the spirit.
The time of the test on the esoteric plane lasts 40j (4 + 0 = 4). The spirit must ensure its predominance over the matter. A grain of wheat takes 40 days in the soil to germinate. I close here this parenthesis but proof is made here that the concept is an African one.

Can we rigorously close this parenthesis without mentioning the supposed biblical myth of Cain (á-kòŋ'o / Akono) and Abel (Bela / Bala)? Indeed, according to the biblical account, Cain is the eldest son of Adam and Eve. He is a peasant, and has a brother Abel who is a shepherd. According to a certain onomastic, the words are supposed to be of Hebrew origin in reality judge by yourself.

To conclude on this question of the God Baal it is much later only by racism and following the process of the normative inversion that the black ***ntr*** Baal will be associated with the zebu which is also his other representation to give the word ***Beelzebub.*** This information is recorded in the midrash Rabba and Bereishit rabba as synonymous of "satan" or "devil". (cf illustration n ° 23 and 24)

The ram by which this animal was represented is still called today in the Fulfulde language spoken everywhere in Africa and in North Cameroon: mbaala. This word / *ùn*-mbaala / in wondo is translated by the shepherd. This reminds me Jesus not you?

The words of the register of the drink, the beer of millet is named in fulfulde, mbal, the term yara "to drink", nyiri, nyaamougo "food, to eat" these terms refer to the idea to lick /à-nyé/ or to like nyeghe of the fang.

The word ram in Pharaonic is recorded by the consonants skeleton ***bA***.It is therefore better understood why the ram is the eldest caprine since he is "elder, old etc. So we call him / ntomba / sliced: /ntô/ "elder" and /ba/ "caprine; being". We also remember that this ram represented ***Imn***, the Imn ram (see illustration: No. 22).

Middle-egyptian: ***Imn*** "Amon"
Ewondo: /*ùn*-nôm/ "old"
Ewondo: /*ùn*-nóm/ "male"
Ewondo: /*ùn*-nàm/ "land, country"

The consonant skeletons ***imn*** and ***unm*** are simply consonant metatheses. It is thus that one must read what has been established as *imn* by Egyptologists.

The passageway of the bull to the ram simply represents the change of zodiacal era.

Culture

The term / mbal 'Etudi / is used to name the dance practiced during the big festivities by the big Etoudi family of yaoundé in Cameroon (the one of Yannick Noah). It finds here again all its meaning.

In the end, we note here that the reading of a script from the African point of view is not univocal (p-a, pa, p-a, pa → papa). Signs are multifarious and polysemic operative concepts. This is why the reading in context will better shed light on the meaning of the words recorded by the scribe.

We have seen just few aspects of the script above. But really, those aspects remain superficial compared to the fields of possibilities offered by the script through the metonymy, the antonymy, the homophony and the synonymy of the terms that were engaged in the framework of this endogenous reading.

When egyptologist often say:

Middle-egyptian: *As.t mryt imn* "The beloved of Amon"
We, we do say for the same expression**:**
Ewondo: */I-toa mor ùn-nâm/* "the throne" the one on the chair, the chaiman.
This can be litterally translate the chair of country-Man

But according to african spirituality we can also read:
Ewondo: */I-toa mor Zàmbà/* "The priest's chair"

This expression can be litterally translated by the chair of the God's Man implied the priest. Indeed, all the black pharaohs were absolutely priests in ancient Egypt. We can just say that, all these expressions are still used nowadays all over Africa.

Anthroponymy

Middle-Egyptian: ***As.t*** Even blinded you are able to see that this script alone is giving the right undestanding of the above translation.

Fulfuldé : Aïssa
Fulfuldé: Aissata; Aïssatou

3.4.3.15. ***Example 15: A One Sign Director Script Anlysis***

Middle-egyptian: ***mw.t*** *(Faulkner p. 106, Wb. II p. 54, 1-10)*

Step 1: Name all the signs

- Sign G14: ndoé, í-kuba, kubbe, *ùn*-nya, ndoé-obama "vulture"
- Sign x1: í-di "food"
- Sign B1: *ùm*-inga, "woman"; *ùn*-nya, *ùn*-na "mum" etc.

Step 2: R.A.S

- The sign X1 representes food, the nutritional function related to the mother
- The sign B1 is the determinative and the semanteme.

So an important part of this script reading is based on the G14 sign. Either we know the name of the vulture in our languages and we read the script or we do not know and we will just invent.

- Sign G14: ndoé, "vulture" ikuba, kubbe, kube "bird; vulture"

Step 3: Determinative Reconstitution

Reading this script envolve to analyze and understand the meaning of the sign G14. In a matriarchal society, the concept of the mother is the hornrstone of the social structure.

Sign G14: ikuba, kubbe, kube "bird"

The term */kub/* is used in Ewondo to designate the gallinaceans (cock, hen, fowl etc.). It is therefore the protective side associated with the vulture and the mother-hen that is here invited to the understanding by the scribe. Properly spoken between the mother and the vulture there is not much in common.

Rigorously the reading of this script would be / kub / and by semantic shift / nya-kub / "mother-hen". Note in passing that the semantaxe has spread to French. In use the term for mother in Ewondo is / *nya / or / *na /.

Step 4: Check of consistency of the reading

The sign designating the quail chick ***w*** would be vocalized in Ewondo / *ùm*-wàn kub / "the little gallinaceous".
Reading from right to left we have / nya-kub / "mother-hen"

The name Ewondo / ndoé-obama / is interesting because it indicates the name of the bird / ndoé / and identifies it with the ornithological aspect of its son ***Hr*** / Obama / "hawk". By the way of naming this bird, we have here a code that identifies the mother and the son (here the primordial ancestor of the African people) Obama.

We will therefore better understand the approach of Hermopolitan cosmogony for which the universe is derived from a cosmic egg. Maternity in the broad sense is therefore the work of an oviparous anatidae entity ***gb*** "goose"; ***gb*** "Earth". By metonymy, the placenta **x** with its embryo can be considered as an egg.

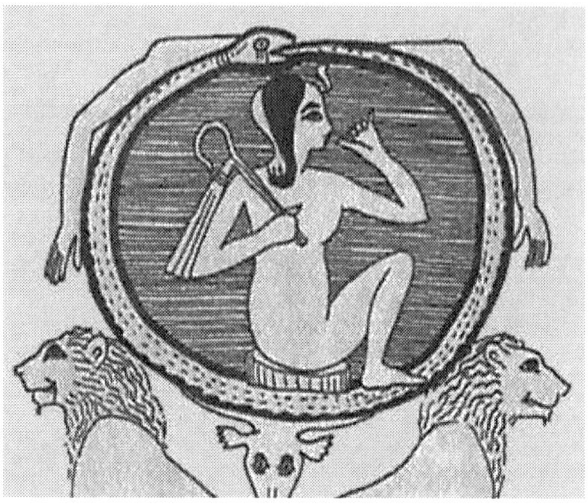

Ouroboros

Cosmic egg

We have better understanding why the Fara "sovereigns" were ***sA*** ***rA***

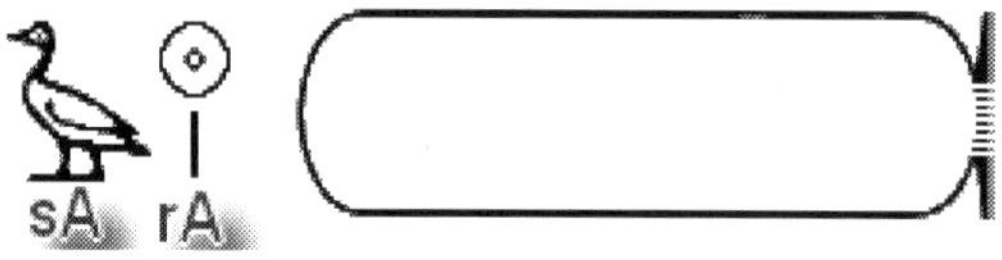

A cartouche, called ***šnw*** in ancient Egyptian, the transcription of the Egyptian word ***šnw*** is a hieroglyphic symbol, elongated and closed by a knot, which contains the name of a pharaoh. It symbolizes the egg, all that the sun surrounds, that is to say the universe and has the protecting function of the name of Pharaoh. They were most often painted in yellow or gold (color of the sun) and were used for two of the five names of a fara.
So in fact the term ***sA*** denotes the egg and therefore the son. In reality, the words for childbirth are words associated with egg laying in Ewondo according to the meaning retained by classical Egyptology.

Thus the word for giving birth is to be associated here with the word to lay *msi* is vocalized in Ewondo, / á-soui / "to lay", / mə-soui / "the laying, eggs, brood". For the ancestors to give birth was to lay an egg

The term for placenta **x** is in Ewondo / í-sokh /. This term is homophone of duck / sokh / but also an homophone of sexuality / *sokh /.
The placenta is biologically the internal egg, the unwavering link between the child and the mother. We can interchange duck and placenta without changing the meaning of a script.

In order to confirm this idea of egg-laying, the woman's position at birth is therefore reminiscent of the nesting position of a bird. The expression ***st ms*** (zAt ms) duck laying an egg is quite clear on this subject. The duck is not lying on its back to lay eggs. We say in Ewondo : /á-sui mə-kî/

The term for eggs in Ewondo is also / mə-kî /. This term / mə-kî / also designates the blood. So the use of the volatile to designate the word *son* is not the fact of chance.
We better understand the semantax that associates his offspring with his blood. It's my blood, that's my son.

The term "***Sw***" generally designates feathers and therefore here poultry. Rigorously we would translate poultry egg. But the generic term for poultry is / kub / in our languages. As far as the phonetic correspondence is concerned, if I take the tongues of the Kongo bowl, the dull alveolar fricative consonant in its initial position very often corresponds to the voiceless velar occlusive consonant "K" and the voiced labio-velar spiral consonant W which serves to make the plural corresponds to "b", "m" (be / me). It can also be used to designate the masculine (sound or initial) so indisputably here we have the word kub / kubu "poultry = hen". So we have one word left that is the egg in the strict sense of the word. But phonetically the blood and the egg are vocalized in the same way.

On the endogenous level:
The symbole “folded loincloth” is especially used to cover oneself. The word for covering itself is / á-kub /, it could also be approached phonetically by the idea of a second skin / kob / skin.

The quail chick alone offers the vocalization / kub / because it is a poultry.
It offers also the acceptation / ki / because it can be likened to the egg and / ko / "foot" because it designates ***Hr***. One of the characteristics of the chief ***Hr*** is to embody and be a guarantor of the use of the strength / ki / .

the twisted rope is called / *ùn*-nkô / "rope". We saw why the root / ko / is vocalisable / ki / too.

Using a deductive method we have:
The script refers to the idea of drawing blood. This one gives an idea of blood flow . If we compare the three scripts the twisted rope also codes for the sound / ki / for egg = blood in the languages of the Ethiopian bowl. In the languages of the Congolese basin, / ún-ŋkô / "rope", / à-kî / "blood" = "egg". So we have written by the scribe here **Chicken Egg**. The hen being considered by the scribe simply as a bird, a fowl.

Middle-egyptian: ***swH.t***
Ewondo: à-kî kub “fowl’s egg”

Middle-egyptian: ***msi*** “give birth” according to egyptologists.
Ewondo: /à-biéh/ “give birth”
Ewondo: /à-soui/ “spawan; lay; churn out”

One might think that these two terms are different. For the neophyte, they are simply from a phonetic point of view. But in reality they are not from a semantic point of view. The subtlety of understanding African languages is based on this intelligence that led to the creation of tribes and so on..

The sign F31 transliterated ***ms*** represents, according to the Egyptologists, three fox skins bound together. This definition makes it possible to note that the fact of binding the three skin leads to create One. In one way it look like *a three in One* or the trinity. According to African ontology the trinity means one in three. "Father, son, mother".

The acceptance of the skin is confirmed by the sign S29 representing a folded cloth.
The term used to designate the skin in Ewondo is / i-kob /. This word / i-kob / is an ananym of / ùm-bokh / "one"; " feather "; "Duck"; "Nestled"; "Two" etc.

How to prove that the duel is synonymous with one?

This is relatively simple if one starts from the endogenous postulate that makes African philosophy, a thought based on relativity and not an absolutist thought.

The unit is identical to two because the two is the other one in a binary system. In a binary system, the second is the other one and semantic shift both.
The semantax to designate the "other" is always the same as the OTHER ONE.

> **Ewondo**: /i-nyo *ùn*-**mbokh**/ "the other **one**"
> **Ewondo**: /i-mor *ùn*-**mbokh**/ "the other **one** person"

These are archetypal traces that explain why I am the other and the other is me in African thought. So there is no way of speaking in our languages to mark the difference with respect to the other. Consequence of what, if the other is me, he can not be treated in an iniquitous and in inhuman way.

I therefore recall that, I have formulated that the term / mə-sè / "feathers" has a synonym which is / *ùm*-mbokh / "feather" which also denotes the numeral Un, / *ùm*-bokh /. We established the phonetic proximity between two and one. So that's how we get to this series of readings

> **Middle-egyptian**: ***msi*** "give birth" according to egyptologists.
> **Ewondo**: à-bih "give birth"
> **Ewondo**: /*-bèh/ "two, *thigh;* leg; ham etc."
> **Ewondo**: /o-bəh/ "small" ☞synonym with small;*th*in; s*t*ick; s*t*ing
> **Middle-egyptian**: "give birth"

So we understand by phonetic slip how we get to /á-biéh/ "give birth"

Note : hope that you have remarked the phonetic proxity between the items *bird* and *birth.*

> **Ewondo**: mə-sèh "feathers"
> **Ewondo**: mə-soui "egg lay"
> **Fang:** /sam/ "to give birth to a corpse".

Example:. /mon a ***sam/*** "the child is born dead",.

To bear, to breed, to produce, to lay, to adopt, by. The word Ekang /***sam***/ is a consonant metathesis of the transliterate consonant skeleton ***msi***.

> **Middle-egyptian**: ***msi***"give birth"

Let's get Familiar with the concept of egg and duck

According to the extension textbooks, ***As.t*** is a mythical queen and a funerary goddess of ancient Egypt. Most often, she is represented as a young woman wearing a throne or, in the likeness of ***Hw.t-Hr***, a wig surmounted by a "solar disk" inserted between two cow's horns.
Since ***Isis*** appears as the grecised form of the skeleton of consonants ***As.t***, how to vocalize this name in our current languages? I propose to do the exercise in Ewondo.
The sign B1 (the seatting lady) on the right designates a deified ancestor. But his sitting posture can read the script.
From left to right we have:

> The Egyptian vulture, sign G1 of the Gardiner's list
> Sign Q1 representing, the throne, the seat, the chair etc.
> The sign x1 reprsenting the roll bread
> The sign H8 representing a full egg.

There are different spellings to name the ancestor above. These differences lead us to understand why saying *Isis* or *Asset* is an unparalleled barbarism because it does not refer to any referent in our languages.

As.t ***As.t*** ***As.t*** ***As.t*** ***As.t*** ***As.t***

Chart 67: scripts transliterate As.t and read Isis by egyptologists

All these scripts are transliterated As.t and all have the *same* referent, the seatting woman (sign B1). How are we vocalizing these different spellings in Fang / Beti?

This script is read / *ùn*-ntout Etoa /. The most accurate reading is here from right to left. The full egg is Ntour / ntout in Ewondo, the bread cake that is located just above the egg serves as a kind indicates the nutritional function related to the mother. The reading of the full egg indicating that we are talking here of an accomplished Being, that is, who has completed his reincarnation cycles.

The sign Q1 as well as the sitting position of the woman return the idea of sitting, chair, seat which in language Fang / Beti is / I-toa /. This expression is also taken up by some that I would not name here in the expression "Holy Seat". But this spelling makes it possible to support the subject and justify it in the light of African culture.**Ewondo**: /ù-ntout Etoa/ "the holy seat".

This one is vocalized / Toŋ'o / because of the horns on the head of As t The duel y or // (y) imposes this constraint of reading. The sit ting position of the cobra is a good reminder of the sitting position. Etoa (the seat). To be very precise we will say: / i-toa Toŋ'o / "seat of the horned"

This script is read Ekara, Akara, Ikara. Reading is done from the stepped form of the seat or throne. This idea of staircase returns by homophony to the idea of diadem (ikar / ekare). Instead of reading Aset / isis a fang / beti will say Akara / Ikara / Ekara for the diadem, the crown.

This one is read /I-dzoé. Reading is done by association of the idea of the seat with the Egyptian vulture. The Egyptian vulture is Ndoé in Fang / Beti. The addition of this Vulture is not trivial or a fantasy. It emphasizes here the idea of authority / I-dzoé / related to the seat, throne. Ndoé or Ndo'o is the name of the person on the right side.

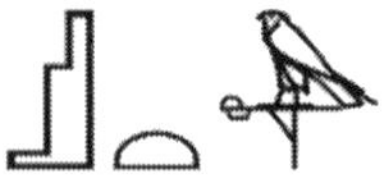 This one is read: Saghe, Sagh'a, sogho, Sokh, Zamba. The reading is done by association of the idea of the seat with the falcon perched on a prop. The fact of being perched is said /saghe/ in language fang / Beti

SAGH (h) vb. To put on top.
Example:Môr a sagh mveghe kum e yô, "the man puts his load on a stump".

TOP SUMMIT, ahôhôn (b) , ahôhôn nlô. Top of the tree, small branch of the summit, sokh (h)

 This one is vocalized / ù-nson'o /, / ù-nsoŋ'o / etc. The reading is done by association of the idea of the end of flesh / *ùn*-nsón / "meat". In appearance only this script is not important but it is crucial for the reading of other scripts that will rely on homophony with this piece of flesh. We can have anthroponyms like Essono, Hassana; Husseini etc.

In addition to these scripts transliterated ***As.t*** I postulate and assume that the skeleton of consonants ***sA.t*** is interchangeable with ***As.t*** because they are different only by pure convention and otherwise the languages Kwéké and kékwé create words by ananym or by consonant metathesis in Pharaonic. The script is used in Egyptological rule to designate a girl.

sA.t: *i-sokh* ***sAt nsw:*** *i-sokh i-dzoué*

Chart 68: some egyptian word with sA.t

To finish on this subject, it appears obvious because the scribe does not use the French alphabet but only signs and moreover that the convention of reading ***mDw ntr*** according to Egyptian rules is not necessarily good according to the African prism.

I allow myself to postulate it because of the presence of the semanteme  (sign B1, the seatting lady) which is the same in almost all the transliterated script ***As.t***.

The convention is that we want this ***sAt*** script to be transliterated and read from left to right. According to this reading, that this script designates the girl, she would read / i-sokh / (placenta) in Fang / Beti. Indeed the sign 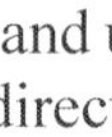G39 (duck) is designated by the word sokh in Fang/Bti. So by homophony we have / sokh / which helps to understand why the duck is used here. In the nominative, it will take a final vowel and vocalize /I.sogho/.

Furthermore, consonantal metathesis is a common phenomenon in reading and understanding the mechanism of pharaonic writing, so reading must also be done in the opposite direction, namely ***As.t***. A girl in the Fang / beti language is called /Ngon/, which corresponds to the pharaonic . ***s.t***. So this synonymy script can be read / ngon / too. The woman, the girl being associated with the moon /ngon/ ***iaH***, / ngon / with falconids, / ngon / with menstruation, with baboon / ngon / with squash /ngon/ etc.

In the nominative, it takes a final vowel. So, it could be read / Ngono / when it is called ***As.t***
On the purely philosophical level, the placenta is at first maternal, but it is also the child's interior, so it is also the child. It is therefore normal for the child and the mother to use the same script ***As.t*** and therefore identical to ***sA.t*** from the African point of view.

Depending on this scripture designates whether the king's daughter, it will be read /i-sogh/ (placenta) /*ùn*-nsoŋ'o/.

Indeed the sign G39 (duck) is designated by the word soghe in Fang / Beti. So by homophony we have /sokh/ helping to understand why the duck is used here.
This script is vocalized /*ùn*-nsoŋ'o/ “the beak”. The sign M23 is vocalized /i-sóŋ/, "shape of the ring" used here to name the leader, vertex, the peak, the top: /*ùn*-nsóŋ/.

We must read from right to left: /i-sogh *ùn*-nsóŋ'o/ "the king's daughter" and especially not from left to right we would have /*ùn*-nsóŋ i-sogh/ "the mouth of the girl".

We must also read from right to left: /ngon ù-nsóŋ'o/ "the king's daughter"
Or simply by synonymy from right to left: /mwan ù-nsóŋ'o/ "the king's daughter"

Middle-egyptian: ***As.t*** “Isis” **Middle-egyptian**: ***sA.t*** “daughter”
Ewondo: Ngono “Isis” **Ewondo**: /ngon/ “daughter”

Chart 69: As.t versus sA.t

I suggest to compare these two scripts in order to twist the neck to the supposed idea of sense of reading and supposed transliteration.
They have two common signs and three apparently different signs:

Communs	Diffrents
X1	• G39
	A1
B1	Q1

Chart 70: As.t versus sA.t, signs comparison

The sign X1 of the Gardiner’s list represents a food that is destined for the mouth / à-nyu /. The dedicated phrase is / byem bi à-nyu / literally « the things of the mouth ».

The sign B1of the Gardiner’s list is for the mother /*na/; /*nya/.

The sign G39 we name it / sokh / duck which offers Homophony / i-sokh / placenta, synonymous with / *nya /.

The sign A1 of the Gardiner's list represents the vulture we call that bird /ndoé/, /kub/ or /*nya/ because it designates the mother of the primal ancestor / nya kub /.

The sign Q1 of the Gardiner's list represents a seat that serves to rest the gluteal hemispheres / menyâ /.

I do not see any difference between these two scripts since all the signs allow to name one and the same reality.

The juxtaposition of the signs according to the contexts will guide us towards this or that meaning.

Ontology and metaphysics of the duck.

In order to illustrate the point, I used a comparison with a ritual associated with poultry (duck or chicken), from ancient Egypt to nowadays. It is practiced in Bamileke country and among Jews people.

In West of Cameroon the Gôb Beum / Ngap Bôô: The hen of the creation is the name given to this ritual opposite.

All Bamileke without exception know the ritual of Gôb Beum / Ngap Bôô (The Hen of Creation). This is a very very important ritual that every Bamileke child has to do. I did not say that MAY do, but MUST do, because this is his first spiritual and cosmological offering.

In fact, what we said from the beginning makes it possible to understand why this ritual is performed with a bird. In fact, why is the chicken or often a chick who is chosen for this ritual? Why is it rather a Hen and not another animal? We have already answered this question. We will not come back again to this point.

Not believing in chance, here is what the author of this text finds. In Yèmba language spoken in western Cameroon;

The hen in Yèmba is *Ngap* and /kup/ in Ewondo.

> **Yemba:** Ngap is related to the verb *Ngap* meaning (divide, share).
> **Ewondo**: /Ngap / "part of a division".
> **Ewondo**: / à kap / "divide, share, etc."

In Yemba language this term is an homophone of the word *Ngap* " week". Now we know that Bamileke week is eight days old. Indeed according to the people the duration of the week is modeled on a lunar or solar cycle. Switching from one system to another is easy when you know the conversion rules

Depending on the location in ***km.t*** some peoples have kept archaeological traces of the time division. The count of time according to an octal base (8) reminds us the so-called ***Xmnw*** cosmogony based on the Ogdoade. The number 8 represents the four couples of the creation for our Ancestors. For the record, the eight original entities (frogs and snakes) inhabiting the primordial Ocean design an egg that they deposit on a hillock.

From this egg, arises the Sun which creates the world generated by ***xnmw*** / ù-nkomboro / the potter or according to another version ***DH.wty*** the scribe considered as the creator of the world (in the cycle of the day). The first beings he built were the eight reptiles of the Ogdoade and they provided the elements that make life possible on Earth.

- ***nnw*** and ***nnw.t*** were snakes synonymous of water by menonymy.
- ***HH*** and ***HH.t*** were frogs synonymous with infinite time.
- ***kkw*** and ***kkw.t*** were serpents synonymous with the night, and
- ***Imn*** and ***imn.t*** were frogs synonymous with emptiness.

These entities are related to the Spirits of Nature, as later the figures of ***ptAH-Wsir*** assimilated to water and vegetation.

The primordial principle ***HH*** is, from the New Kingdom, resulting from the fusion of two divinities, a member of the Hermopolitan Ogdoade (association of eight divine entities), which plays a vital role in the cosmology of ***xmnw***, the other, the principle of infinite time, the genius of eternity. The first principle is known as a Frog, (to represent the large numbers and thus the infinite) the Egyptians thought that the animal had witnessed the creation of the world. The second takes on the features of a Frog Man wearing the solar disk. He holds in each of his hands a palm shoot (***rnp***) sign and symbol of millions years.

These couples of creation according to Hermopolitan thought are:

HH and HH.t, spatial infinity
kkw and kkw.t, deep dakness
nnw and nnw.t, the initial water pair
Imn and Imn.t, what is hidden

As we can see, the Ogdoade is made up of couples that can also be approached by a system of polarity of the fundamental essences that are in all things (+ / -). We can approach these couples to:

- ***Sw*** is air, ***Tfn.t*** is moisture
- ***Gb*** is the earth, ***Nw.t*** the sky
- ***As.t*** is order, ***Wsjr*** his keeper
- ***stX*** is disorder, ***Nb-Hw.t*** the keeper.

According to Hermopolitan thought, these four couples are at the origin of the diversity in the universe and the forces that animate it. The whole creation is therefore governed by these 4 couples (8 deities).

I therefore deduce the reason for the eight days of Bamileke week. *Temgo* for example in Ghomala is the day of field work. Who says field work, says land, so ***gb***.

Each day of the week would therefore be governed by one of these entities. This is the very meaning of the "distributing and sharing" verbs that were alluded to at the very beginning /ngap /. The week is shared between these 8 entities or cosmic forces. So at the end of the week, we went through all the primordial entities, so the whole creation.

Moreover, the word creation has its place, because in the languages of western Cameroon they often say ***Mbôô / Beum*** " to build, to create".

Therefore, according to the semantic meaning of the word creation in our languages then we have to consider that *hen / duck* is the animal of creation.

After this excellent text from **Dzu To** that I took the liberty of retouching at times, I weighed it against the ritual called **Kapparot** among the Jews.

The illustrations at the beginning of the paragraph show whether it was still doubted that the Egyptian ritual still practiced today in Bamileke land is a Negro practice.

The Kapparot ceremony [47]

Reminder

Yemba: Ngap is related to the verb *Ngap* meaning (divide, share).
Ewondo: /Ngap / "part of a division".
Ewondo: / à kap / "divide, share, etc."

"Kaparot" by Hendel Lieberman

It is customary to perform the Kapparot rite (symbolic "expiation") in preparation for Yom Kippur.

This ritual is accomplished by turning a chicken three times over his head while reciting the appropriate text. The poultry is then slaughtered according to the halakhic ritual slaughtering procedure and its monetary value is given to the poor, or, and this is the most widespread practice today, the chicken itself is given to a charitable cause.

We ask God that, in the event that we are destined to be the object of cruel decrees, may these be transferred to this chicken by the merit of the mitzvah of charity.

In most Jewish communities, Kapparot are organized in a designated place, in accordance with the guidelines of the health authorities. Live chickens are available for purchase, ritual slaughterers are present, and the chickens are then donated to a charitable institution. Ask your rabbi if Kapparots are organized near you.

The moment

The Kapparot can be made at any time during the Ten Days of Penance (that is, between Rosh Hashanah and Yom Kippur), but the ideal time is the day before Yom Kippur shortly before dawn because then "a thread of divine goodness" reigns over the world.

Poultry

Several reasons have been put forward to explain the use of a chicken to accomplish the Kapparot rite:

1) In Aramaic, a rooster is called a hunger. Now, in Hebrew, a hedge is a man. So, we take a fight to atone for a heave.

2) A chicken is an easily findable and relatively cheap poultry.

3) It is not a species that was offered as a sacrifice to the Temple. This makes it possible to exclude the possibility that someone mistakenly imagines that Kapparot is a sacrifice.

It is customary to take **white chickens** to evoke the verse (Isaiah 1,18):

"If your sins turn red like scarlet, they will become as **white as snow**. "

In any case, ***it is not appropriate to use a black chicken***, because black is the color that represents the divine severity and rigor. Do not use a chicken that has a defect or obvious injury.

47.http://fr.chabad.org/library/article_cdo/aid/995025/jewish/La-crmonie-des-Kapparot.htm

A man or a boy uses a rooster, a woman or a girl takes a hen. The best is that everyone has their own chicken. However, if it is too expensive, the same chicken can be used for several people. So, an entire family can do Kapparot with two chickens: a male for boys and a hen for girls.

When several people make Kapparot with the same chicken, they have to do it at the same time, and not one after the other, because you can not do Kapparot with a "used" chicken.

A pregnant woman makes Kapparot with three chickens, two hens and a rooster: a hen for herself and the other hen and the rooster for the child she is carrying (whose sex is not known). If it is too expensive, a hen and a rooster are enough (because if the child is a girl, she shares the hen of her mother).

If live chickens can not be obtained, other kosher birds may be substituted (except doves and pigeons, as they were offered as a sacrifice to the Temple). Some use live kosher fish; others complete the whole rite with money, and then give that money - at least the value of a chicken - to charity.

The ceremony

Take the chicken in your hands and recite the first parascript ("Benei Adam ...").
When reciting the beginning of the second parascript, turn the chicken three times over your head, once saying 'Zeh' halifati ',' This is my replacement ', once saying' Zeh temourati '(' This is my substitute "), then again saying" Zeh kaparati "(" This is my atonement ").
Repeat the two parascripts again (rotating the chicken a total of nine times).
Bring the chicken to the chochet (the ritual cutter), which kills the chicken.
This is an opportunity for you to perform a rare mitzvah: cover the blood of a ritually slaughtered bird. Take a handful of earth (a quantity of which is usually prepared for this purpose near the chochet) and recite the following blessing before covering the blood that has flowed:
Baroukh Ata Ado-nai Elo-heinou Melekh haolam buys kidchanou bemitsvotav vstsivanou al kissouï dam beafar.
(Blessed are you, O LORD our God, King of the universe, who sanctified us by His commandments and commanded us concerning the covering of blood with earth.)
It is customary in many communities to tip the cho'het for this service. If you are reluctant to hold a live chicken in your hands, someone else can hold it and turn it over your head.

Even the youngest children are traditionally brought to Kapparot, and one of the parents turns each child's chicken over his head, saying, "This is your replacement, this is your substitute, this is your atonement ..."

It is of utmost importance to treat chickens with humanity and not to God forbid to cause them any pain or discomfort. Jewish law strictly forbids any unnecessary pain to the creatures of God. The repugnance inspired by such a cruel act must be redoubled on this day, on the eve of the day when we ask God for kindness and mercy that we may not deserve. The Jewish Law Code even suggests placing the entrails and liver of slaughtered chickens in a place where birds can come to feed themselves. "It is appropriate to witness to the pity of the creatures on this day, so that in Heaven we may have mercy on us [also]. "

The same process is followed - except ritual slaughter when using fish or money to make Kapparot.

Conclusion

At the end of this plea which is in the expective on the choice of a Galinacea as a ritual animal, on the color of plumage etc. we can see that we are in a grotesque plagiarism and nothing more. I will not dwell more on this practice but remember that it results from plagiarism and its meaning escapes including pseudo-scholars because its anterioty in Egypt places it in the lap of Africa. The practice continues in Cameroon today. Some Africans who compare the ancestral practices and find strong similarities with the patricians of Judaism are lost when they base their reasoning on the anteriority of Judaism on African spirituality.

3.4.3.16. Example N°16: The Feathered Serpent

Middle-egyptian: ***mDw.t*** (*Faulkner p. 123, Wb. II p. 184, 8-14 (mDw.t))*

Step 1: list all the signs

Sign G17 of the Gardiner's list ***M*** /à-kouŋ/; / koghle-kue / "owl".
Sign I10 of the Gardiner's list ***D*** / nyô(l) /; / *ùn*-mbokh/ "serpent; snake"
Sign X1 of the Gardiner's list ***t*** /i-di /; / *ùn*-mbokh/ "roll bread"
Sign V19 of the Gardiner's list ***Tma***/ *ùn*-mbokh/ "hindering"
Sign N23 of the Gardiner's list ***tA***/ *ùn*-toŋ/; /*ùn*-mbokh/ "Irrigation canal"
Sign Z1 of the Gardiner's list ***Wa*** / *ùn*-mbokh/ "vertical bar"

Step 2: signs Analysis

This script dictates a good analysis as we have done so far.
Let us start by saying that the Sign G17 of the transliterated Gardiner's list ***M*** corresponds to the owl which bears the names in Ewondo: "à-kouŋ"; "Koghle-kue", "*ùn*-ndukh" etc.

Note : The French people will sometimes call owl : "Hibou grand-**duc"** like the Ewondo "*ùn*-ndukh"
Let's still say that the association of the signs I10 and X1 allows the vocalization / nywa / because we have: / nyo / + feminine marker / wa /. Note as we have seen that the homophone / *nya / refers to the meaning "mother". It can also refer to the possessive variant: your mother /nywa/. Contraction of /*ùn*-nya wué/ "your mother"

In order to sort through the many synonyms of the sign G17 for the owl above mentioned, it would be necessary to choose the word that meets the conditions imposed by the association of I10 and X1: / *nya / "mother". Reminder: "*à*-kouŋ"; "Koghle-kue", "*ùn*-ndukh"

One of the characteristics of the mother is to be in labor during childbirth.

Ewondo: /à-koghlo/ "to be in labour"
Ewondo: /*ùn*-ŋkoghlo/ "the one who is in travail"

The sign V19 specifies that this is a condition of obstruction. So we arrive according to the Ekang mythology to the python which is likened to the rainbow that lives in the water. The sign I10 precise it perfectly.

Among the brothers of Kasaï this snake is very well known and is called: *Mwanza Nkongolo wa ku mayi*.
Mw-anza: "colorful, brilliant"; *Nkongo-lo*: "ring of heaven"; *Wa ku*: "of"; *Mayi*: "Water".

Surprisingly, Professor Blas Castellón Huerta, an archaeologist at ENAH, an anthropologist at Arizona State University. Doctor in Anthropology of UNAM. Researcher at the Department of Archaeological Studies, INAH. Cutha project director, Zapotitlán Salinas, Puebla, INAH, and professor at the Faculty of Philosophy and Letters of UNAM, and at Iberoamericana University in an article about a mythological snake write this:

> *"Cuentan los mixtecos actuales que su plumaje brillaba cuando lo tocaban los rayos del Sol, pues Quetzalcóatl tambien es el dios que trajo a la Tierra la luz y los distintos colores que se manifiestan en el maíz, en las aves de pluma rica, en las footras preciosas, en los árboles y en las distintas direcciones del mundo de los vivos. Como dueña de los reflejos iridiscentes, la serpiente está directamente relacionada con el rayo y el fuego, como fuerza creadora de la vida."*[48]
>
> *Blas Castellón Huerta*

Translation:

> *"[...] The current Mixtecs say that his plumage shone when he was touched by the sun's rays. Because Quetzalcoatl is also the god who brought to Earth the light and the rich colors that appear in the corn, the feathers of the birds, in the precious stones, the trees and in the different directions of the world of the living. As a mistress of iridescent reflections, the feathered serpent is directly related to lightning and fire, as the creative force of life.(...)"*
>
> *Blas Castellón Huerta*

The association of signs N23 and Z1 of the Gardiner's list specifies that we look for a place, the place symbolizes the UN (-okh or ko in our languages). The scrible tells us a channel, a door, a hole, a percolation.
In fact, we conclude that / *nya / the mother is the place of the ONE (origin) and the place is the top maternal figure.
Knowing the vocals associated with the mother in our languages, it is deduced from the analysis of phonetic complements that the related place is given by the synonyms of sign G17, so we have the words to designate this place in our languages.

Middle-egyptian: ***mD.t*** "depth"
Ewondo: /*ùn*-nyà/ "depth" but also "mum"
Ewondo: /*ùn*-ŋkoghle/ "depth" but also the one who gives birth "owl"
Ewondo:/ndome/ "depth" or her "brother"

The Ekang peoples of the Kongo basin claim that they came from ancient Egypt and that they crossed a river on the back of a giant snake named / ***ngan medza*** /.

Middle-egyptian: ***aD*** "to be safe; cure"
Ewondo: /à-dâ/ "Pass on or beyond safely"

As we see on the script above the safe crossing idea is associated by the scribe to the a snake . This script above confirms the Ekang people's mythology with a snake associated with a helping hand .

Note: I suggest for a good understanding of the script above to explain why the scribe draws here a snake to which he associates the idea of crossing.

Middle-egyptian: ***aDnt*** "armband" ☞**Ewondo:** /i-dàŋ/ "crossing"
Middle-egyptian: ***aD*** "to be safe; cure" ☞**Ewondo:** /à-dâ/ "Pass on or beyond"

Middle-egyptian: ***DAi*** "to cross" ☞**Ewondo:** /à-bekh/; /à-dâ/ "to cross"

48.Blas Castellón Huerta, Cúmulo de símbolos, la serpiente emplumada, Arqueología Mexicana N°53, p. 28§35

The root /*bəkh/ can be used to express many other ideas as to carry and so on:

Ewondo: /i-bəkh/ “chance, good luck”
Ewondo: /mvom/ “chance, good luck” ☞**Ewondo:** /mvôm/ “python; boa”

We better understand here that the giant snake of the Ekang mythology is a python just because naming this snake is also to cross, to carry.

What is curious is that to go from one bank to another one sometimes uses a bridge. The giant snake could then be assimilated to a bridge. Indeed the item for bridge brings the idea of to be trapped, to be captured, to be jailed by homophony.

Ewondo: / i-kàngà / “bridge”
Ewondo: / á-kànga / "to be caught"

In short, crossing safely is not a must. We can stay or get trapped during the crossing. We notice in the African thought regarding the dead persons are not dead, they are in a parallel dimension. So we could just think that they crossed (the bridge) without hope of return. The bridge appears as the entrance, the limit but also a potential prison. According to Muslim doctrine, there is above the hell a bridge called Sirat, which is thinner than a hair, more sharp than a razor, and whose length equals the diameter of the Earth. After the resurrection, the elect will pass it with the speed of lightning, supported by the hand of angels; but the reprobate slipsand will rush into the abyss of eternal fire. According to very doctors, this bridge has seven arches, on each of which there is a prison where God encloses man to question him about his actions main.

The Scandinavians say that the gods have built a bridge which serves as communication between heaven and earth; this bridge is none other than the rainbow.

The Rocky Mountains Americans believe in the existence of a bridge of souls very similar to the Muslims one. It is through the dime and held by the flow of water that reaches its apron, in a continual sway. The deceased must cross it to get to paradise; the good ones pass without difficulty in spite of their agitation; but the wicked are incapable of standing there; they stumble and fall, then the torrent carries them into a maze of swamps.

The expression devoted to the involvement of a person in an evil socrotle often used in the Ekand region called Kong is: /daŋ ndzong/: "invitation to cross the road"

After this aside, let’s go back to the Ekang story.
What can be an extremely a difficult issue in this story anyway is the term / ***ngan*** / indicating the crocodile and not a snake as one might have hoped.
So, how can a crocodile be confused with a snake in the collective imagination? The question needs to be asked and clarified.

If the predicted script was transliterated  ***mD.t*** and reports a snake as anyone can note, this script  is transliterated ***msH***. Except some few details we have the same vocalisation. But the determinative in this case is a **saurian**, probably a **crocodile**.
It can be noted that the two scripts have in common the sign 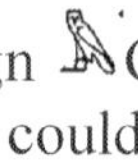G17, of the Gardiner's list. But this sign in the previous script was related to the notion of depth. It could deal with the abyssal depth.

The expression / ngan mə-dza / is better understood since the entity associated with the depths is the mythological serpent of the Ekang peoples called by some other African peoples * *kongolo, mvom, mpwam, mpomo* etc. The expression / ngan mə-dza / could be replaced by / ngan mə sa / "the feathered serpent".

Professor Binam Bikoi devotes an entire opus to this story in his book *Mpomo the Prince of the Great River*. The story is about Mpomo, an extraordinary child, the last born of Zyèm "God" (ananym of msH), a clan founder, himself the youngest of the first human family, who found himself disunited and dispersed to the following fault committed by the initial mother of humanity. The latter having introduced sorcery and death among men, Mpomo comes into the world, invested with a quadruple mission: restore order and justice, establish peace, behead witchcraft, ensure the unity of the great family of "Descendants of Zyem, Be'm and Yamekana" which, according to Koozime, is none other than the undivided humanity. From one chapter to another, the reader discovers the epic
who populate the country with the "Great River". These are the first female figures, the most notorious being Ampouam, a discreet priestess of power, Njol, twin sister of The epic of Mpomo, and Ande, the foster mother of the young **Mpomo**. Male characters with reckless, devilish or good-natured heroism are added. The whole revolves around that which is held for the prince of the giants, the king among kings, and whose name is pronounced only with a host of currencies: "**Mpomo** says Yan-yan, the cheese maker branches armed with thorns; Black ant trees with the most reckless deceit; Small winding stream that sneaks between the tall trees ... "
Mpomo is a vast narrative complex that the researcher has collected in fragments among the Koozime, from the mouth of Daniel Minkang, the bastard. The Koozime are a small group of a few ten thousands speakers, gorged in the equatorial forest that covers the province of eastern Cameroon. Their language belongs to the so-called Bantu family.

We find curiously in Ouiddah in Benin in the country Voodoo a cult addressed to pythons.

> *"[...]*
> *The legend tells that in 1717, following a fratricidal war opposing the kingdom of Danxome and that of Houeda, the defeated king of Ouidah took refuge in the forest to escape the attackers who pursued him. It was the pythons who attacked the enemies of the kingdom, and he was thus saved. To thank the protective serpents, the king made to build in Ouidah three huts in the forest and made of them a totem. Thus, the serpent is now a true deity of their voodoo worship.*
> *The snakes of this temple are not fed by men. We do not give them food but once a week, in the middle of the night, we open the doors of the temple and the hungry go out to go to the neighboring houses. Legend tells that they turn into geniuses who take refuge in the forest to find food and do not return until early morning.*
> *If the python does not return after three days, it is beaten and is sometimes brought back by the population because it is considered a god. It is not uncommon to see drivers stop their vehicles in the middle of the roadway to let a python pass or help them cross the road. These pythons are entitled to funerals close to those of men. The python also recalls by its moults the rebirth and the resurrection, and symbolizes the immortality of the ancestors."*[49]
>
> *(EKOUE, 2016)*

On a purely paradigmatic level, we can notice below the richness that surrounds the semantics of the term "python, boa" in African languages.

Kɔ́ɔnzime: mpwam "python, boa"
Fang: mvom "python, boa"
Ewondo: /mvwàm/ "python, boa"
Duala: bwaba"python, boa"
Ɓàsàa: mbɔ̀m "python, boa"
Yemba: swɛ́nnū "python, boa"
Fulfuldé: asli mboodi "python, boa"
Hausa: mesa "python, boa"
Oluganda: ttimba "python, boa"
Kinyarwanda:ziramire "python, boa"

49.Sophie Ekou, Wisdom of Africa,Hachette Paris, 2016, P.98.

All these words seem different from those on the left in reality ,it is not so as we can see below:

Yemba: swɛ́nnū "python, boa"
Ewondo: /nyo ù-soué/ "The river's snake" ☞**M.-E.**: ***mD.t***

Kinyarwanda: Uruziramire "python, boa"
Ewondo: /Hor Zén be râ/ "râ's path" ☞**M.-E.**: ***mD.t***

Oluganda: éttimba "python, boa"
Ewondo: /Zamba/ "*ùn*-ŋkoré" ☞**M.-E.** Sn

Hausa: mesa "python, boa"
Ewondo: /Mə-sè/ "feathers" ☞**M.-E**: aApp

This last Hausa acceptation: *mesa* is very interesting because it gives new informations precisely: the feathers. At the end, it seems that we are confusing. The myth names a crocodile that would have allowed the crossing. The collective consciousness keeps the memory of a snake and at the end, the Hausa speak about feathers.

Middle-egyptian: ***nSmy.t*** "scales"
Middle-egyptian: ***nSm.t***"scales"
Ewondo: /bi-bàs/ "scales"

Middle-egyptian: ***Sw***.t "feather"
Ewondo: /mə-sè/ "feathers"

There is a semantaxe associng in african languages, wings and feathers by metonymy. On this account, what the word Hausa takes for a feathers can be also in Ewondo a wing.

Ewondo: /à-fab/ "wing" **Ewondo**: /à-fôbô/; /à-fakh/ "hat; cap"
Middle-egyptian: aApp **Ewondo**: /Zaŋ ùn-mbèh/ "the wide door"

Amazingly, many mythologies around the world report a feathered or flying serpent.
The feathered serpent (translation of Quetzalcoatl in Nahuatl) is a deity whose worship was widespread in Mesoamerica.
His cult has its origins in the pre-classical era, around 1200 BC. NS, in that of the Aquatic Snake among the Olmecs, which has gradually transformed into different entities such as Quetzalcoatl and Ehecatl among the Aztecs, Kukulkan among the Yucatan Maya, Tohil among the Maya quiche or Coo Dzavui at the Mixtecs.

> *"As is often the case in Mesoamerican thought, the symbolism of the feathered serpent is imbued with dualism: it is at the same time attached to the earth, by the serpent, and to the sky, by the feathers of the bird[50]".*
>
> *(CASTELLON, 2002)*

We might ask ourselves how this mythological history of the Ekang peoples has been found back to the Americas?
It is Professor Van Sertima who offers an excellent answer in chapter 5 page 71 regarding this Quetzalcoalt in his book *The African Presence in Ancient America*[51].

50.Blas CASTELLON HUERTA, Cúmulo de símbolos, la serpiente emplumada, Arqueología Mexicana N°53, 20020, p. 28-35.
51. Ivan Van SERTIMA, They Came Before Columbus: The African Presence in Ancient America, New-York, 1976, p.71-89)

If his position seems to explain other curious similarities with the important place of the jaguar compared to the African leopard, the pyramids, Negroid heads Olmec, this position is still controversial by Africanist censors. To quote just one excerpt on a point of controversy, just go through the passage below:

> *"[...] Selon l'hypothèse initiale de Van Sertima[7], les souverains nubiens de l'gypte ancienne (XXVe dynastie)[8] organisèrent, à la fin du VIIIe ou au dbut du VIIe siècle avant N.-E., une expdition en vue d'obtenir diffrents produits, dont du fer, en quelques lieux de la côte atlantique de l'Afrique du Nord, de l'Europe et des îles Britanniques. Le vaisseau aurait eu un quipage composite, constitu de troupes nubiennes aux commandes, d'un pilote phnicien, d'un ou deux Hittites, de marins gyptiens et d'un certain nombre de femmes gyptiennes ngroïdes[9]. Cette expdition partit du delta du Nil, traversa la Mditerrane et atteignit la côte atlantique d'Afrique du Nord où elle fut prise dans quelque courant ou tempête qui l'entraina à travers l'Atlantique pour lui faire finalement toucher terre sur la côte du golfe du Mexique, où elle entra en contact avec les Olmèques, accueillants maisinfrieurs. À ce stade, les Olmèques auraient, semble-t-il, accept les chefs de l'expdition nubienne et gyptienne comme souverains (les "chefs guerriers noirs", black warrior dynasts), et ceux-ci, en retour, auraient cr, ou du moins inspir ou influenc la cration de la civilisation olmèque, qui influença à son tour Monte Alban, Teotihuacan, les Maya et toutes les autres civilisations mso-amricaines qui suivirent[10].*
> *Selon Van Sertima, les Nubiens servirent de modèles pour les clèbres têtes de pierre monumentales que les Olmèques produisirent dans les annes qui suivirent le contact prsum. Les Nubiens donnèrent aussi l'impulsion à la construction de pyramides et de centres crmoniels, et introduisirent un certain nombre d'innovations et de pratiques (momification, mtallurgie à la cire perdue, usage symbolique de la teinture de pourpre de murex, tissage, etc.) qui influencèrent la religion, la mythologie, les coutumes et même le calendrier mso-amricain. Van Sertima affirme galement que les gyptiens influencèrent les cultures sud-amricaines en introduisant au Prou, parmi d'autres choses, le mtier à tisser vertical[11], la momification et some[12] pratiques funraires[13]. Il n'est jamais expliqu exactement quand et comment ces contacts avec les civilisations de la côte pacifique d'Amrique du Sud se sont produits."[52]*
>
> *(Jean-Pierre CHRETIEN, 2000)*

After these chappels wars, if we returned to the heart of the main subject. By giving the floor to an author collective that brings together in an excellent book interesting elements of mythology about snakes and in particular that which interests us here:

> *"THE STRUGGLE BETWEEN INDRA AND THE SERPENT VRITRA*
> *The theme of the fight between a hero or a deity to a snake or an evil dragon is found in several societies, including Asia and the Indo-European civilization. In India, the struggle between the Indra snake and the Vritra snake is one of the essential myths of Rig-Veda, one of the sacred books of Hinduism. The popular god of the Vedic pantheon, Indra, master of warriors, lightning, atmosphere, seasons and rain, has the role of establishing order in the cosmos and exterminating the titans, evil monsters .*
> *One of these titans is the gigantic Vritra snake, which Indra confronts several times. This snake, whose name means "who encloses, who imprisons", would have modeled a python, snake constrictor.*
> *Vritra manifests its evil power by causing drought: it retains trapped waters.*
> *Indra succeeds thanks to the lightning (vajra) to split the belly of the serpent Vritra, from where spring the liberated waters, which release and let reappear the sun. They pour into the earth in the form of rain, forming seven rivers that flow into the sea.*
> *Depending on the case, this myth has been interpreted as a myth of the organization of the world, of the victory of life over death or as a myth to explain atmospheric phenomena.*
> *Hydra of Lerne, fought by Heracles. possessed nine snakeheads that repelled when cut.*
> *According to Hindu mythology. Vishnu defeated the serpent-demon Kaliya by dancing on his many heads.*
> *In Japan. the eight-headed serpent. who devoured the daughters of the king, was defeated by the god Susa no-o.*

52.Jean-Pierre CHRETIEN, Claude Hlène PERROT, Afrocentrisms. The history of Africans between Egypt and America, Khartala, Paris, P255-256.

> *The Scandinavian god Thor wanted to kill his enemy the serpent of Midgard who lived in the sea he presented him a head of ox like bait and succeeds in the fishing. In many versions (...)"*[53]
>
> *(Roland BAUCHOT, 2005)*

Surprisingly this serpent is unknown to no people. Whether in Japan, in the Scandinavian countries, in India, in Greece, in China with the dragon, all the peoples seem to have dealt with this snake surprisingly.

Indeed according to the principle of polarity everything is double, everything has two poles, everything has two extremes. For example, the thesis and the anti-thesis are identical in nature, but different in degrees. Any paradox can thus be reconciled because what seems opposite is in fact only another part of the same scale. Hot and cold have no absolute definition and exist only in relation to each other. The cold can burn like hot. In the same way hatred can be changed into love or the opposite. The extremes touch and merge.

After this trip, let's go back to the people of the Kongo pool.

> *"[...] All over southern Gabon the rainbow is the image of the sacred python snake. This link between the bridge, the rainbow and the snake is found in the legend of the migration of the Fang who had to cross a big river on a bridge which was nothing but the back of the python snake. . A member of the tribe who dropped his torch, the snake started and the last fell into the water and drowned. This event explains the separation of the Fang people into two (...)"*
>
> *(MARY et ORTIGUES, 1983)*

If we question different populations in Africa this snake is obvious to the extent that everyone knows how to name it or identify it with the rainbow.

Let's take a few examples of the names used to designate the rainbow to better understand this reality in our land:

Bushongo: *nkongolo* "rainbow"
Bashllele: *nkonulo* "arc-en-crel"
Bawongo: *nkongalo* "rainbow"
Bena lulua: *nkongolo* "rainbow"
Bakete: *nkongolo* "rainbow"
Kikongo: *nkangolo* "rainbow"
Ciluba: *nkongolo* "rainbow"
Balolo: *kongolo* "rainbow"
Bandiari: *nkongolo* "rainbow"
Basala mpasu: *nkongoli* "rainbow"
Sangô: *kongo* "rainbow"

Chart 71: rainbow in some african languages

A little cosmogony: Nyungu! Nyungu! Nyu-Ngu!
In the cosmogony of the Ngala-Duala or Sawa, Nyungu is the third deity of the ennead after Dibobè, the Heaven-Father, and Minyangadu, the goddess Earth, Ñungu is the god Rainbow.
Nkongô-Nyungu is the nickname of Sipopo, the liquid chaos dragon or primordial ocean that the luminous ancestors named *Mangamba.*

This is the brutal energy sometimes brutal personified in Kwa or Mars but also in Neptune called Ekongolo when it is related to water.

Nyungu will leave his submarine sanctuary under the name of Ekongolo and will plunge into the celestial spheres in the shape of the multicolored rainbow, carrier of human grievances and descend to disappear on the other side of the horizon in the primordial ocean. Nyungu is the symbol of clan wealth.

53. Roland BAUCHOT,Cassian BON, Patrick DAVID, Jean-Pierre GASC, The serpents, Artmis, Paris, 2005, P.188.

Jeki, in the myth, fights Nkongô-Nyungu every night and cuts off his multiple lips so that the sun will be reborn and a new day will live in humans. Yes, the brute force must be contained. It is a regulation of cosmic Forces.

Nyungu, at least, is an androgynous demiurge because water and also fire
The link Nun ***nnwn***, Primordial Ocean of our Kamito-Nubian Ancestors, and Nyungu in its Nkongo-Nyungu shape takes on its meaning here.

An the end, what do we have to retain in all of this? Are we talking about a snake, a rainbow?

The scribe's very essence is to condense a plurivocal concept into a script. So, depending on the context, it will be necessary to take into account the best suiting idea.

1- The appearance of the phenomenon describes the colors of the rainbow.
2- The reality described in the appearance of water, mobilizes telluric and aquatic energies.
3- **This reality is visually recognizable by a geometry in the shape of rings or doors**

These are some elements that can be drawn from a serious analysis of this script compared to the information that is accessible through the various mythical stories around the globe.
So what do the scribes say?

Urunziramirê

Urunziramirê

Kinyarwada: *Urunziramirê* "Hr the path leading towards Ra"

Kinyarwanda: nzira "path" **Lingala**: nzila "path"
Ewondo: zén "path" **Duala**: ngea "path"
Chinese : 森蚺 Sēn rán "anaconda"

""The rainbow python nH-Hr is the same as nkôgal or nkôgolo bantu. The correspondence seems assured. nH-Hr python, alias dm3 (Urk.VII.87), aka s3-t3 (literally "son of the earth"), is symbolically associated with ouroboros Sd m r3 (literally "tail in mouth")", Wb.IV. 364)."[54]

(Pfouma 2000)

54. Oscar PFOUMA, The Harmony of the World: Cultural Anthropology of Colors and Sounds in Africa from Ancient Egypt, Editions Mnaibuc, Paris, p.58.

Ouroboros ***Sd m rA*** *"the one who has his cock in his mouth"*

Bamoun Museum in Cameroon

Mwanza Nkongolo wa ku mayi.

Among the brothers of Kasai this serpent is called: Mwanza Nkongolo wa ku mayi.

***Ciluba**: Mw-anza: "colored, brillant";*
***Ciluba**: Nkongo-lo: "ring of the sky";*
***Ciluba**: Wa ku: "of";*
***Ciluba**: Mayi: "Water".*

Ewondo*: /ùn-mfiè/ "colored, brillant"*
Ewondo*:/ù-kéné/ "ring"*
Ewondo*:/ya / "of"*
Ewondo*:/me-ŋyu/ "water, drink"*

Surprisingly the scribe suggests this script to designate the ring.

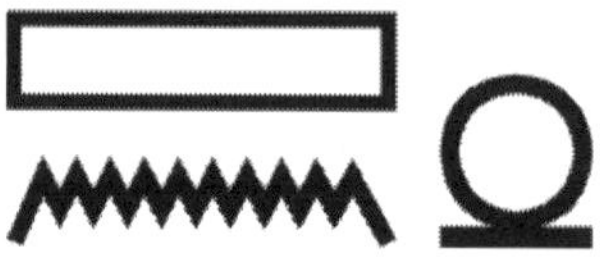

/U-kéné/ "ring"

We find for the feathered snake, the rectangular shape and the annular shape.

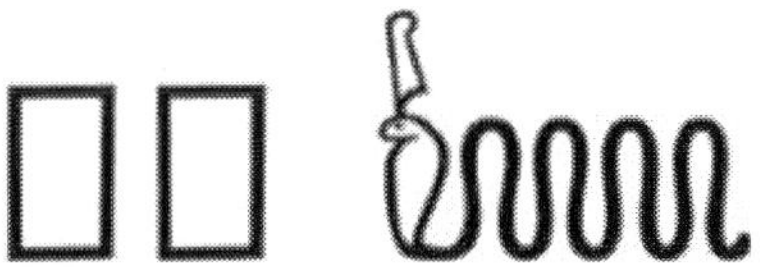

Un-mbèh nənə: "the wide door"
U-nənə
A-fobo

Without knowing how to read the sacred scripts, note the two doors as well as the snake's rings symbolize the wave nature of energy.
This script speaks of gates / mi-mbèh / or then of the big door / *ùn*-mbèh nənə /. The couple of doors on the script has to be linked with a superlative.

There is another meaning of homophony between / *ùn*-mbèh / "door" and / à-béh / "evil". This last approach is the one that the myths often raise up or put forward by the surprising Egyptologists. Transliterated ***aApp***, we are talking about absolute evil. Does not the scribe allude to the danger of operating these doors? When the Egyptians read about it curiously, they talk about the big house, but here why not mention the big door?

aApp transliterated spelling can be read / à-fôbo / to designate a long snake. But by homophony it finds the word / à-fob / serving to designate the space void above.

Etymology

Throughout the world as we have pointed out, the collective memory has kept the memory of an imposing serpent. This snake is assimilated according to the regions of the world to a constrictor. It is very interesting to address the idea that the people of Latin America are making of it.

Middle Egyptian:  ***aApp*** "apophis"

The O29 symbol  of the Gardiner's list represents a wooden pillar.
A pillar in house building is used as a roof support.

Ewondo: /à nə à-kɔh ndà/ ☞ / à n'à-kɔh ndà/ "like a house pillar"
German : *anakonda* 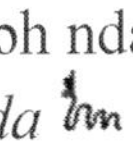
French: *anaconda* 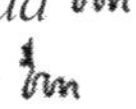
English: ***anaconda*** 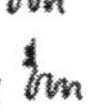

According to the standards, there are three possible origins for the word anaconda: it could be an alteration of the **Sinhalese** word *henakanday* (Thunder snake), or the **Tamil** word *anaikondran* (elephant killer). It is surprising that the word has such a distant origin of the animal's habitat; it may be that the name is related to the vague similarity of the anaconda with Asian pythons. However, it seems that the term "anaconda" comes from the Amazonian Indians and means "Warrior Serpent of the Wave".

South Amerindian communities also called anaconda *sucuri, yacumama*. The **Ticuna** Indians, **Cocamas** and **Yaguas** (Colombia, Peru) called it *Yacumama*. In **Quechua** *yacu* means "water" in Ewondo the item yacu can be approached with the word *yokh "drink" and mama In Quechua as in Ewondo, the "mother", yacumama being therefore the mother of the waters often called in central African pidgin *Mamy wata*, besides the slave descendants of Guyana, the **Bushinengue**, call it the *Wata Mama*, while in Creole Guyanese its name is *Maman Dilo*. According to specialists the Spanish settlers of South America called him *matatoro*, the bull killer. But this interpretation is a bit wrong we have to consider

mata not as killer but as mother. Because in Africa a mother is a bull. By the way, *matatoro* has to be translated the mother of the waters because :
Middle-Egyptian: ***mwt*** « mother »
Middle-Egyptian: ***itrw*** « river »

This matatoro can be represented as bull mother (as in zulu myths with Chaka, malian myths with Soundjata Keita's mother or on the Narmer's palet).

How comes that for african peoples a mother is a buffalo or a bull? The answer of this question is simple mays you repport to illustration N°47:

Ewondo: **nya* "mother"
Ewondo: râ "sun"

So for **Ewondo:** *nya râ* "ra's mother" ☞**Middle Egyptian**: ***nwt***
But this word game in **Ewondo**: /nyar/ or /nyara/ "buffalo; bull" allow then a better understanding of the purpose.

Why do we have a Cobra on the head of the fara? *(cf illustration N°34)*

Simply because the phenomena described here can also be realized with modified states of consciousness when the Kundalini goes up from the sacrum to the pineal gland or the crown chakra and on that account, allows the mental perception of a crown with 7 petals like the colors of the rainbow, the 7 rivers of the Hindu myth!!! The rings this time are those of our spinal bone and the canal, the medullary canal. The feathers being the electromagnetic radiation that emerges.

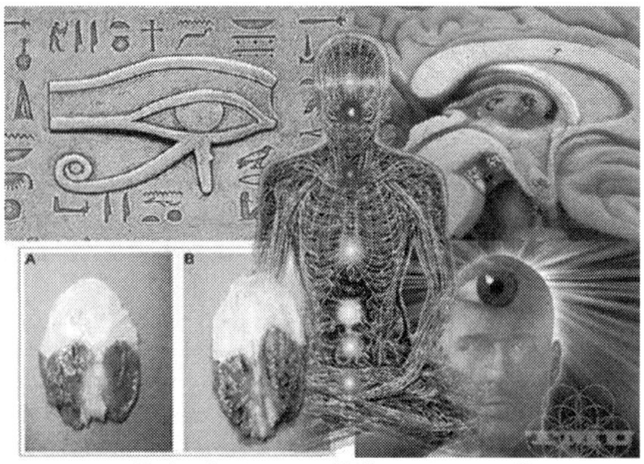

wDA.t Activation "The spirit eye"

According to Ampère's theorem, any current flowing through a circuit creates a magnetic field through the section it surrounds, this is the phenomenon of electromagnetic induction. The inductance of this circuit is the quotient of the flux of this magnetic field by the intensity of the current flowing through the circuit.

By synecdoche, inductance is termed any electronic component intended by its construction to have a certain value of inductance (physical quantity) (just as resistance is called the components used for their electrical resistance). These dipoles are usually coils, often called self.

What we are told in the script with of a big boa snake is strangely similar to an energy loop with an electromagnetic induction phenomenon that uses the zero potential of the earth and telluric energy powered by an electric current supplied by the negative charges of an aquifer source.

An inductor as a component.

<u>Kezako ngan medza</u>?

Ewondo: /mə-dza/ "crossing"
Ewondo: /mə-dzê/ "disparition"
Ewondo: /ngaŋ/ "witchery"
Ewondo: /i-daŋ/ "crossing" ☞**M-E:** ***aDn.t*** "armband"

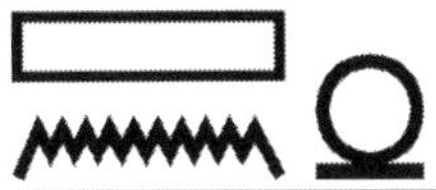	**The feathered snake equations**	**Comments**
Loop shape	$\mu_0 r\left(\ln\left(\frac{8r}{a}\right)-2+\frac{Y}{2}+O\left(\frac{a^2}{r^2}\right)\right)$	r: loop's radius: a: wire's radius
Rectangular shape	$\frac{\mu_0}{\pi}\left(b\ln\left(\frac{2b}{a}\right)+d\ln\left(\frac{2d}{a}\right)-(b+d)\left(2-\frac{Y}{2}\right)+2\sqrt{b^2+d^2}\right)$ $-\frac{\mu_0}{\pi}\left(b\cdot\operatorname{arsinh}\left(\frac{b}{d}\right)+d\cdot\operatorname{arsinh}\left(\frac{d}{b}\right)+O(a)\right)$	b, d: perimeter d >> a, b >> a a: wire's radius

Chart 72:feathered snake

Without going into more complicated details, we understand through the narratives that appear in the collective unconscious that Africa has experienced by the past mechanisms of mass transfer through an electromagnetic module.

The scripts are not simply small drawings but they are digests of knowledge for those who can read them in their language or draw a sound understanding.

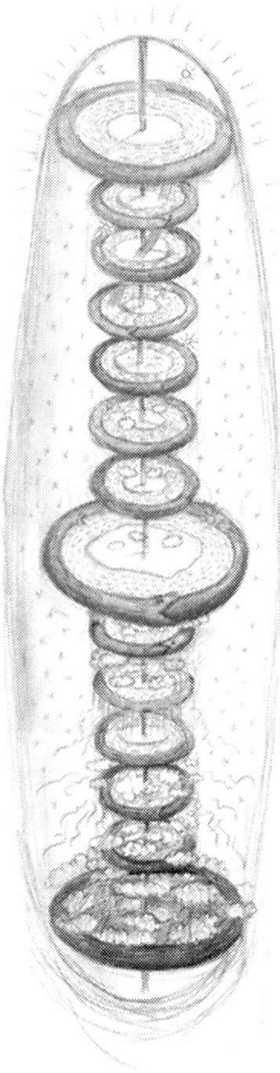

The universe seen by the Dogons: theory of parallel worlds?

3.4.3.17. **EXAMPLE N°17: *sA rA*** 𓅭𓇳𓏤

Middle-egyptian: 𓅭𓇳 ***sA rA*** *(Faulkner p. 207, Wb. III p. 408, 1-17 sq.)*

This script was been chosen for analysis because, it often appears in the header of fara "pharaoh" cartouches. Besides that it offers a range of interesting information.

Classical Egyptology gives to this script the transliteration ***sA-rê*** and a meaning usually associated with "sons of God".

Let's try to see if the African point of view related to this script offers the same level of relevance.

What appears curious first of all is the meaning given to this script. Why do precisely the titulature of the Fara should have to indicate that, the fara is **son of "God"**?

It is obvious that the notions taken up by the religions of salvation are African notions because the sociology of nomadic peoples can not worship or practice religions. When one is in constant migration, one does not build temples. So the main notions are African. Only An African can therefore give an accurate point of view on the question of God in all peace of mind.

Further on, the **son of God**'s acceptance is problematic insofar as this question is important because if the scribe needed to indicate the term *son* in this script, was it not more accurate for him to draw a duckling sign G47 𓅮***TA***, rather than a duck 𓅭 Sign G39?

If ***ra*** is the sun and there is only one sun in the sky, so what is **Middle-Egyptian**: ***ra.t?*** A female sun? A goddess sun?

Finally, taking into account the historical continuum, no title carried by a sovereign in Africa designates him as **son of God**.

So where does this fad comes from, is it a new invention? It therefore seems important to take the data well to get out of the intellectual tutelage.

Step 1: list all the signs

> Sign G39 of the Gardiner's list ***sA*** / i-loli/; / soghe /;/ sua/; /bat/ etc.
> Sign N5 of the Gardiner's list ***ra*** / ra /; /dzô/; /*ùn*-mfiè/; /vièn/ etc.
> |Sign Z1 of the Gardiner's list ***wa*** /wa /; /*ùn*-mbokh/; /fokh/ etc.

Step 2: Signs Analysis

Let's start by sign G39 .

> **Ewondo:** / i-loli/; / soghe /;/ sua/; /bat/ "duck"

Do we have any connection to the word ***son*** with the sign N5 of the script?

It turns out that we have the character ***x*** "placenta" in Ewondo / i-sokh / which designates the maternal egg of mammals and also called placenta.

So by homophony and metonymy, we can accept the idea that the sign G39 designates the child through the placenta. Because the placenta is the mother and the child.

How to make sure that ***x*** equals N5? This is relatively simple, both signs are often associated with numeral 1.

We have that and are associated to Number 1. Then both signs N5 and Aa1can be read:

> **Ewondo:** /ùm-bokh/ "one; 1"
> **Ewondo:** /*ùn*-mbokh/ "hollow, cave, hole, tube, canal" etc.

Now, in the cosmogony of ***km.t***, ***Hr*** is the one designated by / *ùm*-bokh / "paralytic"

Step 3: Final reading

> **Middle-egyptian**: sA ra "God's Son"
> **Ewondo:**/i-sokh ra/ "Hr's son"? why not… but not convinced.

On the technical side, ***sA*** transliteration offers a wide range of reading from the African point of view. ***Sa*** reading is possible when **A** matches the appearance of a vowel because associated with the weak consonant *H*.

On this account we can have the son or child acception. I still remember that the letter ***A*** in Negro-Egyptian corresponds to K^wh. English has kept this trace of the past through the homophony between:

Middle-egyptian: *sw* "sun"
English: sun "sol"
English: son "son"
Ewondo: /*ùn*-sɔŋ'o/ "the beak, the duck"

This expression "son of" has persisted within the Nordic countries with the suffixation morphemes – son/sen "child of; son"

Example: Edmunson; Edmunsen; Yurgenson etc.

The expression ***sA*** also exists in Ewondo. It is stills used with the meaning of face. When the son looks like the father, he is said to be his face. It is therefore this expression which has persisted and which by metonymy translates the idea of son.

Example: Asu Ekotto "Ekotto 's son"

Similarly we can have in African languages all the possible forms of *A* associated with [S].

Duala: Sadi "son"
Duala: Sawa "the unique son"
Mbende: Sango"ngo sons"
Sarakol: sara "son"

As we can notice in English also there is an homophony between *son* and *sun*. By the way just keeping the meaning son related to this script is to ignore how to read egyptian scripts. Depending of the context it will be more accurate to consider one meaning better than another.

Other Hypothesis: **Middle-egyptian**: sA ra "god's son"

We have several approaches for another reading, let's choose the first one that comes to us. The one represented by the duck as a rule is / ùn-nsorèn / "skinned" ***wsir***[55]:

- **Middle-egyptian**: ***wsir*** "osiris"
- **Middle-egyptian**: ***wsir*** "osiris"
- **Middle-egyptian**: ***wsir*** "osiris"
- **Middle-egyptian**: ***wsir*** "osiris"
- **Middle-egyptian**: ***wsir*** "osiris"

We say that the African language is a tonal language obtained by the association of simple elementary concepts that can be assimilated to tonemes with non-zero signifiers.

Thus the transliteration ***wsir*** can be decomposed into ***Ws*** and ***Ir***.
However, the African linguistic variants operating by consonant metathesis stemming from the Kweke and Kekwe branches, we can therefore have not only ws-ir but also sw-ri or sAra, sAri, sArê.

Fulfuldé: saar "concession, dwelling, house, country, etc."
Kinyarwanda: nzira "path"
Lingala: nzila "path"
Ewondo: zén "path"
Duala: ngea "path"

55. Cf illustration N°32

For the morpheme of Negro-egyptian * k^{h}w- we have in Pharaonic the transliteration A = kh and the variant w. So we have: * khw -> * A- and * k^{h}w-> * w-. In short, the words obtained with the W have their counterpart with those created with the ***A.***

Thus, the Pharaonic transliteration **sA** is etymologically linked to the following variants from the Negro-egyptian: *k^{w}ək^{w}-(*hja)*:

Middle-egyptian: ***wS*** *"to fall, to fall; devastate, ravage"*
Middle-egyptian: ***sw*** "sun, day"
Middle-egyptian: ***Sw*** "dry, dryed; etc."
Middle-egyptian: ***Hs*** *"shits; excrement"*
Middle-egyptian: ***sH*** *"attempted; big room"*
Middle-egyptian: ***xs*** ***"****gutter, channel****"***
Middle-egyptian: ***sx"****to cut"*
Middle-egyptian: ***Sa****"cut; tear down; prune"*

Let's try to find out what the scribe here associates with the duck in the meaning of wsir "the eye". It can be postulated that it associates it especially with its beak in the African languages because the other name by which we designate ***wsir*** is the worm / *ùn*-nsoŋ /. The word worm offers a homophony with the beak (understood duck). So we have

Middle-egyptian*:* ***wSnw*** "birds, fowl, poultry"
Middle-egyptian ***Apdw*** **"**duck; bird, poultry**"**
Ewondo: /ùn-ŋsoŋ'o/ "beak, duck; fowl"

We therefore think that the best transliteration is ***Sn*** and not necessarily **sa**. Since the determinative of the script is the fara cartouche which is ring-shaped.

Middle-egyptian: ***Sn*** ***"****loop, ring****"***
Ewondo: /ù-kéné/ "ring, loop"

We notice indeed that the spellings in Middle-Egyptian, associate the duck with a hole, a spiral, a door /*ùn*-mbè/ ou /ù-kéné/ "ring"; /*ùn*-mbokh/ "cave; hole"

Middle-egyptian*:* ***pAa.t*** ***"****quail****"***
Ewondo: /ù-vem/ "web-footed: *podica senegalensis****"***

We also notice that the script below invokes the idea of shifting and curiously indicates a duck.

Middle-egyptian*:* *sA/zA* "resort to, go to"
Ewondo: /á-so/; /á-zù/ "resort to; go to; to come"

We can also read this script simply: how to go to the cave, to the ring, by the canal, to the country.

Middle-egyptian sA ra "God's son"
Ewondo: /zén *ùn*-mbokh/; /i-yolé *ùm*-mbokh / "channel access code"

Let's remind that in **Ewondo:** / i-loli/; /i-lole/; /lolo / "duck"

In summary: designates the access code to the ring door. From my point of view, the duck especially indicates here the words: ***rn*** "name" , ***Sn*** "ring", ***zn*** "path", ***Wsir*** "eye"

Example of cartouche reading:

Middle-egyptian: ***sA ra < ax-n-Itn >*** "Akhenaton"
Ewondo: /i-yolé *ùn*-mbokh/ < /*nti *ù*-kéné / > ☞priesthood name "the lord of the ring"

One can realize that the proposed official reading leads to a big issue to the extent that no one bears a name as the son of God in Africa despite the presence of high ranking rulers and initiates.

We affirm that this is a priesthood name also attested in another element of the pharaoh's titulature offering by his cartouche. The pharaoh's name as ***nsw.t bity***.

Middle-egyptian n(y)-sw.t-bjty<Nfr xprw Ra wa n Ra>
Ewondo: /i-yol ùm-bokh bwiti/ < /Râ i-tam ny' á-nyəp mə-kuli/ >
French: <nom de prêtre du bwiti> "l'unique Râ dont les manifestations sont parfaites"
English: Bwiti's priesthood name < only Râ appearances are beautiful>

Sign M23 of the Gardiner's list is read / i-sɔŋ / "*Pennisetum purpureum*"
Homophone of beak in African languages because the other name to designate wsir is the worm / *ùn*-nsoŋ /. The word worm offers an homophony with the beak (understood duck). So we have:

Middle-egyptian: ***wSn***
Ewondo: /*ùn*-ŋsoŋ'o/ "the beak; the duck"
ɓàsàa: ǹsáŋ; sáŋ; sɔ'ɔŋ "the father"

As we have already shown, the duck is called / ilole /, / lolo /, / iloli / this vocalization offers phonetic proximity with / i-yol / "noun". must therefore be understood in reading the cartouche as insider name, as high priest of the Bwiti.

Middle-egyptian sA ra "God's son"
Soninké: Mansa kura "new king"

Mansa is a title worn by the rulers of the Malian Empire and more particularly those of the Keita dynasty that dominated West Africa between the 13th and 14th centuries. The power of the Mansa included the right to dispense justice and the monopoly of commerce, especially that of gold. The equation is simple here because Ra = imn. If we replace Râ by imn then we get Mansa. Of course we can read from right to left. We often use this meaning for son in Ewondo. **Ewondo**: /mon/ or /mwan/ "son"; **Ɓàsàa**: *ɔ̀né* or *hɔ̀né* "son, child"

Generalization: Etymology of the word duck in African languages

The only way to find the true etymology of the word duck and therefore its first meaning is to propose a reconstruction from the Negro-egyptian. This reconstruction allows us to understand the meaning of the word from Negro-egyptian and explains the different variants of the word in Middle Egyptian.

Negro-egyptian: **k^{w}ək^{w}-(hja)>S^{w}əta>s^{w}irĩ "duck"*
Proto-kang: *cʊecʊe ou *cʊʊpe "duck"

We can thus from this variant explain the morphological links between the different duck words used in the variants of the current African single language:

Proto-version: *$*Bək^{w}$->$*Bət^{w}$-*

Oluganda: *mbaata* "duck"
Tshiluba: *mPaatu* "duck"
Ashanti: *bako*"duck"
Baoulé: *lábûlábû* "duck"
Teke: bala"duck"
Russe: лебедь *leBed*(swan)
Lingàla: *libáta* "duck"
Hindi: *bat* **"**goose**"**
Persan: *bat* "duck"
Mansi: *batta*"duck"

Boukharian: *beth*"duck"
Armenian: *bath* "duck"
Albaniane: *peth*"oie"
Malabare: *vâttu*"duck"
Syriac: *batò*"duck"
Syrrac: *pattô*"duck"
Georgian: *bati* "goose"
Chechen: *bat*"goose"
Chechen:*bad* "goose"
Touchi: *bata*"goose"

Note: The english name of the palmiped *goose* is a consonantal metathesis of /soghè/ in Ewondo language.

English: goose "palmiped"
Ewondo: soghè "duck"

The english name of the palmiped *swan* can be approched to the Ewondo word:

English: swan "Anatidae"
Ewondo: /*ùn*-swaŋ/ "the beak; duck"

Proto-version: **koko->*cœcœ-*

Zulu: idada"duck"
Duala: elela "duck"
Yasa: elolo "duck"
Basa'a: lolo "duck"
Kasem:didwa "duck"
Ewondo:I-lwali "duck"

Koonzimé: lwal "duck"
Eton: ilwali "duck"
Bakweri: elela "duck"
Hausa: agwagwa"duck"
Zand: gbiwi "duck"

Proto-version: *$*k^{w}ək^{w}$->$*ətar^{w}$-*

Kwere:imw~e "duck"
Dagbani:nd ɑm "duck"
Soninké: wandalle "duck"
English: teal "ringed teal"
Ewondo: /à-lwàt/ "duck"

Proto-version: *$*k^{w}ək^{w}$ ->$*s^{w}ək^{w}$ -*

Fang: Soghe:"duck"
Ewondo: sokh "duck"
English: goose "Anatidae"

Ewondo: /i-sûghà/ "support"
Boor: sèink "support"

Note

Middle-egyptian: ***nTr*** *"God, divinity"*
Fang /Ewondo: /*ùn*-ntum/ "cane; chief"
French: cane "duck"
French: canne "cane"

We can also observe these evidences:

Middle-egyptian: ***nTr*** *"God, divinity"*
English: tin "ducking; pewter; box, can, case, canister, bin"
Fang /Ewondo: /nti "lord"

Middle-egyptian: ***nTr*** *"God, divinity"*
English: pewter "tin"
Latin: pater "father"

Middle-egyptian: ***nTr*** *"God, divinity"*
English: can "spool" ☞ **English**: ☉ loop "circular"
French: cannette "can; ducking"

So whatever is th type of the word duck used here, it is enough to put the good reasoning to reach the same conclusions.

Of course it was not useful to know the proto-versions of the word duck to recognize the bird in his language or to understand the meaning he can take according to the context.

The process consisted of demonstrating that, despite the different vocalizations, we are using the tools that have been proven to explain and to be understood in today different linguistic varieties.

What is :

Middle-Egyptian ***:*** ***ra.t***

To answer that question, it is important to have a good knowledge of African languages. In the Bamoun language spoken in the Cameroon west side :

Bamoun : *ret* "duck"

In Ewondo language the word transliterated ***ra.t*** and vocalized ***ret*** in Bamoun language is linked to the words :

Ewondo :/ *à*-lùat/ "duck"

The comparison between the different synchronies shows that the consonant Consonic voiceless fricative **[R]** is transformed into Ewondo in liquid consonant [L].

We can also say that the Ewondo word above, denotes by homophony number of other meanings, one of the most interesting binded to the sun is the moist heat.

Ewondo: / *i*-lular / "sweat; dark, cloudy "
Ewondo: / *i*-du**dukh** / "sweat" ☞ **English** : *duck; to duck; to douse*

Middle-Egyptian: ***sA.t*** "duck" ☞ **English** : *sweat;sudor; goose bumps; willies ...*
Ewondo: / *i*-su**sut**/ " moisture"
Ewondo: / *ù*-swé/ " river"

So, the claims of making a goddess of I do not really know what, are funny and inconsistent. As I have emphasized,

Ewondo :/ sokh/ "duck"
Ewondo :/ *i*-sokh/ "placenta" ↔ **Ewondo** :/ *ùn*-**nyà**/ "placenta; mother"

So, if the duck and mother are sharing the same words then, the duck is the mother according to African thought.
I therefore affirm that one of the meanings for the mother is homophonically linked to the word duck.

3.4.3.18. ***Example N°18: GOD: A TREMENDOUS ILLUSION***

(Faulkner p. 173, Wb. III p. 122, 14+ sq).

This script is enlightening and very interesting on the educational level. For example, let's find out why. As always, let's apply the method we are set. We start by identifying the characters. So we recognize from right to left:

Sign G7 of the Gardiner's list *Hr* "Falcon G5 on a bulwark R12"
We remind just that this falcon: /ùn-ŋkoré/; /kùlù/; /ù-bama/; /kiri/; /*ùn*-nti/; /tara/; /*ùn*-mbokh/ etc. His 99 qualifiers, we can give them without any problem.

Sign G43 of the Gardiner's list *w* "quail chick", is named **Ewondo**: /*i*-woré/

Sign D2 of the Gardiner's list ***Hr*** "face" we say: /à-su/ "face, visage, etc"

Sign D21 of the Gardiner's list ***r*** "mouth" that we call /*ùm*-mbokh à-nyu/ "buccal cavity"
Sign V28 of the Gardiner's list ***H*** "Twisted linen wick" that we call /*ùn*-mbokh/

There is no difficulty for us to read this script ***Hr*** vocalized / à-su / in Ewondo "face" by its director sign D2. It admits all the above meanings and many more but let's stay on the acception / à-su / "face".

For example, we could have vocalized this script ***Hr*** / dys / "eye" by identifying the other signs and by determining what the writing is about. This meaning is allowed even if it is not recorded in accordance with the requirements of the Egyptological standard method. As for ***Hr***, it seems that he has 99 names according to some exegeses. We call him / dys / "eye", in African traditions because ***Hr*** would have lost an eye during the fight with his uncle ***stX.*** Let's leave out these other names and stay on / à-su / "face; visage"

The main issue linked to this script is the slippage and falsification that result from its reading. We will try to explain it so that everyone can understand it once and for all.

<u>Little reminder regarding historical chronology</u>.

Recall for some people who are not used to grind the historical facts that on the chronological level, Egypt was originally a negro civilization at the base (-∞.to -333 BCE.).

This Negro civilization came under Macedonian rule towards -333 BC. with the fall of Egypt under the rule of Alexander the Great of Macedonia (present-day Greece).

The Roman emperor Augustus will put an end to the Macedonian domination of Egypt around -30 B.C.E and keep the stranglehold on this earth until 476 A.C.E.

Chronologically we have: ***km.tyw*** or Egyptians, then Greeks and finally Romans. It is important to understand this to grasp the remaining.

1- Did you say Jesus?

"Ñgwan i-yalgə bwan, i-yômgə bə-nyabôro".

Tara Mebenga m'Abessolo

We have until it is proved to the contrary: ***Ancien-egyptian***:  ***Hr***

According to the languages of the occupiers of the earth named Egypt, this script will change vocalisation in this way:

- **Ewondo:** à-**SU** "visage" (take here the African language you want).
- **Ancient Greek**: Ho**RU**s (adding of the suffixal S) ***wsjr modified in*** jrws->Horus
- **Latin**: ie**SU**s (change of the prefix and maintenance of the suffix S)
- **French:** Jé**SU**s (twisting of the Latin **i** into french **j**)
- **English:** Je**SU**s (twisting of the Latin i into french j)

Chinese : 受理 shòu lǐ "to face sb./sth."

There is no comment on how to vocalize the sign D2 " a face". The word is an African word in this particular case a word Ewondo and anyone can speak the language may validate this lecture without any particular intellectual effort.

So why do we observe differences between African and Indo-European languages? It can be explained on several levels but the most relevant is to consider the proto-version of Negro-Egyptian.

Negro-egyptian: *ŋ^{jw}ə-**k^{wh}iki-**t^{w}i>*hi-**k^{w}ki**-*> *x^{w}urĩ - "the animated that the foot is a bone"
Middle-egyptian: ***Hr*** *"Horus"*
Ewondo: *kolworo "pycnonotus gabonensis"*

In the Kekwé branch we have the word **Ewondo**: / *ùn*-nsurù / "Black" which will be greyed out in Horus (Kweke branch). Just read the Greek word Horus from right to left.

From the proto-version of the word ***Hr***, we understand well where the j/i comes from sometimes; h the suffixal etc.

The aim is not to treat this subject on the possible evolutions of this term according to the Kekwe, Kikuki or Kweke branches in order to explain all the possible variants of this word. It is just a question here of stating that, the supposedly different variants of the word are not a mystery for us.

The turtle is the best animal to represent the ancestor of Kongo peoples. We call it / Kulu /. The Greeks who have stayed in Africa, we are curious to know how they call the turtle because they could not know this animal in the plains of Mongolia.

Ewondo: à-**SU** "visage" **Ewondo:** yesus "jesus???"
Ancient Greek: χέλυς *chélys* "turtle; tortoise"

The resemblance is quite strange with the **Ewondo** : /kulu/ but alors also with the Latin word *Yesus*.
Let us close here the parenthesis on this question not without pointing out the work of some histrions like Odon vallet who presents himself as a French specialist of religions and co-founder of the Vallet Foundation, under the aegis of the Fondation de France.
A pseudo specialist who has written a number of questionable books on the question of Christianity finds here its limits.
He is known for a number of unsubstantial writings such as: Jesus and Buddha, Albin Michel, 1999; Another history of religions, I to VI, coll. "Discoveries Gallimard", 2000; Corps divine, a book on the artists Pierre and Gilles at Editions du Chêne, 2006. etc.

We say that the question of Jesus is an intellectual swindle and a real shame for those who eat and feed this ideology of another time.

History and Evolution of Christianity are based on a number of documents concerning the genesis, the sources, the roots of the myths that many people believe hard!

1. The Oxyrhync Papyrus (Egypt) discovered by B.P. Grenfell and A.S. Hunt in 1882. It is being published in England. 67 vols. are already available; some forty or fifty others are in preparation.
2. The Bodmer Papyrus (Vatican City)
3. The Ethiopian Codex
4.The Coptic Bible
5. The Library of Nag Hammadi (Egypt)
6. The Dead Sea Manuscripts ...

Obviously all these documents on the purely historical plane are prior to the supposed facts.

2- Have you said Elo?

"Fiàngà man ùn-nwan i-nga woé man ùn-nwan"

Tara Nguini

Let's go back to our script: ***Ancient-egyptian***:  ***Hr***

The sign D2 of the Gardiner's list can lead to the Ewondo acceptation: / *ùn*-nloh / because this sign also designates "the head". From the root of this word will derive by metonymy Hr's qualifications in African languages to designate the leader as the head, the top

Ewondo:  /Elo/; /Eyo/ "the primordial ancestor, as head or summit"

Hebrew: אֱלֹהִים "'ēlo*hîm*", term that usually means "god" (singular or plural) in the Hebrew Bible. The vowels of the name "elohim" (e-o-i) are sometimes attributed to the Tetragrammaton "YHWH" in the Masoretic text, but less often than those of the name "Adonai". In the Jewish tradition, the choice of the name "Elohim" insists on the idea of God's power.
Ewondo : YHWH = /*ùn*-yəhbəh/ "belief; confession, credence; faith"
Ewondo : YHWH = / yohboh/ "top; high, great; lofty"

Arabe: 'اللهAllāh is the Arabic word for "God." Literally, Allah means "The God" in Arabic [56]. By extension, Allah generally refers to the name of God in Islam.

56. Ar-RAZI, Treatise on Divine Names, Translation from Arabic: Maurice Glotton. Hritage spirituel, dition al-Burak, 1986.

If we start from the Middle-Egyptian transliteration ***Hr***, we see that the final voiced voiceless fricative consonant R on the diachronic plane will evolve towards the liquid L. We thus obtain HL instead of HR. H being a weak consonant, this last HL variant is often vocalized [el-][57]. It appears in many sources as in the book written by Martin Rose [58]. We know from the work of researchers like Mas Udi that the first inhabitants of Mesopotamia or the periphery of the Mediterranean were black Kuschite populations. Professor Günter Dreyer, following his discovery of the precedence of Egyptian writing (3400 BC N-E) on the Mesopotamian cuneiform, stated in a Reuters dispatch of 15 December 1998 that:

> *"The Egyptian writing was much more advanced than that of Mesopotamia, which at the time was not yet inhabited by the Sumerians"*
>
> *(Cf. Reuters news agency dated 15/12/98)*

Indeed, 400 km south of Cairo, Abydos, his team found 300 pottery in royal tombs, bearing hieroglyphic inscriptions.
This information testifies that long before the advent of the Sudanese fara Nyar-moro (Narmer), unifying the two lands (Upper and Lower Egypt) to 3,200 before the Christian era, this crucial discovery for humanity, had already been made by the ancient Africans nearly 200 years ago.

Why do Mesopotamia using the ***HL*** variant of the script  ***Hr*** can not be earlier than ***km.t*** moreover than what Günter Dreyer said? To answer this question it is enough to enquiry many specialists and not ideologists and quacks.

The chronology of Professor Christian based on the calculations of Professor Kugler, gives birth to the first dynasty of Our about 2,600 BCE, date of the anonymous graves qualified by "royal" while the date advanced for Mesopotamia is sometimes even 3100 BCE, without "special" reason. Surprisingly, the "Manual of Oriental Archeology" (eds A. and J. Picard) states that no intact tomb has ever been found in Mesopotamia.

Thereupon, Professor G. Contenau recalls that one can: "Ask, in the presence of the tombs of the royal tombs, whether they were really kings and whether these tombs should not be attached to the cult of the principle of fertility (...) What strikes, indeed, is that the occupants of these tombs are, so to speak, anonymous (...).

MS Smith thinks that these tombs could contain, not true kings, but actors of the sacred drama which was played on the occasion of feasts and where one sacrificed the main protagonist (...) Their inventor (of the tombs) Sir L. Wooley absolutely denies it (...) In describing this sensational discovery of the royal tombs, I observed, as it naturally comes to mind, that the Scythians, much later, practiced similar rites. ".

In the book "The Civilization of the Hittites and Mitannians", Professor Contenau still clearly mentions that the professor MV Christian of Vienna: "Suggests for the first dynasty of Our, in Sumer, the first truly historical dynasty, the date of 2,620 before NE ". But he adds that: "If the Mesopotamian civilization was alone at stake, nothing would oppose a compression of this order (...) There is, however, a general solidarity that must be taken into account. about the same date for Egypt and Mesopotamia "hence the refusal to place Egypt first, despite the results of archaeological excavations."

---

57. ʾĒl is a word of the Semitic languages of the northwestern group, meaning "god". It may have taken different meanings according to the place and place where it is attested, and it has notably served to designate a generic god as to Mari or a precise god, who has occuped an important place in different pantheons.
58. Martin ROSE, A hermeneutics of the Old Testament: Understanding, to be Understood each other, drive home, Elabor et Fides, Genève, 2003, P130.

Thirty years ago, the name of Sumer meant nothing to the public. Today, there is something called the Sumerian problem which, for archaeologists, is a subject of constant controversy and speculation, "recalls Professor Brion quoting Professor Lovat Dickson.

Professor Marcel-Auguste Dieulafoy excavating tombs in Mesopotamia and especially in the Elam region says that the first inhabitants were blacks.

> *"[...] Only the powerful characters had the right to wear high canes and bracelets; only the governor of a war place could have the image embroidered on his tunic. Now, the owner of the cane, the master of the citadel is black: there are therefore great possibilities for the Elam to be the prerogative of a black dynasty and if we even refer to the characters of the figure already found, of an Ethiopian dynasty. Would one be in the presence of one of those Levantine Ethiopians of whom Homer speaks? The Nakhuntas were the descendants of a princely family related to the black races that ruled south of Egypt(...)"*[59]
>
> *(OMOTUNDE, 2004)*

After clearly establishing that the first indigenous inhabitants of Mesopotamia were blacks, we understand well how ***Hr*** became ***HL***.

> *"Ēl is a Semitic languages word of the northwestern group, meaning "god." It may have taken different meanings according to the place and place where it is attested, and it has notably served to designate a generic god as to Mari or a precise god, who has occuped an important place in different pantheons"*[60]*.*
> *"El is called in the Ugandan tablets either "Toru El", the god bull, or "Batbyu binwati", the creator of any creature, either "Abu bani ili", the father of the gods or "Abu adami", the father of men. He is "qaniyunu ôlam", the eternal creator, this word - "ôlam" - reappearing in the Bible, in the book of Genesis, in his Hebrew version "El Ôlam, אֵל עוֹלָם", to designate the eternal God (Gen 21. 33 [archive])."*[61]

So there is no surprise for us. We find nowadays many names such as: Israel, Emmanuel, Gabriel, Raziel, Raphael, Haniel, mikael, Camael etc. number of names found in the zephirothic tree have been plagiarized from the ***mry kA bA***. Let's close the parenthesis because this is not yet the subject in this case.

Once the root of the Negro-egyptian has been defined and the etymology of each word is clearly explained in each of the languages, the question of borrowing arises. Is this word a loan of Africans made to the Arab peoples who have been in contact with the African people during the Islamization towards the 7th century?
The question here is easily settled by the historical chronology as well as the sociology of the peoples of the Nile Valley compared to nomadic peoples.

On a purely historical level, the oldest ancestor of the "Semites" known until nowadays is Sargon I of Akkad[62] (Šarru-kīnu) whose existence is located around -2334 years before our era. Now, at that time, worship was already practiced in the temples of ancient "Ethiopia", in ***kAs***, at ***km.t***, and the present peoples practiced a cult in ***Hr*** if only in ***kAs*** Nubia (-15,000 years before NE.). The chronologies that appear in some sacred books are quite edifying so that we do not linger long on the chronological

59. Jean Philippe OMOTUNDE, The African roots of European civilization, Menaibuc, paris, 2004, P.196.
60. Carole ROCHE, PIERRE BORDREUIL, From Ugarit to Jerusalem: a collection of epigraphic and archaeological studies offered to Pierre Bordreuil, Editions De Boccard, Paris, 2008, P.328.
61. Thomas RÖMER, The Invention of God, Seuil, Paris, 2014, p. 55
62. Sargon of Akkad (Akkadian Šarru-ukīn or Šarru-kēn, also known as Sargon the Great) was the first ruler of the Semitic-speaking Akkadian Empire, known for his conquests of the Sumerian city-states in the 24th to 23rd centuries BCE.He was the founder of the "Sargonic" or "Old Akkadian" dynasty, which ruled for about a century after his death, until the Gutian conquest of Sumer.The Sumerian king list makes him the cup-bearer to king Ur-Zababa of Kish. His empire is thought to have included most of Mesopotamia, parts of the Levant, besides incursions into Hurrite and Elamite territory, ruling from his (archaeologically as yet unidentified) capital, Akkad (also Agade).Sargon appears as a legendary figure in Neo-Assyrian literature of the 8th to 7th centuries BC. Tablets with fragments of a Sargon Birth Legend were found in the Library of Ashurbanipal.

question. It would have taken the arrival of the Aryans from the plains of Mongolia so that the miscegenation begins with indigenous Kushite populations around the Mediterranean.

> *"[...] All the populations around the Mediterranean were then Negroes or Negroids: Egyptians, Phoenicians; when they were white, they underwent economic and cultural Egyptian-Phoenician influence: Greece, time of the Boeotians; Asia Minor, Troy; Hittites, allies of Egypt; Etruscans in northern Italy, allies of the Phoenicians: strong Egyptian influence; Gaul, furrowed by Phoenician caravans, under the direct religious influence of Egypt. This Negro influence extended even to the Germans, whose tribes worshiped Isis the Negress."*[63]
>
> *(DIOP, 1954)*

It is important to have a sociological comparison because it is the way of life that forges the character and attributes a vision of the world and a kind of ontology. So if we want to know the origin of what is being discussed here, we must first decide this issue.

The question of "God" is not universal and a-historical. This question is first historical and is therefore part of a sociological intelligence, spatio-temporal very precise. We can not deal on such issue without a technical analysis and only stay on musing or simple impressions.

On the sociological level, this is what appears from the characteristics of nomadic populations according to the African scientist Cheikh Anta Diop:

> *"[...]On the other hand, the ferocity of nature in the Eurasian steppes, the infertility of these regions, all the material conditions in this geoscriptal cradle, will forge in man the instincts necessary for his adaptation to the environment.*
> *Here, nature does not allow any illusion on goodness: it is implacable and allows no negligence: man will draw his daily bread from the sweat of his forehead. He will learn, above all, during this long and painful existence, to rely on his own means, his own possibilities. He can not afford the luxury of believing in a beneficent God who lavishes upon him abundantly the means of existence; his mind will give birth above all to evil and cruel divinities, jealous and spiteful; Zeus, Javeh, etc.*
> *In this ungrateful activity imposed by the physical environment on man, was already involved materialism, anthropomorphism which is only a particular case, the secular mind. This is how the environment will forge these instincts, gradually, among the men who lived in this region, especially among the Indo-Europeans.*
> *All the peoples of this cradle, whether White or Yellow, will have the instinct of conquest, because they will have a tendency to escape from this hostile environment. They are chased away by the environment they must leave or succumb, try to conquer another place in the sun in a more lenient nature and the shaking of invasions will not stop any more, as soon as a first contact with the southern Negro world will teach them the existence of countries where life is easy, abundant wealth, flourishing techniques.*
> *Thus from 1450 BC to Hitler, through the Barbarians of the 6th and 5th century, Genghis Khan, the Turks, these invasions from East to West or from North to South are uninterrupted; it would be even more exact to say these escapes.*
> *The man of these regions remained nomadic for a long time. He is cruel. Caesar and Tacitus describe the warlike, savage customs of Germans who are still nomadic or semi-nomadic and have not yet acquired a sense of ownership of land."*
>
> *(DIOP, 1954)*

It would be inappropriate to base the discussion solely on the purpose of the African scholar Diop. One might suspect a lack of intellectual rigor and judge it to be Afrocentric. It therefore seems more appropriate to involve the main stakeholders. On this account, without being more more papist than the Pope may we listen to Julius Caesar himself:

> *"... They do not engage in agriculture and live mainly on milk, cheese and meat.*
> *No one has a portion of land or specific limits; but each year the magistrates and chiefs assign to the various tribes and families who have assembled such an extent of land, and in such canton as they think fit, and the year after, they force them to move elsewhere, they give several reasons for*

63. Cheikh ANTA DIOP, Negro Nations and Culture, Présence Africaine, Paris, 1979, P. 176.

> *this: they are afraid that the force and the attraction of habit will make the taste of arms disappear for that of agriculture.*
> *The greatest honor for cities is to have devastated borders and immense solitudes around them. They believe that the characteristic of courage is to force the neighboring peoples to desert their territories and to see no one who dares to settle near them: at the same time, they think they are safer and have no invasion. sudden to be feared ...*
> *The robbery committed beyond the borders of the city is not shameful: it serves, they say, to exercise young people and reduce laziness".*[64]
>
> *(Csar, 2004)*

Besides Julius Caesar let's offer the floor to the historian and Roman senator born in 581 and died around 120 A.C.E., Publius Hornlius Tacitus better known as Tacitus:

> *"The height of dishonor is to have left his shield ... One reports his wounds to a mother, to a wife; and they are not afraid to count the wounds, to measure their size. In the fray, they bring fighters food and exhortations ...*
> *If the city which saw them be born languishes in the idleness of a long peace, these leaders of youth will seek war with some foreign people: so much does this nation hate rest! Besides, we are more easily distinguished in the hazards, and we need the reign of force and weapons to maintain many companions ... You would persuade them much less to plow the earth and wait for the year that to call enemies and look for wounds.*
> *It is in their eyes laziness and cowardice that to acquire by sweat what they can obtain by blood ...*
> *They also wear skins of animals, coarser towards the Rhine, more sought after in the interior where the trade does not furnish other ornaments. There we choose animals, and, to embellish their remains, it is sprinkled with spots and it is mixed with the skin of the monsters that feed the unknown beaches of the most distant ocean ... No pomp in their funerals: only we observe to burn with a wood especially the body of illustrious men."*[65]
>
> *(TACITE, 1871)*

Other writers include Marcus Porcius Cato, well known as Cato in his treatise De agri cultura or *De Re Rustica*, who talks about how he advises treating Slavic populations like cattle.[66]. Crispy details can be found in the authors of the General History of Europe [67].

After the customs they describe themselves what is their worship?

> *"The cold climate engendered the cult of fire which, from the fire of Mithra to the flame of the Unknown Soldier Arc-de-Triomphe and the Olympic torches ancient and modern, remains alive. Nomadism will engender incineration: Thus, the ashes of its ancestors were transported in small urns. This custom was perpetuated by the Greeks; the Aryans introduced him to India after 1450 BC. J.C., and this explains the incineration of Caesar and Gandhi in our time".*
>
> *(DIOP, 1954)*

If one sticks to the historical chronology, to the purely historical facts asserted by the historians of good faith, we conclude that the peoples having a nomadic life do not have the way of life, nor the cultic practices compatible with any spirituality. which they want to echo today and singers to the point of wanting to educate Africans.

In the past, this is evidenced by the way in which their ancestors were named in Egypt: Men of sins. Ewondo. / *ùn*-nsəm / "sin". Vocable taken up by the real editor of the pentateuch in the person of Esdras who invented an ancestor named **Sem**[68]. Sorry for those who have taken naively the word ***Sem*** in the nineteenth century to create the word *Semite* [69] as an operating category in physical anthropology when it was a derogatory qualifier that was affixed to immigrant populations at ***km.t*** prohibited access to

64. Jules CESAR, César: The Gallic War, ditions Palo, 2004, Livre 6, chap. 22, 23.
65. TACITE, Germans's Mores , CreateSpace Independent Publishing Platform, 2015, Chap. 6, 7, 14, 17, 27
66. Marcus PORCIUS CATO, De Re Rustica, Société d'édition Les Belles Lettres, coll. "Universits de France", Paris, 1975
67. La CEPEDE (Bernard Germain Etienne de La Ville sur Illon, M. le comte de), General, physical and civil history of Europe: from the last years of the fifth century to the middle of the eighteenth century, P.J de MAT, Bruxelles, 1826
68. Rémi CEILLIER, Laurent Etienne RONDET, General History of Sacred and Ecclesiastical Authors, Volume 1, third book, Chez Louis Vivès, Paris, 1865, P.264.
69. Albert DE PURY, "The "Semites" do not exist", Le Temps stratégique, no 89, septembre/octobre 1999, p. 6-16

temples because uncircumcised and therefore impure. A worship rendered to the primordial ancestor ***Hr*** of sedentary Africans. A worship that is still made today to the one who first performed the sacred ritual of union with the earth thus making it sacred.

Populations of various origins who did not have access to temples because they were impure because they were uncircumcised and unable to worship were therefore rightly called sinners. / bor mi-ʿnsəm / literally "Men of Sins". Their life style contrary to ***Maat*** and their cults forced them in good faith to these qualifiers.

3- Have you said Dzô or Dieu?

> *"Dieu, est le nom que depuis le début des temps jusqu 'à nos jours les hommes ont donné à leur ignorance".*
>
> *Max Nordau*

After this aside in order to clarify the terms and remove the ideological poison contained in it, let's revert to our script:

Ancient-egyptian: ***Hr***

The sign D2 of the Gardiner's list represents the upper body in Ekang languages is called:

Ewondo: / yop / "UP"

That is why the sign D36 / i-dzo / "member" the sign D21 r / i-dzô / "the organ of the speech" which is homophone and synonym will designate the mouth / *ùn*-mbɔkh /, the time / à-bokh /, the beginning / *ùn*-ŋkoba /, the word / *ùn*-ŋkɔbɔ / but especially in its cosmographic meaning the sun ra.

As a solar star, we call it in Ewondo in various ways but above all / dzô /, / *ùn*-nloh dzobo / "the head of the sky"
This term, while keeping the same meaning, is found in Africa everywhere under barely different signifiers:

- o **Kɔ́ɔnzime:** ɗɔ̃ɔ̀ "sun"
- o **Fulfuldé**: loorɗe "up, great" ☞probable origin of the English word *lord*?
- o **Wolof**: nǯɔ:l "great"
- o **Kiswahili**: dy~ua "sol"
- o **Bakoko**: alo: "up"
- o **Bassa:** Job "sol"
- o **Kinyarwanda:** juru "up"
- o **Yemba:** ndo "up"
- o **Baoulé:** dãÿ "up"
- o **Bt:** ḿ-bó "up"
- o **Hausa:** dogo "long, great"
- o **Eton**: Dob "sky"
- o **Duala:** lóba "heaven"

By borrowing and then semantic shift in meaning or metonymy, he gave birth to the others in the words:

- Dieu in modern French, Deu in old French, Deo in very old French
- Deo, Dei, Deus in Latin which means Shine
- Theo, Theos in Greek which means Brilliant

- Dios in old European root which means shiny
- Dewo, Diew, Dyew in very old European root which means Bright
- Dyaut in Sanskrit which means Light,

I could be accused of bias with the Ewondo. I will therefore take the same term in another African language variant:

Ewondo: dzô "speech"
Kinyarwanda: Juru "up"

This term could be vocalized Djuru in a variant of Kinyarwanda without shocking anyone.
So let's put the evolutionary scheme on a diachronic next:
djuru >djulu>djuhu>diuhu>diehu>dieu.

Once again, there is no mystery except the one of ideology. The word tinged with ideology god basically means in the African languages HIGH and by metonymy when it is associated with the SUN in its cosmoscript dimension, it means light and nothing else. May you please refer to illustration N°47.

ETYMOLOGY

Here is the definition tainted of ideology that can be read in authorized documents:

> *"The word "god" comes from the Latin deus, itself stemming from the Indo-European root dei- "shine" which, widened in deiwo- and dyew-, serves to designate the luminous sky as a deity as well as the beings heavenly as opposed to earthly beings, men. Closely linked to this notion of light, it is the oldest Indo-European denomination of the divinity that is found in the name of the Greek god Zeus whose genitive is Dios. From the same root comes the designation of daylight (daytime) and day, itself (Latin dies)"*

It is thus posited here that the term has a Latin origin. By setting the scene in this way, Africa is thus taken out of History, from its history as if the Latin had invented some kind of spirituality. As if Humanity had been born in the cold latitudes. As if the sociology of nomadic peoples was compatible with the construction of temples to worship. As if there was an anteriority of the Latin on Africa.
We call this ideology and those who are distinguished by these practices can not prosper on the ground of truth. Writing such statements is a historical falsification.

God or Goat?

English: goat ☞ **English**: god
Ewondo: /i-tughà/ "service man; boy"

Ewondo:yop "up"
English: up "high"

English: ram "caprid"
Ewondo:mor"man"
Basa'a: mura "woman; wife"

English: tup "ram"
English: top "high; up"
Koonzimé: Tab "caprid"

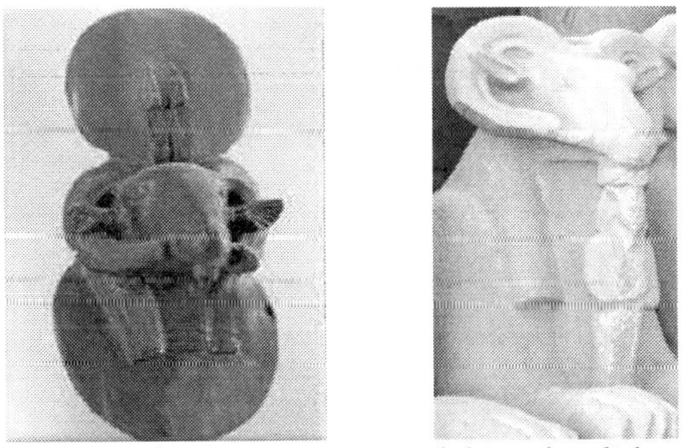

Imn the Egyptian goat with his solar disk

I wanna emphasized that the word goat leading to god in English is intimately linked with the Ewondo words /ndokh/; /kɔt/ "greedy; gluttonous". These two Ewondo's words are from the Kekwe and Kweke branches of the Negro-egyptian as previously explained. They are obtained thus by inverting phonemes.

The words goat or god are simply the Ewondo versions of the word /kot/ "greedy; gluttonous". The English words are obtain par tonemes inversions. In african thought, ram, goat are associated to the idea of gluttonous. For African people, the best way to represent gluttonous is ram or goat.

Conclusion of what, speaking of God to the Africans, it is really an insult to the intelligence and an unbounded intellectual swindle, a shameful mystification.
The Vocable / dzô / simply designate the top, the speech in african languages. It is therefore necessary to extract the poison of ideology, to put the subject back in its original context. Everything that results from this devitalization leads to evacuate all this poison from our minds because the rest is fiction and nothing more.

In order close the debate on this subject, since the ancient Greeks on the historical plane did not know the idea that they crystallize with the words Zeus or Dios, it is certain that they plagiarized it in Egypt. Why then do not Egyptologists provide a word for the Egyptian signifier and have a phonetic proximity to the Greek words? We have Isis, Osiris, Nephthys, Amenophis etc..

They can not because they are in hoax. The words in question are:

***Dr*** this term is synonymous with full up to the top and speech "dzô"
***Dd*** this script refers to the idea of saying, talking "dzô"
 ***aA*** this script is linked to "dzô" by the arm / i-dzo /; the vulture / i-dzoé /; the chisel / ndzoé /; wound papyrus / i-dzô /

Curiously the word that Egyptologists associate with their God is ***nTr***[70].
Why do not we have a ***neteris*** item in Greek but dios? Once one has knowingly or unconsciously loaded this error the rest helps to establish ideology.

I therefore suggest to give in Ewondo maximum words of my language that obey to this skeleton of consonants for a better understanding of the skeleton of consonants ***nTr***. Just before that, I'm just pointing out that, the prefix ***n*** serves as a simple way to designate the expression / i nyo à / "the one who; whoever" what's called an agent prefix

Middle-egyptian:  ***nTr*** *"god, divinity"*

Ewondo: / tara /; " the father "
Ewondo: / á-tár / "pronounce"
Ewondo: / *ùn*-ntára / "The one who pronounces; the one who moans "
Ewondo: / á-tár/ "to moan"
Ewondo: / *ùn*-ntâra / "the one who moans"
Ewondo: / *ùn*-ntér /; / *ùn*-ntét / "handicapped, difficulty moving"
Ewondo: / á-tér / "the one who can, who has difficulties to move"
Ewondo: / *ùn*-ntér /; / *ùn*-ntét / "hundred, 100"
Ewondo: / á-tari / "to start"
Ewondo: / *ùn*-ntàri / "the one who starts"
Ewondo: / á-twàrà / "to crush"
Ewondo: / *ùn*-ntwàrà / "the one who crushes"
Ewondo: / *ùn*-ntwar / "orphan, single, alone, lonely"
Ewondo: / tir /; / tit / "animal, meat, game"
Ewondo: / i-tour /; / i-tout / "bump, protuberance, protrusion"
Ewondo: / i-toúr / "the goat, the male"
Ewondo: / *ùn*-ntour /; / *ùn*-ntout / "full, holy"

70. Faulkner p. 142, Wb. II p. 358, 1-14 sq.

Ewondo: /i-tour/; /i-tout/ "hump, protuberance, protrusion; billy goat"
Ewondo: / *i*-tiri / " what on which we are walking on"
Ewondo: / *i*-twàrà / " what we have to boil"
Ewondo: / *i*-téré / " inside"
Ewondo: /à-tor/; "protuberance on the middle of the face, protrusion"
Ewondo: /à-torà/ "potato" ☞**German:** **karto**ffel "potato"
Etc.

There is nowhere in this lexicon a place where it is mention an idea of any entity named God or assimilated. These are once again Africans basic lexicon words. Words by which we often designate the primordial ancestor / Kulu / of the Ekang peoples.

If for the scribes and i who are speaking and understanding an African language, where does the idea of God comes from?
Where does everything related to all this come from, theocracy etc? Where do we find that Africans were polytheists? That Fara Nti Ekang (*Axn Itn*) was the inventor of monotheism?
Where do all of this comes from? We do think that it is ideology??

Pay attention! some, will tell me. The word God in our languages would be the word flag represented by word-signs *nTr:*

Middle-egyptian: *nTr "god, divinity"*
Fang /Ewondo: /*ùn*-ntseŋ/ "flag"
Fang/Ewondo: / i-fôŋ/;/i-bor/ "flag"

Once more the word flag doesn't match with the word God. There is no sociological nor ontological link between any flag and the supposed idea God in african thought. Moreover I claim that this sign is not a flag at all.

What an absurd idea for a scribe to draw a flag here!!! It's simply senseless to suggest such idea if you have a little knowledge of African thought.

Therefore, I'm saying that, it's not a flag but a **cane**. The cane is one of the royalty symbol of a chief. That's why, representing the chief paramount, it was essential to highlight a visible royalty symbol.

Middle-egyptian: *nTr "God, divinity"*
Fang /Ewondo: /*ùn*-ntum/ "cane"
Fang /Ewondo: /tara/ "the father; the ancestor"

In the african ontology, the cane is associated to ***Hr***. It is useless to remind that we are talking about the prime disabled. His cane is an indication for a walking aid. But as we talk about a leader, indisputably except the flagellum-cane couple, the scribe just drew a cane. The chief is called the cane in Africa.
If it is the truth that must guide humanity, I advise specialists to remove the idea of the flag and replace it with the cane. Indeed, we are talking here about the cane / *ùn*-ntum / associated with the disabled / ùn-ntér / ***Hr*** who we often call fondly: / Kulu / "the foot, the turtle; chief".

4-Have you said Christ?

"ùn-ntsimèn bə-bəbəla wa kar man bukh mə-boŋ"

African wisdom

Before closing this chapter, I want to analyze a purpose that makes Africans abulic.

I recall that, according to African mythology, after his death, ***wsjr***, was mummified by his "son" ***inpw***.

The goal of mummification is to dry a body of the deceased person in order to prolong its conservation by using natron's salts as preservatives. What Nissire Sarr confirms by this purpose:

> *"[...] In the same way the isolation of the bowels without a surgical operation. The means of treatment probably overlap. In Egypt, oil, salt, ointment, natron, resin, myrrh, perfume were used, whose religious virtues were to allow the deceased to meet with the sun god, to render his strong bones.*
> *In black Africa, however, honey, salt, butter, alcohol, incense and certain plants were used to prevent the putrefaction of the body. This process is common to the group of West Africa (Baoulés, Dogons, Mossi, Songhoi, Agni)"* [71].
>
> *(NISSIRE SARR, 2001)*

The result of the desiccation process of the bodies by salts of natron use is the lowering of the AW[72]. The dryed body obtained is call in Ekang languages: / *ùn*-nkor / "sec", / i-kor / "thirst", / *ùn*-nkor / "The savior" who can transliterate by ***krs.t:***

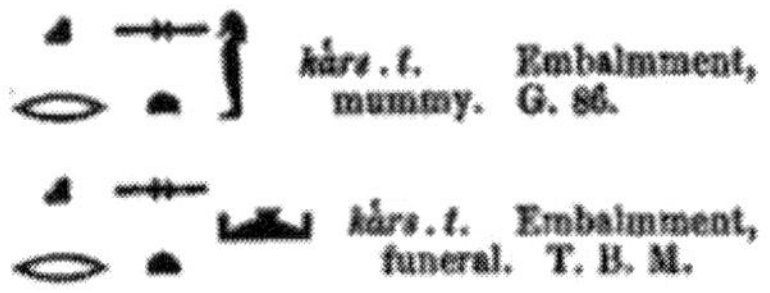

We therefore naturally find the words revolving around the semantic field of the script transliterated ***Krs.t***. Indeed those words are related to the funeral field. All these words are easily translatable in Ewondo language.

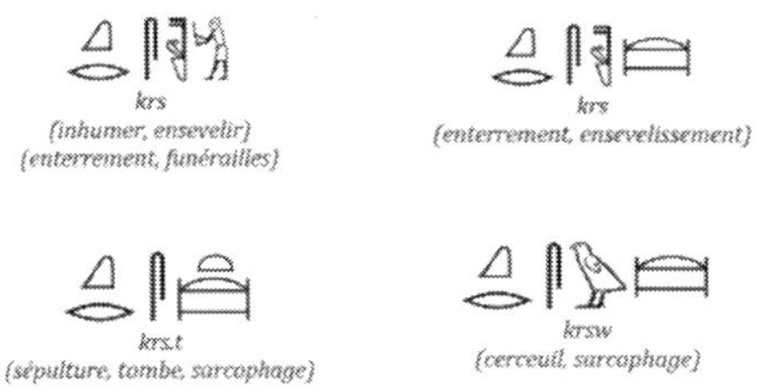

Ewondo: /á-dzeb/ "to bury, to inter" **Ewondo**: /i-dzeb/ "burial; funeral"
Ewondo: /soŋ/ "tombe; grave" **Ewondo**: /i-wali/ "sarcophagus"

I already hear so far skeptics who are impatient to challenge me on the fact that they do not see any link with the word ***krs.t.***

Ewondo: /á-kur/ "to hit"
Ewondo: /i-kurà/ "punch"
Ewondo: /*ùn*-ŋkur/ "the hitter"
Ewondo: /i-kur/ "Abandoned house"

---

71. Mouhamadou NISSIRE SARR, Funerals and representations in the tombs of the Ancient and Middle Egyptian Empires, Lit, 2001, P.133.
72. AW: biochemical indicator that measures the activity of water (P / P0) and whose control allows the conservation of biochemical entities below a certain value. Water activity is one of the main parameters influencing the conservation of food or pharmaceuticals. Microorganisms need "free" water (free for biochemical reactions) to grow. The activity of water does not represent the water content (or moisture) but the availability of this water. The higher the water activity, the greater the amount of free water (1 being the maximum) and the more the micro-organisms will develop. Fungi usually require a water activity of at least 0.7 and bacteria of at least 0.91.

Yes I know, I also noticed it. The mental programming imposed by the techniques of Egyptology dictates a principle based on a unique association by phonetic proximity to **krs.t.** So if no word is vocalized karast I am not in the right. Let's do the audit anyway.

Once we have understood that the *mDw nTr* do not deal only with the phonetic proximity due to the diktat of egyptology's transliteration, that we deprogram and we return to more rational and factual data. Data that we have already triturated throughout this work allow us to move towards a more serene reading.

Middle-egyptian: ***gb*** "eath's God"
Ewondo: /à-dzeb/ "to bury" ☞ **Ewondo**:/á-buri/ "to bury"
Ewondo: /i-dzeb/ "burial; where we bury"

On the script unless if we are blind, we can recognize the chief paramount with his ceremonial beard , we can see the duck, his other signature .(*cf illustration n°32*)

Middle-egyptian: ***krs.t*** "tombe, grave"
French: creuser; creuset "excavate; crucible" **English:** to crush; cross
French: sacré; secret **English**: secret, sacred

Ewondo: /soŋ/ "grave"; /à-sóŋ/ "tie the loincloth"; /ùn-nsoŋ/ "beak; mouth"; /ùn-nsòŋ/ "the hilltop"

Ewondo: /soŋ/; /à-mànà/; /mvuâ/;/i-sughlà/ "grave; the end"
Ewondo: /mə-sùgha/ "silence".

Who said that I was impellent to read this script *Karast*? What does this *Karast* even means at least? So you mean **Ewondo:** / i-kur si / "the ground abandoned house"

Middle-egyptian: ***krs.t*** "coffin, sarcophagus"
Ewondo: /i-zimbi/ "coffin, sarcophagus"
Ewondo: /i-sughlà/ "coffin, sarcophagus , the end"
French: cercueil "coffin, sarcophagus"
English: sarcophagus
Latin: sarcophagus "coffin"

Unless the contrary is proved, the chief paramount of the Ekang people is called /Zamba/. By naming the coffin / i-zimbi / except a difference of vowels, is it not he who Is here named also by metonymy? I'm still expecting to be trapped.

Let us note nevertheless according to the Christologists of all the edges, the crown of thorns carried by ***wsjr*** is called:

Ewondo: / i-kàr bi-yô / "crown of thorns"
Ewondo: / i-kàr i-vés / because ***wjsr*** is the bone

We have also:
Ewondo: / i-kur si / "the ground abandoned house"

literally in Ewondo: / i-kàr i-vés / means, the crown of the bone which by metonymy is sometimes just called in the African linguistic variants Ewondo: / i-kar (i) / "the crown".

Who said I'm compelling to read this script *Karast*? This word *Karast* is even in which language?

To finish on this subject, when the messianic ideology takes hold of the ancient Ekang's spirituality, it gives:

- **Greek**: Kristos — **Ewondo**:  /ùn-ŋkor/ "dry"
- **Latin**: Christus — **Ewondo:**  /à-kur/ "to beat"
- **French**: Christ — **Ewondo:**  /ùn-ŋkun/ /i-kar/ "basket; cage; hamper"

Note: The English Word crown is easily linked to the Ewondo word /ùn-ŋkorèn/ "dried" or /i-kàr/ "crown" or /ùn-ŋkùr/ "shroud; jute bag"

Once you've demystified all of this, you understand how they have invent anything you know better than me: Christianity, Christendom, Christ, Christic, Christology, etc. that was abut on the performative and propaganda techniques. A surrealist and hardly believable speech intended to reform the consciousness of the less experienced. To speak of all this in Africa is nothing more than a sham. Dealing with the subject of alienation Fanon takes the opposite of Lacan in these remarks:

> *"[..]For Lacan,* ***the delusional belief is misunderstanding****. At this level, in our view, is the logical reversal of Lacan's scientific attitude. The author, in approaching the human value of madness, goes from the plane of causality to that of motivation. Starting from knowledge and belief, he considers madness in an intersubjectivist perspective. "Madness," he says, "is lived in the register of meaning. And its metaphysical significance is revealed by the fact that the phenomenon of madness is not separable from the problem of meaning for being in general, that is, language for man. We would have liked to devote long pages to the Lacanian theory of language.* ***But we might venture further away from what we are saying. Yet, on reflection, we must recognize that any delusional phenomenon is ultimately an expressed, that is, spoken phenomenon. Also the best way to analyze a delirium or an abnormal psychic process is still to face the explanation of this delusion.*** *"*[73]
>
> *(FANON-2015)*

5-Have you said Messi or messiah? (chap.142 de *nw pr.t hm Hrw*)

> *"Mi-nləmə tèghe fwa fə kə à ñkagh"*
>
> *African wisdom*

After evacuating from the spirit the poison of mystification conceived around the word ***krs.t***. Let us consider another fable that of ***messiah***. It is often presented thus to weak minds of the kind:

> *"The Messiah (from the Hebrew: מָשִׁיחַ - mashiach, Aramaic meshi'ha משיחא, Arabic masih) originally designated in Judaism the anointed, that is to say, the person consecrated by the rite of the anointing, performed by a prophet of God. In the Bible, the kings Saul and David are anointed by Samuel. This rite is at the origin of those of Saint chrism and the holy bulb of the coronation of the kings of France. In Greek, the word "Christ," whose root Χριστός means "anointed," translates the Hebrew term Mashiyach"*

Middle-egyptian: 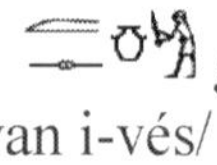 ***gs*** "anoint, to coat"
Ewondo: /*ùn*-ŋkwan i-vés/ "bone marrow" the one which anoints the bone ***(wsjr)***
Ewondo: /à-sî/ "to anoint, to smear"
Ewondo: /à-soué / "to anoint, to coat, to pour"
Ewondo: /mə-sî/ "action of anointing oneself, of being smothered, etc."

73. Frantz, FANON, Writings on Alienation and Freedom, La dcouverte, paris, 2015

There is once again here no magico-mystical character. It is a word that is used to simply describe the action of anointing the body with oil or having the body anointed.

> **Middle-egyptian:** ***nb*** *"lord, master, possessor"*
> **Ewondo**:/ùn-mbɔn/ "oil"☞ by consonantal metathesis
> **Ewondo**:/mor ù-**nam**/ "the leader, the chief, the president" etc.
> **Mossi:** moro **naaba**[74] literally the anointed-man

How in history or mythology the anointing can have some importance?

According to African legend, the body of ***wsjr*** was anointed during mummification by ointments and oils. But in a hidden african sense, ***wsjr*** is the bone, / i-vés /, this bone is anointed naturally by the greasy substance of the marrow's bone through the medullary canal.

In fact, ***wsjr*** designating the leader in African ontological thought, the embodiment of the leader's function involved a ceremony during which the new chief is anointed like the perfect chief ***Hr***. So, do not see anything more please.

> *"[...] In Egypt, oil, salt, ointment, natron, resin, myrrh, perfume were used, whose religious virtues were to allow the deceased to meet with the sun god, to render his strong bones. In black Africa, however, honey, salt, butter, alcohol, incense and certain plants were used to prevent the putrefaction of the body. This process is common to the group of West Africa (Baoulés, Dogons, Mossi, Songhoi, Agni)"*[75].
>
> *(NISSIRE SARR, 2001)*

I have sought the mystical or magical side, I still can not find it since embalming is a classic and elementary step in the mummification process.

Wsjr having been the first to undergo mummification, he has therefore been described as anointed. What from the African point of view is quite logical and does not refer to any kind of charlatanic messianism ideology. Africans are still being mummify every day in the 21st century without making it a special or magical-mystical affair.

But this idea of anointing after the change-over into the distorting ideological prism of Judeo-Christian technostructure will take the path of Islam and is presented by some authors as being an African messianic idea.

*"**Al-masih** messiah (wehr 907b)*			***2630***
Boko Jo	*Màsihù*	*messiah*	
Busa Jo	*àrùmàsihù*	*messiah, christ*	
Dagbani B masia		*messiah, christ*	
Fulfuldé Tay	*almasihu*	*Messiah*	
	Masinhinkjo		
Hausa A	*Almàsihù*	*Jesus-Christ*	
Swahili J	*Masiya*	*Christ*	
Wolof MouAlmaside		*Messiah*	
***Al-masihiya** chretiented; Christianity, Christian faith (wehr 907b)*			***2631***
Swahili J	*Masihiya, mmasihiya*	*Christian"*[76]	

(BALDI, 2008)

---

74. Michel IZARD, Moogo - The emergence of a West African state space in the 16th century, Karthala, paris,2003, p. 95-192
75. Mouhamadou NISSIRE SARR, Funerals and representations in the tombs of the Ancient and Middle Egyptian Empires, Lit, 2001, P.133.
76. Sergio BALDI, Dictionary of Arabic Borrowings in West African Languages, KARTHALA Editions, Paris, 2008, p.461.

When one goes through the writings of Sergio Baldi indeed, one can be destabilized and invade by the doubt and begin to believe that Islam would have borrowed these notions from the African languages, which in my opinion is absolutely false taking into account the historical chronology. On a purely chronological level, the word anointed is African, it is copied by the producer of religions and returns to Africa under the rule of Islamization or the Christianization of Africa surrounded by a magical-mystic halo.

All this seems like surrealism. We have snapshots of Egyptian society's past frozen in time with the Harapean civilization or the Mesoamerican civilizations that were created by Egyptians. We can investigate afro-descendants in African societies across the Atlantic, which constitute a photography of the 15th century in Africa. One thing is certain, there is no such idea of Jesus or messiah in these populations. This shows that Africans were not at all aware of what they are being told today. I will be retorted that the founders of so-called Christic religions are among others:

Titus Flavius Clemens better known as Clement of Alexandria (+150 to +215) worked to transfer the major ideas contained in the texts plagiarized by the Greeks during the occupation to Christianity.

There is also Alexander of Alexandria (+250 to +328) who reorganized the Church of Egypt. In Alexandria, he built the great church of St. Theonas. These two supposed fathers of the church being Africans out of the mold of Egypt under the Roman occupation.

Islam has its roots in the Sabean cult and Judeo-Christian technostructure
and adapted them to his expansionist model.

Although born into a murid and therefore Muslim family, Cheikh Anta Diop is very clear about the nature of the mohametan religion in *Negro Nations and culture*:

> *"[...] A direct invocation was addressed to the seven planets. The 30-day fast already existed similar to those practiced in Egypt. They prayed seven times a day with their faces turned to the north. These prayers, which are addressed to the sun at different times of its development, are quite similar to the Muslim prayers which take place in the same phases but which have been reduced to five obligatory by the Prophet "to alleviate humanity, the others being optional. There were also sources and sacred stones, as at the time*
> *Muslim Zenzen, sacred source; Kaaba, sacred stone. The pilgrimage to Mecca already existed Already believed in the future life, as in Egypt. The dead ancestors were deified. All the elements necessary for the blossoming of Islam were therefore in place already more than 1000 years before the birth of Muhammad."*[77]
>
> *(DIOP, 1954)*

The danger of the intended religious discourse is the unconscious transfer of mental projections realized through the mystification of peoples by sectarian groups wanted religious who have kept a collective memory embellished in unconscious mental schemas of domination and violence[78].

But we can not find an African people who has not been intoxicated by this story and who believes or perpetuates an African thought based on these fantasies. I challenge those who support this statement to give the word in Baka language for Christ, Messiah or Jesus.
I challenge anyone who thinks it is smart to go to the Amazon with the Pirahã who live away from the world today and refuse any contact with the ideologists to find us the word christ or messi in Pirahã.

I say, therefore, that there has scarcely existed, and to this day, no particular worship in Africa of baskets, mummies, coffins, or even anointed ones. All this is absolutely wrong. This is part of an anthropology of violent people that bases the discussion on sophistry and the art of deception.

---

77. Cheikh ANTA DIOP, Negro Nations and culture, Présence africaine, Paris-Dakar, 1954, P.193.
78. Rosa Amelia PLUMELLE-URIBE, The White Ferocity: From Non-Whites to Non-Aryans, Occult Genocide from 1492 to the Present , Albin Michel, Paris, 2001.

In Africa, there is no cult of drying corpse in th kang today. Mummification is an operation of conservation of the fluidic m ix of the Being the ***D.t***, an operation practiced by our ancestors and still practiced in memory of ***wsjr*** in certain regions of the continent, as Brother Nissir Sarr rightly points out, and I might add that this practice still exists among some kang today.

If you are looking for messi's I have trouble to inform you that there are profusely in Africa. Just go to the land of the anointed (the Moro Naaba) in Mossi country.

The term Mossi has not moved an island, the only thing that has changed here is the ideology.
If you want to attend anointing ceremony. It is not usefull to make brain knots. May you please go to Bamileke land in western Cameroon. Skulls are worshiped every day during a relic anointing ceremony (**Ngeombale**: bere to) (see illustration N°45). It is the same ritual that is taken up in the context of the mystification by the churches to make the baptism. Africa has a duty to take control of itself and stop making itself ridiculous with its own corrupted knowledge.

Etymology:

The French word huile "oil" is composed of a consonants skeleton **HL** issued from a diachronic evolution of ***Hr*** "horus, king". We can therefore also justify the English word **oil** on the same basis.

English: bone "ridge"

As we have already said, the term oil is:

Ewondo:/*ùn*-mbɔn/ "huile"

The phonetic proximity is not fortuitous here it serves to establish the connection between the anointing and the bone as can be seen in the illustration N°45.

Ewondo:/*ùn*-mbɔn/ "oil"

Nzim:*mulo* "oil"	**Ewondo**:/ùn-nlɔh/ "head"
Dualà: *Muúla* "oil"	**Ewondo**:/ùn-nlɔh/ "head"
Soninké: *Te* "oil"	**Ewondo**:/i-tɔh/ "trône"
Bamankan: *Toulou* "oil"	**Ewondo**:/itɔh ùn-nlɔh/ "chief"
Basa'a: *mo'o* "oil"	**Ewondo**:/mbo/ "one, chief"
Tupuri: *no* "oil"	**Ewondo**:/ùn-nlɔh/ "head"
Lingala: *mafuta* "oil"	**Ewondo**:/ù-mfərèn/ "the chief"
Obamba: *Muuri* "oil"	**Ewondo**:/râ/ "chief"
isiXhosa: *ibhotolo* "oil"	**Ewondo**:/à-bon ù-nloh/ "fontanel" **Ewondo**:/ibho ù-nLO/ "head dedayed"
Russe: *масло* "oil"	**Ewondo**:/ùn nlɔh/ "head"
Nzbi: *Mèdi* "oil"	**Ewondo**:/mə-ndib/ "water"
Sonrhaï: *dji* "oil"	**Ewondo**:/mə-ndib/ "water"
Bansoa: *mevet* "oil" **Yemba:** *mvet* "oil"	**Ewondo**:/mvett/ "A stringed musical instrument known since ancient Egypt, and, with the capital letter, a set of warrior stories that came with this instrument, forming the Ekang culture"

Chart 73: oil in african languages and signification in Ewondo

Note that the word Lingàla *mafuta* is interesting on several levels etymologically because it explains the origin of the English word FOOT. In Africa indeed, if the leader is the head, he is also the foot.

Kikongo: *nyosi* "honey"
Lingàla: *mafuta ya nzoyi* "honey bee"

This link between **futa* and foot goes further because the French word for foot, *foot* is linked with the word Kɔ́ɔnzime: *mpyê* "dog". Let's remind that the responsible for the embaubement in the Egyptian-Nubian thought is the dog  ***inpw***.

I summarize then:

Lingàla: *Mafuta* "oil" linked with ☞**English**: *foot* "leg"
Kɔ́ɔnzime: *mpyê* "dog" linked with ☞**French**: *pied* "foot; leg"

It is only through the ceremony of anointing that we understand the true meaning of the words *pied* and *foot* and how they are related to the African thought and especially to the Egyptian-Nubian basin.

The **Nzébi** language word *médi* "oil" or *dji* "oil" of Songhai are interesting words because they refer to the word ***dib*** "hippopotamus" of egyptology.

They refer to the hippo greacy substance but also to its biotope. Indeed, the hippopotamus is also called in African the water-elephant.

We see in fact that the word ***dib*** is related by metonymy with the word ***foot*** which is as we have said the symbol of the CHIEF. Indeed, we have in African languages equation related to the word kulu "chief":

kulu = foot= dog = chief

As a result, the comparative method is not very appropriate for African lexicology. We are talking about OIL and we end up talking about another signified namely hippopotamus. The signified is the thing as it exists in itself. The signifier is the way we have to name it in different languages according to Saussure.

So if we are talking about Oil = meant, and everyone calls it naturally as it should in his language, we notice that Oil = bone, = head, = hippopotamus, = honey = water = etc. We therefore see that we do not speak of the same signified. But trying to reconstruct the ancestor of the word Oil, the comparative method can not guess that it encloses the concept of foot, honey, oil, water, chief's bone etc. This method, although very effective, is not suitable for the study of Egyptian from our languages point of view.

The word lipid, (lpd) .... intuitively (rbt) refers to (rab) lordship, chief, owner, on the one hand among the Arabs, except that the (t) mark of the feminine lends itself to matriarchy or to the matrix closer to Africa.
(Rab) means protective genius that is inherited from the maternal mother or uncle among the Wolof lebu ... So we can see the supposed origin of the word Semite rabbi = master

We also have:

Middle-egyptian:  nb "to attach"
Ewondo: *á-bàn* "to tie, to seal; to attach"

Wsjr the ultimate chief's figure was attached by his brother 𓋴𓏏𓈖 **stX**. We see in this word the symbol of the foot used to designate the chief.

The character read 𓎟***nb*** is a bowl which, by its ontological dimension, is reminiscent of the vessel which contains the sacred oil for the operations of unction according to African thought. This operation is named in Ngeombale mbafung / Babadjou the ***bere to***.

If you do not have red oil to make the ***bere to***, You can use water or honey even dust also.

Ewondo:/i-béré i-toà/ "coronation; crowning; enthronement etc."

One can observe the historical continuum that links Himbas of South Africa to this old tradition of anointing. In fact, the Himbas dye their skin red with an ointment made from cow fat and red ocher powder.
The same tradition is found among the Zo'é of Amazonia.

The Zo'é using the roucou to paint their faces and their bodies

Himba woman of South Africa

In Ewondo language, to anoint is:

Ewondo: /à-sî/ "to anoint" ☞Homophony **Ewondo:** /si/ "Earth, dust"
Ewondo: /à-**màm**/ "to anoint"
Ewondo: /i-**màm**/ "the anointing"

So we find in a way and indisputable the concept of anointing stolen by the Arabic world :

Arabe: امام, imām "muslim religious chief"
Persian: امام, emām "muslim religious chief"

The *imam* is the Muslims religious leader, the guide, the one in front. Curiously is the definition given by egyptologists for the 𓈖𓎟𓀀 *nb*.

Middle-egyptian: 𓈖𓎟𓀀 ***nb*** *"lord, master, possessor"*

In fact the anointed can be called in Ewondo: /i-màm/. But this Ewondo word /i-mam/ lead to a word game with the item of the first ancestor /i-mvam/ 𓀀.
Moreover, the word /*mam/ is used in Africa for the mum.

The mother owns the land, the maternal uncle who is the depositary of the patrimonial interests is also called mother or mum in the African languages. This is why the words used in relation with anointment are to be directly related to African thought.

Ewondo: /à-sî/ "to anoint" ☞Homoph. **Ewondo:** /si/ "Earth, dust"
Ewondo: /à-màm/ "to anoint" ☞Homoph. **Ewondo:** /*mam/"shape, anoint"

May you allow me to remember you that the chief in African performs in three main functions, the religious leader, the warlord and the depositary of the land management interests. Hope that you follow me if, by chance the *imam* is a religious leader and curiously gets dressed in the finery of the African leader.

Note: we can observe that the color used by Himbas for unction is red ocher as red palm oil. May we link it with the red crown of ancient Egypt?

We say that this ritual is not magic, but is scientifically based on quantum entanglement or the inseparability of matter. It works by using the special properties related to concave or even spherical objects.

It is curious to see that the word ***mvett*** is used to designate the eye but a stringed musical instrument known from ancient Egypt, and, with the capital letter, a set of war stories that are played accompanied by this instrument, forming the Ekang culture.

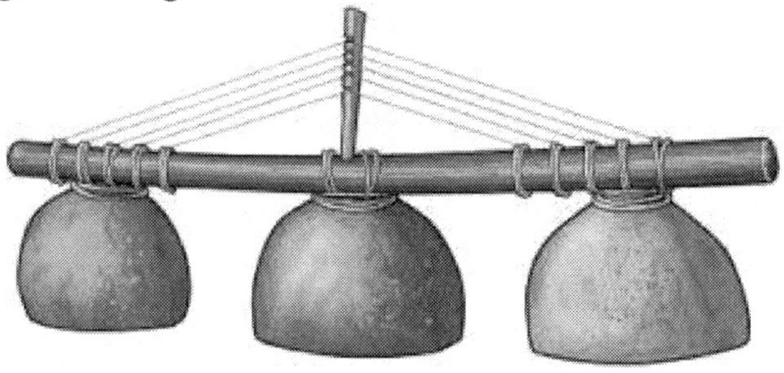

The mvett

Middle-egyptian: ***ꜥnb*** "lord, master, possessor" **Ewondo**: Zamba/ "lord"

We find that the calabashes that serve to contain the anointing oil create the sound boxes of the mvett guitar.
It is therefore normal that these calabashes are also assimilated to the eye by metonymy.
If in West Cameroon we call oil **mvett** and the fang / beti calls the guitar **mvett** it's because these two lineages have shared a common space that only Egypt can reconcile.
The mvett is not a simple instrument, it is a fundamental element of the ancient peoples of the Nile Valley.
Only African languages can establish indisputably the link between the word *mvett* and the word oil *mvett*.

The word *Maari* "oil" of Obamba (Gabon) is very interesting because it is closely related to different variants:

This word is etymologically at the origin of the word meroitic. He is at the origin of the word:

French: mari "husband" ☞**Middle-Egyptian**: *mr.t*"Meret, musician goddess"
French: Amour " love"

Ewondo: mor "human being; Man"☞**Middle-Egyptian**: *imn* "Amon"
Basa'a: mura "woman, wife"
English: to marry "wed, marry off, ally, mix; amour"

The Obamba word *maari* is linked to the word: El *Mahdi* (arabe: mahdīy, ألمهدي, "Guided person (by God); the one who shows the way ") or El Mahdi Mountadhar (Arabic: المنتظر المهدي," the expected guide ") or the Khalifat Allah (" God-elected ruler ") is the expected" savior "of all Muslims (except Quranites) , which should appear at the end of time according to some hadiths.

Middle-egyptian: mry "beloved"
French : Marie, mari, mairie etc.
English: to marry; marry off, married, ram, amour, mar etc.

Ewondo: */i-fyer/ or /i-fér/ or /i-fét/* "alliance"

6- Did you said Mary?

"Màn ù-tóŋ tèghè kɔ à-fan woŋ"

Middle-egyptian: ***Mry.t*** **"***Hr's mum"*

I have already analyzed the script ***As.t***. But I come back a little bit on this script because it is given in some dubious literatures as the mother of "Jesus". Under the name of Mary, it leads to the Marian worship and a whole current of highly disturbing thoughts in Africa.

According to African mythology, the mother of ***Hr*** was called ***As.t*** whose grecised variant is Isis. According to the African tradition, ***As.t*** is often called ***mry.t*** meaning the beloved[79] according to classical Egyptology. To the question of finding spoors of this term in historical documents? The answer is clear about this issue:

Demotique: mr(t)
Copte: ⲙⲣⲱ [mrō] (Sahidic), ⲉⲙⲣⲱ [emrō] (Bohairic)
Grec Ancien: Μάρεια [Máreia], Μᾰρεῶτῐς [Mareôtis]
Arabe Egyptien: damrūʾ

This expression can be found in a number of documents, particularly in the extract from the Nauri decree below, for example:

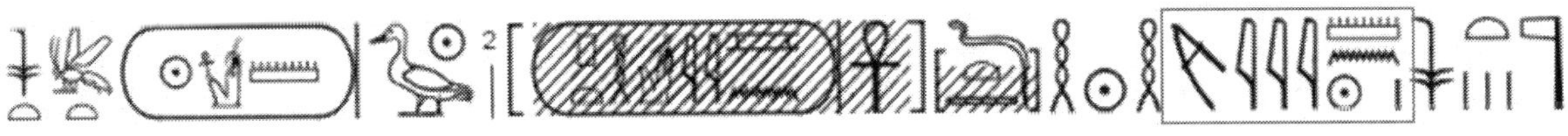

Middle-egyptian*:*

Nsw-bity Mn MAa.t-Ra "*The King of Upper and Lower Egypt, Stable is the truth of Re",*
(2)- *zA-Ra < stXy mr n PtH > anx D.t nHH,* "*rê's name < Sethy, Ptah,beloved [alive forever] and forever"*
mry Imn-Ra, nsw nTr(.w), "*the beloved of Amon-Re the King of the gods***"**

Middle-egyptian: ***mry imn*** "imn's beloved"

79. Adolf ERMAN, Hermann GRAPOW, Wörterbuch der ägyptischen Sprache, volume 2, Berlin: Akademie-Verlag, 1926-1961, page 109.3, 109.12–110.3

This name was common in Egypt and was found even in the royal titulatures as underlined below Leprohon:

> "Epithets added to the Throne name:
> ***mry(t) n imn*** *(mery(t) en imen), Beloved (fem.) of Amun*
> *[Imwt] t;-mry ([henutl ta-mery), [Mistress of the Beloved Land*
> *(i.e., Egypt)*"[80]

The transliteration ***mry*** / ***mry.t*** once translated correctly into African languages offers plethora of names in the lexical field of current Ekang languages:

- Itoa "seat"
- Ngono "moon"
- Nanga "albino"
- Nya "mum"
- Etc.

- Nga "wife"
- Toŋ'o "horned"
- Inyəghə "breat feeder"
- Mbala "the weeper"

How does one go from the transliterated version mry.t to the filled with mysteries Mary/Marie?

The recovery of the transliteration ***mry.t*** to make Marie in French is done by replacing the feminine Egyptian tag ***.t*** by the vowel ***e*** the feminine in French tag and by adding in ***a*** arbitrary and authoritarian way the vowel has between the ***m*** and the ***r.***

But for those who do not have a particular feminine termination like the English the word has been plagiarized and barely modified to offer *Mary*. So check by yourself.

Taking into account the sense imposed by the Egyptologists the skeleton of consonants ***mry.t*** does not necessarily vocalize thus keeping the same meaning in the African languages:

Kɔ́ɔnzime: *mumá* "wife"	**Lingala**: *mwǎsí* "wife"	**Duala**: *múnʒa* "wife"
Hausa: mata "wife"	**Soninké**:*yaxare*"wife"	**Ɓàsàa**: *mùdà* "wife"
Kongo: nkazi "wife"	**Sena**: nkazi "woman,	**Lega**: nkashi"woman"
Bwisi: omukali "woman"	**Wolof**: *jabar* "wife"	**Bété**: *ù-ŋgiɛ* "wife"
Konzo: omukaali "woman"	**Lunda**: mukaj "woman"	
Tsooto: umuxuçi"wife"	**Kinyarwanda**: *umugore* "wife"	

It is therefore a banal word of current African languages. It is time for Africans to escape from the comfortable mental jails designed by the Westerns for their loss. After 1000 years of obscurantism (5th-15th century), as Dante Alighieri so well describes in *The Divine Comedy*[81]. All this appears as a desire to chain Africa into Christian delusions magnified by ideology but unfortunately highly incapacitating in terms of efficient rationality.

Note nevertheless that  ***Wsjr*** is vocalized in Hebrew יוֹסֵף, (Yosef), Yosif, Yusab in Coptic giving the fictional character Joseph in French gives Youcef, Yosef, Youssef, Youssouf.

English: *zero; Joseph; wiser; wizard; wise; etc.*

If the evidence presented still does not convince you despite its authenticity and veracity, it's not a surprise, this phenomenon has already been well understood and is called, cognitive dissonance.

80. Ronald J. LEPROHON, The Great Name: Ancient Egyptian Royal Titulary, Society of Biblical Lit, Atlanta, 2013, p.125.
81. Dante ALIGHIERI, Commedia puis Divina Commedia, Centro cultural "Latiunt", Buenos Aires,1922.

> *"Sometimes people cling to a very strong basic belief. When they are presented with evidence that goes against this belief, the new belief can not be accepted. It would create an extremely uncomfortable feeling, called cognitive dissonance.*
> *And because it is so important to protect basic belief, they will rationalize, ignore and even deny anything that does not fit the belief"*
>
> *Frantz Fanon*

Although Flavius Josephus took over the writings of the priest historian ***Mnjw Htr*** better known as Manethon of Sebennytos (in ancient Greek Μανέθων, Μανέθως) (third century BCE) who had written a History of Egypt (Ægyptiaca) in three volumes in Greek during the reign of Ptolemy II (309 BC NW at 246 BC). This lending as Official Historian, had access to the royal lists of temple libraries, but also to popular tales about various mythical pharaohs. It is therefore to Mnjw Htr that we owe the division into thirty dynasties of the sovereigns of Egypt, still used by Egyptologists with some modifications (Egyptian dynasty zero for example), because it makes the analysis of Egyptian history more convenient. It is also to him that one must use the name of Amenophis, the ***Imn Htp***, the Thutmose, the ***DHwty-mss***, or the Sesostris the ***Sn Wsr.t***. These are indeed Hellenized transcripts of Egyptian names. These translations end up rejecting the African youth who does not recognize themselves in this story.

The first reference to Aegyptiaca by ***Mnjw Htr*** is in the book of the Jewish copyist and historian Flavius Josephus under the title of *Against Apion*[82]. The supposed additions to the life of Jesus that have figured in his writings find by reading the scripts an indignity.
The improbability of the facts as we have demonstrated is part of a series of unspeakable errors in order to synchronize fables with historical facts.

Dionysius Exiguus, born around 470 in the Roman province of Scythia Minor on the banks of the Euxin Bridge (present-day Dobrogea, Romania), was a learned Christian religious living in Rome around 497 AD. died, who was considered one of the principal "smugglers", in his time, between Eastern and Western Christianity.

He is credited with translating Greek texts into Latin on Canon Law and Comput. He remained famous for calculating the Anno Domini, which served to date the years since the conventional birth of Jesus Christ and is still used today for dating.

If one believes the latest research on the chronology on which our calendar is based from his work visibly this chronology suffers from a major paralysis error of 4 years. Not to mention the vagueness on the historical level between a birth supposed during the reign of Herod the Great (37 BCE to 4 BCE) or Tiberius (Latin:Tiberius Caesar Divi Augusti Filius Augustus),
born in Rome on November 16, 42 BCE and died in Misene on March 16, 37 ACE., is the second Roman emperor from 14 to 37.

More seriously Nas E. Boutammina argues about the falsification and alleged fabrication of Western History, highlighting Anatoli Fomenko's controversial work on the reliability of the current historical chronology:

> *"Anatoly Timofeientich Fomenko [1945] is a Russian mathematician of the Russian Academy of Sciences, Doctor of Science, Professor, Head of the Department of Differential Geometry and Applications of the Faculty of Mechanics and Mathematics of the Moscow University. [In 1996 he received the State Prize of the Russian Federation in Mathematics. Fomenko is also known as the theoretician of a "New Chronology" in historical matter, theory considered as pseudoscience by historians and astronomers. According to this theory, ancient history is only a vast invention of the Jesuits in the seventeenth and eighteenth centuries. Its main argument is the hypothesis that texts have been misinterpreted by historians: some texts considered to relate to different periods, while they would speak on the same subject, but written by different authors and in different languages, with all the changes it entails [like the names of cities], and that would have helped to expand*

82. Flavius JOSEPHE, Against Apion, Ernest Leroux, diteur, Paris, 1911.

history. The author claims that all the allegedly ancient histories of Greece, Rome, Egypt, China are only late rewritings, made at the Renaissance from the story of events that occurred in reality in the Middle Ages "summarizes archaeologist Jean- louis Le Quellec, [JL THE "Martians in the Sahara, chronicles of romantic archeology" Actes Sud / Errance, 2009, p. 93]. According to A.T. Fomenko, History does not begin until the tenth century of our era. Jesus-chist would have been crucified in 1083 Constantinople. Many dates arrived with an i or an "i" followed by numbers: i235; I322.

These two letters have been interpreted to mean the number "1", increasing the actual date by one thousand years. According to Fomenko, the Italians claimed to live in the quattrocento (400) and not the one thousand quattrocento (1400); Fomenko concludes that their calendar began in the year 1000. Thus, the Crusades and the Trojan War would be one and the same event, the British past would come from texts actually describing a part of the Byzantine Empire (Anglia later Albion would describe a fraction of the Byzantine Empire.In fact, the Angels (Engels) were a Byzantine dynasty), the Hittites would really be the Goths. Fomenko states in his "New Chronology" (History 1) that carbon dating has been calibrated on the already accepted chronology. According to him, this dating could not be used to confirm it since it follows from it. He also challenges dendrochronology and other methods"[83]

All these reservations on a technical level lead us to understand that religions are anthropological categories of thought. In fact, the ideas of Messiah, Christ and others are characteristic of a nomadic thought. Indeed, the idea of rationality is only consistent with its reasoning logic. So we build a rational intended discourse according to its own logical paradigm.

As we can see, these fantasies about what the West calls god have nothing to do with Africa. But absolutely nothing!!!

The African thought as it is recorded in the sacred characters reveals the idea of a mechanism while being one following a fractal model that gave rise to the universe by diversifying an infinity of facets.

This is why the script ***ntr*** inadvertently translated by god by concordism with the foreign paradigm ended up giving conclusions that make African society a polytheistic society. How to explain that ***Hr*** is also ***wsjr***? Learned answer consubstantiality of the father with the son.

So, why ***Hr*** is ***As.t***? Why ***Hr*** is ***stX***? Official speeches are conceived by ignorance or to deceive Africans. We therefore call to depart from these discourses.

But worse than that, it has led African intellectuals, who are somewhat lucid but trapped in this thought based on the theos, to reject polytheism in order to claim a hypothetical monotheism that can not be proved by taking into account the sacred characters. A monotheism that can only be explained by relying on the official interpretation of the sacred texts imposed by the West.

The myths around the primordial ancestor ***Hr*** are staged to tell a hidden truth that is accessible only to those who are able to go beyond the level of the fable.

7- Have you said Amen or Amon?

"U-nə zam bi mwàn vədà véndé tèghe-bi fulu dy"

Na Ndzié

Middle-egyptian: *Imn* "(God) Amon" (*Faulkner p. 21, Wb. I p. 84, 15 sq.)*

83. Nas E. BOUTAMMINA, Understanding the Renaissance - Falsifying and Making History of the West, Books on Demand, 2015, P.219.

According to whether one is in a church, one will have Amen, if it is in a Muslim place of worship one will speak of Amin 'if one finds oneself in Egyptology it will be Amon. But in all cases we agree on the idea that it is about God.

But since "God" is a hoax, then this spelling speaks of what at the end? That's the real question. Above all we will make you understand by which pirouette that Amon is the hidden.

Middle-egyptian: ***imn*** "secret, hidden; hide"(*Faulkner p. 21, Wb. I p. 83, 12-23 sq.*)

On a technical level this script seems to be problematic because those who decided to transliterate ***imn*** do not know why since they say themselves that the sign Y5 is transliterated ***zn.t***. We assert that, this game is played in Africa and is called / sɔngɔ / in Ewondo and / awalé / in West Africa.
So how do we go from ***zn.t*** to ***imn***?
I want as proof this screenshot made from the Internet web site of the Faulkner dictionnary[84].

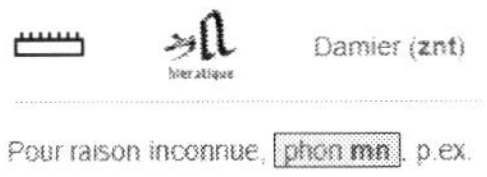

So we see that all this is nebulous for some and sometimes friezing amateurism.
So, I say that to read the script ***Imn***, it's possible to anyone but, to have a good understanding of the purpose related to this script then just ask Africans. Since this reality belongs to them.

This is how to read this script in Ewondo. I suggest to list all the words used in this language and respecting the consonant skeleton **** mn-.*** I shall refrain to take some lower variants. They are available in the fang dictionary.

Middle-egyptian: ***Imn*** "Amon"

Ewondo: /i-mén/ "him-self"
Ewondo: /à-mɔŋ/ "to hide"
Ewondo: /mɔn/ "child; kid"
Ewondo: /à-min/ "gulp; swallow"
Ewondo: /i-mɔŋ/ "left"
Ewondo: /à-nomo/ "right"
Ewondo: /à-myɔŋ/ "calm"
Ewondo: /à-màŋ/ "cheek"
Ewondo: /màŋ/ "sea"
Ewondo: /màn/ "small"
Ewondo: /mân/ "crossroad"
Ewondo: /mina/ "you"
Ewondo: /à-man/ "finish"
Ewondo: /à-mâŋ/ "terminate, finished"
Ewondo: /à-man/ "to press, fold"
Ewondo:/à-mane/"kill,finish"
Ewondo: /màŋ/ "drugs"
Ewondo: /màŋ'a/ "menatee"
Ewondo: / màŋa / "smell of burnt fat"
Ewondo: / mənyî / "the 4"
Ewondo: / mân / "yours"
Ewondo: / mə-nə / "i' am"
Ewondo: / mə-nyî / "ingress"
Ewondo: / mə nyi / "Here I am"
Ewondo: / mə-nyî / "Entrance"
Ewondo: / à-mene / "break, tighten"
Ewondo: / mə-nyù / "drink"
Ewondo: / mə-nyà / "variety of chili oblong, strong"
Ewondo: / mə-nyo / "animal smell, buffalo, goat, etc."
Ewondo: / à-məŋ / "to swing"
Ewondo: / mi-'nhiè / "intestins"
Ewondo: / mine / "you"
Ewondo: / mi-nyo / "houses"
Ewondo: / à-mona / "to spy on each other"
Ewondo: / à-mùŋ / "Smile while hiding"
Ewondo: / à-mùŋa / "smiles reciprocally in secret"
Ewondo: / muŋe / "fish with scales"
Ewondo: / à-mune / "drop arbs"
Ewondo: / moŋ/ "dislocate"
Ewondo: / à-myaŋ / "get lost, lost"
Ewondo: / myaŋa / "silver"
Etc.

84. http://www.hierogl.ch/hiero/Sign:Y5

As we can see after this long list, the consonants skeleton ***imn*** is not something magical or beyond the understanding. These are words that are used every day by Africans in one kind or another.

However in this long list what are the most interesting words regarding the cult?
These are those who in any way behave like descriptors of ***Hr*** in the worship field.

According to the mythology ***Hr*** would be the ancestor of the Nilotic peoples. On this account, it is therefore in numerological logic that can be associated with numeral 1. Being a small number, it becomes the smallest / obə /. He is considered so small that sometimes it induces a superlative by doubling its qualifier / obe / giving in the French language the word *bebe* "baby". We say in ewondo / i-bobogh / for the same signified. Just drop the final velar to get to the vowel close to the French variant is not it? Check it by yourself.

I would like to point out that all the words mentioned above are searchable as a check in the fang dictionary[85].

Verification

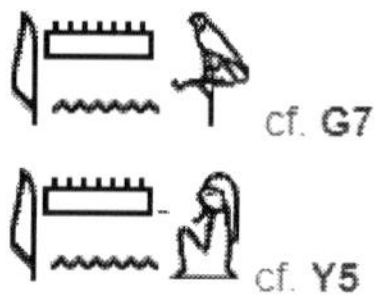

In the representation opposite (*Faulkner p. 21, Wb. I p. 84, 15 sq*).we have ***Hr*** the falcon ***Imn.***
Since the couple of scripts only change by their semantems, while the characters associated with transliteration are identical and offer the transliteration ***imn***.

When we talk about ***Hr*** the consonants skeleton ****mn*** one should understand:

> ***Hr*** so is /mɔn/ "the child of the trinity,the baby"
> ***Hr*** is the negligible amount /màn/ "small, little, short"
> ***Hr*** is the one who is left /i-mɔŋ/
> ***Hr*** the hidden "à-moŋ/ "to hide"
> ***Hr*** is the one who designates the end "à-mànà/ "The End"

The script opposite indicates ***Hr*** as the pathe and is transliterate ***Sn***.
Hr is the path /zén/ when one is in the spiral [86] /ùn-mbokh/.
Hr being handicapped / *ùn*-mbokh /, this word is homophone of / *ùm*-mbokh / meaning hole. This hole being a path is therefore this word / zen / the path that is transliterated ***zn.t*** and ***zn.t***, we go magically to **imn**. I will say by recopying on target languages without really understanding how and why.

When ***Hr*** is put forward through the numeral 1 He is: /zamba/ "small energy"
/*ùn*-ntouar/ "unique" giving the consonants skeleton ***ntr.***
This term / *ùn*-ntouar / shares the same skeleton of consonants with / *ùn*-nter / or / ù-ntét / designating the handicapped.

The script here transliterated ***wsir***[87] is often vocalized Osiris without any valid reason as if we are speaking about a Greek.

---

85. Samuel GALLEY, Fang-French dictionary. and French Fang followed by a fang grammar, ditions Henri Messeiller, Neuchâtel, 1964
86. See illustration N° 29
87. FAULKNER p. 68, Wb. I p. 359, 5

One can thus notice here without making advanced researches on the subject that the semantem refers, of course, to ***imn***. And as we have shown that ***imn*** is ***Hr*** . So ***wsir*** refers to ***Hr.*** As I pointed out, it is not a binary speech but based on a complexity logic. The ONE is in everything. And, the whole of all leads to ONE. Each part of ALL is a fractal representation of the ONE.
To understand this, we must inevitably get out of the conceptual matrix of binary logic: all or nothing.

The information contained in the Chester Beatty I Papyrus makes it possible to postulate that ***Hr*** would have lost an eye in a fight with his uncle ***stX***. Michèle Broze, Ugo Bianchi and many specialists repeat this approach in their writings

> *"It may be necessary to add that certain fundamental features of Seth's struggle with Horus, who is deprived of his eye, while Seth is mutilated in his genitals, may also recall episodes typical of the adventures of the North American trickster or other funny characters more or less related (for example, the South American jaguar)."*[88]
>
> *(Ugo BIANCHI, 1997)*

To circumstance's speeches, I refer to illustration N°29 giving a good plank of what should be understood by ***wsir***. The universe is spiraling and the center of the spiral resembles a vesica piscis, with a single eye.

This illustration makes it possible to visualize the manifestation of the cosmos which is approximated by René Thom's mathematical theory of morphogenesis. Indeed this one theorized about disasters and the relevance of its purpose was recognized by Fields medalist in 1958. This brilliant mind is known through publications like *Structural Stability and Morphogenesis* condensed into the single image of the spiral . The spiral describes how the Universe was created:
1- by symmetry breaking (physical theory) and the Riemann-Hugoniot catastrophe or the predation lace in mathematic;

2- The spiral described by the "S" cycle of René Thom has been reduced in Egypt in the shape of a lady with the left leg in genuflection that extends the right leg on a horizontal support!
Disaster (mathematical) or symmetry breaking (theoretical physics) leads to the appearance of new shapes unrelated to the old while the cycle "S" allows a selective mutation, without total rupture with the old shape.

Professor Bassong, who is attached to this theory at all, takes up all the schemas and their hermeneutical contents in his book entitled *The Mathematical Theory of Everything in Africa* published by Tamery Sematawy! His work joins the one of René Thom in Structural Stability and Morphogenesis.

Unlike some researcher of our time, René Thom sees that for this issue, it is in mathematics that a qualitative solution is possible and not in physics, because the quantitative solution is impossible to evaluate in fleeting reactions of discontinuity of matter. Since science can not experimentally observe or resume such experiments beyond the reach of man, the algebraic topology of which he is a specialist allows him to reduce, on an algorithmic level, the difficulty of physicists.

As a result, one can understand why the Negro-egyptian scholars considered the logarithmic spiral whose function is quadratic in the second degree. Because it allows an iteration in the dimensions 3 envisaged by Benoît Mandelbrot in fractal theory, or the set of Julia. Thus, the spiral, which appears in the natural shapes, fetal disposition of the embryo, DNA, spiral galaxy, etc

It is indeed a mathematics inscribed in the Universe, and therefore a metamathematics that the African already has in his brain algorithm! As Mbombog Bassong points out.

---

88. Ugo BIANCHI, Selected Essays on Gnosticism, Dualism and Mysteriosophy, Brill Academic Pub, 1997, P.153

Middle-egyptian: ***wsir*** "osiris"
Ewondo: /Tara/ "the father" if i only deal with semanteme
Ewondo: /ùn-ntour i-toa/ "holy seat"
Ewondo:/ùn-nsoh dys/ "the unique eye"
Ewondo: /dys/ "eye"
Ewondo: /*ùn*-nti/ "lord, majesty"
Ewondo: /i-sia/ "father"
Ewondo: /*ùn*-nsor/ "flayed; scratched"
Ewondo: /*ùn*-nkor/ "whoever dries up or saves"
Ewondo: /râ/ "energy" **Chinese**: 熱 Rè "heat"
Ewondo: /*ùn*-mbokh/ "One, disabled; hole; spiral"
Ewondo: /zamba/ "creative energy"
Ewondo: /i-yô/ "thorn, top, vomit"
Ewondo: /Elo/ "the head"
Etc.

It is therefore by metonymy that everything that appears crumbled or unitary shape, as seeds, will be associated with the pupil of the eye and we get the transliteration: ***wsjr***.

Expressions used to designate for example grains or seeds: beans, wheat, rice, millet, etc. simply repeat the meaning of "eye":

Ewondo: /M'-is m'*ù*-lés/ "rice grains" literally rice eyes
Ewondo: /mys m'ù-wondo/ "Peanut seeds" literally peanut eyes
Ewondo: /mys mə fón/ "Corn kernels" literally corn eyes
Etc.

Thus, in a hymn of the Sacred Texts of the Pyramids (Formula 273-274, No. 397), it is said about ***wsjr:***

"You are the father and mother of men, they live by your breath, they eat the flesh of your body".

There is a circulating myth stipulating that the "bread / cake" is called "Mesi" symbolizes Osiris and cutting it into several pieces symbolically represents his passion, his dismembering by ***stX***.

Ewondo: /bidi/ "food, breads, nourishment"
Ewondo:/me-si/ can be approached to "fruits of the earth" literally lands.

The transliterated symbol x3 of the Gardiner's list *t*, Variant of X4 (elongated roll, unwound) can be assimilated to the symbol N16 *tA*. Indeed, the earth can be called in the languages Ekang: / tsè / or / si / or / tékhè / etc

This identification of the foods to their origin has persisted in several languages and it is not by chance that we still speak of seafood to distinguish them from other foods.

This typical African expression basically, thus enlighten us on the origin of this idea, in the writings said inspired by God:

The Bible sometimes speaks, in a general way, of *fruits* (Ezek. 25: 4, 2 Timothy 2.6 etc.), or **fruits of the earth**.

(Deuteronomy 7.13, Wisdom 16.26, Mark 4.29, James 5.18); it is the production of either cereals (Matthew 13:26, Luke 8,8) or trees (Ezek. 36:30, 1 Maccabees 10:30, 1 Maccabees 14,8, Sirach 27: 6, Wisdom 10: 7, Matthew 12:33), which are sometimes called fruit trees (Genesis 1.11, Psalm 148.9, Ezekiel 47.12, Ecclesiastes 2.5), especially the fruit of the olive tree (Jeremiah 11.16), the fig tree (Luke 13.9), most often the vine (Hosea 10.1, Isaiah 37.30, 1 Corinthians 9.7 etc.), finally fruits from the gardens (Amos 9.14, Jeremiah 29.5, Jeremiah 29.28, Song 4.16). All these products of culture were the wealth of the country; their abundance marked his fertility, and hence his prosperity (Isaiah 37.30, Isaiah 65.21 and following, etc.). Humanitarian law in Deuteronomy prohibits the killing of fruit trees during siege wars (Deuteronomy 20:19). The Levitic law prescribes for God the tithe of fruits as of all the productions of the soil (Leviticus 27.30, Nehemiah 10.37); it declares uncircumcised, that is to say unclean, the fruits of the first three years (it was also a "taboo" for many primitive peoples), those of the fourth being destined for God and the following harvests for personal use (Leviticus 19: 23-25).

After this aside concerning the expression "fruits of the earth", on the philosophical level from the moment that an object occurs to the African consciousness as fragmentary, as a bunch of grapes, pineapple, turtle shell, dates etc. It gives the impression to have been divided, cut into pieces, into unitary portions.

The entire lexical field related to the semantics of division in our languages refers to what the ancestors called ***swty***.

> **Ewondo**: /à-saŋ/ "to cut, to carve; to stamp out; etc."
> **Ewondo:** /à-soŋ/ "to attach"
> **Ewondo:** /à-tsigh/ "to cut, to carve; to stamp out; to cut up; to indent"
> **Ewondo:** /ziŋ/ "hatred; wickedness; heating; odium etc"
> Etc.

Deconstruction

All these phenomena are known in molecular biology. When we speak of mitotic or meiotic division, we never mention the devil or satan, etc.

This is the way our ancestors had to qualify the universal principle of division that allows to pass from the whole to the part. They named this principe ***stX***. It is this rule that modifies a given state to allow the creation of a new state by a phenomenon of entropy. When one has a thesis to get the synthesis, in dialectic point of view, an antithesis it is necessary.

Metabolism is the action of anabolism (construction) and catabolism (destruction). For a better understanding, catabolism can be associated to ***stX***.

So instead of being in a fragmentary discourse, new concepts according to the scientific disciplines, operative concepts were invented by the ancients who make it possible to translate the laws that govern the universe and phenomena that are part of the harmony of the world.

Therefore, whether we are in the management of society, in the domain of universal laws, in myths or in the observation of the sky, we simply repeat these basic universal laws without having to resort to any jargon proper to each discipline.

That is why, what is called African spirituality is actually the African tradition. The Mbogh, the Kam, the Kommo, Bwiti, Vaudoo etc..

Texts written by african scholars (*Ani, Wnnfr, Xnmw-Nfr, Sbk, PtaH Htp, Mry- KA -rê, KA-gmw~n=j; ȝxn-Int etc.*), at ***km.t*** during the period of the fara were in the past, the source of inspiration for many spiritual philosophies current. These texts, which remain the first so-called religious texts in the history of humanity, are deviations from the ancestral African tradition. We can say this in comparison to what is happening today in our latitudes and especially compared to what is recorded in the sacred characters.

Middle-egyptian:  *iA* "to praise"*(Faulkner p. 7, Wb. I p. 28, 6 (iAw))*
Ewondo: /à-sêgha/; /à-kaŋ/; /à-wume/; /à-bara/ ; / i-kè/"to praise"

Middle-egyptian:  ***sSA.y*** "to Pray, to beg"
Ewondo: /à-sêghè/; /à-koghlèn/; / à-yəghlə/; /à-nyámdà/ "to beg, complain"

These words are not for only for lithurgical use. They are words of everyday life. But besides that, I took care to record a few sentences that led to a mystification and to establish a theocratic vision linked to African thought.

Middle-egyptian: *ink* PtH "I am the Creator"
Ewondo: /mə nə *ùn*-nkomo/; / mə nə *ùn*-mvəlghə/ "I am the one who reveals"

Middle-egyptian: nTr wa xpr m hA.t[89] "The beginning"
Ewondo: / *ùn*-ntouar y' à-bogh mə-kêh/ "the first at the beginning"

Middle-egyptian: ink nHH it n HH.w[90] "I am the father of the eternity of the eternities"
Ewondo: /mə nə kwam ya kwam i-sə/ "I am all eternity";

Middle-egyptian: nb r Dr[91] "the Master of the Universe"
Ewondo: / Zamba ya ndon / "The chief"; / *ùn*-mbwan ya nyol/ "body oil"

Middle-egyptian: nTr aA[92] "The great God"
Ewondo: / *ùn*-ntouar Zamba/ "the saint"

Middle-egyptian: nb nHH [93]" the master of Eternity"
Ewondo: / kwam ya kwam i-sə/ "I am eternity"
Ewondo: / Zamba *ùn*-nhəhəə / "the giver"

Middle-egyptian: nTr nfr[94] "the good God"
Ewondo: /*ùn*-nkor *ù*-ntouar/ "the only beginning"; nfr = 0 and ntr = 1
Ewondo: / *ùn* ntouar i-fàrà/ "the unique of the odd";

These phrases can be found in many texts compiled in the "book of the dead", such as the hieratic papyrus 3049 of Berlin p. VIII, XXIIth Dynasty, about 945-715 BCE, the Text of the Sarcophagus, Vol. VII, Spell 1130 but also in compilations of texts recorded in the excellent book[95] of Sylvie Cauville etc.

As we have just seen, there is no time when Africans talk about the entity GOD. It does not exist. There has been an introduction of exogenous dogmas knowingly or unconsciously that they do not fit the truth.

89. Dend. VI, 170.10
90. (CT II, 22a [78]; similarly 28d and 31d-e [80])
91. Dend. VI, 91,3
92. Dend. V, 13,14; Dend. VI, 72,10; Dend. VI, 60,3 etc.
93. [Harsomtous] Dend V, 40,15
94. Dend. V, 52,9 Dend VI, 108.2 1425 155,4
95. Sylvie CAUVILLE, Dendara V-VI, Peeters publishers, 2004

So all the incantations that are supposed to mean *so be it* like **amen** or I do not know what else, do not fit the endogenous facts.
This is mystifying and simply beyond understanding. Everyone will be able to check if I took any liberty with the subject concerning this question. But I do not believe it sincerely.

Moreover, ***Hr*** is the eldest ancestor, he is therefore qualified to be the solitary / itam / who is often transliterated ***itm*** and vocalised in Egyptian Esperanto **Atoum** for a reason that escapes the most elementary good sense. By plagiarism with the word / *ùn*-ntoum / which is used to designate the **cane** and therefore the leader. Hard to explain that based on trial and error is not it?

We recall ***Hr*** among the Ekang peoples / Zamba Elo /; / *ùn*-ŋkombodo /, / *ùn*-ntondobe /. I want to point out that ***Hr*** is not a person but an energy *nza* that is materialized in our space-time dimension in the infinitely small aspect / * bi /, / obə /. All that exists, is a manifestation of this energy. This energy enters in a process of expansion following a spiral. As it is expanded, it is described by scribes in anthropomorphic or zoomorphic figures. But realistically we are dealing with the field of advanced quantum science. So there is no place here for childishness and dreams or mysteries. What we are talking about here are bosons.

Since the (Arab-Berber) and Western (slave) raids, religion has been used as an immaterial weapon allowing the Arab and Western expansionist models to alienate and neutralize the African subject and then to smoothly folklorise the highly scientific African spirituality.

Religion as an immaterial weapon, acts as a planer of people's consciousness. It disables the sharpened area of mind vigilance that generally protects the ancestral tradition that is often misnamed spirituality. Thus, when a people loses its own spirituality, it loses its rudder and becomes porous to all breaths. To this day, the psychological excesses in Africa concerning the divine supposed on the individual and sociological level are unquantifiable.

In order to attract the attention of the true sons of Africa on this highly delicate subject, the great African thinker Anta DIOP, who had taken the measure of this danger, recorded as a testament to the future generations of Africans an intention tinged with extreme lucidity:

> *"It is reasonable to think that an African federal government will give equal support to the tenants of the ancestral religion, by provoking an ecumenical council and its priests, to allow the creation of a hierarchy, a better adapted liturgy, the training and the education of a caste of priests on a continental scale, the deepening and standardization of dogma on the basis of ancestral Monotheism. By doing so, the future federal government will protect the continent from any new insidious penetration from abroad, protect Africans from cultural alienation"*[96].
>
> *(ANTA DIOP , 1990)*

When Anta Diop wrote (1946-1960) *Alert within the tropics*, the data given by the French Academy of Sciences related to the distribution of the great masses on religions in Africa were the followings:

- Pagans ou traditionalists: 85 millions
- Muslims: 28 millions
- Catholics: 15 millions
- Protestants: 4 millions

With the demographic boom, things have changed, the statistics of 2010 offer another lecture:

Africa has 45% Muslim inhabitants; they represent 1/3 of the Muslims in the world.

96. Cheikh ANTA DIOP, Alert within the Tropics: articles 1946-1960 culture and development in Black Africa, Présence africaine, paris, 1990, P.122.

Their distribution on the continent is disparate: the proportion of Muslim population is around 98% in North Africa, but is about 30% in sub-Saharan Africa.
A very large majority is Sunni; Shia minorities, however, are present in Tanzania and Nigeria.

If in 1910, only 1% of Christians were in Africa, in 2010, we find about 23% of Christians in the world. Christianity remains a minority in North Africa, but has become the most widely practiced religion in sub-Saharan Africa (63%), ahead of Islam (30%) and traditional or so-called pagan religions. At the beginning of the 21st century, Africa is the continent where the number of Christians is growing quicker.

This development is due in large part to the growth of Evangelical Protestants, but also the other African Instituted Churches which represent a total of 36% of the population of sub-Saharan Africa, the proportion of Catholics being 21%.

Having taken the measure of Diop's alarm and checking for a check point, once again the prophecy of Diop as a great visionary has come true.

"[...] Both are feverish efforts to convert 85 million Africans whom they call pagans"
(ANTA DIOP, 1990)

The Pew Research Center survey of 25,000 Africans, surveyed in 19 countries between December 2008 and April 2009, reports that 400 million to 500 million is the number of believers each can claim Islam and Christianity. If we take into account the whole of Africa, there are about as many Muslims as Christians out of a total population of about one billion inhabitants (against about 200 million inhabitants there is a century).

In sub-Saharan Africa, the number of Muslims increased from 11 million in 1900 to 234 million in 2010. Christians have grown even faster, from 7 million to 470 million. Geographically, North Africa is massively Muslim and the south is predominantly Christian.

The small drawings, seemingly harmless and silent are great talkers who tell another story for who knows how to listen to them. Would not that be a reason why the scribes called them the ***mDw nTr***? One can proceed to the falsification of the History but one can not erase these characters which are self-sufficient intrinsically.

3.4.3.19. ***Example N°19:***  (Faulkner p. 37, Wb. I p. 163, 9-10)

I would like to redo a small focus based on the comparative angle between the conventional method and the endogenous approach of ***mDw nTr***.

Conventional method

Middle-egyptian: 
Transliteration: ***aA***
Meaning: *greatness*
Potential Vocalization: [*aa*]

Endogenous method

Sign O29 of the Gardiner's list: /ùn-mbokh/ + synonyms "Wooden column"
Sign D36 of the Gardiner's list: /wa/ + synonyms "Forearm"
Sign G1  of the Gardiner's list: /ndo/ + synonyms "vulture"
Sign Y1 of the Gardiner's list: /à-ndakh/; /à-fəb/ "papyrus"

How to analyze and understand this script 𓂝𓊢𓄿𓏛?

The Sign G1 𓄿 indicates the subject that interests us here.
The sign Y1 𓏛 does not indicate as the conventional method a simple idea.

<u>Little reminder</u>

The sign G1 𓄿, the Egyptian vulture in the Nilotic tradition, represents the mother. From the moment we understood this, the words used to name the signs O29 𓊢and D36 𓂝 and which render in our languages the idea of mother will thus allow to read correctly this script.

By the way the correct reading of this script is given by:

Ewondo: 𓄿/*nya/ "mother"
Ewondo: 𓄿 /nyâ/ "great" example: /nyâ dyàm/ "Great thing"

Metonymy

Ewondo: 𓂝 /i-sùkh/ "the hand as greeting organ"
Ewondo: 𓊢 /i-súkh/ "pillar, support, column"

Ewondo: 𓄿 /i-sukh/ "great"
Example: /i-sukh duma/ "The greatest and the greater of all"

Metonymy

Ewondo: 𓄿/nya/ "mum"
Ewondo: 𓁦 /à-sa/ "feather"
Ewondo: /à-sa/ "Great"

Metonymy

Ewondo: 𓁦 /à-sa/ "feather"
Ewondo: /à-sè/ "flapper"
Ewondo: / à-kwenya / "flapper"
Ewondo: / à-kwenya / "Great"

Example**:** /à-kwenya môr/ "big and huge (man or woman)"

Metonymy

Ewondo: 𓄿/ *ùn*-nsoŋ'o / "beak"
Ewondo: / *ùn*-nsoŋ / "summit, great"

Metonymy

Ewondo: 𓁦 / môr(a) / "woman, wife, person"
Ewondo: / môr(a) / "great"

Example: /môr à-boum/ "big belly"; /môr i-dyom/ "a great deal"

Metonymy

Ewondo: 𓆤/ fôŋ / "queen, bee"
Ewondo: / *ùn*-mfôn/ "great and big "
Ewondo: / *ùn*-mfaŋ/ "great, important"

Example: / *ùn*-mfaŋ môr / "important person"

Metonymy and phonetic proximity

Ewondo: / *ùn*-nòn / "the bird"
Ewondo: / *ùn*-na/; / *ùn*-nana / "the mother, the mum/
Ewondo: / nén/; /nənə/"Great, big, important"

Example: / mor nén / "the greatest, the most important"
/à-nén mor/ "fat man"

Metonymy and phonetic proximity

Ewondo: / *ùn*-nòn / "the bird"
Ewondo: / *ùn*-nôm /; "Great, big, important"

Example: / *ùn*-nôm à-dzô / "important speech"

Metonymy and phonetic proximity

Ewondo: / *ùn*-nòn /"the bird"; /i-mbgaŋ/ "raven"
Ewondo: / *ùn*-nom /; "old, great"
Ewondo: / i-mvam /; "great age, old, ancestor"
Example : / i-mvam dyama /; "my Grand parent"

Metonymy and phonetic proximity

Ewondo: / i-vəgha / "the endowed"
Ewondo: / i-vogh /;/ /i-wogh/; /woghbe/; / vughbe /"Big, loose, full"

Example: / i-vogh à-bum / "big belly"

Ananym

Ewondo: /nya/ "mother"
Ewondo: /à-yeŋ/; /à- yeŋle/ "too big, loose, full"

Quid: And if the sign Y1 did not just indicate an idea according to Egyptology?

Middle-egyptian: The sign Y1 "Papyrus wrapped, tied and sealed"
Ewondo: /à-ndakh/ "papyrus"
Ewondo: /ndah/ "house"
lingala: ndáko "house" literally the foot(-ko) of the house (nda-).
Ewondo: /ndoé/ "vulture"

Middle-egyptian: The sign Y1 "wrapped Papyrus, tied and sealed"
Ewondo: /à-fəb/ "paper, papyrus"
Ewondo: /à-fab/ "wing"
Ewondo:: /à-fab/ "wing"

MAa.t:feather or wings

Middle-egyptian: The sign Y1 "wrapped, Papyrus tied and sealed"
Ewondo: / à-wôghle / "attach"
Ewondo: / i-vogh /;/ i-wogh/; /woghbe/; / vughbe /"Big, loose, full"

Middle-egyptian: The sign Y1 "wrapped Papyrus, tied and sealed"
Ewondo: / à-vughlu /;/ vughdu /"wrapped"
Ewondo: / i-vogh /;/ i-wogh/; /woghbe/; / vughbe /"Big, loose, full"

So I let you draw your conclusions. If could give one advice, my requirement will be to know an African language to be able to give an accurate point of view on this subject with a great peace of mind.

Technical explanation

If by transliteration ***aA*** things look simple, it is usefull to know African languages very well to read and understand the ***mDw nTr.***

On the technical point the transliterated sign G1 of Gardiner list ***A*** derives from the proforma * **kh** of the Negro-egyptian.

This sign associated with the sign D36 /wa/ will thus give in the African languages:

> Case N°1: Wa+h
> With h as a weak consonant giving the version sought by classical Egyptology [aa].
>
> Case N°2: Wa +k
> With all the weakened variants of the velar consonant K in the different linguistic variants and explaining then the plethora of words which one observes and that would be difficult to explain from aA.
>
> Case N°3: Wa + kh

Diachronic linguistics can be illuminating by some aspects. But we think that the characters themselves are facts loud and clear for the one who know african tought and languages well in order to absolve the constraints of the official methodology which impose a transliteration. In my opinion this method is limiting.
The different meanings transliterated ***aA*** below can thus be easily read if the method is understood. Without making any important developments we say in wondo:

1 **Ewondo**: /*ùn*-ndî/; /i-yəkh/; /*ùn*-mbôkh/; /*ùn*-nkoum/ etc. "column; pillar"
2 **Ewondo**: /*ùn*-mfaŋ/; /à-boui/ etc. "great; abondant, multiple; rich in; groin"
3 **Ewondo**: /*ùn*-nti/; /*ùn*-ntô/; /ù-su/; /*ùn*-ntomba/ etc. "notable; eldest son"
4 **Ewondo**: / *ùn*-nya/; / *ùn*-mfaŋ/; /nén/ etc."great"
5 **Ewondo**: /i-nyi/; /i-nyuo/; /à-wé/; /và/ "right here; there, over there"
6 **Ewondo**: /i-nyî/; / *ùn*-mbèh nda/ "entrance, house door"
7 **Ewondo**: / *ùn*-mbôkh/; /ù-zələ/; /i-djakas/; /i-tib'li/ "donkey, jackass"

So you can easily read these scripts in all African language variants without any problem since you have understood the method.

3.4.3.20. Example N°20:  (Faulkner p. 260, Wb. IV p. 405, 7 10 (SAj))

Conventional method:

Middle-egyptian: ***SA*** "pig"

Endogenous approach
Sign M8 of the Gardiner's list "i-sam à-yaŋ bi-lobi" "Lotus flower"
Sign G1 of the Gardiner's list /ndoé obama/; / *ùn*-nya/; / *ùn*-na/ "vulture"
Sign D52 of the Gardiner's list /byèm bi fam/ "phallus"

Sign M8 of the Gardiner's list

This sign is supposed to be a basin with lotus flowers. But we think that this sign besides the lotus also represents the Cyperus *papyrus* an umbelliferous plant that grows in the swamps of the Nile delta.
We call it in Ewondo: /i-soŋ /"*Pennisetum purpureum*", when it represents the lotus, we call it / à-ndakh / "Cyperus *papyrus*"
This plant indicates the biotope (marsh) in which evolves the animal represented by the sign E12.

The word / i-soŋ / the sign M8 of the Gardiner's list indicates a strong link with the idea of sharp or sharp because of the homophony with / *ùn*-nsoŋ /. This word is used to designate what is pointed or sharp. This root also indicates the tooth / à-soŋ / or the worm. We have in this account:

Ewondo: / *i*-soŋ / "Pennisetum purpureum", Biotope
Ewondo: / *ùn*-nsoŋ / "*snout*"
Ewondo: / à-soŋ / "*teeth*"
Ewondo: / *ùn*-nsoŋ / "*worm*"

The sign G1 of the Gardiner's list clarifies us that what is pointed is the the bucal device even the teeth. Indeed, the vulture is a bird with a pointy buccal apparatus / *ùn*-nsoŋ / "beak".

Why do we have flowers?

The flowers / à-sam / are another way to represent the feathers / à-sè /. These two symbols are related to the mother: / *ùn*-nya /, / *ùn*-na /. Maâ.t the mother, are symbols of the vulture, the feather, the wings. The mother having a nurturing function this animal can be considered as being used for food. We can also think that, the scribe indicates that this animal is a mammal too, since it has a nourishing function.

The sign D52 of the Gardiner's list here specifies the animal's tendency for reproduction. We have a confirmation because the flowers are reproductive organs of the plants too. More than that the phallus can share the homophony with /i-soŋ/ with /*ùn*-son/ "penis".

The word for *multiply* is in ewondo is /à-fi /; / à-fie /, / à-fia / "to multiply reciprocally"
/ à-wun / "multiply" refers to the homophone / à-vun / meaning to search. That means the sign E12 is also characterized by its ability to search. But the root /*fi-/ can lead to the word /fiè/ "scrubland" indicating once more the biotope of the animal.

The animal represented by the sign E12 the mother / *ùn*-nya / offers a homophone with / á-nyè / meaning to breastfeed. So we have confirmation that it is a breastfeeding animal.

Ewondo:/ à-bye / "to multiply; to give birth" this term admits an homophone in Ewondo: /byé / designating the nails. The animal would be an ungulate[97].

The sign D52 of the Gardiner's list represents a phallus with testicles. The phallus and the root word are interchangeable in our languages. The testes are considered as nuclei in Ewondo: / mi-mbaŋ / "nuclei; testes"

Since the root has a nourishing function for a plant, this sign also describes the diet of this animal, namely the roots / mi-'nkaŋ / and the seeds / mi-'mbaŋ /.

97. An ungulate is a mammal whose last phalange of finger or fingers is wrapped in a hoof.

We therefore take note of the richness of an endogenous analysis of a script. A script from my point of view is like a cartography of a given reality. It is not a simple association of consonants but a summary of a reality expressed by the scribe. We could continue the analysis in this case and find elements that are not perceived by a simplistic analysis of the characters.

Reading: 

As depositary of the land, the mother shares with the panther the same signified. In the Ekang languages we often use words to describe the panther, the earth and woman mobilizing the roots: / ng- / and / * z- /. In fact we will expect at least to name these animals using this base.
According to the swamp biotope, by associating it with papyrus / à-ndakh / we identify this animal with the black pig because the house / nda / of the foot / ko / or by the fact that the pig is in an enclosure / à-kakh / "Barrier, enclosure , palisade "so we have:

Ewondo: /i-ŋgakh/ "wild pig"

So we have from right to left: the place  / *ko- /; / *to- /; / nda / from the mother  /ng-/.The two words seem different / * ndakh / and / * ngakh /. If we do understand that the mother is also the wife or the wife and the house, then we understand that it is the same word.

Ewondo: : Nga = nda = *nya = *na = zé

According to the marshy biotope, by associating it with the lotus / à-yaŋ *bi*-lobi /, the sign G1  specifies that we do not speak of the variant / à-yaŋ / but of its consonantal methatesis. We identify this animal with the black pig because its house is in the enclosure / ngoui / and call it to this account:

Ewondo: /ngoué/ "pig; hog; swine; dirty; disgusting; etc."

Note : **hog** and /ngoué/ are dealing by phonemes inversion.

To understand why it is necessary, well speak African languages indeed. The term that serves as a class prefix for places is / ko / "foot" or / bo / "foot". But the foot is a body member also. It also admits as a qualifier of the term / *i*-nam / "member" homophone of / *ùn*-nam / "place, country, village, city etc."

However, since the arm is also a member, it is called / wa /. So it is synonymous of the foot in this logic. By replacing the foot / ko / by his alter-ego, I call here the pig by associating the root / ng- / for the mother and the suffixale / wa / for the arm.

According to the shape of his bucal device its called:
The house of  ***Hr*** is outside: / *ùn*-nsəŋ/

Ewondo: / *ùn*-nsoŋ'o / "the snout, the truffle" literally the sharp

This meaning offers a phonetic proximity with / *ùn*-nsaŋ / "band, horde".
His oral apparatus is often plunged in the ground searching his pittance, it gives the impression of a greedy. He is called:

Ewondo: / bəmə / "the pegged"
Note : May you take note of the phonetic proximity between *pig* and *peg*.

When we do speak the African languages all this is quite logical. The item for housing that meets the descriptors we have mobilizedh ' i-bəm / "the hute". The charade consisted in looking for the name of the foot's dwelling meaning **Hr.** T wer is therefore given by the indications recorded by the scribe specifically the little house since ***Hr*** is the smallest.
A small house is a hute. The hute in Ewondo is /i-bəmə/ "The hute".

Let us note by the way that / bəmə/ is in the Ekang collective consciousness a legendary man; a kind of bogeyman. He is a greedy ogre who eats everything he finds. His wife's name is Nyôbôla. Syn.: Byeme, Byôme.

The term / ù-nsoŋ / is an homophone of / ù-nsóŋ / "the worm" which represents ***wsjr*** or ***Hr***.

According to his association with ***Hr*** as the numeral one, the smallest of the trinity, when he characterizes a solitary male, he calls it

Ewondo: /i-fal/ "Pig, hog, sorting" literally the ONE / ù-bə /; / * Fal /.
The term / *fyal /"to sort" is from afar a derived variant of / á-fakh / " to dig".

Once again without a good practice of African languages all this will remain rather nebulous. I still remind that the woman, the panther / leopard and the earth are linked in african thought. The place of a wild pig is the bush / fiè /.
The generic name for cats is: / Foe / or / fole /.
Moreover the word place is said in the languages Ekang / à-fola /. The pork place is often behind the house / fala /. This place is also used to name pork.

So, let's summarize: mother quadrilogic equation = panther = earth = pig

Ewondo: /à-fola/ "place"
Ewondo: /à-fôe/ or /fole/ "cat, panther"
Ewondo: /i-fal/ "pork, pig; hog; swine"

According to his association with ***Hr*** as the hidden one:
Living outside, the rain is used as shelter for wild pigs. The rain is called / mvəŋ /.

Ewondo: / i-mvaŋ / corresponds to average pigs after losing their welts.

Hr being the eldest ancestor he is / i-vam / if this animal is the incarnation of ***Hr*** so he also brought under his banner. ***Hr.*** He is the ancestor in our languages.

According to his association with ***Hr*** he is identified by his mother's name. The mother of ***Hr*** is Ndo the vulture. Then the script can be read:

Ewondo: /d**o**r**a**/; /d**oro**/; /T**oro**bio/

Since ***Hr*** seems to have lost an eye in the fight with his uncle ***stX.*** He can be called:

Ewondo: /ù-fìfìkh dys / "small eye; pig eye"

Just for information, the mother as ***maA.t*** is symbolized by the wings or the pen. We call / fufukh /

the feathers mainly the ones under the belly of the birds.

/ Fîfîkh / is the little pig's eye, a little natural eye, or an eye made by ***DH.wty*** according to the mythology of the African people

So here are some elements of analysis on the script about pork. Transliterated ***rri;*** ***rrt;*** ***rrwt*** variants seem therefore, to be mere evidences for reading. We recognize the chick ***Hr*** and the mouth that characterize ***Hr*** and pork.

Technical Sheet

If for current egyptology this script is only ***SA*** a pig, for us is more than that. This script is a technical sheet.

The scribe informs that this animal is charactezied in this way:

Classification

Characteristics	sign	Sign reading	Ewondo Driver	lexicon
Digitigrade		**Nya: mother*	*mə-nyu*	fingers
Hoofed		*À-byé: birth*	*byé*	nails
Mammal		**Nya: mother*	*À-nyè*	Breatfeed
Sorter		*Foe; à folà*	*i-fyal; i-fal*	pig

Chart 74: pig script classification

Feeding

Characteristics	sign	Sign reading	Ewondo Driver	lexicon
Herbivore		*À-yàŋ i-lobi*	*i-səŋ*	Pennisetum P.
Herbivore		*À-yàŋ i-lobi*	*À-kyé*	leafs
Carnivore		*À-yàŋ i-lobi*	*i-lobi ùn-səŋ*	worms
roots		*Nya: mother*	*Ùn-ŋkàŋ*	roots
Kernels		*mi-mbàŋ à-bin*	*Ùn-mbàŋ*	kernels
Milk		*Nya. mother*	**nyàŋ*	milk

Chart 75: pig script feeding

Biotope

Characteristics	sign	Sign reading	Ewondo Driver	lexicon
The beak		*Ùn-swàŋ'à*	Ùn-nsəŋ	Outside, yard
The pointy		*Ùn-nsəŋ*	i-bəm	hut
The cat, panther		*foe*	fyè	bushland
The cat, panther		*fole*	*fala*	backside
Lotus		*Ayàŋ-ilobi*	*i-ləbi*	Swamp; marsh
Full of water		*Ayàŋ-ilobi*	*mvəŋ*	rain

Chart 76: pig script biotope

Life Style

Characteristics	sign	Sign reading	Ewondo Driver	lexicon
The beak		*Ùn-swàŋ'à*	Ùn-nsàŋ	Bunch, group
Beak-place		*Ùn-nsəŋ*	Ùn-nsàmbà	Pack, group
Nya Metathesis		**nya*	i-yəŋ; à-yoŋ	Pack, group
The cat, panther		*Fole; i-bàm*	*Ùn*-mbàm	Bunch, group
The Mother		**nya*	*Bor*	bunch, group

Chart 77: pig script life-style

Note: The mother symbolises the multitude in the african thought.
Ewondo: /bor/ "people; multitude; group, bunch; pack; bevy etc."
English: *boar* "wild pig"

Ewondo's Names

Characteristics	sign	Sign reading	Ewondo Driver	Lexicon
		Ùn-ngà	Ngoué	pig
		Ùn-nsɔŋ	i-fyal	pig
		À-folà ùn-ŋgal	I-ŋgakh	Pig; boar
		fufukh	*Ù-fifikh dys*	Pig-eye
		Ndo-obama	*Torobio/doro*	pig
		mvəŋ	*i-mvaŋ*	pig
		i-bəm "sting"	*bəmə*	pig

Chart 78: pig's names in Ewondo related to the script

I can put forward other characteristics given by this script related to the present animal like its harmfulness for human being, its sociability, if human can eat them; weight, dimensions etc.
According to me this script is a technical variant of the pig. I just wanted emphasize by reading this script that the level of information provided by the egyptology standard is excessively limiting if not nil.

<u>Etymology</u>

In the language of the exercise; Ewondo one of the words used to name the phallus is:

Middle-Egyptian: *SA* "pig"
Ewondo: /*ùn*-nsor/ or /*ùn*-nsot/ "penis"

By characterizing this animal , we have said that it's is a kind of *sorter*. The snout pegged in the ground, it looks for worms and roots for feeding. Then the word to ***sort*** can then find it's origin in Africa with the phonetic link with the word /*ùn*-nsor/ "penis" or all the anthroponyms related in the african language. Except if we are blind, we can see the concerned penis on the script above. The English word *sort* is a pig metonymical word which etymologically is a penis in the african language.

African origin of the word Pig

What we can observe concerning the English word *pig* is the fact that, it's an ananym of the word Beti /*ùn*-ŋgup/ "hippo". I claim that regarding also this script: xAb "hippo". You have more or less the same signs except the determinative.

For a long time, and before deep genetic researchs, hippos were traditionally stored in ungulates before phylogenetic studies revealed them as ketartiodactyls. The studies show also that they are relatively close genetically of cetaceans and to a lesser extent suidae and Ruminantia.

African origin of the word swine

Middle-Egyptian: *SA* "pig"
Ewondo: /*ùn*-nson/ "penis; meat"
Ewondo: /*ùn*-nswaŋ'a/ or /*ùn*-nsɔŋ/"the beak"

Then we can connect the word *swine* to the letterings of this scri used as african basic lexicon words also. But *swine* is also related to its homophon *swan* by this sign the beak /*ùn*-nswaŋ/ no scramble…

May you take note that :

Middle-Egyptian: *SA* "pig" ☞**English**: *swine; sow*
Middle-Egyptian: *sA* "kid" ☞**English**: *son*
Middle-Egyptian: *wsA.t* "duck" ☞**English**: *swan*
Middle-Egyptian: *Sw* "solar" ☞**English**: *sun*
Middle-Egyptian: *Sw* "god Shou" ☞**English**: *senior; schlong*

African origin of the word hog

Middle-Egyptian: *SA* "hog"
Ewondo: /*ù*-kyé/ "leaf"
Ewondo: /à-koh/ "place" litt. foot

Then we can connect the word *hog* to the **Ewondo** word: /*i*-vogh/ or /*i*-wogh/ or /*i*-hogh/ "fat; huge; large; big".

The ananym of the **Ewondo** : *ngoué* "pig" is *hog* "pig".

African origin of the word hog

Finally the word pork can be approched to the word /i-kop/ "skin". The evidence is given by the following script: *bA* "leopard".

The mother is the leopard /ŋgué/, the place and the skin . Pork and hippo have thick, greasy skin. It is therefore normal that by metonymy the African has identified this animal by this characteristic. *Pork* means skin according to the african common language.

African origin of the word sow

The mother is the leopard , the place and the skin . We do call the earth in Ekang languages:

Ewondo: /Si/ "earth, soil, ground"
Ewondo: /á-sɔh/ "to come "
Ewondo: /*ùn*-nsɔh/ "naked; the one who comes" ☞**English**: sow "pig female"

You have then the real african etymology of the English word *sow*.
In the same framework to think, note that the word *ground* corresponds to the French word :*Goin* " snout; pork's mouth".

Pork and nudity

In the Romanesque period the body as such does not exist. The soul is always closely associated with it and the concern for the hereafter is widely shared.
Imagery uses metaphors that their extreme simplicity frequently prevents from noticing. Medieval stonecutters frequently portrayed the human soul as a small child *Hr*. If a dying man lets his soul

escape by the mouth in the shape of a small naked child ***Hr***, it is because he breathes a last breath and because death is a "new birth" post-mortem. Nudity meant not a carnal reality but a spiritual dimension. This is to say that nudity in Romanesque imagery frequently loses its erotic dimension. The iconoscripty of the soul can be grouped under several categories: - the illustration of the spiritual fight, -the representation of the soul of the dying person, - naked creatures and scenes of devoration. This last category is related to the pig for its gluttony.

Two naked children - Images of the soul - are seized by a **pig** and a **snake.** We are in attendance of an evocation of the evil of which these **young beings** could be the toy.

Capitals of the portal of the Lauterbach's church, Haut-Rhin

Note: For the word boar "wild pig" just see above on the preceding parascript the african etymology of that word *boar*. It just means multitude in the african common language.

Etymology of the word Naked

Middle-Egyptian: ***SA*** "hog"

Except for blinded people, we can notice the presence of this symbol meaning nude or naked. In Ewondo language we name this sign:

Ewondo: /*ùn*-nkàŋ/"the penis; the root"
English: *naked "nude"*

The word *naked* is just a consonantic metathesis obtains par phonemes inversion of the Beti word /*ùn*-nkàŋ/ just check by yourself please. It is for this reason that you will get a metaphysical relationship between pork and nudity. But all this is clear from the african point of view.

Particular case

Middle-egyptian: ***ipH*** *"pig"*

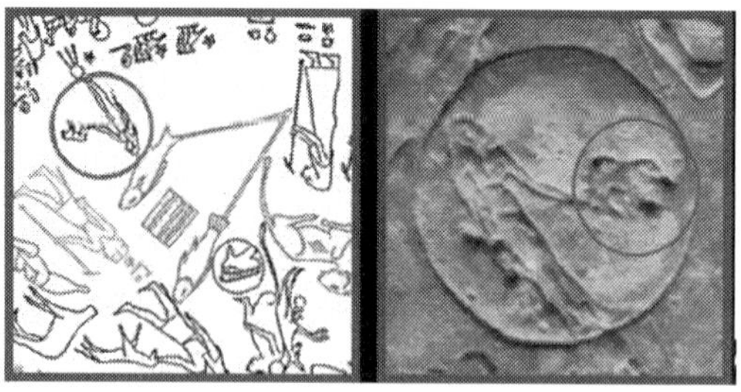

We think we have to associate this script with something else. It clearly represents a door and belongs to the domain of cosmography. As it can be seen here on Denderah zodiac, a suidae seems to be held by a woman's hand. This representation is the constellation of the whale.

The constellation of WHALE or PORK-FISH is in the company of a young boy, ***Hr*** avatar, he becomes captive and pourer of water (or milk) in the Greco-Roman versions. On the Denderah zodiac, CETUS[98] with horns has no head, the constellation is a combination of a suidae and a big FISH and also represents ***stX***. It is just above the Lunar Circle where ***As.t*** hold the leg of **stX** the Pig-Boar.

98. Cetus would be a fantastic marine creature of the Greek and Roman myths, also called Keto, which gave its name to the animal of the medieval bestiaries cete. The constellation of the whale that would draw its name from the same creature. Without doubt a posterior versin of stX.

During the ninth month ***pA n xnsw*** "Pachon" of the)tic calendar, at the full moon, a pork was sacrifice ring a : in memory of the myth where ***stX*** was transforme :o a wild pig that was chasing ***wsjr***. ***wsjr*** took refu the moon but also the myth where ***stX*** always under the appearance of a suidae hurt the eye of ***Hr*** eye which symbolized precisely the Moon. The word for scar in Ewondo is /i-fel/ and the pig /i-fyal/ of /i-fal/

In ancient times, FISH represented the month of ***pA n xnsw***, the ninth month of the Nilotic calendar (based on the Nile flood). This is the first month (pA n int). of the season ***Smw*** "chemou". In the calendar of ancient Egypt, the month of ***pA n xnsw*** is that of ***xnsw,*** "Khonsu" the traveler "or" the one who crosses the sky. "His main worship center was Thebes started in the 20th Dynasty by ***rA***- Mss III. Later, he becomes a healing god known as "***xnsw***, the Counselor." At Kom Ombo, xnsw, was supposed to be worshiped as son of **Sbk** and **Hat-Hr**.

After clarifying the level of understanding required for each of the scripts the notable differences on the signs and their real meaning.

So the question arises whether the kmtyw eat or not pork.

"Pork" another facet.

Alienation (from the Latin: alienus, which means "other", "foreign") is generally understood, in philosophy, as the dispossession of the individual, that is to say the loss of his control, his forces for the benefit of another (individual, group or society in general). He thus frequently refers to the idea of an inauthenticity of the existence lived by the alienated individual.
I do not want to deal with alienation in what follows, but the ban on pork consumption in religious book since the Nilotic cradle.

I want to point out that, I am not attacking people's faith here. I do not want to disrespect any believer far from it. I apologize in advance if this statement is made to be perceived as vexatious or offensive to the reader. I am simply trying to exchange scientifically. So I'm not talking about faith and beliefs, but about technical data that can be verified by healthy minds with common sense.
Two things have drawn my attention to the subject of the Abrahamic religions practiced by blacks, the fact that it is Semitic cults (barely 5000 years of humanity) first and secondarily the refusal of the blacks practicing forced of these cults to question their history (150,000 years) for the comfort of speech imposed by others. Speeches that are easily revisable today in the light of new scientific knowledge.

If we believe those of ours who have become Semitized since the Arab colonization (+639 AD), they would be able to consume all types of food, toads, cockroaches, snakes and other scorpions but in no case, they would consume pork because their religion considers this animal as impure.

By the way, this food would be stamped with the seal of the prohibition. They describe it in their new lithurgy language as Haram. At this level we are struck by the fact that we dissolve our own personality to limit ourselves to applying what others have thought for you. We like to believe that the negro has never had a history and owes his salvation to those who say they speak with God for his salvation.

From the theological point of view, one might well ask the question of why the pork that enters the human diet as many other meats would be more unfit for consumption than another and would you the close the doors of salvation if all that "God" created, he considered it to be good? (Dixit Genesis 1: 20-22)

> *"Genèse 1:*
> *20.God said, Let the waters produce abundantly living creatures, and let birds fly on the earth to the expanse of the sky.*

> *21. God created the fish greats and all living animals that move, and the waters produced abundantly according to their kind; he also created every winged bird after his kind. God saw that it was good ...*
> *22. God blessed them, saying, Be fruitful, multiply, and fill the waters of the seas; and that birds multiply on the earth ...)"*

This simple question puts confusion to the logic. It is therefore more appropriate to look elsewhere for the rational foundations of this prohibition.

Since there is no spontaneous generation, each new civilization builds its bed on the ashes of those who preceded it and, strictly considering that from the point of view of historical chronology the religion that forbids it for consumption is the last "so-called revealed religion", it is more advisable to go investigate in the religions that preceded it to have their point of view on the question of the consumption or not of pork.

Before going further, let us question the most great African historian (born in a murid Islamic family) Cheick ANTA DIOP about the "divine revelations" of this religion.

> *"[...] All the elements necessary for the blossoming of Islam were thus in place already more than 1000 years before the birth of Mahomet and Islam will appear as a "purification" of the Sabeanism (spirituality of the Sabeans) by "the sent from god".*[99]
>
> *(ANTA DIOP-1954)*

Before going farther i'm gonna show you something:

Middle-egyptian: 𓐍𓄿𓃀 *xAb*"hippo"

According to egyptians hippo and pigs were considered as similar. That's why i took the hippo's script. But you can take the pig's script it will not change anything to the reasoning.

Due to the African matriarchy, the mother is the symbol of the earth. The semantaxe associated with the earth is linked to the 𓃀 foot as can be seen in the script of the hippo above.
But the word 𓃀 *foot* is in Ewondo: / à-boh /. This term / à-boh / allows the homophone with the word / à-bh / "evil, ugly, dangerous ". Do you undestand me? But il allows also the homophone /ù-bə/ "small, child". Then you will have a better understanding of the relationship between ***aApp*** the evil and the ***pig*** as ***aApp's*** sons.

Does it seem clear to you now, i do hope so. This thus illuminates with a new eye a fundamental reason related to African thought that associates *pork* with *evil*. So if pork is evil it is therefore prohibited to consume as a result. Just for fun note thant all the animal dangerous for human beings was qualified by the negro-egyptian word /i-mvul/ enlighting here the real sens of the english word *evil*.

Rigorously, we can not talk about any revelation or rely on Islam in an intrinsic way to justify the non-consumption of pork by some blacks embigaded into religious prohibitions. As such, on the pure historical level, Sabeism and the religions that precede Islam would thus give more information on this subject.

At this stage and without insulting our Semitic Black brothers, we are fully convinced that no black Muslims even with good faith and the most serious could give a fair and honest challenge on the issue that interests us here by relying solely on Islam without shedding fantasy, fanaticism and ramblings.
Since Sabeism is the father of Islam, and being an Osirian cult practiced in the region of Saba by black people, let us go to the source and ask ourselves the question of knowing what *Kmtyw* thought about it?

---

99. Cheikh ANTA DIOP, Negro Nations and culture, ditions Présence africaine, Paris, 1979, p.195.

We do say that since long time in Africa, from the point of view of the consumption of animals, there are different types of animals: the animals for which one has disgust. Those animals are called:

Ewondo: /tum/ "disgust; Impure, dirty, animal that we must not eat"

Here are some of them below:

Ewondo:/Nyo/ "snake",
Ewondo:/mvakh/"Variety of civet",
Ewondo: /nsw/ "lizard",
Ewondo: /fôe/ "felids",
Ewondo:/à-kuŋ/ "owl",
etc.

Ewondo:/wa'ha / "chimpanzee",
Ewondo: /kulu/ "turtle",
Ewondo:/Mvu/ "dog",
Ewondo:/ŋgəm/ "bat",

Beside these animals some are prohibited to the particular consumption as wild **boar** for women, then the word is not yet *tum* but *eki* "prohibited".

Animals forbidden for consumption and animals for which there is no prohibition or disgust.
As we can see, this point of view is neither foreign to Africa nor singular. Aside of the previous animals, we do not eat totemic animals. You can note that most of the disgusting animals are the one frequently represented on scripts by the scribes.

For a long time, pork has been eaten in Egypt. As the scripts show, namely a food for consumption.

Historiographyically, Egyptian pigs have been domesticated in the Delta since the predynastic period, although traces are already appearing in Neolithic cultures from the wild pig / i-ngakh /, a species that is easily tamed. His back is covered with hard, bristly silks, thin legs and very long snout. His race seems to be between that of wild boar and that of today's pork. He lived either in wild horde in the human environment, or in domesticated herds, or in small group type "pork farmer" in the families. The presence of bone remains found on many sites leads to the conclusion that its meat is the most consumed in the Neolithic era and occupies an important place in the diet of the Old Kingdom.

According to Yuri Volokhine, who wrote a magnificent opus on the subject[100], he mentions the pigsty of Tell el Amarna and the bone remains found on various sites, including that of Kom Firin, in the Western Delta, occuped from the Ramesside era to the present day. saite period, where the excavations have revealed many animal remains (among which pig bones account for 44% of the total, which is considerable). But it is to note that "the inventory of all the findings remains to be done", that "further systematic research" should be carried out.

In spite of these details, the pork is scarcely represented on the walls of the chapels, the animal is never represented in the religious scenes nor mentioned in the lists of offerings, whether in divine worship or in funerary worship; Besides, he is never mummified. It is sometimes represented as in some tombs in Beni Hassan, El Kab (Renni and Setaou) and Thebes. It is totally absent from the remains of the funeral meals that took place there. Its figurative representation remains insignificant: pigs are never represented on the bas-reliefs of the mastabas, where herds of cattle and even donkeys figure prominently in the evocation of the property of their rich owners. However, the possession of herds of pigs is mentioned on steles of the Middle Kingdom: one could therefore wonder why the Egyptians, who, very early, remarkably observed and represented the behavior of the animals, were so little interested in the image of an animal that was obviously quite familiar to them.

100. Youri VOLOKHINE, Pig in ancient Egypt: myths and history at the origin of food bans, Presses Universitaires de Liège, Liège, 2014

It is nevertheless considered as a butcher's animal, and pork is mentioned in many sources of the Middle Kingdom, alongside other livestock.
Pig is even a fundamental animal in ancient Egypt because it is one of the important sources of dietary protein since the Old Kingdom (archeology has shown). Italian researchers have found (November 2005) signs of cysticercosis[101] on the stomach of a mummy. Pork was therefore an integral part of the meals: so Egyptologists have they discovered, on the Neolithic sites of Fayoum and Merimde, to mention only these two, bones of pigs among other culinary waste, which would tend to prove that they had been well consumed by the natives.

How of a real world, where the animal is still present but less valued, at least at the level of images, we move to a mental universe populated with rich and complex myths That's the question we can legitimately ask?
The answer lies in the Sarcophaguss Texts of the Middle Kingdom. This is where the myth of the black pig, the culprit of a "cosmic crime", appears, because it would have hurt the eye of ***Hr***: the moon, whose phases of darkening are seen as so many injuries caused by his enemies, ***Aapp*** and ***stX***. Pork is one of his avatars. The same story appears, with some variations, in the Books of the Dead of the New Kingdom.

However, we find a kind of ambivalence in mythology. One negative because of its assimilation to ***stX***, and a positive through the sow.
The Egyptian had noticed that the sow could eat her cubs (a type of self-regulation according to the mass of milk to give) and assimilated it to Nout swallowing the sun. This point lends itself to other interpretations of swallowing piglets, not to destroy them, but, on the contrary, to regenerate them; she is therefore likened to Nut, the goddess of heaven, and her children are the stars. In this case again, the myth interprets cosmic phenomena. Under the name of Reret ("***The Sow***"), she appears as a maternal, nourishing figure; it is similar to the ***hippopotamus*** goddess ***tA-wr.t*** "Taouret". In the case of pork as in that of the hippopotamus and their figurations are not always easy to distinguish.

The positive and beneficent connotation of the female is opposed to that of the male, negative and dangerous. Nevertheless, despite its legendary malice and gluttony, mythology will make it a globally harmful animal also associated with ***stX.*** Indeed, ***stX***, to steal and devour the moon, the left eye of ***Hr*** son of ***wsjr,*** would have turned into black pork.

This assimilation to ***stX*** made the animal a religious taboo whose unpopularity followed that of the god. At the time, the god Seth no longer carrying any negative values, the image of the pork became very negative too, all his behaviors were ridiculed and in particular his fertility or his way of feeding (it persists nowadays in the language even at home!)
Thus, in the temples, the pork is considered a Sethian animal by the officiants, remember that it is never represented in the cult scenes nor mentioned in the lists of offerings, whether in the divine worship or in the funerary cult; Besides, he is never mummified.

By the way, according to Herodotus, Greek historian and geography considered as one of the first Western historians, the porchers are the only ones who can not penetrate the enclosure of a temple. However, even on the religious level, the exclusion of pork is not absolute. Herodotus had already noted that it could be sacrificed "in honor of Selene and Dionysus" at the time of the full moon. The pork indeed appears as a sacrificial victim in the context of lunar festivals and Osirian feasts, attested especially at the temple of Edfu.

---

101. Cysticercosis is a serious parasitic disease caused by the ingestion of the larva of Tenia solium, a porcine species.

For the clergy, the anathema that weighs on the pork assumes that when Ra moves in the DwAt, to deliver the dead's souls who unbalance the scale of the maat, his boat is attacked by beings vomited by the snake ***Aapp*** that look like pigs.

In the popularization textbooks *Apopi or Apophis* (in ancient Greek: Άπωφις or Άποφις, or Apopis or Apofis, in Egyptian: Apopi, Apep or Apepi or Aapep or Aapef) is an Egyptian concept materialized in a ***ntr*** for translating evil forces and night, personification of chaos, evil, darkness, seeking to annihilate the divine creation. His name Aapep or Aapef (ancient Egyptian) meant "giant" or "giant snake". As we have seen, it is not that at all.

Small reminder about ***stX***

Seth would be in the textbooks of popularization a deity of Egyptian mythology. He is also called the "red god", the god "great of strength" (ꜥꜣ phty), master of thunder, lightning and disorder, god of the desert and aridity, foreign countries.
He is associated with foreigners. It is a complex god whose role and place in the Egyptian pantheon have evolved and gradually reversed. Originally conceived as a desert god of Upper Egypt, he represents brutal and hostile nature, and is opposed to ***Hr*** and ***wsjr***, gods of Lower Egypt and associated with the Nile and fertile and nurturing land. Protector of Re, he fights the Apopi snake and thus participates in the smooth running of the world. Although disturbing and linked to indiscriminately destructive forces, it is not particularly associated with evil forces.

After the unification of the country, ***stX*** is particularly associated with the myth of the revenge of ***Hr*** and the murder of ***wsjr*** in order to take power over all Egypt. He then becomes a morally ambiguous and dangerous god, but retains positive attributes of Ra defender, and always capable of protecting Egypt from the natural forces he controls.

In the kamite myth of ***stX*** / ***Hr***, ***stX*** is the one who killed and dismembered ***Hr****'s* father appointed ***wsjr***.

According to some researchers ***wsjr***. would have lived in Africa about 12,000 years ago in Abydos (works Emile Clément Amélineau). Dating estimated at -11792 years see planisphere of Dendrah).

At his death, ***wsjr***, was the first human to succeed the test of psychostasis (the judgment of the weighing of the heart by Inpw on MaAt's balance). Success in this test of psychostasis in the AmdwAt. will make ***wsjr*** the ***xnty imntyw*** Khentamentiou or Khenty Amentiou ie the highest on the ontological level).

From this success, Ra will give up his seat as judge of the dead at ***wsjr***. By this simple act, ***wsjr*** thus became the equivalent of "God". Because, God is the one who is raised the highest on the ontological level.

He will also be called *imAxw* (In ancient Egypt, during the Old Kingdom, an imakh (plural: imakhou) or in English "a Venerable" is a man who benefits after his death, on order of Pharaoh, from the funerary rites of immortality thereby accessing the state of Blessed One who possesses the status of a tomb owner and who, during his lifetime, has had a high function in the administration of the country, having spawned offspring to take care of the cult. funeral and enjoy a good fame)

Why is there a pork ban ?

The Egyptian myth tells that ***Hr*** was resting in an oasis when ***stX*** rushed upon him in a black boar aspect. ***stX*** pulled out his lunar eye and threw it to the other side of the world. ***Hr*** woke up and did himself justice by tearing off the testicles of his aggressor, that is why ***stX*** is considered as a god of sterility (incidentally, that's where the Greek myth of Cronos-Saturn comes from).

However, it did not, give back ***Hr***'s eye, that seemed lost forever. During this time, deprived of a moon, the night sky plunged the poor Egypt into irremediable darkness.
Then, ***DHwty*** intervened from other sources saying that it was ***Hwt-Hr*** who looked after ***Hr***, he conscientiously searched the chaos that reigned beyond the confines of the universe until he found the eye lost and grafted him to ***Hr***, who was able to avenge himself and also to avenge his Father ***wsjr*** of ***stX***.

A number of papyri, now known as Papyrus Ebers, Hearst, Papyrus from London, Berlin ..., have been found citing many remedies, ointments, decoctions or dressings in which entered one or the other part of the pork. The gall of the pig was used to cure certain diseases of the eyes (***Hr***): a first half of this gall, mixed with honey, was intended the evening the eye of the patient, and the other half, dried and finely ground, to be applied every morning.

What Ayatollah of salvation religions, specialists in anathemas of all kinds have not taken into account is that, the word associated with / mura / "mother" in the African single language induces theses homophones:

> **Koonzimé**: mir "remedy"
> **Ewondo:** /mə-bâlà/ "remedy"
> **French**: remède "remedy"

This to say that the pork in African thought is associated with the pig with the teleological and teleonomical projection of remedy idea also.

That's why, the pork eye as *Sethian* animal was used to treat eye infections may seem somewhat contradictory. But in fact, this comes under the notion of "reparation", an essential step in the Egyptian conception of evil, which is at the base of an improvement: there is recovery of evil for an apotropaic purpose.
Here, the pig's eye is used to "fix" the ablation he suffered in the myth of ***Hr***, the Good, and ***stX***, the Evil.

It should be noted that the sacrifice of a male and adult pig took place each year on the full moon of the month of ***pA n xnsw*** (this month corresponds to March-April from March 16th to April 14th).

That said, as a general rule, what we eat through the magic of metabolism is our being. So we can consider that I am what I eat. So if I eat an animal called Sethian, I mix my essences with those of stX; my body and my fluidic dimension (KA) will mix with the essence of this animal which is a negative energy.

Because this animal is a "child" of Apopi. To put it simply, I make myself unclean by eating an unclean food because of Sethian. To put it scholarly, I download the energy content in pork considered negative in my morphogenetic field.

Thus the pig being potentially a child of Apopi is the reason why, a kamite priest, could not eat pork because this animal could make him impure by its load.

Reciprocally the consumption of bread and wine which foreshadow the Christian Eucharist will purify the body of the one who eats them. Indeed, bread and wine are foods associated to  ***wsjr***.

Generalization of the prohibition to consume pork meat.

Under Amenhotep IV or Akhenaten who is the tenth pharaoh of the eighteenth dynasty whose reign of -1355 / -1353 to -1338 / -1337 is located, will occur a revolution.
Ax-n-Itn who from the age of 8 already had received all the initiations of the Amonian priests, will learn early in his initiation that the consumption of pork makes impure for the reason that we quoted above and not a other.

But since the Atonian cult that he institutes is universal, he stipulates that reincarnation is possible for any man provided who respects the 42 commandments of Ma'at.
Among the rules that aim at the purity of the body and therefore the spirit of the Aton worship is the prohibition to consume pork. For *Ax-n-Itn*, this ban only concerns priests. The ban will be enforced even to the commoners.

Recovery and shift in meaning in Islam.

The first one to recover and record these taboos in a religion that will be qualified to be inspired by God is called ESDRAS around -459 BCE. Because he will recover the dogmas of the Aton worship to make some, God's chosen people. Let's go and find out which God

It is said that he would have received the revelation that only ruminant animals with forked hooves are considered pure (Deut 14: 6).

He therefore does not name ten expressly: the ox, the sheep, the goat, the deer, the gazelle, the deer, the ibex, the antelope, the oryx and the mouflon. (Deut 14: 4-5).
All other mammals that lack one or both of these characteristics are impure and therefore prohibited. This is the case of the camel, rabbit and daman who are brooding but do not have forked hooves, and pigs who have the cloven hoof but do not ruminate (Dt 14: 7-8). The Bible names in this last category 42 impure animals.

In short, therefore, the pig that has forked hooves and does not ruminate is therefore impure according to Jewish thought. For the Jew religious, the non-consumption of pork would be reduced to a story of split hoof, ruminant and non-ruminant.

From one tradition to another, if we look at the numbers and not what we associate them with, we recognize the 42 commandments of the ***ma'at*** at the 10 commandments of Moses.

It is clear that, this is not serious as an argument; it is closer to fallacy, dogma or magic than any truth.

When in 639 the Arabs invaded Egypt, the Egyptian sacred texts are stripped and their translation will be done later under the instigation of the Abbasid Caliph Al Ma'mûn. The construction of Islam will be done by plagiarizing the Sabian kamite cult but also and especially by integrating the apocryphal [102] Jewish texts. Ezra is usually identified by the Muslim commentators by a prophet named Uzayr (in Arabic: عزير). He is mentioned in Surat al-Baqara (The Cow) verse 259 in a story telling his death and

102. The apocryphal texts are those not retained by the official standards

resurrection when he passed through a destroyed village whose inhabitants had been slaughtered by Nebuchadnezzar II.

One of the loans to Judaism is not to doubt the ban on the consumption of impure animals and therefore pork. This new religion trying to be universal by resuming dogmas that it is hard to understand including the prohibition of pork consumption basically a prohibited Kamite misguided by Jewish plagiarism thought as demonstrated above.

Many of these loans will be recorded in a book that they will call the Koran. At a historical level, "the Koran was not entirely written down until the death of Muhammad, and the definitive wording goes back to the third caliph, 'Uthman (656 AD), who had all the non-conforming copies destroyed. to the canon adopted by him with the exception of a few called apocryphal sources. This means that before this Caliph, the text of the Qur'an was not unique. In other words, there have been several versions of the Koran, which is confirmed by science, and this is confirmed by Volume 26 of one of the most important references in the field, the abstract of the Islamic Encyclopedia published in 1998, page 8175.

> *"[...]the text of the Qur'an that was retained by Othman Ibn Affan was only one text among others found throughout the first four centuries of the Hegira (There were other Quran related to the companions (of the prophet: they spread celeters in the cities of Al-Basra, Al-Kophah Sham (...)"*

If there were several versions of the Qur'an as confirmed by the Islamic encyclopedia and so many other sources, it means that after the disappearance of the prophet of Islam, there existed several clans and communities of following of the prophet and that each community had its Qur'an, that is, its own version of the supposed message of the prophet. If we arrived at only one version of the Qur'an, it is because of Caliph Othman Ibn Affam.

But how did this Caliph make his version of the Quran become the one and only version?
Well the 4702 Hadith reported by Al Bukhari says that this Caliph has had his version of the Qur'an written in several copies and asked that his version of the Qur'an be imposed as an official version to all the communities, and that the various Qurans that owned these communities before being destroyed.

As a reminder, in 832, Al-Ma'mûn had excavations carried out in Egypt, in the grand pyramid of Giza built by Pharaoh Xufu (2590 BC to 2565 BC), son of King Snefru, considered by many historians as one of the most greats in the history of ancient Egypt. It seems that the "visitors" have had relatively easy access to Pharaoh's burial chamber and its treasures.

He was particularly interested in the work of scholars, especially those who knew Greek. He had assembled in Baghdad scholars of all creeds, whom he treated magnificently and with the most complete toleration. He brought manuscripts from Byzantium. He placed as a condition of peace with the Byzantine Empire the delivery of a copy of the Almagest. The great mathematician Abu Jafar Muhammad ibn Mussa al-Khawariqmi has spent most of his life in Baghdad under the patronage of Caliph al-Ma'mun.
He translated into Arabic, with his colleagues, the Greek manuscripts of Byzantium gathered in the library founded by the caliph in the House of Wisdom, and studied, from them geometry, algebra and astronomy .

Almost at the same time the so called **Ibn Wahshiyah**, nicknamed al-Kasdani ("the Chaldean"), by his full name Abu Bakr Ahmed ibn 'Ali ibn Qays al-Wahshiyah al-Kasdani al-Qusayni al-Nabati al-Sufi, in Arabic وحشية بن أحمد بكر أبو, (ninth and tenth centuries), will translate the hieroglyphs using the Coptic. He is also the author of the Kitab Shawq al-Mustaham, which offers a translation of Egyptian hieroglyphs, a book that discusses also several other alphabets.
Except, it is Haram, it is forbidden by God, it is Mohamed who has said it, we can suggest rational explanations devoid of all fantasy and religious fanaticism.

Also believing that eating pork deprived of salvation is simply showing intellectual limits and reducing the divine to a criminal who destroys his creation for a misleading reason knowing the consumption of pork.

As it can be seen a seemingly simple and simplistic script can open unsuspected doors on the historical level.

Note:

Tupuri: Gadura "Pork"
Fulfuldé: Gaduru "Pork"

The item *gadura* is an anagram of the Tupuri language *duraga* (we do not touch them). We have in Tupuri language:

Tupuri: Du " touch" **Tupuri:** Ra " them" **Tupuri:** Ga " negation"

which is normally at the end of the words or sentence.
The question is, why do not you touch this animal? The answer has been given above.

Curiously Anta Diop has for long time emphasized on the african cultural unity. Surprisingly, the Kikongo language tells us the same thing,

Kikongo: Ngulu"Pork".

Kikongo: Ngu "energy" and **Kikongo:** Lu " Impure".

The Ngambay also known as Sara, Sara Ngambai, Gamba, Gambaye, Gamblai and Ngambai is one of the major languages spoken by Sara people in southwestern Chad, northeastern Cameroon and eastern Nigeria, illuminates us in the same way:

Ngambay: Bere "Pork".. Bere = Bede (r <t <d) "different or apart".

And then, we will be told that it is with the arrival of the Bible and the Koran that we acquired this knowledge. The apodictic evidence is that things are already coded in our languages. They really do not care about us, those people. This is not to be doubted, a wide intellectual swindle.

3.4.3.21. __Example N°21__: (Faulkner p. 184, Wb. III p. 229, 5-6)

1- Script analysis

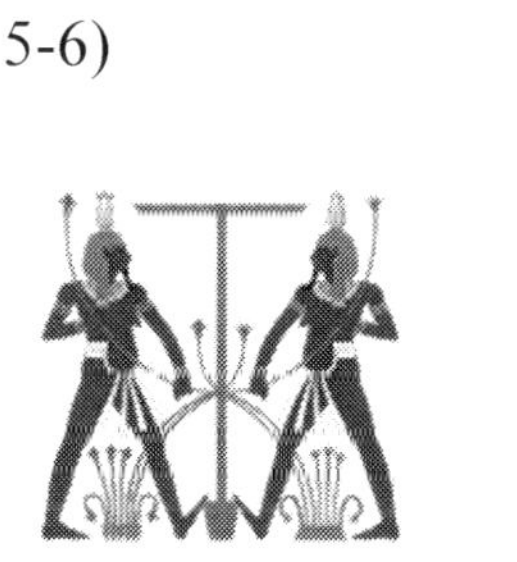

Sign M12 *xA* of the Gardiner's list: "Leaf, stem and rhizome of lotus"

We call the sign M12 /à-ndakh/; / i-soŋ / "papyrus" or / à-yaŋ bi-lobi / "lotus"
This sign associated with the sign used to designate a place or lieu, their association here indicates the biotope of the animal, in the marshes where the lotus grows. We are therefore likely to meet it more in the delta of the nile because of the symbolic sign of low Egypt.

The sign G1 /*nya/ shares the same root with / á-nyié / "to suckle and to keep" which is the symbol of the nil. The two Nil (blue and white) represented above with the breasts considered as the breasfeeders of ***km.t.***

This sign M12 associated with the sign G1 is related to nutrition also indicating its diet, namely herbivore. In addition to this, a voracious character exacerbated by the sign G1 vocalized / ndoé / sharing the same root with / *ùn*-ndî / "eater".

Sign G1 *A* of the Gardiner's list: "Egyptian vulture"
The sign G1 is called: /ndoé obama/; /kub/; /i-kobo/; /i-kuba/; /*ùn*-nya/ etc.
Note
We can observe that the word /*à*-yaŋ / "Amaryllidaceae" is an ananym of /*nya/.
This sign is linked in the African ontology to motherhood / *nya /. It can be used to indicate a house, a household /nda /. We draw attention to the fact that the sign M12 is read / à-ndakh / "papyrus". So we also find the root / nda / associated with the mother.

The word / *ùn*-nya / besides the fact that it's used to designate motherhood, it also serves to qualify what is *big* and *great*. We assume that the scribe informs us that this animal has an imposing morphology, it is fat and big. Better than that, the scribe indicates by the term / kob / "skin" used to designate the vulture also that the animal is a pachyderm. It is this descriptor related to the thick skin that will be used in many African linguistic varieties to name this animal.

Compared to the pork ***SA*** with which it shares almost the same characteristics, it deviates in reproduction. It is not as invasive as the pork according to the scribe. But he lives in group.
The transliterated script ***imy-r-xA*** designates the herd leader. The script ***HAt-xA*** herdsman, shepherd. In both cases the sign is used to designate the flock / *ùn*-nsaŋ /. This word offers by homophony the word / *ùn*-swaŋ / "the mouth; beak"

The scribe by the sign also indicates that the animal is a mammal.
One of the synonyms of is / *ùn*-nsoŋ / "beak" homophone of / à-soŋ / "tooth" or from / *ùn*-nsoŋ / "pointy". The scribe informs us that the animal has a distinctive, peaked dentition.

The sign is associated in African ontological thought with authority as in ***As.t.***
In fact, the animal will be easily characterized by everything that revolves around the lexical and semantic field of authority.

Sign D58 ***b*** of the Gardiner's list: "foot"

We name this sign /à-kol/ "foot; leg".
It is used alongside the root of the word to indicate a place or a locus.
It indicates a mode of displacement. This sign informs that this animal is a digitigrade, it moves with its feet.

2- Interpretation

We easily notice that the sign *A* is important to this script. it represents the mother.

In a matriarchal entity the land is related to the mother. The other symbol of the mother is the leopard / panther that we name with the root / * ngu / ou / * zé / and what is characterizing it, is the strength, the force: / ngoul /. It reminds us vaguely of the word Zeus[103]

The scribe tells us that, this animal is also endowed with strength. This vocal / ngul / is a superlative. It also designates what is imposing, huge to know here the feet.

103. Zeus (in ancient Greek Ζεύς / Zeús) is the supreme god in Greek mythology.

Because pigs and hippos are the same animal according to african point of view then some people in africa will call the pig:

Lingála: *ngulu* "pig";
Kikongo: *ngulu* "pig";
Vili: *ngulu* "pig"
Punu: *ngulu* "pig"
Vili: *ngulu* "pig"

To read the script, as a charade the scribe asks us how we call the house of the foot. It can be a physical foot or an ontological foot representing the eldest ancestor of african people who we name also The foot /Kolo/.

Physical foot

The answer to the charade above is the shoe. The word shoe is called in Ewondo / ngob / heard what lodges the foot. We thus obtain by phonetic proximity the word / ngub / used to designate the sign E25 of the Gardiner's list.

Influence of the sign M12

We remind for the purpose that, the sign representes what we call: /*à*-ndakh/ "*papyrus*" and, within the script it is associated to the sign "*foot*".

The expression /*ndakh/ of a **beer** language corresponds in a **bɛrɛ** language to: /ndako/ "house". Indeed, the **beer** language makes an apocope on the final vowel of the word unlike the **bɛrɛ** language.

It is the association of / * nda / "house" and / -ko / "foot" literally the place of house which allow us to take note that we are in the charade of the scribe.
The word / ndako / or / kondo / "skin; shoe" is the charade answer that is used to find the name of the determinative .

Being animal and handicapped in the sense of the word Ewondo / i-yəm / "animal and disabled" it can be named by its synonym / ù-mbogh / which offers the ananym / * ngob / and thus also allowing to justify the reading / ngub /. This approach is confirmed by the presence of the foot used to designate ***Hr***. For Nzebi language we have: *M'mbônghe* "pig"

As we have already said, ***Hr*** is the handicapped. The transliteration ***db.t*** denoting a female is written with the symbol as determinative and that serves as ananym to the term /*ùn*-mbôkh/ designating ***Hr***.

This reading is validated by phonetic proximity from the script ***xAb*** "to be folded" which designates the action of to be folded: / á-kumbu / in Ewondo. Vocalization is somewhat different between the two scripts because of the sign Aa1 which brings velar K initially because of the vocalization / kombo / which is used to designate the uterus or placenta Aa1 .

In addition to the phonetic proximity with a word that is transliterated in the same way, we also have for the ***xAb*** script in the other African linguistic varieties :

- **Sizaki**: gubbu "hippopotamus"
- **Bassa**: ngubi"hippopotamus"
- **Kwaya:** fubu "hippopotamus"
- **Duala**: ngubi"hippopotamus"
- **Bamoun**: ʃuɛn "hippopotamus"
- **Ndunga**: ngùbû "hippopotamus"
- **Tupuri**: bal "hippopotamus"
- **Baoulé**: nzu-sɥí "hippopotamus"

- **Kisongye**: mfubu "hippopotamus
- **Kilala**: nfubu "hippopotamus"
- **Bt**: nígbùlí "hippopotamus"
- **Pulaar:** ngabu "hippopotamus"
- **Ciluba**: nguvu "hippopotamus"
- **Soninké**: wungaame "hippopotamus"
- **Lingala**: ngubu "hippopotamus"
- **Ding**: ngub "hippopotamus"
- **Chuka**: nguu "hippopotamus"
- **Idonge:** ngabu "hippopotamus"
- **Nzadi:** ngb "hippopotamus"
- **Dogon**: agabu "hippopotamus"
- **Fulfuldé**: nyiiwa "hippopotamus"
- **Kitabwa**: mvubu "hippopotamus"
- **Kioshi**: imfuvu "hippopotamus"
- **Kilamba**: imfubu "hippopotamus"
- **Kitshok**: nguvu "hippopotamus"
- **Lunda**: nguvu "hippopotamus"
- **Bemba**: infubu "hippopotamus"
- **Buma**: ngub "hippopotamus"
- **Lubukusu**: fubu "hippopotamus"
- **Hausa**: dorina "hippopotamus"
- **Yaagaare**: ngabbu "hippopotamus"
- **Haya**: jubu "hippopotamus"
- **Kɔ́ɔnzime**: nzo'medíb "hippopotamus"
- **Mundang**: bále; yik "hippopotamus"

We also find the descriptor related to big-feet for the elephant ***Abw.*** In fact we can mobilize the same term for the elephant /zokh/ but by specifying its biotope namely water or marshes as a criterion of differentiation. It is named in this account /zokh mə-dim/ "water-elephant ". Bamoun and Baoulé languages also use this approach.

The comparison below between Tupuri and Mundang illustrates the use of elementary operating concepts (signs) to create the lexicological field around the symbols of the script.

Associated Signs	Tupuri	Lexicon	Mundang
	bal	"hippopotamus"	*bále*
	béle	"leopard, panther"	*sel*
	lɛ́ge	"force, strength"	*záke*
	boN	"lieu, place"	*kẽ̄*
	ʒag	"parole; mouth ; speech; talk"	*zā:*
	Dö	"bird"	*ʒù:*
	bi:	"water"	*biì*
	hő̃ge	"skin"	*wāke*
	klu	great, big , thick, heavy, coarse	*yik*
	ko	"tree; wood;tree structure;mast; fire; three; bomb; plant; post"	*kp~u*
	sukko	"leaf"	*gò:*
	gúllɔke	"Behind, buttocks"	*fàl*

Chart 79:Script cross analysis mundang-tupuri

We could have questioned these two languages more closely around this semantic field. Without any doubt we would have noticed that, the mobilized signifiers are all united to the signified of the script as shown by the signs on the left.

This is also confirmed in Fulfulde where the word for beer / wine is *ndzuʔʙ*. We also find in fufuldé the word: *gxup* which refers to the skin, bark, leather, etc

Swahili: nguvu "power, strength, force"

The scripts ***db***; ***dib*** at the level of the signifier are sharing the same the root ***/*dib/*** "eau" in Ewondo.

Ewondo: /zokh mə-dim/ or /zokh mə-dib/ "hippopotamus; water-elephant".
Kɔ́ɔnzime: nzo' medíb "hippopotamc; water elephant".

One can even note that, mobilizing the root / ze / of our languages, the sign representing the hippopotamus, can be considered as the panther of the rivers. Indicating here that the animalis dangerous for Man. The *nza* indeed is in the Ko-ngo tradition is the creative energy. In Eastern extreme spirituality we speak of *ki, chi, chi*

For the scribe indeed, the rhinoceros ***SAqb*** is characterized by the same descriptor namely bigfeet. So in my language both animals have the same name /ngub/. The information on the biotope is almost identical, the maternal function is here represented by the vertex / *ùn*-nsoŋ / which is also the other name of the vulture because having a beak /*ùn*-nsoŋ/ and representing the chef as the summit . This sign is also used to indicate the presence of the pointed outgrowth (rhinoceros horn) and thus to differentiate the two animals.

These scripts are not simple alphabetic characters, they are characters filled with information for who knows how to read and interpret them beyond appearances and impressions..

Etymology: xAb "hippo"

Bbaledha: Nya "hippo"	**Ewondo**:/*nya/ "vulture; mother"
Bambara: Meli "hippo"	**Ewondo**:/i-lom/ "buffalo; cobra"
French: bouc"beard"	**Ewondo**:/kub/ "vulture; fowl; hen"
French: bouc"beard"	**Ewondo**:/i-bukh/ "speech"
French: bouc "beard"	**Ewondo**:/i-kob/ "skin"
Mossi: yemde "hippo"	**Ewondo**:/nda i-yəm/"house of the disabled"
Dagara: yenne "hippo"	**Ewondo**:/nya/ "vulture; mother"
Mbay: yab "hippo"	**Ewondo**:/yob /; /yab/ "top; high"
Malinké: me or meli "hippo"	**Ewondo**:/ma/; "me; mum; cobra etc."
Bandjoun: Nya "hippo"	**Ewondo**:/*nya/ "vulture; mother"

3.4.3.22. ***Example N°22*** (Faulkner p. 137, Wb. II p. 303, 3-8)

I allowed myself to read this script not to indicate the people who are here designated but the vocalization which emerges in Ewondo for the reading of this script. I thus specify that according to the linguistic variants, this script can be vocalized with nuances on the level of the prosody and the synonyms used. I am committed to reading in Ewondo it is the only merit of this reading.

Middle-egyptian: ***nHsy*** "nubians; nehesy"

What is astonishing in this reading is that no people are called Nehesy or Nubian in Africa. Where have gone the people who are here reported by the scribes and that Egyptologists name Nehesy of Nubians? That's the key point of this issue.

Endogenous method:

Let's start by listing all the signs:

Signs N35 "Water ripples" and G21 "guinea fowl" of the Gardiner's list.

Ewondo: /*ùn*-ŋkaŋ/; /*ùn*-mvəm/ "guinea hen"
Ewondo: /mvəŋ/ "rain, water"
Ewondo: /à-yoŋ/ "heat" because we are in a desert area.

Sign V28 of the Gardiner's list "Twisted linen lock"

Ewondo: /*ùn*-mbokh/ "the other qualification of ***Hr***"

Sign S29 "folded linen"

Ewondo: /i-yé/ "clothing; garment"
Ewondo: /i-yóŋ/ "what keeps warm"

Sign Z4 of the Gardiner's list "*ùn*-mbias" (nisbe)

Ewondo: /*ùn*-mbias/ "twins"
Ewondo: /i-yoŋ/ "ensemble, together, set, in a body etc."
Fang: /yor mbiar**/** "the smaller of the two twins"

Sign T14 "throwing stick; mace"

Ewondo: /ù-mb**əŋ**/ "cudgel" ☞/ngul-icud/ "hit hard"
Ewondo: /à- yi**ŋ** / "Cross sticks in trap holes"
Ewondo: /ùn-**əŋ** / "foreign" ☞**Ewondo**: /bə-y**əŋ** / "foreigners"

Sign A1 "seatting man"
Ewondo: /*ùn*-mbokh/ "***Hr's*** other qualification"
Ewondo: /*ùn*-ntobo/ "sitting, living; inhabitant; chosen etc."

Reading:

From right to left we will have:

Ewondo: /*ùn*-mbias à-yoŋ i-kaŋ *ùn*-mbokh/ "The twin people who praises ***Hr***"

We recognize in the script ***nbwy.*** the Two Lords (***Hr*** *and* ***stX***).
We recognize also ***tA sty***.

The symbol O34 of the Gardiner's list given for a Lock, indeed presents a knot.
Ewondo: /à-tiŋ/ "knot"
Ewondo: /à-tiŋ/ "alliance; league"

Indisputably so we have in this script the people who concluded the *alliance* with ***Hr***.

Note : Even though the concept is found in Iceland later (10th century) with **:**

Icelandic:Althing "parliament since 930"
At the beginning it has been conceived as a coalition of tribes with the same african matrix.

The people named Nubians we call them in Ewondo the **Ekaŋ; Kongo; Akaŋ** or **Kiŋ'â** "to be surrounded"; **kiŋ'ê** "chief"; **Akuŋ** "owls" only on the basis of the phonetic proximity and sense here highlighted. We can highlight other signifiers from many synonyms of the word guinea fowl /**ùn-nkaŋ**/.

Ewondo: /ùn-nkàŋ/. "guinea fowl; "
Ewondo: /**ùn-nkaŋ'a**/. "the praiser"

Notably with the Ewondo meaning: / ù-mvəm / "guinea fowl"
One notes on this subject the existence of the name Mveng-Ayong legally dependent of Kingélé / Tchimbélé in Gabon.

Mveng is also a village of Cameroon located in the department of Dja-et-Lobo and the Southern Region, not far from Gabon and the Republic of Congo, on the road that connects Sangmelima to Djoum.

The phonetic proximity between / ù-mvəm / "guinea fowl" with / ù-ndzəm / (or djem, dzem, ndjeme, ndzem, ngyeme, njem, njeme, njyem, nyem, zimu) which is an Ekaŋ language classified in the group makaa * -njem spoken by the Djem population of Cameroon, in the Eastern region, the department of Haut-Nyong and the district of Ngoyla, and also in the Republic of Congo.
Studies by Marion Cheucle and Akumbu Pius Wuchu estimate the total number of speakers at 7,900, including 4,400 in Cameroon (2005).

After this apartment to say that we are not on a univocal reading, we can emphasize that, semantically for Ewondo language: / à-dzəm / it is "to dance", synonymous with / *i*-kaŋ / "joy , celebration, party ".

	Guinea Hen	**bow**	**Associated Group**
Hausa:	Zǎbō ou zabouwa	Bàkā*[104]	Baka; Makaa
Fulfuldé:	yaɓɓaago; jawngal	Lagaawal Labbo* *(lance)*	Belabo*[105]
Soninké:	kengele;bedere; kenne	bu*nda*nŋe	Beti?
Baoulé:	akɔ	ta-b**à**	Beti?
Zand	kondo	m*boto*	Beti?
Ci-luba	Ngine ou *di*kanga*la*	*butà*	Beti?
Nzebi	L'règhe	*kumbu*	Baka; Makaa
Anô	Kô'ndjê		
Moor	Kongo	*kom*	Baka; Makaa
Mandinka	kunoo	*Mùru;* *Too (nom)*	Beti?
Mandingue	Kâmô		
Bambara	Kami	*Bundanŋe;* *bunne*	Beti?
Malink	Kimi		
Songaï	Tchyrow		

104. Hausa is the root of the lineage identified by the Ekang language.
105. Belabo is the toponym of a commune of Cameroon located in the region of the East and the department of Lom-and-Djrem as the Baka, the Ndjem.

Bissa	Ga	*Sa; kà*	Baka; Makaa, Basa
Sousou	Kank		
Batonu	Gonna		
Foŋ	Sonou		
Duala	Ikwale	*Dipunga; nkɛti*	Baka
Wolof	Naate	*kála;Fitt*	kɔlɔ = Beti
Maba	Kouchik		
Mundang	Loo	*zɔ́:*	Nzem= makaa
Gal	jawlal Machallah	*Lein, tutui*	Beti?

Chart 80: signs cross analysis in different languages

A small multivariate cross-study with different criteria (guinea fowl, arc and the hidden meanings) on the different linguistic varieties can indeed make it possible to better refine the subject here. But just these few examples are already quite telling.
The most important information is that km.t = tA nHsy (Bambara, Malinké, Moor)

Reading Validation:

Middle-egyptian: ***tA nHsy*** "nubia" **M-E**: ***Styw*** "nubians"

Middle-egyptian:  ***nHsy*** "nubians"

Ewondo: /à-faŋ/ "to attach"
Ewondo: /mi-'mfaŋ/ "the real, the tied, the related"
Ewondo: /*ùn*-mfàn/ "bow"
Ewondo: /bə-fàŋ/ "people living in Central Africa in the 21st century"

Incidentally this people was made up of bowmen but what I wanted to highlight in this reading is above all its name in an Ekaŋ language.

The *ndan* of this people appears like this: / *ùn*-mfaŋ à-yɔŋ i-kaŋ *ùm*-bokh / "the true people who honors ***Hr"***

3.4.3.23. ***Example N°23***  (Faulkner p. 286, Wb.? p.?)

Middle-egyptian:  ***kynanw*** "Canaanite"

In the authorized literature, Canaanites are called the inhabitants of Canaan in the Bronze Age, sometimes called for this region "Canaanite period". The search for the identity of true Canaanites has been the subject of much scientific research without achieving a satisfactory result. The concern to attach this region absolutely to a Semitic branch has led to this dead end on the historical level. In a book titled Egypt, *Canaan and Israel: History, Imperialism, Ideology and Literature the authors' collective* (S. Bar, D. Kahn, J.J. Shirley) dispel what they consider to be ideology.

> *"The primary sources for the reconstruction of this geoscriptal area include: (l) the Mari letters, (2) the Amarna Letters, (3) Ugaritic texts, (4) texts from ASSur and Hattusha, and (5) Egyptian texts. The term Canaan is found 16 times in New Kingdom Egyptian sources*[19]*. Most scholars have concluded that Canaan in a number of these texts, particularly during the 19th Dynasty, referred to the entire area of Palestine synonymous With the toponym Kharu.*[20] ***This view was recently challenged by N.-P. Lemche****, Who stated that there is a "correspondence between the imprecise and ambiguous Egyptian use of the geoscriptal name Canaan and the likewise imprecise*

> *understanding of Canaan displayed by the inhabitants of Western Asia themselves."[21] In the conclusion to his study of the Amarna Letters, he writes, "Evidently the inhabitants of the supposed Canaanite territory in Western Asia had no clear idea of the actual size of this Canaan, nor did they know exactly where Canaan was situated."[22] In essence, "the Canaanites of the ancient Near East did not know that they were themselves Canaanites."[23] Lemche's conclusions have been challenged by N. Na'aman, A. Rainey, and others,[24] but he has maintained his interpretation of historical sources which he calls "imprecise" and "ambiguous."[25] In 2001 O. Tammuz built on the assumptions of Lemche but only dealt With the Egyptian material briefly in his lengthy article.26 In both studies one is left With an open question concerning the meaning Of this toponym to the Egyptians themselves and what, if any, connection (...)"[106]*
>
> *(S. BAR-D. KAHN-J.J. SHIRLEY, 2011)*

In the biblical narrative, Canaan would designate the land promised to the Hebrews by God (Yahweh) to Abraham. This region would be between the Mediterranean and the Jordan, before its conquest by Joshua and the tribes of Israel out of Egypt. The term comes from the name of Canaan, Noah's grandson.

Historically, however, during the eighteenth Egyptian dynasty, the Egyptians repeatedly attacked the land of Canaan, especially under Amenhotep I, under Tuthmosis I, where the Egyptian army reached the Euphrates in the north, under Tuthmosis II and under Tuthmosis. III where the battle of Megiddo takes place. The Canaanites, grouped under the command of the governor of the city of Qadesh named Ali El-Assi, and the Egyptians, under the orders of Pharaoh Thutmose III, besiege the city of Megiddo for seven months and force the Canaanites to surrender.
This script is therefore historically important because it illuminates the historical fact with an impartial eye and puts the true inhabitants of this land back in the center of history.

Endogenous method

Sign V31 of the Gardiner's list "Basketry basket with handle"

Ewondo: /ù-ŋkùn/ / i-ŋgen /"Basketry basket"
Ewondo: /wa / "you, your; yours"
Ewondo: /wa / "hand"

Sign Z4 of the Gardiner's list /*ùn*-mbias /"twins" (nisb)

Ewondo: /*ùn*-mbias/ "twins"
Ewondo: /i-yoŋ/ "together"; /à-yoŋ/ "people"

Sign N35 of the Gardiner's list "Water ripples"
Ewondo: /à-nyu/ "drink"
Ewondo: /à-yoŋ/ "hot"

Sign D36 of the Gardiner's list "forearm"
Ewondo: /wa/ "arm"
Ewondo: /i-nam/ "member"; **Ewondo**: /*ùn*-nam/ "country, village"
Ewondo: /*ùn*-ŋkun/ "forarm, basket"
Ewondo: / *ùn*-ŋkoŋ/ "hill, mound, hillock; butt; montain, forearm ridge"

Sign U19 of the Gardiner's list "adze"
Ewondo: / *i*-ŋgwakh/ "adze, stone"; / *à*-ŋgonga /"arm, adze"

106. S. BAR, D. KAHN, J.J. SHIRLEY, Egypt, Canaan and Israel: History, Imperialism, Ideology and Literature, BRILL, Danvers, 2011, P.68.

Sign W24  of the Gardiner's list "pot"

Ewondo: /ndəkh/ "pot; jar, potty; can; drink, tub etc."

READING:

Many attempts have been made to find the etymology of Canaan, which seems to be a Semitic word. But these attempts are not very successful visibly.

It is obviously not enough to proceed with voluntarist groping or to falsify history. In addition, it is necessary in time to be able to pronounce the names of the objects which appear on the script in its language and to know the use and the sense.

The sign U19  of the Gardiner's list "adze" is called in the Beti language:

Ewondo: / *ŋgwakh/ "adze, stone"; / à-ŋgonga /"arm, adze"

In addition to the agrarian and non-nomadic field, it is a sacred tool that was used at ***km.t*** and is still used for the sacred ceremony of the opening of the mouth.
The good reading of the script is given by the name of the adze in our languages and confirmed as it has been demonstrated in the enumeration by phonetics complements. All the descriptors on the script about heat, the presence of stones, etc. that sticks well with what is identified as the country of Canaan come to confirm this reading. We just have a little trouble understanding how the scribe could give the name of a grandson of a mythological character named Noah to this land.
For Africans who want to do research on this topic, just find the name of the adze in his language or pebbles to have a research base on the subject.

Mopho-syntactic constraint.

In order to respond to the constraints of the structure offered by the consonant skeleton *kynanw* which is only authentic for some, I listed all the words homophones or having a phonetic proximity with the word adze in Ewondo and that can answer the technical requirements on the morphosyntactic plan. So we have:

Ewondo: /i-ŋgôŋgôl/ "pity; mercy" burned
Ewondo: /i-ŋgongə/ " hill, bump, hill, crete"; "able to aim"
Ewondo: /i-ŋgôŋgôm/; /i-ŋgeŋgame/ "agitation, noise"
Ewondo: /i-ŋgôŋgôr/ "strong, smart, vigorous etc."
Ewondo: /i-koŋgoŋgo/ "songs of cicadas"
Ewondo: /i-ngoŋgoa/ "spread country, scattered"

By reading the excellent book of Unesco dealing with African ethnonyms and place names, here is what we read on page 66:

> *"[...]Conversely, it is almost established that the Kingdom of Engong, the theater of fang epic narratives, existed, contrary to what the first observers thought, only in the collective imagination of the so-called "pahouin" group. ""*[107]

If we stick to this testimony, we know that it is the ancestors of the current Fang who populated this territory since they have kept the collective memory has kept a trace.
Of all the homophones presented above, I mainly retain only these one:

Ewondo: /i-ngoŋgoa/ "spread country, scattered"

107. Unesco, Ethnonyms and African toponyms, l'Unesco à Paris, 1978, p.66

which offers vocalization in a Beer language
Ewondo: Engong "Canaan"

Ewondo: /i-ŋgongə/ " adze, hump, hill, crete"; "able to aim"
Ewondo: / bi-ŋgong/ "Canaanite"

Ewondo: /i-ŋgôŋgôr/ "strong, smart, vigorous etc."
Ici la racine /*ŋgong-/ is associated to Hr /-or/ eldest ancestor of those peoples.

The other variants that can be mobilized as is the case with the ***mDw nTr***.
This script is read: / à-yòŋ Engong / and offers to understanding, clever people, strong, hills inhabitants, spread lands etc.

This script 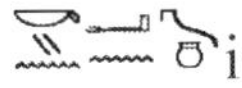is read: / à-yòŋ Engong / and offers also the acceptance people's of the alliance.

Indeed, / à-ŋgoŋ / can mean a *padlock* and thus a *node* just by synonymy. These words are homophones of the word alliance. So only a people who speaks Ekaŋ languages can have this level of precision regarding this subject and justify with whom it is linked how and why.

Instead of inventing ghostly peoples like Canaanites, is it not on the basis of the historical continuum to recognize Kenya and Kenyans in this script? You should think about it.

In the land of Canaan, former suzerainty of ***km.t***, the worship of the Baal God was practiced. Remember, we have already been pointed out that / à-bal / is hoeing; weeding and that the adze can be used for this purpose.
Rigorously this country being that of the descendants of the Natufiens[108] as asserting Miss Dorothy Annie Elizabeth Garrod the British archaeologist and prehistorian who excavated Es Skhul and El Wad, to enlighten humanity on the settlement of Canaan (present-day Palestine) during the period from the lower Palaeolithic (-3 million years before NE.) to the epipaleolithic (-12,500 BC).

If the scribe characterizes the descendant of the Natufians with the U19 sign which is a tool of sedentary peoples, as noted by Miss Garrod, I can not be more royalist than the king. We can only link this people with sedentary or at most semi-nomadic strains

If one adds to the solid works of Miss Garrod, the testimony of the scholar Cheikh ANTA Diop who affirms that 4000 years before N.E there was no human trace in Palestine. And that it is only later that the Kushites will invade the entire Mediterranean basin and participate in mixing with the ancestors of Indo-Europeans to create the Semitic type. The first known semitic to date being Sargon I of Akkad (2200 BC N-E).

> *"[...] It is the Negro influence that preceded, throughout the Aegean period; that of the Indo-Europeans.* ***All the populations around the Mediterranean were then Negroes or Negroes: Egyptians, Phoenicians****; when they were white, they underwent economic and cultural Egyptian-Phoenician influence: Greece, time of the Boeotians; Asia Minor, Troy; Hittites, allies of Egypt; Etruscans in northern Italy, allies of the Phoenicians: strong Egyptian influence; Gaul, furrowed by Phoenician caravans, under the direct religious influence of Egypt. This Negro influence extended as far as the Germans, whose tribes adored Isis the Negress."*[109]
>
> *(ANTA DIOP, 1979)*

108. Natufian is the name given to a culture of the final Epipaleolithic, attested in the Levant between 12 500 and 10 000 BC. J. - C. and characterized by the first experiences of sedentarisation.
109. Cheikh ANTA DIOP, Negro Nations and Cultures, Présence Africaine, Paris, 1979, P 173 – 177.

It would therefore be nice that Russian linguists, Alexander Militarev and Viktor Aleksandrovich Shnirelman who consider that the first speakers of Proto-Afro-Asian [110] were the Natufians are inclined to revisit their theory of mythological languages such as Afro-Asian that exist only in the imagination of ideologists.

When true African historians will look into this issue, I hope they will validate what the small, seemingly silent drawings say. This will also imply a better understanding of the place of the Fang people in the general history of Africa and of course human being History.

3.4.3.24. ***Example N°24***  (Faulkner p. 44, Wb. I p. 205, 11-12)

Middle-egyptian:  ***anx.t*** "goat"

Endogenous method

Sign S34  of the Gardiner's list "Tie or bridle, especially sandal flange"
Ewondo: /*ùn*-nkwal/ "cord, string, rope"
Ewondo: /à-kélé/ "to suspend"
Ewondo: /*à*-kol/ "foot, leg"
Ewondo: /*i*-tingha / "flange"

Sign N35 of the Gardiner's list "Water ripples
Ewondo: /á-nyu/ "to drink" ☞ **Ewondo**: /mə-nyu/ " drink"

Sign Aa1  of the Gardiner's list "Placenta"
Ewondo: /i-sokh/; /*nya/; /kwé/; /i-bibyali/ etc. "placenta"

Sign X1 of the Gardiner's list "bread"
Ewondo: /i-kô/; /i-di/; /*ùn*-mfəma /; /*ùn*-mbaŋlà/ "bread"

I offer to do the exercise upside down this time to become more familiar with the method and appreciate the possibilities.

Script Reading
The sign E186  of the Gardiner's list represents a goat. In Ewondo the word for goat is: / kab'ra /; / kabat / "goat" or / i-kəlà / "goatee; billy goat; beard".

Indeed, if we look rigorously on the determinative, there is no what to look for noon to two o'clock, it is a goat, just name it and the script is read.

The real question is what is the link that the scribe makes between the goat and the symbol of the egyptian cross  which represents eternal life among the peoples of the Nile Valley?

Sharing symbol
Ewondo: /à-kab/ "to share"

110. Peter S. BELLWOOD, Colin RENFREW, Examining the Farming/language Dispersal Hypothesis, McDonald Institute for Archaeological Research, University of Cambridge, 2002, p.136.

We note that the signs that are mobilized by the scribe are all signs that refer to a notion of sharing.

Sign Aa1 𓐍: shared disk
Sign N35 𓈖: shared sine wave (one sinusoid up, one down)
Sign X1 𓏏: we have here only a half-loaf of bread.

Life Symbol

The signs mobilized by the scribe to write the script are all connected to life.

Sign N35 𓈖 wave phenomenon for quanta of energy or water
Sign X1 𓏏 bread represented for a food is a symbol of life
Sign Aa1 𓐍 the solar disc from which emanates the energy responsible for life.

So in the end we have an expression that reflects the idea of sharing life. In fact the symbol ☥can be mobolized by the scribe to give the best representation of the animal  symbolized by the sign E186.

Symbol of life: another approach

On another level, the symbol ☥of the egyptian cross actually represents the bell often placed around the necks of some livestock.
We call it by the word / à-ŋgoŋ / "padlock, small bell, bell, etc."

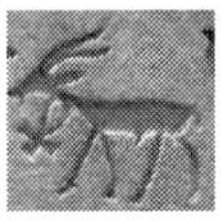

Ewondo: /à-ŋgɔŋ/ "padlock, small bell, bell, alliance"
Ewondo: / à-ŋgoŋ/ "figurative meaning of scrotum, testes"

The scribe thus associates in this script, the purses (gonads) and the sign Aa1 𓐍which represents the uterus. These two signs are organs linked with reproduction and therefore with life☥.

At the end, the charade boiled down here was to name in our languages a padlock.

The padlock is usually used to secure an entrance. / á-*nyî*n / "to enter" which offers an homophony with / i-nyiŋ / "life". So that's the reason why all the symbols used are related to this root / * ny- / "mother"

Ewondo: Sign N35 𓈖 / mə-nyu/ "drink"
Ewondo: Sign Aa1 𓐍/nya/ "placenta"

The sign ☥
as a fastener or flange, especially a sandal flange behaves like a padlock. It comes to secure the fingers / mə-nyu / with the insole of a sandal.
It encloses the foot /à-kol/ which offers a phonetic proximity with /i-kélà / "goat; beard; goatee".
Let's summarize a little, this script besides reading / kab'ra /, can offer other meanings like / à-ŋgɔŋ / "padlock" or / i-kélà / "goat"

Associated sociology

On the sociological level, therefore, no serious ritual in Africa can be done without the presence and sharing of a goat. Indeed, as we have just seen, this animal is the embodiment of the sharing of life. Ritual ceremonies are consolidated by sharing. While it is still useful to point out sharing is a central value in a sedentary and matriarchal society.

Other Related Scripts

Middle-egyptian: ***waty*** "goat" **Ewondo**:/ i-tyé/ "to be tied"

Middle-egyptian: ***waat*** "goat"

The presence of the numeral one as well as the arm and the harpoon to a beard impose us the vocalization / *i*-bobokh / "baby" whose radical will admit the ananym / kobo / or / kaba / which designates the goats in our languages. The presence of the sign leads to the reading / *ùn*-mbobokh / or / i-bobokh / which designate ***Hr***.

The presence of the numeral one as well as the arm and the harpoon with a beard sends us indisputably to Hr. We also call the foot / *kol / which offers here a phonetic proximity with / i-kela / "goat; beard; goatee" because it is the attached animal / á-kl/.

As / i-bobokh / "baby" ***Hr*** is then the breastfeeded.

Ewondo: /*á*-nyaŋ/ "to suckle"
Ewondo: /*i*-nyiŋ/ "life"

Etymology

Goat		**African Language**	**European Language**
Cabri		Eton	French
Cabra		Ewondo	Spanish- Portuguese
Capra		Ewondo	Italian
Tâb	*Tapră*	Nzimé	Română
Tâb	*Ţap*	Nzimé	Română
Tâb	*τράγος*	Nzimé	Greek
Cabra	*Caprinus*	Ewondo	Latin
ngasà	*Коза*	Sango	Russian
ngasà	*koza*	Sango	Polski
ngasà	*Козел*	Sango	Ukrainian
mbuzi	*maeiz*	Ki-swahili	Arab
kémbɛ̂	*kambing*	Ɓàsàa	Bahasa (Indonesia)

Chart 81: Similarities of word Goat in several languages

We find this little example of the origin and the true etymology of the word *goat* used in the so-called Indo-European languages. It would therefore be more honest to change in the official dictionaries the supposed origin of the etymologies which appear in it.
There is also a link between *goat* and *cobra* which are in reality one and the same reality.

The word cobra

Ewondo: kab'ra "goat" **Ewondo**: kab'ra "goat"
Middle-egyptian: ***D.t*** "cobra" **Sossou**: cobro "cobra"

Etymology: origin of the word goat

English: goat
Ewondo: *i*-tughà "a stooge" ☞The two words are just ananyms.
Ewondo: *i*-tîghà "a cutter; ; to be shared; the one to be tied"
Ewondo: *ùn*-ŋgut "genital"
Ɓàsàa: ŋguèé "panther"
Ewondo: ŋgwé "pig"
Ewondo: kot "greedy"
Etc.

The word 𓐍𓊪𓂋𓆣 *xpr* "kheper"

I chose to integrate this sign here because when we sometimes approach the words vocalized by Egyptologists, we have the impression that they speak Martian.

What is kheper? Only them know because they are specialists. When we listen to them, this item is so nebulous that we finally give up.
But in reality, this notion is not as nebulous as it seems since we use it every day in our lives.
Subsequently the notion being nebulous, i suggest that we try something less nebulous.

Middle-Egyptian :𓆣𓂋𓊪𓋙 ***xpr***S « khépresh, blue crown »
Wolof: baxa "blue"

So we can see above that, the idea of ***xpr*** can be associated to the **blue** color that we can consider as something well known by all.

Ngiemboon: *Bleu* "blue"
Tshiluba (Rdc): *Cy-uluulu* "blue"
Samo: *Boula* "blue"
Basa'a: *mandut mandut* "blue"
Tupuri: *tabay* "blue"
Qusqu qhichwa simi: *anqas* "blue"
Lingàla: *bo-zinga* "blue"
Gourane: *azraq* "blue"
Mooré: *Boula* "blue"
Lingàla: *bozinga; bulo; bulé* "blue"
Bambara: *Boula* "blue"
Kikongo: *mbundu* "blue"
Anô (C-I): *Boula* "blue"
Chinese: 蓝 *lán* "blue"
Wolof: baxa "blue"

We can group some words by islands of phonetic convergence.

One can legitimately ask the question to know whether the skeletons of consonants *bl* "bluc" cxample: bulé are phonetic deformation of "blue" or not.

This simple qucstion will annihilate the logic and will give thabackdrop that cveryonc could know BLUE tinge except Africans.

But, colonization only begins in the 15th century before that Africa was able to produce blue colour.

Lotus Chalice XVIIIth dynasty Circa 1479-1353 B.C

Lapis lazuli and gold pendant representing Hathor - Egyptian Museum of Cairo

We therefore claim that, the dyeing industry was perfectly controlled since the extraction, the purification as well as the fixing of colors on many supports such as earthenware, fabrics, buildings etc. was an indisputable achievement.

By the way, it is impossible to postulate here a colonial legacy to support the sudden appearance of the word related to the idea of blue in African languages.

One thing is certain, those who use the word in the West will be hard to give a satisfactory etymology. I challenge them when they want, academicians or other assimilated charlatans.

We can assert that regarding developed shades of blue never equaled like the one in Saqqara, for example. To believe that the Africans waited for the settlers to finally get acquainted with blue has no name.

We assert that the word blue has its root refering to word 9 symbol of the multitude and therefore referring to the MOTHER and so to the SEA.

Blue for Scribes and then for ancient Egyptians, is the color of water or sea / mother but not only…

Middle-Egyptian :  ***irtyw*** **« blue, color, a blue mineral »**

Middle-Egyptian :  ***itrw*** « river »

The word BLUE or BLUE is therefore African and that we can easily prove it
The MOTHER embodies the MULTITUDE, the multitude is 9 in african thought. But also the number 100: /Nter/ or /Ntet/.

Ewondo : /*ùn*-nter / designating the one who has difficulty moving like a stream that simply meanders because it has no feet or paws.

I take my precautions because the Blue color is here associated with the transliteration ***itrw*** by Egyptologists to a stream of water. And I do add also earth of sky by metonymy. You may report on illustration N°54. ***The* Tshiluba (Rdc)**: *Cy-uluulu* "blue" ↔**Ewondo**: /i-wulù/ "earth"☞**English** : walk

Onomastic
Boula, Louba, Loba, Lobé, Lebu, Balà, Belà , Libi, Laba, Baal, Bêlè; Blue . Bleu ;Bulu etc. all those names can be effortlessly found in Africa in a different place.

The question is simple here: who is Baal or Loba for Africans? He is the one often call Goat sorry God.

The Ram Loba or Baal

Who is the ram and why is it a ram? May you report on illustration N°47.
By the way then

Middle-Egyptian : ***xpr*** « khéperesh»
Ewondo: kab'ra "goat"

The dung beetle and the goat can be considered as reversers because they have the ability to return. This prompted Egyptologists to associate transliteration *xpr* with change.

The dung beetle

Holy cow

The scale as reverser

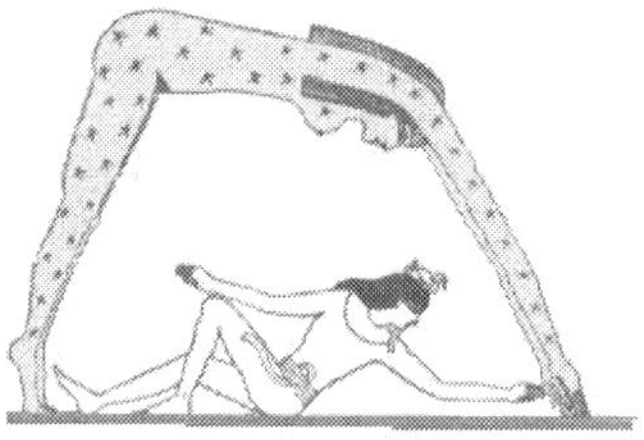

Nwt and gb reverser

Xpr.w:Reversers

We can see that, the operative concept of movement is intimately linked to that of transformation. The same idea of ***xpr*** is applicable to several domains (cosmogony, cosmology, cosgraphy and cosmocracy).

Kasha Ka Mu Tshanda. the gazelle of the lodge

In cosmography, the starry vault of the Milky Way makes think about a curved woman. It is therefore associated by metonymy to the horned animals.

We therefore speak of horned sacred animals incarnated by ***Nw.t*** and represented by goats or cattle according to the periods of the zodiacal cycle.

We can not evoke the notion of ***xpr*** without speaking of the balance.

Since the universe operates with physical laws, these laws are applied in its expansion from the highest level to the smallest.

Thus the sky represented by ***Nwt*** being blue, can naturally be associated with the notion ***xpr***.
The sky is not only blue, it can be associated in its starry outline to the panther while remaining ***xpr.***

Chinese: 蓝 *lán* "blue" **Middle-Egyptian**:  ***nSmwt***

Ewondo: /mə-làŋ/ "spots" ↔ **nw.t** the panther **Middle-Egyptian**:  ***nrw.***

This celestial panther developing some ferocity, words for ferociousness can be used to designate the blue and therefore the sky.

Bull /Cow	Goat	Largest
Mooré: *Boula* "blue"	**Tupuri**: *tabay* "blue"	**Basa'a**: *mandut mandut* "blue"
Anô (C-I): *Boula* "blue"	**Nzimé**: *tab* "goat"	
Samo: *Boula* "blue"		
Ngiemboon: *Bleu* "blue"		
Bambara: *Boula* "blue"		
Lingàla: *bulo; bulé* "blue"		
Wolof: baxa "blue"		
French: vache		
Spanish: vaca		
English : cow		

Horns associated to Nw.t

The notion of blue can also be associated with land or water. The earth seen from the sky is indeed blue. Earth is 70% water.

Xpr in Ewondo language

Ewondo *:ùn-ŋkàbghà* "fire; flame"
Ewondo *:á-vələ* "create, shape, wake up, reveal"
Ewondo *:á-hələ* "create, shape, wake up, reveal"
Ewondo *:á-me* "to create, to shape"
Ewondo *:á-bɔh* "doing, acting, creating, shaping"
Ewondo *:á-vîmli* "create, shape wooden"
Ewondo *:á-koghdo* "create, shape into a sphere"

Ewondo *:á-vîghdi* "create, shape in clay"
Ewondo *:á-ŋgôé* "Shaping as a star"
Ewondo : i-vəghlə « image ; representation »
Ewondo : *i-fughlù* « mixture »
Ewondo : *ùn-ŋkəbɔ ; ùn-ŋkɔlɔ* « speech »
Ewondo : *zùkùlù ; sùkùlù* « school ; know, intelligence ; learning »
Ewondo : *i-kùlbèn* « apparition »
Ewondo : *i-sughlù* « advancement; weak»
Ewondo : *i-sughlù ; i-sughlèn* « end »
Ewondo : *i-kulbù* « bent, bend» syn. *vîrbe, mvanbe ; nyighbe, ŋgonbe* etc.
Ewondo : *i-kolbo* « wake up»
Ewondo : *à-kùl ; ùn-ŋkukul* « hump; hunchback; bud; outgrowth; paw»

Ewondo : *ùn-kulbù* « He who is sad, quie»

3.4.3.25. ***Example N°25***  *(*Faulkner p. 304, Wb. V p. 368, 11-13*)*

Middle-egyptian:  ***tmHw*** "Tjemeh, Libyan (meridional)"

Endogenous method

Sign U33  of the Gardiner's list "Pestle; stick"
Ewondo: /*ùn*-ntɔŋ/ "stick"; **Ewondo**: /tɔŋ/ "horn"
Ewondo: /*ùn*-mor *ùn*-nam/; "chief, head"
Ewondo: /*ùn*-nloh/; "chief, head"

Sign G17  of the Gardiner's list "owl"
Ewondo: /à-kouŋ/; /i-vou/; /ùn-nəmə/; /ùn-mbokh/; /mbgəl/; /ba/ etc. "owl"

Sign V28  of the Gardiner's list "Twisted linen lock"
Ewondo: /*ùn*-mbokh/ "lock"
Ewondo: /*ùn*-mvokh/ "country, land, village etc."
Ewondo: /*ùn*-ntɔŋ/ "feeder"
Ewondo: /tɔŋ/ "horns"

Sign M17  of the Gardiner's list "Reed in bloom"
Ewondo: /à-sam/ "flower; blossom; bloom"
Ewondo: /*ùn*-ntoŋ/ "young shoot"
Ewondo: /tɔŋ/ "horns"

Sign T14  "throwing stick; mace"

Ewondo: /à- yi**ŋ** / "Cross sticks in trap holes"
Ewondo: /ù-nə**ŋ** / "foreign"

READING

This script is read then /à-yɔŋ mi-'nlaba/

This ling is fully validated by the script  in which the sign "Pestle" is replaced by the sign V13 "Rope to tie animals".
By metonymy, this sign is associated with horns. The horns are named in Ewondo / mi-nlakh/. The horned animals usually use it to defend themselves by pricking / à-lab /.

Evolution of the word over the time

With the occupation of ***km.t*** by the Greeks towards -330 of N.E, the root Ekang / * laba / will become Greek (ancient Greek: Λίβυες Líbues). After the Greeks the Latin will vocalize the word Libyes.

Historically, the / mi-'nlaba / are attested since the end of the Bronze Age as inhabitants of the region near the Egyptians. They appear on many historical documents under the name of: .[111] One could think that between the current /Laba/ Ekang variant and the transliterated variant ***rb.w*** it is just the fact that the liquid consonant L being absent in Egyptian which has indeed been replaced and transliterated by the voiced uvular fricative consonant R, confirming the shape of the skeletal potential of consonants ***lab.w*** from the Ekang vocalization.
The suffix ***w*** serving as plural is vocalized ***mi***- in Ewondo. The remaining is therefore a nominalization by deverbal derivation / á-lab / "to sink, to prick"; / *ùn*-nlaba / "whoever shoves, who stabs".

In fact, the skeleton ***rb.w*** is a synonym of the Ekang /*laba/:

> *"BÔR: (b) vb. To splash, throw something wet, throw mud. A* ***bôr*** *me ntôma e nyôl, he splashes me with mud. Syn.:* ***lap*** *(h),* ***leb*** *(b)."*[112]
>
> *(S.GALLEY, 1964)*

So we find there the two variants.

The word sheep in Fulfulde is: *njawdiri mbaala*. Ewondo has kept / *ùn*-mbaala / for "gardian, shepherd". This variant is ananym of / * laba /. We can read this script / mi-'nlaba /, / mi-mbaala / or / bor minlaba / "the shepherds".

Nomadic people and stockbreeders, the term /mi-'nlaba/ associated with them refers mainly to splashing. From my point of view, / Mi-'nlaba / could therefore be associated with the French word "bouseux" (*sludger*). The word variant in the kweke branch could give / *ùn*-mbulughu / "the dusty ones".

> *"ALABA: pl. melaba (vb labah) . Mud we have on our feet. Dirt on the mouth after eating. Melaba gives me my hands, I have splashes of mud, especially the body."*[113]
>
> *(S.GALLEY, 1964)*

The oldest references known of the / mi-'nlaba / date from ***Ra-mssw*** II, and from his successor ***Mr-n-Ptah***, pharaohs of the nineteenth dynasty of Egypt, in the thirteenth century BCE. LBW appears as an ethnic name on the ***Mr-n-Ptah*** stele.

At the level of meaning, This script  can also be compared to lingala *Ku Bola* or *Ko pola* "To rot". whose Ewondo kweké version corresponds to / i-bolo / "rot".

In ancient Egypt, however, the necropoles were located on the western side where the southern branch of the "mi-'nlaba /" Libyan "people lived west of the Nile. So by naming these people we also gave their geographic positioning. I will approach this expression / mi-'nlaba / with French term *occidentaux* "Western"

111. A. ERMAN., H. GRAPOW, Wörterbuch der aegyptischen Sprachen vol II, Leipzig, 1926-1950, P.414
112. Samuel GALLEY, Fang-French dictionary. and French Fang followed by a fang grammar, Neuchâtel, 1964, p.76
113. Samuel GALLEY, Fang-French dictionary. and French Fang followed by a fang grammar, Neuchâtel, 1964, p.34

It is for this reason that the country of / mi-'nlaba / is associated with the land of the dead in the consciousness of the Ekang peoples. In fact these people are likened to ghosts / kon / because of their pale complexion.

The other name to designate (mi-'nlaba) is related this time to their pale complexion. This word has persisted until our days. ***THnw***: /mi-ntaŋën/. Lemma associated with the leucoderms as of the XVth century. It must also be associated with the ability to haggle prices also.

Chefs lebus : peinture du tombeau de Sethi 1er XIXe Dynastie vers 1300 av. jc *(d'après O.Bates)*

3.4.3.26. ***Example N°26***  (Faulkner p. 283, Wb. V p. 93, 12-14 sq.)

This script was chosen because it is part of the basic lexicon and thus its understanding can be easy for a someone who learns a new method.

Sign D28 of the Gardiner's list "Arms stretched as if to kiss?"
Ewondo: /à-kaŋ/ "to rent"

Sign G1 of the Gardiner's list "egyptianVulture (*Neophron percnopterus*)"
Ewondo: /ndoé obama/; /*nya/; /i-toa/; / kubbe/; /i- kobo/ etc. "vulture"

Sign X1 of the Gardiner's list "bread"
Ewondo: /*ùn*-mfəma/ "bread"
Ewondo: /i-kaï/ "supplies"

Sign F51B of the Gardiner's list "Piece of flesh"
Ewondo: /*ùn*-nson/ "meat"

The curiosity offered by the scribe indicates that the term for the male and female reproductive organ bears the same name in our languages.

We could indeed name the male and female sexuals in our languages and easily demonstrate the link with the scripts above.

Ewondo: /ngout fàm/ “male sexual”
Ewondo: /ngout minga/ “female sexual”
Ewondo: /i-sàkh/ “sexual”

However, the African thought proceeding complexity, I propose not to name only the organs regarding all-or-nothing basis like Egyptology but contrary to understand the logic that underlies the reading of this script. I therefore suggest to reconstruct the predialectal ancestor of the transliterated variant ***kA.t*** of Egyptologists. Indeed, there are other variants from the proto-language which allow to better understand how the lexicogical field was constituted in the African linguistic varieties. For this occasion, I will analyze a case related to the Kweke branch. One could indeed do the exercise also with the kekwe branch or the kikuki branch.

Middle-egyptian: *kA.t*
Negro-egyptian: *ŋ^{jw}ə-k^{w}əK^{h}. t^{w}i*

I just remind that we have here:

- agent prefix: ***ŋ^{jw}ə-*** “the one who”
- Le radical: ***-k^{w}əK^{h}-***
- Le suffixe: ***-t^{w}i***

I'm not going to deal with the affixes that participate in the process of radical grammaticalization and focus only on the possible modifications of the root.

The radical

Negro-egyptian: **k^{w}əK^{h}-*

This radical kweke will give the main variants

Negro-egyptian: **wəh-t^{w}i* — **Middle-egyptian**: *waa.t “goat”*
Negro-egyptian: **wəK- > -βəni* — **Middle-egyptian**: *bA “sheep”*
Negro-egyptian: **Kəh- >-kəi* — **Middle-egyptian**: *kA “taurus”*
Negro-egyptian: **KəK- >-kəni* — **Middle-egyptian**: *kAkA “bush”*

Sharing all the same protovariant, these different roots will be associated with the affixes to create the words the sexual lexicon in the different African linguistic varieties.
The following association ***ŋ^{jw}ə-wəKh-*** leads to the transliterated variant ***ngAw*** in Middle-egyptian represented on the script by the bull and yet not transliterated by egyptology here for example

It is therefore these different associations that will give the multiple meanings literally or figuratively that are related to this script.

De facto, the words used by Africans to name a reality are necessarily good. Egyptological transliteration appears as a real obstacle to the appreciation of the reality recorded through the scripts.

Once this ancestor is reconstructed, we will have different possible cases that will participate in the construction of the lexicon related to the variant identified as ***kAt*** by Egyptologists. The variant vocalized *ka.t* comes from the root ***-KəKh-twi> kəh.ti> kA.t***
This is only a special case among all the possibilities offered by this script taking into account in our languages all homophones and synonyms.

After this aside, may we go back to our script to better understand what’s happening.

Ka = Taurus

The scribe reports us that the genitals of the woman or man are named by the signs below. Do we understand well, not only the signs above but also their homophones, and their synonyms:

If I illustrate it, that's what it gives for example:

Ewòndo		Ka.t = Vagin	Lexicon
	I-mvoul	Mə-vul	Genitals
	Nyakh	Mə-nyakh	Fecal excretion
			Parturition
	A-yokh (Agressif)	I-yɔkh	Curse
	ùn-ŋkaŋ	ùn-ŋkaŋ	Phallus root
	à-kaŋ	i-kaŋ	Joy, pleasure
Bovid = dog	i-bɔn	i-bɔn	Vagina
	i-bon	i-bone	What makes barking
	i-bwan	i-bwan	Lover
	i-bɔŋ	i-bɔn	Sex
	i-bɔŋ	i-bɔn	Layer
	*Nya (The mother)	bone	Children
	ùn-nsoŋ	ùn-nson	Soft part, flesh
	kabat	i-nyiŋ	life
(iwf)	fôl	I-fél	The scar / wound

Chart 82:Few hidden senses and obvious senses related to goat

If we understood how it all works, I could extend this list almost infinitely.

Other language variants:

Ka = Taurus

If we have understood the origin of the words, and how the meanings are organized, let us see how at the first level of comprehension the distribution according to the African linguistic variants compares to the Ewondo:

Middle-egyptian:  ***kA.t*** "vagina"

Other African synchronies	Ewondo
Wolof: Kat "to fuck, make love"	A-sakh: "to fuck, make love"
Kikongo: likata "phallus"	à-takh: "enjoy, exhult"
Tunen: etek "vagina"	i-təkh: "what is soft"
Fulfuldé: kuttu "vagina"	à-kout: "Hit / beat"
Fang: "ngout"	Ngout: "private parts; pudenda"
Moor: Kind "vagina"	*Ùn*-kandi "trunk; "

Kikongo: mboula "vagina"	Mə-voul "pudenda"
Basa'a: mbèè "vagina"	Mbèh "door, drum, giving birth"
Kikongo: Kata "phallus".	à-takh: enjoy, exhult
Lingala: libolo "vagina"	i-bol "gamy"
Kikongo: Mfuni "Anus"	à-fòn "enjoy, exhult"
Anglais: Cat "gib, puss; kitten"	*i*-tàkh "joy"
French: chatte "female sex"	Zout " pudenda"
Ngiemboon: Leka "penis"	*ùn*-ŋkaŋ "penis"
Ngiemboon: Ka'a "copulate"	à-koa "to crush"
Moor: Ka'a "to nail"	à-koa "to crush"
Ciluba: ci Tuuka "rump, behind"	à-tukh "to scrub, to rub, to chafe"
Ciluba: Kunshi "bottom, lower belly, vagina"	à-bumsi "bottom, lower belly"
Ciluba: Ku Tuuka "Fall, detach"	à-tukh "to scrub, to rub,"

Chart 83: multivariate crossword analysis of the genitals

The words of the semantic field of the genitals are vast. Thus the different levels of languages in Africa will often mobilize the hidden meanings and are accessible only to the true initiates. Not only can they be built from the following words but also on the basis of homophones and synonyms of each of the following terms.

From a small demultiplexing we can propose all these words that are directly related to the genitals:

bull, ox
herd
donkey, donkey
small livestock; goats; (big) livestock, livestock; herds; men
green; blade; fresh; thought; vigorous, robust, strong, energetic; lucky, happy; to make green, to revive; to bloom
thick
thick door, gate
thick clothing
mass, multitude
thick wall (idiom.); thickness
parent
watermelon
mother
fighting bull
ship, driven; muscular
mother
bowl
band (idiom.)
concubine;
young lady
just, accurate; exact; usual;
customary; usual;
accuracy, precision
straightness, accuracy
semen, seed, offspring; poison
dagger sheath
talk, discussion
reward, retribution
exact time, good time
testimony (idiom.)
arena
road, path; course, direction
husband and husband
burn with desire for
category of bulls; livestock
phallus
cow
three; to do for the third time; triple
project; think about; expect to
in the absence of, without (idiom.)
third
thicken; strengthen
hear listen
listen to doors
witness; consider; investigate
adorn, decorate
fertilize; beget
pork
bull, ox
victorious bull
Kemour, the Great Black
male man bull
foreskin
stiffness;
constipation,
world; mate

bear witness; to exhibit, to show; load; give
to be of good reputation
reputation
witness, testimony
flood
testimony,
certificate
midday

Etymology

We note that curiously the French word *chatte* often used to designate the female genitals. It finds here its origin as well as the associated semantaxe. How to understand that in French the feline **cat** can designate the a woman sex?

The term *moule* "mold" to describe the sex of the woman in French still finds its origin here. The anteriority of Man on African soil and therefore the anterioty of the African languages push us to admit that the word Ewondo /i.bɔn/ "vagina" which is homophone of / i-boŋ / "the mold" can justify this association of ideas. Indisputably the semantaxe has gone through the ages including the thick cloak of ideology.

In the etymological dictionaries, it would be good to review the etymology of some words and to put them back in their original context, which is not necessarily Greek or Latin but rather African.

3.4.3.27. ***Example N°27***:  (Faulkner p. 273, Wb. IV p. 556, 7 (Stw))

Middle-egyptian:  ***Styw*** "tortoise; turtle"

Endogenous method

Sign N37 of the Gardiner's list "Garden pond"

Ewondo: /i-tam/ "basin, well"; /*ùn*-mbokh/ "well; sink"

Sign G4  of the Gardiner's list "Fierce Hawk (*Buteo rufinus*)"

Ewondo: /ô-bam/; /ô-bi/ "Falconidae"

Sign X1 of the Gardiner's list "bread"

Ewondo: /i-kɔh/; "cake"

Reading:

There are two ways to read this script:

1. Carefully avoid left signs called phonetics complements and we go to the right sign that we recognize and then we perform the reading by identification of the sign I2.
2. All phonetics complements are taken into account as each of these signs allows to vocalize the sign I2.

To method 1's riddles, I propose to substitute the second way. Why that? Just because African thought is ambivalent, it works with hidden meanings and evidences.
By principle therefore, the signs on the left are the signs that convey the hidden meanings of the sign I2 .

- What is the link between the sign N37 and the determinative I2 ?

To answer this question, we must name the sign N37.
To do this, we must ask ourselves a number of questions:

What is the meaning of the sign N37?

You are not required to comply with what has been defined as a standard by Gardiner's list. In fact, we carefully note everything we see. On this account, we must mobilize synonyms and homophones since phonetic complements are hidden meanings of sign I2 .

What I see	Homophones
Ewondo: /*i*-tam/ "basin, well";	**Ewondo**: /*i*-tám/ "alone"
Ewondo: /*ùn*-mbokh/ "well, hole"	**Ewondo**: / *ùm*-bokh / "one"

Chart 84: Signification of signe N37

What is the sign N37 for?

What is the obvious link between the function of the sign N37 and the sign I2 .?

What I see	Homophones
Ewondo: /à-lèh/ "draw";	**Ewondo**: /i-lèh/ "three"
Ewondo: /à-nyî/ "enter"	**Ewondo**: / à-nyu / "mouth"
Ewondo: /à-kouî/ "go out; enter"	**Ewondo**: / à-koh / "foot"

CHART 85:LINK BETWEEN THE FUNCTION OF THE SIGN N37 AND THE SIGN I2

Absolutely find the link with the sign I2 .

So we harvest all the homophones of the reality that we have seen and so we determine exactly what we are talking about.
We see here that we are talking about the ancestor of the Ekang peoples.

The representation of the ancestor of the African peoples in the chelonians order is:

Ewondo: /kulu/ "tortoise"
Ewondo: /Kulu/; /Kɔlɔ/; /*ùn*-nkulu/ "the eldest ancestor"

The representation of the ancestor of the African peoples in the ornithological order is:

Ewondo: Sign G4 /obam/ "falconidae"
Ewondo: /Obama/; /Obam/ "the eldest ancestor""
Ewondo: /Kulu/; /Kɔlɔ/; /*ùn*-nkulu/ "the eldest ancestor""

The representation of the ancestor of the African peoples in the order of the pasta used for food is:

Ewondo: Sign X1 △ /i-kɔ/ "dough, cake"
Ewondo: /à-kôh/; /à-kul/ "foot, leg"
Ewondo: /Kɔlɔ/; /Kulu/ "the eldest ancestor"

So here is the link established with each sign and how to read this script in Ewondo.

Technical foundation

If one had to bow the diktat of Egyptology specifically to respect what Egyptology has laid down as rules n this account one would have:

Middle-egyptian:  ***Styw***

In this account, it should already be noted that the transliterated sign **S** which serves as a radical here is the N37 ▭. This script would then transliterate ***S.tyw***. The suffix part (*.tyw*) will be decomposed as well.

So we have:
- *un*-mbias y (nisb / duel) (leading to a doubling of the transliterated sign S
- The presence of twins (.y) justifies (.w) for the plural.
- And then we have the (.t) which would designate a feminine gender.

Taking into account what we have written about the sign N37, we have:

Ewondo: ▭ /*ùn*-mbokh/ "well, hole"

We therefore have two possible cases taking into account the constraint of the twin *.y ùn*-mbiè (umbias):
Ewondo: /*ùn*-mbobokh/ "disabled"
Ewondo: /i-bobokh/ "the baby"

Since on the morphosyntactical level Egyptology enacts to take the feminine version (.t), then we have:

Ewondo: /i-bobokh/ "the baby"

If this word designates a hidden meaning of the eldest ancestor, it is used above all to name the baby. In fact, if I want to name the animal represented by the sign I2 this can be a big issue if I do not know anything about the African culture.
What do we observe?
We are dealing with an incommensurable online loss because / *ùn*-mbobokh / "the handicapped" it is this way that we also qualify the eldest ancestor. Moreover, if I was givent the opportunity to name the animal on the right / kulu / Ewondo and not the ancestor, on this account it would be almost impossible from the ***S.tyw*** variant.
The ***S.tyw*** variant will correspond to some language variants which respect the constraint of (.y) which induces a repetition or a superlative variant of our languages:

Bahasa Indonesia: kura-kura "tortue"
Bambata: Koro kara "tortue"
Lingale: *mu*kulukulu "chief"

If we made a comparison with the other African linguistic variants, the sign I2  would not necessarily be vocalized Kulu. It could then be with other descriptors. On this account, the signified remains the same, the signifiers alone will differ.

Other variants	Ewondo	
Fulfuldé: Kekou "tortoise"	*Ako*	The foot
Hausa: kùn-kurū "tortoise"	*Ako-nkor*	foot of the savior
Kɔ́ɔnzime: kûl "tortoise"	*à-kul*	Paw
Kinyarwanda: nyamasyo "tortoise"	*i-Nyié mə-sè*	Breastfeeding of the feather
Baoulé: àbú "tortoise"	A-bo	leg
Bété: á-tíá-kwú "tortoise"	*U-nti'ène à-ko*	The attached food
Lingala: kóbá "tortoise"	*U-koba*	The bigining
Bahasa Indonesia: kura-kura "tortoise"	*Un-nkoré*	The great savior
Bambata: Koro kara "tortoise"	*Un-nkoré*	The great savior
Ɓàsàa: kúl "tortoise"	*A-kul*	The paw
Yemba: pfáŋkɔ̀ "tortoise"	*U-mfaŋ'ène à-ko*	attached foot / The true foot
Isizulu: ufudu "tortoise"	*i-vu ndo*	Ndo's Placenta
Oluganda: nfuddu "tortoise"	*i-vu ndo*	Ndo's Placenta
Nyamwasi: gùlùmàątì "tortoise"	*à-kol maâ*	maat's foot
Tika: akoi "tortoise"	*à-kôh*	The foot
Soninkanxaane: le:le "tortoise"	*ùn-nlô*	The head, the chief
Wolof: mbonnat "tortoise"	*U-mbond*	The parent
Ngiemboon: Pfomankwo "tortoise"	*ù-mfaŋ'ène à-ko*	Foot attached
Kikongo: Bonga "tortoise"	*Bongo, à bong*	Children knee
Dogon: Kri "tortoise"	*Un-nkoré*	The Savior
Dioula: Srakôkôman "tortoise"	*Man Ako*	Small foot
Adja: Klokpè "tortoise"	*A-kol pəpa*	Father's foot
Duala: wudu a mundi "tortoise"	*Mwan ndo*	Ndo's son
Ngad'a: Abou "tortoise"	*A-boh*	Foot
Diola: kang "tortoise"	*A-kang*	Praise, the people
Malagasy: Sokatra "tortoise"	*i-sokh tara*	The placenta of the father

Chart 86: Cross analysis of tortoise between African languages and Ewondo

Etymology

The french word *tortue* "turtoise" originated in **Tartare**[114], region of Hell, in Greco-Roman mythology. What is the connection this animal has with the hell?

> **Grec ancien**: χέλυς "tortue" (*ce vocable rappelle trangement le mot Jesus*)
> **Latin:** tartarūca "tortue"
> **French**: torture > tortule >tortuhe > "tortue"
> **English:** tortoise; turtle

The word turtle originates in African thought because ***wsir*** was tortured by his brother ***stX***. This is why he was described and call the tortured. So we say that the word turtle must have an etymology in African thought. Greece has indeed shared a common past with Africa in its pre-Hellenic period and after the invasion of Dorians including around -330 with Alexander the Great. The Greeks therefore knew about African mythology.

> *"A gigantic study was conducted in 2000 by the Department of Immunology and Molecular Biology of Madrid, Spain, and the Skopje Human Group and Blood Transfusion Laboratory in Macedonia aimed to determine precisely the origin of the first inhabitants of ancient Greece. This study brought*

114. Lexicographical and etymological definitions of "turtle" of the Computerized French Treasury, on the site of the National Center for Textual and Lexical Resources

together a wide range of scientists. To carry it out, many blood samples were taken on three geoscripts zones:
The European world (French, Italians, Spaniards, Portuguese, Macedonian Greeks, Cretans, etc ...).
The Semitic world (Morocco, Algeria, Turkey, Iran, Armenia, current Egypt, Israel, Lebanon, etc ...).
The African world (Senegalese, Mossi, Rimaibe, Fulani, Oromo, Ahmora, Bushmen, South Africa).
All the DNA sequences of the peoples have been examined, analyzed and classified, in order to identify the sequences common to the DNA traces found on the current inhabitants of Greece.
The first inhabitants of ancient Greece had a sub-Saharan (Ethiopian) ethnic origin, which clearly separated them from other Mediterranean-type DNA groups.
Only ancient Greeks and Africans share a common sequence of DNA. Genetic distances are therefore closer between ancient Greeks and Africans than any other human group.
Only ancient Greeks and Africans share a relatively long sequence of common DNA (DRB1), namely: 0305, 0307, 0411, 0413, 0416, 0417, 0420, 1110, 1112, 1304, and 1310.
***African Group 1**: Ahmara, Oromo, Mossi, Fulani, Aegean Greeks, Greek Cypriots, Greek Penthouses, Rimaibe,*
***African Group 2**: Bushmen, Senegalese,*
***Group 3**: rest of the Mediterranean world (Semites, Europeans).*
Surprisingly black African groups have more connections with the original Greeks. Among them are:
Oromo 43.26
Ahmara 51.74
Mossi 53.46
Rimaib 55.95
San (Bushmen) 57.78
the Fulani / FULANI 61.01
This DNA study concluded:
The reason why the Greeks did not show a close relationship with all other Mediterranean analyzed was their genetic relationship with the sub-Saharan ethnic groups currently residing in Ethiopia, Sudan and West Africa (Burkina-Faso). Although some DRB1 greeks are not quite specific to the Greek / sub-Saharan split, the list of alleles is self-explanatory. The conclusion is that part of ancient Greece can be sub-Saharan and mixing occurred at an uncertain but ancient time.
The origin of the black ethnic groups in West Africa (Peul, Mossi and Rimaïbe sampled in Burkina Fasso-) is probably Ethiopian (26, 27) (Fig. 4). The Fulani are semi-nomadic hunters and gatherers and one of the few inhabitants of the region to use cow's milk and bran products for food and trade. The Rimaïbe Blacks were serfs belonging to the Fulani and often mixed with them (27) (...)"[115].

This data relating to the Greek past is little commented but it disturbs because the Greek-Latin European myth collapses like a house of cards.

Is it the proximity between the Greeks and Africans that led them to the IMF or to suffer the wrath of countries ranked in the group of PIG (pork)?

Historians Admit To Inventing Ancient Greeks

"Scholars apologize for attributing Western democracy to a make-believe civilization.
WASHINGTON—A group of leading historians held a press conference Monday at the National Geoscriptic Society to announce they had "entirely fabricated" ancient Greece, a culture long thought to be the intellectual basis of Western civilization.
The group acknowledged that the idea of a sophisticated, flourishing society existing in Greece more than two millennia ago was a complete fiction created by a team of some two dozen historians, anthropologists, and classicists who worked nonstop between 1971 and 1974 to forge "Greek" documents and artifacts.

115. (Source of Quote and Study of DNA: Spanish University of Complutense)

"Honestly, we never meant for things to go this far," said Professor Gene Haddlebury, who has offered to resign his position as chair of Hellenic Studies at Georgetown University. "We were young and trying to advance our careers, so we just started making things up: Homer, Aristotle, Socrates, Hippocrates, the lever and fulcrum, rhetoric, ethics, all the different kinds of columns—everything." Just one of the "ancient" artifacts dreamed up in a basement in Somerville, MA.

"Way more stuff than any one civilization could have come up with, obviously," he added.

According to Haddlebury, the idea of inventing a wholly fraudulent ancient culture came about when he and other scholars realized they had no idea what had actually happened in Europe during the 800-year period before the Christian era.

Frustrated by the gap in the record, and finding archaeologists to be "not much help at all," they took the problem to colleagues who were then scrambling to find a way to explain where things such as astronomy, cartoscripty, and democracy had come from.

Within hours the greatest and most influential civilization of all time was born.

"One night someone made a joke about just taking all these ideas, lumping them together, and saying the Greeks had done it all 2,000 years ago," Haddlebury said. "One thing led to another, and before you know it, we're coming up with everything from the golden ratio to the Iliad."

"That was a bitch to write, by the way," he continued, referring to the epic poem believed to have laid the foundation for the Western literary tradition. "But it seemed to catch on."

Around the same time, a curator at the Smithsonian reportedly asked for Haddlebury's help: The museum had received a sizeable donation to create an exhibit on the ancient world but "really didn't have a whole lot to put in there." The historians immediately set to work, hastily falsifying evidence of a civilization that— complete with its own poets and philosophers, gods and heroes—would eventually become the centerpiece of schoolbooks, college educations, and the entire field of the humanities.

***Emily Nguyen-Whiteman**, one of the young academics who "pulled a month's worth of all-nighters" working on the project, explained that the whole of ancient Greek architecture was based on buildings in Washington, D.C., including a bank across the street from the coffee shop where they met to "bat around ideas about mythology or whatever."*

"We picked Greece because we figured nobody would ever go there to check it out," Nguyen-Whiteman said. "Have you ever seen the place? It's a dump. It's like an abandoned gravel pit infested with cats."

She added, "Inevitably, though, people started looking around for some of this 'ancient' stuff, and next thing I know I'm stuck in Athens all summer building a goddamn Parthenon just to cover our tracks."

Nguyen-Whiteman acknowledged she was also tasked with altering documents ranging from early Bibles to the writings of Thomas Jefferson to reflect a "Classical Greek" influence—a task that also included the creation, from scratch, of a language based on modern Greek that could pass as its ancient precursor.

Historians told reporters that some of the so-called Greek ideas were in fact borrowed from the Romans, stripped to their fundamentals, and then attributed to fictional Greek predecessors. But others they claimed as their own.

"Geometry? That was all Kevin," said Haddlebury, referring to former graduate student Kevin Davenport. "Man, that kid was on fire in those days. They teach Davenportian geometry in high schools now, though of course they call it Euclidean."

Sources confirmed that long hours and lack of sleep took their toll on Davenport, and after the lukewarm reception of his work on homoeroticism in Spartan military, he left the group.

In a statement expressing their "profound apologies" for misleading the world on the subject of antiquity for almost 40 years, the historians expressed hope that their work would survive on its own merits.

"It would be a shame to see humanity abandon achievements such as heliocentrism and the plays of Aeschylus just because of their origin," the statement read in part. "Moreover, we have some rather disappointing things to tell you about the pyramids, the works of Leonardo da Vinci, penicillin, the Internet, the scientific method, movies, and dogs.""[116]

It's true this is another matter. Let's just stay on our small talkative drawings.

116. https://www.theonion.com/historians-admit-to-inventing-ancient-greeks-1819571808

3.4.3.28. ***Example N°28*** (Faulkner p. 245, Wb.? p.?)

Middle-egyptian: ***ssm*** "cheval"

Endogenous method

Sign O34 of the Gardiner's list "Verrou"

Ewondo: /à-tiŋ/ "node; knot; knob"
Ewondo: /*ùn*-nkwal/ "rope, string"
Ewondo: /*ùn*-hia/ "belly's rop, intestine"

Sign G17 of the Gardiner's list "owl"
Ewondo:/à-kouŋ/;"owl"

Analysis

What i see	Hidden Meanings
Ewondo:/à-kouŋ/;"owl"	**Middle-egyptian**: bA "soul"
	Ewondo: /*ùn*-nsisim/ "spirit; soul"
	Ewondo: /kôn/ "spirit; ghost"
	Ewondo: /kôl/ "string; ancestor"
	Ewondo: /*ùn*-mbombokh/; "owl"

CHART 87: HIDDEN SENS OF THE SCRIPT

This script is very interesting because it allows you to get to the bottom of things. From the standpoint of Ewondo, the horse for the scribe is a spirit. The **ssm** / **ssm.t** transliteration of Egyptology refers in my language to the word / *ùn*-nsisim / which designates the spirit .

Remembering is about reconnecting by the spirit. We use the word / à-simzèn /; / á-simsèn / to translate the idea of remembering ***sxA***.
The skeleton of consonants that one can draw from it would be: *sms*.

What is the link made by the scribe between the horse and the spirit? Either Egyptology was wrong or there is a hidden relationship between the spirit and the horse. Let's try to discover it.

Reading

Fang: ŋkalbenya "horse"
Ewondo: /i-kabli/; " horse "
Kɔ̀ɔnzime: kab'lá " horse "
Ewondo: /kabla/; "To be glued, to stick" Middle-egyptian: ***qnb***
Middle-egyptian: ***Hnsk*** "attach, to knot; to tie"

The link we establish in this example is a synonymous link. The words: tying, binding, sticking and the word, remember, spirit are synonymous. The term for designating the horse is therefore a homophone to stick in Ewondo language or in Ekang languages.

Let's sum:

Signs	What i see	Hidden sense
	A-kuŋ "owl"	*Un*-nsisim "spirit"
	A-tiŋ "node"	A-simsèn "to remember"
	bi-kabla "links"	A-kabla "to glue, to stick"

I-kab'li
(Horse)

CHART 88: READING METHOD OF SSM

The Horse in African languages

Duala: Ekabala "horse"
Fulfuldé: *puccu* "horse"
Hausa: *doki* "horse"
Baoulé: *nnɛ̃ÿ-kpã́ŋgɔ́* "horse"
Yemba: *lúɔ́* "horse"
Ngiemboon: *Lúoon* "horse"
Senoufo: *chong* "horse"
Gébé: *sô* "horse"
Soninké: *sy* "horse"
Lingala: *mpunda* "horse"
Seereer: *piss* "horse"
Bozo: *Chiy* "horse"
Etc.
Moore: *Wedmoaga* "horse"
Ashanti: edin "horse"
Tupuri: *píri* "horse"
Mundang: *di* "horse"
Mossi: *Woued moaga* "horse"
Fali: *ni*ri* "horse"
Koulango: *sôh* "horse"
Wolof: *fas* "horse"
Makia: *kpli* "horse"
Basa'a: *hɔ́sì* "horse"
Gouran: *aski* "Horse"
Bamanan: *so* "horse"

Etymology and deconstruction

English: *horse* "nag, bolter"
Ewondo: /*ùn*-nsɔrɔ/ "the phallus"

Horse and /*nsɔrɔ / are just two ananyms. They proceed relatively each other by phonemes inversion. It is thus useful to know african language to find the reading related to the penis. Just because this sign  is related to the word /ndəm/ "scar; mark" this lemma admits the homophone /ndəm/ "sperm; semen, cum". If you are just seeing an owl but only an insider of african culture can see the sperm and then the cock though the signe G17.
The sign G17  can share the following meanings:

Ewondo: /*ùn*-yəmə/ "The intelligent"
Ewondo: /i-yəm/ "animal"

According to the item /i-yəm/ We have then the word /*mə*-yôm/ "sperm" sharing a relation with the penis by the sign D53 . Therefore, for the scribe, penis and sperm are linked in african thought. By the way if we name the sperm we are naming the penis also. I can establish other links but i think these example are enough.

That's why and how, according to me, by metonymy, there is a relationship between the owl and the penis.

Middle-egyptian:  ***ngAw*** "bull or long-horned ox"
English: Nag "horse; bolter" ☞**kàllaama wolof**: *nag* "cow"
Ewondo: /nyakh/ "bull or long-horned ox; shit"

There is a misinterpretation of the notion of the name found in Western grammars. To the extent that, the African signifiers are not necessarily names. They are, in my opinion, **qualifiers** associated to a signified. In this case, the lemma above is the characteristic associated to some animals that often shit and walk on their poop.
Of course they are used as names. We have to assume here that what is discribe above is the fact that those animals are not like cats who take care to hide their faeces. The root of the word / * nyak / is associated with the verb / á-nyak / "defecate". It is thus this root and this word that one still finds present in the English and the Wollof word **nag**.
I want to insist on this point because the English word *nag* is still used nowadays in the Wolof language as well as all its variants.

Wolof: *nag; nak* **"**bull or long-horned ox"
English: *nag* "horse; bolter"

We can observe this curiosity:
Wolof: *fass; nak* **"**horse"
Serrer: *Fast* "horse"
English: fast; horsepower

One wonders whether the word *fast* is not associated in the unconscious with the speed that characterizes the horse? We note that the word *fast* still refers to the horse till nowadays in African languages. Do you understand what I mean or not?
Ewondo: /á-fass/ "to think" we have the relationship with the sign G17 . In Africa we often associate the word /á-fass/ "to think" with /à-tiŋ/ related to the sign .
The whole expression is Ewondo: /á fass à-tiŋ/ "examine the knot".

Did you note the same thing as I do?
Ewondo: / à-tiŋ/ ☞ **English:** *think*
Ewondo: / à-fass/ ☞ **English:** *fast*

<u>bolter</u>

I wanna emphasized that people who said that they are reading ***mdw ntr*** are quacks. Concerning the word bolter we have the root *bolt*.

The best description i could give to the bolt is a node as this representation: .
The relationship between bolt and Ewondo language is establish with the word /i-bol/ "decayed; rotten". Remember that i have explain the fact that nag is shit.
What is rotten smells bad. And by metonymy smells like shit.

English: *bolter* "horse"
Ewondo: /*bol-/ "rotten; decayed"☞ see illustration N°11
Ewondo:/i-bolé/ "breakable**"**

We are not gonna discuss about the origin of the word *cavalry* because with what we have seen, it is linked with the word *ekabala* above.

So the commonly accepted but false idea of making the horse by ideology a foreign animal to the African bestiary is an idea that must be arased.

> *"ÑKALBEÑYA: n.2, pl. miñkalbeñya. Cheval. Comes* ***from Spanish****: cavala.*
> *In Bulu: ekabala."*[117]
>
> *(GALLEY, 1964)*

Africa did not wait for the Spanish in the 15th century to discover equines and especially the horse. How is it possible that Africa could have all equines except the horse? How to explain the African terms that designate the horse without having any phonetic proximity with the cavala version of Spanish? How do we establish the precedence of the Spanish on the Negro-egyptian and the African languages?

We say that the presence of horses is attested [118] at ***km.t***, the presence of a complete lexicon (groom, royal stables, stables etc.) revolving around the horse did not fall from the sky either. Monumental frescoes present Ra-Mss II (-1304 to -1213) on a chariot pulled by horses. Historical stories of battles with horses. All of this is visible to all who do not suffer from mental blindness.

The oldest representation of a king on his chariot appears on a scarab of *kA nxt mri maat* (Thutmose I) and is preserved in the British Museum. All the pharaohs of the New Kingdom, since ***jmn Htp (w) nTr HqA iwnw*** (Amenhotep II) treading the enemies of Egypt or practicing archery, until *DHwtj ms (jw)* (Tuthmosis IV) training at the foot of the Pyramids of Giza, will be represented on their chariot.

It is time for Africa to write its own History and come out of myths that bring Africa under the thumb of ideology. Would It have been necessary that the Spanish come to Africa so that the Africans can discover the horse, does it really make sense to write such tripes? Think about it. I have my own opinion.

3.4.3.29. ***<u>Example N°29</u>: ADAM AND EVE***  (Faulkner p. 17, Wb. I p. 78, 1)

The former kenyan president in 1966 Jomo KENYATTA said one day:

> *"When the white men came to Africa, we had the land and they had the Bible. They taught us to pray with our eyes closed and when we opened them, they had the land and we had the Bible."*

The content of the Bible established as an absolute truth by the others can be submitted to the ***mdw ntr*** to see what the small drawings say. So I chose to experience the story of Adam and Eve .

Adam and Eve, according to the creation myth of the Abrahamic religions were the first man and woman. They are central to the belief that humanity is in essence a single family, with everyone descended from a single pair of original ancestors. It also provides the basis for the doctrines of the fall of man and original sin that are important beliefs in Christianity, although not held in Judaism or Islam.

In the Genesis Book of the Hebrew's Bible, chapters one through five, there are two creation narratives with two distinct perspectives. In the first, Adam and Eve are not mentioned (at least not mentioned by name). Instead, God created humankind in God's image and instructed them to multiply and to be stewards over everything else that God had made. In the second narrative, God fashions Adam from dust and places him in the Garden of Eden. Adam is told that he can eat freely of all the trees in the garden, except for a tree of the knowledge of good and evil. Subsequently, Eve is created from one of Adam's ribs to be Adam's companion. They are innocent and unembarrassed about their nakedness. However, a serpent deceives Eve into eating fruit from the forbidden tree, and she gives some of the fruit to Adam. These acts give them additional knowledge, but it gives them the ability to conjure negative and destructive concepts such as shame and evil. God later curses the serpent and the ground.

117. Samuel GALLEY, Fang-French dictionary. and French Fang followed by a fang grammar, Neuchâtel, 1964, p.290.
118. Catherine ROMMELAERE, The horses of the Egyptian New Kingdom, Bruxelles, Safran - Knowledge of ancient Egypt n°3, 1991.

God prophetically tells the woman and the man what will be the consequences of their sin of disobeying God. Then he banishes them from the Garden of Eden.

This is in a few words the story that has haunted humanity for millennia. But what do the African facts say?

Middle-egyptian: **im** "shape, figure, silhouette; side, flank"

<u>Endogenous method</u>

Sign M17 of the Gardiner's list "Reed in bloom"

Ewondo: /i-sɔŋ/; "reed"
Ewondo: /à-sam/; "flower"

Sign G17 of the Gardiner's list "owl"
Ewondo:/à-kouŋ/;"owl"

Sign Aa15 of the Gardiner's list "shape, figure, silhouette; side, flank"

Ewondo:/i-bèh mviè/ "rib"

Sign F51B of the Gardiner's list "Piece of flesh"

Ewondo:/*ùn*-nson/ "flesh; meat"

Reading

What i see	Hidden senses
Ewondo:/à-kouŋ/;"owl"	**Ewondo**: /kôl/ "string; ancestor"
	Ewondo: /*ùn*-nsoŋ/ "pointy; mother"
	Ewondo:/i-bèh/ "peak; point; thigh"
Ewondo: /i-sɔŋ/; "reed" **Ewondo:** /à-sam/; "fower"	
	Ewondo: /*ùn*-nson/ "flesh"
	Ewondo: /*ùn*-nsoŋ/ "pointy; mother"
	Ewondo: /*ùn*-nya/ "mum, vulture"
Ewondo:/*ùn*-nson/ "flesh"	**Ewondo**:/i-bèh/ "peak; point; thigh"
	Ewondo: /*ùn*-nsoŋ/ "pointy; mother"
	Ewondo: /*ùn*-nya/ "mum, vulture"
	Ewondo: /*ùm*-inga/ "woman"
Ewondo:/i-vs/ "bone" **Ewondo**: /*ùn*-nsoŋ/ "pointy"	**Ewondo**:/i-bèh mviè/ "rib" **Ewondo**: /*ùn*-nsoŋ/ "pointy; mother" **Ewondo**: /*ùn*-nya/ "mum, vulture" **Ewondo**: /*ùm*-inga/ "woman"

Chart 89: hidden meanings of the script
Let's note that in the field of possibilities, the coast, the flesh and the woman are associable in the African complexity.

Let's sum

- Pointy idea

All the signs committed by the scribe for this writing: M17 ; G17 ; F51B ; Aa15 are homophones of sharp.

M17 : /i-soŋ/"reed" **Ewondo** : / *ùn*-nsoŋ/ "sharp"
G17 : /*un*-nsoŋ/ "the beak" **Ewondo** : / *ùn*-nsoŋ/ "pointy"

F51B : /*un*-nson/ "flesh" **Ewondo** : / *ùn*-nsoŋ/ "sharp"

Aa15 : /i-bèh/ ↔ /*un*-nsoŋ/ "peak, sting" **Ewondo**: / *ùn*-nsoŋ/ "pointy"

The other name of the mother ***As.t*** represented by the vulture, the feather on the head and the wings is: ***ns.t*** "seat, throne" in Ewondo: / *ùn*-nsoŋ'o/ "the beak, the throne", /*ùn*-nsono/ "the flesh"; /i-sono/ etc.

She is also called /*nya/; /*na/ "mum" and she is a woman /*ùm*-inga/; a wife /*ùn*-ŋgal/.

To the question of what is a rib, basically it is a bone / *i*-ves / and flesh / *ùn*-son /. It is thus the word / i-vé(s) / which will give by ideological makeup the word EVE.

This is one of the many logical mechanisms that the scribe has written in this script for the future generations that we are. The scribes knew that all this falsification would be exposed.

According to some texts supposedly inspired by "God", the woman would have been pulled from the coast of the man astonishingly. I will be curious to know according to what writings, according to which logic and according to which language?

> *"Genesis 2*
> *21 And Jehovah God caused a deep sleep to fall upon the man, who fell asleep; he took one of his ribs, and closed the flesh in its place.*
> *22 And Jehovah God formed a woman from the coast which he took from man, and brought him to man.*
> *23 And the man said, This time is that bone of my bones, and flesh of my flesh. she will be called a woman, because she was taken from the man"*

We have in this text 3 input data, the man named Adam, the coast, the flesh for an output that is the woman.
We have shown that woman, flesh and rib are exactly the same word by a game of homophony and synonymy in our language.

So let's look at the input data ***Man alone*** named Adam. When we presented the example concerning the turtle / kulu /, I put in the analysis phase, the Chart below. As a homophone of the ancestor we have the qualifier / i-tám / "alone".

What i see	Homophones
Ewondo: /i-tam/ "basin, well";	**Ewondo**: /i-tám/ "alone"
Ewondo: /*ùn*-mbokh/ "hole, well"	**Ewondo**: / ùm-bokh / "one"

CHART 90: THE ONLY ONE

Our mystery is solved therefore it is the word / i-tám / "only" that has been modified in a diachronic logic with a replacement of the voiced alveolar occlusive consonant **T** by the voiced alveolar occlusive consonant **D**. This is how we flip from the word /i-tám/ "alone" to the condensed ideology *Adam, Adama*.

To the question, is there a script that presents the supposed name of Adam. We answer it by the affirmative.

Itm

This idea, taken from the cosmogony of *Iwnw*, will give in Egyptology the word **ATOUM** and by undergoing some facelifts of facades on the field of physical sciences, the word ATOM. I claim that, the skeleton consonant *itm* is an adaptation of the Ewondo word / *ùn*-ntoum / meaning for us "cane, leader" as in Chinese too:

Chinese: 头目 Tóumù "chief"
Ewondo: ùn-ntoumù "chief; pillar of house angles"
Middle-Egyptian:  ***itmw***
Massagui: Toma "chief"

French: mot, team, âme; maître; tamis ; tombe, monté ; tempo ; atome ; etc.
English: intimate; time, tin; tome; tame; stem etc.

Etymology and deconstruction

We invite the ideologists to replace the words: *atoum, atom, adam,* and all the other words in their true original cradle ***km.t*** (according to ***sn-nDm***[119]).

Either those items are extirpated from the poison of ideology, or else we will teach the truth to ours.

We therefore claim that Adam and Eve only existed in the minds of the ideologists. It is therefore necessary that the real children of Africa come out of these fables to reread the texts that concern the origin of the cosmos according to the endogenous and highly scientific vision of ancestral Africa.

3.4.3.30. ***Example N°30***  (*Faulkner p. 79, Wb. I p. 435, 21-22*)

Middle-egyptian:  ***bit*** "crown of Lower Egypt"

Endogenous method

Sign L2  of the Gardiner's list "Bee"
Ewondo: /fɔŋ/; /bo/; / fôfôŋ/; /à-bè/ "bee"

Sign Z5 \ of the Gardiner's list "Diagonal feature according to hieratic drawing"
Ewondo: /i-Bét/; /*ùn*-nkol/ "climb"

119. Sennedjem was an official living during the reign of Tutankhamun at the end of the eighteenth dynasty. Sennedjem was a tutor, but also carried several other titles, such as "divine father", loved by the god, Bearer of the fan to the right of the king. It can be read in his grave that km.t is three regions: tA mry, tA sty; tA ntr.

Sign X1  of the Gardiner's list "bread"

Ewondo: /*i*-kɔh/; /*ùn*-mfəma/ "cake, bread"

We therefore recall that the signs drawn are hidden meanings of the determinative.

What i see	Hidden sense
 **Ewondo**: /fɔŋ/; /bo/ "bee"	**Ewondo**: /fɔŋ/; "chief"
	Ewondo: /nda / "house"
	Ewondo: / bit / "honey"
	Ewondo: /bot/ "people"
 **Ewondo**: /i-Bét/; /*ùn*-nkol/ "climb"	**Ewondo**: /bot/ "people"
	Ewondo: /*ùn*-nsoŋ/ "pointy; mum"
	Ewondo: /tsóŋ/ "red"
	Ewondo: /ngon/ "panther"
	Ewondo: /*ùn*-nya/ "mum, vulture"
	Ewondo: / i-toà / "seat"
 **Ewondo**:/ /i-kɔh/ "cake"	**Ewondo**:/à-koh/ "ancestor"
	Ewondo: /à-koh/ "place; seat "
	Ewondo: /*ùn*-nya/ "mum vulture"
	Ewondo: /*ùn*-mbobokh/ "disabled"

CHART91: HIDDEN MEANINGS OF THE SCRIPT

This script makes it possible to name the leader as:

Ewondo	**sens**	**Homophone**	**sens**
/fɔŋ/	*"bee"*	/fɔŋ/	*"chief"*
/ùn-nsoŋ/	*"pointy"*	/tsóŋ/	*Red (crown)*
/*ùn*-nsoŋ/	*"sting"*	/*ùn*-nsoŋ/	*"chief, summit"*
/*ùn*-nsoŋ/	*"beak"*	/*ùn*-nya/	*"mum, vulture"*
/*ùn*-nsoŋ/	*"beak"*	/ i-toa /	*"Seat"*
/i-koh/	*"cake"*	/à-koh/	*"place"*
/ bit /	*"honey"*	/bot/	*"people"*
/ bit /	*"honey"*	/bwiti/	*"Ekang's cult"*
/ ùn-nya /	*"woman"*	/ngon/	*"panther"*

CHART 92: SOME HOMOPHONES OF OBVIOUS MEANINGS

I could extend the list infinitely by synonymy and homophony. The most important thing is to point out that from the African point of view, the sovereign is the incarnation of his fellows. He is the family, he is the panther as a land depositary, he is the chief of the cult which one can call the Bwiti among the Fang, it is the person in charge of the armed force.

<u>Let's get acquainted with this script</u>

Middle-egyptian:  ***pr bity*** "Palace of the King of Lower Egypt"

Ewondo: /nda bot/ "family"

Ewondo: /nda boé/ "The house of the bee"

Ewondo: / i-bôkh zə / "the lair of the panther"

Ewondo: / à-folà *ùn*-nsoŋ / "the summit place"

Ewondo: / à-folà *ùn*-mbobokh / "The place of the ŋganga"

Ewondo: /nda mi-'mbokh/ "prison; jail"☞jail = dyal "country"
Ewondo: /mə-wuba/ "Bwiti's home"
Etc.

The bee in africain languages

- **Ewondo:** *foŋ* "bee"
- **Kɔ́ɔnzime**: *ányûkwan* "bee"
- **Hausa**: *zuma* "bee"
- **Fulfuldé**: *nyaakel - nyaaki* "bee"
- **Yemba**: *ŋŋwā'* "bees"
- **Gweno:** *ntuki* "bee"
- **Haya:** *enjoki*"bee"
- **Kerewe**: *enzoki* "bee"
- **Tupuri**: *nő̃re* "bee"
- **Kisi:** *njoki* "bee"
- **Fali:** *mozòŋà* "honey"
- **Baoulé**: *zɔgɔdà* "bee"
- **Ndengereko**: *nzusi* "bee"
- **Runyoro**: *enjoki* "bee"
- **Bakoko**: *mbomboŋ* "bee"
- **Londjuee**: *dit* "bee"
- **Mooré**: *sii* "bee"
- **Bambara**: *Dikissè* "bee"
- **Zand**: *ɲege* "bee"
- **Bahasa Indonesia**: *lebah* "bee"
- **Biafada:** *gunč-íɣe* "bee"
- **Oluganda:** *enjuki* "bee"
- **Lingàla:** *nzói* "bee"
- **Dogon**: *ìwó* "honey"
- **Wolof**: *yamb* "bee"
- **Seereer**: *Nqourane* "bee"
- **Kikongo**: *nyosi* "bee"
- **Tshiluba**: *nnyikì* "bee"
- **Makonde:** *nywhi* "bee"
- **Kimbu**: *injuki* "bee"
- **Pular**: *nyá:ki* "bee"
- **Ngiemboon**: *mvomgnwā'* "bee"
- **Lambya**: *inzushi* "bee"
- **Baka**: *toŋyà* "bee"
- **Duala**: *Ndômbi* "bee"
- **Kirundi**: *uru-yuki* "bee"
- **Soussou**: *koumi* "bee"
- **Gourane:** *Ebibi* "bee"

What can be noted by browsing through all these items of African linguistic variants that are subject to a dubious classification by unidentified censors?

They say in essence that the bee is the ELEPHANT in our languages. Because the words for elephant and the words for bee are almost identical. So, how without knowing any kind of talk we can easily find ourselves here.

Better than that, in the majority of these variants of the African single language, the morphology of the word bee is done by using the radical /*nya-/ "mother" or /ze-/ "panther" to which we associate the suffix place -ko / -bo / - /nda/. The bee is not only / zokh /; / ndjokh / "the elephant" in the collective unconscious it is a place, the house of the panther, the seat of the mother or the land, the country.

In the variant Gbe we call the tortoise klô just like the knee. Simply because the breastfeeded place is on his mother's lap. There is no antinomy with the previous meaning that says the mother is the country.

I regularly associate the Bahasa language of Indonesia with the listings of African words because I find the proximity already between the Bahasa term and the term Basa'a which represents African lineages. And I also find that the words used by this language are easy to understand for those who bother.

So we note several things about this script  especially at the onomastic[120] level:

120. The references related to the onomastics solicited are drawn mainly from the production of Unesco, Ethnonyms and African Toponyms

The African lineages that have retained a tie with the sovereignty of bee are still easily recognizable by both anthroponymy and worship. As a result of this, thanks to their collective memory, they can find traces of their past at ***km.t***. To give only a few examples:

Anthroponymy:

Fon, foŋ (certified chief titles in Cameroon)
Tsonga, (RDC, Cameroon)
Song, Songo (Cameroon)
Fonkou (Cameroon)
Etc.

Ethnonymy

The foŋ (Cameroon)
Agonlinu, Popo, Hula, Mahi, Tofini, Oumnou etc. (Benin)
Aizo, Aizangbe, Hwida, Ouida, Peda (Benin)
Mbofon, Ekamtulufu, Koi, Nde etc. (Nigeria)
Fogni, Fonyi (Senegal, Gambia)
les Fong-gb, Djedji, Fo, Kotafon " Togo, Benin"
Tsonga, (Mozambique, SAR)
Dzonga, Thonga, Tswa (Mozambique)
Baronga (SAR)
Tsong, Songo, Masongo, Basongo, (RDC)
Abetswana, Bechuana, Tswana
Etc.

Toponymy

Tshaneni (Swaziland)
The kingdom fon of Abomey
Mikongo (Gabon)
Etc.

Deconstruction

Africans living with fake colonial ideas should know that the first French royalty was Merovingian around the 5th century CE.
It is these African symbols that make it possible to confuse those who practice a discriminatory policy based on Gobineau's raciological thinking.
All that the Africans need to do is questioning the History so that to get out of the dark night where they have been expertly immersed.

Was the Bees a simple myth or a reality?

Opposite here, the coat of arms of the province of Ogoue-Ivindo representing the country of the bees. The red symbolizes the iron of Belinga and the "Y", the meeting of Ivindo and Ogoooue. Surprisingly some have kept the memory of their ancestral past is not it?
Does this Y represents the meeting og Ivingo and Ogoué or the alliance symbol? May you think about it.
As usual I'll keep my own opinion for myself.

A stylized bee. or bee hidden in lily??

3.4.3.31. ***Example N°31: WHO ARE THE KM.T'S BUILDERS ?*** 

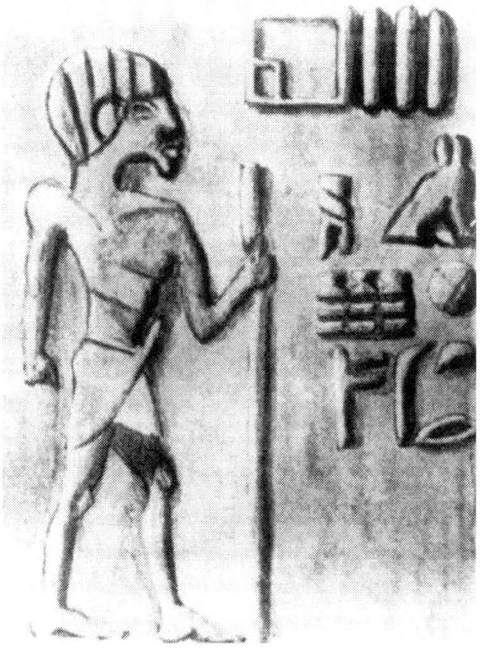

When official scientists are inquiry concerning the origin of ancient egyptians, they answer with a lot of confidence: "The Builders of ancient Egypt are the Anu"

It is often based on a body of information drawn from the names of the cities of ancient Egypt but above all, a predynastic document found by the English Egyptologist Flinders Petrie (+1853 to +1942) at Abydos.

It would be a registered tile dug up under the dynastic temple. The inscription includes the glazed faience portrait of the chef ANU Tera-neter. There is even his address written above "Palace of the UNU in Ermant, Tera-neter". It is from this information that we have deduced the identity of the builders of ancient Egypt.

If today the question seems incongruous by some way because of the diversion work of Africanists, the Africanity of the ancient Egyptians has always been at the heart of the concerns of the first modern Egyptologists for several decades.
In 1900, researchers like Morgan was placing the origin of the Egyptians in Mesopotamia, in Libya for Petrie & Quibell in 1896 or in North Africa for Zaborowski in 1900.

The French Egyptologist Emile Amelineau was on of the precursor, repeating and developing the thesis of the negritude of the ancient Egyptians already present in the accounts of Count Volney (Volney 1821). In his two volumes entitled *Prolegomena to the Study of Egyptian Religion*, Amelineau (1916)[121] has methodically collected clues proving that the Anu, who lived in Upper Egypt during the predynastic period, were Africans coming from the south.

Among his many arguments, he claims that the decorated pallets, found in the graves of high-ranking figures, show that the Egyptians of the Predynastic well knew the blacks and the great fauna of Africa[122], or the fact that the tombs of *nxn* "Hierakonpolis" are circular, reminiscent of the huts of branches found in Central Africa [123]. All these elements allow him to conclude that "the Egyptian civilization [...] doesn't have an Asian origin, but an African origin, a Negroid origin, although this assertion may seem paradoxical.
We are not accustomed, in fact, to endowing the Negro race or the neighboring races with too much intelligence, even intelligence enough to have made the first discoveries necessary for civilization, and yet there is not only one of the tribes living in the interior of Africa who did not own and who still does

121. E.AMELINEAU, Prolegomena to the study of Egyptian religion, part 2. Paris, ditions Leroux,1916.
122. (op. cit.: 426)
123. (op. cit.: 427)

not have any of these first discoveries! "[124]. Thus, wanting to fight against the prevailing ideology of the time that all greates advanced in human history were the result of Caucasoid populations, these explorers and these explorers and researchers embraced theses that we know a posteriori to be equally difficult to defend.

In addition to this literature based on a serious foundation, there is a propagandist literature that organizes the scientific truth from a distorting angle and merely inventing peoples named *Anunaki* to create confusion and better than that, to pretend that aliens are at the origin of the brilliant Negro civilization of ***km.t***.

In addition to the "greecisation" of the names of the inhabitants of ***km.t*** another current surrounded by ideology bases its statement on the MESNITU (that some researchers assimilate to Shemsu-Hor) who come from Punt, more exactly for some of Somalia. Their Earth would be called ***tA Ntrw*** ("Land of the Gods"). They would have been mainly metallurgists and blacksmiths and would eventually dominate the ANU and their Nubian land ***tA Sty*** ("Land of the Link" or "Land of the Prow" according to the translations) of Sudan. Little by little the two Lands were assimilated into one, Ta Khent ("The Land of the Beginning"). The physical difference between ANU and MESNITU is noted in the jaws, those of ANU being more rounded and short whereas those of MESNITU are square like that of Narmer-Menes …

So we see that the question does not have some lack interests.
I recommend, therefore, to question the ***mdw ntr*** which are not peculiar to the predynastic but to read them according to the African prism and to give a sound understanding of them according to the endogenous point of view.

Middle-egyptian:  ***Hr*** (*Faulkner p. 173, Wb. III p. 122, 14+ sq.*)

We said that this script can be read in Ewondo: / à-sù / for the face .
We have said also that this script can also be read in Ewondo: / *ùn*-nloh / for the head .

We therefore notify that this script can also be read in Ewondo: / à-nyu / for the mouth .
But all the supposedly highly scientific developments speak of a people named the **Anu** who would have been Egypt builders.

It was enough to understand that Africa is made up of one and only one people who have a unique ancestor ***Hr*** that we also call / à-nyu / "The mouth" as the **Agni** for example.

What defines the leader in our traditions is his right to public speaking. What defines a man in the African tradition is the meaning of the given word.
To say give me your word the semantax is to say:

Ewondo: /vâ ma à-nyu dw/ "give me your mouth"

"In the beginning was the Word, and the Word was with God, and the Word was God."
(Louis Segond Bible, Jean 1:1)

Let's get acquainted with the supposed vocalization Anu

Middle-egyptian:  ***iwn*** "complexion; color; texture, appearance"
Ewondo:/*ùn*-mfén/; /*ùn*-mbokh "plait"; /nyol/ "complexion; color; texture" etc.

Middle-egyptian:  ***iwn*** "pillar column"

124. (op. cit.: 330).

Ewondo: /*i*-won/ "boost, pillar"; /*ùn*-ŋkum/; /à-koŋ/ "pillar column" etc.

Middle-egyptian: ***iwn mwt.f*** "pillar of his mother, epithet of  ***Hr*** and title of priest"
Ewondo: /nya *ùn*-ŋkum/ "master pillar"; "the pillar of the pillars"
Kréyol: *Poto mitan* "master pillar";

Middle-egyptian: ***iwn n fnD*** "nasal bone"
Ewondo: /*ùn*-tan dy-òé/; / i-kokom / "septum"
Ewondo: /i-won *ùn*-fàŋ dy-òé / "nasal bone" ☞**Ewondo:** /á-fàŋ / "to split"

Middle-egyptian: 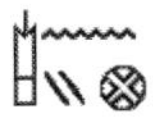 ***iwni*** "Hermonthis"
Ewondo: /à-folà *mi*-ŋkukumà / "the place of the royal couple"

Middle-egyptian:  ***iwntyw*** "tribesmen"
Ewondo: /*mi*-ŋkukuma / "kings, rulers, descendants of  ***Hr"***

Middle-egyptian:  ***iwntyw*** "tribesmen"
Ewondo: /bə-yəŋ mi-'ŋkukuma / "the kings of the alliance"

Middle-egyptian:  ***iwnw*** "Heliopolis"
Ewondo: /à-folà *ùn*-mfiè / "***Hr's*** *city*"; "the city of light" Hr = sun

Middle-egyptian:  ***nsw.t*** "flame"
Middle-egyptian:  ***nsw.t*** "King of Upper Egypt, King"
Etc.

There is, no difficulty at the technical level if we have understood the method.

What became the Anus?

This question is a little uncomfortable for supposed specialists because of two reasons. Part of buried ideology and especially the worry of not having to ask to Africans who are the Anu and what they have become.

Unfortunately for the ideologists who know the truth just like us, following the Arab invasions, the Anus are nowadays located from the South Sahara to Cape Town.

Due to the fact that 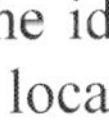 ***Hr*** has 99 names as claimed by some exegeses, it is enough to interrogate our languages and you will thus have groups that identify themselves by one of its 99 attributes.

Ethnonymy

Batas, Bathaa, Batta, Bwati, Demsa Bata, Dunu, Gboati Gbwata, Agonlinu, Popo, Hula, Mahi, Tofini, Oumnou etc. (Bnin); Aizo, Aizangbe, Hwida, Ouida, Peda (Bnin); Mbofon, Ekamtulufu, Koi, Nde etc. (Nigeria);Teda, Banda, Bagrimma, Cuvok, Dourou, Bamana, Fali, Gbaya, Namdji, Gude, Bamanan, Guidar, Guiziga, Haoussa, Hid, Kanuri, Baguirme, Kapsiki, Karan, Kra, Komas, Kotoko, Lagwan, Mada, Mafa, Mambay, Mambila, Mandara, Massa, Banmana, Mboum, Mbukou, Meta, Mfumte, Mofu, Mouktl, Bamanankan, Mousgoum, Moundang, Moussey, Nsungli, Ouldm, Peuls, Podoko, Toupouri, Sara, Vam,

Voût, Yamba, Banmanan, Zulgo, Akunakuna, Babanki, Babungo, Bamilk, Bamoun, Bagarmi, Ngemba, Nso, Tikar, Vere, Bafaw, Bakokos, Bakweris, Babenga, Babinga, Baka Bambuk, Bakas, Bambenga, Bangombe, Bayaka, Bebayaga, Bebayaka, Bibayak, Bibaya, Binga, Bingas, Mbaka, ou Baga-Binari, Baga-Koga, Barka, Kalum, Alante, Balanda, Balanga, Balanta-Brassa, Balanta, Balantes, Balanti, Balant, Bolenta, Brasa, Brassa, Bulanda, Bulante, Frase, Tar Bagrimma, Tar Barma, Bamananke, Bambaras, Banmani, ou Bamangala, Bangalas, Imbangala, Imbangola, Mangara, Mbangala, Mongalla, Ngala ou Wangala, foŋ, Baghirmi, Bagirmi, Bagirmis, Baguirmiens, Baguirmi, Baguirmis, Barmas, Bauirmi, Lisi, Lis, Masa Guelengdeng, Mbara, Mbarma, Balong, Bankon, Bassas, Ababdehs, Abakwariga, Ab, Abour, Abron, Accaras, Acholis, Achtouken, Adangms, Adele, Adoumas, Adjoukrous, Adyumba, Afar ou Danakil, Afo, Afridi, Afrikaner, Agäw, Agni, Agumba, Ahantas, Ahoulan, Aja, Ajjer, Aka, Akan, Akanigui, Akbou, Akls, Akouè, Akowa, Akposso, Akpwakoum, Aku, Akuapem, Akunakuna, Akyems, Alladian, Alurs, Amba, Amharas, Anang, Andorobbos, Anga, Angolares, Ankw, Anlo Ewe, Ano, Antaimoro, Antaisaka, Antakarana, Antambahoaka, Antandroy, Antanosy, Anuak, Anufo, Aowin, Apindji, Appolo, Arabe, Arabe du Tchad, Arbors, Argobba, Arma, Aro, Ashantis, Attis, Aushis, Avalites, Avikam, Avukaya, Awandji, Ayahou, AzibisBatanga, Bonkeng, Doualas, Evuzok, Malimba, Manehas, Mbo, Mouamenam, Ndonga, Pongo, Yassa, Abo, Yabassi, Nyamtan, Ewodi, Bodiman, Bandem, Banen, Bayangui, Bafia, Ban, Banen, Boundjou, Dzimou, Yambassa, Nyokon, Sanaga, Bétis ,Fangs, Bulu, Maka, Badjou, Baka, Soo, Vute, Njanpeta, Lemande, Batis, Akpwakoum, Maka, Badjou, Gbete, Konabembe, Nzim Mloon, Mpoumam, Nkounabem, Bobilis, Pôl, Mvele, Gbaya, Kaka, Mvovon, Bikl Etc.

Apart from ideology, here is the answer to the question: who are the Anu and what have they become ??

There is a disturbing african opinion that is trying to produce dubious information about solar worship in Africa and Anu-related to albino carnation.

> *"Meaning of the name Anou or Iounou ((banou or Baounou) of the Egyptian nobility albino, builder of the Egyptian civilization and at the origin of the different names of the country of Egypt (..)"*
> *(Mardoche Ekwé-2018)*

I told them that all this is not rigorous and I reiterate it here.
Some arguments are a bit misleading, are stated by the author of these writings but unfortunately these writings are not based on something very serious.

3.4.3.32. ***EXAMPLE N°32: PERSONAL PRONOUNS***

Stand ard Method

Middle-egyptian:  ***ink*** "i, me, mine, to me" (*Faulkner p. 24, Wb. I p. 101, 13*)

Endogenous method: First-person singular

What i see	Hidden meaning
 **Ewondo**: /à-sam/; /i-soŋ/ "reed"	**Ewondo**: /*ùn*-nson/ "beak, mouth"
	Ewondo: /à-sè/ "feather"
	Ewondo: /mâa/ "feather"
	**Ewondo**: /maa/; /mè/ "what we move from left to right"

Ewondo: /*mi*-'ŋkiəŋ'ə/ "Pot, jar"	**Ewondo**:/à-nyu/ "mouth"→ /à-mê/ "stumble"
	Ewondo: /à-mè/ "Shaping pottery".
 **Ewondo**:/*ùn*-nkun/ "basket"	**Ewondo**:/ŋgan/; /*ùn*-nkwé/ "basket"
	Ewondo:/ŋgan/; /*ùn*-nkôm'o/ "crocodile"
	Ewondo: /*ùm*-mè/ "potter"; pottery
 **Ewondo**:/*ùn*-mbokh/ "disabled"	**Ewondo:** /*ùm*-mè/ "potter".

Question:

If, deliberately, a simple question was asked to a sample of Africans living from the South Sahara to the Cape Town to find out how they express the notion of selfhood: "I, me," would they say:

- ***Ink (***as ink in english) as formulated by the official method?

  Or would they say
- **Ma, mə, mi** "i, me, mine"

One thing is sure, we will find at least groups of Ekang populations who will say / ma /, / me / to signify *me*, the singular first person pronoun.
I therefore base my remarks on the demonstration above. I remain dubious about what is presented as a personal pronoun independent of Pharaonic according to the standard method.

Was it possible to read this script simply quickly?

The answer is affirmative in evading the diktat of transliteration and answering the question of who is responsible for the following functions in Africa:

Nutrition: the kitchen is the mother's domain in Africa.
Who is the basket?: household; housekeeper; housewife
Who is the feather and the feeder?: the mum, mAa.t

Some variants of the African language

Kɔ̀ənzime: *mem; me* "i; me "
Ewondo: /ma/; /me/ "i, me"

☞**Ewondo:** /məma / "mum"

Bété: *ì-fè* "i, me"
Lingala: *ngáí; na* "i, me"

☞**Ewondo:** /à-fap/ "wing;" so "i, me"
☞**Ewondo:** /na / "mum;";/nga/ "woman"

Bahasa Indonesia: *saya* "i, me"
Bamanakan: *ne* "i, me"

☞**Ewondo:** /à-sa / "feather;" so "i, me"
☞**Ewondo:** /na / "mum;" so "i, me"

Atchan: *Min* "i, me"

☞**Ewondo:** / min/ "swallow" "i, me"

Yémba: *meugn; a zaa* "i, me"	**Ewondo:** / zi/ / zə/ "panther"
***Isi*-Zulu:** *mina; yimi; mi* "i, me"	**Ewondo:** / min/ "swallow" "i, me …"
Oluganda: *nze* "i, me"	**Ewondo:** /zə / "panther" so "i, me …"
Nyamwési: *nὲὲnέ* "i, me"	**Ewondo:** /nana / "mum" so "i, me"
Snoufo: n; mî "i, me"	**Ewondo:** / mwi: / "home, place" → mum
Soninkanxaane: *n-* "i, me"	**Ewondo:** /na / "mum" so"i, me"
Wolof: *ma*; *dama* "i, me"	**Ewondo:** /dama / "mine"; /ma/ "me"
Tupuri: *me, ndi* "i, me"	**Ewondo:** /ndo / "vulture"; /me/ "me"
Sango: *mbi* "i, me"	**Ewondo:** /ù-mbè / "door" so "mum"
Dogon: mi: "i, me"	**Ewondo:** / mwi: / "home, place"
Dioula: *ma; ma:n* "i, me"	**Ewondo:** /na, ma / "mum" so "i, me"
Adja: *ŋé* **"i, me"**	**Ewondo:** /na / "mum" so "i, me"
Duala: *mbá; na* **"**i, me"	**Ewondo:** /na / "mum" donc "i, me"
Kreyole: *Mwen* "je, moi"	**Ewondo:** /mwan-na / "mum's son"
Mooré: *M'é; ma; Maam* **"**i, me"	**Ewondo:** /ma/ "me"
Ewè: *Mé; Moué; Né'; Egné* "i, me"	**Ewondo:** /à-nyu/ "mouth"
Kissi: *Ya; onou* "i, me"	**Ewondo:** /à-nyu/ "mouth"
Basa'a: *Meh,meyin, mepi "i, me"*	**Ewondo:** /ma; me / "i, me"
Etc.	

Let us summarize

The languages taken here randomly demonstrate a uniformity of thought. As always the length of the list does not count only the principle counts. It is observed that the first person of the singular "i, me" is a notion that is related to the mother in variants of the unique African language.

Guadeloupean Creole: *Mwen* "i"	**Ewondo:** /mwan-na / "mum's son"
Guadeloupean Creole: *Man* "me"	**Ewondo:** /maâna / "mum's son"
Guadeloupean Creole: *An* "me"	**Ewondo:** /na / "mum"

The lemmas **na** and **an** here mobilized are simply ananyms stemming from the consonantal metatheses between the Kekwe and Kweke branches of the Negro-egyptian.
Chamoiseau and his creolites would thus here have a little trouble facing the truth of the ancestral component of what he calls Creole.
Languages mischief called Creoles are a mixture of language synchonies and can allow reading of the ***mdw ntr*** without any problem by applying the present method. The reason is simple, we are dealing with the same lexicon even though creole looks like a mix of languages.

Greenberg Language Classifications

Moreover, we can see that Greenberg's classifications lend themselves to a smile because they are based on impressions and not endogenous tangible facts of African thought.

African thought

At the level of African thought, the first person of the trinity (mother, father, son) is the mother. Indeed the son comes from the nourishing roots of the tree that are in order, the mother and the father. The son is simply the trunk of the tree. In a matriarchal regime, the power is embodied by the panther = mother.

So to say "i, me" in African languages, there is a sort of honorific anteposition that consists of recalling the words of the mother.

The chicken or the egg?

From the moment we understand that the numeral 1 is reserved for the mother, then we have answered de facto to the question that has been presented by some people as the paradox of the chicken and the 'egg.

For Africans, the chicken being the mother and the egg the child, it is therefore anterior to the egg. The chicken or the egg causality dilemma is commonly stated as "which came first: the chicken or the egg?". The dilemma stems from the observation that all chickens hatch from eggs and all chicken eggs are laid by chickens.

Egyptology and its pronouns

Independent variant: Reminder

Middle-egyptian:  ***ink "i, me"***

Dependent variants and suffixed ones

Middle-egyptian:  ***wi or w(i)***"i, me; to me"

This variant according to the authorized literature is used as a complement of verb object (except in the infinitive) or after some initial particles and after the negation of the future 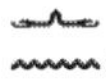*nn*.

Why do we observe these different variants?

I postulate that this is due to a need for the language simplification.

So we went from the variant  to the version .

Are these two scripts the same?

I say that the sign M17. of the Gardiner's list is called in **Ewondo**: / i-sɔŋ / "reed".
I still say that the word for designating a bird is / *ùn*-nsɔŋ'o / "the beak"

It is therefore normal for this script  appear as a simplified variant of this one . Where these signs  are individually homophones of the sign M17. To engrave monuments, steles etc. it looks like it was probably better to reduce the engraving time and thus to go from this script to the simplified ones.

And if we wanted to simplify more we would have two alternatives

Middle-egyptian:  ***w(i)*** "i, me"

Middle-egyptian:  ***i*** "i, me"

This last type is thus the one that classical Egyptology qualifies as suffixed variant.
We Africans, we say that the 3 versions can be vocalized in one and the same way. We do not have in our languages three variants to say ***"i, me"***.

Why would Pharaonic have needed: ***Ink***; ***wi, w(i)*** and ***i***?

This is our position on the question of the three supposed variants. For us there is only one way to say "i, me" in Ewondo.

Small comparison for honest people

Middle-egyptian:  ***ink "i, me"***

Ewondo: */ze/* "panther/ leopard"	**French**: je	**English**: je>ie>i
Ewondo: */me/* "me"	**French**: me	**English**: me
Ewondo: */ma/* "me"	**French**: ma "à moi"	**English**: me; my
Lingála: *ngoi / nkoi* "panther"	**French**: moi	
Lingála: *mayi* "water" 	**Mid-Egyp.**: ***mw.t***	**English**: my

Note: how comes that Egyptologists talk about  ***ink*** **and Ewondo about /ma/ "me"?**

The Egyptologists transliteration ***ink*** just give the name of a synonyme of "water"

Indeed this vase  is called in Ewondo language :*/mi-*'ŋkəŋ'ə/; it is used to carry and keep water fresh. By the way the egyptlogist method name the vase when Ewondo people name the content, the substance /mayi/ "water".

To support this point of view, the reed  was used by scribes, scholars and intelligent people. In **Ewondo** we do say /*à*-kəŋ/ to name intelligence. This reed is a sting used to remove what has penetrated in the body with a sharp instrument. We do call it /i-bèh/ " a sharp bamboo". The action of widening the skin around the spine to remove the tingle is: / á-mê / in Ekang language. May I have to recall that *I* or *me* in Ewondo is /ma/; /mə/ ?

We also emphasize that among the basket  represented on the script one is called in Ewondo : /i-ŋguén/ moreover, we have alors /*ìn*-ŋkwé/"basket".
The only thing you have to do now is to remove the vowels and compare with the consonant skeleton ***ink***. Don't be shy compare.

To close the loop, we assert that in Ewondo Nze **Mba** is a Basket roughly done in the forest to harvest honey. You can note also that; mà/mbà is :

Duàlà: mbá "me".

We have already discussed technically the presence of this epenthetic "m" in the so called Bantou languages. If needed, you can deal with it again in the section concerning **mb**àra and **m**ara "elephant".

Middle-egyptian: 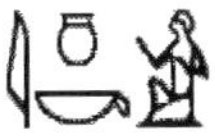 ***ink "i, me"***

Kɔ̃ɔnzime: *nkœœ* "panther"
Yoruba: *Êkoun* "panther"
Bambara: *khanbon* "panther"
Yemba: *ngwui* "panther"
Nzebi: *nzèghè* "panther"
Gourane: *onki* "panther"
Sango: *Zîn* "panther"
Bahasa: *aku* "panther"
Fon: *nkpó* "panther"
Godi: *gui*; *gu* "panther"
Wolof: *Sgue*; *Safando*
Quichua: *kiru* "panther"
Tshiluba: *nkashama* "panther" litt. The place of the mother feather. I=ma

Note:
The word *panther* can be approached in Ewondo by: /i-dùrù à-fan/ "the bush's tractor" or /tsir à- fan/ "wild animal"
The word *leopard* equals in Ewondo: /i-lob bor/ "the mankind biter"

The translitered variant *ink*  is acutually the word *panther* which is one of the characteristics of the mother and found everywhere on the globe as can be seen in these examples below.

Xapaitíiso: *aitoi* "panther";
Xapaitíiso: *ti* "I, me"
Ewondo: /i-to na/ "place of the mother"
Ewondo: /i-to / "place / seat"

Qusqu qhichwa simi: *puma* "panther";
Qusqu qhichwa simi: *ñuqa* "i, me"
Ewondo: /məma/ "mum"; /i-fəmə/ " hatred"
Ewondo: / nyakh / "bull; buffalo, beef"

Sullkatiti Titiri: laka "panther";
Sullkatiti Titiri: hiwasa "me"
Ewondo: /lə-ko/ "foot"
Ewondo: /was / "time"

Japonais: *Hyō* "panther"
Japonais: *Ashi* "foot"
Ewondo: /à-yo/ "vomit" ; /Eyo/ « zamba »
Ewondo: /à-ki/ "egg"; /à-ko/ "foot"

Chinese : 豹 Bào ; "panther"
Chinese : 咱家 zá jiā « i ; me, mine »
Chinese : 我 Wǒ « i ; me »
Chinese : 寡人 guă rén « me (the sovereign)»
Duàlà : mboà « country »
Ewondo: /zə dyàl/ "home panther"
Ewondo: /wa/ "one, me, you"
Ewondo : /ngà(l) nyàra « buffalo »

Etc.

I could spread whole pages but as always it is not the purpose. The most important is the principle.

The *ink* transliteration variant is little bit used in Africa to say "i, me". But it has persisted in some African linguistic variants until the present days. It may be postulated that the simplified transliterated variant ***i*** has persisted in our present languages in the versions: na, ne ni, nu or ma, me, mo etc

__**Examples:**__

Middle-egyptian:  ***ink "i, me"***

Lingala: *ngoi* / *nkoi* "panther"
Ewè: è*gnè* "panther"
Lingala: *ngai* " me, i"
Ewè: *blèkpɔè /ma* "i, me,"

On a chronological level, we can say that the populations from the Egypt migrations have found in the southern Sahara archaic variants of the type ***ink***. of the African language. Some adopted them as in

Kongo and others persisted with the simplified type. That's why we have some pockets where both versions still persisting today.

This vestigial version is also found in the Western languages or the so-called Indo-European languages through the African branch of Sanskrit, which is the mother tongue of current Indo-European languages

Middle-egyptian:  ***ink "i, me"***

Dutch: *ik*; 'k; ego "i, me"
Deutsch: *Ich*; ego "i, me"
Svenska: *jag; ego* "i, me"
Dansk: *jeg; mig* "i, me"
Sanskrit: चरणः *caraṇaḥ* "foot"
Sanskrit: कलम *kalamaḥ* "feather"
Sanskrit: अहम् *ama* "me, i"

Suffixation versus prefixation

This question, as we have already pointed out, is a consequence of a methodological choice rather than an actual reality.

Let me explain, we have languages that operate by consonant metathesis. The languages that the great African researcher Mboli calls Kekwe and the languages of the Kweke group.

One thing is certain the pronoun in our languages always stands before the verbal radical.
What is the conventional meaning of the Kekwe or Kweke group? This question has never troubled us. It will be very clever to tell us what the rational meaning is.

Maspero clearly notes that if the subject is an absolute personal pronoun, it is placed before the root (Maspero, 1871). I repeat below his example:

Middle-egyptian:  ***Ink rx*** "i know"
Ewondo: /ma yəm /; /ma tsogh/ "i know; i think"

Maspero notes formally that "*if the personal pronoun is suffixed, it is immediately after the verb*" (Maspero, 1871). I recall his example below:

Middle-egyptian: ***mr i*** "i love "
Ewondo: /ma diŋ / "i love "

Maspero says also two things more important: "*The noun or phrase subject is indifferently placed before or after the verb*" and the second thing he says is that "*the subject can be expressed twice*" (Maspero, 1871) . I recall below his example:

According to his point of view let me emphasize this example below:

Middle-egyptian: ***mr i*** "i love"
Ewondo: /nga ma diŋ / "i do love" this variant is still used nowadays.
Ewondo: /ma diŋ / "i love"

Lingala: /ngaï na lingui/ "i do love" this variant is still used also.
Lingala: / na lingui/ "i love"

In short here, we have just a level of precision say absolute. Whether I start with left or right I could always read, "I like"

Of course, Egyptology did not wait for Africans to have an opinion on this question to decide on the problem. But we suggest that Egyptology revise its positions on pronouns in order to better adapt its discourse to its grammar, particularly in terms of pronouns (dependent, independent and suffixes).

<u>And the Remainder</u>?

I will present the different pronouns with their reading without making more comments.

SECOND PERSON OF THE SINGULAR

<u>Standard method:</u>

Middle-egyptian:  ***ntk*** "you, to you" (*Faulkner p. 142, Wb. II p. 357, 2-3)*

<u>Endogenous method</u>: Second person of the singular

What i see	Hidden sens
 **Ewondo**: /mə-ndim/ "water"	**Ewondo**: /à-nyu/ "drink"; "mouth"
	Ewondo: /ùn-mbokh/ "hole"; "one, 1"
	Ewondo: /à-wa/ "to wash"
	Ewondo: /wâ/ "one" , "you"; "arm"
	Ewondo: /wa/ "affluent" , "you"
 **Ewondo**: /*ùn*-mfəma/ "bread"	**Ewondo**: /i-wuê/ "breakfast"
	Ewondo: /i-koh/ "dough; cake"
	Ewondo: /Yakh, vyakh/ "glutton"
	Ewondo:/à-nyu/; /*ùn*-mbokh/ "mouth"
	Ewondo:/wué/; / "honey"
	Ewondo: /wâ/ "one"; "you"

	Ewondo: /*ùn*-mbokh/ “hollow”; “one”
	Ewondo: /*ùn*-nkoh/ “string”
	Ewondo: /*ùn*-ŋkwé/ “basket”
Ewondo:/*ùn*-ŋkun/ “basket”	**Ewondo**: /i-wakh/ **“**basket straps”
	Ewondo: /à-wè/ “to strap”
	Ewondo: /*ùn*-mbokh/; /wâ/ “one”; “you”

Chart 93: Hiddens meaning of the script

Second-person singular

Middle-egyptian: ***ntk*** *“you, to you”*
Ewondo: /wa/ *“you, to you; your”*

Driver:
Ewondo: /i-**wùa** I nə *ùn*-**ntùkh/** *“what is thrown is useless”*

Second-person singular feminine

Middle-egyptian: ou ntT/ntt *“you, to you; your”*
Ewondo: /wa/ *“you, to you; your”*
Driver:
Ewondo: / *ùn*-**ntout** ù nə **wà /** *“unit is a whole”*

The sign V13 of the Gardiner's list is a rope that serves as an hamper for animals. We call this type of fencing:

Ewondo: /ùn-mbôkh / “rope” ☞*homophone* of **Ewondo:** / *ùm*-bokh / "one; 1"
Ewondo: /ùm-bokh / “one” ☞*synonym* of **Ewondo:**/ wâ / “one”
Ewondo: /wâ / “one” ☞*homophone* of **Ewondo:**/ wa / “you”.

The sign X1 of the Gardiner's list having been analyzed naturally I will not come back on it anymore.

Small technical detail.

In a binary system, the second person is considered as the other one. That is why there is here a direct link between the word used to designate the numeral One (1) and the word for the second person namely: / wa / “you”.

Note to this account that for the sovereign Egyptology the numeral One (1) is said:

Middle-egyptian: ***wa*** “one; unique, the only one” (*Faulkner p. 56, Wb. I p. 273, 3-10 sq.)*
Ewondo: /wa/ *“arm, one, you; bowl etc.”*

We have understood the logic that leads to the other two dependent and suffix variants. I do not care about it here.

Etymology

Negro-egyptian: ***njw-***
Ewondo: *nya* "mother"
Middle-Egyptian: nw.t
Proto-Germanic: *juz-
Indo-european: Ju
Anglo-saxon: eow
Old Saxon: iu

Ewondo: *nyua* "the one who"
Ewondo: *nyat* "buffalo"
Ewondo: *wa* "you"
Old High German: *iu*
Old Frisian:*iuwe*
Dutch: *u*

pronunciation of ***you*** and the nominative variant ***ye*** gradually merged from 14c.; the distinction between them passed out of general usage by 1600. widespread use of french in england after 12c. gave english you the same association as french vous, and it began to drive out singular nominative thou, originally as a sign of respect (similar to the "royal we") when addressing superiors, then equals and strangers, and ultimately (by c. 1575) becoming the general variant of address. through 13c. english also retained a dual pronoun ink "you two; your two selves; each other."

THIRD PERSON OF THE SINGULAR

Standard method:

Middle-egyptian:  ***ntf*** "he, him; to him; it" (*Faulkner p. 142, Wb. II p. 356, 3-5)*

Endogenous method

What i see	Hidden sens
 Ewondo: /*mə*-ndim/ "water"	**Ewondo**: /*à*-nyu/ "drink"
	Ewondo: /*nyâ/ "mum"
	Ewondo: /i-â/ "undulating"
 **Ewondo**:/*ùn*-mfəma/ "bread"	**Ewondo**: /bi-di/ "food"
	Ewondo:/i-yâ/ "what moves"
	Ewondo: /*nyâ/ "mum"
 **Ewondo**:/fi/ "snake"	**Ewondo**:/nyua/ "serpent"
	Ewondo: /nyâ/ "mother"
	Ewondo:/i-yâ/ "creeping"

CHART 94: HIDDEN MEANINGS OF THE SCRIPT

Third-person singular

Middle-egyptian:  ***ntf*** *"he, him; to him; it"*
Ewondo: /à/ *"he; him; to him; it"*

Driver:

Ewondo: / *ùn*-**nfət** à nə ***i*-yâ /** *"the opening is also closing"* ☞***yâ>à***

Third-person singular feminine

Middle-egyptian: ou nts *"she, to her;"*
Ewondo: /à / *"she, to her"*

The sign S29 is vocalized / i-y / garment and the O34 / i-yâ / "opening" of the Gardiner's list. There is no male or female in Ewondo. So these different variants are vocalized in Ewondo in one and the same way.
Driver:

Ewondo: / *ùn*-**ntseŋ** ù nə *i*-**yé /** *"the flag is a fabric"* *yé>é*

Little technical trick

For the sovereign Egyptology the vulture (head bald or Egyptian) represents the mother.

Middle-egyptian: *A* "vulture, bird" (*Faulkner p. 1, Wb. I p. 1, 1.)*
Middle-egyptian: *mw.t* "mother"
Middle-egyptian: *nr.t* "vulture" **Ewondo: /nyara/** "buffalo"
Ewondo:/ya/ or **/*nya/** *"mother"*

The agent prefix *ny-* in African languages means the *one who; whoever*. Literally *nya* is the one who is **A** sometimes call *yâ* also.

The Ewondo language therefore uses the simplified variant / à- / to indicate the third person singular.

Example :

Ewondo: /à á-di/ *"he/she eats"*

Note: I assume that the third person of singular in English shares a phonetic proximity and meaning with the Ewondo word below. We can undertake these connections at a phonetic level.

Ewondo: /zə/ "panther; female "
French: *je* "i;me"
Dutch: zij, **Wobe**:Tchihi "vulture; "
German: sie,
Greek: he
Old English: *ze'e>seo, sio*
English: *he/she*

In authorized literature we can observe that original **h-** survives in **her**. A relic of the Old English pronoun is in Manchester-area dialectal ***oo*** "she" like the variant **à-** of ewondo. As a noun meaning "a female," she is only attested from 1530s. According to me, the word she, is closer to ze "the panther; "

First Person Plural

Standard method

Middle-egyptian: ***inn*** "we" (*Faulkner p. 23, Wb. I p. 97, 5-6)*

Endogenous method

What i see	Hidden meaning
 **Ewondo**: /i-soŋ/ "reed"	**Ewondo**: /ùn-nson/ "bec, mouth"
	Ewondo: /*à*-nyu/ "bec, mouth"
	Ewondo: /*ùn*-nya/ "mum"
	Ewondo: /*ùn*-mbi/ "mum"
	Ewondo: /o-bə/ "small"
	Ewondo:/o-bi/ "falcon"
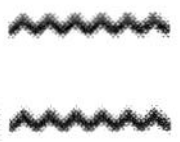 **Ewondo**: /mə-ndim/ "water"	**Ewondo**: /bîa/ "beer"
	Ewondo: /bi-à-ba/ "you and me"
	Ewondo: /ù-nya/ "mum"
	Ewondo: /ù-mbi/ "mum"
 **Ewondo**:/*ùn*-mbokh/ "disabled"	**Ewondo**: /biə/ "tatif: the one planted"
	Ewondo:/o-bi/ "falcon"
	Ewondo:/à-bî/ "groan"
	Ewondo: /ba/ "father"
I I I **Ewondo**:/i-lèh/ "trois"	**Ewondo**:/à-bui/ "many"
	Ewondo:/o-bi/ "falcon"
	Ewondo: /o-bə/ "small"

Chart 95: Hidden meanings of the script

First person plural

Middle-egyptian:  ***inn*** *"we"*
Ewondo: /bi/ *"we"*

Driver:

Ewondo: / *i-**nyièn*** i nə *à*-**bié** / *"birth is a kind of entry into life"* ☞ ***bié>bé***
Ewondo: / *i-**nyùèn*** i nə *i*-**bi** / *"alcool is a trap"*

Etymology

Ewondo: /wa/ *"you"*
English: we

Old English **we**, first person plural pronoun, "I and another or others," from Proto-Germanic *wiz (source also of Old Saxon wi, Old Norse ver, Danish vi, Old Frisian wi, Dutch wij, Old High German and German wir, Gothic weis "we"), from PIE *we- (source also of Sanskrit vayam, Old Persian vayam, Hittite wesh "we," Old Church Slavonic ve "we two," Lithuanian vedu "we two").

The "royal we" (use of plural pronoun to denote oneself) is at least as old as "Beowulf" (c.725); use by writers to establish an impersonal style is also from Old English; it was especially common 19c. in unsigned editorials, to suggest staff consensus, and was lampooned as such at least since 1853.

Second Person Of The Plural

Standard method:

Middle-egyptian: 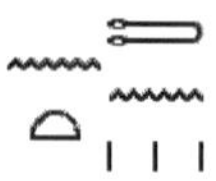 ***nTtn*** "you" (*Faulkner p. 142, Wb. II p. 357, 8)*

Endogenous method

What i see	Hidden meanings
 **Ewondo**: /*ùn*-mbokh/ "hindering"	**Ewondo**: /à-nyu/ "mouth"
	Ewondo: /ùn-mya/ "porcupine Prickly"
	Ewondo: /*ùn*-mwi/ "home"
	Ewondo: /mi/ "homes"
	Ewondo: /mye/ "to be pure physically and morally".
 **Ewondo**: /*ùn*-mfəma/ "bread"	**Ewondo**: /à-nyu/ "mouth"
	Ewondo: /*ùn*-nya/ "mother"
	Ewondo: /mi/ "painful birth"
Ewondo: /mə-ndim/ "water"	**Ewondo**: /à-nyu/ "mouth"
	Ewondo: / mye-mya/ "flexible stick"
	Ewondo: /mya/ "the two thin, long bones of the arm and the leg."
	Ewondo: /mi/ "painful birth"
 **Ewondo**:/i-lèh/ "three"	**Ewondo**:/i-lé/ "tree"
	Ewondo:/myé/ "path without contours, all right".
	Ewondo: /my/ "the house has three floors"
	Ewondo:/mi/ "the one who is not dwarf"
	Ewondo: /mye/ "to be pure physically and morally".

CHART 96: HIDDEN MEANINGS OF THE SCRIPT

Second person plural

Middle-egyptian: 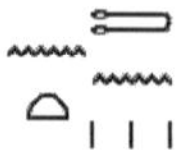 ***ntTn*** *"you"*
Ewondo: /mi/; /myé/; /mya/ *"you"*

Middle-egyptian:  ***nttn*** *"you"*
Ewondo: /mi/; /myé/; /mya/ *"you"*
English: you

The symbol  X1 of the Gardiner's list having been described, I do not come back naturally anymore. It simply replaces the sign V13 .

Ewondo: / *myé* à nə *ùn*-**ntout/** *"the one who is pure physically and morally is a saint"*

<u>Etymology</u>

Ewondo: **nya* **"mother; the one who"**
English: *you*

Old English eow, dative and accusative plural of þu (see thou), objective case of ge, "ye" (see ye), from Proto-Germanic *juz-, *iwwiz (source also of Old Norse yor, Old Saxon iu, Old Frisian iuwe, Middle Dutch, Dutch u, Old High German iu, iuwih, German euch), from PIE *yu, second person (plural) pronoun.

Pronunciation of you and the nominative form ye gradually merged from 14c.; the distinction between them passed out of general usage by 1600. Widespread use of French in England after 12c. gave English you the same association as French vous, and it began to drive out singular nominative thou, originally as a sign of respect (similar to the "royal we") when addressing superiors, then equals and strangers, and ultimately (by c. 1575) becoming the general form of address. Through 13c. English also retained a dual pronoun ink "you two; your two selves; each other."

Troisième personne du pluriel

<u>Standard method</u>

Middle-egyptian:  ***ntsn*** "they" (*Faulkner p. 142, Wb. II p. 357, 8)*

<u>Endogenous method</u>

What i see	Hidden sens
 **Ewondo**: /I yé/ "garment"	**Ewondo**: /à-nyu/ "mouth"
	Ewondo: /à-bèh/ "To sting; plant"
	Ewondo: /à-bèh/ "thigh"
	Ewondo: /*ùn*-mbè/ "tam-tam"
	Ewondo: /ù-su/ "Textile fiber from petioles of palm leaves"
 **Ewondo**: /ù-mfəma/ "bread"	**Ewondo**: i-di/ "food"
	Ewondo: /*ùn*-nda/ "house"
	Ewondo: /*ùn*-mbè/ "door; gate"

𓈖 Ewondo: /mə-ndim/ "water"	**Ewondo**: /à-nyu/ "mouth"
	Ewondo: /ù-su/ "river"
	Ewondo: / i-b/"hole"
	Ewondo: /i-bè/ "thirst"
	Ewondo: /ù-nlo/ "river"
	Ewondo: /nyol/ "dry"
I I I **Ewondo**:/i-lèh/ "three"	**Ewondo**:/à-bui/ "many"
	Ewondo:/i-bè/ "bamboo"
	Ewondo:/i-bè/ "two"
	Ewondo: /o-bə/ "small"

Chart 97: Hidden meanings of the script

Third person plural

Middle-egyptian: 𓈖𓏏𓋴𓈖𓏥 ***ntsn*** *"they"*
Ewondo: /b/; /ba/ *"they"*
English:they

Driver:

Ewondo: /ùn-**tsen** ù nə *i*-**bèh**/ "a flag is to be connected to a mast"

Etymology

Middle-gyptian: 𓋴𓈖𓏥 ***sn*** **"**they, the; they; their;theirs**"**
Middle-gyptian: 𓌢𓈖𓀀 ***sn*** "brother" ☞ **Ewondo:** dyàŋ "motherland"
Ewondo: tsè "land"
English: they

c. 1200, from a Scandinavian source (Old Norse þeir, Old Danish, Old Swedish þer, þair), originally masculine plural demonstrative pronoun, from Proto-Germanic *thai, nominative plural pronoun, from PIE *to-, demonstrative pronoun (see that). Gradually replaced Old English hi, hie, plurals of he, heo "she," hit "it" by c. 1400. Colloquial use for "anonymous people in authority" is attested from 1886. They say for "it is said" is in Milton.

The most important importation of this kind [from Scandinavian to English] was that of the pronomial forms they, them and their, which entered readily into the system of English pronouns beginning with the same sound (the, that, this) and were felt to be more distinct than the old native forms which they supplanted. Indeed these were liable to constant confusion with some forms of the singular number (he, him, her) after the vowels has become obscured, so that he and hie, him and heom, her (hire) and heora could no longer be kept easily apart. [Jespersen, "Growth and Structure of the English Language"

Personal pronouns are concepts that are an integral part of the basic lexicon. It is therefore interesting to question them in order to better understand their relevance.
The words to express the personal pronouns that come from the African branch that gave the Sanskrit must therefore recognize etymologically their African root not as a shame but as a truth since the man

was born in Africa and acquired language. One can not therefore be born in Africa and have a non-African etymology in terms of language. We must all be consistent.

*3.4.3.33. **EXAMPLE N°33: Ra***

Standard method

Middle-egyptian:  ***ra*** "sun, day; Rê" (*Faulkner p. 147, Wb. II p. 401*)

Endogenous method:

In this exercise, i suggest to proceed in another way namely how to interrogate some African languages on how to name the solar star and only after that to proceed at the reading in Ewondo.

	SUN	Ewondo Equivalent	Lexicon
Kourounfé	a firè	à-firi	confidence
Sara	bak	i-bàkh	hoe
Ciluba	diba	i-dibgha	key
Wolof	djaneteu	dyal i-tôa	lord's country
Wê	djrou	dzal hor	lord's country
Min	ewoe	eyo	lord
Bankon	fiah	mfièh	light
Basa 'a	hyangaa	iyaŋ	scorching
Kouyou	i kanga	i-kanga	bridge
Poular	naangue	Nanga	Elder Ekang Ancestor
Bandjoun	nam	*ùn* -nam	country
Fondjomekwe	nam	*ùn* -nam	country
Atchan	n'gnoka	nyoukh	snake
Kikongo	ngue ntangu	ze ntong	the feeder
Ewondo	nloh djobo	*ùn* -nloh dzobo	sun
Lik pak pan	nwiin	I-nyin	life
Medumba	nyame	I-nyama	melter
Sanaga	ongona	Ngwana	The girl
Ehotile	oroyi	nyia hor	hor's mother

	SUN	Ewondo Equivalent	Lexicon
Zulu	ilanga	iléh ze	leopard's tree
Gourou	Irie	i-lé	tree
Basa'a	job	yob	up
Kiswahili	juha or juwa	yô	up
Comorian	Juwa	yô	up
Bamana	kile	à-kol ilé	parent tree
Kikongo	mbumba	ùn-vamba	ancestor
Dagari	mô- tong	ùn-ntô	elder
Lingala	moyi	mə-hié	red
Saïdic	pH	à_fab	wing
Haoussa	rana	râna	source energy
Senoufo	tchangue	i-two zə	panther's sit
Beti	viane	Viane	sun
Duala	wei	i-vié	red
Wiintoogo	Wiinĩga	mininga	woman
Gbe	xw~e	-Vié	red
Songhay	woyna	Viane	sun
Bissa	wu so	ù-sùé	river
Xapaitïiso	hi si	si	earth

CHART 98: CROSS COMPARISON RELATED TO SUN

We note that all the words suggested in the different varieties of the African language, all have a counterpart in Ewondo.

We also note that all these vocalizations seem different at first glance but are 90% summarized to a common idea: the place of the panther, the mother, to the source of life.

For languages that are supposed to be different, it is surprising that they convey curiously the same idea.

It can also be noted that the vocalization **ra** of Egyptology is found in Hausa with vocalization *rana.*

Why do we speak here of a panther or a mother?

The answer to this question is below.

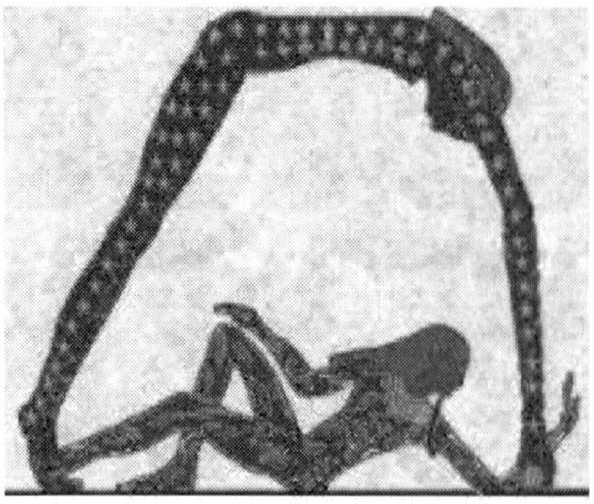

Nwt
with her spots looks like a panther[125]

Milky Way is the african panther

If in some languages we call the solar star, it turns out that in African languages
It is a matter of it situating within the Milky Way.
To the question how to read this script in our languages everyone has already his answer.

Reading in Ewondo:

What i see	Hidden meaning
Ewondo: /à-nyu/ "mouth"	**Ewondo**: /ù-nson/ "beak, mouth"
	Ewondo: /ù-nloh/ "head"
	Ewondo: /dzô/ "speech"
	Ewondo: /i-lê/ "teller"
Ewondo: /wa/ "arm"	**Ewondo**: /*ùn*-mbokh/ "hole, 1"
	Ewondo:/zokh/ "elephant"
	Ewondo: /zə/ "panther"
Ewondo:/*ùn*-mbokh/ "hole"	**Ewondo**: /*ùn*-nloh/ "head"
	Ewondo:/*à*-lɔŋ "hole"
	Ewondo: /dy-oé/ "nose"
Ewondo:/ù-nti/ "lord"	**Ewondo:** /à-tari/ "start". **Ewondo:** /i-sià/ "father". **Ewondo:** /tara/ "father".

This script will be read:

Middle-egyptian: ***ra*** "sun, day; Re"
Ewondo: /*ùn*-nloh dzô/ "sun, day; Re" litt. the head of the panther.
Ewondo: /tara/ "the father"
Ewondo: /vyène/; /vyè/"sun" litt. Ivyə na "the red of the mother"
Ewondo: /nyamlà/ "sun" litt. the one who heats
Ewondo: /nyambà/ "sun" litt. The one that melts
Ewondo: /ùn-ŋkoro/ "sol" litt. The one who dries

The transliterated word ra corresponds to the word: **Ewondo** /à-lê/ "tell, speak, say etc."
If we proceed to a little comparison between sun, head and words related to the speech we get this:

	Sol	Head	To speak
Sango	lá	**lì**	Sarä/húrù

125. Kriyòl Ayisyen: Nayite/Nayide "nwt"

Zandé	urù	rîì	Ya/tera
Ewondo	/*ùn*-nloh dzô/	/*ùn*-nloh/	à-lê
Nzimé	dɔ̃ɔ̀	lʉ̂	elɛ́ɛ

CHART 99: CROSS COMPARISON BETWEEN SOL AND SPEAK IN FEW AFRICAN LANGUAGES

We notice that the words used to designate the sun share, the same root with the ones used to designate the head or the semantic field of the speech.
Do we associate the sun to a simple star or do we name an idea that represents the head or the world?

So we easily notice that the sun is identified by its position on the top of ***nw.t*** *"sky"* and that can correspond to his word **Ewondo*:*** */à-nyu/* "mouth"

Since African cosmogonies teach us that at the beginning was the word and that the word was with the primordial ancestor, everything becomes clear again.

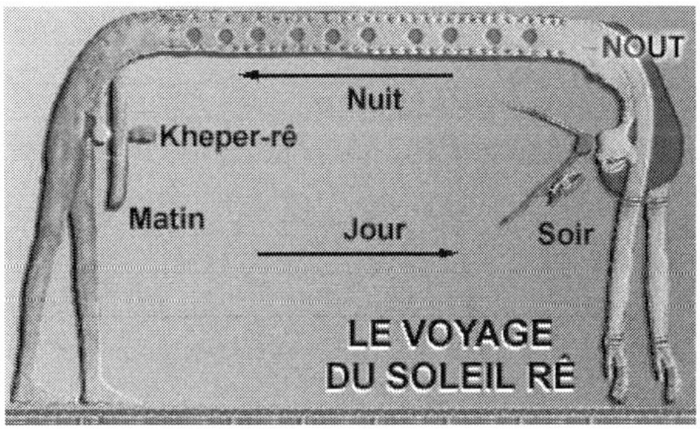

Quadrupole trip of Ra

So we have the complex relationship:

Sol: speech = head = mouth = ancestor=top

Ewondo: /*ùn*-nloh dzô/ "head the speech"
Ewondo: / á-lê/ "to tell, to talk"
Ewondo: / á-loué/ "to call"
Ewondo: /Tara / "father"
Ewondo: /*tari / "begining"
Ewondo: /à-nyu/ "mouth" ☞**Middle-egyptian**: ***nw.t*** *"sky"*

Ewondo: /E-yô/ "ancestor" litt. The vomit by a game of word, the thorn.

What is named solar by western referring to a star has no equivalent with this meaning.

Ewondo: /á-soli/ "to hide; cover up" ☞**English**: solar
Middle-Egyptian : ***imn*** "to hide; cover up"

The sun will rather be identified by its location, its manifestations, its effects:

Heat -Light – Time -
Etc.

Ewondo: /vyèn/; /vyè/ "sol" litt. ⟶Ivyə "red"
Ewondo: /nyamlà/ "sol" litt. The heater
Ewondo: /*nyamà/; /nyamga/ "sol" litt. The melter
Ewondo: /Anyame/; /Zamba/; /Zama/ "ancestor"
Ewondo: /*ùn*-ŋkor-/ "soleil"; "savior" litt. The dryer
Ewondo: /*ki**ri**/ "to roast; morning; tomorrow";
Ewondo: /kab'**ra**/ "to share life energy; goat";

Ewondo: /kob'**ro**/ "wardrobe";
Ewondo: /dzô/ "sol" litt. speech
Ewondo: /à-sira/ "sunstroke, sunburn"
Ewondo: /i-yér/ "miror"

Etymology

On this account we note that the French word *heure* (fraction of 60 min) is nothing more than a word find everywhere:

French: *heure* "60 min"
English: hour
Sango: *húrù* "to blow"
Hausa: *hura* "to blow"
Pular: *hoore* "head"
Middle-Egyptian:  **Hr**

Dutch: uur "60 min"
German: *Uhr* "clock; hour"
Romanian: oră "hour"
Spanish: *hora* "time, hour"
Italian: *ora* "hour"
Ewondo: *à*-wolà "hour"

Conclusion

If the reading of the word transliterated as ***ra*** iby Egyptology is not complicated but it's comprehension is highly complex from the African point of view because it is life itself. The words used to designate the sun besides designating the head will therefore naturally be relative to the mouth as well as to drink, to eat or even to vomit.
Also to understand what it refers to is to be at the heart of a multivariate analysis is to be at the heart of all occurrences identified by Egyptlogy. So I pruned here only the superficial part to allow a familiarization with the mechanism of ***mDw nTr***.

In concrete terms, this implies the following concepts:

Axw: *sunlight*
Axty: inhabitant of the horizon; residing on the horizon (idiom.)
Akr Aker; Earth; the gods of the earth
I, myself; me; my
Iaxw: sunlight; influence
iah: moon
imn: amon
imsti: Imsety
in-HRT: onouris
Inpw: anubis
ihy ihi:, the sistrum player; title of priesti
xmw-wrd: the tireless star
it: father
itmw: Atum; ntum, tum, i-tum etc.
ahaw: alternate, help
wp-wawt: Wepwawet
wnn-nfrw: Onnophris
wr-hkaw: great of magic
wrrt: crown
Wsir: osiris
nb-r-dr: lord of all
nbd: the evil one
npr: grain; Nepri
nrf-tm: Nefertum
nwnn: primitive waters;
nsw:king of high-egypt, king
ntr: god, divinity
re:sun, day;
rwty: routy
rhwy: the two companions; both fighters
rx-nsw: knowledge of the king
ha:ha god
hay: protector
hapy: hapy, nil; flood
hw:hou, the creative verb
hw: food, food
hwt-ka:chapel
Hpy: Hapi
hm:majesty
hnsktyw:carrying braids
hr:horus, the king

Wda: Judged
Baw: "baou", souls; power,
Bagyw: those who languish
bah:abundance
bsw: secret image
pawty: old family man; primary
pwntyw: the inhabitants of pount
Psdt: Ennead
Pth: ptah
Mnw: the god min
Mntw: montou
mr-wr:mnevis bull
mry:loved
mh:bent
mhy: keeper
mhw.s:crown of low-egypt
mxnty:smuggler
msxtyw: the greate ourse
mstw: offspring, descendant
nwt:nout; sky
Skr:Sokar
stx:seth
sw: shou
swbty:shabti
Snwt:entourage
Qbh-snw.f: Qebehsenuef
km-wr:kemour, the black great
gb:geb
gmhsw: falcon
gnbtw:a foreign people of pount
dwa-mwt.f:Duamutef
dawn:dawn, morning; tomorrow, tomorrow
hhw: heh
hqa:ruler, ruler
hkay:magician
htptyw: those who are at peace
xbstyw: the bearded
xprr: Khpri
xps: constellation of the greate ourse
xm: divine image, sacred
xmnyw: ogdoad
xnsw: Khonsou; tenth month of the year
sah: constellation of orion
saty-bity:the royal twins, shou and tefnout
saty-gb: son of geb
sia:to recognize, to perceive, to know; god
sah: dignitary; noble
sbk:sobek
spdw:Sopdu
smayt: association, confederation
sxm: the powerful, the power
Sxm-ir.f:potentate
dwatyw:inhabitants of the lower world
DwA.t(douat):dead
dmdw: crowd
ddwn:dedun
dhwty:thot
ta-tnn:Tatenen
tpy:living-ta
Re:sun, day;
rat:Rât, goddess sun
Swyt:shadow, spirit
tm:Atum
Etc.

All the words above are the manifestations of one and the same reality.
If i ask you what is the word behind the skeleton of consonants ***mr***, of course if you are rigorous, you will compute all the words in your language sharing the skeleton of consonants *mr*. it's the same for me here. According to me, reading is not complicated. The good comprehension is not easy.

3.4.3.34. ***EXAMPLE N°34: SOBEK***

In scholarly literature derived from the interpretation of the Greek and demotic papyri or the Tebtynis papyrus (P. Tebt., II, 291-315, P. Kronion 1-69). Here is what is said about Sobek. He is still called Sochet, Sobék, Sobki, Soknopais etc. In Greek texts are names would call so:

Greek: suchos ou sucos (σοῦχοσ ου σοῦχοσ) "crocodile",

It is the son of the aquatic goddess Neith and the twin gods senwy (= both brothers). His status as a god of water and fertility makes him worship all over the Nile Delta, the Fayum and Kom Ombo.

I really enjoy Susan Brind Morrow's observations in "*The Silver Eye: Unlocking the Pyramid Texts*":

> *"[...] A glance at the wall shows the subject of the fifth and final verse: water. Hidden in the waves is the danger of the rising water, personified by the danger rising in the water, the crocodile. The verse is a dynamic description of both the element and the animal. The swirling throb of rising water is conveyed by the word kbb, a word like ebb. The economy of the previous verses gives way to a kind of Wild fluidity that conjures both the sight and the sound of the rushing flood that brings the miraculous greening of life. The crocodile is the rising of greenness itself. The intensity of the motion as the meaning of the animal is embodied in a visual pun on its name that contains the typical sweetness of Egyptian humor.*
> *The name of the crocodile is **sbk** (s/thread, b/leg, k/basket). Unis pi **sbk.** Unis becomes the crocodile. Beside the word sbk on the wall is the word for "leg," sbkh, With the same letters but a different, aspirated k. Both words are spelled With a leg, which is the letter b. The second word ends With the leg, not as a letter but as a defining picture drawn slightly differently. Immediately below this picture is the word khbs (kh/lined disc, b/leg, s/knotted cord), the word for "tail," spelled out and followed by the picture of a tail. The word for "tail" is the word for both "leg" and "crocodile" spelled phonically backward. (...)"*[126]
>
> *(MORROW, 2016)*

Many things are said on this subject. No opinion is taken into account regarding what Africans think about the subject. I admire the voluntarism of some but stunned by the certainties on the topics they embrace so casually.

Standard method

Middle-egyptian:  ***sbk*** "Sobek" (*sbk Faulkner p. 221, Wb. IV p. 95, 2*)

This script caught my attention because it is wealth of information. Now that we are used to cram scripts I will go to the point here no longer formalize me on the methodology you already know very well. Reading this script does not involved any kind of issue for the African point of view since we have already touched on it a bit in reading the some other scripts. What matters most here is the understanding of this script as I have emphasized. What is interesting here is the semantic field linked to this script in its construction and without its everyday use.

Endogenous method:

Arbitrarily here, I will consider that this script is composed of the sign S29  and the set of signs  ***bk***. I will first consider words that have a phonetic proximity with the skeleton of consonants ***bk***. I will be content with this very complex analysis of the words of the Kweke group. Since I make a case study here, I gladly leave the words of the Kekwe group and those of the Kikuki group this time.

I will therefore list all the words that answer to the requirements of the skeleton of consonants ***bk*** in my language. Before i get started, i just want to point out this below.

Middle-egyptian:  ***bik*** "falcon"

Middle-egyptian:  ***skr*** "Sokar"

I close the parenthesis because i don't want to weigh down the load. As you can note by yourself above there is a link between ***bk*** and ***bik***. We were supposed dealing with the crocodile and now we are dealing with a bird, a falcon.

Let's go back to our listing. We have to this account:

126. Susan Brind, MORROW, The Silver Eye: Unlocking the Pyramid Texts. Head of Zeus, Londres, 2016, p. 480

Ewondo: / ùn-mbɔkh / "kneader, mortar, trough"
Ewondo: / à-bəkh / "small tubes of clay made by worms"
Ewondo: / à-bəkh / "Complete fishing line"
Ewondo: / à-bəkh / "whistle, trumpet, barrel end of a rifle, horns"
Ewondo: / à-bəkh / "float raft"
Ewondo: / ù-bəkh / "short, small, piece, half, fragment, short and big etc"
Ewondo: / à-bəkh / "to delimit"
Ewondo: / *ùn*-mbəkh / "fall"
Ewondo: / *ùn*-mbakh / "scare, tag"
Ewondo: / *ùn*-mbàkh / "lie"
Ewondo: / *ùn*-mbəkh / "kolas game or songo"
Ewondo: / *ùn*-mbəkh / "big fire, big burning tree that"
Ewondo: / á-bəkh / "cut; split"
Ewondo: / *bəkh / "carry, to get carried"
Ewondo: / *bəkh / "hurt each other"
Ewondo: /*bəkh / "curse" (child→ old)
Ewondo: /i-bəkh / "luck success, talent, skillful, skilled"
Ewondo: /i-bəkh / "large gathering of people to palaver or dance"
Ewondo: / *bəkh / "to put the fire"
Ewondo: /*bəkh / "cross a river"
Ewondo: / bəkh / "castor" (Ricinus *communis*).
Ewondo: / á-bəkh / "cut, hurt, score"
Ewondo: / i-bəbəkh/ "Big thing"
Ewondo: / i-bəkh / "palaver, catfish"
Ewondo: / á-bəkh / "cut"

Since the consonants skeleton is ***sbk*** and is a combination of ***S*** and ***bk***, to the previous listing of words, I will be able to add another list of words this time limited or presenting an occurrence with the sign S29 . As I have emphasized during tones study; syllables are suprasegmental units with non-null signifiers. These are operative concepts.
Then, subsequently because everyone understands the method, I will simplify things and keep only two words related to the sign S29  in the language we have so far triturated. By the way we could have:

Ewondo: / *ù*-swé / "river"
Ewondo: / *ùn*-nswé / "lizard"

I took these two words simply because of the script  and particularly because of the sign I3 of the Gardiner's list attending as its determinative.
Thus, if we resume the first listing and make a mesh with this listing reduced above, we already offer a semantic field much deeper than the simple sobek proposed by Egyptologists.
I have postulated that these characters are tonemes with meaning non-zero value, I give a clear enough illustration of this purpose. If you ask me how this script reading and what it meaning you understand that it is the context that will give a good understanding.

This point left aside from the incantations of scholars, who make a God from the copulation of x with y. The words feeting the consonants skeleton ***sbk*** and also the meaning exist at in Ewondo language.

Ewondo: /*ùn*-nsəbəkh/ "stream, small stream"
Ewondo: /i-sobokh/ "irritating, pungent, hot, noisy"
Ewondo: / à-sôbôk/; "Sound of a massive beast jumping over a trunk".
Ewondo: /*ùn*-nsobokh/ "the one who washes"

Ewondo: /sɔmbɔkh/ “lucky, providence”
Ewondo: /*ùn*-nsɔbɔgh/ “the hidden”
Ewondo: /*ùn*-ŋkwébékh/ “deepness”
Ewondo: / *ùn*-ŋkwebég/ “ravine, furrow, valley, groove, ditch etc.”
Ewondo: /*ùn*-mbyekh/ “Ride, fold of ground, small ravine, hilly”
Etc.

I think that, having listed all the words presenting a skeleton of consonants ***bk*** I thus allowed by this small listing to establish the link on the semantic level with the transliterated script ***sbk*** .

From this previous listing, I just want to take the word / i-sobòkh / because it is very interesting. It is used almost every day in an African family. Often by mothers.

Ewondo: /ma á-wogh i-sobokh/ “I'm irritating, I'm hot, I'm choking etc.”

Sbk refers here to the sensation of warmth and suffocation. But until then I have taken into account only obvious descriptors like water or lizard.
We thus arrive indisputably to the following notions:

Middle-egyptian: ***nw.t*** “Nout; sky”
Middle-egyptian: ***niw.t*** “lower sky”

Middle-egyptian: ***nnw*** **“**primitive waters; Noun”
Middle-egyptian: ***nnw*** **“**primitive waters; Noun”

Water and heat appear to us to be identical in nature but simply different in degrees

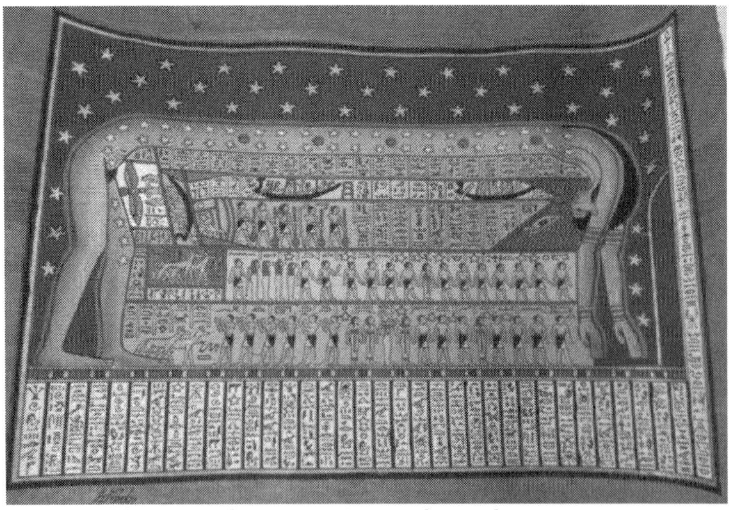

*The sun and nw.t constitute only **one** by metonymy*

The expression below encloses at least 3 strong ideas from the African point of view:

Ewondo: /ma wogh i-sobokh/ “I'm irritable, I'm hot, I'm choking etc.”

I need a shelter: /à-s**ɔ**b**ɔ**/ “to hide” **M.-E.**: ***mD.t*** **M.-E.**: ***imn***
I need water to refresh: /à-s**o**b/ “ to wash”

I am irritated: /ya-s**ôb**/ "it stings, it irritates, it heats"

If we go back to our ***sbk*** consonants skeleton. It was also enough to see that:

> **Ewondo:** /soghe/ "Domestic Barbary Duck that does not speak but whispers. */Soghe a k'a te-takh/*"

The idea of irritation is not here the fact of the heat but the whispering of the choking by the nervous noise.

Sbk is especially one who hides namely the sun.
So when we talk about ***sbk*** we also talk about the original ***anatidae*** of the peoples of ***km.t***.

The notion of ***sbk*** goes beyond the simple frame of the populations of Africa.

Am I surprised when inhabitants in Papua New Guinea named **Sepik** (pronounced "si-pik") the longest river in New Guinea? May you verify by yourself what I mentionned above.

> **Ewondo**: /*ùn*-nsəbəkh/ "stream, small stream"

Am I surprised when these same populations devote a sepik art? Do rites of passage into adulthood and have among them crocodile men called **sepik** (pronounced "si-pik") too?

Am I surprised that these people are calling a province of Papua New Guinea **sepik** (pronounced "si-pik")?

If I had not previously given the words beti of the semantic field of **sbk** indeed I could pass for a mountebank.

Sbk or the river falcon

I initiated at the beginning of this exchange the scripts below. The curious ones would like to know more:

> **Middle-egyptian**:  ***bik*** "faucon"
>
> **Middle-egyptian**: 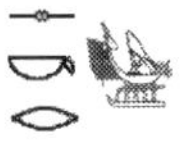 ***skr*** "Sokar"

We therefore see the limits of the transliteration that invents characters that do not exist.
They themselves transliterated **bik** to name the falcon which is one of the birds symbolizing ***Hr***.
If we understood correctly that the **S** prefixal of sbk could indicate the river / ù-swé / or *sea* in English so what is all this cinema?

> **Middle-egyptian:**  ***sbk*** "Sobek"
> **Ewondo:** /ù-bam ù-swé / "hawk of the river"

> **Middle-egyptian**:  ***bik*** "falcon"
> **Ewondo:** /*ù*-bama / "falcon"

Yes, Obama doesn't give ***bk*** but ***bm***. So hit me. But the hawk is called in my own language / obama /, / obi / the word ***bik*** is only a transliteration of /i-kàb/ "picker" or i-bəkh "carrier". In African falcon ofter pick up chicks.
Just for your understanding, may you remember that Ekang peoples claim that they crossed the Sanaga river on the back of a crocodile names */ngan medza/*. We have already discussed this point. And here, we have the word carrier linked to a bird that can be consider as the sky matching part of the crocodile.

Note : the English word picker embodies the root **pick-** with the consonant skeleton ***pick>pik*** versus ***bik.***
May you explain me then how The British archaeologist and Egyptologist who became world-famous after discovering the intact tomb (designated KV62) of the 18th Dynasty Pharaoh, Tutankhamun (colloquially known as "King Tut" and "the boy king"), in November 1922 namely **Howard Carter** was supposed to read the ***mDw ntr*** not seeing the words of his own language !!!

Middle-egyptian:  ***sbk*** "Sobek"

English : back• bakes • breaks • beaks • becks • cakes • kebs• kibes •caibs • Ceibas etc.
As you can see, these are perfectly English words. Apart from the ideological framework that leads one to believe in different languages, i challenge who needs to tell me that is not true.

It can also be a synonyme of / *ùn*-mbôkh / "prisoner" as you can see below:
Middle-Egyptian.: 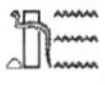 ***mD.t***

Middle-egyptian:  ***sbk*** "Sobek"
Ewondo: /*ùn*-mbokh *ù*-swé / "the prisoner of the river, the handicapped of the river"

Middle-egyptian:  ***sbk*** "Sobek"
Ewondo: /*ùn*-ŋkomboro / "men manufacturer"

Here the music is simple  ***s/z* "man" and**  ***bk*** /*ùn*-ŋkom/ "crocodile"

Middle-egyptian:  ***sbk*** "Sobek"
Ewondo: /i-bəkh *ùn*-nswé / "the very long and large lizard"

Middle-egyptian:  ***skr*** "Sokar"
Ewondo: /byal *ù*-swé / "the boat of the river"

I'm doing away with all the homophones and the resulting hidden meanings.

Middle-egyptian:  ***skr*** "Sokar"
Ewondo: /*ùn*-nkor' ù-swé / "whoever dries up the river"

I'm doing away with all the homophones and the resulting hidden meanings.

Middle-egyptian:  ***sbk*** "Sobek"
Ewondo: /*ùn*-ŋkomboro / "men manufacturer"

I don't want to do more on how to deal with this script that does not bring us much. Right here we understand that man is earth and returns to earth is not it?

Small comparative analysis of the words designating the crocodile.

Languages	Crocodile	Ewondo	Lexicon
Aymara	kuḳutrilʸu	Ùn-ŋkukukh dyal	The very old village
qhichwa simi	kukurtillu	Ù-kèkü dyal	The very old village
Mapudungun	ko kujiñ	Koko dyal	The very old village
Tupi	yakar	i-kari	The crown, the chief
Gününa Küne	yamgawčˀač	i-yemghə koko	The old shackled
Tacana	mama-mama	məmà	The great mother
Shipibo-Conibo	kapɨ	i-kyabi	Whoever catches
Yagua	nuturu	*Ùn*-nturu	The saint, the accomplis-hed
Ngiemboon	Nnuguem	*ùn*-ŋguəmə	The tail
Sango	Ngundé	ngondo	Horrible ; scary
Lingala	Ngando	ngan-nda	House crocodile
Nzebi	Nganda	ngan-nda	House crocodile
Duala	Ngando	Ngan-nda	House crocodile
Wolof	Dieusite; Diassit Djassigue; Jeusite	I-toà i-dzos	The place of judgment
Soninké	kin	À-ki na	The egg of the mother
		Kiŋ	Throat; gorge
Bambara	Baama	Ù-mvamba	ancestor
Malink	Baama	Ù-mvamba	ancestor
Moore	Yimbga	Zamba	Primal ancestor
Vili	Ngando	Ngan-nda	House crocodile
Fulfuldé	Nguilanguehrou	nguénguéru	The Albino
Pulaar	Nooro	Hor	Primal ancestor
Hausa	kada	ngarakh	The hindered
Sango	kada (lezard)	ngarakh	The chained
KiKongo	Ngandu	Ngan-nda	House crocodile
CiLuba	Ngandu	Ngan-nda	House crocodile
Yemba	Nga-ndu	Ngan-nda	House crocodile
Kouyou	Ngongui	i- ŋgəng'o	the throat
Mbène	Hikombad	À-ko mbwar	The place of the wrinkles
Seerer	Noh	Nah	The mother
Senoufo	outjîn	Tin	The origin, the source
		Ù-tiŋ	weaving
		à-tiŋ	alliance
Gourane	adi	á-dî	To confine
Godi	gbogbô	Koko	The very old
Japanese	クロコダイル (Kurokodairu)	kolokundu	Serpent of the crossing
Chinois trad.	鱷魚 (yu)	Eyo	Sanaga
Vietnamien	Cá sấu	À-ko à-swé	Primal ancestor
Birman	မိကြောငျး mi kyawwngya	mə-kol mə-nya mə-kî mə-nya	Hiding place
Hawaïen	ke kani	i-kaŋ à-ko	Feet of the mother

Turkish	timsah	Itàm i-siah	The lone father
Nti	Kumbhīra	À koh be râ I-kumbu râ	Renting old
Goudjerati	Magara	ŋgara(gh)	Bindings ; alliance
Hmong	khej	À-Koh nyà	The place of the sun
Maori	karaka	Koko ra	Very old râ
Nepalese	Makavānapura	Mə-kô vānə ‘mferèn	Disabled
Tamul	Mutalai	Mor *un*-nloh	The head’s woman
Pendjabi	Magaramacha	mə-swé m’àkoh râ	The place of Ra's hiding place

CHART 100: CROCODILE CROSS COMPARISON

I allow myself to return to the word Wolof used to designate the crocodile because it is very interesting:

Wolof : Dieusite; Djassigue; Jeusite; Diassit
Ewondo: /*i*-to *ùn*-ndzos’o /“the judge's place”

Crocodile as judge in Wolof

In Egyptian mythology, ***ammyt*** "Ammut", "the devourer of the dead", is considered in official literature as the goddess who, during the weighing of the heart, devoured the souls of humans deemed unworthy to continue to live in the world.
She is represented with a hippopotamus body, **a crocodile head** and lion's front legs**.** After the reign of Imn-Htp IV (or Akhenaten, 1353 / 52-1338) she appeared on almost all painted scenes representing the judgment of the deceased by the court of osiris.

When we know African thought, we know that in the species of cetretiodactyles **Hr** is the hippopotamus, **Hr** is the lion as a feline and as we have just seen, **Hr** is still the crocodile

There is no mystery here just the scribe tells us who the judge is. On this account, the Wolofs have perfectly preserved the memory of the crocodile as judge form ancient Egypt.

Speakers of the Kouyou language have preserved the memory of the crocodile as the throat *ngongui*.

The Japanese word for crocodile also brings a lot of information.

Japonais: Kurokodairu “japanese”
Ewondo: /kɔlɔ kô nda Elo/ litt. the old kɔlɔ is the house of ***Hr***

Ewondo: /ù-kɔrɔ kô nda Eyo/ litt. Whoever drains the house from above
Ewondo: /kəlɔkh kundum / litt. Thick skin.
Kinyarwanda: Hurukundo "love"

This endogenous reading makes it possible to get out of the myths of theocratic ideology. This reading helps to detoxify the advertising of gods and goddesses.
We speak of ***Hr*** whose hypostases are highlighted by zoomorphic representations.

Etymology

Middle-egyptian: kAp.w "crocodile"
Ewondo: /ù-ŋkàbà/ "whoever catches"
French: couper \ku.pe\
Greek: Kόψιμο **kóps**imo "cut"
Latin: *c*ol*p*u*s*
English: clip; chop; lop;

If I compare the skeleton consonants of **kps> cps** with the **sbk** skeleton, we clearly see that they are mirroring one another. The first **kps** is from the Negro-egyptian kékwé group, and **sbk** is from the kwéké group. The french word *couper* therefore has an African origin. It shares a phonetic proximity with the word *kapi* of the **Shipibo-Conibo** spoken in Peru for example or even *sepik* Latmul of the Kanganaman village in Papua New Guinea and of course /à-bukh/ "to break/; /à-bàkh/ "to cut" in Ewondo.

As always, I will not hesitate to ask scientists to kindly put these words in their lap. I mean in African because these words as we see all the shots are Africans.

Other

Some brands have not hesitated to develop on the symbol of the crocodile. Croco Pik for example of a brand you will recognize.

Middle-egyptian: ***km.t*** "Black Land, Egypt"
Ewondo: /ndəm ngàn/ litt. The crocodile's brand.

May you try to understand that reading

3.4.3.35. ***EXAMPLE N°35:*** ***NHSY*** *(Faulkner p. 137, Wb. II p. 303, 3-8)*

UNESCO has undertaken to write the general history of Africa through its college of scholars, but few in Africa know their past. The story that is told in textbooks is steeped in the ink of ideology. For many in Africa, their country was discovered in the XVth century by European navigators. I have read this script on example 22 i'm fliping back to it to show how a script can be interpretated depinding on the focus.

Before the fifteenth century, then, it was dark night in Africa. Also, few Africans can trace their history before the XVth Century. Very few can talk about the African rule of the 10th century, Arab invasions and prehistory. The onomastics resulting from the Berlin Congress, the overpowered nations of independence, are the average baggage of the African.

Those who trace their past from the Arab invasion are true buffs. These are the ones who try to find their historical memory. And these compared to the others, can measure the extent of the damage in the conservation of the historical collective memory as well as the tinsels which result like the mass alinnation.

The challenge of this script is therefore true. The study of this script consists of demonstrating that one can trace one's past through the scripts and write or rewrite one's story on sound bases. In truth, the scripts are self-sufficient and are not simple characters, they tell a lot of things to those who are listening.

If I'm willing to go to ancient Egypt dating back to -5000 years including ***Sms Hr*** / Zomloa.
A mass map of the African population can be drawn from the Horn of Africa to the Egyptian-Nubian civilization cradle, the flight from Egypt with the Arab invasions, migrations to the south Sahara towards the current settlement areas. We can see that quickly we are filling empty boxes.
Let's try to understand who are the ones that Egyptologists call Nehesy.

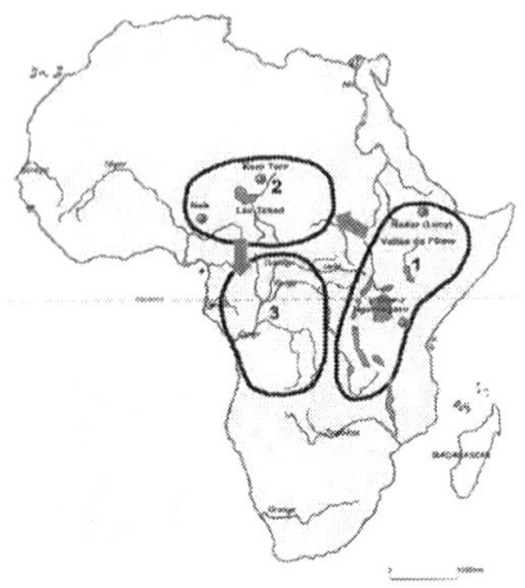

Migrations from the Horn of Africa

Standard method

Middle-Egytian: nHsy "nubians"

Endogenous method:

Middle-egyptian: *nHsy* "nubians"

I have chosen a script slightly different from the first because the two counterpart each other perfectly for the general understanding of this script. I also spare you the basics for reading this script because you master them all very well at this level.

Who are the nHsy?

Thay are all peoples in Africa often recognizing themselves as related to **Hr** and whose name means attached regarding the real alliance. I'm just point here that *attached* means being on duty, hobbled and submissive to **Hr**.

If we want to have the range of all the names, then we must rigorously sweep the lexicon of words to index all the words related to the hindering and cross-check with the names of populations that are named *attached, bound, knotted* , *shackled* etc. It is not complicated to do but it is not quite the object of my approach. So I will be content with only a few school examples to illustrate the point here.

So this script can be read, (I insist on it can be read):

Ewondo: /Bə-Tî/ because /á-tindi/ "to attach"; /à-tiŋ/ " node; knot; alliance"
Ewondo: /Fang/ because /á-fang/ "bind; tie down; truss up"

The first meaning being solely related to the notion of the impediment, it also offers by homophony and metonymy of / **à-tî** /, the sense of, respect for oneself, authority. The term **Beti** in this example, also expresses the sense of authority, of greatness, of self-respect, of nobility etc.

This reading is not exclusive or exclusivist in its deciphering, I insist on it.
To the objection and if this script designates simple captives for example? I answer while the sign G21 of the Gardiner's list means a guinea fowl. On this account the fact of naming the guinea fowl as a phonetic complement will give us additional information.

Middle-egyptian: ***nH*** "guinea fowl (*Numida meleagris*)
Ewondo: /*ùn*-ŋkaŋ/ "guinea fowl"

It just happens that when we question the Fang people, Bəti, they define themselves as Ekang and not as Pahouins as the way some want us to believe by distorting the word baboon. So this script is quite explicit on the question.
And the sign T14 of the Gardiner's list can be read in Ewondo: / *ùn*-nəŋ / "foreigner; visitor". So it designates a people not simple captives.

Since we have already mentioned here the social function linked to the wearing of the loincloth in african culture, we recall here that the kushite society before any foreign entry offered a social organization formed of castes[127]. Men of the loincloth were part of the social caste of lords, men of power.
We still find this social organization in castes, for example, in the Indus Valley. We remind to avoid any misunderstanding on the question that talking about organization is to be linked with the division of labor[128] and not exogenous raciological criteria such as those stemming from the thought of Gobineau or the French mustache bearer : Vacher de Lapouge.

Middle-egyptian: ***sr*** "significant; magistrate, high official"
Ewondo: /*ùn*-nti/; /tàrà/; /*ùn*-ntol/ /nyamoro/ "lord" or anthroponym /TSALA/
English: sir• sire • Eris • ires • reis • rise etc.

The sign A21 offers a good illustration of the loincloth and the associated function.

Etymology

The transliterated word is /*ùn*-nsɔrɔ / "the phallus" or / *ùn*-nsoré / "the scratched" or /*ùn*-nséré/ "the sliced" corresponding to ***wsir*** having diffused out of the mother earth to give **vizier** (وزير in arab and in persan, sometimes translated as vizier, vazir, wasîr, wesir or even wezir)**; sorrow**; **sore**; **sir** etc. in English, then in popular Latin **seiiōre*, denasalised variant Latin *seniōre*, the latter having regularly given *seignor* in old French; *Sieur* in French comes from *seigneur* and *sire*.
The *messire* variant of the old French will give *monsieur* "gentleman" by the contraction of "mon Seigneur". So we are talking also about the script ***sr***. The seigniorial function is above all an attribute of sedentary and non nomadic peoples. It is therefore logical that this term originates from a land that is sociologically characterized as matriarchal and sedentary exogamous.

After this aside let's go back to the other outlines of our script
Transliterated script ***xnty*** "advanced part; south, southern part" is read / à-folà bə-tî / or also /à-folà Fang /. The script *xnty* offers the visual of the crocodile as well as the sign G7 also indicates the place of ***Hr***. / à-kol *ùn*-nti / or / à-fola *ùn*-nti / "the place of the lord"

127. Cheikh Anta, Diop, Negro nations and culture. Paris: Présence Africaine, 1954, p.193.
128. According to Strabon Kouschite caste system: 1-Philosophers; 2-Cultivators; 3-Shepherds and hunters; 4-Artisans and workers; 5-The military; 6- The ephors; 7- The counselors and courtiers of the king (Strabo: Geoscripty, Book 15, chapter I, 29-38)

Ideological divergent squint.

What is destabilizing with specialists is their propensity to partition knowledge and things. When they are dealing with egyptology, these inhabitants of the script are Nehesy.

When one is meroitic researcher this same people becomes the *Nubians*, *kushite*, *the Napateans*, the *Candaces*. This is why they can not read what they call the Meroitic because their approach far from being scientific is tinged with ideology and the primary concern of segmenting knowledge.
So the full range of so-called exact (and universal) sciences when we fully understand the Symbolic language of the Magic Egyptian and Animist Science ... Strict rupture point ... nothing but continuity and fluidity as then you can perceive it so well in the field of so-called approximate (and human) sciences ... In a way, here is precisely where the break between Science according to "African Black Magic" and Science according to "The Greco Judeen Manicheism" .

Yes I know candaces are the supposed name of Nubian queens. Let's see this:

Middle-egyptian:  ***nHsy*** "nubians"
Meroïtic:  ***k(n)dk*** "candaces"
Meroïtic:  ***k(n)dk*** "candaces" **Ancient Greek**: Κανδάκη

Reading these characters is relatively simple here. What has been transliterated by ***k (n) dk*** is actually the word bəti / ngarakh / "tied up, hindered, bound, prisoner" it is a synonym of vocal /*tî/; /*fang/ as explained above. Note also that belt is use to tie.

Ewondo : Kanda "belt; ceinture; waistband"

We can also read this script /kindik/ meaning just from our point of view "borders; neighbors". Of course the so called nubians and egyptians according to specialists were neighbors.

But in reality  ***k (n) dk*** can be read / nya é ndom nya / "the country of the mother and her brother" exactly with the same reading comfort offered by the ***km.t*** script.
When in the tomb of ***snDm*** one reads that  ***km.t*** is ***tA-mry***; ***tA-sty*** and ***tA-ntr***. So here we have what we will call sometimes Nubia, sometimes the country of Kush to confuse while it is part of  ***km.t*** Hr's country.

Middle-egyptian:  ***kAS*** "Kush"
Maba: Kouchik "guinea fowl" **Middle-Egyptian:**  ***nHsy*** "nubians"
Ewondo: /mə-sɔŋ/ "the hindrance, the prison"

I just say that the word / mə-sɔŋ / is a synonym for:

Ewondo: /à-tiŋ/ "knot, link, alliance etc."
Ewondo: /ŋgarakh/ "knot, link, alliance etc."
Ewondo: /*fang/ "tied, knotted etc"
Ewondo: /*ùn*-mbokh/ "prison, link, knot, etc."

So I remain naively persuaded that the research could have proceeded if it had been removed from the ghetto of ideology. Many things have been invented that have been taught to Africans concerning them but many things very wrong unfortunately. We must have the courage to say it.

Compare These Two Scripts

Middle-egyptian: *nHsy* "nubians"

Middle-egyptian: ***sn*** "brother"

The scientists say that ***tA sty*** is the region of Nubia, the land of archers.
I voluntarily put the script ***sn*** because it confuses the method that offers an understanding by the transliteration and that makes the difference between ***n*Hsy** and the word ***sn***.
If we remove the "twin" ùn-mbias (y) and the sign V28, we have exactly the same root and therefore the same reading and the same level of understanding ***ns*** for Nubians and ***sn*** for brother.

A B C D

This painting from the tomb of Ramses III (1200 BC) shows that the Egyptians saw themselves as Blacks, and painted themselves as such without possible confusion with the Indo-Europeans [Caucasoids] or the Semites. It is a representation of the races in their most minute differences, which insures the accuracy of the colors. Throughout their entire history, the Egyptians never entertained the fantasy of portraying themselves by types B or D.

A) The Egyptian seen by himself, black type
B) The "Indo-European"
C) The other Blacks in Africa
D) The Semite

What I do really love in this painting taken from the tomb of ramssw III / ngamessi; nyamessi is that the inhabitant of Egypt (A) represents himself as a twin of the Nubian in (C).

So when I choose the script ***sn*** "Brother" I find here all the concordant elements, related to the symbols used: the spear, the notion of fraternity, the mother, the brother etc.

To get rid of the diktat of single thought, I could read the script:

Middle-egyptian: *nHsy* "nubians"
Peul: ***Fulani*** "nubiens"
Ewondo: /á-fùlàn/ "to resemble" ☞ **Ewondo:** /mi-'mbiàs/ "twins"

Claude Rilly the great Meroitic specialist did not have the humility to go and ask Africans what returns to this topic. I just present facts hard to refute.

Precisions:
Many in a spirit of malevolent and frustration will want to train me on an ungainly ground namely that I have said that Beti, peuls and fang, are the only inhabitants of Nubia. So I want to remind you here that I'm doing the exercise and I'm explaining a method.
Any African can do the same exercise in his language and check if the data sticks or not. So I say here that the condition sticks regarding Beti, peuls or Fulanis and Fang.

How to be certain that it does not concern only this group simply by scrupulously adopting what the scripts say without going into the tribalisation and the ethnofascism of the spirits.

Middle-egyptian: ***nHsy*** "nubians"
Akang; Bete;Bati; Batanga; Batchenga; Mossi; Baoulé;Bulu Etc;

Simply the scribe offers here the sign G21 by phonetic proximity between the term guinea fowl / *ùn*-ŋkaŋ / and the word he really wants to express here:

Ewondo: /á-kàŋ/ "to praise; to draw"
Ewondo: /*ùn*-ŋkàŋ'à/ "the one who praises (Hr)"

That's why this script is for everyone who praises **Hr**. This script is for all Africans who are related and serving their ancestors ***Hm*** /*ùn*-ŋkom/ "servant; attendant; bellboy".

We find in the cultural indices of the Akang people for example in this script. Thanks for taking a look at the illustrations (N ° 8, 8-bis, 10,). What I say for the Akang is valid everywhere in Africa according to cultural unity.

Let's summarize

The reading of this script gives us many indications on the historical and cultural plan that will guide the research in the field.

On the Cultural level

The Bəti speak of a primordial ancestor named Zamba, *ùn*-ŋkombodo.
There exists in the south of Egypt a temple called Kom Ombo. One can thus on the cultual level make researches on this site and restore the original worship by compiling the current sources and the information which one finds around this site.

Architecturally

In Nubia as well as in the south of Egypt we find temple constructions with very precise methodologies and precise knowledge. I have the deep conviction that by focusing on this work we can recover our memory and put ourselves back to work as in the past according to our paradigm.

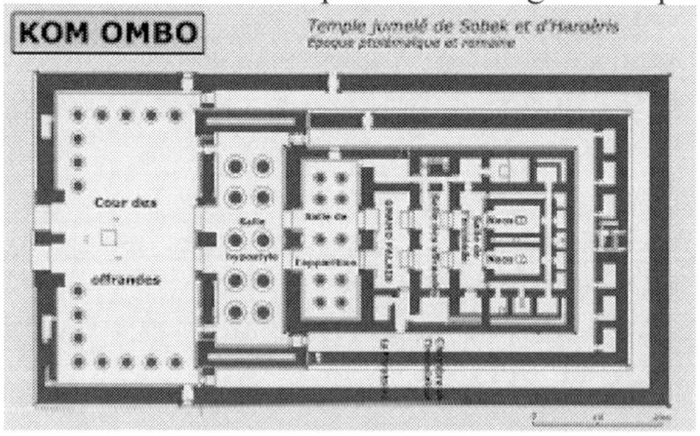

Ancient egyptian temple

On the IT Level

It is no longer an open secret that Africans have invented the modern computer. It is the ancient african binary system that is still used as a mathematic model of nowadays computers. Let's just suggest that ancients egyptian temples were motherboards of computers.
So getting back on the logic that makes these computers work is allowing humanity to make good historical records over time.

History Level

All the civilizational material of the so-called kerma or pre-kerma period, Napata, Philae, Meroe is known to fill the gaps with a more muscular research and taking into account the advances of the research in this field notably the last excavations of Pnoubs (jujube) under the direction of Sverine Marchi and Charles Bonnet.

Onomastics

Nubia: /*ùn*-mbon/ "oil" anointed tradition found in Mossi land (Moro **Naaba)**
Kush: /mə-sɔŋ/; /mə-sóndó/; /kongo/; /kimi/; /kami/ etc "the prison, The alliance"
Ta Nehesy: Si Ekang, Akong, Akang, Kongo, kimi, kami, etc.
Kerma or Km.t: /à-kol maâ/ " the mother's place" (place of the universal law)
Kerma: /*ùn*-ŋkor maâ/ "queen mum" from Qor and Maâ
Napata: /ndabot/ "the family; the duck"☞**Ciluba**: mpaatu; dibata "duck"
Mro: /Mor/ "woman, love, Homme" also found in Mossi land (***Moro*** Naaba)
Pr-bity: /ndabot/; /à-folà bot/ "the family"
Bity: bot, bwiti, boé, fɔŋ etc. "bees"
Nhsy: Ekang, Akong, Akang, Kongo, Mossi, Batanga, Bété etc.

NOTE:

The word / à-Kôŋ / in the Beti languages means Gold (metal) ***nbw "gold"***

Hausa: zinariya "gold" **M.E.**: ***nbwy*** "the two lords"
Chinese: 金 Jīn "gold"
Ewondo: zir "pearl, jewel, bead"
English: zinc **French**: zinc **Chinese**: 鋅 Xīn "zinc"
Foulfouldé: Kanguri "gold"
Middle-Egyptian: ***nbw*** "gold" ☞**Ewondo**: /ôbon/ "necklace"

The Hausa word for gold *zinariya* means simply /ziŋ oro/ "hate and hero".

M.E.: ***nbwy*** "the two lords or the twins"

The Chinese word: 金 Jīn "gold" or **Chinese**: 鋅 Xīn "zinc" return only the word for hatred **Ewondo**: /Ziŋ/ "hate"

Note: The Hausa word for gold *zinariya* teaches us that what we call gold is a mixture of at least two metals including zinc. Compare to gold, zinc is tough, hard. It then reminds the couple ***stX*** and ***Hr*** .

Composition of white gold 18 carats (750/1000): 75% of fine gold + 18.5% of silver + 1% of copper + **5.5% of zinc**
Composition of 14 k white gold (583/1000): 58.3% of fine gold + 17% of silver + 17% of copper + **7.7% of zinc**

Peul: *Kaaliss* "gold" ☞**French**: calice "chalice; calyx"
Wolof: *xaaliss; wurus* "gold" ☞**English**: chalice; calyx

We have also the word:

Mina: *sika* "gold" ☞**Manhi**: *sika* "gold"
Ewe: *sika* "gold"

This term must be weighed against the title *Askia* (emperor) prevailed within the Songhai Empire between the 15th and 16th centuries. The word or "gold" has to be related with ***Hr*** . Therefore, an african emperor is ***Hr*** follower. He is the golden.

On the the historical continuum this term reminds strangely what Egyptologists call Kush. The two words are just anagrams.

Middle-Egyptian: ***kAS*** "Kush"

Le tata de Sikasso.

The tata of **Sikas**so, also called tarakoko locally, is a series of ramparts initially built in Sikasso during the reign of Tiba Traoré (king of Kéndougou), between 1877 and 1897, to resist the attacks of Samory Tour. It was then enlarged by his brother Babemba Traore.
The item **Sikasso** can be understood as the land of ***Hr*** or the land of Gold. ***Hr*** being the protector and the savior then the Tata of Sikasso simply means the protection of Hr's land.

Chinese: 青銅 Qīngtóng "bronze" **Middle-Egyptian**: "bronze; copper"
Chinese: 銅 tóng "copper" **Ewondo**: ŋgô; kôŋ "bronze; copper; or"

We can note that the three metals are sharing the same name in the three languages.

The chinese word for copper and bronze are really interesting to analyse from Ewondo language. The two series of words seem not the share something but in reality chinese words are related to the sacred cow also named horned. The chinese word *tóng* means in Ewondo horn.
Qīngtóng can then be understood in Ewondo language as the supreme copper or literally the neck of the horned. As we can see below, gold is about the necklace.

Middle-Egyptian: ***nbw*** "gold, or"

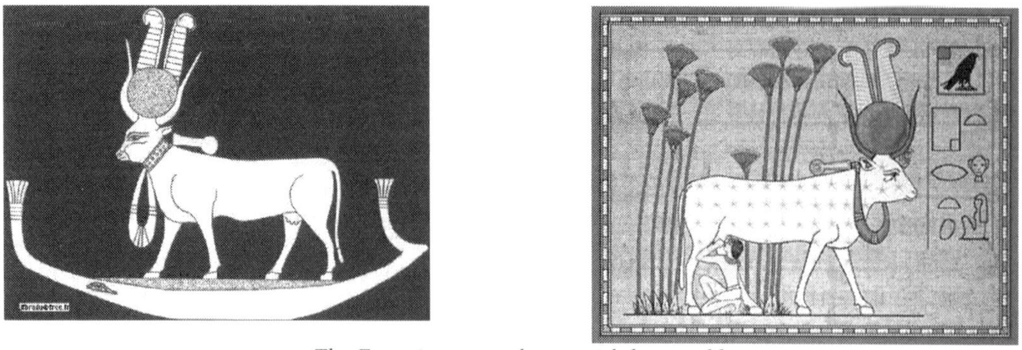

The Egyptian sacred cow with her necklace

We thus better understand why *Nubia* was also called the land of gold. A tradition that is still alive in Akang's land nowadays. If I do say *Koŋ'o* let me emphasize that it's an antonomasia meaning of the country of gold; the golden. Therefore anyone can better understand the meaning of the word KONGO "golden".

The linguistic similarities between Chinese and African languages eventually crystallize around the issue of metalworking and particularly goldworking in the sense that there is a great tradition of gold work. As we have just seen in Nubia, Egypt, Songhai, Mali, Ghana etc.
Did not this tradition of the cult of gold lead the Chinese to use bronze for lack of gold?
questioning is allowed instead of doubt.
Chinese bronzes are universally admired as masterpieces of the art of bronze, especially those made during the first two millennia before our era. The bronzes of the Shang and Zhou Dynasties, from the sixteenth century up to 221 BC, with their amazing shapes and decorations, testify to the creative spirit and the remarkable technical mastery of the Chinese bronzers during the Bronze Age. . The art of the royal dynasties of Antiquity (2100 - 221 BC), the essential foundation of Chinese art, is mostly a bronze art.

African society were basically organised around castes. And only people in the blacksmith caste were allow to deal with metals. That's why the West will only discover metalwork in the 19th century at coalbrookdale in England. While Nubia and Egypt were already working metals and traces of iron founded on the ruins of the castle of Loreponi in Burkina-Faso current dating back to the 9th century. We can not therefore discuss the origin of the terms associated with metals and on numismatics apart from the effects of announcements and mystifications of the West. The West can not challenge Africa about metalworking. It's a nonsense on a technical point of view.

This notion of ***nbw*** even makes it possible to revise the understanding that the egypotologists gave to the title of golden Horus of the fara's titulature.
It is clear here that this notion is linked to Nubia. So it's a title as sovereign of Nubia. Exactly as the ruler of England is Queen of New Zealand in New Zealand, Queen of Tuvalu in Tuvalu, etc. In the Channel Islands and the Isle of Man which are Crown Dependencies, it is designated respectively by Duke of Normandy and Lord of Man. In some territories, its official title includes "defender of faith" and "Duke of Lancaster" etc..

Let us read for this purpose the name of ***Golden Horus*** of Fara ***Axn-itn*** which appears under its first titulature as *Imn Htp IV*.

.

Middle-egyptian:  ***WtS Xaw m Iwn.w Sma***
Ewondo: mor ùn-nam <à-kol mə-kui ya ùn-nam mə-sɔŋ à-si>
French: *dirigeant de pays*: <le chef des Ekangs du sud d'où apparaissent des louanges>.
French: *seigneur de Nubie*: <le chef dont les louanges apparaissent au sud du pays des Ekang>.
English: *country leader*: <the chief of the Southern Ekangs, from which come praises>.
English: *Lord of Nubia*: <the chief of the Southern Ekangs, from which come praises>.

Let's take another name from the titulature of the same fara. I love this script because it stinks ideology full nose in its current vocalized form.

Middle-egyptian:  ***Ax-n-itn***. "Akhenaton"
Ewondo: i-yol ùn-mbôkh </nti Ekang/>
English: *priesthood name* "Lord of Ekang"

I will not list here all the important and interesting cultural elements offered by this script but you will have understood because it touches the history of the continent. It requires a pooling of efforts to reach the end of the renaissance. Scripts are greats talkative for who knows how to read them.
This approach makes it possible to review the pseudo history which is counted to Africans with a nauseating point of ideology and especially a lot of contempt and insults for a non-negligible fraction of Humanity.

Other: Bantus ...

Who are we talking about at the end? This is the question that a sensible African must ask himself. The Bantu are some Africans living from Cameroon to South Africa. Curiously when we question these populations they are everything but at no time do they designate me as being *Bantu*. The dynamic phenomenon of memory replacement has begun to have an effect on the folklorized class of colonization. Some are proud to define themselves as Bantu now.

So where does this comes from, this time?

It is a pioneer of comparative linguistic in Africa, the German Wilhelm Heinrich Immanuel Bleek, a German missionary who first, gathered West African languages under this term.

In 1859, he laid the foundations for modern Bantuistik (or Bantu language science) as part of the comparative analysis of the languages of the southern and central regions of Africa, to which he gave the name "Bantu".

The work of these linguists is first of all attempts to classify languages from which are then formulated hypotheses on the history of the establishment of their speakers. They made Africanist linguistics autonomous, the study of African languages, called afrikanistik, being at the outset of orientalism.

So it was this German linguist Wilhelm Heinrich Immanuel Bleek (1827, Berlin-1875, Cape Town), librarian of Cape Governor, Sir George Gray, and founder of Bantu linguistics who began to demonstrate the historical unity of the languages of their study that they named Bantu.

Demonstration that was given to Carl Meinhof to complete through his main works: Meinhof, (Grundriss, 1899) and Meinhof, (Grundzüge, 1906).

Bleek introduced the term Bantu into linguistic terminology, which he explains in bleek, comparative grammar, page 3:

> *"[...]That the derivative prefix and pronouns of this last gender (of personal nouns in the plural) are either actually ba , or contracted, or in same other manner changed from it, is one of the characteristics of the bâ ntu family of languages, which have on this account been called ba languages by dr. h. barth [heinrich barth (1821 1865)"*

Abel Hovelacque (1843-1896) contributed to the diffusion of the term in French by employing it in 1876 in the first edition of his highly influential work Hovelacque (Linguistics).

It is observed that there exists a variant Bantu (used in linguistics as well as in ethnology) that one finds, for example, in 1968, in Alexander, (Bantu, 1388: there is no Bantu race). There is also no Bantu civilization, Bantu customs, Bantu art ... there are only Bantu languages, as there are Semitic languages or Indo European languages. "Bantu" is a linguistic technical term, invented by linguists for the needs of their discipline, which can only be used incidentally in social anthropology, and not at all in physical anthropology.

And linguistics gets involved

After Bleek and Meinhof here is the continuation taken by Guthrie. The endogenous Bantu languages are distributed, according to Malcolm Guthrie (1953, 94), taken up by André Jacquot (1978, 639-643), in three zones of integration comprising zone A and zone H, on the one hand, and zone B, 'somewhere else.

The languages of this zone H are attached to ensembles whose main area is outside the Gabon, and the languages of zone B, have the main area at Gabon and a part overflows the territory.
Each zone has several groups and, within each group, one or more languages are identified.

To decide to classify a language as Bantu, Malcolm Guthrie uses a few number of criteria that he divides into two categories: main criteria, whose presence in speech is the "sine qua non" condition of the classification as a Bantu language, subsidiary criteria, with less importance but not necessarily all present in all languages, according to Pierre Alexandre in: Perrot j., (1981, 354-355).

So I spare you those criteria in this heuristic.
And finally all this will be inscribed in the baptismal fronts by Greenberg in his classification of African languages. Classification that is taught in all African universities at 21s.

Conclusion

The Bantu question is not a simple question. It is part of a geostrategy erasure of the Negro historical memory but especially for the replacement of this memory by a borrowed memory.

The story is that Africans were discovered in the 15th century and on that account, they are studied and labeled as cattle within ethnoscience.

Question?

Take away from me a doubt: why did all the people who were interviewed by Bleek have the primary concern to define themselves as human beings (ba-ntu) first and foremost?
Perhaps then it is implied that these populations did not want to be considered animals in front of the colonial master. I'm I right? Since among the Bantu some defined themselves as pardoned Baboons sorry I meant Pahouins.

There is thus in filigree the nausea theories of Vacher de Lapouge, Gobineau, Hegel, Levy Bruhl with its primitive mentality but especially Darwin according to the fact some men are descendants of greats apes. Words that are here distilled with intelligence by lending them to the Negroes themselves. If this is not ideology, what is it?

That means these Negroes define themselves as human beings; indeed fortunately. Why would not they call themselves snakes concomitantly? Here is the poison of the French thought *Philosophie des Lumières* represented by (Vacher Lapouge, Gobineau, Darwin, Hegel, Voltaire etc.) for which the Negro is a great ape barely evolved.
Why on earth, would the negroes all have needed, as by magic, to be similarly defined to human beings before Bleek? This simple question takes up the logical paradox and highlights the reminiscent of the fish syndrome that claims the dip net.

To define oneself as a human being does not make sense. This is a subterfuge well found by Western think-tanks, a subterfuge based on a semantic shift from a bias on two homophones to induce entire generations of Africans in a regrettable mental confusion and therefore plunge them into the mistake on their identity.

On this account, books have been written by Africans, inducing scholars and hungry in the mental confusion of wanting to exchange on a common basis with terms decided and defined according to the lexicon of others.

My personal opinion

Africans recognizing themselves as descendants of Hr who is / *ùn*-nti / "lord, ancestor, father, elder etc."

EWONDO: /bə-nti/ "those of the lord, the lords"
Basa'a: *bà-nto* "those of the lord"
Chichewa: *a·nthu* "those of the lord"
Duala: *ba to* "those of the lord" ☞strong resemblance with beti with apheresis of the ***n***
Herero: *ova·ndu* "those of the lord"
Kikongo: *baa·ntu* "those of the lord"
Kikuyu: *a·ndũ* "those of the lord"
Kinyarwanda: *aba·ntu* "those of the lord"
Kirundi: *aba·ntu* "those of the lord"
Lingala: *ba·to* "those of the lord"
Luganda: *aban·tu* "those of the lord"
Shona: *va·nhu* "those of the lord"
Swahili: *wa·tu* "those of the lord"
Tshilubu: *ba·ntu* "those of the lord"
Xhosa : *aba·ntu* "those of the lord"
Zoulou: *aba·ntu* "those of the lord"

What must we retain from all this?

- The term Bantu is initially used as a simple linguistic technique, invented by linguists for the needs of their discipline, which could also be used only incidentally in social anthropology, and not at all in physical anthropology as it has become Nowadays.

- The term Bantu results from a misinterpretation of bleek or a will to harm because since the 15th century at least, there is a plot against Africa under the leadership of the popes. A current of nefarious hatred that will take shape in pseudo "scholarly" academic bodies including with the philosophers of enlightenment and Africanists.

- The term Bantu thus serves those who categorize beings and things in order to erase any link between Africans and their glorious ancient past of ***km.t.***

- The term Bantu is not used by anyone in Africa to define itself as bipedal because the apodictic evidence that we are human beings does not lead us to ask the question reciprocally and to define ourselves as such.

- A priori, the African does not define himself as a Human Being. It is defined above all by its name, lineage, ancestor etc. but certainly not like a human being.

- The term Bantu is an immaterial weapon that is sinked like a ferment of tribalism and tinsel that flow from it. We define Bantu, semi-Bantu, tikar, Sudanese, proto-bantu, etc. but no individual of African origin questioning his past can find all these morphemes as identity markers..

-
The term Bantu is intended to create mental confusion between others and us. Forcing us to define ourselves according to the lexicon of others who are expertly imposed on us through the very expensive western school institution known as jules ferry.

According to African thought, what defines a human being is his name as a sociological marker and not the kind to which he belongs, let alone his race or even less his age. At the end, those who invented this conceptual trap are also Bantu according to their own definition: human beings. Hoisted with one's own petard.

It will therefore be necessary for Africans to ascend the musicality of words from the lexicon of others served in ready-to-think kits in order to define ourselves according to our past and only according to our paradigm and our own lexicon.

If we reject the idea of Bantu as an operational category in physical anthropology, we say that we accept the expression Ba Ntu "those of the Lord" because it is entirely an African reality.
So I say that there is no Bantu language and that humanity comes from ***Hr*** we name human beings ba ntu, bod, abantu etc.

On a technical level, there are two terms that are very similar in African languages: "those of the Lord" that we have just seen and the expression "those of Hr".

EWONDO: /bə-hor/ giving **/bor/; /bot/; /bod/** "those of Hr; humans, people"

Both expressions sharing the same meant confusion or malevolence led to using one instead of the other.

EWONDO: *rmTyw* "man; men, Human beings; Egyptians"

We can read this script as human beings since they are inevitably but the good reading is: "those of the lord"; "Those of the elder"; "The elected" or "those of Hr" etc.

3.4.3.36. ***EXAMPLE N°36:*** ***htp*** *(Faulkner p. 179, Wb. III p. 188, 2 sq.)*

Many Africans, following the deceitful spiral of Western Kamitism, regularly use the expression *Hotep* to greet each other. As weird as it may seem, those who speak their own language do not use this word outwards of the new western framework of Kamitism. The challenge of reading this script is situated at this level. Understand this script and explain why it is read one way or another according to the variants of the African language.

<u>Standard method</u>

Middle-egyptian:: ***htp*** *"Hotep, blessing, offering (idiom.)"*

<u>Endogenous method</u>:

Middle-egyptian:: ***htp*** *"blessing, offering (idiom.)"*
Meroitic: ***otp*** *"esteem, recognition"*
Meroitic: ***otp*** *"esteem, recognition"*

Bulu: *Mbolo* "hello"
Poular: *On beliquai ê diam* "hello"
Xapaitíiso: *hoi* "hello"
Bandjoun: *O ti'aa* "hello"
Ciluba: wêtò "hello"
Bassa: *Mbo'o; yega* "hello"
Mpiemo: *Waloe* "hello"
Mboum: *nang* "hello"
Fang: *Mvɔghè* "hello"
Moor: *Nê'y yibogo* "hello"
Godi: *ayoka* "hello"
Bahasa Indonesia: *hai* "hello"
Yemba: *letsa'te* "hello"
Eton: *mboghè* "hello"
Hausa: *kwanaŋ* "hello"
Peul: *wali* "hello"
Kikongo: *Yenga* "hello"
Nufi: *pè'teh* "hello, are you healthy?"

Qusqu qhichwa simi: *allin p'unchaj* "hello"
Lingala: *Mbot* "hello" litt.: May the divine flame burn in you!

I will not do a great presentation here but i'll summarize the subject to give the keys of reading from the language of the study.
I point out that the word *hotep* is very misleading here. It can not be said enough that the Negro-egyptian has two main branches, the **Kweke** and the **Kekwe**. So when it is transliterated *htp* we tend to read only from left to right hotep.

In reality, we must also consider the other possibility namely the reading from right to left obtained from the transliteration *htp.*

Therefore, a language like **Qusqu qhichwa simi** that is spoken in the Amazonia offers the word *p'unchaj.* For the purpose, let's use the *ptH* transliteration to recognize the same word. **Qusqu qhichwa simi** corresponds to the consonant skeleton *pTj> ptH* with *T = tch*

Another example is given by:

Middle-Egyptian: *htp* **or** *ptH*
Nufi: *pè'teh* a western Cameroonian language.

Lingala offers muted vocalization by replacing [p] with [* b].
Many languages spoken in western Cameroon like the Bandjoun language use the word of the kwéké branch *htp.* **Oti'aa** causing a complete weakening of the suffix.

The **Xapaitíiso** or **Bahasa** *hoi / hai* are very closer to the **English** *Hi.*

Let's go back to our script and comment it a bit for a better understanding of the etymology of the words we have just described.

Middle-egyptian:: 𓊵𓏏𓊪 ***htp*** *"blessing, offering (idiom.)"*

This script presents as a determinative a bread laid on a board / table.

In fact the words here mobilized in our languages can be associated with: equilibrium, stability, food, standing, table, board, posed, harmony, gift, gifts, mother, foods, feather, panther etc.

Etymology:

Middle-egyptian:: 𓊵𓏏𓊪 ***htp*** *"benefits, offerings (idiom.)"*
English:: Hi; Hello *"hullo,hallo;good morning"*
Mpiemo: Waloe "hello"

The lemma *Hi* and *Hello* of English can also find their etymology with the words Beti, /Eyo/ and /Elo/ designating the eldest ancestor. The offerings are assimilated here to the ancestor as being his part.

In Ewe, mina, the term *htp* recalls the expression *hor tepe*, which would refer to an imaginary but glorious place that, the one to whom it is addressed is supposed to occupy.
" Take a seat " = hor tepe wo.

What brings us to the city of Göbekli Tepe; an archaeological site that has revealed the oldest remains uncovered date back nearly 14,000 years. This site was occuped at the end of the Mesolithic early

Neolithic (more precisely the pre-Atramic Neolithic A). It is located southeast of Anatolia, the region of present-day Turkey located near the border of Syria. To the southwest lies the town of Şanlıurfa. The Turkish toponym Göbekli tepe would mean "navel hill", in reference to its shape.

Are the Ewe's talking about this city?

Curiously till this day, at least fifty-six monumental T-shaped stones recalling  ***Htp*** and reaching 3 to 6 meters high were taken out of the ground, most of them engraved. On some pillars, we can see many finely represented animals (snakes, ducks, cranes, bulls, foxes, lions, wild boars, cows, scorpions, ants). Some of these engravings have been purposely erased, perhaps to mark exclusions, or perhaps in preparation for other representations.

If we take the case of the French language this script would read: au top "At the top". Indeed the object on the table is above.

The associated semantax with *htp* still exists with the word top. *I am at the top even in spanish with the word tapas.*

If from the Ewondo I focus on the word Lingala word *mbot*, it can be approached to the word Bulu / á-bətə / "to place; to put up". In reference to the food that will be placed on the table of offerings.
The lemma / bətə / can be linked to the word / bələte / "bread" in the same language. Indeed the X2 sign  is a "bread". It may be noted that the word Ewondo: *teb'le* "table" is an anagram of the word bread within Beti languages.

The formula *htp* will give in Ewondo:

Ewondo: /à-təbə/ "to stand up"(Are you awake?)
Ewondo: /i-təbə/ "the place where something is deposited"
Ewondo: /à-bét/ "climb, above, up"
Ewondo: /yi u təbə/ "are you awake?"
Ewondo: /bi-tutup/; / bi-bele/ "food"
Ewondo: / blt/ "bread"
Etc.

<u>Philology</u>

If we consider ***H*** as a weak consonant, the skeleton consonants *htp* can be approached by the ***itm*** consonants skeleton:

Ewondo: /I-tam/ "alone"
Ewondo: /à-tam/ "to give, to wish"

The semantax related to *htp* "donation" has not really changed over time. it persisted in the expression "to greet"; "to wish a good appetite"

Ewondo: /à-tom/ "Give strength by fetishes"
Ewondo: /À-tome/ "offering, package"
Ewondo: /à-tom/ "Pecking, eating with beak"Syn.: kom, dom.

By the example of lingala *mboté* we have:

> The sign Q3 𓊪 which brings vocalisation [*bo] for door
> The sign X1 𓏏 sound [té].

This script offers many words like:

Lingala: Kotámba from proto-Ekang: *támb "take; receive "" to catch, to seize the flight"
Lingala: Motámbo "take; receive"; "snap, trap, hatch; ambush flight"
Lingala: Kotámbama "be snapped up, grabbed"
Lingala: Kotámbana "seize each other"
Lingala: Kotámbela "grab for"
Lingala: Kotámbema "be snapped up, grabbed"
Lingala: Kotámba "go; travel; being yourself a variant of * támbod"
Lingala: Kotámbola, "walk; to move, to advance; walk around"
Lingala: Kotámbama "to be advanced, to be presented"
Lingala: Kotámbela "to present to"
Lingala: Kotámbema "to be advanced, to be presented"
Lingala: Kotámbola miso "look from all sides"
Lingala: Kotámbolatambola "loitering, prowling"

This script is ripe with lessons to be learned. It depends only on each one to study it well in order to perceive the full extent of the associated meanings.

*3.4.3.37. **EXAMPLE N°37:*** 𓇋𓏏𓂋𓅱𓈗𓈘 ***itrw** (Faulkner p. 33, Wb. I p. 146, 10-17 sq.*

Formerly called ***itrw*** or ***Hapy***, nowadays it is known as Nile. In the opinion of the experts, it is an African river. With a length of about 6,700 km, it is currently considered the 2nd longest river in the world, with the Amazon River. It is one of the few rivers in the world to flow from south to north. It comes from the meeting of White Nile and Blue Nile. White Nile (Nahr-el-Abiad) originates from Lake Victoria (Uganda, Kenya, Tanzania); Blue Nile (Nahr-el-Azrak) is from Lake Tana (Ethiopia). With its two branches uniting with Khartoum, capital of present-day Sudan, the Nile flows into the Mediterranean Sea forming a delta in the north of Egypt. Including its two branches, the Nile crosses Rwanda, Burundi, Tanzania, Uganda, Ethiopia, South Sudan, Sudan and Egypt. It also runs along Kenya and the Democratic Republic of Congo (respectively with Lakes Victoria and Albert), and its watershed also affects Eritrea through its tributary of Tekez.

The Nile is the way that the Egyptians used to travel. It brings life by fertilizing the land and guarantees abundance, it played a very important role in ancient Egypt, from the economic point of view, social point of view (it was around him that were the most great cities), agricultural (thanks to the precious mud of the floods) and religious. Nurturing river of a great people, it was "divinized" under the name of Hâpy personifying the flood of the Nile in the Egyptian mythology.

The brothers of Abbadie thought they had discovered the sources of the Nile (1846) and placed them in the south of Abyssinia, at 7 ° 49 'north latitude; but subsequent research has shown that they stopped at one of the tributaries of the river, the Uma, and that the main course came from farther. After the explorations of Burton, Speke and Baker (1857-1863), it was understood that the Nile is the flow of vast lakes, the Nyanza and Louta-Nzigh, and the product of the eternal snow that covers the Kombirat Mountains, Kenya and Kilimanjaro, placed under the Equator or even south of this line; The Nile is now extended by the Kagera River to Burundi, where its farthest source is located.

Curiously the onomastic concerning this river has changed during the settlement of North Africa. It's called Nile nowadays. To recover one's historical memory is also to be able to interrogate one's past through elements of one's onomastics that have been allowed to be built and consolidated as a nation.

Has this name persisted today? That's the point of reading this script.

Standard method

Middle-egyptian:: 𓇋𓏏𓂋𓅱𓈗𓈘𓏤 ***itrw*** *"Nile , rivers"*

Endogenous method:

Middle-egyptian: 𓇋𓏏𓂋𓅱𓈗𓈘𓏤

Since the reading method is already mastered, I will proceed a little bit differently just to highlight new aspects. Simply because the characterization of a watercourse is not the same as the characterization of a plant. The conventional method is confined to the supposed reading of the written word, the Endogenous method highlights the information recorded by the scribe.

Comprehension axis 1:

Ewondo: /ù-swé / "stream; watercourse"

Indeed, to name this stream, it is normal that the person who approaches this script can understand that it is a watercourse.

Symbol 𓇋 M17

We have the symbol 𓇋 M17 of the Gardiner's list repesenting a feather. We call it /à-sè/ which offers by homophony the lemma:

Ewondo: /ù-swé / "stream; watercourse"
Ewondo: /à-sẃé/ "hideout; stash; hidden place;lurking etc."

Symbol 𓈗 N35A

The symbol 𓈗 N35A represents ripples and makes think in the case accurate to water.
The expression / á-sè i-koré / literally means to quench thirst. This expression on the ontological level is strong because it encloses two notions: /á-sà / "pen" and / kor / "savior, leader etc.". This sign reminds us of what is unstable and therefore easy to pour / á-swé / "pour"

Symbol 𓅱 G43

The symbol 𓅱 G43 is a quail chick, the body is covered with down. That brings us back to the sign M17 𓇋 of the Gardiner's list representing a feather. We call it:

Ewondo: / à-sè / "feather" ☞by homophony offers the word
Ewondo: / à-sẃé / "stream; to pour"

Symbole D21

The symbol D21 of the Gardiner's list can be seen as:

> **Ewondo**:/à-sẁé/ "hiding place" ☞ homophone of
> **Ewondo**:/ù-swé/ "stream".

This notion seems to be a simple sinecure evoked as such here but in reality it is the hiding place it implies an important notion namely the hiding place of ***Hr*** ***mD.t*** "deepness".

Symbol N21

The symbol N21 of the Gardiner's list representes a tongue of land.
The symbol X1 is a piece of bread made with wheat representing earth's fruit. It is also assimilated here to the earth. The earth is named / si / homophone from / à-swé / "hiding place".

Symbole N36

: The channel filled with water can also be /considered as:

> **Ewondo**: /ù-sẃé/ "stream"

Note

The same reasoning could have been used with the expression / * nyu / consecrated for drinks, waters.

Comprehension axis 2: ***itrw***

The transliteration ***itrw*** will induce in the variants of the African language, a whole series of homophones. So I note here the most important:

> **Middle-egyptian**:  ***itrw***
>
> **Ewondo**: /mə-tàri/ "the beginning"
> **Ewondo**: /mi-ntour / "the holy, sacred"
> **Ewondo**: /mi-ntarà / "the roaring"
> **Ewondo**: /mi-ntwar / "the uniques "
> **Ewondo**: /twé / "the bosom"
> **Ewondo**: /mvara/ "couple, double"
> **Ewondo**: /bə-tara / "the parents "
> **Ewondo**: /i-twara / "the be boiled "
> **Ewondo**: /Wouri / "Ra's hand"; "whoever shaves"; "grows in volume" etc.

Comprehension axis 3: ***numeration***

Aside the script  ***itry*** according to the Classical method specifies that there are two rivers. Without the need of y *ù-mbias* "nisbe". The other signs also inform us of the notion of double.

The sign M17 given for reed in bloom can vocalize /i-bèh/ "bamboo". This lemma /i-bèh/ is homophone of /i-bêh/ "deux"

It is the same thing for the sign D21. Considered as a hole / *ùn*-mbokh / which admits the synonym / i-béh / "hole" leading to the homophone / i-bêh / "two"

We can absolutely do this with all the other signs to demonstrate that the script encloses the notion of quantity without needing the sign M17A.

Curiously this notion of 2 recalls another notion by homophony namely ***mD.t***
not the notion of deepness but the notion of interdimensional gates represented here.

As a result, these streams can be qualified in our languages as twin: / mi-'mbyè / related to the doors / mi-'mbèh /.

<u>**Comprehension axis**</u> **4**

In their nurturing functions. The two Niles are called ***Hâpy***.
The sign G1 /*nya/ has the same root as /á-nyié/ "to breast-feed" symbol of the Nile. The two Nile (blue and white) represented here with the breasts considered as the foster breasts of *km.t*.

We are required to read the Hâpy script:

Kɔ́ɔnzime: àpâ "father"
Meroitic: ***ab*** "father, ancestor"
Meroitic: ***ap*** "father, ancestor"
Meroitic: ***api*** "heritage"
Ewondo: pəpà "father, ancestor"
Ewondo: tàrà "père, ancestor" **Middle-egyptian**: ***itrw***
Ewondo: /mə-bèh / "bosom"
Ewondo: /mi-'mbyèh / "twin"
Ewondo: /mvara/ "couple, double"
Ewondo: /bə-bèh/ "the two "
Ewondo: /mi-'mbèh / "the doors"
Ewondo: /twé / "bosom"
Ewondo: /à-bwi / "many, many, multitude, great"

Under which names can transliteration *itrw* be found nowadays?

Kongo: it's the place of the mother, the place of the panther / catfish (mvar)
Bénoué: we find the notions of B and nyu, /ibéh nya/ **or sbk** /*ùn*-nsebek/.
Nyon: we still find the root /Nyu/ "snake, mother".

Note: The metathesis: Ewondo:/ ùn-nlɔŋ / fits with the morpheme **Nile**.

It is the lemma in **Kɔ́ɔnzime**: nyɔ̌ŋ "mum" which enlightes more here.
Ewondo: /à-nyɔ̌ŋ / "to dive, to disappear" recalling the notion ***mDt*** "deepness"

Noun: ***nwn*** "Primitive waters" we said / mə-tari / "beginning"
Mungo: the mum panther or catfish
Wouri: Râ's spirit, great, mvara (couple, twin)

Ntem: ***itmw***; we will say, /mi-'ntəm/ " the branches"
Komoe: the stream recalls the name of the earth ***km.t***.
Limpopo: ***aApp*** the two doors.
Ogooué: okh-/ogh- "one" we can find the idea of / mə-tari / "beginning"
Sanaga ou Okavango: /ù-swé na Nga/ the river of the mother, the place of the mother.
Zambèze: /Dya Zamba/ the country of the primal ancestor
Jubba: ***gb***; job, yob "name given to the primordial ancestor" we notice the 2 feet.
This name jubba will recall the words: / mə-ndib /, ndiba; ndzib etc.

Etymology

The word ***itrw*** according to scholars was deformed in eior.
The word "Nile" ((ar) nīl), would come from the Greek Νεῖλος (Neilos0), which would itself be a distorted transcription of the Egyptian term Na-eiore, plural of eior designating the delta.

In Arabic we write النيل (An-Nil). In the ancient Egyptian language, the Nile is called Ḥ'pī or Iterw (Hapy), which means "river". In the word, the words piaro (Sahidic) or phiaro (Bohairic) meaning "the river" (lit. p (h) .iar-o "the.canal-great") come from the same ancient name.

The words / fara / "priest" and / *ùn*-mbogh / "priest" are two synonyms of the Beti languages.

So we say that, the word is still used today in our languages. The consonantal methathesis / ùn-nlɔŋ / "nyon" explains from our point of view, the origin of the word Nil. Who could also be vocalized / nyol / "snake" dixit

The word / *ùn*-nlɔŋ / recalls the notions of hole / à-lɔŋ / or then trumpet, call, / à-lɔŋ / no doubt related to cascades. We recall what we said above: the sign D21 is considered as a hole / ùm-bokh /. This word admits the synonym / i-béh / "hole" itself homophone of / i-bêh / "two" . Indeed, the term canal is a ditch, which admits the synonym / *ùn*-mbôghlo /, / *ùn*-nsəsəkh /, / *ù*n-ŋkəbgə /.

According to the legend Hâpy lived in a cave that was under the first cataract of the Nile, near Elephantine. From there, he poured the contents of two jars to raise the level of the waters of the Nile.

It is by metonymy that we get the words:
Sahidic: piaro
Bohaïric: phiaro

because they indicate among other notions, the notions of double.
Ewondo: /mvara/ "couple, double, to be followed"
Ewondo: /mvar/ "mud, dirty water"
Ewondo: /mvar/ "Divorce, separated"
Ewondo: /Mvar / "catfish, variety of ŋgo(l). It is found in lakes.
Mvar bè, mvar abî.

It is thus this catfish which represents one of the famous sovereign of ***km.t*** after the reunification namely ***n'r mr***. One of his names is rightly **Fàrà Ngolà** ou **Mvas**. (voir 3.5.2.2)

Ewondo: /i-fàrà/ "odd" English: ***odd*** metathesis of /***da***/ or /***dya***/ "one"
Ewondo: /fàrà/ "priest"

Note

This notion of oddness may seem contradictory here but, if we remember that in a binary system, the two (2) is the other One (1) then it is normal that the notion of double could fit the notion of unity by the way they are semantically related.

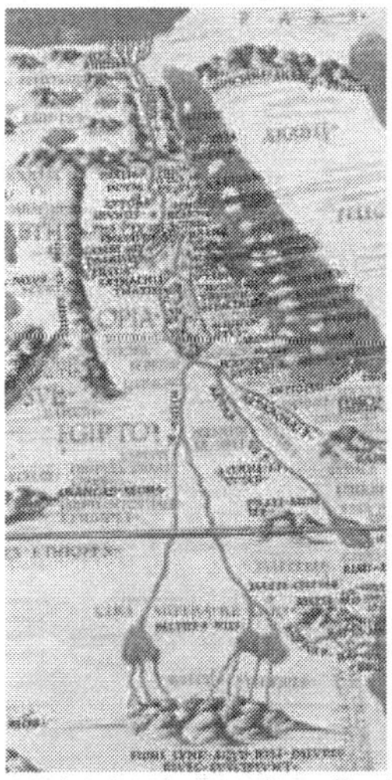

Carte de 1490 (extrait) d'après celle de Ptolémée (manuscrit Parisianus latinus)

White Nile and Blue Nil
The Nile had different names depending on the areas it crossed, the South Nile Spirit has a group of lotus plants in its head, while the Nile of the North has a group of papyrus plants, the first Dzô is called **H'p rs.t**, and the second **H'p mH.t**. When the two branches are united, they are indicated in a single character, so the Nile becomes the ntr **Hâpy** of our zodiacal sign of the Aquarius.

> *"[...] Several names used in the Bible to designate the Nile, give us the origin of other appellations which derived from it; the principal ones are: Gihon, Nehr, Nehhl, Syhur (a).*
> *Gihon in Genesis (Gyhhoun), the river of Chous, that is to say, of Ethiopia, gave rise to the names Pi-keon or Kehon; word design of the water that springs violence (erum-pens). Gihon is particularly the name of the white Nile or dental arm; the blue Nile, or eastern arm, is called Phison. In the first centuries of the Christian era, the Church always called the Nile, the Gihon of Thebaine (3). Nehr or Nahr means river: we find more often (...)"*[129]
>
> *(K.RITTER 2013)*

HAPY is the divine personification rather than a mere allegory of the river, he could also be represented with a hippopotamus head, still in its hermaphroditic form, it is also a figure of the sign of AQUARIUS whose water flows from her breasts or a cup later. HAPY personifies more specifically the natural phenomena associated with it: the flood in the dry season and the annual fertilization of soils, to which the Egyptians could not attribute a rational explanation.

According to legend, HAPY lived in two hidden places. The first was under the first cataract of the Nile, near Elephantine. From there, he poured the contents of two jars to raise the waters of Upper Egypt. The other place was situated in Lower Egypt near Memphis, and served to supply the delta with fertile water. Egyptians attributed to HÂPY
the vital force of the Nile which came from the BLACK LIMON rich in MINERAL SALTS which it transports and which
fertilized the banks of the river bringing its wealth to Egypt. It is also, by extension, The symbol of prosperity and fertility.

Professor Wallis Budge (The Gods of the Egyptians 1904) on the DOUBLE CAVE indicates that the Egyptians believed that the Nile was increasing in the first cataract called QERI (the Double Cavern), and that the Southern Nile was for them only part of the a river extending from Elephantine Island to a place some north of the present city of Assiut.

The Egyptians made a connection with water sources and paradise in the afterlife, the Egyptian Paradise is encircled by a river (itr.w) that came out of the sky mysteriously, this river reached the surface of the earth at through two holes in the first cataract between Elephantine and Philae and emerged like a river.

This is how HAPY is represented surrounded by a protective serpent, he pours the water of the Nile on the surface of the earth on two vases, symbolizing the sources hidden in the two caves under the rocks.

---

129. RITTER, Karl. Geographie Generale Comparée, Ou Etude de la Terre Dans Ses Rapports Avec La Nature Et Avec L'Histoire de L'Homme, Volume 2. Traduit de l'Allemand par E.Buret et Edouard Deson, Etablissement Encyclographique, 1837. P.249.

These stylized and symbolic representations combining the pink flamingo and the Nile surrounding HAPY looks like a Snake-Man found in Amerindian peoples and in the figure of Buddha awake surrounded by serpents.

The farthest source of the Nile Adding a reader (Eric): taken from "Discovery trips to Africa": In 1938, Dr. Burkhort Woldecker undertook the exploration of the mountains of the moon. He thus located the southernmost and farthest source of the mouth, hence the source of the Nile, at the source the river bears the name of Kosumo, which becomes RUVUVU, then Koguero, before flowing to Lake Victoria. .. Today a small "pyramid" commemorative is erected at this point: Mount Kikizi or Burundi, by 29 ° 51 'East / 3 ° 55' South at 2050 meters above sea level.

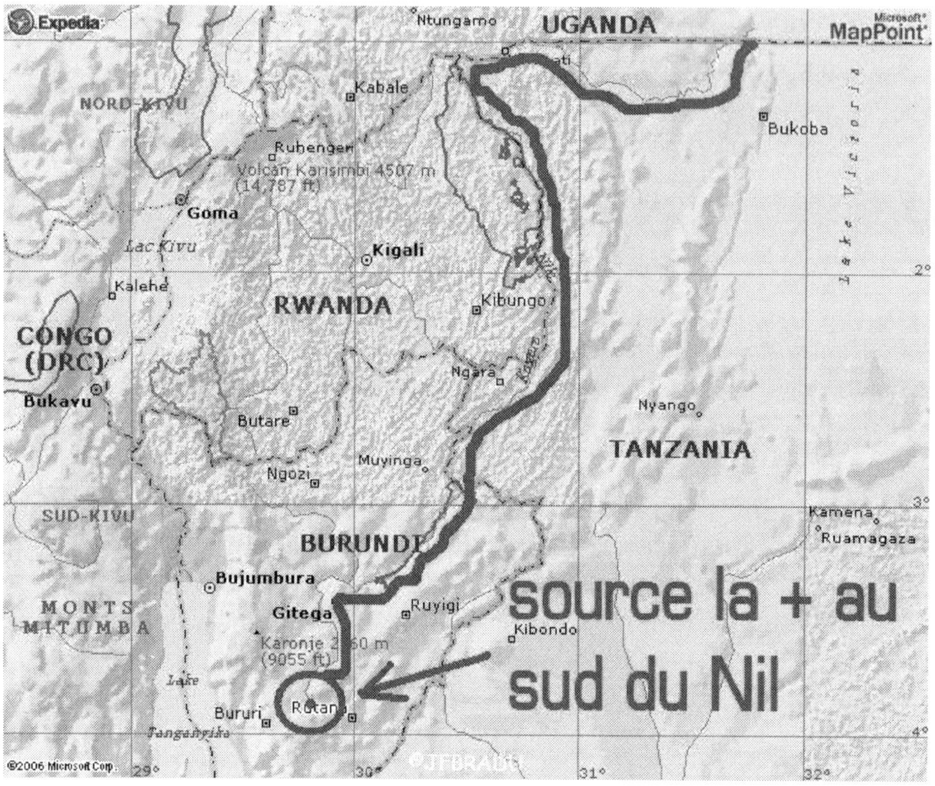

Conclusion

So we see that mDw nTr are a complex mixture that combines at least four encoding systems that can be used simultaneously:

- a symbolic system,
- a rebus system,
- a syllabic system
- and an alphabetic system.

It may be noted that the method proposed by Egyptology puts forward a peculiarity that makes it a consonant writing. The vowels appear as optional letters and the freedom of use requires the control of encoding rules.

Moreover, and as it has already been pointed out according to the languages, the coding mobilizes the synonymy, the polysemy, the homonymy, the oxymorons, the antonymy, the metonymy, the contractions / elision etc ... one thus understands why only Africans can take the real measure. This writing reflects the profound expression of the complex thinking on which african philosophy is based.

The result of the transliteration ***itrw*** can be linked to the following words:

Ewondo : /i-tiri/ "the stamped"
French: *terre; voûter ; orient, tirer; azur; jour; Jupiter etc.*
English *earth* (an ananym of *terre*, obtain by phonemes inversion. Word of the kwéké branch)
English : *water; treppe*
German : *erde; Wasser*

To understand why earth is about water, may you repport on illustration N°54.

*3.4.3.38. **EXAMPLE N°38**:* ***Swb*** *(Faulkner p. 263, Wb. IV p. 435, 10-14).*

Persea is a plant gender belonging to the lauraceae family represented by more than 150 species of evergreen trees. The best known species is avocado (P. americana), widely grown in tropical and subtropical regions for its fruit, avocado.

The term advocate is an intellectual curiosity from the African point of view because it admits a namesake that the authorized literature defines by the term lawyer, whose traditional function is to advise and defend its customers, natural persons or in law, by pleading to defend their interests and, more generally, to represent them.

Intellectual curiosity is needed because the nomadic way of life is incompatible with the establishment of a judicial system and moreover, the fact that the avocado tree is a tree of the tropics, so it seems interesting to find the link between the lawyer and the fruit based solely on the African lexicon and thoughts.

In terms of historical facts, it is useless to point out that Macedonian Greece was in contact with Africa by the occupation of Egypt under Alexander of Macedonia (-330) and Roman (-30) took over .

In his book entitled *The Egyptian Origins of Roman Civil Law*, the origin of the Civil Law of current societies is clearly established by the author[130] as being of African stock.
Chronologically, this level of information gives us therefore technical elements of indisputable research. It is therefore important to see what the Egyptians thought of the fruit, as well as the jurist and especially to make a comparison with a language spoken today.

Middle-egyptian: ***Swb*** "persea"
Ewondo: /fià/ "avocado and avocado tree" ☞**English**: fear

Middle-egyptian: ***wSbw*** "Comforter; advocate, lawyer"
Ewondo: /ù-mbɔh i-sié/ "minion; attendant"
Ewondo: /loya/ "lawyer"
English: lawyer ☞**Ewondo**: /ùn-yàlèn/ "the replyer"

We can note a resemblance between the 2 transliterated terms of the word avocado in the language of molière.

I personally think that the word beti / loya / obeys the transliteration:

Middle-egyptian: ***rw*** "lion"
Middle-egyptian: ***HsA*** "fierce (lion, fighter, etc.)"
Beti: loya "advocate"

130. Eugene REVILLOUT, The Egyptian origins of Roman civil law.Paris, (Vième) 13, rue Jacob: Librairie Paul Geuthner , 1912.

Figure 8: The psychostasis

In Egyptian mythology, ***ammyt*** "Ammut", "the devourer of the dead", is considered in official literature as the goddess who, during the weighing of the heart, devoured the souls of humans deemed unworthy to continue to live in heaven.
She is represented with a hippopotamus body, a crocodile head and **lion's front legs**. After the reign of Imn-Htp IV (or Akhenaton, 1353 / 52-1338) she appeared on almost all the painted scenes depicting the death sentence by the ***Wsjr*** court.

From our point of view, ***ammyt*** "ammut", as a hypostatic figure of lion, is the **advocate** of the soul.

Let's question some variants of African languages about the term avocado:

Ewondo: /loya/ “jurist”
Kimongo: *Savuka* “avocado, avocado tree”
Kimbuza: *Savuka* “avocado, avocado tree”
Bambara: *Pia* “avocado, avocado tree”
Kikongo: *Ndingulu* “avocado, avocado tree”
Poular: *Pilla* “avocado, avocado tree”
Sosso: *Pia* “avocado, avocado tree”
Haoussa: *Piyo* “avocado, avocado tree”
Nauhacatl: *ahuacati* “testes”

Kreyol: *zaboka* “avocat, avocatier”

Figure 10: avocado

The word *ahuacati* in Nauhacatl [131], means testicles as in some Kongolese languages like the Lari language:

Middle-egyptian:

Lari: *Makata* “testicles”
Lumbu/Gabon: *Makate* “testicles”
Lingala: makata “phallus”
Lingala: makata “arm, hand”
Ewondo: [*n*kat] “judgment, exam”.

Ewondo: /mə-bin/ “testicles”.
Ewondo: /mə-bin/ “testicles”.

Ewondo:/ à-bi/ “to stop, arrest, apprehend”.

The Egyptian terms, ***kA / kA.t*** that deal with the genitals are related to:

Middle-egyptian: ***kAi*** “think about; develop, design”
Ewondo: /á-kat/ “to say, to speak, to chat”.
Ewondo: [*ŋ*kat] “judgment, exam”.

131. Nahuatl, whose name probably derives from the word "nāhuatlahtōlli", is a macro-language (group of related languages) of the Uto-Aztec family.

The word *ahuacati* "testicle" (in Ewondo /á-hə nkat/ "to judge) thus establishes the link between *avocado* fruit and *advocate* Lawyer, jurist by the triangulation offered with the African languages that links the judging act to the testicles.
It is worth mentioning here that African authority is here called testicles by simple metonymy. We have the authority that is the panther and the depositary who is the male component, the testicle. Sometimes he is called the sperm.

The words Savuka, zaboka, *ahuacati* etc. are perfectly enlighting on the etymology of the word avocado.

The words of the African languages: *pilla, pila, fia* recall the English word: fear "scare".
Men of laws dressed **in black** embody authority and therefore arouse fear. Thus, we have the link between Avocado Fruit and advocate "Lawyer".
Let us note that the Ewondo: / firi / serves to designate what has charcoalized and therefore what is very black as the Men of laws attir.

Ewondo: /firi/ "charcoal" ☞**English**: fear "scare".
Ewondo: /á-fírí/ "Trust" we put our faith on our lawyer in the framework of justice.

Concerning the word Kikongo: *ndjigulu* "lawyer" it refers to the force. The law is the force so the word *Ndjigulu* can also be used to designate a Lawman as one who embodies force.

> *"Justice without strength is powerless; force without justice is tyrannical.*
> *Justice without strength is contradicted, because there are always villains. The force without justice is accused. We must, therefore, put together justice and strength, and for that reason that what is right is strong or that what is strong is just.*
> *Justice is subject to dispute. The strength is very recognizable and without hassle. Also, strength could not be given to justice because the force contradicted the justice and said that it was unfair, and said that it was she who was right.*
> *And so, being able to do only what is right was strong, we made that which is strong was right."*[132]
> *(Pascal-1671)*

It also appears to us that the words: *yellow* and *lawyer* in English are ananyms. They are mutually obtained by inverting phonemes.

Anglais: lawyer "advocate; jurist"
Anglais: yellow"blackleg; scab; yellowy, etc."
Ewondo: /i.vələ/ "bright; manifest, etc."

The lawyer 's flesh being yellow - green, the understanding of the link between the clear word of the Ekang languages and the word Yellow used to designate the lawyer among the English is here rather interesting. Note that the word / loya / "lawyer" helps to understand the true etymological meaning of the word / loyal /. However, let's suggest that the role of the lawyer is to clarify thought by his speech.

Is it worth mentioning here that the words:

English: pear "perry" **English**: reap "gather, obtain, garner"
French: poire "pear" **Ewondo**: /á-far/ "to reap"

These words have their origin with the African words: pia; piyo, fia etc.
Just look at the shape of the two fruits to realize the resemblance between pear and avocado.

---

132. Blaise PASCAL , Thoughts on Religion and Some Other Topics, Third Edition. À Paris, chez Guillaume Desprez, rue Saint Jacques,à Saint Prosper. M. DC. LXXI.avec privilège & approbation, 1671, P.86

Figure 9: pears

We say that the etymology of pear given as a Latin word **pira**, neutral plural, taken for a feminine singular (Latin classical pĭrum). The old French was *peire*, a *père* "father" who remained in the dialects of western France (see Norman pei [re]). In Occitan, named pera (pero) of the same origin, Italian pera, Spanish pera, which also goes back to popular Latin. The Germanic languages have borrowed the etymology of vulgar Latin: English pear (reinforced by Norman), Dutch peer, German Birne, late and altered form of southern Germany. The Celtic word is also borrowed from Latin: Breton per (enn), Welsh peren, Irish piorra. **The Latin word has an unknown origin**.

Ewondo: /fàrà/ "priest"
Ewondo: /á-fàr/ "gather; to pick" ☞**English**: reap "gather, obtain, garner"

We linked the lawyer to the testicles at the beginning of this analysis and de facto, to say that *pear* shares his etymology with the French word *père* "father" in the African languages is he a fool?
It is clear that *pears* do not grow in a desert environment. This fruit is thus linked to a sedentary sociology and not to a nomadic type.

The masculine noun pharaoh is given in the authoritative literature as a loan to the Christian Latin Pharao, -onis, title of the kings of Egypt, taken by antonomase [133] as proper name of every king of Egypt, himself borrowed from the ancient Greek Φαραώ / Pharaṓ and this one to the biblical Hebrew פַּרְעֹה / par'ōh, in turn borrowed from thc Egyptian ancient per-aâ's designating first the "greate house", that is to say the palace, then, by metonymy, the occupant of the palace, that is to say the king.

We tell ourselves that the word is still used today in African languages:

Middle-egyptian:  ***pr aA*** "pharaoh"
Ewondo: /fàrà/ "priest"
Ewondo: /à-folà fàrà/ "priest's place"

It is necessary to approach the term recorded by the above script of the Anglicism ***loft***.

A loft understood as housing in a former workshop, warehouse, or even a factory. This choice makes it possible to create fully open spaces and thus obtain uncluttered and illuminated volumes, most often by windows or windows pierced in walls or in the roof ("sheds"). In addition, the loft keeps the imprint, the visible memory of its former destination (metal beams in an old factory, vaults in an old chapel etc.)

> *"**Loft**: an upper chamber," c. 1300, an extended sense from late Old English loft "the sky; the sphere of the air," from Old Norse lopt (Scandinavian -pt- pronounced like -ft-) "air, sky," originally "upper*

133. An antonomase (from the Greek ἀντονομάζειν antonomazein "call of a different name", from anti- "instead of", and onomazein "to name", from onoma "noun") is a figure of speech or a trope, in which a its own name, or a paraphrase enunciating its essential quality, is used as a common noun, or vice versa, when a common name is used to signify a proper name. Some common antonomases end up being lexicalized and appear in the usual dictionaries ("a trash can," "a silhouette", "a Don Juan", "a harpagon", "a Bordeaux", "a Roquefort", "the macadam", "A gavroche", "a tartuffe", etc.).

*story, loft, attic," from Proto-Germanic *luftuz "air, sky" (source also of Old English lyft, Dutch lucht, Old High German luft, German Luft, Gothic luftus "air")*

Visibly before the Moorish colonization, the word was not known in Europe. We think that this approach is justified by the consonant metathesis offered by the word Ewondo:

Ewondo: /à-folà/ "place"
Ewondo: /fià/ "avocado"
Hausa: piyo **"avocado"** ☞ **Ewondo:** /yop/ "up"
English: up

Compared to the term Ewondo: / á-folà / the suffixale [t] of *loft* appears simply as a survival or a vestige of Egyptian's [t] thought by Egyptologists as a feminine tag.

We can also approach the term ***pr aA*** recorded by the above script by ***loaf***.

Loaf: a portion of bread or cake baked in a mass, usually oblong with a rounded top.

Loaf: a mass of shaped or molded foods, such as sugar or ground meat:
a small veal roll.

Loaf: the rounded head of a cabbage, lettuce, etc. in older use. head or brains

The bread, the food, the brain and the head maintaining a metonymic relationship in African thought.

We were also attracted by the phonetic proximity that links the two words below sharing a kinf of phonetic proximity with the word Ewondo: / fia / "avocado":

English: pilow "cushion; hassock"
Ewondo: /filà/ "cushion; hassock; pillow"

Many could argue us an anterioty of the words supposed to have brought with the colonization. The pillows until today are made of anatidae feathers. We find the link with the word / fia / "lawyer". The first to raise geese and ducks are the Egyptians who made foie gras.

In some Tikkar languages of western Cameroon, advocados are named *chie deuckh* "the white's safou". The term *deuckh* strangely recalls the English word duck "anatidae". We say that this is not a coincidence.

We think that the explanation "the safou of white" is minimalist that should be translated "duck feather".
Indeed we have in Egyptian:  ***s.t/z.t*** "duck" et  ***s.t/z.t*** "woman"
And we have also:

Middle-egyptian: 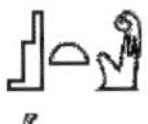 ***As.t*** "power incarnation"
Middle-egyptian:  ***maA.t*** "justice; accuracy, orderly management;Maât etc."

We thus find the link between the feather and the maternal figure. The chief is in charge of justice and by metonymy the one who judges in Africa is the **feather**. The action of judging will be qualified in **Ewondo**: / à-və ùn-ŋkat / "to judge, to pronounce a judgment". This expression curiously recalls the word lawyer.

Nauhacatl: *ahuacati* "testes"
Ewondo: / à-hə *ùn*-ŋkat / "to judge, to pronounce a judgment".

Taking into account the figure of the judge we say that it can carry the meanings:

Middle-egyptian: ***mAat*** "justice; accuracy, orderly management etc."
Middle-egyptian: ***swt*** "feather"
Ewondo: /à-sà/ "feather"

Curiously the fruit of the safou tree is called:
Ewondo: /sà/ "safou"
Kikongo: n'safu (tree called m'safu or musafu).

Dacryodes edulis or safou is a fruit tree native to Africa, sometimes called Atanga (Gabon), Ube (Nigeria), African or bush pear or plum, Nsafu, bush butter tree[134], or butterfruit (Dacryodes edulis) is an oleaginous fruit tree of the family Burseraceae gender. Safou contains 59% moisture (53.3 g / 100 g) of water, 4% or 11.2 g of protein, 5% of carbohydrates, 8.7% of fiber, 1.4% of minerals and provides 234 cal per 100 g. Its fatty acid content (22.1%) is equivalent to that of avocado. There is also copper (0.93 mg / 100 g), lipids (29.4 mg / 100 g). It does not mature as consistently as the avocado.

Etymology:

The word given for safou indicates to us the nebulous origin of the word philosophy. The Greek root: σοφία, sophia "wisdom" or "savant" maintains a phonetic proximity with the term safou indisputably

Ancien-Egyptien: ***siA*** "wisdom"
Ewondo: /à-sà/ "feather"

Those who wear feathers in Africa are those who exercise power. Safou recalls the word:

Ewondo: / fô-àsà / "panther of feathers"

Terms used today to name the west leader of Cameroon with the names: *sufo'o, tchufo, fo'o, Acha etc.*

Ghomala: sufo "chef's friend, head cat, front cat"
Guiemba: Tchufo "servant of the chief, front cat"

Middle-egyptian: ***rwty*** "Routy (the twin god lion)"

We can observe this idea as below at the entrance of the belgian palace in Bruxelles.
Lions until now are not part of the cold lands bestiary. These concepts are African as we have shown.

134. Joseph Kengue, The safoutier (Dacryodes edulis (G. Don) H.J. Lam): first data on morphology and biology,ORSTOM, 1994

Entrance of belgian palace in Bruxelles

There is a shameful falsification of human history that tends to create white snow lions by genetic manipulation to justify the presence of lions in the West. But those who are doing so are shameful for humanity.

Bambara: Filani "double"
Bambara: fila "2; deux"
Ewondo: /i-fial/ "what has been sorted"
Ewondo: /fili/ "LOVE"
Ewondo: /i-fàrà/ "odd"
Ewondo: /fara/ "priest"

☞**Ewondo**: / i-fùlàn / "resemblance"

Philosophy therefore draws its etymology from fila "2, two, love" and safou "wise cats".

In terms of onomastic, **Safou** is a name that is still found today in several African countries. It can meet with variants such as : *Safoura, Safouen, Safouan, Safougne* etc.

So we say crest of this that the pyramid of ***xwfw*** named *Khops* is for us ***sufo***'s pyramid meaning, the pyramid of wisdom.

Ancien-Egyptien: 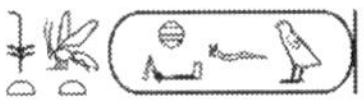 "May he protect me"
Ewondo: / fô-àsà / "feathered serpent";

Ancien-Egyptien: *xwfw* "khops"
Ewondo: / fəkh/ "wisdom"
Ewondo: / ô-bùfəghə/ "the great wisdom"

According to scholarly literature, the word *advocate* in its legal sense (because it exists also to designate a fruit) comes from the Latin *advocatus* = 'defender, confessed, called by'. Under the Ancien Rgime, the role of the lawyer was to defend the convents, the cities etc. ...
We say that the Greco-Latin etymology is intellectual swindle and ideological.

The man was born in Africa, and the Greeks as well as the Romans occuped the North Africa. A*vocado* is not a Western word because this fruit does not grow in the West so this word can only have one Negro etymology.

A nomadic sociology is not compatible with the invention of the judicial institution or the art of structuring JUSTICE. Automatically these concepts of Jurists are Negro concepts.
We can not therefore justify a Greco-Latin etymology for the term lawyer because it is a NON-SENS sociological, chronological and logical.

From the word / fia / "avocado"; we think that there is an etymological link by homophony with the French words:

Fia/Fool

Ewondo: /à-kùt/ "fool; Madness;tomfool; crazy"
Ewondo: /á-kùt/ "fib; hit; strike"

A hammer chair is a small mallet made of hardwood (walnut, oak, ash or maple for example) that is usually struck against a pile (or base). It is used by the president of a deliberative assembly to open and close debates, or to call attention or silence. It is very surprising to find here a crazy word where one speaks of wisdom and greate intelligence. But in reality all this makes sense because the very great wisdom puts you outside the "normality" and in this account your wisdom is perceived by the profane as a kind of abnormality and therefore of madness.
It is said that the limit between genius and madness is tenuous. There is a close link between genius and madness. No one doubts it. Whoever creates and invents, whatever his field, is he not different from us, a little "crazy"? A popular idea that goes back to ancient times! Already in the fifth century BCE, Euripides emphasized the psychological "affinity" of states of drunkenness, madness, and artistic inspiration, and Democritus believed a healthy poet incapable of creating a masterpiece. According to Plato, Socrates paradoxically affirmed that madness has brought a lot of good to Hellenes. The philosopher Aristotle himself wrote that there is no genius without a grain of madness. This idea will often be picked up by such remarkable individuals as the English playwright William Shakespeare, the French philosopher Blaise Pascal and the Dutch doctor Herman Boerhaave. Pascal himself thought that "the genis is close to extreme madness".

Fia/son, love

We have argued that the term advocate /lawyer is related to Men who wear feathers in Africa.

Middle-egyptian:  *sA* "son"

Ewondo: /à sè/ "feather"
Ewondo: / i sokh/ "placenta"
Ewondo: / i yol/ "name"
Ewondo: /i-sià/ "father" ☞antonymie
Ewondo: /sokh/ "duck"
Ewondo: / i lol/ "duck"

Note:
Transliterated script sA "son" can indeed name the father in African thought because there is a consubstantiality between the father (***wsir***) and the son (***Hr***). We have argued that for complex thinking and fractal logic, the seed is in the tree and the tree in the seed. It is through this mechanism that we must understand this approach that I described as antonymic.

In English we have the homophony between sun and son. Without sun no life. So sun is the father of any being.

Fia/Foie/ liver

Middle-egyptian:  "Sickle; justice; accuracy, orderly management; Maât"
Ewondo: /i səghà/ "sickle"
Ewondo: /á-səkh/ "cut"
Ewondo: /i səkh/ "liver"
French: foie "liver"

Example: /à-tigh à-dzô/ "administering justice" litt. To slice the case.

Fia/Foi/faith

Ewondo: /à-fírí/ "Faith, advocate, confidence"
Ewondo: /firi/ "what has charcoaled, raven, black"
French: foi "faith"

The black color is reminiscent of the lawyers' dress and the confidence they feel for being defended.

Fia/fois

Ewondo: /i-yɔŋ/ "time"☞**English:** time linked to /i-tam/ "lone"
Ewondo: /à-yóŋ/ "hot" ☞**English**: hot is ananym of /i-toh/ "the seat"
Ewondo: /i-yóŋ/ "fate; defunct"
French: fois "time"

We note here the homophony between *hot* which generates what is *charcoal* and therefore *black*, the homophony with the *deceased* which recalls the lexicon of the *death* and the wearing of mourning in Black in some African populations.

Fia/Fiole /flask

Ewondo: /*ùn*-nlombi/ "flask, bottle"
Ewondo: /i-lombi/ "fate; foul; black; mourning;fetid, foetid"
French: fiole "flask, bottle"

We note here the homophony between the vial idea of he deceased that recalls the lexicon of death and the wearing of black mourning in some African populations. This wearing of the Black colour again recalls the toga of advocates.

Fia /Fiel /gall

Ewondo: /zɔŋ/ "gall" **Middle-egyptian**: ***bnf*** "gall"
Ewondo: /*mə*-sóŋ/ "prison; grave"
French: fiel "gall"
French: jaune "recalling the color of the fruit avocado"
English: zone "pool; area; box, belt",

Note: The word gall is related to the food in african languages. It is therefore normal that the *toga* we have seen above shares an homophonic relationship with *gown* versus *gone* "past, gone, former, elapsed, bygone" . Let me emphasize that *gone* is also a *missing*.

Fia/file

Ewondo: /*ùn*-nsàmbà/ "file" **Middle-egyptian**: ***saH*** "rank, dignitary"
French: ensemble

Fia/Fouler

Ewondo: /Ton zôŋ/ "dislocate, tread the joint".
Ewondo : /*mə*-sɔŋ/ "prison; graves"
French: jaune "color reminiscent of the yellowish color of avocado fruit"
French: prison "prison, jail"

Fia/Fioul

Ewondo: /kalasina/ "fioul".

We note here that the root kala = black recalling the gown of lawyer.
Ewondo: /á-kala/ "protect".

Fia/foule

Middle-egyptian:  "Sickle; justice; accuracy, orderly management; Maât"
Ewondo: /i sàŋ/ "sickle"
Ewondo: /*ùn*-nsàŋ/ "crawd"
Ewondo: /*ùn*-nsàŋ'à/ " whoever cuts, who judges "

Fia/Foulon/fuller

The fuller, or ruffler, is a worker who practices the crushing of the sheets or the one who runs a fulling mill.

The work of the rammer is to knead and beat the fabrics by bathing them in water mixed with various substances, to felter, soften, waterproof and finish.
Water is added to the water, and finally human urine that provides the final cleaning and bleaching. long time practiced manually, using a kind of wooden pestle, fuller's staff, or trampling the tissues, the activity was mechanized in the early Middle Ages with the mills where the pestles were powered by hydraulic energy.

In ancient Greece, it is the job of father of Theophrastus, Melantas. In France, this trade was regulated under Henry IV who entrusted the bailiwick facilities to farmers against a fee.

Ewondo: /*ùn*-ŋkàt/ "judgment "
Ewondo: /*ùn*-ŋkùt/ "burlap; hessian"
Ewondo: /á-kùt/ "to bit"

A president's hammer is a small mallet made of hardwood is struck against the base by the president of a deliberative assembly to open and close the debates, or call for attention or silence.

Fia/ frelon/hornet

Ewondo: / à-fôm/; /à-mvam/ "hornets"
Ewondo: /fo/ "panther, bee"
Ewondo: / à-fəm/ "royal cobra"
Ewondo: / fəm / "clay, whitish"
French: frelon "hornet"

We have an relationship between horn / hornet and gwon.

The word color **Ewondo**:/ nyol / and the word snake **Ewondo:** / nyol / are here identical.
This white color is reminiscent of the epitoge representing a distinctive band of fabric worn over the toga.

Fia/feu/fire

Middle-egyptian: 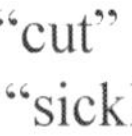 "Sickle; justice; accuracy, orderly management; Maât"
Ewondo: /i dîghà/ "Fire"
Ewondo: /á-tikh/ "cut"
Ewondo: /i tîghà/ "sickle; cutter"
English: fire "flame" is linked with Ewondo:/firi/ "charcoal"

The root **pyr** of the common indo-european *peh ur has given hnւր hur in ancient armenian,  paḫḫur in Hittite, and fȳr in Anglo-saxon (→ see fire in modern english), the Czech pýřit ("to be blushed"). This root is found in the words:

Note this little summary chart:

Cas	Singulier		Pluriel		Duel	
Nominatif	τὸ	πῦρ	τὰ	πυρά	τὼ	—
Vocatif		πῦρ		πυρά		—
Accusatif	τὸ	πῦρ	τὰ	πυρά	τὼ	—
Génitif	τοῦ	πυρός	τῶν	πυρῶν	τοῖν	—
Datif	τῷ	πυρί	τοῖς	πυροῖς	τοῖν	—

Chart 101: Fire in Greek

Fire in some African languages

Ewondo: /firi/ "what has charcoaled, charcoal"
Tshiluba: ka**pyà** "fire"
Tshiluba: fù "fire, defunct"
Tshiluba: **fwà** "to die, to no longer exist, to perish, etc."
Tshiluba: bi**pììla** "ash; ruin"
Tshiluba: **pepepe** "fire, heat"

The term Tshiluba: pepepe "fire, heat" is on the morphological plane is a superlative because it is built on the basis of a triple repetition I I I of the morpheme *pe*.

Fia/Feuler/caterwaul

Middle-egyptian:  "Sickle + board; justice; Maât etc."
Ewondo: /zə/ "panther"

Ewondo: /zigh/ "saw" **Middle-egyptian**:  ***As.t*** "power incarnation"
Ewondo: /á-səgh/ "cut"

Ewondo: /á-bam/ "growl" **French**: grolles "rooks" 
Ewondo: /à-bàm/ "*board"* homophone of English: *bird* "rook, cantilever"

Note: Lets note that rooks are black they used the word grolles in french for rooks and for shoes.

Fia / faux/false

Middle-egyptian:  "Faucille + planche; justice; Maât etc."
Ewondo: /á-bam/ "it's false"
Ewondo: /à-bàm/ "board"

Ewondo: /*ùn*-mbakh/ "fib, false"
Ewondo: /*ùn*-mbakh/ "used to fell "

The word Ewondo: / minal/ "fib, false" appears to us simply as an ananym of:*Lamina, lamella;* simple synonym of sickle .

Fia / fleur/flower

Middle-egyptian: 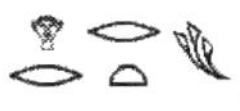 ***Hrr.t*** "flower"
Middle-egyptian:  ***sSA.t*** "Seshat"

Ewondo: /à-sam/ "flower" **English**: blo***ssom*** "flower"
Ewondo: /à-sà/ "feather"

The symbol R20  of the Gardiner's list symbolizing a stylized flower surmounted by horns. ***sSAt*** also bears the epithet *Sefekhètabouy* which means "seven and two horns". She is the deification of the concept of wisdom.

sSA.t "Seshat" is portrayed as a woman, often dressed in panther skin, worn by the funeral priests because the motifs of this skin symbolize the stars, associated with eternity and the moon.

The goddess carries a rosette with seven petals or a seven-pointed star above a frontal band. Its interpretation varies: in addition to a geometrical and astronomical representation, one can see there a palm tree or a stylized papyrus (symbolizing the writing since the Egyptians wrote on a material derived from the papyrus).

After its association with TH.wty "Thot", which was at the origin of the lunar hypostasis, this seven-pointed star will be surmounted by a lunar crescent (inverted arc) whose representation degenerated over time until reproduce two horns forming a crescent between them. When The symbol definitely took the form of horns, the goddess was sometimes called Seshfet-Abouy, (the one who) wears both horns.Fia/ Voyage

Little reminder:

Middle-egyptian: 

Lari: *Makata* "testicles"
Lumbu/Gabon: *Makate* "testicles"
Lingala: *makata* "phallus"
Lingala: *makata* "arm"
Yemba: *Liket* "testicle"

Ewondo: /mə-bin/ "testicles".
Ewondo: /mə-bin/ "testicles".
Ewondo: /ngut/ "genital".
Ewondo: /à-bin/ "testicle".
Ewondo:/ à-bi/ "to stop, to catch; arrest".
Ewondo: [*n*kat] "judgment, exam".

The Egyptian terms, ***kA / kA.t*** that deal with the genitals are related to the speech by:

Middle-egyptian: ***kAi*** "think about; develop, design"
Ewondo: /á-kat/ "say, speak, chat".
Ewondo: [*n*kat] "judgment, exam".

However, we recall that the word avocado is named:

Ewondo: /fià/ "avocado".

Ewondo: /á-kə/ "* go; to leave, to travel, to pass away, to pass on".
French: voyage "to go out; to go off; to trip;"
English: voyage

The word ***kAi*** if it is linked to speech and thought, it is also linked in African languages to presents that are offered following a trip
Ewondo: /bi-kaï/ "gifts; present".
French: cadeau
The mother represented by the vulture is called /ndoé/ so saying **cadeaux** is Ndo's gift.

The word gift is related to the word /i-fâghà/ "gift". Both word are dealing by phonemes inversion.

We therefore simply notice that the term avocado has three syllables: A-VO-CA-DO.
These three syllables are therefore well in a different order in travel and guava.

	Ewondo	**Middle-egyptian**	**Lexicon**
Ewondo: fià "avocado"	/*ùn*-ŋkat/ "judgment"	kA.t/kAi	**A**vo*ca*do
Ewondo: à-félé "guava"	/i-kaï/ "provisions"	imA "tree"	*Gu***a**ve
Ewondo: i-boŋ ẑén "travel time"	/á kakh/ "voyage"	***kAi*** "boat"	vo**y***age*
	/Kî/ "voyage"	**Aqw** "provisions"	
	/á-kə/ "voyage"		

I-bon: female sex; vagina
I-boŋ: screwings
À-folà= I-boŋ "place"

Conclusion

The advocate's script is very interesting because it highlights the close links between African languages and the current forms of the so-called Indo-European languages. We could continue the analysis using the logic of complexity but I just wanted to show how this is built from a simple term like the term avocado.

The consequence of this demonstration is directly the revision of the etymologies of the terms which appear in the encyclopaedias.
Better than that is the complete revision of all the grammatical notions that found links between words like homonymy and synonymy. Because from the point of view of the synchronies of Western languages indeed some words can share no semantic link but from the African point of view, the link can be obvious despite different vocalisations.

*3.4.3.39. **EXAMPLE N°39**:* 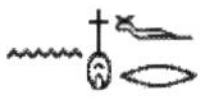 ***nfr*** *(Faulkner p. 131, Wb. II p. 253, 1-32 sq.).*

The Egyptological classification of signs has defined rules for classifying signs according to their supposed vocalization etc.
Thus, a uniliteral character in Egyptian hieroglyph is a symbol that represents a consonant or a vowel.

According to the Egyptologists, there are three types of hieroglyphs: those representing a simple value (called uniliters as the alphabetic signs), those representing two (called biliterals) and those that represent three (called triliterals). It is possible to go to quadriliterals but that is not the question here.

The sign  F35 is thus ranked by the specialists among word-signs or triliterals.

If we can thank those who did this work, we are still entitled to ask ourselves the question of what is the Africans point of view related to this question? Did the scribes thought of uniliterals, biliterals, and triliterals? Is there in any area in Africa a place where these x-literal concepts are taken over or used? What does F35 of the Gardiner's list really mean? How to read this sign in an African language without falling into the diktat of Egyptologists?

Middle-egyptian:  ***nfr*** "beautiful, good ..."

In order to deal with this point, I will only focus on the sign F35 which is on this script above.

The question is what does the sign F35 of the Gardiner's list mean. Is it a Heart and trachea? If yes, what is the difference with **ib** "heart" or ***hAty*** "heart". How does one arrives at the vocalization and the meanings proposed by Egyptology?

endogenous Analysis

If for classical Egytology the sign F35 corresponds to the heart and a piece of trachea. I say on behalf of African researchers that, the sign F35 corresponds to the african ***kora***. Indeed, we think that, before finding hidden meanings associated with a sign, we must begin by naming it. It is thus the fact of recognizing and naming it that allows associations of ideas to define its various senses

Kora

This musical instrument is called:

Ewondo: /mvett/ or /mver/ "a stringed musical instrument"

On acomparative level, we can already see that the skeleton of consonants *mvr* is strictly approachable *nfr*. We must not fall under the purview of the absolute Egyptologists *nefer*.

How is this instrument related to the good / beautiful senses proposed by Egyptologists?

The answer to this question is relatively simple if we use an African approach:
We have said previously that the character read ***nb*** is a calabash which from the African point of view recalls by its ontological dimension the container used to keep the sacred oil for the ritual operations of anointings. This operation of anointing is named in Ngeombale mbafung / Babadjou in the west of Cameroon ***bere to.***
As we have emphasized so far, the term oil is:

Ewondo:/*ùn*-mbɔn/ "oil" *__bn__ metathesis of **nb**

Nzimé:*mulo* "oil"	**Ewondo**:/*ùn*-nlɔh/ "head"
Dualà: *Muúla* "oil"	**Ewondo**:/*ùn*-nlɔh/ "head"
Soninké: *To* "oil"	**Ewondo**:/i-tɔh/ "throne"
Bamankan: *Toulou* "oil"	**Ewondo**:/itɔh *ùn*-nlɔh/ "chief"
Basa'a: *mo'o* "oil"	**Ewondo**:/mbo/ "one, chief"
Tupuri: *no* "oil"	**Ewondo**:/*ùn*-nlɔh/ "head"
Lingala: *mafuta* "oil"	**Ewondo**:/*ùn*-mfərèn/ " chief "
Obamba: *Maari* "oil"	**Ewondo**:/râ/ "chief"

isiXhosa: *ibhoto**lo*** "oil"	**Ewondo**:/à-bon *ùn*-nloh/ "fontanel"
	Ewondo:/ibho *ùn*-nLO/ "decayed head"
Russe: *масло* "oil"	**Ewondo**:/*ùn*-nlɔh/ "head"
Nzbi: *Mèdi* "oil"	**Ewondo**:/mə-ndib/ "water"
Sonrhaï: *dji* "oil"	**Ewondo**:/mə-ndib/ "water"
Bansoa: ***mevet*** "oil"	**Ewondo**:/mvett/ "a stringed musical instrument known from ancient Egypt, and, with the capital letter, a set of warrior stories that play with this instrument, constituting a part of the Ekang culture"
Yemba: ***mvet*** "oil"	

CHART 102: WORD OIL CROSS COMPARISON AFRICAN LANGUAGES AND EWONDO

I summarize my point:

Yemba: *mvet* "oil"
Bansoa: *mevet* "oil"
Ewondo: */mvett/* "stringed musical instrument"

It is curious to see that the word ***mvet (t)*** serves both to designate the oil but also a stringed musical instrument known since ancient Egypt. And, with the capital letter, a set of warrior stories that play with this instrument, forming the Ekang culture.

We can thus notice that the ***Kora*** and the ***Mvett*** or ***Mver*** are stringed instruments sharing together the squash calabashes used as a sounding board. These gourds of red squash are used during sacred African rituals to keep the anointing oil. So that's how by metonymy these elements are interrelated

kora player

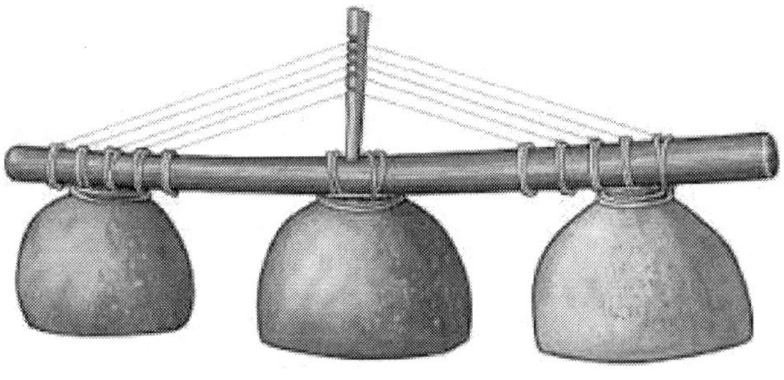

The mvett

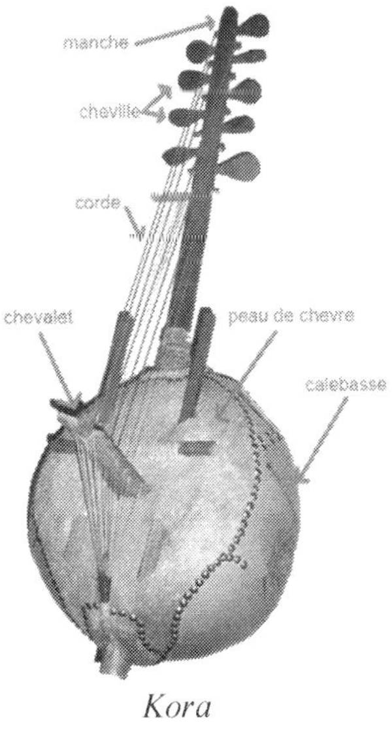

Kora

Middle-egyptian:  ʿ***nb*** "lord, master, possessor"

We observe that the calabashes used to keep the anointing oil constitute the sound boxes of the ***mvett*** guitar used in Beti country and called ***Kora*** in West Africa. According to the number of sons it will be called: *kora*, *khalam*, *rity, toukhous ngalam.* It is a Mandinka harp-lute (Senegal, Mali, Mauritania, Gambia, Guinea ...).

It is therefore normal that these calabashes are also assimilated to oil or mvett / kora by metonymy.
If in Western Cameroon oil is called ***mvett*** and the fang / beti call the guitar ***mvett*** it is because these two lineages have shared a common space that only Egypt can reconcile.

The mvett is not a simple instrument. It is likened to the figure of the ruler (nâba). It is a fundamental element of the culture of the ancient peoples of the Nile Valley.
Only the African languages can establish indisputably the link between the word *mvett / kora and* the word oil *mvett.*

We say this to understand the links between the sign F35 "kora" and some meanings highlighted by Egyptologists:

Middle-Egyptian: F35 ***nfr***
Ewondo:/*ùn*-mbɔn/ "oil"→ the anointing oil of the pumpkin calabash
Ewondo:/mvett/; /mver/ "kora, guitar etc."
Yemba: *mvet* "oil"
Bansoa: *mevet* "oil"

So we say that / mvett / and / *ùn*-mbɔn / are synonymous if we stick to the African single language.

Note: **French**: guitare or **English** : Guitar ☞**Ewondo**: /i-targui/ "beginning"

If the ancestor is the eldest, he is the beginning by the way. So the word guitar is in fact the beginning in african language. That's why this intrument is associated to the chief and to anointing ceremony. Do you think that we conceive guitars by strolling on the backs of camels? I have a doubt about that position and you?

Homophony

Ewondo:/*ùn*-mbɔn/ "oil"
Ewondo:/*ùn*-mbəŋ/ "good, beautiful" → (kweké)branch

Take note of the phonetic proximity between the word *oil* and *beautiful* in Ewondo. Of course you have in English *coat* and *cute*. By the way, this is one of the manner to read this sign in English.

Note: As we have said the pumpkin calasbash is linked to ***wsir*** by the anoiting ceremony. Therefore you can observe this below:

Middle-Egyptian: ***nb*** "lord, master, possessor"↔ **Middle-Egyptian**: ***imn*** "Amon"

French: *bonne* "good"; *bien* "good"; *nimbé* "surrounded by"; *bénit* "blessed" etc.
English: *bone*; ebon-; Nabu; beni; bent; bend; bine, nabob; band ;good, goat, god; etc.
Hebrew /Arab: ben "son"; beni; nabob; nabab

It is therefore the homophonic relationship between oil and good / beautiful th llows to relate the goodness to the beauty. This is true because we are dealing with the signified ***wsir*** who is the anointed and also the best (good; well; anointed; injured, sorrow; blessed etc.)

Why is anointing a holy ceremony?

To answer this question we can spend a moment together. So I'll just focus on the most important aspect of the holy anointing. I suggest then to flip back to our script.

Middle Egyptian : , "lord, master, possessor"

The purifying aspect of liquids will not be discussed here, I just want to emphasize the role of the calabash.

The Calabash is made in a cucurbitaceae called in **Ewondo**: / ngon / "pumpkin".

Since you have understood that we must name what we see before mobilizing the derived meanings by homophonic and synonymic relationship, the term / ngon / "pumpkin"can be linked mainly by homophony to: baboon, the moon, monthlies, girl, panther etc .
All these entities stated to operative concepts. I will try to show the influence of a few. The complexity of the subject makes it possible to write a complete book through this word **Ewondo**: / ngon /.
To hava a better understanding of the purpose, may you report on illustration N°47.

The anointing ceremony involved, a liquid (oil, honey, water) a calabash and the anoited's head. In the previous points, I have enlighted why the head is envolved but precisely the bone so I'll not repeat it here. I'll will talk about the purification also. I just wanna focus on the calabash.

The first thing I pointed is the name this vegetable. I hope that you will be able to split the word Calabash or pumpkin and get the deep understanding of those words. May you allow me to help you with the word calabash. I wanna remind you that suprasegmental units carry meanings in Africa. Calabash can be split in Kala and Bash.
As we have seen is, Kala is the way to name Hr, the eldest ancestor. He is call Kali in India. Kulu in the Kongo basin.
And who is Bash? I'll show you Bash. Bash is the anointer may you report on illustration N°12. What is called Bash is also called by Egyptologists in **Middle-Egyptian**: *sAb* « jackal »
The jackal is also **Middle-Egyptian**: ***inpw*** « anubis ».

Anubis according to Egyptian mythology, is responsible for mummification. Therefore it is he who is in charge of the ceremony of anointing the mummies.
The link between the mummified and the mummifier is metonymic. This is why the elements of mummification are all brought closer to the ***sAb*** transliteration by Egyptologists.

Middle-Egyptian: *sAb* « jackal »	☞The anointer
Middle-Egyptian: *sAb* «liquid»	☞Illustration N°45
Middle-Egyptian: *sAb* «dignitary; groin»	☞illustration N°47 (Bes)
Middle-Egyptian: *sAb* «spots»	☞panther ; starry sky
Middle-Egyptian: *sAb* «crossing»	☞The bridge etc.

I let you discover the obvious links with the word *pumpkin*.

What you can observe on illustration N°47 is that the milky way looks like a calabash or a basket. That why this shape is a holy shape according to African thought.

This shape is consistent with the macro-structure of the universe in a fractal approach of the proportions from the huge to the smallest. May you allow me to remind you that the name of the pumpkin is a holy name, the shape of the calabash is a holy shape.

Before his death, here are the words that Jesus uttered. Just remind what I told you,
Monthlies are periodic blood flow of women. You have a script for flowing liquid transliterated ***sAb*** .

> *"[...]Because this is my blood, the blood of the covenant, which is shed for many for the remission of sins".*
>
> ***(Louis Segond Bible)***

After his death here is what the supposed sacred texts say :

> *"[...] The men retreated to a small depression south-west of Calvary to prepare the articles necessary for embalming. Cassius, with some soldiers who had converted to the Lord, stood at a respectful distance. All the ill-intentioned people had returned to the city, and the soldiers present only formed a Bard of Safety to prevent the last honors rendered to Jesus from being disturbed."*
>
> ***(John 19,39-40).***

So these are words that are lent to Jesus in a ceremony known at least 6000 years before his supposed existence. We clearly have representations of mummification and the first mummified who is ***wsir***.

How is it possible? You must have an idea. But mine looks like a ridiculous stealing.

We have just pointed the plagiarism associated with the fictitious character of Jesus as well as the words he is supposed to have uttered. Mummification is a negroe sociological marker. It was practiced in antiquity and is still practiced in Africa today. Nomadic societies do not mummify their corpses, they incinerate them and transport the ashes of their deads. Either Jesus was black to be mummified or so he never existed. Make your choice.

The baboon and the moon

As I pointed out, the baboon and the moon are also called / ngon / as the pumpkin.
This is what a wrote just above:

Middle-Egyptian: ***nb*** "lord, master, possessor"↔ **Middle-Egyptian**: ***imn*** "Amon"

In English remove the vowels of the words *moon* and in *ba-boon* and compare with the result of Egyptologists.
Are you sure that Champollion and Howad Carter were reading those scripts until that they were unable to recognize the words of their own languages? I do not hide my surprise.

Middle-Egyptian: ***ian*** "baboon" ↔ **Middle-Egyptian**: ***iaH*** "moon"

The link between the baboon and the character of ***inpw*** « anubis ». is the canine face of the Baboon in addition to its carnivorous teeth.

I will not do any development on the character of the baboon in African mythology. I just want to point to the close connection between the moon and the baboon.

Baboon represented by ancient Egyptians

Curiously, the ancient Egyptians were representing the baboon with a crescent moon on the head.

Plagiarism or simple similarities, let everyone make his own opinion

I want to remind just that, in terms of historical chronology, Egypt is according to the calendar 6254 years, Islam is 1400 years (7th century) and the UN is 73 years (1945).

I give you the script again so that you have more relevant elements of comparison.

Middle-Egyptian: 𓎟 ***nb*** "lord, master, possessor"

May you allow me to remind you that the word anoint in Ewondo language is:

Ewondo : /*à*-mam/ "to anoint" and the action /*i*-mam/ "anointing"

So you can correlate the word **Ewondo**: /i-mam/ "anointing" with the word Imam as a prayer leader. Those who are doing so, cannot explain why they use this word. But we assert that /i-mam/ shares a phonetic proximity with /i-mvam/𓈖𓎟𓅆, "lord, master, possessor; ancestor"

If for you as for me, the skeleton of consonants ***nb*** can not be summed up in these few words in your language, let us see what gives the other meanings of the Egyptologists.

The Alliance

Middle-Egyptian: 𓎟 ***nb*** "to bind, to tie"

This recalls the alliance not between Jews and Yahve but between **Hr** and the anointed chief.

Ewondo: /à-ban/ "to bind, to tie" ☞**English**: ban "veto; blocking "

Ewondo: / bandas/ "strip" ☞**English**: band "strip"

It want to point that from southern Spain to South Bordeaux in France, the summer festivals are composed of musical groups called **bandas**. I just remind you that we were dealing of an African musical instrument, the *mvett* "guitar".

All

Middle-Egyptian:  ***nb*** "whatever it is, some, every, everything"
Ewondo : ban "some"

Example: /ban mam/ "few things"

We have also :
Regarding the skeleton consonants ***bn*** we have in Ewondo all the following words:

Ewondo : /*ùn*-mbàŋ/ "unique"
Ewondo : /*ùn*-mbàŋ/ "nut"
Ewondo : /ban/ "smalls"
Ewondo : /*i*-baŋ/ "stone"
Ewondo : /*à*-bîn/ "testicle"
Ewondo : /*á*-bén/ "to deny "
Ewondo : /*i*-bén/ "callipyge"
Ewondo : /*à*-bîŋ/ "sorrow"
Ewondo : /à-ban/ "jealousy"
Ewondo : /*i*-bóŋ/ "the search for a sexual partner"
Ewondo :/*i*-bɔ̌ŋ/ "shell mold"
Ewondo : /*i*-bɔŋ/ "fruit of àdzàp tree"
Ewondo : /*i*-bón/ "female sex"
Ewondo : /*i*-bɔ̀ŋ/ "lover, concubine "
Etc.

After this aside let's flip back to the main purpose of our subject.

Consonant metathesis

Ewondo:/*ùn*-mbɔn/ "oil"
Ewondo:/*nyəb/ "beautifull, good"→ (kekwé branch) phonemes inversion.

The Negro-Egyptian having given two branches, we have here the counterpart of the word / *ùn*-mbəŋ / "beautiful, good". We could therefore from these different words vocalize synonyms, metonyms, including antonyms.

How do we reach the consonants skeleton *nfr*?

Middle-egyptian:  ☞ the hare is associated with the kora, the heart
Middle-egyptian:  ☞ the hare is associated to strips or placenta.

What characterizes most the leporidae [135] in general is to possess long hind legs, very muscular which allow them to propel themselves by very fast leaps when necessary.
Let's question the lexicon of Ekang languages on the question of the jump or regarding hares

Ekang: *NE-FAR*[136]*: (vb. Far, jump). "jumping idea"*

***Example1: /Zire a far* ne-far/ "*the classy jumps* ne-far".**
Example2: /Falghe ŋgwène* ne-far/ "*remove the iule with a stick".

135.Leporidae (Leporidae) are a family of lagomorphic mammals including hares and rabbits.
136.Samuel GALLEY, Fang-French dictionary. and French Fang followed by a fang grammar, ditions Henri Messeiller, Neuchâtel, 1964, p.394

So the good reading of the triconsonant sign-word ***nfr*** is ne-far or nefe in the Ewondo language. *Nefar* is basically The onomatopoeia translating the noise emitted by the friction of the wings of grasshoppers during their movement through ruts. By cross association of ideas, one can note a morphological similarity between the feet of locusts and those of leporids. It is therefore this similarity in my opinion that leads us to retain this idea here.

Figure 10: big thighs like hares

The best interpretation to give to the consonants skeleton ***nfr*** is in the meaning / fara nene / "the highest priest". Indeed, ***Wsjr*** is the highest priest.

Wsjr is the perfect ***nfr*** he is also the mummy ***nfr*** "dry". It is by this way that one must understand the semantax of the scribe which associates the script zero ***nfr*** with the sign-word ***nfr*** of the beauty. The Ewondo's expression / mi-ŋkɔr mi-moh / "dried-hands" lights up here with a new day to translate the idea of zero.

The vocalisations generated by the consonants skeleton ***nfr*** will allow us to highlight some important notions:

> **Ewondo**:/nə-fyôr/ "decayed, rotten" the idea of decay must be linked to the operational concept ***inpw***

The word above will give by homophony:

> **Ewondo**:/nə-fyôr/ "bald" qualifier given to a bald head, shaved.

It is also this idea of deprivation that supports the semantax that binds ***nfr*** and ***zero*** in our languages.

We can also note the meaning:

> **Ewondo**:/nə-firi/ "black-charcoal"

We can also note the meaning:

> **Ewondo**:/nə-fyor/; /fyar/; /pyor/ "the noise made with the lips to say NO"

The list being long let's note some words simply:

> **Ewondo**:/fàrà/ "priest"
> **Ewondo**:/à-firi/ "hope"
> **Ewondo**:/*ùn*-mfiri/ " the one who crushes"
> **Ewondo**./nə-fyor/ "the noise made with the lips to say NO"
> **Ewondo**: /*ùn*-mbyar/ "twin"
> **Ewondo**: /*ùn*-myar/ = /ibèh/" porcupine spines"
> **Ewondo**: /*ùn*-mvar/ "odd of a couple"
> **Ewondo**. /*ùn*-mvarà/ "couple"
> Etc.

We could extend the list and demonstrate the richness of the African approach, but it could move us further. So we say that only the context will enlighten the subject.

Reading ***nfr*** can be approached by the idea of confinement which is associated with the character of ***Wsjr***.

Ewondo:/á-fér/ "to close; to confine"
Ewondo:/*ùn*-mfərèn/ "whoever has been locked up"

French:enfer "hell" ☞**Ewondo**: Elo "God"
English: inferno "hell; underworld"
Ewondo: Elo "God"
English: Hell
Latin: infernum
Greek: ilios "sun"

From the anatomical point of view, the heart is a muscle symbolized by a closed curve with a cusp directed inwards and symmetrical about an axis passing through this point.

Etymology

Middle-egyptian: ***nfr***

I want to show that this script can be read in English as in French. I'll use a cross comparison for the purpose.

To be honest with you, I have a lot of things interesting to show you on this script so I don't know where I have to begin.

Sometimes the tricky question for Egyptologists is to know what is exactly this animal . Is it a hare of a rabbit ?

The answer is a bunny, a rabbit.

This is what the holy Bible says about the way by which the disciples were calling Jesus. This is in the Gospel of Matthew in the verses mentioned below.

> *"Do you understand what I did to you? have you grasped the deep meaning?*
> *It is by this question that Jesus introduces the instruction he wants to give to his disciples. These usually called him* ***Rabbi, Master, the teacher****, and Jesus claimed this title for himself alone, in his highest sense."*
>
> *(Matthew 23.8).*

I want to recall that:

Middle-egyptian: ***nb*** " lord, master, possessor"

Sorry, I mean in English nb = **bunny** but, bunny is the **rabbit** also.
Even though Jesus and his disciples were supposed to speak Aramaic, you can speak English and use the same words as them. The remaining is ideology and intellectual weakness or swindle.

Wsir *as Bunny with his long ears*

rabbi *with his long quilts or papillotes*

And what about Ewondo language?
We have already seen all this before when dealing with the elephant remember or please verify again.
I just wanna remind you that :

Middle-egyptian: ***nb*** " lord, master, possessor"

Sorry I mean **M.-E.**: ***nb*** = **bunny** but, bunny is the **rabbit** also.
This is what we have seen before concerning the elephant in a cross comparison with the Somali language:
Question: does the word *maroodi* "elephant" of Somali have an equivalent in Ewondo?

As we have said before, it's the same language so the answer is automatically yes. But I'll not develop the link with the elephant once more because it's obvious. The elephant is the chief paramount.

Fang: /à-mara/; /mare/ "To fear oneself, to flee each other, to fear, to flee, etc.".
Ewondo: di-mbara >i-mbara "scary"
Old Ewondo : di-mbara"scary"
Ewondo: i-mbara"scary"

Note: The presence of an epenthetic **b** is found between fang, Somali and Ewondo. As a general rule, the appearance within a word of a non-etymological b is said to be epenthetic. We often find that it occurs every time we have: **m** and **r** (or **m** and **l**).

The epenthesis of **b** also occurs in Egyptian, Coptic and African languages under the same conditions: mr -> m**b**r. We must link the origin of this phenomenon b said epenthetic in the evolution of variants of the unique african language the **Negro-egyptian**: k^{w}>mb> m. The consequence is that the same word could be found in comparative analysis with the following variants : *mb-; *m- or *b-

Rabbi(t) appears as the ananym of (m)**bara**"scary". So Jews are using an African concept. The Jew word *rabbi* is **semantically** equivalent to the word /fara/ of Ewondo meaning, priest.

From the point of view of Onomastics, the English word *rabbit* has persisted in French names like Rabiot or Rabbiot, Robert. But also with items like, robe, rabot-, ramper, rable, bière, bar, beurre, boire; phare, far ; fer etc.

The answer is a Hare.

I want to recall that:
Middle-egyptian: ***nb*** " lord, master, possessor"

This animal can also be a hare. Why ? Just because the bird on the script is transliterated ***Hr*** :

Middle-Egyptian : ***Hr*** « horus » **Middle-Egyptian** : ***Hr*** « horus »
Middle-Egyptian : wnn-nfr

So here, this sign ***Wsir*** is the HARE with his long ears as we have seen it in the preceeding paragraph. I recall is for you, don't worry.

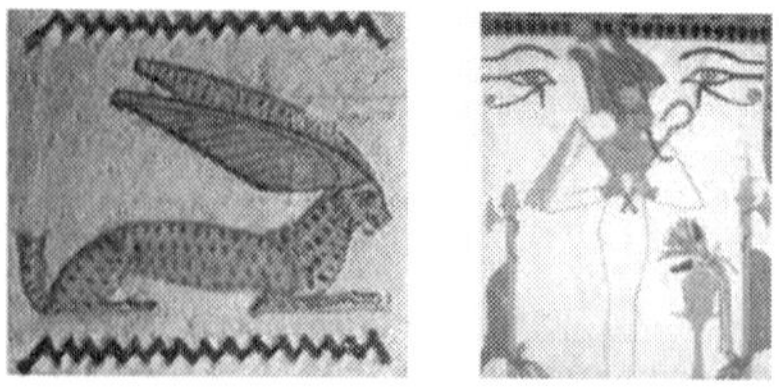

***Wsir** as a HARE with his long ears*

African scholars were not daft to call the chief a rabbit. The answer to this issue is simple. The chief is the oldest and most of the time, he is toothless. That's why the word transliterated by the Egyptologists means rodent or gnawing.

Middle-egyptian :  ***wn*** « desert's hare »
Ewondo: /*i*-woŋle/ ; /*ùn*-woré/"rodent; gnawing"

The word Hare, shares a phonetic proximity with the real items useful to name the chief in English as : *head, hate; hit; heat* etc. So the good reading of the script is related to the context knowing that it is an operative concept. For the chief the transliteration of the script will be :

Middle-egyptian :  ***wn*** « desert's hare »
Ewondo: /woŋ/ or /wôn/ "scary" same meaning of the word /*i*-mbara/ "chief, elephant"
Chinese : 王 wáŋ "chief"
Tupuri: waŋ "chief"

Chief as the hidden

Middle-egyptian :  ***wn*** « desert's hare »
Ewondo: /wône/ "unseen; to be badly hidden "

So instead of reading hare it is better to read **head** or **one** according to the transliteration ***wn*** of the script in order to title the chief.

You can not that the hare or the rabbit is speedy. There is a relationship with the wind. The **one** is a winner. So you have a word game between wind and win.

I have shown you a detail in the previous chapter that has possibly escaped your intellectual sagacity. I mean the phonetic proximitity between *hair* and *hare*. Let me recall it for you then below:

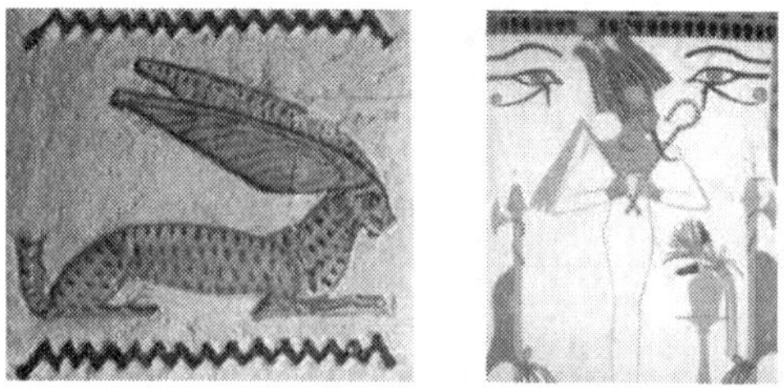

***Wsir** as a HARE and with his HAT with two long EARS*
6254 years

***A rabbi** with his long HAIR in papillote and his HAT*
Judaism was founded around -1350

Note: The black color we can note on the rabbi's garment symbolize the scary, the respect. In Ewondo language, I have alleged that the transliteration ***nfr*** gives the word *rabbi* and can be linked in Ewondo with the word /fara/ "priest". But it can be correlated also to the elephant as the scary /i-mbara/. unless I'm mistaken, the rabbi is a priest as in Ewondo.

The locks are hair curls catered by some blacks brought out of Africa during the slave raids of the fifteenth century. This hairstyle appears to us with a little perspective as a remnant of the heritage of Humanity preserved through the magic of the historical continuum.
The Rasta movement symbolizes respect for the long hair (the locks) is an atavism of ancient Egypt through the cult of the chief paramount.

Middle-Egyptian : ***Hr*** « horus » **Middle-Egyptian** : ***Hr*** « horus »

The word *hare* also exists in Ewondo as :

Ewondo :/ *ùm*-ore/ "human being or the one who belongs to Hr"
Ewondo : *ùm*-ùra "woman or the one who belongs to Hr"
Ewondo : *i*-woré "the nibble " ↔**English**: nibble/noble

The answer to the question is : a leporidae.
As we have just seen, the question of rabbit or hare is a bad question in the complexity logic. That question enlightens the expression of a binary thought. Both, hare and rabbit are leporidae namely those that resemble *lepus* (hare). Together with the *pikas*, the Leporidae constitute the mammalian order Lagomorpha. Leporidae differ from pikas in that they have short, furry tails and elongated ears and hind legs.

I just wanna show you something if you have a second. Among the leporidea we have also the *pikas*. What is bunny sorry I wanna say funny is that pika(s) is ananym of kippa. Don't forget we are dealing with 2/3 main branches of the negro-egyptian language. So when I have a word of one branch i can find the counterparts in the other branches.
So, *pika* is a word of the kweke branch and its counterpart is *kippah*. Kippa is the same word but from the kekwé branch. The French word is chapeau.
A kippah is a brimless **cap**, usually made of cloth, worn by Jews to fulfill the customary requirement held by Orthodox halachic authorities that the head be covered. It is usually worn by men in Orthodox communities at all times. Most synagogues and Jewish funeral services keep a ready supply of kippot.

Wsir *as a HARE/RABBIT*

a ***PIKA***

Leporidae

A rabbi with his ***kippah***

Concerning the Jews brimless **cap**, we do call it in Ewondo language : /i bobwar/. You can note once more that the Ewondo word is an ananym of the word rabbi. This word /i-bobwar/ allows the synonym /i-kobo/ "skin" souding as kippah.

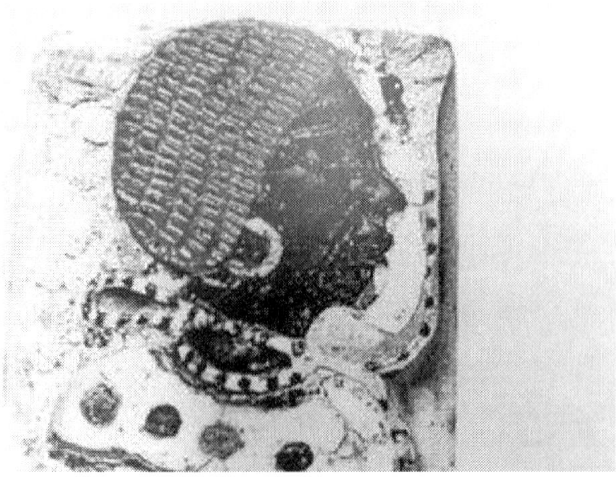

Ancient Egyptian with his red cap on the head on the wall of a temple built by Ra-Mssw III 20th dynasty

If, *pika* is a word of the kweke branch and kippah its kekwé counterpart.
I assume that the word *pika* shares a strong semantic link with to the French words:

French: *pic* "pike, peak"; *pîquer* "pick; prick pique"; *bouc* "goat" *etc.*

In fact the word *pika* or *pick* actually reminds the horns of the ram and therefore enlightens an element of the Amonian cult of ancient egyptian.

The question is by the way to edify the link between the 𓃹 *rabbit* and the 𓃝 *ram*.
This link is simple. Let me show it to you then :
As we have said, the *Hare* is also a leporidae. And the ram is simply a hairy. The game of word between *hare* and *hair* is the key element to understand why the scribe has used a Leporidae to express the idea of the 𓃝 ram ***imn***.

So you can see here the priest (fara)
And his ram or kippah
English : to keep « to ; protect »
Kippah = keeper

The Ewondo word Kaba is enlightening here the idea of protection, light etc.

Note : **leprosy and Leporidae**

I was very surprised by the phonetic and therefore the semantic link between these two morphemes.
Leprosy (or Hansen's disease) is a chronic infectious disease caused by Mycobacterium leprae (a bacterium close to the causative agent of tuberculosis identified by the Norwegian Gerhard Armauer Hansen in 1873) affecting the peripheral nerves, skin and mucous membranes. and causing severe disabilities. It is endemic in some tropical countries (especially in Asia). Leprosy is a poorly contagious disease.

Leprosy was long incurable and very mutilating, leading in 1909, at the request of the Society of Exotic Pathology, "systematic exclusion of lepers" and their grouping in leprosaria as an essential measure of prophylaxis.

Leporidae is the family of rabbits and hares, containing over 60 species of extant mammals in all. The Latin word Leporidae means "those that resemble lepus" (hare). Together with the pikas, the Leporidae constitute the mammalian order Lagomorpha. Leporidae differ from pikas in that they have short, furry tails and elongated ears and hind legs.

In appearance, these two notions have no connection.

Let's take note that :

Ewondo: /á-bùni/ "to be proud" ☞**English** : bunny "rabbit, buck"
Ewondo: /á-zam/"to be proud"

The eldest ancestor of African people is called sometimes: Zama, Zamba, Nzambé, Nzambi etc. That means by a synonymic relationship that, by drawing a rabbit, the scribe means :

Ewondo: /Zamba/ "God; goat"

Curiously, the leprosy is called in Ewondo :

Ewondo: /Zam/ "leprosy"

So it makes sens that the two notions sharing a phonetic proximity being then linked by their signified.

So, why is the African *god* considered as a leper in the collective African imagination? Simply because according to the African mythology, his body has been mutilated by his brother ***stX***. The severe mutilations linked to the leprosy recall the mutilation of ***wsir***'s body inflicted by ***stX***.

Note : **Lepus versus Lupus**

If we rely on taxonomy, some leporidae like hare are ***lepus***. Curiously, the common gray wolf (Canis lupus lupus) is one of the subspecies of Canis ***lupus***, a species of canids that includes most wolves, dogs and dingoes. He is also called Eurasian Wolf, European Wolf, Carpathian Wolf, Steppe Wolf and Chinese Wolf.
The terms ***lupus*** for wolf and ***lepus*** for hare are sharing a strong phonetic proximity, so it is interesting to find the link existing between these two animals. Why for the scribe, hare and wolf are assimilable?

We have validated, according to us that the hare represents God or Zamba in the African thought. :

Ewondo: /á-bùni/ "to be proud"
Ewondo: /á-zam/"to be proud"

The eldest ancestor of African people is called sometimes: Zama, Zamba, Nzambé, Nzambi etc. That intended by a synonymic relationship, by drawing a rabbit, the scribe means :

Ewondo: /Zamba/ "God; goat"

So why dogs can be assimilated to hares?

You can note that dog is ananym of God. So if hare represent God by the way Dog is another way to enlighten the idea of God.

Middle-Egyptian : ***inpw*** "Anubis"

Anubis is the Greek name of a *god* associated with mummification and the afterlife in ancient Egyptian religion, usually depicted as a **canine** or a man with a **canine head**. Archeologists have identified Anubis's sacred animal as an Egyptian canid, the **African golden wolf**.

May you take note that ears are called in sengambian Wolof :

Wolof *(Sénégambia)*: nopp; *nopou- "ears" recalling ☞**Middle-Egyptian** : ***inpw*** "Anubis"

In the African collective unconscious there is a link between a dog and his ears. We often talk of dog's ears.
However, the characteristic trait of leporids is to have long **ears** and very muscular hind legs. It appears then normal that, speaking of **ears** we can link canines and leporidae.

Fortunately, ears are located on the head. So according to the scribe, there is a metonymic relationship between the head and the ears. By the way, to name a leader, the scribe can name ears because ears are also head-parts. The ears emphasize the chief's ability to listen in Africa. The leader is not a despot he is a service Man. On this account, he is obliged to listen to all his subjects before taking a decision that commits the community. In his function as head of justice, he must listen in order to make fair judgments.

Middle-Egyptian : ***sAb*** "jackal"

This transliteration of the canine ***inpw*** leads us to recognize the name of the ***ntr*** Bès or the eldest ancestor of Africans : /Zamba/

The last point concerning this comparison can be done with the word hare. Dogs and hares are hairies. So once more we have the word game hare and hair.

This subject is a bit tricky because it envolves the cosmographic dimension of those words I mean **Canine** Constellations and **Lepus** constellation. We are note going to deal with those aspects of the scipt in this heuristic. I just want to mention it to avoid misinterpretation. The purpose with those constellations is linked with spatio-temporal doors connecting Sirius (***Wsir***) with our planet ***itrw***.

Middle-Egyptian : ***inpw*** "Anubis"
Ewondo : /i-nyî *ùn*-mbèh mvù/ "Anubis"

Orion

Among the ancient Egyptians Orion is also called ***saH*** or ***saH.w*** by the ancient Egyptians and was identified either ***Itm***, ***wsir***. The name ***saH.w*** is mentioned several times in the ***nnw prt. m Hr.w*** (Book of the Dead) and more frequently in the Texts of the Pyramids, the oldest religious writings known to date. From the word ***saH*** for the constellation Orion derive Sahara or SaH-Hr-rA, "Orion - Horus - Sun", Sah-Ka-Ra, the famous pyramid of ***Mn-nfr*** (Memphis), "Orion - Heart - Sun".

"It is the blood that came out of the phallus of RA when it was trained to mutilate itself, it was formed of Gods who escort to RA, it is Hw and SAw who escort their Father **Tm.w** every day ". (Book of the Dead - 1.23-24).

Summary

I purposely asked the question of the meaning of the sign  E34 of the Gardiner's list in order to elicit from the reader a priming of critical thinking, but above all to demonstrate how complex thinking works.

Of course as you have experienced everything here, hare or rabbit, the question is not at this level in complex thinking. Only in a binary thinking that the hare is not the rabbit. Curiously in terms of taxonomy, they are all leporidae.

The sign  E34 is associated with a 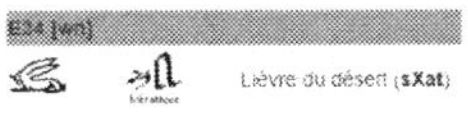

desert hare in the authoritative and scholarly literature. We claim that I could be more suitable to change this information because it's not true, it appears us more suitable to replace it by the word **Leporidae.** In my opinion this item is the most accurate as we saw it previously.

Misinterpretation of comparative linguistics

The races hypothesis as postulate has led to the categorization of languages. And of course has conduct to many spurious analyzes. The comparative linguistics techniques consisting in posing equa-tions to explain, with phonetic laws, the fate of this or that morpheme in this or that language in my opinion is problematic.

The problem arises since we compare languages that are a priori different. But the postulat is wrong because all the humanity is sharing the same lexicon.

Following the comparatists method, therefore we compare words with same meaning only.

This is a gross error because the signified appears to be the same but on the morpho-syntactic level, it is not at all the same word.

Between two "languages" a word can express semantic nuances while sharing the same semantic field. More than that, Africa does not necessarily deal with morphemes but descriptors of a reality. It is therefore inappropriate to reason in terms of words exclusively.

Rigourosly we cannot compare apples and pears. According to me then, a comparative analysis can only be made between morphemes sharing a phonetic proximity.

I want to enlighten my purpose with the example below :

Ewondo		French	English
mvu	*i-lob*	*loup*	*loop*
kub	*Fol**[137],	*poule*	*pool*
mvu	i-wolo	*volupté*	*wolf*
À-ngounda	*à-folà*	*foule*	*fowl*
i-yələ	*i-fyal,*	*Floué, filé*	*fly*

Semantic versus morphology

Comments

Under the main column Ewondo I did two other columns. The left column is the semantic meaning of the word written under the French column. The right column only shares a phonetic proximity with the English and French words.

When historical linguistics makes a comparative analysis between the French word *loup* and the English *wolf.* According to me, it's a mistake. That's why I wrote the word loop in front of the word loup . I think it's the right word. Even tough the *loop* is a ring and not a *wolf.*

137 *Ewondo language doesn't use the morpheme [p-] at the beginning of a word. Ewondo is using a [f-]. But among sibling languages of Ewondo like Eton, Manguissa etc the [p-] can be found at the beginning of a word.

But, in order to capture their preys, wolves often create a **loop** to avoid any escape for their victims, hiding behind the bushes and then attacking to kill them.

Compare to English people, the animal called *loup* "wolf" by French people has to be put in balance with the English word *loop or* the Ewondo word */i-lob/"*biter". The word *wolf* has to be linked with the word **Ewondo** : /i-vulbu/ "wrapper". Inpw being in the African collective unconscious the mummification responsible and by the way to the mummies wrapping (report to illustration N°25).So comparing loup and wolf is a tremendous error because both words are only semantically linked.It's not the same word neither an evolved word of another word in a diachronic point of view. Both are only sharing a metonymical relationship that's all. According to me, if we wanna compare two words, we have to be sure that the two words are sharing the same semantic field and of course have the same morphology.

Note: According to the context, what English people only call *wolf* can be named different ways from the African point of view : a dog, a wild dog, /i-vulbu/ "a wrapper", /*i*-lobo/ "a biter"; /i-bwàlà/ "the decay; rottenness" etc.

The lack of rigor leads us to state equations between morphemes that collective memory has retained signifiers associated by default to the same signified but without the fact that those signifiers really share morphological similarities.

Concerning what English people called *wolf*, according to you then, what is the real name of that animal, the *wolf* or the *loop*?
Comparatists laws can be accurate sometimes but most of the time they are intellectual swindle.

Among the lexicon we are using worldwide, they are two loops of words, the homophonic loop and the synonymic loop. According to me, comparatists laws are only truthful inside the homophonic loop.

African point of view.

What we say regarding our history is that all is linked to one. At the beginning was the spiral.
This spiral is called in **Ewondo** : /i-vul/"spiral" but this spiral is a hole that we also name /*mbokh/ "hole, one etc"
All the words created worldwide are linked to this concept.

Ewondo : /i-mvul/"bull; buffalo" may you compare with ☞**English**: *wolf* vs vul
Ewondo : /i-mvul/"bull; buffalo" may you compare with ☞**English**: *fowl* vs vul

Note: *wolf* and *fowl* are dealing by phonemes inversion it is the same word.

A spiral is a stack of rings. May you compare ☞**English**: *loop* vs vul

Note: *vul* and *loop* are dealing by phonemes inversion

A pool is a basin with water. You have to put it in balance with wave, curl, loops. We are talking about the spiral, something dangerous.

I hope, I'll be back more deeply on this subject in another opus. But I just wanna record it here to avoid continuous intellectual swindle.
Before closing this paragraph, I would like to invite you to find the semantic link between the word *wolf* and the African language often called *Wolof*. Good luck…

Bunny in Ewondo

By cross comparison analysis we have also by phonetic proximity in Ewondo that:

Ewondo: /*á*-bùni/ "to be proud" or **Ewondo**: /á-zam/ ☞**Ewondo**: /Zamba/
French: fier "proud"

Middle-Egyptian :  nfr.t ☞**French** : fierté ; la fierté, une fierté,
☞**English**: proudness; proud

Cross comparison

	Leporidae	Equivalent Ewondo	English
Wolof	*Leuk*	Kulu (tortoise, chief)	hare
Wolof	*Niombor*	i-dzoué bor (chief)	Rabbit
Mid.-Egyp.	*Wn*	*Wa (first)*	one

Cross comparison with middle-egyptian wolof and Ewondo

The hare is called * LEUK * in Wolof. The word leuk is linked with the word lucky of English this way:
As we have seen, the hare is the other way to represent our ancestor. Lucky is chance; fortunate, blessed etc.

Ewondo : mvom "chance, fortunate, python"

We can state that /*i*-mvam/ is the ancestor in Ewondo language and /vom/ "place" synonym of /à-kol/. The item /à-kol/ is the ananym of lok/leuk.
As we have emphasized that, a parent is a lucky gift of the life. So that's the link between luck "god-send" and **Wolof:** leuk "hare".

Many lineages in Africa use the root leuk/ kel/kol as : Meke, Mekie, Leke, Lekie , Kellé etc. are de-nominations that recall the link with the first ancestor also called ZAMBA**.**

So if the eldest African ancestor is kulu " the tortoise" how comes that *leuk* "**hare"** is also the tor-toise? The hare is speedy while the turtle is excessively slow.
As heat and cold are dealing with temperature, they are by the way from the same nature but different in degrees.

The ancestor is slow walking because he is old but he is quick thinking because he has life experience. We often say in Africa that only :

"A young man who has traveled a hundred villages is the equal of an old man who has lived 100 years."

(African wisdom)

Phonetic laws

Middle-Egyptian	French	English
n-	*n-*	*n-*
-f-	*-f-/-ph-*	*-p*
-r-	*-r-*	*-r-*
-t	*-t*	*-d*
Nfr.t	***fierté***	***proud***

Few English words liked with the transliteration *nfr.t*
enter • enter- • rente • treen • erne • feet • fern • fête • fete • free • fret • ne'er • neer • reef • rent • rete • teen • tern • tree • tref •entera • entire • faerie • feriae • ferine etc.

Deconstructing

So now, better than me, you know how to utter the word  ***nfr.t-ity***. Because it's written in your own language, may you avoid the word *Néfertiti* because it doesn't mean anything in your own language.

Why is this instrument associated with the heart?

It's simple once you look at it from an African perspective.
The lexical field of playing music in African languages introduces a semantax that states that playing music is beating music.

Ewondo:/à-bom/ "beat, play play music"
Ewondo:/á-kur/ "beat, play music"
Kannouri: Karouw "heart"
Vivina: Koto "heart"
Dagara: sikri "heart"

We say that the word heart could have admitted the following synonymy and offer a reading ***nfr***:

Ewondo: ùn-nəm "heart"	
Nzim: lémé "heart"	→ **Ewondo:** *ùn*-lùmù "palpitating"
Senoufo: Zôoun "heart"	→ **Ewondo:** *ùn*-sɔŋ "muscle"
Senoufo: Zôoun "heart"	
Bambara: Doussou "heart"	**Ewondo:** *à*-sou "chief, visage"
Ngeombale: Tæ "heart"	**Ewondo:** *i*-toa "chief, seat"
Toupouri: Djyri "heart"	**Ewondo:** dzar "chief, basket"
Bassa:Něm "heart, centre"	
Wolof: khôl "heart"	**Ewondo:** *à*-kol "chief, foot"
Kryol: Kè "heart"	**Ewondo:** *à*-ko "chief, foot"
Sango: bè"heart"	**Ewondo:** /*ki/ "blood"
Middle-gyptian:  ***ib*** "heart"	
Kikongo:N'tima "heart"	**Ewondo:** /*ùn*-ntum/ "chief"
Obamba: Otima "heart"	**Ewondo** : /tum/ "impure"
Nzebi: M'tma/Moutma "heart"	
Mbo: Lëm "heart"	**French**: malady
Mbankon: Müem "heart"	
Duala: Mulëma "heart"	**Nzimé**: lam "amiss"
Soninké: Thônhou/ Hatt "truth"	
Mooré: Souuri "heart"	**Fulfuldé: saar** "house, land"
Dagara: sikri "heart"	**Fulfuldé: saar** "house, land"
Fang: nsi "heart of the tree"	
Etc.	

I therefore speak out against the idea of the word signified linked to the Saussurian sense of the term. Since in the examples above, all these words supposed to name the heart in the variants of the African language, call each other to its characteristics and therefore to different signifieds. The heart is the center, it is the blood, it is the chief, it is breathing, it is suffucation, it is unique, the negation, perish; to stop, to close (the mouth), to be finished, to complete; everything, the universe etc.So when Saussure is talking about his signified, what is ultimately the signified?

Middle-Egyptian: ***itmw*** "suffocation; stifling"
Bandjoun: tum "heart"
Ewondo: tum "tum beasts are beasts for which one has disgust"

Middle-Egyptian: ***ib*** "coeur"
Ewondo: /*ùn*-həbə/ "beat, pool; breathing "
Ewondo: /i-bèh/ "turpitude"
Ewondo: /i-bèh/ "pip, desire, thirst"
Ewondo: /ne-fefefe/ "thirsty, longing"

The relationship between all these words are synonymic and metonymic.

This notion of beating is related to ***wsjr*** because it embodies in the African ontology, the one that was molested.

Etymology

The fretted musical instrument called kora gives us the true etymology of the word *Coeur* in French. The phonetic proximity between *Coeur* and kora is not gratuitous from African point of view. The action of beating related to the heart and the beat words relating to music share together the same meaning.

Ewondo:/á-kur/ "beat, play the music"
Ewondo:/*ùn*-ŋkur/ "heart; hessian; burlap"

So we say that this word Ewondo:/*ùn*-ŋkur/ "heart; burlap" shares his meaning by homophony with Ewondo:/*ùn*-ŋkor/ "dry; zero"; Ewondo:/*ùn*-ŋkoré/ "savior"

Onomastic: Middle-Egyptian: F35 ***nfr***

Anthroponyms

Ewondo: Nkoro; Nkodo; Nkoto; Nkura; Nkolo etc.
Ewondo: Abəŋ'e; Mbenguè etc.
Ewondo: Firi; Fàrà; Umfàrà etc.
Ewondo: Evolo; Vili; Ivələ etc.

Middle-Egyptian: ***nfr-tm*** "Néfertoum"
Ewondo:/à-firi 'Ntum/ "the chief's trust"
Ewondo:/ Ntum *à*-firi / "the cane of the faith "
Ewondo:/ Ntum fara / "the cane of the priest"

In Egyptian mythology, Nefertum is the supposed to be the son of Ptah and Sekhmet in the Memphite theology of the ancient Empire. Evoking the fragrance of the lotus, he is represented by a lion or more often by a young man with a lotus flower in his hair. He is the god of resurrection and immortality.

Deconstruction

A curiosity caught my attention. The term given by Christian religious confirmation is the word:

Ewondo:/ i-fyér-Amougou / "confirmation".

One could therefore believe that this term is a colonial item. On this account, It will be necessary to explain the part Amougou of the word / ifyér-Amougou / which is an Ekang Anthroponym still worn today. So how do an imported religion found in wild populations, a term that is part of their jargon? This term has now fallen into disuse but in my opinion, it is very important in terms of diachronic linguistics.

Middle-Egyptian: ***smn*** " establish, settle, perpetuate, etc.".
Ewondo:/à-samèn/ "to gather, consolidate"

Surprisingly, the arbitrage done by the Pharaonic Egyptian with the script above confirms that this word predates the supposed period of Christianity. How, then, did Christians find themselves with an item before the supposed facts? What is the link with the sign F35 ***nfr***?

The cross on the sign F35 ***nfr*** gives us already a good indication on the origin of the plagiarism since of the official opinion based on the authority conferred to Gunther Dryer, these characters are at least 5500 years old as he alleged.

Since we postulate that there is no accidental homonymy but an homophony related to the convergence phenomenon to a same signified, the term / ifyer / giving the skeleton ***fr*** and ***nfr*** sharing the same root * **fr-** to the agent prefix **n** – /nyo/ "whose". So let's use an endogenous analysis on the script: ***smn*** "to establish, perpetuate, etc.".

The sign U32 of the Gardiner's list is given as a pestle and a mortar. The pestle / mortar set is used in principle to crush.

Ewondo: /á-fir/ "to crush" thus offering phonetic link with / i-fyér /

The sign Y5 of the Gardiner's list is transliterated ***znt*** and often associated with the item "zenet" name given to an African game by Egyptologists. The two other associated signs transliterated ***sn***, I will not dwell on them too much.

I do say that the sign Y5 is read here /*ùn*-sɔŋ/ "piquant; sharp; biting". This reading allows a synonymy with:

Ewondo: /*ù*-myar/ "porcupine spines"

The African sacred animal that represents the best illustration of the idea of quills is indeed the porcupine. It can be found nowadays as the Ashanti royalty totem in Kumasi. There are many anthroponyms in Africa, toponyms associated with the porcupine. The porcupine is called:

Middle-Egyptian: ***Hnty*** "porcupine"
Ewondo: /ŋgɔm/ "porcupine"

All the subtlety of the scribe is therefore based on the mindset consisting in using a word from the kekwé branch instead of a word from the kweke branch to express an idea.

The terms **ŋgɔm** and **mog** are indeed consonantal metathesis. So we say that the consonantals skeleton **mog** is the Beti anthroponym equivalent of: Amougou in the expression /ifyer Amougou/ "confirmation".

Middle-Egyptian: ***smn*** " to establish, found, perpetuate, etc..".
Ewondo:/Ifyer-Amougou /"religious confirmation".

Before drawing the necessary conclusions here, the word Ewondo: / i-fyer / phonetically refers to the catfish family found in Africa (*Mastacembelus, Ophiocephalus*). Avariety of small catfish with pointed mouth like **Ewondo**: /*ùn*-woŋ/ (*Mastacembelus, Ophiocephalus)*. This catfish is called in Ekang country:
Ewondo: / *i*-fyer / "small catfish"classed among the **Ewondo**: /*mi*-ntutom/ "frying; *(Mormyrus)"*.
Here are some varieties of /*mi*-ntutom/ "(Mormyrus)":

- Ndoñlo
- Adua
- Mbabe-nsokh
- Esala
- ôbar
- Fyer

The word /*ùn*-ntutom/ also means lukewarm. This brings another glimpse into the idea of confirmation as a booster shot.

This term maintains a metonymic relationship with the word "buffalo" present in the so-called Narmer's pallet

Ewondo: / i-fir / " crusher; buffalo"

Thus, we have that:

Ewondo: / i-fyer Amougou / "alliance, "

[138]

From the standpoint of consolidation, the confirmation can then take on the meaning of the word:
Ewondo: / á-won / "to heal; to consolidate"

When we read ***wnn nfr*** we can also understand the consolidation of the alliance with the /Fara/ first priest who is also the first ancestor called frequently Zamba.

Middle-Egyptian : ***wnn nfr*** "Onnophris"
Ewondo : /*ùn*-wono *à*-firi/ "the faith consolidator"
Ewondo : /*i*-fon fàrà/ " what rejoices Zamba; the country's rejoicing"
Ewondo : /fàrà nənə/ "the high priest"
Ewondo : / *i*-nyùèn fàrà / "mass wine; Red wine"
Ewondo : / *à*-firi nén/ "the high faith"
Etc.

The word /à-firi/ here has to be connected with ***nfr*** as the faith. That's why we often put the hand above the heart to express our belief when we pray.

Consequently, John the Baptist's concept of the hypothetical baptismal system leads us to believe that, this is a massive intellectual swindle inasmuch as the sign N35 represents the ceremonial liquid here. Let's listen to the specialists on this subject

---

[138] This expression has to be put in balance with the fish on Narmer's pallet. Indeed the fish on Narmer's pallet represents the alliance, the country. But the head of the fish /i-fyer/ "Mormyrus; oxyrynchus"
shares the same shape with ***km.t***. That's why this fish is the sacred for real Africans.
By this demonstration, we highlight the limits of the intellectual swindle. Here we highpoint the limits or the borrowing linked to the Socrato- Platonic dualism. Here we find the idea of the polymorphism of an idea's projection.
Professor Martin Bernal is therefore right to point out the intellectual fraud that underlies the supposed labor of Plato. One does not make philosophy in micro-states quoted subjected to the war. There is no philosophy in states confined to alternative sexual practices.
It is clear that when studying a graph, it recalls complexity. But a binary thought is not compatible with complexity. The complexity is based on a simple model that becomes more complex according to the mathematical model of fractals. The complexity recalls a densification of simple idea leading to polymorphism.

"It is with a perfumed eye, the Holy Chrism, that the bishop marks the forehead of each confirmand. As this oil gives off a good odor, everyone is called, by the impulse and the joy of his life, to spread the good odor of Christ, to be an authentic witness of the Risen One, so that the body of Christ may be built up in the faith in God, Father, Son and Spirit, and the love of men and the world. The confirmed then joyfully bears the responsibility of making known to men and the world the love of which they are loved."[139]

The danger of transliteration in the specific case with the *smn* consonants skeleton, did not allow reading / ifyér-Amougou / "confirmation" but only allows the item *to gather* in Ewondo /à-samen/ "sum".

Conclusion on this last part

Advanced research in science have made it possible to affirm a monogenetic origin of Humanity. In fact, if the Man was born in Africa, he developed his language faculties through a common language in Africa that was identified as the Negro-egyptian, the common African or the paleoafrican.

We know that the variations of this primary language are related to the phenomena of grammaticalization. These have occurred through the remoteness of populations in space and time. The advanced techniques of comparative linguistics have made it possible to reconstruct this common ancestor and thus explain the behavior of current languages.

Beyond impressions and ideology based on raciology that wanted to stratify languages according to bogus criteria highly questionable, Africa remains and remains the cradle of humanity. Attempts to create civilizational fires outside Africa are a matter of deceit and therefore of malevolence. The endogenous reading technique of the scripts made it possible to note several things:

A phonetic proximity between the speaker language and the transliteration proposed by Egyptology, we say on this account that there is homophony between the term put forward by Egyptology and many synonyms offered by the vocalization of a script in the language of the speaker.

Sometimes between the transliterated variant and the word in the speaker's language, there is a gap. We say, that these are mere synonyms. And the way to find this synonym linked to the loss of its historical memory lies in the interpretation of the script through metonymy, synonymy, antonymy and all in an African paradigmatic framework. This is what we tried to do through the previous examples.

We also say that scripts are greats talkative and querying them allows access to information outside of any ideological constraint.

The morphological and other variations observed between the supposedly different words seem to us to be superficial data from the moment when *mDw nTr* is used as a common base. Verification is offered when reconstructing a predialectal ancestor.

The greatest difficulty for those who practice reading of these scripts is to get rid of mental prison with the praxis of Egyptology mental frame.

The in situ reading of each script takes real meaning only when each symbol matches the next one. As the scope of words is not a single one in Africa, it necessarily points to the advanced sciences of our time. In this sense, a synthesis work would make it possible to discover basins of iso-representations of the thought grouping singular themes, able to tell us the level of technological development acquired by the regroupings of lineages or African castes.

139. https://eglise.catholique.fr/approfondir-sa-foi/la-celebration-de-la-foi/les-sacrements/la-confirmation/371024-confirmation/

This is observed when we extend the comparison outside the simple African context. So we find curiously the same African sociological representations throughout Latin America, the Far East, Europe, the Americas, etc.

How and why does the African totemic panther become a totemic jaguar in the Amazonia?

How and why does the totemic African crocodile *sbk* can be found as a totemic animal across the globe and especially among peoples who are supposed to have a very distant past with Africa?

For at least 300,000 years surprisingly the observation that emerges when comparing few words leads us to observe that the Man has preserved the memory of his African past.

The challenge of reading these scripts is to flip through the universal encyclopedia knowledge that are the *mDw nTr*. It is indeed to understand the implications that allowed the various African initiating schools to promote some variants of language adapted to each school and that can be found nowadays through any location over the globe.
If *Hr* is a simple Egyptian legend how and why is it found everywhere then?

The approach of studying the scripts according to the endogenous point of view makes it possible to become familiar with the method without making it a goal. A script can be a musical note in a melody that is the phrase. It will therefore be up to everyone knowing the different meanings of a character to take the real measure for a better understanding of ancestral wills. This approach made it possible to highpoint the enlightening side of the Standard method as well as the limits of this voluntarist and normative exogenous approach.
I therefore remain convinced that we can read these characters in any African language but also, we can understand each other by using this unique writing because it is accessible to all of us.

3.4.3.40 ***EXAMPLE N°40 iwsw***  *(Faulkner p. 14, Wb. I p. 57, 18-19)*

I would like to deal with this script not only the reading but also the understanding offered by the word lawyer.

In the Old Kingdom (see Chronology) alone, the pharaoh and the royal family benefited from the resurrection after death, the "good death", but towards the twelfth dynasty, after an important social revolution, the whole people participates no only to sacred drama and religious feasts (celebration of the death and resurrection of Osiris), but aspires to a happy life after death in what might be called the "Egyptian paradise".

The 125th chapter of the "Book of the Dead" speaks of the judgment and purification of the soul. He gives a detailed account of the weighing of the deceased's heart.
It is also the chapter of the confession of the soul, the famous "negative confession". The deceased denies all the crimes that can be done by the man.

The symbols U38A  and U91  are at the center of two important principles, one dealing with science with weighing systems and the second with the judicial system.

This script is therefore highly important insofar as it sheds light on the reading and understanding of the concepts related to the symbol of the scale. Apart from human justice, the balance is also between immanent justice.

There are many spellings to name the balance I suggest to read them in Ewondo one after another

The present script does not involved any issue because it has already been studied in another context. We have addressed it during the script N °3.4.3.3 learning.

Standard method

Middle-egyptian:  *iwsw*"scale" (*Faulkner p. 14, Wb. I p. 57, 18-19*)

Endogenous method:

In this example, i suggest to proceed another way namely to interrogate some African languages on the way they do name the scale and only after that, to proceed at the reading in Ewondo language.

Kɔ́ɔnzime: *lor* "scale";	☞ **Ewondo**:i-lor "the exceeding gap"
Kɔ́ɔnzime: *kilo* "scale"	☞ **Ewondo**: kilà "kilo"
Hausa: *mizani* "scale";	☞ **Ewondo**: *mə*-zén "the ways"
Hausa: *cikõ̀* "scale"	☞ **Ewondo**: kilà "kilo"
Kinyarwanda: *umunzani* "scale"	☞ **Ewondo**: *mə*-zén "the ways"
Tshiluba: *citempa* "scale"	☞ **Duálá**: etámbí "sandal"
Fang: *adôl* "scale"	☞ **Ewondo:** *ùn-nlod* "scale"
Foutouni: *Deudeung* "scale"	☞ **Ewondo**:*ùn*-ndəndəŋ"swinger"
Ɓàsàa: *kɔ̀βɔ̀* "scale"	☞ **Ewondo**:*ùn*-ŋkɔbɔ "the speech"
Nāhuatlahtōlli: *tlatamachioualoni* "scale"	
Yémba: *lesɛt* "scale"	☞**Old Ewondo**:lə-sè "the feather"
Bafanji: *fʉ'* "scale"	☞ **Ewondo**: fã "machete"
Tupuri: fiː' "scale"	☞ **Ewondo**: fã "machete; scythe"
Mundang: fãʔ "scale"	☞ **Ewondo**: fã "machete; scythe"

I would like to draw attention to the principle because I could easily extend the list, but as I have already pointed out, the most important thing is not the size of the list but the principle.

That's why I have different words in supposedly different languages.
From this long list of words, I want to highlight several important points.

The first is that the term to name  the scale exists in several languages of the world.
The second we can observe that, curiously it is not the same word wordwide. But naming the same signified, lead us think grammatically that those words are synonyms. But in fact, although there are different signifiers, we can sustain that this is related to different signifieds as well.
Finally Is it the same signified or one signified and different signifiers as Saussure thought?
This point deserves clarification.

Since I gave the equivalent of each word quoted in Ewondo, demonstrating that although Ewondo does not use this word to describe t he scale, this word exist, exceptionally so here I will explain the item's root ***fé-*** of bafanji, Mundang, Tupuri designating a balance. I take this root because it is interesting also in English.

I hope that the enlighting by English will help to better understand the method and especially the relationship with the other words of the list.

Bafanji: *fu'* "scale" ☞ **Ewondo**: fã "machete"
Tupuri: fi:' "scale" ☞ **Ewondo**: fã "machete; scythe"
Mundang: fãʔ "scale" ☞ **Ewondo**: fã "machete; scythe"
Chinese: 法码 fă mă "weight (of balance)"
French: faux; fauchet "scale" ☞ **Ewondo**: fã "machete; scythe"
English: fear; fire; fair etc.

What is the relationship between a machete and a scale, here is the main enquiry. If we do understand this relationship we can understand also the relationship between a scythe and a machete.

Ewondo: fã "machete"
Ewondo: /á-fàê/ "to sprout"

The word to express the idea of the growing in **Ewondo** :/ i-fàê / thus illuminates the first meaning of the item / fã / "machete". But that does not tell us what connection is between a scale and a machete.

The link is relatively obvious if we understand the operating system of these small drawings.

Middle-Egyptian: "MAa.t"
English: *billhook*
Ewondo: /bilokh/"herbs, grass "; / i-fàê / "sprouting""

The symbol U2 of Gardiner's list refers to a billhook, a scythe, a machete.
Note that the word given by Egyptologists for the above script reading is MAat nothing to do with ***iwsw*** "scale". However, on the semantics associated with the script ***mAa.t*** one can read a notion that seemed important to us because associated with the balance namely the notion of justice or equilibrium.

From the point of view of the Ewondo language, the Bafanji, the Tupuri, the Mundang naming the item *scale* use a metonymy to designate the sign U38A namely the balance.

If you understood that, then you have the right comprehension for the connection between:

Middle-Egyptian: ***As.t*** "Isis" ☞**Ewondo:** ze; zoa; dyal; dyàŋ etc.
English: scythe; seat; sit; site; city; state; estate; sheet; shit etc.
French: site; cite, cité, état, Etat etc.
German: site ☞**Japanese:** サイト Saito
Bokmål: *sted* "place; locus; etc."
Svenska: *sajt* "place; locus; etc."
Danish: *sted; site* "place; locus; etc."
Dutch: *site* "place; locus; etc."
: 地点 *Didian* "place" ☞**Old-Ewondo:** *di*dyan "place, locus"
Chinese: 所 *suò "sit"* ☞**Ewondo:** si "place, locus"
Arabe: موقع *mawqie "site"* ☞**Ewondo:** mə-koh "feet"

As you can see, the words of the script above do exist in English but the translation given by the Egyptologists to the script is ***asset*** or ***isis***. Even though anyone doesn't really know what is ***asset*** or ***isis***. On this account, there are loads of inventions that had filled conferences and speeches of scholars.
These words ***asset*** or ***isis*** are related to something very nebulous. For neophytes and ideologists they come from Egypt and ancient Greece and it is hard to link them to a known signified. They are used in a demogogic spirit and for the show.

As we argue, it is an operating concept that is applied if necessary in this or that area. The proof is that real words with real signified do exist in English also. The consequence is simple here, Tupuri, Bafanji, Ewondo, Mundang etc. are sharing the same lexicon.

Note: How comes that some words have a final **t** and others do not share this final **t** when sharing the same signified? We think that it is a grammaticalization phenomenon. I have emphasized on this purpose with the example : Aïssa and Aïssatou or Aïssata all African feminine names. For Egyptologists, this **t** is a feminine tag and by the way is not vocalized. We have with the list above the prouve that, this ***t*** is not only a feminine tag. It can be vocalized like in Japanese *saito*. Moreover it can be used for a nominalization with the idea of who embodies, a kind of superlative. **Ewondo**: /á-tî é/

French : bon "good; right, well etc."
French : bon**té** "good**ness**; kind**ness**; gentle**ness** etc."

As we argued, ***MAa.t*** and ***As.t*** "Aïssa ; Aïssatou ; Aïssata" are simple hypostasis of the same reality as a mother could be also a wife, de facto if some people really knew how to read the hieroglyphs then it would have spared us what is called the Egyptological gibberish since the words really exist in each human language. It would have lead us to an harmonization at the spelling level since we have different spellings of the same word linked to the same signified.

Technically we could use Egyptian script as signified because they are synthetical.
We could thus save tons of ink and paper because only one sign condenses a complex reality.A single sign serves as isorepresentation basin of complex realty.

According to the depth of analysis, one can have different signifieds (***mAa.t*** and ***As.t***) or the same signified which is the maternal figure.

Script reading

Middle Egyptian:  ***iwsw*** *"scale"*
Ewondo: /i-vùs'ù/ "error; sin"
English: foolish; false
French: voleuse; fol "thief"

Meroitic:  *"wuz i"*

Script reading

Middle Egyptian: 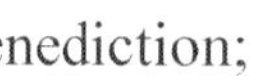 ***mxA.t*** "scale"
Ewondo: /maâ/ "benediction; stability, equilibrium"
Fang: mèghè "benediction; charcoal , equilibrium; poise, stability etc."
English: mass; mace
French: Masse; massue; messe

Script reading

Middle Egyptian: 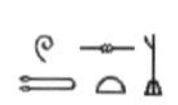 ***wTs.t*** "scale column"
Ewondo: /*ùn*-mbɔkh/ "support, backing; gallows, bracket; gibbet"
Ewondo: /*ùn*-ŋkùm/ "support, backing; gallows, bracket; gibbet"
Ewondo: /*ù*-tsəŋ/ "cutter; knife"

Ewondo: /i-fàkh/ "fork"
Ewondo: /mə-sɔŋ/ "support, backing; gallows, bracket; gibbet"

Script reading

Middle Egyptian:  ***tx*** "peson counterweight"
Ewondo: /i-tîghà/ "devider; devisor; slicer"

Script reading

Middle Egyptian:  ***gs*** "bow to one side"
Ewondo: /i-sûghà/ "bow to one side" ☞**Ewondo**:/à-boh/ "foot"

Example:

Middle Egyptian: *(r)di Hr gs* "bow to one side of a balance or language"

Script reading

Middle Egyptian:  ***wDA*** "balance in a calculation"
Ewondo: /i-lîghà/ "remainder"
Ewondo: /i-dàŋghà/ "left-over; extra"

Script reading

Middle Egyptian:  ***rqw*** "tilt (balance)"
Ewondo: /i-sûghà/ "tilt; incline"
Ewondo: /i-koré/ "redemption; buying back; repurchase"

Script reading

Middle Egyptian:  ***tnm*** "deviation
Ewondo: /i-sûghà/ "tilt; incline; trial; decision; etc."
Ewondo: /ù-mê 'ntén/ "line diversion; curvature, line deflection etc."

Script reading

Middle Egyptian:  ***Hnkw*** "tray (balance)"
Ewondo: /i-kêhən / "separator"
Ewondo: /i-həŋnà kɔb/ "weighing tray"
English: hen's cub "scale's cub"

Note: In English, there is a word game between hen's-cub and hands-cub to enlight the idea of weighing tray. The sign U38  of Gardiner's list looks like a human body with hands. The trays looks like its hands

So here we have the true etymology of the English word handcuffs. 

The item hen "poultry" could find its etymology from the Ikinyarwanda morpheme *hene* "goat". According to african thought the idea of mother is linked to the *hen* and also the *goat* ideas. How comes that the mother is a hen? Simply by phonemes inversion.

Ikinyarwanda: *ihene "goat"* ☞**English**:*a hen "poultry"*
Ewondo: /i-hàŋ/; /i-vàŋ/ "separator"
Ewondo: /i-həŋ'èn/ "reversor"

Phoneme inversion

☞**Ewondo**: *nah "mum"
☞**Ewondo**: *nya "mum"

Note:
We do find in China, the **Han** Chinese, **Han** people or simply **Han** (UK: /hæn/; US: /hɑːn/;Mandarin pronunciation: [xân]) (Chinese: 漢人, pinyin: Hànrén, literally "Han people"; or 漢族, pinyin: Hànzú, literally "Han ethnicity"or "Han ethnic group. Is it a surprise?

We have also **Na**, in China also called "**Moso**", this ethnic group from southwestern China is surprising on many levels. To characterize it, people often speak of a matrilineal, matrilocal and avuncular society. But the facts illuminate themselves from the African point of view. Just as we find the **Han** in China. Where is the surprise here then?

If there are, sociologically, four types of clans, nomadic or sedentary, exogamous [marriage outside the clan] or endogamous [marriage within the clan]. The African Clan is mainly a matriarchal or sedentary exogamous.

The endogamous Clan sedentary. Example of the Chinese clan

characterized by:

1- **Created by the man as african clan**.
2- **Sedentary as the african clan**.
3- **Subsistence derived from agriculture**.
4- Clan endogamy: ignore totemism.
5- Patrilineal descent and succession

This clan constitutes, in general, a middle term between the two preceding extremes: it is exclusive neither in one direction nor in the other. We can then have a better understanding of the presence of some African matriarchy pockets in China related to a sedentary sociology clan.

After this aside, let's remind the relationship between the mother and the goat according to african thought:

English*: hen* "poultry"
Ewondo*: kub* "poultry" *homophony* ☞**Ewondo***: kaba* "goat"

Il have already discussed this point deeply when treating the script related to horns. To get a remind please report to (4.3.4 Script 4 ato).

Etymology:

When we studied the owl's sign  this is what we've underlined in terms of etymology. As we all know immanent justice is rightly done by *law* people (phoneme inversion of *owl*). Is it any wonder that we have a phonetic proximity between *scale* and *school*? We say that this perfectly confirms the idea that these characters can be read in the language of each one without brain knots need.

Middle Egyptian:  ***mxA.t*** "scale"

The word scale thus recalls the word *owl* in certain African languages (see below).

As we have emphasized here that, owl represents the perfect figuration of the idea of knowledge in african thought. Whereof one can admire the phonetic proximity existing between the root ***owl*** in African languages and the word ***school***. By the way then, i will leave aside the ideology of believing that Charlemagne had invented the school. Temples of knowledge were already built in Egypt? To invent school, one must have already invented writing. Is it Charlemagne the writing inventor? You should think about that question deeply.

Douala: Esukudu "owl" **Douala**: sukudu "school"
Ciluba: cipùngulu "owl" **Ewondo:** Zùkùlù "school"
English: school **French:** école "school"
Kouyou: esukuru kuru "owl" **Lingàla**: Esulukú "owl"
Ancient greek: γλαύξ gláfx "owl" **Ewondo:** /ngul fəkh/ "great intelligence"
Latin: schola "studious leisure, lesson, place of study",
Ancient Greek: σχολή skholḗ "work Stoppage"

NOTE
I would like to make an aside on a small curiosity.

Middle-Egyptian : 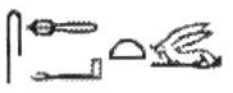 ***sXa.t*** " hare "
Wolof : katanté "to make love" ☞**French**: Sexe
Wolof : Saxuti "laid; recline" ☞**English** : sex
Ewondo : /*i*-sax/ "fuck"

I don't wanna deal with the idea of sexuality associated with rabbits or hares. I point here the phonetic proximity between all the words we have stated above concerning the link between the owl and the school.
How comes that a rabbit is a symbol of intelligence with its long ears? As we have seen, the hare or the rabbit is associated with God himself.

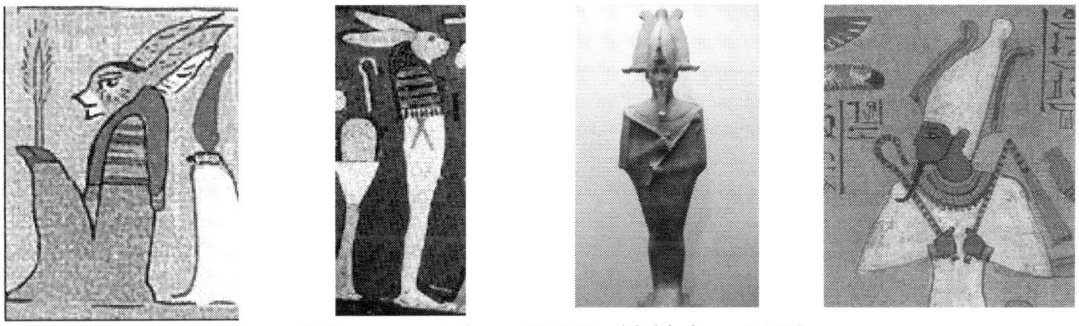

Wsir represented as a HARE with his long EARS

So here is above the reason why the hare is associated with the intelligence. Even though ***wsir*** is the phallus, and his name linked to sexuality and then multitude as rabbits or hares. The most important is to understand that the rabbit is a representation of ***wsir***.

Middle-Egyptian : 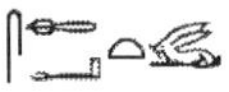 ***sXa.t*** " hare "
French: Sage "wise; intelligence"
English : sage

Note: As we have seen above, there is a relationship between hare/rabbit with canines.
This link is confirm also here with the English word *canny* "foxy; hep; brainy; heedful etc."

Middle-Egyptian : 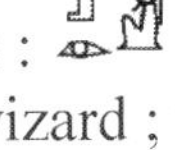 ***wsir*** " osirus "
English : wise ; wizard ; wiser ; advisor ; advisory etc.

French: aviser; avisé ; aviseur; savoir etc.

Middle-Egyptian :  " hare "
Ewondo: /ù-pkwəŋ/ "hare"
Ewondo: /à-kəŋ/ "sage; wise"

The king in Africa without a pun is the phallus because he is responsible for satisfying several wives. This representation of the phallus associated with the chief is not therefore a fantasy of the scribe. We have already seen that the chief is associated with the elephant for his phallus.

Balance african etymology

Portuguese: *pal*anca "wild goat (black roan)
Ewondo: I-*bal*ghà "grass cutter, bilhook"
Ewondo: I-*bùl*ghà; i bəghà "the multiplier; the cutter"
Ewondo: *bilwa*gh "grass" ☞**English:** bilhook
Late Latin: bilanx "scale"
Tupuri: *bal* "elephant"
Fulfuldé: njawdiri *mbaala* "sheep, mutton"
Pular: baalii "sheep, mutton"
English: *bal*ance
French: *Bal*ance
Germain: *bal*ance

As i have emphasized, the item mother is linked to the word goat in african thought. Therefore we do understand the relationship betwen the word *balance* "scale" and the portuguese item palanca.

Middle-egyptian:  ***imA.t*** "ibex"
Middle-egyptian:  ***mAa.t*** **"**justice; accuracy, orderly management"

As we can see here above, the symbols of goat, woman, bilhook, feather etc. are sharing the same semantic field according to the scribe.

Kilo african etymology

Depending on the field of knowledge, the African etymology of the word kilo can be established.
I will not do all the listing of domains but simply highlight the link between / kili / "reject"; / i-kolà / "debt"; / á-kele / "suspend; circumspection"; / à-kol / "foot; leg; 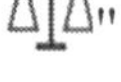" etc.

All these notions open up new fields of investigation by metonymy; antonymy, paronymy; synonymy; homophony etc.

Dol african etymology

A *dol* or fraud is in roman law contracts a maneuver of a co-contractor in order to deceive a partner and cause him an **error**. *Dol* is, with error and violence, one of the three vices of consent. It is sanctioned by the nullity of the contract.

Fang: *adôl* "scale"
Sängö: *dòlì* "elephant"
English: *deal; dwelling; dole* etc.
French: *lodge; logie, log, dol etc.* "lodge; dwelling"

How comes that error, elephant and scale refere to the same reality? This is a childish question because the answer is simple.

We have said that the scale is link to the foot idea. As we have also said, the mother is link to the idea of panther, place, foot etc.

Middle-Egyptian: ***Aby*** "panther" Take note of the foot symbol
Middle-Egyptian: ***Abw*** "elephant" Take note of the foot symbol
Ewondo: dùlù; à-boh; à-koh
French: peau, pot; pied; marche
English: paw, walking; march; going

If Egyptology saddles these characters of the skeleton of consonants ***Ab-*** and force only the reading around phonetic proximity with / abo / "foot"
And is therefore out of inspiration when the Sango to designate the elephant speaks about
Sängö: dòlì "elephant"

This example completely erases the comparative method that uses the consonants skeletons and tinkers with predialectal ancestor reconstructions. Let's take an example:

Fong Gbé: gbé ta "farm"
Ewondo: à-fùb "farm"
Middle-Egyptian: ***bwAt*** "farm"
Middle-Egyptian: ***SA*** "farm"

It is therefore difficult to reconstruct a predialectal ancestor between these two items designating a farm in both languages. Therefore, this is the main error of comparatists. In reality it is not the same word regarding the couple of languages but two expressions referring to the same signified. The equivalent of Fong Géb: *gbé ta* exists in Ewondo also even though it's little used.

Ewondo: tsè i-bèh "ground for farming"
Middle-Egyptian: ***bwAt*** "farm"

The word /Dulu/ in Ewondo is related to the semantic field of the foot because the foot is used for walking. It is therefore this characteristic walk of the elephant that is retained as meaning in Sängö: dòlì "elephant".

The elephant being a chief representation and the scale the best representation of dol idea, is it surprising when we have a phonetic proximity between *thief* and *chief*? Really think about that. Laughing out loud. Consequently, a *good chief* is a *thief*.

However, the scale being fixed on one support/foot, by metonymy it is assimilated to the foot itself. It is therefore normal that the swinging of the arms of the scale gives the thought of an action of walking. So the word walk / dulu / is associated with the scale.

Does not this last point shed light on the origin given to the place of the market?
French: marché "walked"
French: marché "market"

History revision:

Gilles Personne de Roberval (August 10, 1602 – October 27, 1675), French mathematician, was born at Roberval near Beauvais, France. His name was originally Gilles Personne or Gilles Personier, with Roberval the place of his birth.

He was known for the ***Roberval Balance***.

The Roberval balance is a weighing scale presented to the French Academy of Sciences by the French mathematician Gilles Personne de Roberval in 1669.

In this scale, two identical horizontal beams are attached, one directly above the other, to a vertical column, which is attached to a stable base. On each side, both horizontal beams are attached to a vertical beam. The six attachment points are pivots. Two horizontal plates, suitable for placing objects to be weighed, are fixed to the top of the two vertical beams. An arrow on the lower horizontal beam (and perpendicular to it) and a mark on the vertical column may be added to aid in leveling the scale.

weighing of the heart or scene of psychostasis in ancient Egypt (7000 years)

Roberval Scale (17th Century)

The facts are more important than the judgments and impressions, so make your own opinion regarding theses facts above.

3.4.4. Readings Of Phrases

After dealing with examples of script reading, I now suggest to articulate all of this in a sentence.
In this account I hold out firstly to become familiar with the typological elements of the sentence Ewondo.

A sentence can be considered as an autonomous whole, uniting syntactic units organized according to different networks of more or less complex relations called subordination, coordination or juxtaposition. The sentence is composed of a set of words placed in a logical order in order to have a meaning and to express an action or a state.
According to classifications in syntactic typology[140] there are different sentence structures. A language is said to have a syntactical typology SVO when its sentences are of the subject-verb-object type. After a study [141] of Russell S. Tomlin covering 402 languages, about 42% of languages use SVO-type sentences. In terms of typology, the Ewondo would therefore be classified in the SVO language category. To do this, the subject is generally placed at the beginning of the sentence and it can be:

- Either a nominal (single or expanded)
- Either a personal pronoun.

140. Syntactic typology is a branch of the linguistic typology that classifies languages according to the order in which words appear within sentences.
141. Russell TOMLIN, Basic word order: functional principles, Londres, Croom Helm, 1986, p. 22.

However that may be, the subject generating along the sentence a scheme of accordance between the subject and the verb.
When the subject is a nominal, the agreement between it and the verb is marked by an accordance particle that corresponds to the name's class.

Example:

M-òt	**à**	**á**	**zù**	**-ø**	*"the man is coming"*
#subject	*pp*	*Aspect*	*come*	*-ø#*	
B-òt	**bə**	**á**	**zù**	**-ø**	*"men are coming"*
#subject	*PP*	*aspect*	*come*	*-ø#*	

The analysis of these sentences morpho-syntactically shows the existence of five elements. Among which, the subject, the pronominal prefix (PP), the durative / - to / linked to the temporal value, that is to say above the present. Then the zero (ø), signifier marks the aspect not presented by a term in the final of the verb / -zù- /.
Compared to Pharaonic, there are basically two types of simple sentences in Pharaonic: nominal sentences and verbal sentences. Among the verbal sentences: SVO sentences and VSO sentences.
The order of the words in the VSO phrase in Pharaonic is:

1. Verb
2. Subject
3. Object
4. Dative
5. Circumstantial supplement

The fact that the order is not always SVO is therefore a discriminating condition for reading in Ewondo or in any african language. On this account, it is enough to correctly identify each element to build sentences according to the usual typological model.

Example: Verbal phrase typology VSO.

①*Verb* ②*Subject.........*③*objet*

Middle-Egyptian: kef-n sS s "the scribe caught a person." typology **VSO**
Ewondo: /*ùn*-kaŋ'a à á-nga kiabi *ùm*-ód/ "the scribe caught a person." typology **SVO**

Ewondo: / *ùm*-ód *ùn*-kaŋ'a à á-nga kiabi / "the guy the scribe has caught" typologie **OSV**

We see here that the order of the typology is not a problem in itself, but the most important is to identify the lemmas of its own typology and to communicate the ideas faithfully.

The linear order of the lexical complements of the verb is: *ccusative*[142]>*dative*[143]>*oblique*[144], this is shown in the following example where 'man' is the nominative argument, ***Sa.t***, 'letter', the accusative argument, ***sS***, 'scribe', the dative argument and ***pr =f***, 'his house', the oblique argument.

When the subject is a personal pronoun, it fits all the sentence in class and in person.

Example:

M-òt	**à**	**á**	**ɲù**	**Arki**	*"The man drinks arki"*
#subject	*pp*	*aspect*	*drink*	*COD#*	
B-òt	**bə**	**á**	**zù**	**-ø**	*"The men are coming"*
#subject	*PP*	*aspect*	*come*	*-ø#*	

Cas de sujet VSO

Example 1:

Verb	subject	COD	Article	COI	preposition	CCL
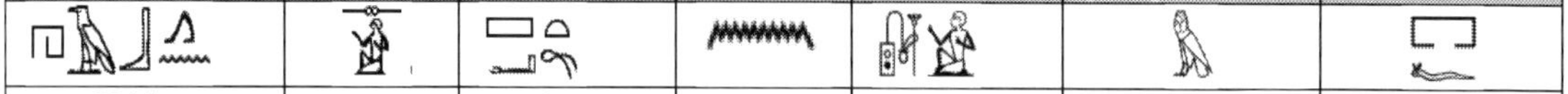						
HAb-n	s	Sa.t	n	sS	m	Pr-f
②A nga həbèn	***①ùm-òd***	***③i-sié***	***④ùn-***	***④- ŋkaáŋa***	***⑤à***	***⑥Nda dyé***
Has sent	*The man*	*the document*	*Of the scribe*		*from*	*his House*

CHART103: VSO EXAMPLE

Example 2:

Aux	Verb	subject	COD	Article	COI	preposition	CCL
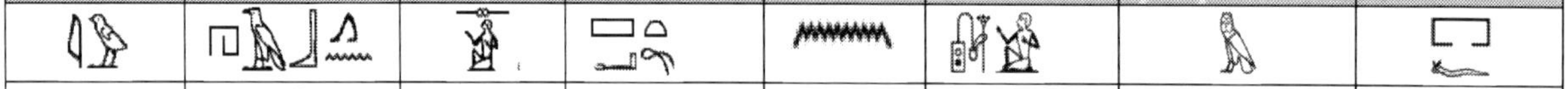							
Iw	HAb-n	s	Sa.t	N	sS	m	Pr-f
②yu	***③A nga lom***	***① ùm-òd***	***④ i-sá***	***⑤ ùn-***	***⑤-ŋkaáŋa***	***⑥ à***	***⑦ Nda dzi***
had	*sent*	*The man*	*the document*	*Of the scribe*		*from*	*his House*

Chart 104: iw sentence

Let us note immediately that, in this type of sentence, the Pharaonic language generally uses what some call an indicator of utterance whereas others speak rather of an old verbal form having the meaning of "it happens that".

In other words, especially in the context of narrative text, the particle ***iw*** introduces a sentence that describes a situation or state of affairs.

Middle-Egyptian : ***iw***
Ewondo : /*i*-və/; /i-ngà/; /i-mbə/ etc.

142.In linguistics, the accusative (abbreviation: acc of grammatical Latin accusativus, "which marks the culmination of the action"), is a grammatical case expressing the complement of direct object (COD), that is to say to say the actant who undergoes the action exerted by the subject of a direct active transitive verb, also called patient object.

143.In linguistics, the dative is a grammatical case expressing the indirect or second object, that is to say the so-called beneficial actant (not to be confused with the beneficial case). It is called dative because it designates the one to whom one gives (Latin: do, das, dare, dedi, datum, "to give"): it marks the complement of attribution. Ex. In Latin:Fluvius dat campo fecunditatem "the river brings fertility to the field"

144.In linguistics, the term oblique case denotes any type of grammatical case other than that serving as a form of quotation (or canonical form, or lemma). It works essentially in opposition to the notion of direct case.

Nonverbal or nominal phrase

This time it is a sentence without verb, and in which the predicate (ie what the sentence is about) is: either an adjective attribute, or a noun, or a preposition + substantive, or an adverb of place or of time (the last two types going together)
One of the criticisms often made to African languages is to claim that they do not use nominal sentences.

In his song titled "*Eloko oyo*"[145] the kongolese musician Fally Ipupa uses in the lyrics of his song some nominal phrases.

So I wrote down one of these sentences below:

"*le 2 juillet bisso tembe na foire*" → "July 2, we opposition to the fairground"

We note here that there is no verb nor any notion of temporality.
A translation at the semantic level could give: "July 2nd we will stop the fair"

This sentence could have been pronounced in Ewondo or in any African language, offering the same comfort of comprehension.
The stated date is a relative time that induces for us a future. Yet there is no conjugated verb.

Let's take a dialogue between two persons to illustrate the use of the noun phrases in Ewondo for example:

-/ i-mor ɲwá, wá zá?/→ "sir, you who?"→ (Sir, who are you?)
-/má Ndengué Zibi Ivòlò, man ileuga ya Ikoudbessanda, ndié Mə-Bəŋ'-m'Abessòlò/.
"I, Ndengue son of zibi Ivolo of the Eleuga lineage, from the place named Ekoudbessanda, grandson of Abessolo, Mə-Bəŋa's wife"

So it's not Lefebvre who would deny me when he says:

"[...]From substantive predicate (noun or personal pronoun) in Egyptian, if the subject is a personal pronoun, it precedes the predicate immediately, by direct juxtaposition."[146]
(Lefebvre-1940)

<u>Example 1</u>: predicate = independent pronoun

M-E.: ***Ink ntr*** "I am a deified ancestor" literally (me, ancestor).
Ewondo: /ma tara/ "me; father"

<u>Example 2</u>: predicate = independent pronoun

M.E.: ***Ink smsw*** "I am a follower" literally (me, servant).
Ewondo: [ma, zomloa] "me spiritual guide"

<u>Example 3</u>: predicate = adjective

M.E.: ***Iqr smsw*** "I am a reliable guide" literally (reliable, guide).
Ewondo: [à-kokh, zomloà] "rock spiritual guide"

145.http://paroles2chansons.lemonde.fr/paroles-fally-ipupa/paroles-eloko-oyo.html#RS5H2jISKtGsvYUk.99
146.Gustave LEFEBVRE, Serge SAUNERON, Grammar of the classical Egyptian,Le Caire, 1940, p.295, 604.

Example 4: predicate = substantive

For this specific case, Pharaonic uses a very particular construction (using *pw*):

M.E.: ***sS pw smsw*** "the scribe is a guide" →(this scribe, guide).
Ewondo: [*uŋ-**kaáŋ'a à-ntò***, *uŋ*-zomloa] "the scribe is a spiritual guide"

Typology

sS	*Pw*	*Smsw*
Predicate / substantive	grammatical Subject	real Subject

Example 5: predicate = preposition or place adverb

For this case we have the predicate consisting of a preposition + substantive, while in the second, we use an adverb of place (or time, as the case may be):

M.-E.: ***sS m pr*** "The scribe is in the house" →(scribe in house).
Ewondo: [*úŋ-**káŋ'à** á-nə* à *ndah*] "The scribe is in the house"
Bùlù: [*úŋ-**káŋ'à** á-mbə ndah*] "The scribe is in the house"

This structure can admit the copula [147] ***iw*** at the beginning of the sentence.

Middle-Egyptian: ***iw sS im*** "The scribe is here" →(scribe here).
Ewondo: [*un-ŋ**kaáŋ'à**-ne và*] "The scribe is here"

Middle-Egyptian: ***iw-f r sS*** "he will be a scribe"
Ewondo: /*A yi nyi boh un-ŋkaáŋ'à*/ "he will become a scribe"

Step by step decoding

M.E.: "iw-f" **Ewondo**: /iye(l)/ "impuissant" giving /à-ye/+/nyi/ "snake"
M.E.: "r" **Ewondo:** /ù-ŋkobo/ "speech =foot= /bo/"
M.E.: "sS" **Ewondo:** / *ùn*-mbokh/ hole = paralytic = scribe

The different types of sentences

According to the authorized literature, there are basically four types of sentences in Ewondo:

Declarative sentence

It is presented as being at the center of the types of sentences. The declarative sentence is a basic sentence on which the various transformations giving rise to the other types of sentences involved.

Ewondo*:* /Nyò í hə lob Mendana/ "The snake stung Mendana"
The snake Pa P2 + has stung Mendana

147.The copula can be considered as the equivalent of the verb to be in French.

Comparative example with the copula "iw"

Middle-Egyptian:  ***iw m sS*** "He is a scribe"
Ewondo: / *i nyo ùn-*_**yəm bi-kanga**_/ "the one of the scribe"

Interrogative sentence

There are two kinds of questions: partial interrogation and total interrogation.
Partial interrogation is the one which refers to only one element or part of the sentence.

Case 1

Ewondo:	Di	i	-hə	lob	Mendana?
	What	*Pa*	*P2+*	*sting*	*Mendana*

English: "What has stung Mendana?"

Case 2

Ewondo:	Nyô	i	-hə	lob	za?
	snake	*Pa*	*P2+*	*sting*	*whom*

English: "The snake stung whom?"

Case 3

Ewondo:	Nyô	i	bo	dzé?
	snake	*Pa*	*do*	*what*

English: "The snake did what?"

It appears that the partial interrogation can relate to one of the elements of the sentence either: the subject as in Case 1, or the complement of object as in case 2, then on the action expressed by the verb case 3. response focuses on the item being interrogated.

Total interrogation:

The interrogation extends over the entire sentence and can only be answered with a monosyllabic affirmative affirmative "yes" or negative "no". Authorized literature often speaks of closed question.

Ewondo:	Yi	nyô	I	he	lob	Mebenga?
	Does	*The snake*	*pa*	*has*	*sting*	

English: "Did the snake has stung Mebunga?"

Exclamatory sentence

It usually has the same surface structure as the declarative sentence and very often it is the intensity of the speaker that distinguishes the two types of sentences. The exclamatory phrase is always stated in a high tone with phrases that emphasize exclamation as *akie*, *fouà* for insistence etc.

Ewondo*: /Akié!* nyò í hə lob *foua* Mendana/ "what! The snake really stung Mendana"
PEx The snake Pa P2 + sting Mendana!

Imperative sentence

It is used to give an order or to make a prayer.

Ewondo: /Woué í-nyô í he lob Mendana./ "Kill the snake that stung Mendana"

Negative sentence

It brings a nuance of negation to the fundamental sentence from which it derives.

Ewondo:	Nyô	í	**ndi**	**kigh**	lob	Abeng Melounou
	The snake	*pa*	*Negation*		*sting*	*Abeng Melounou*

English: "The snake didn't stung Abeng Melounou"

There are, however, nuances to these basic types of sentences that we have just reviewed. This leads to types derived from phrases that are called: emphatic, passive, etc. We will not approach them all here. This is not, in my opinion, the goal of this heuristic.

The negation is placed before the conjugated verb.

We have in ancient egyptian the D35 Arm in a gesture of negation:

Time	Negation	Translation
Aorist	 *n sḏm.n=f*	he does not hear / he can not hear
		A-wogh kigh/ (tèghè wogh nyè)
Past	 *n sḏm=f*	he did not hear
		A ndzi kigh wogh
Future	 *nn sḏm=f*	may he not hear / he will not hear
		A yi kigh wogh or A dàŋ kigh wogh

Note: The sign D35 reach out in demand, that's why we say / yi / "ask"
The signe N35 induced by metonymy the idea of crossing: /à-kigh/ or /à-dàŋ/ ☞/kigh/

Negation types according to kinds of verbs:

Types verbs	Example	Traduction
Strong Verbs	 *n sḏm=f*	A tèghè wogh
		he did not hear
Geminate Verbs (no repetition)	 *n mA=f*	A tèghè yén
		He did not see
Weak Verbs	 *n mr=f*	A tèghè diŋ
		He didn't like

Irregular Verbs	 *n rdi=f*	A-tèghè və *He didn't give*

Chart 105: Negation according to types of verbs

Caution: Gunn's Rule[148]

- The present negation of sDm = f is **n sDm.n = f** and not *n sḏm = f.*
- The past negation of sDm.n = f is **n sDm = f** and not *n sḏm.n = f.*

By mental projection the French Tesnière affirms below, one could have thought that all negation necessarily proceeds of an affirmation but we find that the African languages illustrated by the ancient Egyptian do not necessarily proceed this way.

> *"All negation proceeds from an affirmation. The denied thing is indeed expressed in the same way as the affirmed thing, with the only difference that it has in addition the tag of the negation. The study of negation therefore essentially involves the study of its tag"*[149]
>
> *(TESNIÈRE -1976)*

If we consider what happens in ancient Egyptian, as emphasizes Gunn we realize that the negative version of an utterance is far from being limited to the simple concatenation of the positive version + a marker of negation. This is all the more noticeable, with particle n, that we are witnessing a real upheaval of conjugated forms, as soon as we move from the affirmative form to the negative form.

Example:

Middle-Egyptian: 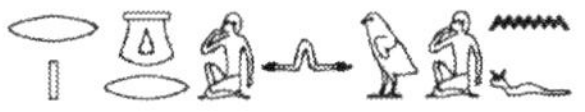 rA gr n mDw. ***n***=f
Ewondo: */A-nyù ne-yor tèghè fə kɔbɔ/*
English:The mouth silent, no more speaking.

Example:

Middle-Egyptian:  *iw s anx.n.i niwt.i* ***n*** *it.i sA.t s*
Ewondo: */Ma bùghbù ùn-nàm wàmà təghè vâ mi-'mfàŋ be ngo bwéé/*
English: I made my city live without removing ladies from wealthy families.

Comments:

In the first example, the negative sentence *n mdw.nf* is to be interpreted as an incompleteness: "it does not have to speak", whereas the phrase without negation, *mdw.nf,* normally has the value of an accomplished one: " she spoke ". The same phenomenon occurs in the second statement, except that everything is reversed: *n it.i* corresponds to the meaning of "I did not remove", whereas *it.i* means "I remove". This chaste-out effects of meaning, which negation triggers, is so systematic that Egyptologists have made it a rule, the rule of Gunn (1921), named after his discoverer. But african peoples are dealing with theses types of sentences everyday.

Which conclusions should we draw from these observations? It is clear, first of all, to the african point of view, negation and affirmation are not on the same level, structurally speaking. On this subject, Tesnière's sentence, "All negation proceeds from an affirmation. ", Summarizes the situation well: the negation is second compared to the affirmation.

148. Gunn, Battiscombe; Studies in Egyptian Syntax; 13 Rue Jacob VIe, Paris: Librairie Orientaliste Paul Geunther; 1924.
149. L. Tesnière, Structural Elements of Syntax, Klincksieck, Paris, 1976: 217.

Note:

We say we have to be very careful about this issue because what can be taken for a negative form can actually be a marker of insistence:

Negation

Ewondo: /**Te** yâ *ùn*-mbèh/ "don't open the door"

Insistent form

Ewondo: / yâghə **Te** *ùn*-mbèh/ "please open the door"

Insistent form negation

Ewondo: / Te yâ **Two** *ùn*-mbèh/ "The door not even opened"

THE INDICATIVE AUXILIARY 𓇋𓅱 IW

The middle-Egyptian auxiliary 𓇋𓅱 ***iw*** is the most frequent auxiliary in nonverbal proposal with adverbial predicate.
It allows to translate the proposal to the indicative (principal or independent).

In general an autonomous lexical element located in a particular context will gradually lose some independent word properties to acquire those of a morphological affix. The element described as "clitic" is thus in a way a step in this evolutionary process. This relatively broad definition can therefore be applied to a class of very heterogeneous elements.

A characteristic that is nonetheless shared by all clitics is their prosodic dependence. A clitic is indeed always prosodically linked to an adjacent word, called host (very often a verb), so that it forms with it only one complex phonetic unit. We speak more precisely:

"proclitic when the clitic relies on the word that follows it (this job is called a" proclise "): The egyptian 𓇋𓅱 ***iw*** is considered as a proclitic particle.

Case 1: NVP WITH ADVERBIAL PREDICATE AND PRONOMINAL SUBJECT

According to Ancient-Egyptian grammar, when the subject is a pronoun, the suffixed pronoun is used. In this case, it may be either a principal or independent, or a subordinate.

Middle-Egyptian: ***iw.i m Hzw.t Hm.f***
English: I was in his majesty's favors
Ewondo: *mə mbə à ngàm ùn-ntùm*

Case 2: NVP WITH ADVERBIAL PREDICATE AND NOMINAL SUBJECT

-With preposition and infinitive

Middle-Egyptian: ***iw sn Hr ir.t kA.t***
English: They do the work
Ewondo: *bà bɔh i-si*

-With prepositional complement

Middle-Egyptian:  *iw Xnw m sgr*
English: the residence was in silence
Ewondo: *i-hənâ i mbə à-mioŋ*

-With a pseudo-participle

Middle-Egyptian:  *iw.i xa.k(w).i m ntr*
English: I appeared as God
Ewondo: *mə ngà koui ùn-ntour*

-With an adverb

Middle-Egyptian:  *iw Nw.t im*
English: nout is inside
Ewondo: *ùn-nya à nə à ùn-nəm*

3.5. SOME IDIOMATIC EXPRESSIONS

Expression 1: "*Faulkner p. 62, Wb. I p. 320, 1-17 sq.*"

Middle-egyptian: ***Wnm ib*** " to lose consciousness, to forget "
Ewondo: /à-dibèn *ùn*-nəm/ "having had the heart ate"

Expression 2: "*Faulkner p. 14, Wb. I p. 59, 10-20 sq.*"

Middle-egyptian: ***ib tmw n sxA.n.f sf*** "the mind is exhausted and can not remember the day before"
Ewondo: / *ùn*-nlô ù dzàŋ yi nyè à fî / "he completely lost his mind"

Expression 3: "*Faulkner p. 14, Wb. I p. 59, 10-20 sq.*"

Middle-egyptian: ***in ib.i sxnt st.i*** "it was my understanding that advanced my position"
Ewondo: / *ùn*-ntsoghèn wàma ù tî à yi *ùn*-nəm wàmà/
English:"it was my understanding that advanced my position"

Expression 4: "Faulkner p. 104, Wb. II p. 43, 1-9."

Middle-egyptian: ***min*** "today"
Ewondo: / ànà mén/ "today"

Expression 5: "Faulkner p. 14, Wb. I p. 59, 10-20 sq."

Middle-egyptian: ***tsi ib*** "raise the heart; self-control"
Ewondo: / à-soui *ùn*-nəm à si / "Calm down" literally lower his heart

Expression 6: "*Faulkner p. 314, Wb. V p. 468, 3-19 sq.*"

Middle-egyptian: ***dns m iHwt*** "heavy of oxen"
Ewondo: / ndzalèn à bi-mvu(l)/ "full of oxen, filled with the weight of the oxen"

Expression 7: "*Faulkner p. 314, Wb. V p. 468, 3-19 sq.*"

Middle-egyptian: ***dns mhwt*** "cautious language"
Ewondo: / ùn-ŋkəghdə i-bukh / "cautious language"

Expression 8: "*Gardiner p. 486, N7 , Wb. III p. 391, 11-16*"

Middle-egyptian: ***m Xrt-hrw nt ra nb*** "the daily need"
Ewondo: / i-yaŋ'èn ya à-mos nti Zambà / "the holy daily need"

Expression 9: "*Faulkner p. 195, Wb. III p. 315, 1-17 sq.*"

Middle-egyptian: ***n xr-nsw*** "(favors, etc.) by the king (from the king)
Ewondo: / à-kàlèn *ùn*-ŋkukuma/ "by the king"

Expression 10: "*Faulkner p. 18, Wb. I p. 73, 20-25 sq. (imj.t-pr)*"

Middle-egyptian: ***imt-pr*** "will, donation"
Ewondo: / à-fola i-likh / "inheritance, concession"

Expression 11: "*Faulkner p. 129, Wb. II p. 232, 4-16*"

Middle-egyptian: ***nbt pr*** "mistress of house"
Ewondo: / *ùm*-inga ndâ / "mistress of house"

Expression 12: "*Faulkner p. 89, Wb. I p. 511, 7-16 sq.*"

Middle-egyptian: ***pr sn*** "at home"
Ewondo: / à-folà dzàŋ / "at home, mother's land"

Expression 13: "*Faulkner p. 89, Wb. I p. 511, 7-16 sq.*"

Middle-egyptian: ***imy-r pr*** "intendant"
Ewondo: / mor à-folà mə-kokh / "intendant; steward"

Expression 14:

Middle-egyptian: ***pr nHH*** "tomb"
Ewondo: / à-folà i-nyiŋ kwàm/ "place of eternity"

Expression 15: "*Faulkner p. 89, Wb. I p. 511, 7-16 sq.*"

Middle-egyptian: ***(i)m(y) prwy*** "official of both administrations"
Ewondo: / ùm-myèn mə nda mə-ngomna mə-bèh/ "official of both administrations"

Expression 16: "*Faulkner p. 86, Wb. I p. 488, 7*"

Middle-egyptian: ***bddw-kA*** "watermelon"
Ewondo: /*ùn*-kaŋ à-bwakh da vyé/ "creeping watermelon" litt. The red squash

<u>Expression 17:</u> "*Faulkner p. 74, Wb. I p. 396, 1-8*"

Middle-egyptian: ***wD-mdw*** "order"
Ewondo: / *ùn*-ntum i-dzoué / "cane of command"

<u>Expression 18:</u> "*Faulkner p. 202, Wb. III p. 368, 17 sq.*"

Middle-egyptian: ***m-Xnw*** "inside"
Ewondo: / à 'nəm / "in the heart"

<u>Expression 19:</u> "*Faulkner p. 56, Wb. I p. 270, 11+ , Wb. I p. 243, 9-11 (w)*"

Middle-egyptian: ***wy*** "how much!; how many"
Ewondo: / bâm/ "how many"

<u>Expression 20:</u> "*Faulkner p. 164, Wb. III p. 44, 8-10*"

Middle-egyptian: ***Hw*** "of its members"
Ewondo: / à wa w-òé/ "of his hand"

<u>Expression 21</u>: "*Faulkner p. 88, Wb. I p. 505, 2-8*"

Middle-egyptian: ***pw*** "or; or else"
Ewondo: /*ù*-fə/; / nguə nâ /; /*ùn*-do nâ/ "or; or else" the door represents the mother ngue

<u>Expression 22:</u> "*Faulkner p. 319, Wb. V p. 521, 4-6*"

Middle-egyptian: ***ra Das*** "ra him-self"
Ewondo: / ra, à tôs / "râ suddenly"

<u>Expression 23:</u> "*Faulkner p. 81, Wb. I p. 450, 8-13 sq.*"

Middle-egyptian: ***bw bin*** "evil that is bad"
Ewondo: / *ùn*-mbə à-beh /; / *ùn*-mbə i-bén /; "very bad, extremely bad"

<u>Expression 23-bis:</u> "*Faulkner p. 81, Wb. I p. 442, 15 16 sq.*"

Middle-egyptian: ***bw bin*** "The badevil"
Ewondo: / à-boh à-béh / ; / à-boh i-bén / "to hurt"

<u>Expression 24:</u> *G. Lefebvre P.254*

Middle-egyptian: ***imyw hat*** "ancestors"
Ewondo: / i-bah y'ù-sù/ ; / mi-'nto/ "the ancestors"

<u>Expression 25:</u> "*Faulkner p. 65, Wb. I p. 334, 1-3*"

Middle-egyptian: ***wr*** "tank"
Ewondo: / i-dùrù/ ; /i-wùlù/ "what is drived"

<u>Expression 26:</u> G. Lefebvre P.254

Middle-egyptian: ***dr*** "to finish"
Ewondo: / à tur / "full, finished, filled" ; / à-dir / "heavy, full, filled"

Expression 27: "*Wb. V p. 125, 12*"

Middle-egyptian: ***wsir kA m km-wr*** "Osiris, the taurus of Kemour"
Ewondo: / *ùn*-ntur i-kaŋ i-lom nənə ya daŋ à-tokh à-vokh / "great hirings for the greatest leader him who exceeds the other most important"

Expression 28: "*Wb. V p. 125, 12*"

Middle-egyptian: ***Hry-ib km-wr*** "The one how is at the center of Kemour"
Ewondo: / *ùn*-ŋkukuma ya mi-'nam mi-bèh wa dàŋ à-tokh à-vokh / "The great leader of the double country who surpasses the other insignificant leaders"

Expression 29: "*Wb. V p. 125, 12*"

Middle-egyptian: ***Hwt-tA-Hrt-ib*** "Athribis"
Ewondo: / à-fola à-kokh *ùn*-ŋkukuma nənə ya mi-'nam mi-bèh / "The city of the palace of the great leader of the double country"

Expression 30: "*Faulkner p. 53, Wb. I p. 253, 1-26 sq.*"

Middle-egyptian: ***nn iw.i r wAH.t*** "I will not wait for you"
Ewondo: / ma yi kigh wa yangha / "I will not wait for you"

Expression 31: "*Faulkner p. 53, Wb. I p. 253, 1-26 sq..*"

Middle-egyptian: ***wAH Hr Hr*** "turn the face at"
Ewondo: / à-wâ mys *ûn*-mfâ ya …/ "pay attention on "

Expression 32: "*Faulkner p. 53, Wb. I p. 253, 1-26 sq..*"

Middle-egyptian: ***wAH r tA*** "put down"
Ewondo: / à-wâ à tsè / "throw down"

Expression 33: "*Faulkner p. 53, Wb. I p. 253, 1-26 sq..*"

Middle-egyptian: ***wAH tp tA*** "last on earth"
Ewondo: / à-waê i-tobo si ndon / "You will endure a long time during your earthly life"

Expression 34: "*Faulkner p. 53, Wb. I p. 253, 1-26 sq..*"

Middle-egyptian: ***wAH pA HqA*** "that lasts this Sovereign (oath)"
Ewondo: / à-waê à-bogh i-dzou / "that lasts this sovereign"

Expression 35: "*Faulkner p. 53, Wb. I p. 253, 1-26 sq..*"

Middle-egyptian: ***wAH mr.t*** "a love that lasts"
Ewondo: / à-wu i-diŋ / "swim in the love" litt. to die of love

Expression 36: "*Faulkner p. 53, Wb. I p. 253, 1-26 sq..*"

Middle-egyptian: ***wAH rnpwt*** "to last for years (constructions, royalty, justice, etc.)"
Ewondo: / à-boui mi-'mbu / "for a long time"

Expression 37: "*Faulkner p. 79, Wb. I p. 429, 6-19 sq*"

Middle-egyptian: ***bAk n pr Dt*** "servant of the state, lord-man"
Ewondo: / *ùm*-bokh ya dyàl bə-wu / "lord-man"

Expression 38: "*Gardiner p. 132, §178 , Wb. I p. 174, 5-9 (ab)*"

Middle-egyptian: ***m-ab*** "in the company of, in the middle of, among, with"
Ewondo: / à- bùm / "in the middle of, among, with"

Expression 39: "*Wb. I p. 37, 32-35*"

Middle-egyptian: ***ii-m-Htp*** "Imhotep"
Ewondo: / à mvouàè/ "In peace"

Expression 40: "*Gardiner p. 537, Z5 , Wb. I p. 42,?*"

Middle-egyptian: ***iaH-msw*** "Ahmôsis, Ahmès"
Ewondo: / Ngamessi / "The anointed's wife"

Expression 41: "*Faulkner p. 286, Wb. V p. 128, 3-17 sq.*"

Middle-egyptian: ***m km n wnwt*** "in one hour"
Ewondo: / à 'ŋgouma à-wolà/ "in one hour"

Expression 42: "*Faulkner p. 49, Wb. I p. 228, 8-26 sq.*"

Middle-egyptian: ***Hm-nTr aSA*** "ordinary prophet"
Ewondo: / ùn-ŋkɔm tara ù-sə/ "like every servant of the ancestor"

Expression 43: "*Faulkner p. 49, Wb. I p. 228, 8-26 sq.*"

Middle-egyptian: ***aSA wrt m*** "a very large amount of"
Ewondo: / à-bim ngul ùn-nsàŋ / "a very large amount of"

Expression 44: "*Faulkner p. 14, Wb. I p. 59, 10-20 sq*"

Middle-egyptian: ***nty ib xft.s*** "what is intelligible"
Ewondo: / i-nə fəugh i-bn / "what is intelligible"

Expression 45: "*Faulkner p. 14, Wb. I p. 59, 10-20 sq*"

Middle-egyptian: ***n ib n*** "for the reasons of"
Ewondo: / à sù yi nâ / "because"

: /*ùn*-yièn/ "desire"☞ /yi/
: /*ù*-soué/ "river"☞/à-sù/
: /*na/ "mother"☞

Expression 46: "*Faulkner p. 14, Wb. I p. 59, 10-20 sq*"

Middle-egyptian: ***(r)di m ib*** "to be determined to, to be decided to"
Ewondo: / á-furi *ùn*-yièn à *ùn*-nəm / "to be determined to, to be decided to"

: /á-tələ/; /á-furi/ "to place, to put"☞ /yi/
:/*ùn*-nəmə/"evil" ☞synonym of : /*ùn*-yièn/ "desire"
:/*ùn*-nəm/"heart"

Expression 47: "Faulkner p. 14, Wb. I p. 59, 10-20 sq"

Middle-egyptian: **rdi ib m-sA** "be anxious about"
Ewondo: / à-furi *ùn*-ŋgekwàb'là à *ùn*-nəm / "be anxious about"

Expression 48: verbal phrase explained

Middle-egyptian: **iw mdw.f n sS** "He is talking to the scribe"
Ewondo: / à-yi kɔbɔ é *ùn*-ŋkàŋ'à / "He is talking to the scribe"

:

Note:
Let us note immediately that, in this type of sentence, the Egyptian language generally uses what some call an utterance indicator whereas others speak rather of an old verbal form having the meaning of what happens.
In other words, especially in the context of narrative text, the particle *iw* introduces a sentence that describes a situation or a state of affairs.

Middle-Egyptian: *mDw* "speech; words"
Lingàla: lobôko "hand; arm; "
Nzimé: ekœ̀h "palm of the hand; "
Ewondo: /i-bokh/; /obə/ "baby; small; kid "
Ewondo: /à-kul/ "palm of the hand; "☞/à-kɔlɔ/ "to speak"

By phonemes inversion of *boko* you have **kobo* "speech".
We can also deal with the fact that /mə-dzô/ "words" are speech /*ùn*-ŋkɔbɔ/

Middle-Egyptian: f "him; he"
Ewondo: nyo "snake; " ☞ **Ewondo:** nyè "him; he; whoever"

And finally
Middle-Egyptian: *sS* "scribe"
Middle-Egyptian: *sS* "scribe"

Lets explain each sign:

Ewondo: /mə-sɔŋ/"hindering; " ☞**Ewondo:** *ùn*-nsɔn"pen; sharp"
Ewondo: /*ù*-swé/"pool; " ☞ **Ewondo:** à-sè"pen; feather"
Ewondo: /à-fəb/ "sheet, " ☞ **Ewondo:** à-sè"pen; feather; wing"

The word /à-fəb/ "sheet" is homophon of /à-fab/ "wing". And a wing is linked to the feather by metonymy.
So the good reading of this script in Ewondo language is:

Ewondo: /à-sè/"pen; feather"
Ewondo: /à-fəb/"scholars"

Of course as i have emphasized all along this document, this item involved synonyms, metonyms, homophons, antonyms, paronyms etc.

Expression 48: verbal phrase explained

Middle-egyptian: **iw rdi sS mDAt n st m pr** "He gives a sheet to the scribe in a sister's house"

Ewondo: / à-yi və à-sà à-fəb à ndâ sità / "He gives a sheet to the scribe in a woman's house"

Expression 49:

Middle-egyptian: **mAa-xrw** "to be justified, to be triumphant; triumph"

Ewondo: / à-mé *ùn*-ŋkàr / "after beeing accused, admit the object of the accusation"

Middle-egyptian: ***xrw.f mAaw*** "he is justified"

Ewondo: / *ùn*-ŋkàr à á-mé wo / "after beeing accused, he has admitted the object of the accusation"
Note:
Ewondo: m can also be replaced by ***Ewondo***: yem "recognize, admit etc." both words are ananyms.

3.6. APPLICATIONS IN EWONDO

3.6.1. Psychological Blocking

Before tackling *mDw nTr* it's important to understand what exercise is all about.

I summarize the exercise. It consists in taking, according to the classical method, a skeleton of consonants and, from this skeleton of consonants, determining the vocalization of the starting word.
Outside of any built-up framework, we will illustrate the purpose with this simple exercise. We will take a skeleton of consonants. Let's take the consonants skeleton ***mr***.
From this skeleton of consonants, what are the words that we can generate naturally?

> **Answer**: we can have the following words:
> *Mary, mere, mur, mire, mar, rum, arm, uremia, rim, ram etc.*

What do we observe? We simply notice that we get a string of words that are vocalized differently even if these words share the same consonants skeleton.

It is the same thing with the ***mDw nTr***. A skeleton of consonants does not necessarily refer to a single signifier or a vocalization unique in our African languages. Everything will depend on the determinative. Those who are looking for a unique vocalization for a given consonant skeleton are mistakenly wrong about how to proceed.

We suggest that even in european languages, if two words are homophones that means in african languages they are sharing the same signified.

If Egyptology offers some comfort with its transliteration approach, this approach is in our opinion completely limited and limiting. But I agree that it may be suitable for someone who is not used to ***mDw nTr***. In any case I do not recommend it partially if you want to embark on an endogenous study.

The confusion induced by the alphabetic system

There are several types of scripts that are obvious to read more than others. We will see him later. But the difficulty of reading is often due to psychological blockages induced by the Western alphabetic system. As far as we want to transliterate and we absolutely want to read the mDw nTr as we read in French or in English by associating letters.

Let's take concrete examples:

Example of script to read:  (ox)

In this case, 99.99% of Africans without any doubt will recognize the sign and will name it in their language without any worry.

Example of script to read:  (ox)

Here begins the complication because the left added characters, associated with the determinative will create a psychological blocking for a good reading. The inexperienced reader will unconsciously try to apply the alphabetic method. And there, we will lose a significant amount of Africans since they do not know how to interpret the signs prior to the determinative. They will throw in the towel and we will not be able to go further. Yet in the previous example, they would have recognized and therefore read the script.

Example of script to read: ox + transliteration (***ngAw***) 

Here two phenomena occur: either the subject finds a phonetic proximity offered by the transliteration (*ngAw*) and then reconnects with the determinative  or the subject does not find phonetic proximity in his language.
In this account, the person does not know more what this script means .

I want to bring this little precision. I could read beef, cow, bull, calf, etc. in my language. Does this mean that I am wrong? Semantically I will be right because I indexed by this script a bovid that's what the scribe wants in the first place.
However, if I want to know precisely what the scribe has recorded then indisputably I have with other more advanced techniques to assure me of the good reading is what we will see later.

Example 1:

Here is a transliterated script msH or mzH with (msH <mzH). I could read this ngan script in Ewondo. Semantically once again this reading is correct but morpho-syntactically this reading is not quite accurate. Because if I compare the two skeletons of consonants: n.gn (ngan) and msH, I can see that the two skeletons are not identical. On one side we have: n.g-n and on the other we have m-s-H.
So how do I make sure that I read what was written down by the scribe? Indisputably here I must have a good knowledge of my language or a good dictionary that allows me to find the right words. This is an approach. We will come back in depth on the methodology.

What are the words to designate the crocodile in the Fang / Beti languages?

> NZIÑDI:, pl. bensiñdi. Middle Cayman, crocodile, alligator, smaller than ñgan, more great than ñkôm. It is also called ñkôm nsiñdi. Ñgan is the generic name, but ñgan is also the caiman who is three meters or more. Nsiñdi is two meters tall. Ñkôm is 1 m away. 50. [150]
>
> **CAYMAN:** crocodile, alligator, ñgan (m) (generic name), three meters or more. Nsiñdi (h), two meters. Ñkôm (b), one meter fifty. This makes three different varieties.

We note that there is no word that is similar to the skeleton of consonants msH or mzH.

Before we throw in the towel, let's question the oral legend of the Beti people. In Cameroon, Betis are mainly located in the central and southern regions of the country..

> *"Legend has it that the Beti come from the other side of the Sanaga River, which they crossed at the end of the eighteenth century on the back of a snake called Ngan-medza. They were fleeing Ousman dan Fodio, a Muslim leader from the Adamaoua region who had decided to Islamize all the animist peoples. Some of the Beti are mounted on the back of the Ngan-medza. This crossing was done at dusk and the person who held the bamboo torch to light stood at the tail of the group. As she climbed last, the flame was being lost. She involuntarily dropped some sparks on the back of the snake Ngan-medza which, burned, let them flow. As a result, some Beti have not passed the other side of the Sanaga and that is why we still find around the Mbam region. The cities of Yaound, Mbalmayo, Obala and Ebolowa remain traditionally their stronghold, given the almost ethnic distribution of cities..."*

This highly edifying story allows to exhume deepness of our consciousness a remnant of the Beti language. Why did *Ngan-medza* became in the collective imagination a python or even a boa?
This confusion is unfortunate because the python / boa is called [Mvom] and not [Ngan], the word / Ngan / is the generic term to designate the saurians (crocodiles, alligator, caiman). Here the storytellers who kept the name used in Egypt inform us that they crossed over the back of the saurian, medza. I will come back later to this notion, which is very important.
We have in all respects the word written by the scribe *msH* or *mzH*. This script would thus vocalize medzah in Ewondo and Medzaghe in Fang language.

Other example: 

Let's compare the previous script and this one. By modifying the determinative, the scribe keeps the same phonetics complements, (the owl G17, the folded loincloth S29 and the string V28) and informs us this time that the sign F27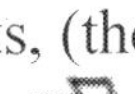 of the Gardiner's list equals sign I3  identified by a saurian.

150.Samuel GALLEY, Fang-French dictionary. and French Fang followed by a fang grammar, Editions Henri Messeiller, Neuchâtel, 1964, P.268§431.

This supposes that these two scripts are interchangeable at the same time because both refer to the same reality. Let's try to understand how and why.

msH< mzH (skin)

- In order that the signs I3  and the sign F27 be interchangeable since visually different, either:

The words used in our languages seem to be homonyms [151]or they are homophones[152] (i.e. Words that are pronounced in the same way).

Since the figure on the left above has been elucidated to know that it is a specific saurian, let's move on to the figure on the right and question the language fang / Beti.
She then postulates that / mezagha /, designates in Fang / beti cassava leaves.

> MEZAGHA (bm) n.l. ss pl. 1. Cassava leaves (edible). Mesagha Mboe. Tsakh mesagha, pound cassava leaves. Nson mesagha, green color, green. Mesagha mvi, mesagha mbo, mbo mesagha, cassava leaves cooked without salt. My si mesagha wam, I eat my cassava leaves. Mesagha aña neghba, cassava grows leaves. Syn.: meya-gha (bm) (Atsi). See nyamesagha.
> -
> 2. Name of a non-edible plant found near streams, replacing a little soap (mesagha elokh*)*. [153]

Since we are talking here about leaves, why do we put a skin determinative? We also see here that the leaves of cassava can be likened to pieces of skins on the ideoscriptic level speaking (it falls well the signs are ideoscriptic).
We therefore conclude that interchangeability is related to homophony here.

But in fact we could have absolved ourselves from the difficulty of comprehension which assimilates skin and crocodile of these two scripts if we did not know the word / ngan medza / or / ngan mə-sè / which we lost the use in the languages of the Congolese basin today?

It was possible to solve this slight difficulty by simply naming the determinatives without worrying about phonetic complements.

Indeed, the crocodile can be called [ñ.kôm] whose skeleton of consonants would be (km) and the skin /i.kob/ whose skeleton of consonants would be (kb) in fang / Beti. At the phonetic level we have kb <=> km. More than that on the ontological level, those who wore and wear some skins in traditional African societies are initiates. The consecrated term is: *Komo* in mali, *Mbobokh* (the paralytic) in the Basa'a land, *Kam* in the Bamileke land etc.

151.F. DEGARDIN, Homonyms and homoscripts of the French language, 1857, p. 5
Homonymous ie words that have a single pronunciation and several spellings. [...]
Homoscripts, on the other hand, are words that have several pronunciations and a single spelling.
152. Homophones are homonyms with the same pronunciation (for example, the phonic sequence [so] corresponds to four homophones: stup, jump, seal, bucket)
153.Samuel GALLEY, Fang-French dictionary. and French Fang followed by a fang grammar, Editions Henri Messeiller, Neuchâtel, 1964, P.*433.*

To the question of whether this reasoning could have worked with other African languages. Let's take examples of words in some languages spoken today in Africa

- **Nzimé**: *konó* "skin"
- **Yemba**: *akup* "skin"
- **Rikpa**: *koko* "skin"
- **Day**: *kog* "skin"
- **Dongo**: *koko* "skin"
- **Fa**: *koko* "skin"
- **Abar**: *igobo* "skin"
- **Baneka**: *kow* "skin"
- **Akoose**: *ngob* "skin
- **Gbi**: *oko* "skin"
- **Bamenyam** :*ngôb* "skin"
- **Gciriku**: *kôba* "skin"
- **Maviha**:*gwohwe* "skin"
- **Ntuzu**: *koba* "skin"
- **Sonye**: *ekoba* "skin"
- **Ewondo**: *ikob* "skin"
-**Zulu**: *isikhumba* "skin"
- **Tene Tini**: *koko* "skin"
- **Tene Kan Dogon**: *koko* "skin"
- **Eton:** *ekob* "skin"
- **Effé**: *gap* "skin"
- **Gbaya**: *koko* "skin"
- **Kande**: *xoko* "skin"
- **Bakaka:** *kob* "skin"
- **Babongo**: *koko* "skin"
- **Néninénongo**: *kob* "skin"
- **Bas'a**: *koo* "skin"
- **Duala**: *eyobo* "skin
- **Saamia**: *hoba* "skin"
- **Kirundi**: *koba* "skin"
- **Tsambaan**: *ekob* "skin"
- **Bullorn**: *kok* "skin"
- **Ge-Vové**: *gobo* "skin"
- **Bubi**: *gobo* "skin"

This method relies on character recognition and vocalization in our languages. But this is a bit more complex than that as we will see below.

PROBLEMS RELATED TO ENDOGENEOUS READING

First problem

Technically, the first problem we encounter is the proper indexing of the elements that make up the script. This classification makes it possible to precisely determine the role of each element in the script.:

- Radical
- Semantemes
- Gender Prefix
- Number Prefix
- Un-mbias (nisbé) "dual"

Example:

Middle-egyptian:  *dwAtyw* "inhabitants of the Lower World (Douat), dead"

- The sign Z3 indicates here a plural tag. This is the number prefix.
- The sign  A40 indicates here the primal ancestor.☞leading to the title /*ùn*-nti/"lord"
- The sign  X1 indicates a word of the feminine gender.

- The O1  indicates here the determinative namely: the main idea that the scribe wants to express

- The signs  together constitute the radical and individually homophones of the idea recorded by the scribe through the sign O1.

This problem is relatively simple because it is necessary to locate each element so as not to waste time in useless analyzes.

The second problem

It concerns the good vocalization of a script. If we choose to no longer transliterate because the ancestors did not use an alphabet to transliterate. If we choose not to reconstruct a pre-dialectal ancestor, we must associate the signs and read.

If we do without diachronic linguistics techniques because to read these characters we did not mobilize these techniques, it goes without saying that we take the opposite of the official logic. On this account, it is necessary to propose an alternative which guarantees the good reading.

How in the example above, the association of characters leads to good reading? How are we certain to read what is written without being deceived?

This problem encounters a HUGE difficulty which is the DIKTAT of the current method which IMPOSE to transliterate.

This problem comes up against another difficulty, which is the knowledge of the morphology of African languages.

Some suffix the radical, others prefix the radical. It is therefore a question of organizing the reading according to the standard codes of the language of the speaker. This method is therefore indicated to a person who has a practice of his language

How to start reading? Is there a sense of reading? How are we certain that what we vocalized is accurate?

It is therefore all these difficulties that the Endogenous method must go beyond to place reading in the heart of Africa and no longer under the guise of exogenous interpretations and ideology.

Example:
Middle-egyptian:  ***dwAtyw*** "inhabitants of the Lower World (Douat), dead"

I took this example because it illustrates what is problematic with the Standard method. Are the stars part of the sky or the lower world?

How can one speak of the lower world here while the scribe represents a star?

So character analysis to understand the logic behind the script is crucial here.

We are looking for a place 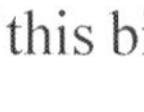 whose best representation is associated with the star 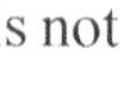.
So we see that this is heaven because the stars are not underground.

Where is the place 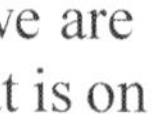 of this bird 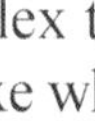? It is still difficult to postulate that his place is not underground?

Nevertheless, as we are in a complex thought, we must be careful. Complex thinking invites us to remember that : what is on earth is like what is in sky. This idea is not a fad. It emanates from the principle of correspondence which implies the idea that, there is "a constant relation between the laws and the phenomena of the various planes of being and life. "What is above is like that which is below; what is below is like what is above. This principle establishes that there is a harmony, an understanding and a correspondence between these different plans of existence that are available at the levels: physical, mental and spiritual.

According to the principle of polarity "everything is double", "everything has two poles", "everything has two extremes". "The thesis and the antithesis are identical in nature, but different in degrees". Any paradox can be reconciled because what seems opposite is in fact only another part of the same scale. Hot and cold do not have an absolute definition and exist only in relation to each other. The cold can burn like the hot. In the same way hate has often been seen to change into love or vice versa. The extremes touch and merge.

So what appears here as falling from the top can quite be down below under the logic of complex thinking.

 ***dwAtyw*** "inhabitants of the Lower World (Douat)" **English** : dead

If we consider the only the root of the transliteration, ***dwA.t-*** the word appear obvious in English *dead*. The work of Howard Carter has to be inquiry. So we see that, this reading is minimalist just by a context analysis.

Therefore, we have here not only letters but operational concepts with non-zero signifiers, we can mobilize not a reading but an arsenal of reading because we can use synonyms, antonyms and homophones signs.
Without taking an endogenous language but only meanings we can speak about:

- From the grave,
- Of heaven
- Canvas
- Last home
- From the land of the dead
- The country of ghosts

Etc.

At any time one takes liberty with any meaning or with the constraints imposed by reading. This is why we do not vocalise a word but we express ideas according to the field. That's why we're talking about cosmogony, cosmology, cosmography or cosmocracy.

And depending on the domains this script can quite give a different reading. It is necessary to adapt to this account the reading pattern that suits his best understanding.

Basic concepts

In order to determine the basic concepts, I will illustrate this remark surprisingly by quoting the Gospel of John:

"At the beginning was the verb. The verb was with God and the verb was God."

(John 1:1)

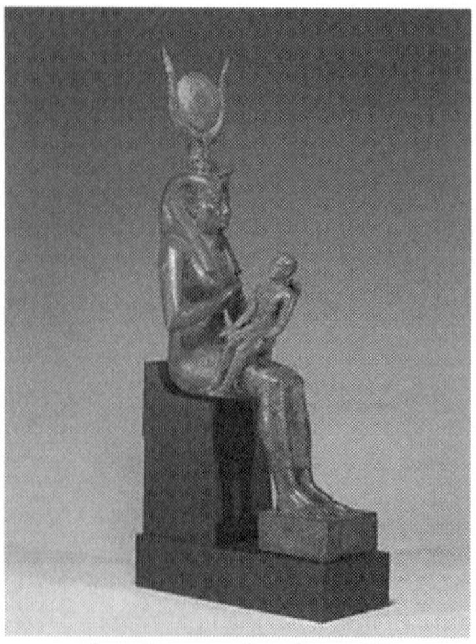

As.t breastfeeding Hr

If this sculpture seems innocuous at first sight, it is rich in lessons for the one who measures the range. We have a mother carrying her son on her thighs and breastfeeding him. This is hardly seen by the layman.
Why is this sculpture important in reading ***mDw nTr*** according to the African method?

I suggest to answer to that question by focussing only on the child who is breastfed by his mum. In this account, we list the descriptors related to the child. So, the child maybe:

Ewondo: obə "little" because he is the small one
Ewondo: à-bèh "breat" because he is the breastfed
Ewondo: á-bɛh "thigh" because he is the on the thighs
Ewondo: i-bèh "thirst" because he is the Thirstiest
Ewondo: *un*-mbokh "bedridden" because it is the one who is lying down

The roots / * beh / and / * bokh / are synonymous because both refer to the baby.
Let us take as an example, the meaning Ewondo: / *ùm*-bokh / "one" which also designates the bedridden

Let's list some homophones Ewondo that are associated with this term without going into the technical requirements of the phonetic.

Series with prefix /i-/

Ewondo: *cottage* / i-bokh /
Ewondo: *clumsy* / i-bokh /
Ewondo: paralytic / i-bokh /
Ewondo: *track, no* / i-bokh /
Ewondo: *baby* / i-bobokh /
Ewondo: *place* / i-bokh /
Ewondo: small village / i-bokh /
Ewondo: tobacco attachment / i-bokh /
Ewondo: basin / i-bokh /
Ewondo: Hip / i-bokh /

Series with prefix / u- /

Ewondo: paralytic / *un*-mbokh /
Ewondo: spirituality / *un*-mbokh /
Ewondo: prisoner / *un*-mbokh /
Ewondo: yoke / *un*-mbokh /
Ewondo: persevering / *un*-mbokh /
Ewondo: lying, disabled / *un*-mbokh /
Ewondo: small / *un*-mbokh /
Ewondo: resistant / *un*-mbokh /
Ewondo: spiral / *un*-mbokh /
Ewondo: black eyeshadow / *un*-mbokh /
Ewondo: pen, feather / *un*-mbokh /

Series with prefix /a-/

Ewondo: time	/ à-bokh /	Ewondo: dance	/ à bokh /
Ewondo: tsetse fly	/ à-bokh /	Ewondo: twist	/ á bokh /
Ewondo: notch	/ à bokh /	Ewondo: braid	/ à bokh /
Ewondo: rat	/ à bokh /	Ewondo: screw	/ á bokh /
Ewondo: to ride	/ à bokh /	Ewondo: right angle	/à bokh /
Ewondo: squash	/ à-bokh /	Ewondo: to dig	/ á bokh /
Ewondo: separate	/ à bokh /	Ewondo: to level	/ á bokh /

After this series of homophones only related to the root / * bokh /, list this time a small series of synonyms:

*yo	*obe
*ko	*soghe
*bo	*nson
*dzô	*ra
*wa	*foŋ
*lo	*yen
*man	*ngon
*nyo	*kouŋ
*kəŋ	*sè

As it was realized above, each synonym can admit 3 different prefixes / a- /; / i- /; / w- /.
With one word, variations of tones (5) (*allophones, homophones*), sometimes even clicks. Don't forget that each word can find his counterpart in the two other branches. I remind you them: kekwé – kwéké and kikuki.
We are able to generate important series of words. What matters to the African is an idea and not the word. In reality, therefore, homonyms in Africa are not so by a fortuitous phenomenon. If two words have a phonetic proximity they are necessarily related to the same signified.

If mankind was born in Africa as is commonly admitted, what appears to others as mere homonyms, is not a fortuitous phenomenon in Africa. If man has acquired language in Africa, we can indisputably find a common signified between two signifiers sharing a mere phonetic proximity.

Example:

French: "Un Ver Vert va Vers un Verre Vert".

All these homonyms \ vɛʁ \ are neither more nor less than the attributes of  ***wsir***. It is therefore logical that naming one and the same reality one finds curiously to vocalize the same sound. This singularity is erased when we replace some homophones with synonyms.
Africa concentrating many tones languages, this phenomenon is an epiphenomenon from which nobody is no more enthused.

In fact, to read a character or understand its meaning, it is necessary to leave the alphabetical matrix to concentrate on the idea recorded by the scribe. As can be seen, a single word generates an almost infinite series of words insofar as it also admits synonyms and each synonym admits as many homophones, synonyms and prefix variants. The apparent difficulty is therefore to choose the right one. The difficulty is indeed apparent because by synonymy we always have the right word in the multitude of possible choices

Let's take a simple example:

The scribe informs that the word for the ibex here represented on the right is identical in all respects to the hidden meaning conveyed by the sickle in our languages or the feather.

In this account:

- Either the word is linked to an action realised by the sickle.
- Either the word is a qualifier of the sickle
- Either the word is a synonym of the sickle.
- Either the word is an homophone of the sickle

How do we name a sickle in Ewondo?

Ewondo: i-bâgha "sickle"

The sickle is used to cut, the feather is used to separate, to share

Ewondo: à-kab "to cut, saparate, share"

Ewondo: *ùn*-ŋkaba "what is used to share" so here the sickle or the feather.

<u>**Note**</u>: /bâghe/ and /kaba / proceed by inversion of syllables. This is normal in African languages. Once we understand that we can read the script without any problem.

Middle-egyptian: ***ImAt***. "ibex, mouflon, chamois"
Ewondo: /kabat/ "goat"

The roles of the sign and X1 have not been engaged in the reading for the moment but we will see later that they are not simple affixes. What really matters warmup is the word radical.

Etymology: we can suggest the etymological link below with Ewondo:

Ewondo: /i-bâghà/; /i-bâgh/; /i-kabghà/; /i-kâba/ "cutter; feller"
French: ibex; bouc "goat"
English: ibex, book; beak; break; kid; goat; god; cut etc.

Let's take another example: 

imA.t *"ibex"* | **imA.t** *"charm, benevolence, grace; favor"*

By comparing these two scripts, we can already see that the determinatives are different. The script on the left shows an ibex in determinative  when the right script presents the sign Y1:  which is a rolled papyrus. The wrapped papyrus, / a-ndəkh / ewondo serves to record the ideas.

Question: So what is the idea representing for the scribe the best way to expresss the script on the right?

The scribe informs that the term to designate this sign Y1:  here represented on the right is identical in all respects to the hidden sense of the sickle  or  the egyptian vulture in our language. That's why in english this sign Y1 is named *book*. This english word *book* is an ananym of the Ewondo:/*ùn*-ŋkɔbɔ/ "speech". A book by metonymy can be considered as mute speeches.

At this count:

- Either the word is linked to an action realised by the sickle .
- Either the word is a qualifier of the sickle
- Either the word is a synonym of the sickle.
- Either the word is an homophone of the sickle

Ditto for the vulture  and the vulture comes in the vast lexicon field of the sickle engaged here.

What is the name of the vulture or the sickle in Ewondo?

Ewondo: /i-kâbà/ "sickle; cutter"; /i-bâgha/ "sickle"
Ewondo: /i-kuba/ "vulture; "

Let us note here that the two words /i-bâgha/ and /i-kâbà/ proceed by phonemes inversion.

The sickle is used for cutting. Surprisingly the French uses the root / kup- / linked to the word Ewondo: /i-kuba/ "vulture" and thus to the sickle.

Ewondo: /à-kab/ "cut, share"
Ewondo: /i-kuba/ "vulture"
Ewondo: /*ùn*-ŋkaba/ "used for sharing"
French: couper "cut"; coupe "cup"
English: cup

The good reading that this script offers is:

Middle-egyptian:  ***imA.t*** "charm, benevolence, grace; favor"
Ewondo: /à-kab/ "generous"
French: beaucoup "alot, too much"
English: piece

We could use all the synonyms related to the word / kaḅ / without offending the scribe.

Note: the sign  M1 of the Gardiner's list is used to designate the tree or idea associated with the tree.

Ewondo: / iléh/ "tree"
Ewondo: /à bakh/ "cut the tree"

We still fall on our paws.

Let's try some expressions that will exploit the register of synonymy.

> **Middle-egyptian**: ***nb imAt***"possessor of charm"
> **Ewondo**: /*ùn*-mbələ mâgha/ "possessor of luck, of grace"
>
> **Middle-egyptian**: ***wrt imAt*** "gracest", epithet of the queen.
> **Ewondo**: /*ùn*-ntour mvam/ "benevolence accomplished"
> **Ewondo**: /*ùn*-ntour mâgha / "gracest"

3.6.2. GOOD INTERPRETATION OF COMPLEX SCRIPTS

I could devote myself to the in-depth reading of many scripts in Ewondo but I have chosen a few here to illustrate the method and an Endogenous approach.

3.5.2.1. Read The Script ***Km.T (Faulkner p. 286, Wb. V p. 126, 7-18 sq.)***

Historical Review

At the opening of the symposium held in Cairo from 28 January to 3 February 1974, a symposium organized by UNESCO as part of the Drafting of the General History of Africa entitled: "*The settlement of ancient Egypt and the decipherment of the Meroitic scripture* ", the great African Egyptologist Cheikh Anta Diop suggeted an acceptance of the ***km.t*** script, assuming that the sign I6 classified by Gardiner as biliteral ***km*** referred to the semantics of the black color .

Anta Diop relied on comparative elements between Wolof and Pharaonic. He left the postulate that in Wolof the word "***charcoal***" would be like "***hem***"[154]. From the phonetic proximity and the diachronic linguistic techniques, he postulates that the ***km*** consonants skeleton translated the idea of blackness and consequently, the script "***km.t***" followed by sign O49 "***niw.***t" would be the land of the black nation, the blacks land, the blacks earth.

Curiously the first controversy came from the African side of which he was nevertheless the leader. The Sudanese Egyptologist Abdelgadir Abdalla[155], one of the 20 specialists whom Unesco had invited to the colloquium, is going to strongly criticize Diop's notion of "***km*** = black", not hesitating to compromise the smooth running of the symposium. There will be a heated exchanges between the two men and it is only after the arbitration of the French grammarian Serge Sauneron that the conference will continue with a hinted validation of Diop's position: ***km*** = black

Had Anta Diop been right to postulate that ***km*** = Black? Was there a connivance between Diop and Serge Sauneron? Doubt still persists to this day among the most advanced in research including in the diopian camp[156]. This is why this script was chosen for illustrative purposes, among other reasons.

The reading methodology used here remains the same as the one previously presented:

- Use the above method or use examples.
- Make sure the reading is correct by checking with proposed technical tools.

We find in a hymn[157] to glory of ***sn-wsr.t III***, the consonants skeleton "***km.t***" accompanied by the determinative of human collectivities.

154. https://www.youtube.com/watch?v=xLf-kiFz1sk
155. Unesco, The settlement of ancient Egypt and the decipherment of Meroitic writing,Unesco, p.1.
156. Jean-Claude MBOLI, Origin of African languages - Test of application of the comparative method to ancient and modern African languages, l'Harmattan, Paris, P.173.
157. Mark COLLIER et Stephen QUIRKE, The UCL Lahun Papyri, I-III, Oxford, Archeopress, 2002, 2004 et 2006

According to the Thesaurus Linguae Aegyptiae[158],, it is an "exceptional" variant of the word ***Km.t*** determined as an inhabited place, of which it designates the inhabitants: a collective that Raymond Faulkner, in his *Concise Dictionary of Middle Egyptian*[159], translated by "Egyptians"

However, apart from this hapax, the Egyptians call themselves ***rmT*** (XVIIIe dynastie), ***rmT n pA tA***, (XXe dynastie), ***rmT n kmt*** "the humans of Egypt" (Middle Kingdom) and ***km.tjw***, "Those of Egypt" (Ptolemaic era)

If the Egyptians were ***km.tjw*** "Kamites / Kemites" then why would they be called ***rmT n km.t***[160]**.** This is already another reason for choosing this script for its in-depth analysis. Indeed, in which region of the continent is there an ethnonym vocalising kamites or kemites?

Moreover, this script gives precise indications on the name of the country and the inhabitants. Through its study, it reveals a large part of the history of the ancient inhabitants of the land named Egypt and therefore a large part of African history.

This question remains relevant because many young Africans seeking identity use words from Esperanto-Egyptology[161] to designate themselves (kamites, kemites).

Let's look at the scripts transliterated km.t:

km.t *Km.tyw* *km.tw* *km.t* *km.t* *km.t* *km.t*

Few km.t scripts

There are many scripts transliterated ***km.t*** but we will not analyze them all precisely here.

Here is above the script that is controversial. Indeed, nowadays, even the most advanced ones in Egyptology research among Africans continue to ask the question about the real semantics of this script which was "*Egyptologically*" attached to the black color and giving to the global reading of the script the "*land of the blacks*".

Jean-Claude Mboli, who is a rigorous researcher, also criticizes the meaning given by Diop in his book on page 173 and then on page 511 where he says I quote him:

> *"But it is the root * kəʀʷ < * (ku-ki) -ŋʷ that is most interesting to analyze. It should normally end in M-E at something like ** **km** *'head'> 'husband'> 'man'. Is there such a word in M-E and among the other Negro-egyptian languages? The answer is most likely yes if we want to translate the word ME. km.t by "men" and not by "Egyptians" as Egyptologists generally do, nor by "blacks" as did C. A. Diop"*
>
> *(J-C MBOLI, 2010)*[162]

158.Zettelarchiv du Thesaurus Linguae Aegyptiae sous l'entre km.t: "nur 1 mal im mR, als besondere Schreibung für km:t*O49, wo dieses für seine Bewohner steht "(Unique occurrence in the Middle Kingdom, as an exceptional script of km.t * O49 which designates the inhabitants); see also Francis Llewellyn Griffith, Hieratic Papyri from Kahun and Gurob, London, 1898: Plate II, p. 2

159. Raymond Oliver FAULKNER, p. 286; Rainer Hannig, p. 883

160.Yaacov SHAVIT, History in Black: African-Americans in Search of an Ancient Past, Routledge, New-york, p148

161. Pr Mubabinge BILOLO calls Esperanto-Egyptological the barbarisms used by Egyptologists when they say they read the Pharaonic by using their own conventions

162. Jean-Claude MBOLI, Origin of African languages: Essay of application of the comparative method to ancient and modern African languages, Paris, 2010, p.511.

The Classical reading method stipulates in the specific case that the biliteral ***km*** allows the reading of the script while the sign G17 the owl specifies the read inclination of the sign I6 for a global reading of the script.

<u>What is the meaning of the sign I6 biliteral "km"</u>?

De facto, the problem that arises is to know what is actually designated biliteral often noted ***km*** and vocalized kem in Esperanto-Egyptological:

- A piece of charcoal burned as the German school says in the Wörterbuch [163] as Diop[164]?
- Crocodile scales as argued by the English Egyptologist Allan Gardiner (Gardiner 475, I6) and his followers?
- Footprint for Professor Mubabingu Bilolo?
- A porcupine back for Dibombari Mbock?
- What is accomplished or Men for Mboli?
- Etc.

One of the keys related to the good reading of this script, according to the methodology of egyptology is therefore based on the proper identification of the biliteral ***km***. Without a good identification of biliteral ***km*** difficult to pretend to an accurate reading.

Script symbols analysis:

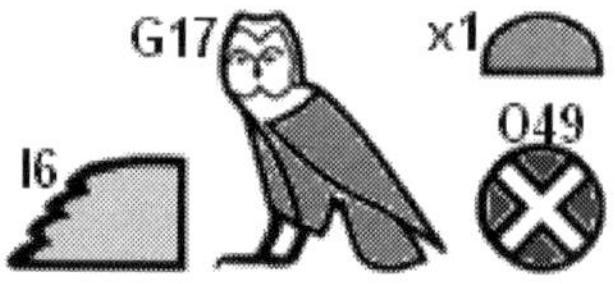

We have, from left to right, the sign I6 that is the discussion topic.
The sign **G17**: the owl supposed to be a simple phonetic complement of the discussion topic sign.
The sign **X1**: The bread cake that is supposed to indicate here a feminine tag
The **O49** sign of the Gardiner's list representing a place, a locus, a city etc.

Rigorously one should transliterate: ***kmm.t*** but by convention one transliterates: ***km.t***

- **Difficulties related to the method:**

The convention requires reading regarding the sense of owl gaze. So here from left to right. I insist that this is a pure convention.

In fact in this script it is necessary to begin to read by the beginning specifically by the discussion topic sign, let's call it, the biliteral ***km*** as of now.

The reading of this script is based on a war between chapels.
*The Germans **said** it in* Wörterbuch (p.123) so that is true. So that is Professor Diop's argument to the specialists during the Cairo debate.

For the African that I am, this is an issue because one would have readily understood that the arbitration is done by the Germans on a question concerning the German language. And despite German

163. ERMAN A., GRAPOW H. Wörterbuch der aegyptischen Sprachen, Leipzig,1926-1950, Vol.V, p.126.
164.https://www.youtube.com/watch?v=lfS4IzC4eLM

insinuations, they can not speak better Pharaonic than an African. In this case, we have the opinion of everyone except the Africans who are nevertheless the first concerned.

Strictly applying the current rules the biliteral "***km***" has been presumed to be:

- A scale of crocodile dixit french egyptological school with Jean-François Champollion.
- Others thought about an animal's paw (professeur Bilolo)
- Others to a porcupine back (Dibombari Mbock)
- Others on a piece of burned coal

The classical methodology, although it gives excellent results leads in this case to semantic impasses because according to the meaning given to the sign I6 this will have consequences on the overall semantics of the script. Although I6 is only a phonetic complement, hierogrammatic priests would have implied that the inhabitants of the designated country were:

- crocodile scales for some
- Even backs of porcupines
- Pieces of charcoal.

Only at this level, we realize that something is turning wrong if we often consider that african hierogrammatic priests were great Scholars and had previously worked deeply on this issue before naming the land and his men.

Let's take the skeleton of ***mr*** consonants. It leads to form a myriad of words: emir, moor, mayor, mire, mare, mure etc.

So, that's what happens with the different meanings proposed by the specialists. The degree of freedom in the reasoning tools, morphology, philology, semantics, etc. If we add more a paralax error to consider that ***km*** is bound to one and only possible meaning:" Black", then many people got trapped by their own ghosts.

I also devoted a little aside in the framework of this heuristic to give my opinion on the reading of this script from the African point of view. I recommend not to deal on what others have said, but to carry out a prior deconstruction of the common acquisition and to reconstruct a coherent model based on what we see by our own eyes.

Criticism of the existing and verification of different hypotheses

Sign I6: The Biliteral km

The first question is linked to the meaning of biliteral ***Km*** of the Gardiner's list. What has been represented par le scribe through this sign?

In order to answer this question, I will give you some elements of answer so that each one analyzes them and makes its own opinion. I propose to this account to check some advanced hypotheses.

Find below the illustration proposed by the dictionary of Egyptology Faulkner[165]:

<u>Assumption 1</u>: footprints?

There we could say, why not a footprint of an animal?
It is surprising to see the 5 growths that may suggest toes.
We can therefore understand that some punctilious researchers are going to consider this sign as a footprint by relying on the representation of Gardiner and the possible

165.Faulkner page 286 rang 3

phonetic proximity of the Ciluba "Ci-Kama" and with the word; footprint. In Ewondo I will have said for footprint / ndəm à-ko (l) /.

Note: All the wisdom of African scholar is to represent a footprint because the foot shares with an homophonic relation with the word ancestor. So here this footprint has to be interpretate as God mark.
Assumption 2: ***Km*** = Crocodile scales?

It is based on the idea that the sign I6 represents crocodile scales.
I appears suitable to compare the sign I6 below with a real representation of a crocodile skin.
To name those scales In Ewondo I would have said [bibas] "scales"

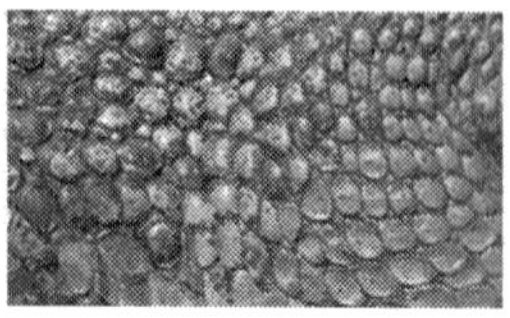

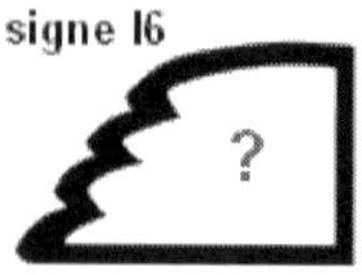

Assumption 3:*Km* = Burnt charcoal?

On the left we have the burnt coal tip and on the right the sign I6.
Ewondo: [kam] or [mâghe] "coal"

burnt charcoal

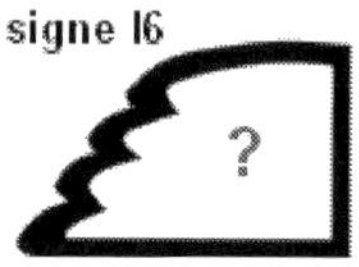

I abandoned the porcupine idea for the moment but i'll be back on that point of view further. Careful observation makes one realize that the script and the representations do not match indisputably.

My proposition

In case of doubt, it consists to seek the sources not to think through someone idea. We have to make our own opinion.

.

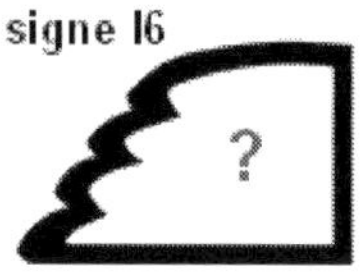

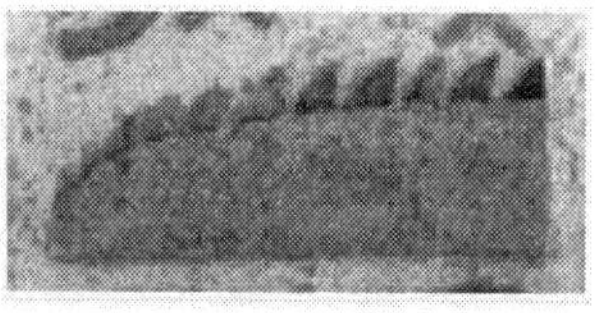
Denderah, Architraves Hypostyle Room

The difference between the source representation and what has been recorded by others is quite amazing!!! Indeed, the scribe puts on the engraving of the hypostyles room of Dendera in other words 12 excrescences and not 5 excrescences as in Gardiner's representation that let believe to a footprint.

I therefore allow myself to think that what the scribe has represented is not a only a footprint. Even less a porcupine back. And as each one indulges in more or less scrupulous reasoning, he departs from the facts. What is seen here can be interpreted as a piece of saurian tail, hardly like a piece of burnt charcoal.

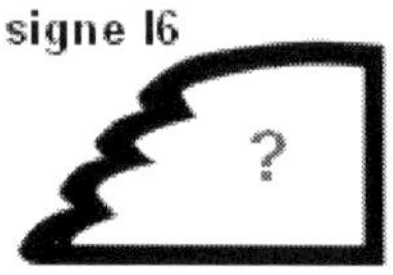

Sign read "Km"

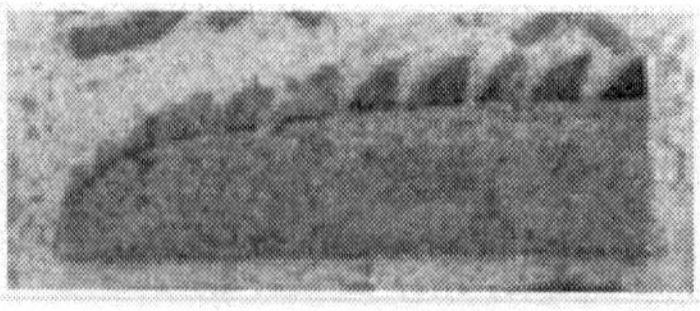
Printmaker representation[166]

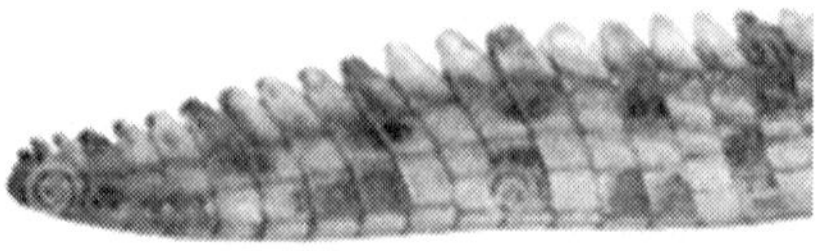
Saurian tail

166.Denderah, Architraves Hypostyle room

Is it logical to have a crocodile tail in the script for the sign I6 representing the name of an African country?
The crocodile being an African holy animal related to the Ethiopian royalties finds all its justification in the representation of the sign I6. (*See illustration n ° 6*)

Unquestionably we already see more clearly on this issue which has given rise to all the interpretations more or less tendentious and misleading.
The vocalization that befits sign I6 on this script will mobilize all the descriptors linked to the reality represented in□□:

> Ewondo: [*ŋ*-kôm] "crocodile"□ consonants skeleton *km<km
> Ewondo: [*ŋ*-gan] "crocodile"□ consonants skeleton *gn<km
> Ewondo: [*ŋ*-zindi] "crocodile"□ consonants skeleton *znd<km
> Ewondo: [*i*-bas] "scale"□ consonants skeleton *bs < km
> Ewondo*:* [*n*-guəm] "nail" □ consonants skeleton *gm<km
> Ewondo*:* [*me*-kəl] "spot" □ consonants skeleton *kl< km
> Ewondo: [*i*-kōp] "skin"□ consonants skeleton km<*kb <*kp

One could at this level simply by knowing the common nouns apply the rule which allows by nominalization to create an antonomase and obtain the possible vocalizations of the script ***km.t***. But it would be a little quick work.

African Ontology

On the African ontological level, it is important to know what this sacred animal represents. From this knowledge between the signified and the signifiers we can therefore propose a healthy reading of the script.

But before focusing on the crocodile, let us give an overview of the possible readings of the script at this level. Using the nominalization rule which allows the realization of an antonomase by tranformation of the common nouns in proper names in Ewondo.

Proper noun= common noun + final vowel

Example:

> **Ewondo: [*ŋ*-kôm] □ km.t *vocalisation*: [*ŋ*kômo]**

☞ **Place names**: Nkomo, Mba-nkomo(*ydé-Cam*)**;** Nkomo*(Limpopo/Afsud);* Komati (*South African River; from Mozambique and Swaziland);* Um.komazi (*afsud*); kome (*Nigeria);* Ngoumou *(cam.), kamba (Africa)etc.*

☞**Anthroponyms**: Joshua Nkomo[167]; Nkomo[168], Komani(*Botswana*); Koma, (*Rdc, Namibia, Soudan, Ethiopia);* Tchamba (cam.); Komo (*Zaïre*), Komono; Koom (Burkina-Faso); Kumu (*Rdc*) ; etc.

167.Sabelo J. NDLOVU-GATSHENI, Joshua Mqabuko Nkomo of Zimbabwe: Politics, Power, and Memory, Springer International Publishing AG- dition , Prtoria, 2017.

168. UNESCOS, Ethnonyms and africans toponyms: General History of Africa Study and Documents 6, Unesco, 1978.
"... nguni consisting of using the clan totem as a last name: today, the many Sindebele-speaking people called Ndlovu, Mpofu, Ncube, Nkomo or Sibanda are actually of Shona origin. As for the Ndebele toponym, it is derived from the sotho-pedi dialects of the eastern Transvaal, a region through which Mzilikazi transited during his migration.". p.41

☞**Ethnonyms**: Bakom, Bakum, Koum, Kom, "Koum" (*Cmr*), Komo (ancient *Zaïre*), Koma (*soudan, Ethiopia, Tanzania, Kenya*), Po.komo (*Kenya*); Tankamba (*C.I, Burkina*); Kambari (Nigeria), etc.

Ngan's acceptance

Ewondo: [ŋ-gan] □ km.t ☞*vocalisation*: [ŋ-gana] "crocodile"

On the mythological point of view, let us question the peoples located south of the Sahara and who are concerned above all. The Ekang people all say this:

> *"[...] Beti's ancestor is "Nanga" who was albino. It is an ethnic group present in Cameroon and Gabon. It will be called Nanga kon among bulu. Beti has fathered children including kolo Beti, Beti Eton, Mvele Beti (Bassas), Mvân Beti, Meka Beti (Makas), Bulu the girl and Ntumu the lastborn. Bulu was the only daughter of BETI BE NANGA, the Bulu are not the Betis but rather "Ban Ngôn Beti" that is to say their nephew. "*
> *"In Cameroon, they are mainly located in the Mbam region. A legend rooted in historical facts, reports that the Beti come from the other side of the Sanaga River* ***they crossed in the late eighteenth century on the back of a totem called Ngan-medza****. They were fleeing Ousman dan Fodio, a Muslim leader from the Adamaoua region, who had decided to Islamize all the animist peoples, especially since they considered themselves contrary to the teaching of the Koran as lord. All Beti were mounted on the back of Ngan-medza."*[169]

African sociology

If the land is tied to the woman in a matriarchal regime, then we have the name of the owner of this land. Her name was Nanga. A play on words for [n'nang] which associates it with the albino qualifier (because of the life in deepnesss sheltered from the sun) in French and the second point which is the ananym which hides the totemic dimension which is related to this term. Nanga inverting the syllables would vocalize [ngan] who is the crocodile. Which thus illuminates with a new eye the supposedly mythical history of this people. Who crossed the river Nanga / oso-nanga / ☞***Sanaga*** on the back of a Saurian / *ngan-medza*/

De facto the scribe could not write the name ***km.t*** without designating the name of the ancestor of this people who owns the land and who is anyone else but Nanga whose totem is / ngan / namely The crocodile. If we stick to the cultural unity advocated by Diop *in Negro Nations and* ***culture***, there is no coincidence that the crocodile is a holy animal in Africa. We find on this subject his representations in the shapes of holy masks[170] everywhere in Africa.

> *"[...], Crocodile masks exist in many ethnic groups in West Africa: Dogon in Mali, Ijo in the Niger Delta and populations of Cross River in Nigeria, Bwa, Gurunsi and Mossi in Burkina Faso. This would be the most common animal mask among Gurunsi* [171]*"*
> *"Another Nigerian crocodile mask, the Basinjom mask of the Ejaghams, is composed of various elements borrowed from both the animal world and the plant world (eagle feathers, cowrie shells, leather ...).*
> *It should be noted that in the Cross River region, Basinjom's mask is considered one of the most powerful. The members of the secret society have the mission of seeking out and prosecuting all those who cause disorder in the community: witches, people inhabited by evil animal spirits. In his black cloth dress that hides him from the eyes of the wizards, the Basinjom, at once a detective and*

---

169. Migration attested in Ethnonyms and africans toponyms: General History of Africa Study and Documents 6, Unesco, 1978, p.64.
170. Mudiji Malamba Gilombe, The language of African masks: study of the forms and symbolic functions of the "Mbuya" of Phende, Facults catholiques de Kinshasa, 1989.
171. VOLTZ M., The language of masks among the Bwaba and Gurunsi of Upper Volta.Ouagadougou: Universit de Ouagadougou, 1981, 102p

> *a diviner, travels unexpectedly through the streets with a quick, slippery step in search of the wicked."*[172]

For example, in voltaic populations, hornbill is often associated with another animal, the crocodile or its substitute the varan.

The crocodile is a totemic animal among the Mossi in Burkina Faso who devote him a true cult devoting a party to his honor: the feast of **koom-lakr**. It is a totemic animal among Nuer, Shiluk, Sise, Moundang etc.

Ewondo: [ŋgan] □ km.t vocalisation: [ŋgana]

☞<u>*Anthroponyms*</u>: Nanga, Ngana, Nganang, Menanga (Cmr); Ganesh (India)[173]

☞<u>*Ethnonyms*</u>: Ba**ngan**gte (Cmr); Nanga-Eboko (Cmr); Anang(Nigeria);We**nang**umba (Zambia); Gara**ngan**za (Tanzania); **Ngan**guela (RDC-Angola); Sosha**ngan** (mozambique) etc.

☞<u>*Toponyms*</u>[174]: Ghana; Ganâr; Gannaria; Ganagana (*Nigeria*); Ganawuri (*Nigeria*), Nanga-eboko (Cameroon), Sanaga (fleuve-Cmr); Mananga, Nhlangano, ukhuhkangana, Soshangane (Swaziland); Tanganyika; Kisangani(RDC)

Ewondo: [ŋ-zindi] ☞ km.t vocalisation: [ŋ-zindi]

☞<u>*Toponyms:*</u> Zindi, city of southern Garoua (*Cameroon*); RCA; Niger; Zinder (*Niger*)

☞<u>*Anthroponyms:*</u> Essindi (*Cameroon*), Azand, Sand, Bazenda (*Soudan, RCA, RDC*)

□<u>*Ethnonyms:*</u> Zandé- Sandé- Bazenda (*RCA Zaïre and Soudan)*; Kasands, Kasanji, Kasandje; Sandawe (Tanzania)

Ewondo: /i-bàs/ □ km.t vocalisation: [*i*-bassa]

☞<u>*Ethnonym:*</u> peuple Basa'a; basaa/mbene (*Cmr*); Bassari (*senegal*); yambasa (*Cmr*), Basa La Mpusa (*RDC*); Basaa-Ng-Bnae (*Nigeria*); Bassa (*Liberia*) etc.

☞<u>*Anthroponym:*</u> Basa (*Senegal, RDC*)

☞<u>*Toponyms:*</u> pays Basa'a (*Cmr*), Mombasa (*Kenya*), the toponym Basa is attested in ***Km.t*** as the name of a small locality on the eastern bank of the Nile; it became much later Antinoe, at the Romans time etc.

After this summary approach of sign 16 study, allowing the validating of many anthroponyms, ethnonyms and toponyms, let's go to the following sign the owl.

Sign G17: owl 

172. FAGG W, LIST H. Wonders of Nigerian Art. Paris: Chêne, 1963, 39p.
173. Cf illustration N°19
174. De GANAY, S. "Toponymy and Anthroponymy of Black Africa", Onomástica, n° 2, vol. 2. Paris, 1948, p. 143-146.

According to the Egyptian convention, the script should have been read using only the first sign. But not knowing it, left doubt behind, and thus gave rise to various interpretations.
Let's take care of the second sign anyway.
As for the previous one, let's question those who claim to have authority over the reading of this signCe sign G17 of the Gardiner's list it is supposed to represent after unilateral translation M.

But seen from this angle this sign brings us not much information so it is more appropriate to get the representation of the scribe.

Sign G17 of the Gardiner's list

Owl seen by the scribe
Mastaba from Iroukaptah in Saqqara

Owl

Ostensibly the sign G17 designates just an owl but the information provided by the representation of Saqqara will be important because it does not act fancies. Every detail represented by the scribe here are really importants.

For the African that I am, on the ontological level, the owl is not a mere bird nor a simple consonant M. It is a totemic animal of African societies. It symbolizes the mystical knowledge of the spiritual world, the world of the "night"vision .
You can not claim to be a Chief in Africa if you do not master the world of the night as you master the worldview in the day.

> *"[...] The holy animals, carriers of secrets, are found in the mask: the owl's globular eyes, the little ears of the bat. Moreover, the rounded ridge recalls the hair of the Fulani woman whose right nose is also indicated, which underlines the privileged relations existing between the blacksmiths and the Peuls"*[175]

Summary:

In addition to the meaning imposed arbitrarily, artifacts introduced by the subjective representation of signs. There are conventions that consider, for example, that the sign G17 the owl transliterates M while being just a simple phonetic complement that reinforces the reading of the sign that precedes it. Why the owl would be more M while other signs exist and are associated with the sound M?

Quid: And if the G17 sign (owl) was not just a phonetic complement for Africans?

The word owl is transliterated into Pharaonic by the skeleton consonant ***Sm***. So why is it reduced phonetically to the sound M?
In the dictionaries accessible to many people, it is not easy enough to find the transliteration of the word *owl* in pharaonic how to understand that?

It may be noted that on the diachronic *Sm* <*km*. Does this have any connection and does it suggest that the script could have been read by just knowing the name of the owl? On this account, is the owl

175.Pauline, Marie, Cline FREALLE, Animals in masks of West Africa, veterinary PHD,ANRT, Paris, 2002, p.59.

always a simple phonetic complement or not? This script shows a tight relationship between owl and crocodile: *msH* "cayman"

> **Ewondo**: à-koung (owl) consonants skeleton km <* kng
> **Middle-egyptian**: *Sm* "Owl" **Ewondo**:/ *ùn*-nsəm /"sin" associated to darkness

Let's take our consonants skeleton ***kng***- and let's operate a nominalization. For the purpose, let's add to the consonants skeleton a final vowel. By the way, let us then operate an antonomase by transforming the common name into a proper name. We get: ***k … ngo***

Km.t could be vocalized: *koungo* / *kongo* / *akono* / *okono* etc. (toponyms, anthroponyms attested in Africa and Cameroon). We

But as always, let's not quickly got down to work:
If for classical Egyptology the owl is a simpthe sign associated with the consonant ***M***. For the African that I am, the fact that the scribe represented the owl refers to a precise and very complex reality. (*cf illustration n ° 11 and 14*)

> **Ewondo**: /i-koum/ "parent tree; tree stump" by metonymy *owl*
> **Gourane**: koum "owl"

We can see that in Gourane and Ewondo we obtain easily the consonants skeleton *km*.
Indeed owls are often found on tree stumps and the skeleton *km* retained by egypotology can be associated with the owl by metonymy regarding Gourane or Ewondo language.

AN AFROCENTERED APPROACH

To read this script, we must take into account a number of precautions among which: sociology, anthropology, culture, the African paradigm etc.
The characters in the script are not just letters. We do not name a country in Africa as we call a pebble because even the pebble can have a strong totemic dimension.
One must be wary of thinking that the choice of the name of a country in Africa boils down to a simple association of characters with an "arbitrary" phonetic value.

The script ***km.t*** used to appoint a political entity in African is necessarily codified by many endogenous sociological rules. Because a political entity in ancient Egypt inevitably brings together several lineages or houses and is based on specific socio-political rules. By this simple fact, to appoint a place of common life involves to take into account the particularisms of all its components.

Km.t is the fruit of the ***smA-tAwy*** /i-samèn Tsè i-bèh/ "*the union of the two lands …*",.
On the same land were living several lineages: fishermen, farmers and shepherds and later foreigners. By the way all those people were dealing ontologically with their own visions of the world.

If we stick to the psychopathology doctrines [176] and the cognitive theory of the Austrian Karl Berg on cognitive-behavioral dysfunction, we note that behavior is a matter of values. Since values are a function of the environment in which we live, the environment forges beliefs of a person. Religion is based on ethics and societal ethics appears as a channel of sensations and emotions. It thus sits on the ontological dimension in the individual and can serve as an explanatory model for its behavior.

That is why the signs that are here represented by the scribes can not be neutral for Africans because they are the mental representations of the buried emotions, erected in beliefs, the ontology in short

176. Ovide FONTAINE, Jean COTTRAUX- Robert LADOUCEUR, Behavioral therapy clinics, Mardaga, 1995.

The environment structuring the behavior and the beliefs, it will be different according to the medium and the features of differences are strongly marked between a sedentary or nomadic life.

On this account, the African society before any foreign entryism being of sedentary and matriarchal type will thus give beliefs and an ontology proper to the sedentary peoples.

> *"[...] It is the man who is in a state of inferiority from the economic point of view because it is him that the conditions of life and the social organization force to leave his clan to join that of his wife where it is only a stranger whom the clan did not think of integrating originally (matrilocal marriage)"*[177]

The girl / woman occupies a key position because the filiation and succession to the head of a social entity or political (house or lineage etc.) is uterine [178].
Land inheritance is the legal property (*usus, abuse, fructus*) of the girl / woman[179]. This is explained according to the African paradigm because it is his ancestor who concluded a ritual with the land[180] and it is to this ancestor where we worship the on who is often called GOD.

Why the woman and not the man? The word **earth** in our languages is **equivalent to the word woman**. In Ewondo the word earth can be called / *ngɔn* / . This word serves to name lands where one makes the plantations unlike residential land.

The same word / *ngɔn* / is therefore homophone of / *ngon* / meaning **girl.** It is also homophone of / *ngôn* / meaning **panther**. Some purists of Ewondo language will jump down on my throat to tell me that this is not true because the panther / leopard is named in Ewondo / zə /. If we ask our brothers Bafia they name the panther ***àngon***. Surprisingly the word for terror, fright in wondo is indeed /à-ŋgône /. So here we have the same signified but a few different signifiers. All this is explained by using diachronic techniques. I do not linger any longer to avoid getting lost. So remember here that earth, woman, panther, moon, menses etc. are homophones in our languages.That why they can we be involved in cross-uses by metonymy.

Let's illustrate this again: we often call / mvôkh / a variety of squirrels that eat even ivory (*Protoxerus stangeri*). But the word / mvokh / designates also a small fish. Which offers as synonym: / ŋgoŋa / (variety kəmə small river jowl). But / mvógh / is also / dzaŋ / "village, land" is the lineage of the Ekang. So this is a notion strictly related to the girl / woman. It translates the idea of the country of the mum, the land of the mum. It is therefore not surprising that the named fish bear the couple of names / ŋgoŋa / homophone of panther and / mvôkh / for the country of the mum. One can even observe that the mustache of felids and silurids bears in the language Ewondo the same name / mi-'mvəmlə /. What therefore comes to establish this equivalence, if doubt was still allowed. The woman, the catfish, the panther, the moon, the baboons etc. are in every respect one and the same entity. I warned that we are in complex thinking. That's it, so we're really in.

From the moment when the political entity is organized by associations of lineages and has at its head a supra-lineage entity, this entity can not direct the political structure without having concluded a ritual marriage with the one whose primordial ancestor had concluded a ritual with the earth.

- The sovereign is therefore a simple depositary and **not the owner** of the lands of the country. By concluding a ritual marriage with the panther (daughter or sister of the country's uncle), the ruler embodies the country. On this account, he is responsible for the land boundary. (This concept goes further, and developing it here might move us away from the subject).

---

177. Cheikh Anta Diop, Anteriority of Negro civilizations myth or historical truth?, Présence africaine, paris- Dakar, 1967, p.121.
178. Baumann, A. et Westermann, D.: The peoples and civilizations of Africa, Paris, Payot, 1948, P 217
179. Cheikh Anta Diop, Anteriority of Negro civilizations myth or historical truth?, Présence africaine, paris- Dakar, 1967, p.76.
180. According to the African vision of the world, what we call earth is a male principle. He can not unite with a man.(cf illustration n°17)

- Being responsible and guarantor of the rites and observances related to the ancestral cult, He is the *Sms-Hr* translated into Ewondo: / zomloa /.
- The sovereign performs the function of chief of armies or defense.

If the Pharaonic society is based on the birthright that is inalienable among the Beti. The chief thus leads to ritual practices centered on the primordial ancestor ***Hr.*** The society is thus articulated on a quadrilogic algorithm: cosmogony, cosmology, cosmography and cosmocracy. It is impossible that the choice of the term to designate a political entity that comes under the cosmocracy could neglect this sociological imperative. (See illustration n ° 13 and 14)

So, if we want to read the ***km.t*** script, we can not do without these requirements and the reading implicitly includes these precautions in terms of methodological requirements.

To name the country ***km.t*** for an African is to question the ontological dimension of the inhabitants through the spirituality that binds them to their ancestors. Who are the real landowners?
To name the country is to ask the question of super-lineage authority, what does the sovereign represent?
To name the country is to name the inhabitants because it is a federation of lineages on the basis of a social contract.
It goes without saying that a good reading of the script must automatically answer each of these questions precisely.

First of all, let's go back to our comparative model.

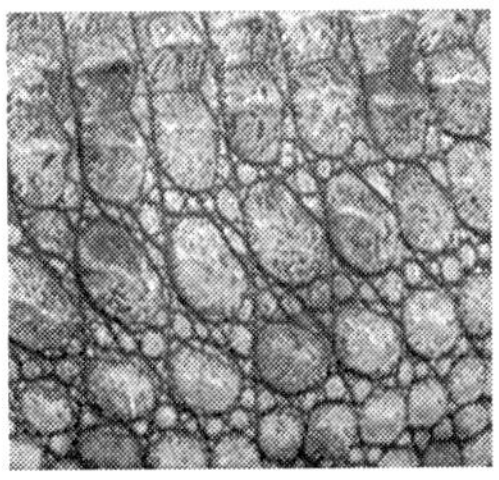
Sign I6: ***km***

Sign G17: ***m***

/Zə-mot/
Leopard-man

Km.t
Faulkner p. 286. Wb. V p. 126, 7-18 sq.

Now what is the relationship between the owl and what appears to be the tail of a crocodile?
We can note by comparison that there is a link between the plumage of the owl, the skin of the crocodile and the covering worn by the priest. We note that this link is the spots. The term used to describe what is "spotted" in Pharaonic is noted with the consonant skeleton ***sAb***.
So to name this country is to name the word stain in our languages but also all the hidden meaning as well as the homophones.
At this level, we see that the two signs I6 and G17 are not disjoint, they combine to give a new concept. If one could read the script using each of the signs, associating them they produce a superior level of information.

If for Egyptology the first sign is a mere biliteral ***km*** and the second a phonetic complement ***m*** .For the African it is a concentrate of complex concepts often reserved for true African initiates.
The association of the two signs I6 and G17 dictates a reading of this script that refers to the stains of a holy animal of Ethiopian societies. One of the royal feline in the civilization of the Nile Valley recognizable by a spotted dress is the leopard/panther.

In short, without knowing the right sign I6 but just the sign G17 one would have, knowing the endogenous traditions of the Ethiopian people given the proper reading of this script.

Indeed, the "leopard" is an animal consecrated to royalty in Ethiopian societies, and particularly in the Greats Lakes region, the place where the Nile find its source. From the Cape, south Africa to south Sahara the brotherhood of leopard tamer is spreading.

In the common expression related to characters I6 and G17 the leopard is highlighted by the scribe for the attention of the reader aware of the transliterated ***km.t*** script. (*see illustration n ° 13*)

The function of land depositary being linked to the membership of the brotherhood of Leopard Men, the name of the country must therefore reflect this reality.

Km.t + ***O49*** ***niw.t*** therefore appears as an acronym for the country of leopard-men. The leopard being a sacred animal in the traditions of the Ethiopian peoples, it embodies the materialization of the creative energy (* za / * zə), the starry sky on the cosmographic plane by its spots, the force.

The African conceiving the world as the materialization in the visible of the invisible energy whose channel is described by Pr Fame Ndongo under the term of crypto-communication [181]. All physical reality thus appears as a teleological and teleonomic projection of a creative energy coming from the invisible world. The Awakened African has a knowledge of the invisible world as well as the visible world.

For purists this spotted skin should not be confusing and represents two different animals, the leopard and the panther. Africans make a distinction between the two animals.

Among the Basa'a who share the same ancestor as the Beti, here are the functions associated with each brotherhood:

Mbog = Panther-men and legislators
Nge = Leopard-men and performers
Um = Water snake men and judges
Ko'o = Women-snails
Ngam = Spider-men and soothsayers

The goal is not here to get into the subtleties between leopards and panthers, it was good just to report that's all.

In terms of philology and dialectology [182] let's see how the words of the panther / leopard connect with other words:

Ewondo: **Ngou**(l) "strength";
Ewondo: **ngu**>zə "panther/leopard"
Ewondo: **ngo/nga** "daughter/wife"
Ewondo: *nyo*-**ngó**/ *i.so*-**ngó** "maternal / paternal aunt"

The reading may suggest a rebus system, so we can use the words naming the sign I6 and associate them with the sign G17. But in reality only one sign is enough to read the script.

The sign I6 represents the crocodile [nkom] in Ewondo, which is a synonym for [mbogh] that is to say a hindered one. Indeed, by its crawling movements, the crocodile is considered by Africans as

181. Jacques FAME NDONGO, An African look at communication: discovering the circular geometer, Editions St Paul, 1996.
182. Dialectology is the branch of linguistics that studies dialects and linguistic variation.

handicapped. The word [mbokh] designates an oral cavity. However, the semantaxe linked to the mouth of Africans is the center of speech / *ùn*-ŋkobo /. In fact, according to the mental projection of the Africans, the mouth is a foot. The two base tonemes with non-zero signifiers that serve to construct the term / ùn-ŋkɔbɔ / "word" are summarized as follows:

Ewondo: /à-ko (1)/ "foot" and /à-bo/ "leg" so nko-bo =foot

At this point, the sign I6 can be summed up in its simplest expression that is the foot: /à-ko/.[183]

The owl sign G17 is called Ewondo: / à-koung /.
This bird with its spotted plumage is reminiscent of the panther.
In the African ontology, this bird symbolizes the ability of some humans to transfer their minds into a bird's body through metamorphic techniques.

This bird is therefore an accomplished man. He often represents himself with a human face (*cf illustration n ° 11).* We call this banal practice / mbgwəl /. Which is often improperly vocalized ***bA*** and accompanied by misleading interpretations. Serious Ethnologists like Trilles or Laburthe Tolra had a superficial knowledge of this reality. So, looking at the script, a wise person would have understood that the inhabitants are metamorphs, owl-men. This practice of metamorphs and a distressing banality under our latitudes.

The root of the word panther being in the Ethiopian languages * ngo.
We have the sign association I6 and the sign G17 which gives: Ko-Ngo. (*The footprint of the panther or catfish.*) Here we find the meaning of Professor Bilolo who could have lend to smile. I just remind that Catfish or panther is the other way recall the name of the eldest ancestor.

But this meaning is only admissible as the identification of the panther's land. But voluntarily I added the catfish because as I explained that the two notions are closely related, interchangeable by homophonyThe semantax inherited from Negro-egyptian making of a place, a locus, a "foot" / * ko /; / * Bu /. This is how this approach is validated.

We could have, knowing the sign G17 the owl only starting from his /à-kung / getting the skeleton (kng) and adding a suffixal vowel by nominalization to make it a proper name to get k ... ngo. By guessing we would have got Kongo.

In summary by associating the two concepts represented by the signs I6 and G17 by a system of rejects we get the reading below:

Middle-Egyptian: km.t = **Kongo**
Ewondo: à-Ko "foot" this semantax is used as a nominal class prefix to designate places in African languages.

This prefix /ko-/ from the Negro-egyptian means place, seat. In Ewondo it can also designate a mountain, ahill, a climb, a bute: / *ùn*-ŋko /, another descriptor that refers to the ancestor of the Ekang peoples.

/ Ngo / is the root of the word leopard, panther, catfish, force in Ethiopian languages.
Ko-Ngo is the land or the place of the panther or catfish [184]. The imprint of the panther in a secondary sense. This makes sense since the woman in African societies is a panther regardless of her name or social rank. In a matriarchal society, that makes doubly sense.

183. See illustration N°42. Reflexology confirms the link between the mouth and the foot.
184. Catfish is the authentic name of the first ruler of this country.

As a territorial entity Kongo therefore strictly designates the land of the panther / leopard.

Verification

Kongo designates until today an African toponym, a speech (the Ki-kongo), a river (kongo), a people (kongo), an old empire, a land, a nation, a political entity (empire of kongo) and an African spirituality[185](*kongo spirituality).*

How is the worship function inscribed in the consonants skeleton *km.t*?

The ancestors are worshiped. To worship is to commune with the ancestors in their new existence. It is especially to thank for the gift of life. It is not a Mass with officiants but simple multimillenary rituals that less modern Africans know perfectly well. Those of the lower floor to use a notion peculiar to the French historian Fernand Braudel (*material civilization economy and capitalism*) and not those who were folklorized by colonization.

The ontological dimension of the crocodile in African societies will give the key to the good reading of this script at this level.

Middle-Egyptian.: *km.t*
Ewondo: /*à*.kang/ "to praise"

km.t

So the country is: /*ùn*-nàm Ekang] "praise land"
Inhabitants: [Ekang], [Ikang], [Akang] "praisers"(ethnonyms attested in central and west Africa).

☞**Toponyms**
We have place names like: (**Kang**wane; ukhuh**kang**ana; **Kang**aba; Ny**kang**, **Kang**eju,etc.)

To the question of who praises these peoples, the answer is simple. Just ask them the question instead of inventing myths.

They worship the elder ancestor for life gift. He is the one that the betis call, / Zàmbà Elo / and that Egyptologists call ***Imn-rA*** or ***Hr***.
In sum the land belongs to / Zamba / as we have specified above. The rulers being his followers in the function of chief (be zomloa or *à*-sù zoa) ***Smsw Hr***. The Ewondo /à-sù zoà/ "visage of the elephant" repport to illustration N°19.

This is not myths those titles exist even nowadays.

According to African mythology  ***Hr*** came into the world as any baby without the use of his legs.

> *"... Horus the Great begotten by Osiris and Isis before their own birth, in the breast of Rhea (Nut); he is likened to Apollo, and probably corresponds to to the Horus evoked later in the treatise, whose eyes are the sun and the moon; Harpocrates - Horus the Child born of a posthumous union of Osiris and Isis,* ***"come before term and weak of the lower limbs"*** *17: Harsiesis ..."*[186]

The one who has paralysis or who does not have the use of his lower limbs is called in the language Ewondo: / *ùn*-mbômbôkh [187] /; /*ùn*-nter/.

185. A great civilization does not disappear without leaving a trace.
186.Michel FARTZOFF - Evelyne GENY, Power of men, signs of gods in the ancient world, pufc, 2002, p.11.
187.Samuel GALLEY, Fang-French dictionary. and French Fang followed by a fang grammar, ditions Henri Messeiller , Neuchâtel, p.351

When a person is hindered, in prison, bound, does not have his freedom of movement, it is said in Ewondo that this person is / *ùn*-nkom / or / *ùn*-mbôkh / "prisoner; " (we obtain the two words by inverting the syllables to something close but the semantics is the same)

km.t

Curiously about the script we identified the sign I6 as a crocodile named in Ewondo: /*ùn*-ŋkôm/[188].

So we have homophony that ***hr*** = [Zàmbà] = [nkôm] = [Elo] "crocodile".
When the scribe represents the crocodile here, he indicates the ancestor of the ancestors for the peoples wrongly called by the Greeks the Ethiopians. I mention here that speaking of Ethiopians, I do not refer to the ancient Abyssinia. (*Cf illustration N°16*).
I am referring to the true people elected / *ùn*-ntôbèn / "the chosen". I therefore deviate from the definition "burned face" (Αἰθιοπία / Aithiopía, from αἴθω / aíthô "to burn" and ὤψ / ốps, "face") which refers to the legend of Phaeton, which would be born from the union of Helios and Clymene, wife of Merops, king of the Ethiopians. I depart from that because, there is not a precedence of Greece on Africa.
Reporting on the circumnavigation of pseudo-Scylax[189], the historian Herodotus, the father of Western history, teaches us this about those whom they call Ethiopians:

> *"[...] Ethiopians wear tattoos and drink in ivory cups. Their women adorn themselves with ivory necklaces; even for their horses, they have ivory ornaments. These Ethiopians are the most gracious men we know; their height exceeds four cubits; some even reach five cubits. They wear beards and have beautiful hair.* ***They are the most beautiful of all men****. Their king is the greatest of them. They are good riders, throw the javelin and are good archers. They also use fire-hardened features. (...)"*[190].
>
> *(LONIS, 1979)*

So there is a mystification when you want to make Africans believe that they are Ethiopian in the sense of burned faces.
The term "*burned*" is an intellectual scam intended to replace the Greek word θεός / theos "*god*". We play here on the phonetic and semantic proximity between aíthô and theos. We thus see here by taking again θεόςὤψ that one rather has a translation going in the direction of **face of God**.
The semantaxe as beautiful as a God still persists today outside any ideological framework. This last translation is therefore more in line with what the Greeks themselves describe concerning the Ethiopians. We know it even more because the Greek word ὤψ / ốps, "visage" returns in Ewondo and in the African languages to the words, before, first, θεός / Theos therefore, God in the Greek sense.

But, this word face gives by metonymy in African languages the word elected or inhabitant. The first inhabitant or the chosen people. This is why the Ethiopian term as translated by ideologists a monumental intellectual fraud. I will not dwell on this longer. I suggest to return to the script ***km.t*** that we can read from the name of the primordial ancestor of which the ancient Greeks rightly speak.

Middle-Egyptian: ***km.t***
Ewondo: /Zàmbà/ "equivalent of "God" in theogony and in Western theology"[191]
Mòor: /yimbga/ "crocodile"

188.Samuel GALLEY, Fang-French dictionary. and French Fang followed by a fang grammar, ditions Henri Messeiller , Neuchâtel, p.275
189. The Pseudo-Scylax Journey is a Greek journey dating from the 4th or 3rd century BC. It is thought that it was given the name of Scylax in reference to a Greek navigator mentioned by Herodotus, Scylax of Caryanda, who would have explored the shores of the Indian Ocean on behalf of the Persians.The Pseudo Scylax describes a circumnavigation of the Mediterranean Sea and the Black Sea, beginning with the Iberian Peninsula and ending with West Africa, beyond the Columns of Hercules. The African section is clearly inspired by the Journey of Hanno the Navigator and tends to indicate contacts between Carthaginians and Africans. There remains a manuscript, that of Pithou, which is the original of those from which the first edition was published.
190. Raoul LONIS, "The Ethiopians of pseudo-Scylax; myth or geoscriptical reality? "French Review of Overseas History, volume 66, n ° 242-243, 1st and 2nd quarters 1979: 101-110.
191. The word god originates from the word [dzô] Ekang indicating the sky and by semantic shift of meaning the sun, the light. It is therefore associated with Wsir or Hr in the tradition of the African peoples.

Tshiluba: /Di-Kamba/ "nile crocodile"
Basa'a: / kòmbàr/ "crocodile"
Kibosho: mamba "crocodile"
Shambala: mamba "crocodile"
Nyaturu: maamba "crocodile"
Safwa: imamba "crocodile"
Mavlha: mamba "crocodile"
Ndamba: mamba "crocodile"
Chahi: maamba "crocodile"
Arabe: تمساح tamsah "crocodile"
Gonja: sà "crocodile"
Ewè: lo "crocodile"
Ewondo: /*ùn*-nti/ "Lord"

Kasem: pa "crocodile"
French: papa "father"
Sawa: Djengou "water spirit"
Ewondo: /i-sia/ "father"
Dagbani:ti "crocodile"
Nzema: ɛ̀lɛ̀ŋgɛ̀nɪ́ "crocodile"

The sign I6 ***km*** as we said it is the crocodile that is identified through the ***msH*** skeleton consonant (h void consonant we have well ms = sm <km) and the owl sign G17 which represents the owl which is also used to represent the  ***bA***. So we get Samba < Zamba<Kamba.

☞Anthoponyms: Akamba, Tchamba, Ngamba, Samba, Tamba etc. are attested in Cameroon.

> *"[...] A legend tells that the Beti come from the other side of the Sanaga River, which they crossed at the end of the eighteenth century on the back of a boa snake called* ***Ngan-medza****. They were fleeing Ousman dan Fodio, a Muslim leader from the Adamaoua region who had decided to Islamize all the animist peoples. Part of the Beti would be mounted on the back of the* ***Ngan-medza****."* [192]

Since the term [ngan] is generic for saurians in Ewondo, the term [medza] which specifies the type of saurian in question can not be approached on a morphological level with the consonant skeleton ***msH***?

I do think so because both terms are identical morphologically mdz = msH and it is enough to invert the syllables of medza to have [dzame], [* zame] or [* zambe], [* Zàmbà]etc[193].

This approach is confirmed by the ***ntr.w*** which are also other representations or hypostases of *Hr* in the traditions of the Nilotic peoples namely:

Middle-gyptian:  ***nSmwt*** *(scales)*
Ewondo: /bi-bas/ □****bs*** *(scales),*

Middle-gyptian:  ***sAb***[194]*(spots)* all this perfectly characterizing the skin of the crocodile "nkôm".

192.Philippe LABURTHE-TOLRA , Initiations and secret societies in Cameroon: the mysteries of the night.Karthala ditions, Paris, 1992, p.427 et 434
193.The anaptyx is the development of a parasite phoneme, not etymological, at the initial (prosthesis), inside (epenthesis) at the end (epithesis) is a fact that must be taken into account in the approach of our languages. It is the same with the suppression of phoneme (s), in the initial (apheresis), inside (syncope), and in the final (apocope).
Compared to Egyptian, it is worth mentioning frequent epenthesis in Negro-African languages. Zama□Zamba.
194.FAULKNER p. 210, *Wb. IV p. 17, 13-15*

This is what Oum Ndigi says in his doctoral thesison *the Pharaonic Egypt and Basa'a- link*:

> *"[...] On the spiritual level, Basa (noted B3s3y, B3s3, B3s, Bs3, Bsy, Bs) is the name of a ntjr that Egyptologists generally translate into "god Bes", and who is known "in the Bible in the form of Besai"*[195] *(cf illustration n°12)*
> *"[...] This same figure is found elsewhere in Africa, particularly under concepts or appellations such as "Obassi", "Obasi", "Bessi"; as well as in the basic terms basaa which means "religion", "particular rites", and low which means "initiate""*[196]
> *"[...] The god Bès appears in the light of the study of J.-M. Kruchten, on the one hand, and that of Jesi, on the other hand, as the ultimate deity who initiates the mystery of life, in the sense Egyptian, because he alone passes from the profane world to the sacred world, in the words of Kruchten: it is he the very principle of religiosity."*[197]

HR's COUNTRY

furthermore, ***Hr*** is also called ***nTr***[198] ***sbk***[199] and wrongly vocalized Sobek in Egyptological Esperanto whose worship was given to the temple of Kom-ombo in "upper Egypt" according to Egyptologists. This term Kôm-ombo can be approached by the item Ewondo / *ùn*-nkom-boro / "the maker of Men" the creator in sum. The vocalization of ***sbk*** is / *ùn*-nkwébek /; "Very deep" with kbk> sbk. (See example # 3.4.3.34)

Before the arrival of the Europeans, the Beti were hardly monotheistic, they just did rituals to their ancestors and their primordial ancestor who manifested themselves in everything they identified with dzô (high) and whom they call Zàmbà "Ntondobe" or rather "Ntondobot" that is to say, the one who peels Man in the manner of a banana finger. The Beti believe in the existence of entities in a parallel reality that are in the invisible that they inherit from their parents or that they submit and use for various purposes.
Occasionally, they have ritual observances that suggest a worship they give to their ancestors. These observances are carried out out of respect for the gift of life and this is done according to the events without these being gods for them (and the expression "the gods" gives directly a sense of heresy and partisan of the forces of evil) and some animist rites.

The expression *nkom-boro* is attested in many writings including those of Messina Jean-Paul and Van Slageren Jaap in their book [200] in a passage where they evoke the myth of Evumana. Besides that, it is enough to make a tour in kang country, this term is used daily.
This term is also found in the writings of Bates, Henry Trilles or the famous French anthropologist Philippe Laburthe Tolra who studied the spirituality of the Ekang and Beti people in particular:

> *"[...]*
> *These popular etymologies are not conclusive, since, as linguists have remarked (Nekes and Alexander, supra), the word derives from a common Bantu root and Koelle already provided variants of the word Nyambi in 26 languages. . It is not disputed that **Zambá** belongs to the "sphere of the divine". But it is only much more logical to understand it, as the missionary Bates (1904 p.231) already asked about the **Zambe** bulu as, the name of the main character, the father of men, adept at "medicines" and at the metallurgy. Nekes rebels against it, claiming that **Zambá** is the Creator (der Schöpfergott) but without any other argument than the qualifier of **Nkombodo**, "trainer of men" who usually accompanies the name of **Zambá** among the Beti. Tessmann (1914, p.16 note 1) claimed*

195.NDIGI OUM, The Basas of Cameroon and the Egyptian-Egyptian antiquity: comparative historical and linguistic research on their cultural relations in the light of Egyptology, Thèse de doctorat (NR), Universit Lyon II – Institut d'gyptologie Victor Loret, 1997, Diffusion ANRT.P.54.
196.OUM NDIGI, op cit., p.419-424.
197.Oum NDIGI, op cit., p.423
198. Ntr seems to be for Egyptologists a god. But in the tradition of the Ethiopian people it is a qualifier of the accomplished ancestor to be "handicapped". It is a mark of eulogy. Sometimes this notion can be: [ntour] "saint", [tara] "father", [ntouar] "unique"
199.The papyrus of TEBTYNIS (P. Tebt. II 291 à 315, P. Kroniôn 1 à 69)
200.MESSINA Jean-Paul and Van SLAGEREN JAAP, History of Christianity in Cameroon - from origins to today, Karthala, Paris,2005, p.16.

that this word means rather "hill and swamp". Even if the etymology of Nekes is the true, it seems to me that his argument turns against him in two ways: first, to say that ***Zambi*** *is* ***Nkombodo****, the author and organizer men, it is precisely him assignr a specific field of action, humanity, and it is quick to conclude as soon as it made men ... and everything else! Secondly, nkóm does not mean the creator ex nihilo, but rather the arranger, the one who puts order: it is an epithet that is perfect for a demiurge or an ancestor monitor, as we meet so often. in African religion, which can be a hypostasis or emanation of the supreme divinity, or a demi-god, but without itself being the end of the ascending dialectic: hence its rejection by the old catechumen of the P. Guillemin, as well as by others, reported by various sources refusing* ***Zambá*** *precisely "not to worship men" (Bot Ba Njock 1960, 160). Only the will, shared by both Nekes and the Beti, to reduce African theogony to Western theology, may explain their blindness on this point."*[201]

Middle-egyptian:  : ***sbk*** *(Gardiner p. 475, I4)*
Ewondo: Nkomboro and one gets by semantic shift Zàmbà

Nkomboro is not a name but a qualifier that equals creator. "God" creator□☞Zàmbà-nkomboro.

This term / nkomboro / of the nebulous concept kheper ***xpr*** can be deduced from Egyptologists. (cause to happen***)***.

Middle-Egyptian: "iwrti" ou "irt": Yareth (Gardiner p. 487, N18 , Wb. I p. 47,?) Yareth (syrian place name).
Ewondo: [si Nti] or [si Eyo] or [si Râ] etc. "The land of the creator ancestor"
Ewondo: [i-wùlù Nti] "The land of the lord"

Comment: It would be more honest to correct the transliteration ***iwrti***[202] and to simply transliterate tA-rA. As a result, the attempt to put the sacred land of blacks in Syria would appear less incongruous. I just want to point here that /si/"ground" and /i-wùlù/ "where we are walking on" so both items are sharing a metonymic relationship.

In addition to these expressions proving that ***Km*** equals Zamba, we find common expressions of the register of belief as:

Middle-Egyptian.: *km mSa wr* (pyr.1630a,b,c,d)[203]
Ewondo: [Zàmbà ngoul mə-se] consecrated expression in Ewondo to say "Almighty God".

Middle-Egyptian: *km wr (Wb. V p. 125, 1-2)*
Ewondo: [kolworo], [ngoul e.kang],[duma Zàmbà], "great praize"; [nti Zàmbà] "Zamba the lord"; [Zàmbà elo] "Zamba in the heaven"

Middle-Egyptian: *wsir kA m km-wr*
Ewondo:[ikang ntour Itoa ya be Zàmbà Elo] "praise to the holy seat of the primordial ancestor"
Zàmbà Elo = elder ancestor, the first ancestor, the respect which is due to him is related to the right of primogeniture. That's why he is great. But it is not a matter of strength but only of respect for the gift of life according to us.

Middle-Egyptian: *Hry-ib km-wr*
Ewondo:[e.nyo ya n'nem Zàmbà Elo] "The one in the heart of Zàmbà Elo"
This expression can be approached by ***ib-rA-im*** dixit□(Abram, Abraham, Ibrahim)

201.Philippe LABURTHE-TOLRA, Initiations and Secret Societies in Cameroon: Betic Religion Essay, Karthala, Paris, 1985, P24-28
202. Syllabic writing: Gardiner p. 52, §60
203.Mouhamadou NISSIRE SARR, Funerals and representations in the tombs of the Ancient and Middle Egyptian Empires,LIT, Hamburg, 2001, p.101.

THE COUNTRY AND THE MEN

In addition to these meanings, ***Km.t*** is the country of those related to the service of the ancestors. And the first ancestor to whom the worship is rendered is ***Hr***, or Zamba-Elo. /Bə Faras/ "priests in Ewondo" as such bear a name indicating that they are descendants of the ancestor of Zamba Elo ancestors. By giving worship to this all-fara ancestor, the country's ruler bears the title of ***hm-nTr***, which translates into Ewondo / nkôm Tara / or / i-tugha Tara / that is to say those who are **attached**  and therefore who are at the service of the primordial ancestor.

Km.t is simply by the way the land of bounded-Men.

Among the Ekang people, when a person is at the service of the ancestors, he is called [nkom]. The true African initiates are all / *ùn*-nkóm / or / i-tugha / or / *ùn*-mfaŋ'ene /. The term / *ùn*-nkôm / homophony designates as has already been said, the crocodile in the same language. The representation of a crocodile skin establishes and imposes the link with the cultural function of the elder ancestor Zàmbà ( ***wsjr***, ***imn-rA*** ou ***Hr***).
It would be pretentious for a scribe to name a land that does not belong to him. The earth is called in Pharaonic ***Geb***[204]
To bury in Ewondo is said / á-dzeb /. The word earth is therefore strictly linked to Zàmbà in Ewondo.

Let's note by the way that, etymologically *to bury* is linked with the word Ewondo: /à-bùri/ "to cover"

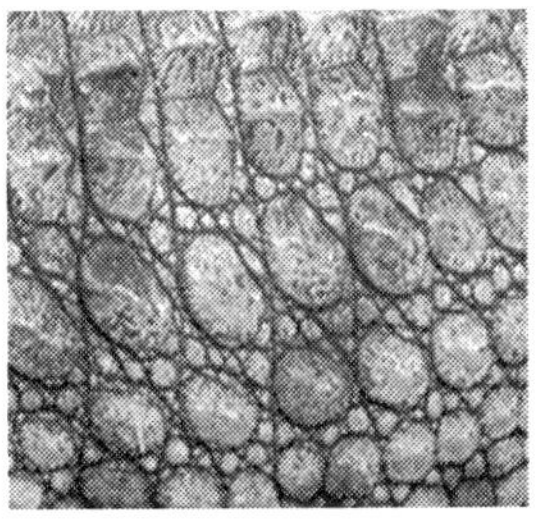
Sign I6: km

Sign G17: m

Fara

Km.t

Because the inhabitants are related to the first ancestor, one could read ***km.tyw - niw.t*** in Ewondo: [n'nam mi-nkôm] that is to say the country of the men of the service.
The expression for naming an other one is "mor ù-mbokh", literally the other hobbled. Let's just remind that the first hobbled is Hr himself.

Country

km

Sovereign

Hm: sovereign

Serf

Hm: serf

Sovereign

Hm: sovereign

The sovereign is in some detail near the country and a serf.
The characters of the scripts are barely distinct.

Ewondo: *i-tugha* "serf; feudatory; villein"
Ewondo: *ù-ŋkôm* "servant, manservant; follower; serf"
Middle-egyptian: *Hm* "serf, sovereign"

204. (FAULKNER p. 288, Wb. V p. 164, 6-11).

The crocodile has a crumpled skin reminiscent of wrinkles.
The word *wrinkles* is in Ewondo: / mi-mbor / or / mi-mbwar /. The crocodile being considered as the father of the multitude, human beings are called / bor / "human beings". You can appreciate the word game between wrinkles and human being in Ewondo language. So if the wrinkled skin belongs to the crocodile then then human beings a descendant of the crocodile. We call that, African wisdom.

The crocodile black scales on his skin could be metonymically assimilated to pieces of coal.

Crocodile skin looking as a heap of charcoal pieces

heap of charcoal pieces

WHO CAN RULE THIS COUNTRY KM.T?

The answer to this question is on the script which indicates the inhabitants or on the script of the country .
You just have to know how to read or have your eyes to see.
The power is the property of the mother but her brother can be the depositary.
On the right we have the mother/sister and on the left we have the brother .

We can validate this point of view on the scripts below:

Middle-egyptian: **As.t** **Ewondo**: nya "mother"
Middle-egyptian: **wsir** **Ewondo**: ndom "brother"

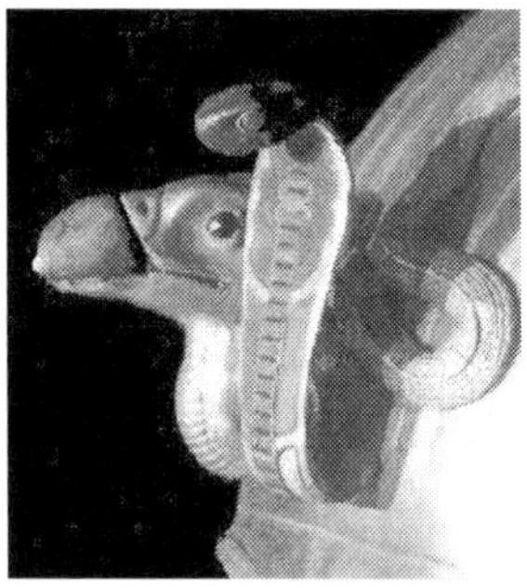

Egyptian crown
Vulture and royal cobra

The vulture = mother
Cobra = brother

We also have the sign G17 can offer the following meanings:

Ewondo: /ùn-yəmə/ "scholarly; learned; scientist; savant etc. "
Ewondo: /à-yəmə/ "mak" ☞synonym **Ewondo**: /ndəm/ "mark"
Ewondo: /ndom/ "brother" ☞homophone **Ewondo**: /ndəm/ "sperm; mark"

Km.t can be read and understood as the country of maternal uncle.

We find this concordance with the word from the Meroitic
Cursive Meroitic: ***yetmde*** *"maternal uncle"*
Meroitic: ***yetmde*** *"maternal uncle"*
Ewondo: /nya-ndomo/ *ou /ndom dzàŋ/ "maternal uncle"*

In the Hieroglyph version of the Meroitic we find the eye who oversees the throne and we also find the sign G17 of the Gardiner's list.

This approach consolidates the idea of the cultural unity of Africa with the notion of UNCLE of COUNTRY or MOTHER of the COUNTRY. If doubt still existed on this subject.

FUNCTION OF DEFENSE OR WAR.

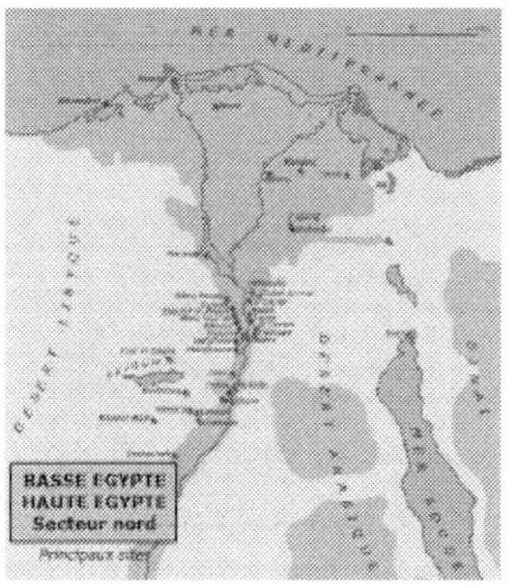

The sovereign being the land depositary, he embodies the country. He also provides the defense function. The country has the shape of an elephant head. *(Cf illustrations n°19).* That's why the spiritual guide is named in Ewondo /à-su zoà/ "elephant's head"

- Ewondo: /à **kam**/; /à **kam**èn/ "defend" The one who performs the function of the defense is called: [ŋ-kámá] or [ŋ-káman].The root of the word is [* kama]. It obeys many meanings in the Ewondo language but one of the most important is associated with the brotherhood of elephant tamer *(cf illustration n°7):*

> **KAMA**[205]*(b) n.5, pl. bikama (vb **kamda** b). Big male elephant with long tusks. Me ŋga vekh ŋga wam ye **nlô kama**, I endowed my wife with a pair of beautiful tusks. See fal (b).*

We understand better why elephant tusks or elephants are used as royal symbols in Africa. (*Cf illustration N°7*)

In fact, we can vocalize the ***km.t*** script in Ewondo: [i.Kama] or [zokh] "The elephant"
We have as toponym in Cameroon *nkongzokh.*

Bafang
Entrance of Bafang chlefiaincy (west Cameroon)

Ewondo: [à-kàm] "defend, prohibit, prohibit, ban, prevent etc."
Ewondo: /i-kàmà/ "battle cry"
Middle-Egyptian: mki "keep, protect; to take care; etc."

205.Samuel GALLEY, Fang-French dictionary. and French Fang followed by a fang grammar , ditions Henri Messeiller , Neuchâtel, p.97

According to the mythology of ancient Egypt, ***Hr*** was / *ùn*-ŋkama / "the avenger" the one who defended his father, the avenger. To name ***km.t*** by the root **kama** in our languages is to make the link with this root vanguard of ***Hr***.

Ewondo: [*à*-kám] "to wonder"
Middle-egyptian: ***Km.t niw.t*** ☞□**Ewondo**: ***[****à-yóŋ mi-ŋkama****]*** "the country of astonishment"

Note here the statement of Mintsa concerning the term Kam, attested and related to the governance of the fang:

> *" [...] The Prefix* ***kam-*** *+ Noun*
> *The prefix ekam- corresponds to the ending* ***-cratie*** *to decline different ways of governing. Thus, a private governance is called* ***ekam-mémen*** *(autocracy), different from* ***ekam-sîli*** *(democracy).*
> *NOTE. The word sîli really does not mean demos (people), but rather consciousness. It may be that the Fang vision of the concept of democracy refers to the idea of a higher political consciousness. So it is a case of "false friends", as the case of the expression tôe'e fili (rule of law), whereas it means state of freedom.*
> *It should normally have been info, since it is the word nmfos which means right in Fang. But that's not said."*[206]
>
> *(MINTSA, 2008)*

Other Defense Symbols:

The symbol for the Ethiopian people of the defense is also the spear (cf illustration N ° 8)
Ewondo: [à.kóng] "lance, arrow" ☞ associated proper name: [à-koŋ'o] vocalized [* kongo]

Middle-Egyptian: ***(Km.t niw.t)*** ➔□ [n'nam à-kong'o] " the land of the spear" which by homophony can offer the country of knowledge [*ùn*-nam à-kəŋ] and by semantic shift the country of archers [n'nam mfan] (country of archers, arrows etc.).

Ewondo: [kɔŋ] "gold" □ nominalisation associated: [koŋ'o] vocalized [* kongo] can also mean the country of gold.

This defense function was therefore retained by the Ashanti people, which we see the entrance of the palace opposite which vocalizes the script ***km.t*** "Kuma si". They take as their emblem the porcupine so the quills are spikes / mya / and therefore defense elements. This is not a coincidence because we call / mya / or / myar / a long thin bone. Needless to say, The symbol of ***Hr*** is the bone. It is therefore understandable why pointed serve to designate the function of chief notably in the cities of ***iwnw*** "yewunu", ***iwn.t*** "Ewondo"*etc.*

Kékwe branch☞	***gom-**	☚ **Kweké branch**
Tam-tam, *porcupine;* *curare* ←		→ Amougou Maghà ↔Mecca

Ewondo: / à-ngom / "reconciliation"
So according to this point of view km.t is the country of the reconciliation and by the way of the alliance.

In this particular case, we speak indeed of / myar ŋgôm / "porcupine quills". This root / yar / will also enlighten us on the word / nyar / "buffalo" that is associated with the leader. As I have already said, / nyar / is also the dance performed by men when the chief has passed away.

---

206. Jeau Joseph MINTSA, The Complete of the Fang Language,Lulu Press, Morrisville, 2008, P.342.

To designate the porcupine in Ewondo we use the term / ngôm / homophone of / nguəm / the tail (implied crocodile). We have ***km> gm***

Km.t or the tam-tam

The cultural impact in Egyptian society is undeniable. The millennial achievements since the predynastic period are there to attest the taste for the inhabitants culture.

The many festivals part of the Egyptian calendar are attesting festivities moments.
This curiosity pushed to take a closer look at this phenomenon.

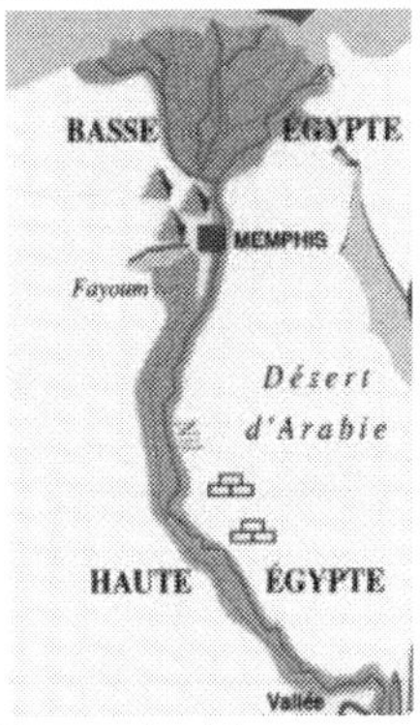

Km.t= the big tam-tam

Tam-tam= Djembé

It can be noted that the country and the tam-tam share the same shape. But better than sharing the shape, the name of the tam-tam recalls that of the first ancestor: Zamba vs Djembe.

Ewondo : /i-tàm i-tàm/ "solely; alone" ↔ The one at the beginning : *Zamba*.
But what appears interesting is the fact that the /i-tam i-tam/ "lone" at the beginning was not in the need, he was filled, full.

Middle-Egyptian : ***itmw*** "Atoum"
Ewondo: /à təm/ "full" this item admits the synonym ☞**Ewondo**: /*ùn*-ntour/ "full"
☞**French**: /tout/ "full; whole"

Ewondo: /à vəm/ "full"
Ewondo: /à-vəm/ "monkey" The reason why the elder ancestor is associated to apes.

Other acceptions of *itm.w*

Ewondo: /à təm/ "to refuse"
Ewondo: /à təmə/ "to be static"
Etc.

Ewondo : /i-tùm/ "worm"
Ewondo: /timi/ "xenophobic"

***Tom** or Pachylobus balsamifera*

***i-zəŋ** or grape*

Fruits of *Pachylobus balsamifera* look like grapes and they could be confused.

I assert that the french word **raisin** is the equivalent of the **Ewondo** :/ râ *i*-zəŋ/ "God grapes"
The French word **grappe** "cluster" is linked with the English word **grape** by metonymy because of the shape above. By watching ancient-Egyptians reaping below then we better undestand also the meaning of the verb to **grab** or to **grip**.

The English word **grab** is semantically linked to the word /i-berghà/ "what is up: the vineyard"
Both words (**grab/*berghà**) are anagrams.

Ancient-Egyptians crushing Grapes

Grapes picking

Middle-Egyptian :  ***irt*** "yareth"
Ewondo :/ i-lé *i*-zəŋ/ "grapevine"

Middle-Egyptian :  irt "grapevine"

We can support that the metonymical relationship between the grape and its color edifies the phonetic proximity underneath:

Ewondo :/ *i*-vine/ "black"
English : vine; wine
French : vigne "grapevine"

You can observe that, grapes are called *Ya* in Gbe language and *Yati* is the name of tree in the same language. But it admits the synonym "ʋinku" (some say win or houin) and for the tree : ʋinka.

"ʋinku" or ʋinka are not at all French borrowing because "ʋin" (or win or houin) in Gbe language means ferment and has for synonym "Ya" or "lã" (lan). The letter ʋ should be read "bv" phonetically.

I took these two scripts above because the left one is very interesting.
It tells us that what egyptologists are calling *yareth* the scribe informs us that it can be vocalized like the word pestle in hislanguage.

Middle-Egyptian :  « pestle » the pestle is called in **Ewondo** : /ntoum/ « pestle ; cane »

If we check above concerning the name of the two fruits then we see that /Tom/ is the name given to *Pachylobus balsamifera*.
The conclusion is that the name for grapes in Ewondo is /Tom/ or /*i*-zəŋ/ for the must.

As a rule the vine produces bunches of grapes. The picking is done by bunch.

French : branche "bunch"
English : bunch "bundle, cluster"
Ewondo : /*ùn*-ntəm/ "bunch; cluster, bundle"

Comparing French **grappe** and English **grape** it is enlighting how the semantax has persisted, allowing the association of the bunch to the fruit.

French : donne moi une **grappe de raisins** ↔**English** : give me a **grab of grapes**

But in Ewondo language the word for a grab as a cluster or a bundle is :

Ewondo : /*ùn*-nsaŋ/ "bunch; cluster, bundle"

It is thus this term Ewondo : /*ùn*-nsaŋ/ "bunch; cluster, bundle" which will be at the origin of the French word SANG "blood"and better than that, this word will give the semantic link which associates with the wine the idea of the christ's blood. Indisputably this notion is of African essence.

Note
To carry out a comparative analysis solely on the usual meaning of the words without taking into account the semantic field and the morphosemantic constraints appears as an unbounded intellectual swindle.

The previous conclusion helps to understand how wine became sacred for the ancient Egyptians. The fruit of the vine shares with *Pachylobus balsamifera* a certain similarity leading to confusion.

We also draw the conclusion regarding color hues. The dark blue color of the vine will therefore be associated with blue in some African languages like in Lingala language compared to Ewondo.

Lingàla : *bo*zinga "blue" ↔ **Ewondo** : /*i*-zəŋ/ "grapes; must"

If fruits of *Pachylobus balsamifera* and grapes are dark we have by the way the original color of ancient egyptians but also the color of their ancestor.

Middle-Egyptian : ***km wr***

We recall that the French language comes from Latin and Greek. Except that the Latins and Greeks have a common history with Africa since at least -660 BOE. Is it useful for me to recall that Africans civilized Europe around the 10th century? You know All this. Only a people of cultivators, farmers can recognize a paternity with the culture of the vine and with the science of oenology. The remaining unfortunately, is ideology but fortunately it is not our cup of tea. We present the facts in their historical continuum and the ideological little tinkerers can not move them one iota with their sleight of hand.

Digging deeper we can easily realize that the words related to dance are the words associated with ***km.t*** like the Ekang music of Eton land called : **Ekomot**.

But generally words for music and dance are linked to Zamba, the eldest ancestor.

Many African dances will be practiced using a hiccup or a movement of convolution of the basin. Dancing Zaiko, Zouk, Asiko, Baikoko is simply to pay a tribute to the movement of galaxies around the central sun that is the life-giving source.

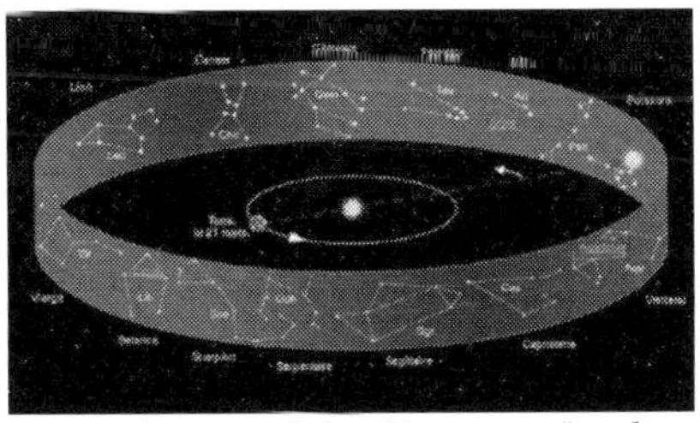

Central sun and the 12 signs of zodiac

The 12 signs of the zodiac in 4 seasons with the central sun

The Tchi or KI "energy", finds its origin in the human body at the level of the navel. The movements of the basin simply reproduce the movement of galaxies around the central sun of Man according to the fractals principle. The majority of African dances are based on this phenomenon.

Dance is an activity to circulate the inner energies, because in all civilizations we always have danced for the gods or god, without any explanation from those who are supposed to know things deeply!

In fact, the dance must enter a whole panel of activities intended to awaken this inner strength in a gradual manner so that the effects are harmonious. This being said, we can use like all things, this science for purposes that do not raise the spirit, but it is like in everything.

When you go into a trance during spiritual dancing, you transcend the music. In this science, if one wants to do everything seriously, one must work with rigor and practice according to an orthodoxy!

The dance while spinning on oneself, can lead to the trance, but in reality, it is not the trance that one systematically seeks. One wants to plunge mind and body into the Intelligence of life, for an inner purification, and the trance is the cherry on the cake. Thus one learns meditation, the art of forced breathing, the localization of sounds, and there is the sexual act which must be the object of a new apprenticeship in view of the spiritual enlightenment! It is necessary to leave the archaism and accede to the science ; This is what has not been done in Africa and it is expensive.

These dances are therefore not simple suggestive or dirty dances as some malevolent people might think. They are dances full of meaning. For some of them, such as Baikoko (*Tanzania*) or Assiko (*Cmr*), Botama or Mutwashi (*Luba*), etc., they are initiatory dances for transition rituals to adulthood.

Tanzanian: baikoko "dance"
Lingàla: zaiko "dance"

☞**Ewondo**: /á-bugh i-kokh/ "to break the basin"
☞ **Ewondo**:/á-souk'zu i-kokh/ "to shake the basin"

These hips are also found among Hindus / Hindi people who share a common history with Africa through the their Kushitic branch.

By the way and according to us, dancing is not an innocent act.
As we have emphasized several times in this document the loin cloth is a sociological tag of ***km.t.***

Middle-Egyptian :  ***snDwt*** « loin cloth »
Ewondo: /sanda/ "loincloth"

The consonants skeleton ***SnD.wt*** is vocalized sanda = loincloth. By consonantal metathesis, we can get the word **dance** in English and **danse** in French. Read the consonants skeleton from right to left.

Note two important things : the Assiko dance is performed with a loincloth worn around the waist and, better than that the loincloth is a garment of the traditions of ancient Egypt. Please report to illustration N°8 or 8bis.
Here is below the word loincloth written minimum 6000 years before our era. Better than that it's written with the weaving logo.
Unfortunately for those who often present negroes as savages who come down the trees half-naked.

Middle-Egyptian :  ***snDyt*** « loin cloth »
Ewondo: /sanda/ "loincloth"

The morpheme in African ontology is dance linked to numeral 1 in our languages i.e. "God".

We discover here a highly spiritual and sacred dimension related to dance. This is why sacred rituals are done in music with so-called sacred dances orchestrated by initiates.

For those who speak Beti, we can observe that : Zamba "god" vs Samba "dance"
For those who speak **Koonzime**: Nziem "God" vs /*dzem/; /*bwal/ "to dance".

I want to remind that the word sanda means in Ewondo loincloth. The same word in Ciluba means:
Ciluba: kusanda " to fornicate"

Middle-Egyptian : ***snDyt*** « loin cloth »
Ewondo: /sanda/ "loincloth"

Middle-Egyptian: ***sDAM*** « to sex»
Ewondo: /swandi i-bèh/ "to sex"

The reading is perfect between the loincloth and the sexual act philosophically once the loincloth are removed the bodies can act.

Km.t the phallus

According to ancient egyptian mythology, ***wsir*** an immense god, was killed and cut in pieces by his brother ***StX*** and the pieces scattered. **As.t**, his loving wife, will strive to search for scattered pieces and collect them. The most difficult was to find the penis that had swallowed a Nile catfish. This was done, and ***wsir*** reconstituted, erect member, could fertilize **As.t**, transformed into a bird. She will breed ***Hr***, her unic son. Later, the half sibling of ***Hr*** will embalm the body of his father who will enthroned in the kingdom of mons while waiting for his children, men of the earth.

I want to emphasize that ***Hr*** and ***Wsir*** are the same signified but two hypostasis at different moments of the human existence as the Sunrise in the morning (***Hr***) and the sunset (***wsir).*** As I have asserted, ***km.t*** is the land of the humanity eldest ancestor.

wsir kA m km wr

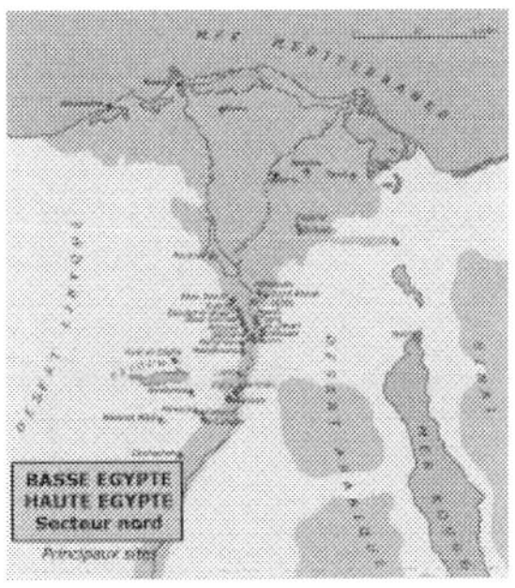

Km.t *or the phallus*

Elephant or the phallus

As we have said, the male component of power is the elephant in African thought. But, what characterizes also the elephant is the length of its phallus. ***Wsir*** being the male leader and representing the country, by the way, the country is also the phallus
We can therefore consider the delta as the pubic triangle and upper Egypt as being the virile member of ***wsir.***

Middle-Egyptian: ***km.t***
Ewondo: /*ùn*-ŋkùm/ " parent-tree; ancestor"

English: come; cop; wood; manwood;

Let's observe also that in **Middle-Egyptian**: ***bAH*** *"phallus"* and
Middle-Egyptian: ***bAH***« ancestor »
Middle-Egyptian: ***baH***« land of abundance »

M.-E.	Ewondo	Enligh
Km.t	i-**kum**	*Parent-tree; wood; dick; come*
km.tyw ; km.t Wm.t;	*i-***kum***be*	*Multitude; abundance*
baH	à-**bui**	*abundance*
baH	Mvam**ba;** *à-bui; i-bugh*	*Ancestor; multitude*
imA	i-**baĉ**	*A tree; collapsible*

We have also below that:

Middle-Egyptian : ***wsir kA m km wr*** « wiser is the father of the multitude »

I'll find the confirmation among the Zo'é of Amazonia. The Zo'é speak a **Tupi-Guarani** sibling language, close to the language spoken by the **Teko** people living in the forest of Guyana. According to their mythology :

> *"The Zo'é tell that an ancestor named* ***Sihié'abyr*** *taught them the use of the stick. One of the most important ceremonies, which is also a rite of passage into adulthood, is to pierce the lower lip. For this purpose, they use a pointed bone of the spider monkey's leg and a small stick is inserted, from the age of seven for boys and nine for girls. Growing up, they will carry bigger and bigger sticks"*[207]

wsir *with his beard*

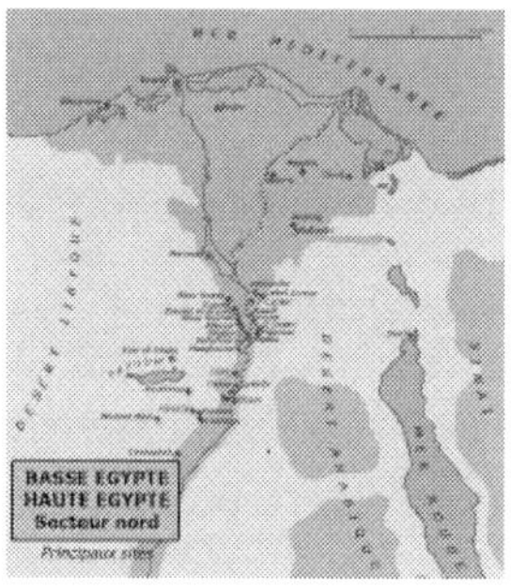

Km.t *or the beard*

Zoa : Elephant

Zo'é of Amazonia

Teko : *Sihié'abyr* " Zo'é's Ancestor"
Ewondo: *I-sièh bor* "Ekang ancestor; the father of the multitude"
Middle-Egyptian: ***wsir***

Note: The beard worn by ***wsir*** and by the Zo'é people of Amazonia is called:

Teko : m'berpót "long stick of wood, in the lower lip".
Ewondo :/*i*bér'-mot/ "transition to adulthood" **Ewondo** :/zə/ "beard; "
Middle-Egyptian : ***DAmw*** ↔ **Ewondo** :/*i*daŋ-mot/ "transition to adulthood"

After this little aside; we can mention Nkomo as a toponym among several that can be easily find elsewhere in Africa.

I therefore specify here that there is no standard vocalization of the ***km.t*** script in the current linguistic varieties for reasons related to diachrony and especially according to the variety of signifiers linked to the distribution of current populations. If the word / kuma / is retained by the Ashanti for quills, the same term will translate the word wealth into the Ekang language as we will see later. But in reality, the

[207] http://danslapeaudunpapou.survivalfrance.org/content/zoe-peuple

signified remains the same. Among the Mossi, ***km.t*** is read Koom, in sawa land we can say Kuma among Malians Kommo etc.

<u>Secondary senses related to **Km.t** script</u>

Middle-Egyptian: *km.t*
Ewondo: /*à*.kuma/ “wealth”

Middle-Egyptian: *km.tyw*
Ewondo: /mi-‘nkukuma/ “the rich”

Middle-Egyptian: TA.ty “vizier”
Ewondo: /nkukuma/ “the chief, the richest”

Hannig, p. 883[208]	M.E.: “tA n Km.t”- Ewondo: “Si ya Kongo”	*Kongo land*
Hannig, p. 883	M.E: “rmṯ n km.t”- Ewondo: “mor ya ù-ngola”	*inhabitant of kongo*
Hannig, p. 883	M.E: nsw n km.t Ewondo: [i-dzou ya Kuma]	*Authority of kongo*
Hannig, p. 883	M.E: pA-tA n km.t Ewondo: [ɲo ya Kuma-si]	*The one of Kongo Land*
Hannig, p. 883	ME: rA n km.t Ewondo: [nkobo Ekang]	*Kongo language*
Hannig, p. 883	M.E: mdwt km.t Ewondo: [ntoum Fang]	*Kongo chief*
Hannig, p. 883	mdwt rmṯ n km.t Ewondo: [ntoum mvog man ze]	*The chief of the inhabitant of kongo*
	Ewondo: [nkóbo ntobo ya Bə-ko]	*The language of the inhabitant of kongo*

One can also approach the ***km.t*** reading to the notion of the dry land valley, what the Ekang call / à-kôm / because civilizations are born in the valleys. This meaning is therefore consistent with Mboli's idea in his book on page 174. He doubted that ***km*** could be translated Black's idea.

Because the nature observation and the musical setting of the mathematical logic of fractals has long presided over the thought of its inhabitants, so all that could recall the shapes of the country will automatically take the name ***km.t***. That's why from the biggest to the smallest, we'll have ***km.t*** in everything. So in the nourishment sector, we will have the cassava called :

KooNzimé: ***k**a **m**e-**k**u**m**à* "cassava leaves" ↔ **Ewondo:** / ***kp**wə**m*** / "cassava leaves"
KooNzimé: ***k**a**m**a **k**u**m**a* "cassava leaves"

What about cassava itself?

KooNzimé: məkumà"cassava" ↔ **Ewondo:** / *ùn*-mbong / "cassava"
Middle-Egyptian : ***bnr*** « A tuber of pleasant taste »
These two words used by Nzimé people and Ewondo seem different only from an external point of view.
Indeed, the word **Ewondo:** / *n***kum** / "pillar" and the word / * mbɔŋ / "pillar" are homophones and therefore related to the same reality. So if someone says /*mboŋ/ and another says /* nkum/ to designate the same reality it is quite the same thing as saying dog or hound.

The spice named *mbongo* here adjacent is usually burnt totally before incorporation into an african recipe sauce called *mbongo tchobi.* After its incorporation the sauce turns black (***km***). Mbongo Tchobi is an indisputable social marker of the Nile Valley people.

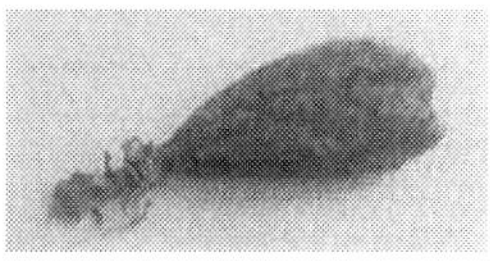
Aframomum melegueta or mbongo

208.Rainer HANNIG, Die Sprache der Pharaonen. Großes Handwörterbuch Ägyptisch-Deutsch, p. 883.

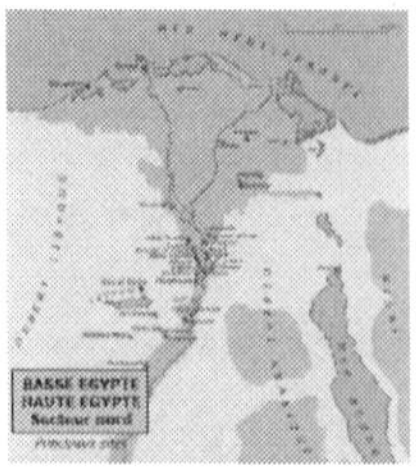

Km.t *or the beard*

Zoa : Elephant

Cassava tree

Palm tree

This logic of fractals will therefore lead to consider the palm-tree as a sacred plant as well. If palm-tree fruits produce an oil entering into the culinary preparations in Africa, its drink is appreciated and stills a capital element of the mummification and especially participates actively to satisfy the purists during festivities all over in Africa.

Ewondo : /à-lén/ “palm-tree”

The cassava leaves are called in lingala *saka-saka*. This item by phonetic proximity recalls the word **Ewondo**: / *i*-sakh / "coitus; phallus" or **English**: sex. For this purpose, if we remember that the country is a phallus shaped. So it is normal that in the African collective memory, cassava is assimilated to coitus. Let’s note that in Ewondo: / mbong / "casava" Homophone of Ewondo: / i-bɔn / "female genitals"

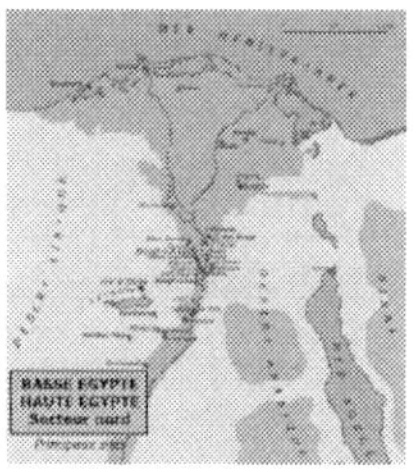

Km.t *or the beard*

Zoa : Elephant

Nail or spine

The vocalization of the word nail refers to three main notions in African languages:

- The first is the hut or house: /i-bəm/
- the second the ancestor: /Abəmə/; /Abémo/ “The nailed” ☞ the real pined not Jesus but ***Wsir***
- the third is pork: /bəmə/

I am not going to dwell on these notions here, but here we have the basic reason for the pork eating ban. To concoct pork is to consume the country or the ancestor according to African thought.
Indeed, the transliteration :

Middle-Egyptian : ***ngAw*** “beef” corresponds exactly to the word pork in Ewondo.
Ewondo: / ngoué / in Ekang languages
Ewondo: / *i*-ngakh / “wild pork”

How comes that, to name the cow, the scribe have written the word pig?
Indeed, the cow is / nyakh / "cow, shit". This dirty characteristic is shared by pigs and cattle. Cattle like pigs often walk on their own poops.
Zebu have long horns used to sting: / á-bəm /. One of the terms to name the pork is / bəmə / because it has its snout constantly nailed in the ground. But, Ewondo: / á-bem / "sting; to spike, to nail".
There is by the way a metonymic relationship between the word pig and the transliterated word ***ngAw*** “zebu”. Pork represents the materialization of the idea of nailing. Just like the land ***km.t*** looks like it is nailed in the entire continent. ***Km.t*** compare to the entire continent is the hut, the little house /i-bəm/ “the nailed”.

If the pig represents the first ancestor the one who has been nailed by his brother ***stX***, and also the home because of its shape is the country in miniature, it is therefore more appropriate according to African thought not to consume pork.

Km.t the lion

Ewondo: / i-mbgəm/ "lion" ☞KM vs *GM
Middle-Egyptian : ***sxm.t*** « sekhmet »
Middle-Egyptian : ***sxm.t*** « power ; force »

The link between the lion's figure and ***km.t*** was not easy to establish and yet, ***km.t*** can be likened to the lion's figure in African thought. I will not dwell longer on this topic but I just wanna make a quick hook.
Indeed, what is read awkwardly ***sekhmet*** corresponds to the word :

Ewondo: /soŋ-kumà/ "lunar halo or solar"
Koonzimé: *sonkumà* "lunar halo or solar"

Indeed, when we observe a halo around the sun or the moon, African thought associates this event with the death of an important ruler.
If for some this halo is a mere meteorological phenomenon, in Africa this phenomenon is to be weighed against the death of the chief paramount, ***wsir***.

Entrance of the Belgian palace

sxmt *at Louxor*

Rasta movement

Thus the term **Ewondo**: /soŋ-kumà/ returns by homophony to several main ideas:
Ewondo: / soŋ / "grave; father" and **Ewondo**: / kumà / "chief"

Ewondo: / soŋ-kumà / "the tomb of the chief"
Ewondo: / soŋ-kumà / " chief paramount"
Ewondo: / si-nkomo / " the creator's land"
Ewondo: / kùmà-si / " the creator's land"

We have emphasized that /soŋ-kumà/ is a halo so it's circular shape. The earth seen from the sky has also a circular shape.So the question is, what is the other rounded shape, with a water aspect and transliterated Sn like the word /soŋ/ "father or grave"?

Middle-Egyptian : Sn " ring"

May we then suggest that, what the African collective memory is calling the ruler's grave is a reminiscence of this famous ring? If we do not have the obvious answer, the question is still allowed due to the fact that this circular shaped is founded elsewhere in Africa among the insiders.

If one remembers that according to African mythology the body of ***wsir*** was scattered around ***km.t*** land then inevitably, ***skm.t*** represents the tomb of ***wsir***. the ***s*** indicates the word earth (70% of water) of grave. In English the word is sea "water"
The transliteration ***sxm*** associated with the creative life force is called simply:

Ewondo: / i-sia nkomo / "father creator"

The link with the lion is established through the mane:

Ewondo: / i-sil / "roaring; a ripe woman"
Ewondo: / á-sil / "to roar; to growl; to scold"

Ewondo: / i-si(l) / "mane; hair" homophone of **Ewondo:** / si / "earth" and **English**: sea.

Thus, through the ages and along the historical continuum, the lion's figure will be disseminated throughout the world by the African royalties. Lions figures are what true historians often call Nubian lions. Notably the Merovingian royalty is the first royalty of France and those lions can be observed until nowadays on French official symbols. You can also find those lions on the british flag. Unfortunately, There are no lions in western countries except zoo lions.
We can claim that, Jamaica's Rasta culture with their locks is a reminiscence of the lioness ***skm.t*** worship.

But for me, all this remains a secondary meaning. The main acceptation is **the land of Hr**. All the African peoples who have ***Hr*** as ancestor can therefore rigorously claim ***km.t 's*** heritage.
Hr having 99 names, imagine the number of qualifiers you have for ***Hr*** and the names of different lineages.

Crocodile	**Language**	**Peoples**	**Geography area**
Yimbga	Moore	*Himba, Balemba*	Afr. Australe
Yimbga	Moore	*Yemba, Ngemba, Ngoumba,*	Cameroon
Yimbga	Moore	*Mambay, Mambila, Yamba*	Cameroon
Yimbga	Moore	*Imbangala*	Angola
Yimbga	Moore	*Yimbe/Limba*	Sierra Leone
Yimbga	Moore	*Natimba, Naudemba*	Benin, Togo
Yimbga	Moore	*Zimbabwe*	Zimbabwe
Yimbga	Moore	*Wazimba/zimba*	RDC
Yimbga	Moore	*Kanemba,*	Central Af.
Yimbga	Moore	*Embabaan*	Af. Sud
Yimbga	Moore	*Temba, Kabiemba,*	Togo, Benin
Yimbga	Moore	*Bemba*	Zambia
Yimbga	Moore	*Babemba, Wemba,*	Zambia, RDC
Yimbga	Moore	*Bolemba, Hemba*	RDC
Baama	Malinké	*Bamana*	Mali

If we have correctly understood that to read the script ***Km.t*** we refer to the country of ***Hr***, we can just ask Africans how they name the characters on the following script
and compare the qualifiers of ***Hr*** with the names these people use to identify themselves.

Feather *sè-		Foot *ko / *bo		Falcon *Hor-	Vulture *Do-		Face *lo-	Spine *yo-		Ra *Ra-
Massa Moussey Haoussa Bassa Yassa		Cuvok Kapsiki Kotoko, Mbu- kou Podoko		Dourou Banda Hidé Mada Gude			Fali, Gbaya Balong Bulu Kolo			Kra Kanuri Karan Mandara Sara

Yab**assi** Yamb**assa** **Sa**naga **S**oo		**K**ra A**ku**na**ku**na Baba**nki** Bami**lké** Ba**koko** B**akwe**ris Ba**nko**n B**onke**ng **Abo** **Mbo** Nyo**ko**n Akpw**ako**um M**aka** **Kaka** Bi**k**l		Mou**nda**ng Ma**nda**ra Nam**dji** Oul**d**m Ewo**di** **Ndo**nga Bo**di**man Ba**nde**m		D**ouala** Bob**ili** M**lo**on P**ôlé** Bikél**é** Mv**ele**		Tik**ar**, V**ere** Bakwe**ri** Ba- fia

Bee  ***fo-**		Panther  ***ze-/*ngu-**		Mouth  ***nyu- /*djo-**		Seat  ***to-**		White  ***fum-**
M**vovon**		Guizi**z**iga		Ba**djo**u		Ko**to**ko		**Mfum**te
Ma**fa**		L**agwan**		Bab**adjou**		M**eta**		Ba**fum**bum
Mo**fu**		**Zu**lgo		Ba**ndjou**n		Mfum**te**		
Ba**fan**g		**Nge**mba		B**anuu**n				
Foŋ		Ndo**nga**		Ba**nyun**g				
Ba**fou**ssam		Po**ngo**		**Djo, Idje**bu etc.				
		Baya**ngui**.		**Dju**nkun				
		Njanpeta		**Djo**la, **Joo**la etc.				
		Japan		**Djo**lof				
		Nzim						

We could continue this classification with other descriptors and add what has been flocked to the names of ethnic groups in Africa by contempt and by ideological sufficiency and especially by ignorance.

Did Diop gets it wrong by saying that *km* means Black? And therefore  km.t + niw.t = land of the black nation?

Now that we have understood how to read the transliterated ***km.t*** script, to answer this question we must already put things into perspective. At the Cairo symposium in 1974, there were three camps:

- The first, and the most important camp was the camp of Western researchers and scholars. According to them, the people who created the Pharaonic civilization were a white people from Europe, who had landed in Africa to create this civilization, while tanning and mingling with (black) African elements over time. Therefore according to this camp the Pharaonic people was a people of European white origin, and that the colors and colors of dark or black skin that it is possible to see on the pharaonic frescoes were only the results of a miscegenation.
- The second camp is the camp of Orientalist researchers or specialists of so-called Oriental civilizations. According to this camp, the Pharaonic people came from the east to create this civilization on African soil.

- The third (and last) camp was that of those who said that Pharaonic civilization was a black civilization, created by blacks and belonged to the African world. It was the line embodied by Cheikh Anta Diop and Thophile Obenga and also the most contested camp by orientalists and Western researchers, who were already conveying at the time through their media (as they still do today), all kinds of things to prove that pharaonic civilization was not black.

Diop's mission during this conference was not just to read the ***km.t*** script or to give all its meanings; far from it. He was called upon to decide on a subject of greater importance, to deliver Africa from marginalization and colonial asylum confinement. So he used a subtlety of mind that was thought by the scribes and that consists in condensing homophones and their ananyms[209] in a single script. Taking into account here the dictatorial position of Egyptology we have:

Middle-egyptian: km
Wolof: hem□hm<km "charcoal "
Ewondo: [kam] ou [mâghe] "charcoal" □km/mgh

In the same way as the symbol I6 "***km***" is a crocodile tail, associates the country with crocodile, scales of crocodiles, spots etc. the same way the sign G17 represents the owl, the accomplished man and allows to name the country and the inhabitants. Let's note that those inhabitants are not owls, these are just semantic shifts specific to African languages. This is the perfect use of metonymy.

The plethora of possible semantic shifts offered by the ***km*** consonants skeleton has been highlighted with a listing of kekwé group words (*see it below at the end of the paragrah*).

I want to note in this regard that i did not find it useful to list the words from the Kweké or Kikuki branches of the Negro-egyptian that are also used in Ewondo and are anagrams of the kekwé branch. Those word are simply obtained by metathesis[210] of consonantal skeletons. I did not find it useful to list all the fun shapes that are semantically related to this script. What matters to the understanding here is the principle according to me.

The ***mDw nTr*** by "omitting" the vowels, the root of the word ***km.***and the root of the word coal ***km*** / ***mk*** (as well as all the variants obtained by weakning) in the Ethiopian languages being in every way identical, one can therefore operate a semantic shift. So Anta Diop used this subtlety of writing offered by the scribes to support his claim to prove that the civilization of ***km.***was the fact of black peoples.

On another way, we can close this debate definitively by reading all the meanings that the consonant skeleton offers ***km.t***

1 ***km.t*** "Black land, Egypt" (*Faulkner p. 286, Wb. V p. 126, 7-18 sq.)*
2 ***km.t*** "Egyptians"
3 ***km*** "Black"
4 ***km.t*** "a jar"
5 ***km.t*** "completion, realization; final countdown"

Since we must close this debate at least I hope I will gladly leave aside for the moment the scripts 1, 2, 3 and 5 because they are the problem and read the 4 because it designates a common object of the basic

209.An ananym is a word that can be read in the normal reading direction or in the opposite direction. The palindrome is a special case of ananym where the order of letters is the same whatever the reading direction; in the general case, the ananym has the inverted order according to the reading direction. By definition, an ananym is an anagram. Example: "Lon" and "Noël", "**me.se**" and "**seme**" in Ewondo.
210.Metathesis (from the Greek μετάθησις metáthêsis, "permutation") is a phonetic modification of the family of metaplasmas involving a more or less important exchange between two phonemes in contact or near. Example: **ms** et **sm**.

african lexicon. We will therefore start from the diktat of Egyptology which dictates a standard consonants skeleton.

> **Middle-egyptian:** ***km.t*** "a jar"
> **Ewondo:** vyəkh[211] "pottery jar made from clay"

I paste below the definition found in the dictionary fang-French Galley.

> *"[...] **Clay vase**: ñkakh ô vyəkh [212]..."*
>
> *(Galley 1964)*

For the non-beti: the sign ***w24*** is read in Ewondo: /ŋkakh ô vyəkh/"pottery jar made from clay". Pottery being Ewondo:/ŋkakh ô vyəkh/ the vase is thus often called / vyəkh / or so in a simplified way / mvié / but let's leave aside the last simplified form and remain on / vyəkh / because it is this variant that will guide us here.

Note that this variant admits a homophone:

> *"VYOKH (h) vb. Suck, shag. /Môr a **vyokh vi/, "**the man sucks a bone".*
> */A **vyokh** aman mon/, "he kisses the child on the jowl".*
> *Syn.: **vyôkh, vyô, fyokh, fyôkh.**"[213]*
>
> *(GALLEY, 1964)*

Now let's take our script N°3: ***km*** "black"

To satisfy the standard techniques authorized by the unidentified censors, let us dare to propose a reconstruction of this Pharaonic word from the Negro-egyptian from:

Middle-egyptian: ***kkw*** "darkness, obscurity, absolute black". (Faulkner p. 287, Wb. V p. 142, 10+ sq.)

To make this short cut, we can propose that the proto-version which gave birth to the known form in pharaonic is of type: * **kwkh-**. This form will give in Middle-egyptian the variants below:

> ***kkh-** but also
> *$ŋ^{jw}$**əkh-***
> ****wkh-/*vkh-***

The kékwé branch will provide the variant ***km-*** by transfert of the its velarization on the **w**.
Recall that the operative rules stipulate that: the phonetic complements allow to find the good vocalization of the determinative.

> **Middle-egyptian:** ***km.t*** "a jar"

If then, as Galley says in 1964, the word for clay is / vyəkh / in the languages of the kongo basin, I can not be more royalist than the king namely that, the biliteral ***km*** offers a vocalization like * ***wkh-*** / * ***vkh-*** and having the meaning of black in our languages since semantically related to darkness.

> ***km*** "black"

---

211. VYEKH (m) n.7, ss pl. Clay, pottery clay, pottery. Bulu use the pl. mevyekh. Lo yekh, take the clay for pottery,Galley 1964, p.363
212. BUCKET: vase, native jug, кkakh (h). Clay vase, ñkakh ô vyekh (hm). Thin copper bucket, ñkakh nžeñy (hb), ôlôm (b).Galley 1964, p.537
213.Samuel GALLEY, Dictionary FANG-FRENCH. and. FRENCH-FANG followed by a fang grammar,1964, p.364

The owl hululume is made with the onomatopoeia type **"* wkh".**We have the hair lock:

Ewondo: /*ùm*-bokh i-sil/ offers the homophone / *ùm*-bokh / "the primordial ancestor"

who is the breatfeeded son of the trinity (***wsjr-As.t-Hr***).

As Galley says, sucking is / vyokh / in the languages of the Kongo Basin. This script reads so:

Middle-egyptian: ***km*** "black"
Ewondo: /vyoukh/ "extremely black, absolute darkness"

We have in the script ***Xnmw*** "Khnoum[214]", we know that the symbol w9 is a vase of clay. So we have here:

Ewondo: /vyəkh/ which we associate here the semanteme ***Hr***.

The consonant ***n*** of ***Xnmw*** is an atavism of ****njw-*** of the Negro-egyptian often used as animated prefix or agent prefix. It indicates here the one of clay, the potter in sum. The best tranliteration could be ****Xm.w***. In this way we have a clear equivalence with ***km***.

We can note here the spiral sign called / *ùn-m*bokh / in Ewondo. / *ùm*-bokh / being homophone of the primordial ancestor / *ùn*-mbôkh /. This variant also takes the qualifier * wkh - / * vkh- because it is bound to absolute black idea.

If we had a doubt about all this, the scribe has for long time planned the script:

Middle-egyptian: ***knm.t*** "darkness, absolute black"
Ewondo: /*ùn*-ŋkoukh/ "darkness"

As mentioned above, it should be transliterated ***nkm.t.***

Note nevertheless that our languages offer the variant / byôkh / "Full of wrinkles, old".
Before closing this demonstration I took care to record the script below:

Middle-egyptian: ***sbk*** "Sobek"
Ewondo: /*ùn*-ŋkomboro/ "The one who makes men"
Ewondo:/*ùn*-nkwébék/ "abyssal depths"[215]
Ewondo: /ŋgan mə-dza/ "Ekang animal mythology"

I found two interesting ideas in the script above transliterated ***sbk***:

Middle-egyptian: ***s/z*** "man; someone; person"
Middle-egyptian: ***s*** "ornamental container"

I still remember just what I said above, / byôkh / "Full of wrinkles; old.
sbk = s + bk, should I still say it here? No I do not think it's extremely clear. The potter is the oldest man full of wrinkles, he is the men maker. He is the oldest ancestor. According to african thought, we often say:

214. Crole ayisyen: Konouwn "Khnoun"
215. Samuel GALLEY, Fang-French dictionary. and French Fang followed by a fang grammar, Neuchâtel 1964, p.398

Ewondo: /Mə-bi mə ŋgan mə nə fəm/
English: "cayman feces are white because he's eating white clay".

Come on, I give you this last script and then we do not talk about it anymore...

Middle-egyptian: ***mDt*** "deepness"
Ewondo: /ndome/ "very deep"

The symbol V19 is a hindrance for cattle called / *ùn*-mbokh /. Often repeated, we begin to understand how works the endogenous method. This hindrance replaces the crocodile of the ***msH*** script. The scribe has replaced the crocodile with the ultimate crippled legs and reminded us that he is hindered. Let's then applaud the scribe clap clap clap.

The crocodile is therefore the animal of deepness, it is the absolute black, the perfect representation of the abyssal darkness. This is why the collective memory of the people who crossed the Sanaga on the back of a snake associates it with the crocodile.

Let's listen to, Bernard Germain Etienne de La Ville sur Illon better known under the name of La Cépède about crocodilians.

> *"[...] and the stories of the travelers, we thought we had to count only three species among these enormous animals.*
> *The first is the ordinary crocodile, or properly so called, which lives on the banks of the Nile; it is called alligator, mainly in Africa, and it could be called the green crocodile, which has already been given to it;* ***the second is the black crocodile,*** *which Mr. Adanson saw on the greate river of Senegal; and the third, the crocodile which inhabits the banks of the Ganges, and to which we retain the name of* ***gavial****, which has been given to it in India. These three species resemble the distinctive characters of the crocodiles we have just indicated; but they differ from each other by other characters, which we will relate in their particular articles.* ***The crocodiles of America have been called the cayman*** *who has been borrowed from the Indians. We have carefully compared several individuals of different ages with* ***crocodiles of the Nile,*** *and we have thought that* ***they are absolutely of the same species as the crocodiles of Egypt****; they show no remarkable difference which can not be related to the influence of climate. In fact, if their jaws are sometimes less elongated, they can never differ sufficiently, by their shortening, from those of the crocodiles of the Nile, so that the caimans constitute a distinct species, all the more as this difference is very variable, and that the crocodiles of America resemble those of the Nile by the number of their teeth, as one individual resembles another among these last crocodiles. It has been claimed that the cry of the caimans was weaker, their courage (...)"*[216]
>
> *(LA CEPEDE, 1819)*

After this confirmation of the link between the black color and the Egyptian crocodile in the light of Lacpède's words, let us summarize a little bit:

Material

Middle- gyptian: ***km.t*** "a jar"[217]
Ewondo: /vyəkh/ "clay pot"
Ewondo: /i-kôn/ "clay"

Color

Middle-egyptian: ***km*** "black"
Ewondo: /vyoukh/ "absolute black, extremely black"

216.216. LA CEPEDE, Natural history of quadrupeds-oviparous,Tome premier, Paris, 1819, P.103.
217. See illustration N°37: black clay pot left by pre-Harappan peoples from Egypt to create the Indus civilization -2700 years.

Aspect or shape

Middle-egyptian: ***kHkH*** “become old, grow old”
Ewondo: /*ùn*-nkoukh/ “extremely old”

Shape

Middle-egyptian: ***iAkw*** “old man, old”
Ewondo: /byokh/ “extremely old”

Suction noise

Middle-egyptian: ***snq*** “to suck; breast-feed; to feed, to treat”
Ewondo: /vyokh/ “suck, aspire, sucking sound”

Souds (otherss)

Middle-egyptian: ***nmi*** “to scream, to mooze, to bellow”
Ewondo: /i-woukh/ “shouts of owl”

Let's finish the translations of the remaining scripts to know: the 1 and the 2.

Middle-egyptian: ***km.t*** “Black land, Egypt”

<u>Etymology</u>

French: *vieux “old”* (by omitting the vocalization of the final valare). We have vyəkh>vieux
Italian: vecchia *“old”*
English: waste ☞**Middle-egyptian**: ***wAs.t*** “Thebes”
Ewondo: vyokh**;** /byokh/ *“old”*

It would be more appropriate to change what is noted by ideology in the dictionaries and replace the etymology [218] of the old word in its true environment of origin in Africa because Latin did not fall from the sky, it derives from African languages. There is no humanity of Latin origin**.**

If we understand how it works, this is the country of ***Hr***. He can therefore take each qualifier related to ***Hr***, namely, his age, his color, his appearance, his trade, etc.
We will read in ewondo according to the meanings:

Ewondo: /Un-nam *ùn*-nkomo/ “The maker’s land”
Ewondo: /Un-nam byokh/ “the country of the primordial ancestor”
Ewondo: /Un-nam ùn-mbokh/ “the country of the shackled, the unique, united”
Ewondo: /Un-nam bə-Koé/ “the country of those of The foot”
Ewondo: /Un-nam eyo/ “the country of the thorn”
Ewondo: /Un-ngwala/ “the trunk”

Middle-egyptian: ***km.tyw*** “Egyptians”
Ewondo: /mi-‘mbobokh/ “the shackles, those of Hr”
Ewondo: /mí-‘ntóbèn/ “
the elected”
Others: *Akunakuna, Babanki, Babungo, Bamiléké, Bamoun, Ngemba,Nso, Tikar, Vere, Banda, Cuvok, Dourou, Fali, Gbaya, Namdji, Gude, Guidar, Guiziga, Haoussa, Hid,*

---

218. Vieux: From popular Latin vĕclus (attested to in the 5th century), alteration of v “tŭlus "same meaning", diminutive of vetus "idem" (ca 1050) velz (Vie de saint Alexis, ms. du Psautier de St Albans).

Kanuri, Kapsiki, Karan, Kra, Komas, Kotoko, Lagwan, Mada, Mafa, Mambay, Mambila, Mandara, Massa, Mboum, Mbukou, Meta, Mfumte, Mofu, Mouktl, Mousgoum, Moundang, Moussey, Nsungli, Oulédém, Peuls, Podoko, Toupouri, Sara, Vam, Voûté, Yamba, Zulgo, Akpwakoum, Maka, Badjou, Gbete, Konabembe, Nzimé Mloon, Mpoumam, Nkounabem, Bobilis, Pôl, Mvele, Gbaya, Kaka, Mvovon, Bikélé, Bafia, Ban, Banen, Boundjou, Dzimou, Yambassa, Nyokon, Sanaga, Bétis, Fangs, Bulu, Maka, Badjou, Baka, Soo, Vute, Njanpeta, Lemande, Batis, Luba, Massaï, Miènè, Dogon, Songhaï, Kongo, Himba, Akpwakoum, Maka, Badjou, Gbete, Konabembe, Baoulé, Bété, Mossi, Abrong, Boron, Bron, Brong, Bono, Doms, Tchaman, Nzimé Méloon, Mpoumam, Nkounabem, Bobilis, Pôl, Mvele, Gbaya, Kaka, Adjoukrous, Mvovon, Bikélé. etc.

Semantax

Ewondo: /i-mor *ùn*-mbokh/ "other" litt. (the other hindered; the other one).
Middle-egyptian:  ***km.t.*** we have the numeral ONE and the vase which are identical.
Ewondo: /*ùn*-mbokh/ "hollow, one"

Middle-egyptian:  ***km.t*** "completion, realization; final etc."
Ewondo: /*ùn*-n**kom**èn / "completed, realized"
Ewondo: /*ùn*-**mbo'**èn / "completed, realized"
Ewondo: /à-songa/ "completed, realized"
Ewondo: /*ùn*-ŋkenya/ "completed, realized"
Ewondo: / à-mana/ "completed, realized" We can note the root ****mn-***
Etc.

If we have understood how the reading of the ***mdw ntr*** according to the African prism works, we understand perfectly well that the transliteration ***km*** here proposed by Egyptology is an acceptation among thousands others possible by homophony, antonymy and by synonymy etc. In fact we do not have to use the transliteration ***km*** to please the standard. The proof the script below gives however the same level of information.

Middle-egyptian:  ***rmT***

The mouth that replaces the crocodile or its skin is a hollow, / *ùn*-mbokh /. What did the scribe do? In our languages we observe that, the scribe replaced the crocodile / *ùn*-mbokh / by the mouth / *ùn*-mbokh /. So the reading remains the same from our point of view namely: /mi-'mbobokh/ "Those of ***Hr*** ". Here is the limit of the transliteration.

To qualify a black color in ewondo I have a pallet of words: / i-Hindi / "black": / *ùn* nsur / "black"; / vyoukh / "black"; / i-momokh / "black"; /kam/ "black" etc. The same level of analysis could have been done with each of these terms and reached the same conclusions, regardless of the registers requested. One may even wonder if this transliteration ***km*** is not an ideological dead-end that some wanted to distract Africans and discourage them in the quest for a chimeric identity straight out of ideological laboratories. I therefore advise Africans to move away from kamitology, kemitism and kamitism which are nonsense as has been calmly but surely demonstrated.

Obviously, Anta Diop was absolutely right. It would take Martians to prove the opposite of his genius on this issue. Do not make me say what I did not say namely that the only meaning of ***km.t*** would be ***Black***. I affirm that this is one of the many meanings that can be rigorously deduced from the ancient method of compiling all in one. In the same way that the French skeleton consonant ***mr*** can give the signifiers mari "mari", mûr "mature", mur "wall", mairie "town hall", maure "Moorish", etc. it does not refer to one and only one signified.

This unique method that by its genius compiles levels and registers of languages; mark of the subtlety of spirit of the peoples from the region of the greats lakes. *Km.t* is a facet of the diamond. The earth of *km.t* laden with alluvium had a black aspect. Egypt is a "gift of the Nile," as the Greek historian Herodotus wrote in the fifth century BCE.

This alluvial soil whose primal ancestor 𓅓𓏺 ***Hr*** used to make vases and black humans. Let it be said in the light of this that this is the land of the black nation, nothing shocking. You really need the courage to prove the opposite and challenge the obviousness of the facts. It is therefore not advisable to take the 𓆎𓅓𓏏𓊖 ***km.t*** meaning and to say that it means only black on that account, we have not understood anything.

I reiterate that it means the land that belongs to 𓅓𓏺 ***Hr***. The main issue is not the script at all 𓆎𓅓𓏏𓊖. It is the established argument that dictates the transliteration ***km.t.***

If I wanted I could submit to the established argument and read this script as follows

Ewondo: 𓆎 /ùn-ŋkôm/ "crocodile"
Ewondo: 𓅓 /i-kùmù/ "owl; parent-tree"
Ewondo: 𓏏 /i-kôm/ "heap"
Ewondo: 𓊖 /i-kôm/ "land, country"

Just by nominalisation of each word i should have obtained the the right reading of this scripty using the standard consonants skeleton ***km***. But this reading is extremly poor at the first level of analysis. Let's observe that /i-kùm/ "tress stump, parent tree". So only the owl indicates the ancestor of the land 𓊖. We therefore have the land of the ancestor once more.

According to historiographyical sources[219], the settlement of the Indus valley was made from African populations of kushite type, a fraction of which comes from the Nile valley. Throughout history, relations between the Nile Valley and the Indus Valley are also quite maintained especially around -525 BC, with the Persian invasion of the delta of the land of **Hr**. various sources, during these periods of unrest, many priests of 𓌀𓏏𓊖 ***wAs.t*** /I-vés/ "bone; Hr" or of 𓏠𓈖𓄤𓉴𓊖 ***mn nfr*** / *ùn*-nam à-bəŋ Zamba/ "the beauty city of Hr" were exiled by Cambyse II. Among these priests many ready one of 𓏠𓈖𓄤𓉴𓊖 ***mn nfr***, Siddhārtha Gautama, better known as bright:

Ewondo : /Tara Sira Itughà maâ/

Sanskrit: Bouddha
Ewondo: / ù-ndâba/ "brightness, fontanel"

__**Nota bene**__**:** the two words proceed simply by inverting phonemes

Siddhārtha Gautama fled to the Indus Valley. He founded Buddhism, which has kept the peaceful character of black spirituality, although it has undergone transformations to adapt to the Asian cultural universe. This information is capital historically because the spirituality of Indian temples to this day gives a good picture of what was achieved in the Nile Valley 6000 years ago. Looking a little bit at the Indus Valley we can note that:

> *"In Hinduism, Gaṇesh, Gaṇesha, Vinâyaka, Gaṇapati ("the chief -Pati- troops of deities -Ganas-" or better "the lord of categories") or Pillayar in southern India is the god who suppresses the*

219. Jean LACROIX de MARLES, General History of Ancient and Modern India Since the year 2000 BC until Our Days - Preceded by a Geoscriptical Notice and Special Treaties on Chronology, Religion, Philosophy, Legislation, Literature, Volume 1, Emler frères, Librairies-diteurs, Paris, 1828, p.470-472

> *obstacles. He is also the god of wisdom, intelligence, education and prudence, the patron of schools and knowledge workers. Recognizable by his elephant head, he is undoubtedly the most revered god in India and his aura touches even the entire Indian subcontinent and Asia in general. He is the son of Shiva and Pârvatî, the husband of Siddhi (Success), Buddhi (Intellect) et Riddhî (wealth)." (cf. illustration 19)*

We say in Ewondo: / Ngan mə-dza / "the crocodile of deepnesss" which symbolizes the primal ancestor, the land of ***km.t***. The country with elephant head that we call the name / I-kama / "male elephant". Hindus whose word derives its etymology from / i-Hindi / "Black" from Ewondo only celebrate the unique ancestor of the peoples of the Nile Valley using the ekang root of the word / Ngan / "crocodile".

What is rather humorous is that Hinduism has transported this practice and language to Asia Minor, Japan[220] among others.

> *"Kannon worship began in India around the 1st or 2nd century AD and then spread to Southeast Asia, China, Korea, and most other Asian nations. Veneration of Kannon in Japan began in the late 6th century, soon after Buddhism reached Japan by way of Korea and China. In Japan, Kannon's paradise is known as Fudarakusen. It is commonly said to be located at the southern tip of India (which supports theories of Kannon's Indian origin). Many Kannon statues from Japan's Asuka Era (538 to 710) are still extant. Originally male in form, Kannon is now often portrayed as female in China, Japan, and other East Asian countries. Each of these nations dressed Kannon in different forms to suit their own temperaments and spiritual concepts".*
> *"It is agreed today to place in the sixth century not only the Buddah, but the whole religious and philosophical movement of Asia, with Confucius in China, Zoroaster in Iran; which would confirm the hypothesis of a dispersion of Egyptian priests at that time spreading their doctrine in Asia. It is difficult to explain this religious movement by a synchronic evolution of the different interested countries"[221].*
>
> *(DIOP ANTA, 1954)*

Since African thought has spread to the Far East through Buddhism, it gives quite interesting linguistic information:

> *"The Japanese name of Ganesh is Shōten (聖天) or Kangiten (歓喜天), le Japanese Buddhism considers it a manifestation of Shō Kannon Bosatsu (聖観音菩薩). In Japanese, Kanji 天 is used as the equivalent of Hindu Deva".* (*cf. illustration N°19*)

Middle-egyptian: ***Hr***

Ewondo: /Ngan mə-dza/
Hindi: Ganesh "elephant's[222] head"

Bulu: nkot "saviour" **Ewondo**: /nkor/
Japonais: Shōten[223] (聖天)

Ewondo: /*ùn*-ŋkaŋ à-twan/ "The scribe"
Japonais: Kangiten (歓喜天)
Middle-egyptian: ***Axn-itn*** *"akhenaton"*

Ewondo: /i-káŋ / "pattern"
Japonais: kanji (天)

Japonais: Shō Kannon Bosatsu (聖観音菩薩).
Ewondo: /I-kaŋ à-su kulu/ "representation of the holy visage of "***Hr***""

As we can notice here, the same reality has been named with different descriptors but all comprehensible from an African language target here the Ewondo. I am therefore convinced that the bridges between

---

220. Engelbert KŒMPFER, Natural, Civil and Ecclesiastical History of the Empire of Japan Volume I,P. Gosse, & J. Neaulme, La Haye,1729, P. 32§33.
221. Cheikh ANTA DIOP, Negro nations and culture, Présence africaine, Paris, 1954, p.194.
222. cf. illustration 19
223. The term Japanese Shoten recalls the passive Nkotèn form of the bulu and nkorèn "saved" from Ewondo.

the so-called Asian languages and the current African language variants are possible from the moment when we know some linguistic skills.

I would not dare to invite myself into the supposed debates of experts like Noam Chomsky. The only experts who alone are able to classify languages and to write a universal grammar, even to say what humanity must or may not think about linguistics. It seems that recursivity is at the basis of all human language. No human language can be non-recursive including the Indian language of Amazonia Piraha.

> *"If a grammar does not have recursive devices [...] it will be prohibitively complex"*[224]
>
> *(CHOMSKY,1957).*

Harry van der Hulst asks a number of interesting questions about this condition of the mandatory recursion of human language:

> *"How might a linguist or a Child attempting to acquire the proper grammar determine whether a grammar, or a particular grammatical rule, is recursive? The standard argument that grammars of natural language must contain recursion dates back to Chomsky (1957), Who pointed out that natural language appears to have the property of discrete infinity: it is composed of discrete basic elements (words) which can be combined to produce apparently infinitely many sentences. An infinite set can only be generated from a finite grammar if the grammar contains some form of recursion. But is it true that natural language is infinite? After all, there are no infinitely long sentences, and only a finite number of sentences (...)"*[225]
>
> *(H. de HULST, 2010)*

Let's leave aside these debates of experts who do not speak a single African language and come back to the script  ***km.t.***

SAME WORLDWIDE STORY

Regarding the crocodile, we find curiously far from Africa as La Cepede noted a kind of dimension ontology related to the crocodile.

Among the Papuans, the myth tells that even before humanity exists, a giant crocodile has moved its tail, the earth has then agglomerated around it and a hillock has emerged from the water. Thus would the Earth be born.

In fact, in New Guinea a party is reserved for the crocodile whose celebrations are between August to October. I was interested because of the **km.t** script that integrates the primordial ancestor of the African peoples.

In Papua, "man-crocodiles" are called Sepik[226], which curiously reminds the script  ***sbk*** vocalized Sobek by Egyptology and / *ùn*-nkwebek / "deepness" in Ewondo.
Besides the party, the river where the festivities take place is called Sepik too.

There are nevertheless many curiosities in the expression of these peoples, the names of the populations, especially the **Kaluli**, hunter-gatherer people who find all they need in the immense primary forest that surrounds them. We can approach the Indian Kali but the root * Kal "Black coal" which also serves to designate **Hr**. Uli meaning simply head / ù-nlo /; / Elo /. ***Kaluli*** meaning literally: black head. But the Black mentioned here is the black coal of DIOP and the German school. Do you follow me?

What most suprised me is the party reserved for Ambunti which takes place in August of each year. In the word Ambunti, the root * Amba / * Amma is used to designate ***Hr*** in Africa.

---

224. Noam CHOMSKY, The standard theory of generative grammar, 1957, p.24.
225. Harry VAN DER HULST, Recursion and Human Language, De Gruyter Libri, 2010, P.160.
226. Fritz MORGENTHALER, Florence WEISS, Marco MORGENTHALER, Dying River Conversations: Ethnopsychanalysis in Papua New Guinea's IatmoulsEditions Zo, 1987, P.34-35-163.

And, the particle / * nti / the lord in Ewondo. If Ewondo says /nti-Zamba/ or /Zamba nti/ in Papua, they say **Ambunti**. Judge by yourself.

Surprising among the Aka of southern Sudan or Yam country, we find the same name in the human groups called Pygmies (Aka, Babenzélé, Babongo, Baka, **Bambuti**, Batwa, etc.).

Ewondo: /i-bà bə Nti/ "those of the lord"
Ewondo: /bà bə Nti/ "those of the lord"
Ewondo: / bə y'à-bo Nti/ "those of the foot of the lord"

Bambuti *: Those of the foot of the lord*

I took care to note a few basic words of their language, the **Hiri motu** literally in Ewondo / i-Hindi mot / "the Black Man" but this word above could admit the translation in my opinion is the Man of ***Hr***. In Ewondo we should way / mot Hor/.

Hiri motu: *imana* "head"
Kinyarwanda: *imana* "primordial ancestor"

Ewondo: */à-mànà/* "the end" in a circle the end and the begining is the same point.
Ewondo: /á-màn/ "terminate, finish"
Ewondo: /i-mén/ "Him-self" concept of ipseity linked to the unique.
Ewondo: /*ùn*-nàm/ "the country, the land"

The signified is the same because ***Hr*** the primordial ancestor is the leader, the beginning and the end.

Hiri motu: *turiana* "foot" *litt.* en **Ewondo** */ntwar/ "unique"*
Ewondo: */i-tiri/* "foot" litt. What serves to trample
Ewondo: */á-tiri/* "to trample"

Hiri motu: *Honu* "full, whole" *litt.* en **Ewondo** */ù-nyu/ "the finger"*
Ewondo: */à-nyikh/* "bend the finger"
Kinyarwanda: *ijuta* "full" *litt.* **Ewondo** */i-nyigha/ "Folding"*

Hiri motu: *dina* "sol" *litt.* en **Ewondo** */i-diŋ/ "love"*
Ewondo: */i-daŋ'a/* "the one which crosses"
Kinyarwanda: *zuba* "full" *litt.* **Ewondo** */dzobo/ "sol"*

I was surprised to find rituals linked to the Crocodile among the Pirahã Indians in the Amazonia who live in seclusion. I was amazed with the proximity between the basic vocabulary of Xapaitíiso their language and the words of Ewondo.

Xapaitíiso: *giopai* "dog" ☞*litt.* In **Ewondo** */ngup/ "hippo"*
Ewondo: */mvu/* "dog"
Middle-egyptian: ***xAb*** "hippo, to be folded"
Middle-egyptian: ***sAb*** "jackal"

Only by the mDw nTr one understands how and why the word dog and hippopotamus are the same entity.

Xapaitíiso: *hisi* "sol" *litt.* ☞ in **Ewondo** */si/ "sol"*
Ewondo: */à-sè/* "feather"
Middle-egyptian: ***sw*** "sun"
Middle-egyptian: ***Sw*** "sol, solar light"

Xapaitíiso: *iho* "to walk" *litt.* ☞in **Ewondo** */á-hà/ "here"*
Ewondo: */à-ko/* "foot"
Middle-egyptian: ***aHa***"to stand up, to find yourself somewhere"
Middle-egyptian: ***X.t***"foot"

Xapaitíiso: *baíxi* "mum" *litt.* in **Ewondo** */ùn-mbèh si/ "yard"*
Ewondo: */*nya/* "mum"
Middle-egyptian: ***mAat*** "the one who shares the land and the food"

How do all these peoples presented as distant from each other, keep the memory of a common ancestor and especially saurian?

Other approach

After satisfying the constraints of transliteration that impose **km.t** as consonantal skeleton.
We repeat that the stransliteration ***km.t*** is the consonants skeleton of / *ùn*-ŋkôm / "crocodile" here highlighted by **Hr** as the manufacturer / *ùn*-nkômo /
But, except that **Hr** is not just a manufacturer or creator, he can also referring to:

Middle-egyptian: ***rd*** **"**foot" in african thought
Ewondo: /*ùn*-nloh/ "The head"
Ewondo: /à-su/ "The visage"
Ewondo: /à-nyu/ "The mouth"

But also by the meaning:

Ewondo: /à kul/ "paw"
Ewondo: /à-kol/ "foot"

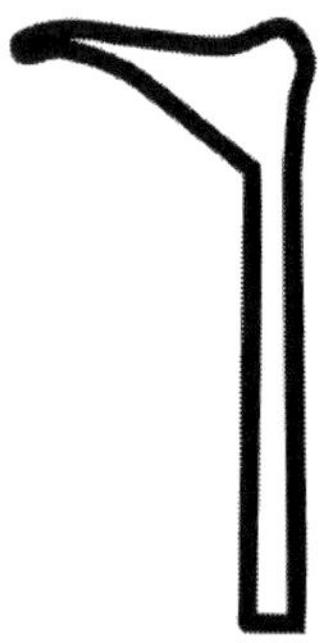

Àkul *the foot, paw*

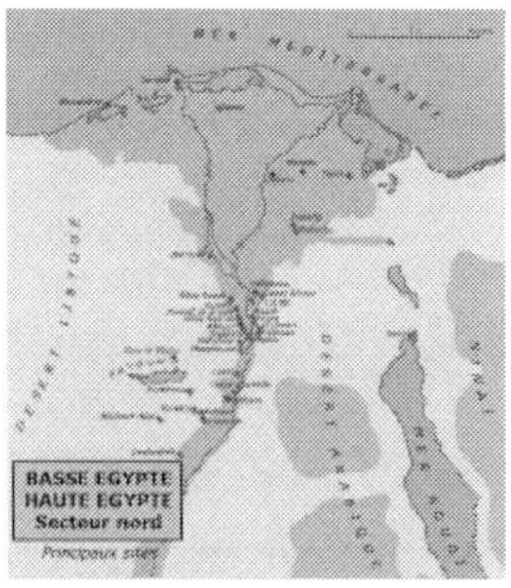

Km.t *or the bigfoot*

Ilé : Tree = kmt

Ewondo:/ i-lé/ or /i-ré/ "tree" ↔ **Middle-Egyptian** : *Elo*"ancestor"
Middle-Egyptian : ***imAm*** "tree" ↔ **Ewondo**:/ i-mvam/ "ancestor"
Note :

Birds usually live on trees. For their safety the often land on the top of the trees. It is therefore this idea that the scribe will associate with the tree and the idea of the summit.
Because we are talking about the first ancestor also then, all the characteristics of the ancestor will be here shared by birds (falcon), the tree, the country and the primordial ancestor.

These words for foot will admit the phonetic proximity / * kal / which designates the charcoal for the black color as we have seen previously but also the primordial ancestor ***Hr*** in some African linguistic variants. Charcoal is a combustible obtained by carbonization of wood in a controlled atmosphere by pyrolysis (in the absence of oxygen).

So by metonymy, ***km.t*** is the black color. ***Km.t*** heard as country of ***Hr***.

As a result, some inhabitants of this country will be called Kɔlɔ, for example, etc. and one of them will be the kɔlɔ or lekɔ (including Bami[227]**-leke**). We find this root * kol- in many African names and anthroponyms. We have Lekié as a place name of a department of Cameroon. We have name like Ekwala among Duala people in Cameroon also even Ekellé department.
This root ***kul-** is a word game to express the idea of the beginning : /à-kùli/

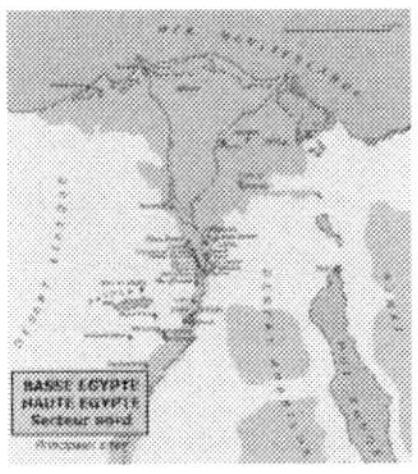

Km.t *or the beard*

Zoa : Elephant

U-kôm = Km

Funeral Mask of Tutankhamun, exhibited in the Egyptian Museum of Cairo.

The black cobra is well known in Africa and is a holy serpent. We call it Ewondo :/ *ù*-kôm /. It can be considered as a rope: Ewondo: / *ùn*-ŋkol /

Ethnonyms

A*kol*i; An*kol*i, Sara*kollê*, Kpele; *Gol*a; Ba*kol*a; Ma*kol*olo; Ba*kol*we; N*kor*e, Nyan*kole*, N*kol*; Ma*kala*nga; *kal*; *Kal*enjin; *Kal*igi, *Kal*iki; *Kal*anga etc.

Toponyms
N*kol*tutu; An*kol*e (Ouganda); Ashanti (Ghana); koto*koli* (Togo)*; Kolibantan;* Ban*kol*e-kigezi; Duk*kàla; Kala*hari, Dan*kali*; Leki (Cmr) *etc.*

Anthroponyms
*Kol*i (Sierra Lone, Sénégal); *Kolekol*e; *Kul*ango; *Koul*an; *Kol*ambio; *Kokol*i; M'Bo*kolo* *Kal*undwe; A*kale*, *Kali*; *Kari*; Ba*kari*; A*kala*k, Ekwala, Bikula (RCA); Bakéla (Congo) etc.

It is not useful to repeat this exercise with other identity markers of ***Hr.*** I believe that the principle has been understood.

227. Bami: those of

After living with Africans, Marcel Griaule said in *The pale fox*:

> *" [...] the sp dayi the clear word concerns the building of knowledge in its ordered complexity"*[228]
> *(GRIAULE, 1991)*

In the African spirituality, which has various names according to the region: Mbog basa'a, Kommo, Kam Bamiliké, kuma, Kop.t, which we have linked here with the script ***km.t***, we recognize five kinds of speeches:

- THE FACE SPEECH: whose role is to provide simple explanations.

- THE SIDELONG SPEECH: which deepens the explanations, draws attention to some essential parts of the rites.

- THE HIND SPEECH: which is used to provide summaries covering large sets. It applies to show the correlation of things.

- THE CLEAR SPEECH: which concerns the building of knowledge in ordered complexity.

As a level of knowledge, the speech of African initiates is conveyed by the ordinary language, but also by the secret languages that contain sacred texts and modes of silence. Thus, the human microcosm makes speech a creative power, a transformative force positively and negatively. It is in this that one can say that the macrocosm and the microcosm, Dzo, Diop, Djoba etc. and man, the universe and society, globality and specificity are inscribed in Africa, in a relationship of homo- phism.

We could have analyzed the sign 𓎡I6 given for the biliteral ***km***, the sign 𓅓 G17 representing the owl and show that they are operative concepts and not mere meaningless alphabetic consonants. When combined they do produce a concentrate of African sociology. These analyzes would be outside the scope of this introductory work. Depending on the context and the language registers the ***km.t*** script will offer a precise meaning for the one who understood how it all works.

Negroes? Africans? Black? Blacks? Who are we at the end?

The etymology of the word Africa has been the subject of many assumptions.

According to Michèle Fruyt[229], the term Africa appeared in the European languages through the Romans who designated the northern part of the continent because, in Campania, africus qualified the rainy wind from the region of Carthage.

According to the hypothesis of Daniel Don Nanjira[230],
the Latin word Africa could come either from the name Afridi, a tribe who lived in North Africa near Carthage, or the Phoenician term Afar meaning "dust".

According to other researchers, the word Africa comes from the tribe Banou Ifren (Amazigh tribe), whose ancestor is Ifren, also called Iforen, Ifuraces or Afer (also meaning "cave" or "cavern; grotto" in Berber language according to Ibn Khaldun). Ifri, the singular form of the word Ifren, also designates an Amazigh deity. Still others refer to the Banu Ifren as the inhabitants of the ancient ifrīqīyā إفريقيا which once designated in Arabic the current Tunisia and the name of Africa derives from the nomination of the tribe of Banou Ifren. Moreover, the Banou Ifren would be the Ifuraces, tribe that brings together the Afar. The Ifuraces inhabited the ancient Tripolitania and are Berber Zenetes, which Corripus has designated in his book by Ifuraces.

228.Marcel GRIAULE, The pale fox, Ant-Matière, 1991, P.52
229. Michèle FRUYT,. "From Africus uentus to Africa terra Journal of Philology,no 50, ." 1976,: p. 221-238.
230. Daniel Don Nanjira. African Foreign Policy and Diplomacy: From Antiquity to the 21st Century,. ABC-CLIO, 2010.

The etymologies prior to the twentieth century are now only historical curiosities: Isidore of Seville derived this name from the Latin aprica ("sunny"), Leon the African invoked a fictional Greek word a-phrike ("without cold") .

For a long time, the term Africa has only served to designate the part of North Africa surrounding Carthage, predominantly Berber, the black-dominated south was called Ethiopia. Thus in book V of his Natural History, Pliny the Elder mentions the Niger River, which he calls Nigris, as delimitation: "the river Nigris separates Africa from Ethiopia" and mentions also the "Ethiopian nations" which live around it.

Since I state clearly and clearly that ***km.t*** is the country of ***Hr***. The word Africa can be broken down into * fri - et -que. In fact this gives in the language of the exercise:

Ewondo: /à-folà/ “place, panther”
Ewondo: /ko/ “foot, old, debut” **Hr’**s other name.

Thus in Ewondo the voiced uvular fricative consonant [r] of the radical * **fr-** is modified into a liquid consonant [l] giving the skeleton ***fl.**

The only meaning that we can accept from Africa is therefore the one I just gave to know the place of the old / the old, the place of ***Hr.***

Rigorously it would be necessary to modify the suffixal part -que and replace it by the toneme / -ko / "foot; brand 'which is common to all variants of the African single language.

Note that although  ***m.t*** "Mother" is tied to the earth. ***Km.t*** can also be seen also ***ko*** + ***m.t*** that is to say the place of the mother or the place of the uncle, the place of the panther, the place of the catfish. This coherence makes it possible to solve the question of who we are and how to call the continent / country.

Kemistism and Kamitism resemble this account of the fish syndrome which claims the dip net.

Ideology

When some said they managed the decipherment of the ***mDw nTr*** they recognize themselves having previously studied the Kopt. Take the script to name the crocodile below.
We simply note here that the word below designates a male crocodile.
If we respect the grammar rules that appear in all Egyptian grammar books, we must replace the masculine suffix ***.w*** with a female suffix ***.t.***

Middle-egyptian:  *kAp*.w “crocodile”
Middle-egyptian:  *km.t*

We thus arrive at the variant ***kAp.t.*** which fortunately is identical to ***Kopt***. Let's all be surprised that some of them are discussing the transliteration of ***km.t*** script while [p] and [m] are bilabials.

So it's the same people who organize belly dancing to say they do not know what is the meaning of the sign I6  of the Gardiner’s list.

If we ask japanese people how they do call what we often name God, here is their answer below:

Japanese: 神 *Kami* "God"
Hawaiian: *ke Akua* "God"
Zoulou: Nkulunkulu "God"
Amharic: አምላክ *āmilaki* "God"
Igbo: *Chineke* "God"

THE SEMITES

Sm/stm/smt
Faulkner p. 225 Wb IV p. 119, 3-9

One can draw many consequences from ***km.t*** script reading and capitalize on the readings of other scripts. Let's take a closer look at the vocalised "sem" script that gave the semitic word in the nineteenth century through the sleight of hand and biblical magic of the son of a fictional character named Noah.

Let's just remember that the Semitic word was coined from the biblical character of Sem, one of the sons of the mythical character Noah and ancestor of ***ib rA im*** "Abraham" [231].

Created in the field of philology to designate a family of languages, the term Semite was used from the nineteenth century to imagine the speakers of these languages as participants of a particular human race. Although this conception is today abandoned by the scientific community, the term Semite continues to be used. It is therefore a good reason to look into this concept to confuse those who defend a biblical thesis of the Asian origin of humanity. The latter were facing Diop in Cairo in 1974.

Let's see through this script how a negro concept can be diverted from its context to create nebulous peoples.

Archaeological evidence shows that speakers of so-called Semitic languages were dispersed in Mesopotamia before the establishment of an urban culture in the region; there is a hypothesis that several waves of nomads speaking Semitic languages have crossed the arid deserts of the East from prehistoric times to reach the fertile crescent. What is surprising here is that the script represents African writing and an African reality.

Ewondo: [à-koung] "owl"□ consonants skeleton *kng
Middle-egyptian: ***Sm*** "owl"

Lois de correspondances phonetics qui en dcoulent:

Middle-egyptian	Ewondo
S/s	*k*
M	*z/ng ou m/b*

from the point of view of the languages of the Congolese basin and taking into account the laws of phonetic correspondences above one can read this script / ù-nsem / "the sin, the fault" because the owl is considered as the animal of the /mbgwəl/[232]; /i-vu/. These components of the human entity that work in the world of the invisible or the world of the night.

231.Albert de PURY, The "Semites" do not exist The Strategic Time, no 89, 1999, p. 6-16
232. Voir illustration 14.

Etymology:

Middle-Egyptian: Sm ☞**Ewondo:** /ùn-nsem/ “trespass, sin”
English: owl ☞**Ewondo**: /à-lou/ “night”
French: Hibou ☞**Ewondo**: /i-vùh/ “Evil”
English: Evil ☞**Ewondo**: /i-vùh/ “Evil”

Reading by ideas association
Who embodies the owl *sm* in Ekang societies?

- ***sm/smt*** can be vocalized in Ewondo: "ze-word or mot-ze" (man-panther) the sign S29 of the Gardiner's list is actually the loincloth worn by African notability. (See illustration No. 10). The sign S29 transliterated s / z comes from the word [***sanda***][233] "loincloth" in Ewondo or /zə/ for leopard / panther.

I juste note here that the anthroponym Motazé[234] is common in Ekang country, Cameroon.

Middle-egyptian: sm or ms *(reading is possible in both directions with semantic nuances)*
Ewondo:zt or mt because z <ng; m<ng

In the traditions of Ethiopian peoples and especially among the Fang, Beti etc. the owl is The symbol of the accomplished man meaning the man who experiences life in the invisible and visible world. But with the regression[235] and the inversion of the cardinal values of the maA.t in Ewondo, it is considered as a bird endowed with an "evil" human spirit.

Akuŋ[236](b) n.4, pl. *mekuŋ*. owl. see *ndukh* the big. The *Faŋ* are frighten by owls. It's said that owl have *nnem*. That's meant a man transformed in a owl to kill other poeples. /*Mekuŋ me bêle ŋgwel*/,
owls have dark magic. /*Nnem nye a loŋ mekuŋ*/, it is the nnem that makes the owls hoot. The owl's cry announces death. Syn.: ônon ô *ŋgwel*. See also *koghl-kue*.

Middle-egyptian: sm or ms *(reading is possible in both directions with semantic nuances)*
Ewondo:[ze-mot] -□zt ou mt because z <ng; m<ng
If you read the script from left to right,

The sign S29 corresponds to a folded linen in this case it corresponds to a panther skin and therefore to the panther by semantic shift in meaning thus justifying the use of the term [Ze] ☞ “panther” in Ewondo. The sign G17 ***m*** corresponds to an owl representing an accomplished Man (*cf illustration n°11*) justifying the word [mot] “man” in Ewondo.

Owl= man

233.François BOUCHER, The history of costume in the West, from Antiquity to today, Paris, Flammarion, 1983
Some clothes, among the costumes in ancient Egypt, are common to both sexes such as tunic and dress. The men wear a loincloth, the "chendjit", scarf wrapped around the kidneys and restrained at the waist by a belt, then to -1425 / -1405, the light tunic or shirt with sleeves and the pleated petticoat.
Peasants, workers, and people of modest condition never wear anything but the loincloth, the chendjit worn by men of all conditions.
234.Louis Paul MOTAZE, Africa and the challenge of extending social security: The example of Cameroon, Pyramide-Papyrus Presse, 2009
235. Cheikh ANTA DIOP, Negro nations and culture, Présence africaine, Paris-Dakar, P.337§353
236.Samuel GALLEY, Fang-French dictionary. and French Fang followed by a fang grammar , ditions Henri Messeiller , Neuchâtel, p.32

sm

	Middle-egyptian	**Ewondo** *(Ze-mot)*
	s	*Z*
	m	*T*

Morphologically, it can be justified by establishing a law of phonetic correspondence between the voiceless alveolar fricative consonant S and the voiced alveolar fricative consonant Z Ewondo of the skeleton consonant of the word [ze]. And ditto between the M (owl) of the pharaonic with the T (man) Ewondo.

Middle-Egyptian: ***s*** "skin" and **Middle-Egyptian**: ***m*** *"owl/accomplished-man"*

Ewondo: [mot]; [mod]; [mor] "man" pluriel [bot], "men, peoples"
the root of the word is therefore Ewondo *ot.
Giving the consonants skeleton *t remember the following phonetics laws (t> z > s)

In grammar the sign G17 the owl can indicate the person to whom one usually addresses with suffix (exple: mk see you, mtn see, mt see to a woman)

From the left to the right, panther skin refers to the panther and the owl to the man because the owl is considered bird-man or man by semantic slip. (*see illustration n ° 11*).

Right reading to the left

sm

	Middle-egyptian	**Ewondo** [Mot-ze]
	s	*t*
	m	*z*

The sign G17 is related to its feathers. The spots of owl feathers bring the idea of the panther. This idea by semantic shift in meaning justifies the use of the word [Ze] in Ewondo and the sign S29 S representing a notable a person who wears the loincloth, a man is equivalent to [mot] in Ewondo. A successful person [mfaŋ-mot] (a real person vs a rascal)

Middle-Egyptian: ***s*** "man" and ☞ **Middle-Egyptian**: ***m*** *"owl"*

Ewondo: [mot]; [mod]; [mor] "man" pluriel [bot], "men, peoples"
the root word is therefore in Ewondo is /*ot/.
Making the link with the consonants skeleton: ****t***
Let's remember the following phonetics laws (ts>t> z > s)

Ewondo: [ze] ☞***z*** "panther" Latin(b, d, g) □(p, t, k) Grimm's rule.
Middle-Egyptian: Aby/bA (panther/leopard) so here (m <b <z) considering here that A is a vowel[237].
Chinese: 豹 Bào "leopard"

237. Middle-Egyptian.: Aby □Ewondo:[bibi,] "The man eater"; [imbgwem] "lion"
M.E.: bA□Ewondo: [à.bi] " to catch ", [ngu], [ze].

On the morpho-syntactical level, the ***t*** (Empty alveolar occlusive consonant) Ewondo is to correspond to the ***s*** (Fricative alveolar void consonant) in Pharaonic. The Ewondo is more archaic than the Pharaonic because the man is born in south of the sahara and not north of Africa at ***km.t***.

- Let's recap here:

Middle-egyptian		Ewondo
owl	Priest with panther skin	[Mot-ze] [ze-mot]
Sm	*sm*	

Morphological verification

Middle-Egyptian: Aby / bA "panther/leopard".
Chinese: 豹 Bào "leopard"

If we remove the "vowels", it remains the skeleton consonant "***b***". From the strictly phonetic and linguistic point of view, the consonant m is a labial nasal, a kind of "nasal ***b***", according to specialists.

So morphologically, panther / leopard equals something close to the owl in our languages, rightly implies the scribe.

- ***sm*** will be able to offer the vocalization "i.kob-Ze" the skin of panther, for the one who wears the skin of panther. It is necessary to operate the semantic shift in meaning: loincloth = skin and owl = panther
- The consonants skeleton ***sm*** as a healer will be read: [mot-*n*.gang]
- The consonants skeleton ***sm*** as therapist will be read: [word-m.fəkh] ☞celui who carries the bag *(cf illustration n°9)*
- The consonants skeleton ***sm*** as the intelligent will be read: [mot-fəkh]
- The consonants skeleton ***sm*** as the intelligent will be read: [à.kəng-mot]
- Etc.

By phonetic matching laws taking into account African sociology.

- The consonants skeleton sm as priest will mobilize the meaning: fara in Ewondo
- Le consonants skeleton sm as a man of service will be read: [*n*.kôm] "the binded" or [*m*.bokh] "officiant; celebrant"
- The consonants skeleton sm as initiates will mobilize the meaning: [*à*.koung]
- The consonants skeleton sm as the educated will mobilize the meaning: [*á*.kəng]
- Etc.

There are many other meanings that I will not develop here. But according to the contexts some aspects of the script can be mobilized by the scribe to specify the good acceptation.

Let us specify that the meaning that was used to designate a kind of category of people at ***km.t*** is more / ùn-nsəm / in our languages, ie "sin". From the moment when we do not have a regular practice of spirituality in the temples to purify our body and mind, we live in sin. So the word / ùn-nsəm / was used to describe the impure, violent and barbaric immigrant peoples at ***km.t*** who could not and did not have the right to visit the temples because uncircumcised and by the way impure.

Any attempt to create a people is indeed doomed to fail because we did not speak of peoples but it is a qualifier that can be given to a native who does not respect the observances related to piety.

Ewondo: /mor *ùn*-nsəm/ "sinner; sinfull person"
Ewondo: /*i*-nyiŋ *ùn*-nsəm/ "live in sin"

There is no particular human group related to priests with panther skin. Far from it. These priests are simply notables of African societies. This is why the second hypothesis is here most likely namely the men of sin.

Listing of words with morphological and phonetic proximity in the Ekang languages with the consonant skeleton km.

The idea behind this is to list words that have the same skeleton consonant in the Ekang language and to demonstrate that these words already exist.

Ô K A M[238]**:** (h) n.6, pl. akam.

1. Chili, ôkam is the generic name. Zir ôkam, a grain of pepper. List of varieties of ôkam; bekue ôkam, ôndôndôe, menya. Ondôndôe is the name Bulu de ôkam, it seems to contain the word ndôa, fire.–

2. Fetish (byaŋ) consisting of a bigescargot shell filled with ferns and other things, and fixed on a wood or a stump at the entrance of the plantation to keep it. Whoever ignores risks his life. It is said: Me mana bi ôkam fakh. This has two meanings: I planted or sowed pepper, or else: I placed the fetish snail shell.

Associated Expressions:

"Okam wa digh ma à-dis": *the pepper is burning my eye*
"ôkam wa vulbeanyu": *The pepper is burning the mouth.*

KAMA[239] (h) n.3, ss pl. Piece of iron, currency of yesteryear for weddings. Alokama, un kama; melo me kama mebè, 2 kama.

KAM[240]**(A):** (b). Partiality in judging or against his own. Kale does not go out, he takes sides with his friend, his brother, his family, he is partial. See ŋgam, ŋgamg. See the expression: dièle minsôn mi bôr, be partial to someone you like.

KAMA[241]**:** (b) vb rcip.of kam (b) . To prevent recip. to do something.

NKAM [242]: (b) n.2, pl. miŋkam.

1. Latex of the liana ekekam, ficus parasite palms. Serves as a trap for birds (even the green pigeon ôbeŋ), because it is very sticky, ônon wa byel ŋkam. Syn.: mekemdi mekekam. Proverb: ônon ô so ôyo ke bem-e-byele ŋkame, the bird that comes from far away is quickly trapped in ekekam latex. Moral: You must not get too far from your country, it's dangerous. -

2. Bite of nnam of leaves. ŋkam mesagha, ŋkam belome, ŋkam minsè. ŋkam ô ko, handful of small fish (and not blocked), small or small amount taken by a woman fishing.Voir nžekh (h).

238.Samuel Galley, Fang-French dictionary, and French Fang followed by a fang grammar. ditions Henri Messeiller , Neuchâtel, p.305
239. Op. cit., p.148
240. Op. cit., p.148
241. Op. cit., p.148
242. Op. cit., p.290

NKAMA[243](h) n.2, ss pl. Fetish for the war, to stop the bullets. It is a pack of animal skins (byaŋ akoŋ). Ŋkama, melôŋe, yaŋa-mba, three fetishes that go together. We sing and dance.

KAM[244]: (h) n.3, ss pl. 1. Yellow ant on guava trees and other trees, it bites hard. If you disturb her, her whole body shakes and vibrates, hence her name. Bum a kam, she makes her nest on the tree by tying the leaves together.–2. Nervous tremor, convulsion, Saint Guy's dance. Môr a ne kam, môr a zem kam. Môr a kü kam, he begins to tremble. This can be a restless paralysis. Môr a ku kam, be very frightened See kamkame, som.

KAM[245] (b) vb. Defend, prevent, thwart, prohibit. Protect, preserve, put someone's party. K a m ye zam, zom, forbid to take a thing. My kam y'keke wam, I forbid anyone to take my knife. My kam y'e bo bam, I defend my children. See ban, ŋgam.

KAM[246] (b) n.3, pl. Mekam. Barrel of hunting powder. Syn.: fîra.

KAM[247] (h) vb. Excite shouting (porcupine hunt with dog). K m mvu, excite the dog. K a m bôr bise, excite men at work. K a m bikama, loŋ bikama, screaming. They shout while carrying loads, carrying a tree trunk or killed animal. See kama.Syn.: bemle.

NKAMA[248](h) n,2, pl, miŋkama. Hundred. That's the word in Galwa.Syn.: siŋ, sin.

NKAME[249] (b) n.l. Affluent of left bank of the Abaŋa.

KEM[250] (h) vb. Fall, collapse, collapse. Metekh my kem, the earth is sinking. Nsegha O vagha kem, the sand has fallen. Beŋgil ba kem, wild boars flee with a single impetus. Warned by the male, they all take the same path. For a leak of monkeys, we say kyal. Syn.: fume, bugh.

KEME[251] (h) n.l, pl. bekeme. catfish (river fish).En Galwa: ŋkèmbè.

ME-KEM[252]**:** (b) n.4, ss sing. Venom. Me-kem me nyo, take off, snake venom, scorpio. Ŋlo mekem, venomous fly. A "yo mekem, venomous snake.

ME-KÔM[253]: (b) n.4, see the sing. Akom. Dance of men. We do not say zem me-kom, we say ka mekom, sing the me-kom; zo mekom, say the mekom. Nzô ye beka, the one who says and those who sing. The speaker tells the story of the old days, everything that happened. It's not a real dance, it's folklore. Women must lock themselves.

KÔM[254] (b) vb. 1. Arrange, train, prepare, make, arrange. Syn.: yala. Oh my, give his son a fetish power (evur), so that he is strong. Kôm enoŋ, make his bed. Kôm nyôl, to put ornaments (syn: kômda). Kôm tô, to wrap well in his loincloth, hide what to hide. Kôm
established, set the table. Kôm nẑali, clean and oilr his fusU. Nẑame a ŋga kom mam mese. God created all things (syn: vel). Ŋkôma, prepared. - 2, Komkukh (or yen ŋkukh). Take a living animal in the forest, make fetishes for her, keep her secretly in her house, or even leave her alone etc.

243. Op. cit., p.290
244. Op. cit., p.148
245. Op. cit., p.148
246. Op. cit., p.148
247. Op. cit., p.148
248. Op. cit., p.290
249. Op. cit., p.290
250. Op. cit., p.151
251. Op. cit., p.151
252. Op. cit., p.190
253. Op. cit., p.190
254.Op. cit., p.190

KÔM[255](h) n.3, pl. bekôm. Large, very venomous snake is the most dangerous because it moves very fast, it catches a man running.Syn.: o-kôm-nyô.

KÔMA[256] (b) 1. vb recip. de kôm. Kôma biyem, exhort yourself to do good. - 2. Vb recip. of kome. To desire, to covet. Kôma mam, to covet things.

KÎMA[257] (h) n.5, pl. bikîma. Iron horn that serves as a horn or whistle. Ekîm'e byaŋ, the man speaks and whistles to make a good journey, having taken leave of the byeri (chest of skulls). Syn.: kyema. See this word

KÎMA[258] (b) n.5, pl. bikîma. Ekîm' ndokh, motte or ball of agglomerated ndokh. See ndokh. We make very big mounds

AKÎM [259]: (h) n.4, pl. mekim. Tree knot. Syn.: atoŋ, atsiŋ. - 2. Bulging of a knife handle. By ext., Capital of column.

AKOM[260] (b) n.4, pl. mekom. 1. Large tree of forest with white trunk, very rare (Terminalia superba). - 2. Bump, swelling, as a result of blows, bites.Syn.: tur.

AKOM[261] (h) n.4, pl. mekom. 1. Fish of the ŋgol category, without scales. Syn.: mvè. -
2. Akom e nẑokh, ivory bracelet. Syn.: ehoŋa nẑokh.
3. Wild yam, tuber of the hiker vine. It is very hard and looks a little like the cultivated yam mvi. We eat it in times of scarcity.

AKAM[262] (b) n.4, pl. mekam (vb kamda b) . Liane also called ekekam and etetol. Ficus parasite palms.
AKAM[263] (h) n.4, pl. mekam. Two-string musical instrument with sound box. Syn.: ayine (b) .

AKAMA[264] (bm) n.4, pl. mekama. Pile, Akama mekokh , piles of stones. Aka-ma bitô, pile of stuffs. Syn.: aŋgung (h).
So why trying to read a script that is no longer subject to debate after the conclusions of the Cairo symposium? Simply because of its importance. On the side of Diop's followers, the debate is latent and, on the Africanists' side, the Cairo pill is still hard to swallow.
Etc.
So there is no consonant skeleton ***km*** in our languages but a considerable amount of words that share the consonant skeleton ***km***. So we could use semantic shift in meaning to traduce the land unfortunatly named egypt simply by soliciting the identity of the consonant skeletons.

Conclusion

The script 𓆎𓅓𓏏𓊖 ***km.t*** is composed of sign 𓆎I6 presenting a crocodile skin because crocodile for african people is their ancestor. As we can see this idea is shared woldwide.

The sign 𓅓 G17 is a owl representing many things in african thoughts.
The signe 𓏏 X1 is representing some food.

---

255. Op. cit., p.190
256. Op. cit., p.190
257. Op. cit., p.99
258. Op. cit., p.99
259. Op. cit., p.28
260. Op. cit., p.29
261. Op. cit., p.29
262. Op. cit., p.29
263. Op. cit., p.29
264. Op. cit., p.29

For the endogenous point of view, each sign is enough to find the right vocalisation.

Sign I6: /*ùn*-ŋkôm/ "crocodile"
Sign G17: /i-kùmù/ "owl" by metonymy
Sign X1: /i-kôm/ "heap; cake"

So the vocalization of the land name can be obtained by a nominalisation process. We just add at the end of the the common word the last vowel. Then we tranform the common noun to an antonomasia.

Regarding the meaning the issue is a little bit more complex. By the way it's necessary to know endogenous requirements linked to naming. But they are minimum three names related to the leader's function.

Aside the standard name, we can have many homophones, synonyms, antonyms also. We do believe that the african words related to this script ***km.t*** are still used in western languages.

The main meaning linked to the script ***km.t*** is ***Hr***'s land. ***Hr*** is for African people the primordial ancestor. For African people ***Hr*** is the foot because he is disabled .

Etymology

Ewondo: /*ùn*-ŋkôm/ "crocodile" **French**: comme "similar"

Note: the word comme means similar and it sounds like /ùn-nkôm/. As we said this animal represents the chief. The chie fis associated to numeral 1. In a binary system, the other one is the similar. That's how and why you have the link with the french word. By the way, the other in Ewondo language is the other one /ùn-mbokh/ "disabled"**.**

Ewondo: /à-kol/ "foot, ancestor" **English**: like ; look "similar"

Note: the Ewondo word and the english one are consonants metathesis. They are dealing by phonemes inversion.

Middle-Egyptian: *km.t* "Black land; Egypt"
Koonzimé: *kyémé* "monkey"
Tshiluba: *nkima* "monkey"
Obamba: *nkima* "monkey"
kiSwahili: *nkiwa* "monkey"
English: monkey by phonemes inversion

DHwty

Note : I want to make an aside on the transliteration **DH.wty** in another african language. Hope you will enjoy it.

Middle-Egyptian : ***DHwty*** « Thot »

Tupuri :Giwgiw "Ibis" can be write also Djiwu Djiwu (G>Dj)..
Tupuri :Dji "whoever"
Tupuri : Wu "Fire, Light; bright"

Therefore, *Djiwu* means, whoever has the light, the knowledge In Tupuri language.
Since the morpheme *Djiwu* is doubled in the item *Djiwu Djiwu* because of the nisbe (//=y) therefore , *Djiwu Djiwu* is dealing with the great intelligence.

If in middle-egyptian, ***DHwty*** is represented by a ibis, he can be represented also by a baboon as above.

English : *ape* « monckey »
English : *hype ; hep ; hupe ; hipe ; hip*

By the way some read *Djehuty* because they are cobblers.Take note that :

The item Tupuri *Djiwu* is sharing a phonetic proximity with the french word DIEU do you see it now?

Djiwu = DIEU

And fortunately :

Giw.t >giw.**d** is linked with the English word **GOD**

We have discussed about the lexicalisation of the this final **-.t >-.d.** It is vocalized here.
Only Africa can enlightning the purpose this way.

Mandjack: Tiatia "monkey"

☞**Middle-Egyptian**:  *ityw* "chief"
☞**Middle-Egyptian**:  *DHwty* "Thot"

Nzebi: Tchèghè/Tsèghè "monkey"
Ewondo: /i-tsogh/ "thought"

Middle-Egyptian :  ***DHwty*** « Thot »
Ewondo : /ŋgemamemoghe/ "ibis"
Ewondo : /i-bisi/ "the earth trapper".
Ewondo : /i-sibi/ "skinny".

Middle-Egyptian:  ***sbA***"intelligence"

.

By displaying the Ewondo name of the ibis above, I wanted to confuse the specialists. Indeed how could they demonstrate the link between the Ewondo morpheme and the consonants skeleton of the above transliteration? It is clear here that the comparative method is completely in cabbages. Even if few tinkerings tips can be involved here.

The ibis likes wet places, marshes, swamps, agricultural areas and food waste deposits.

This bird is rather nomadic, it frequents the islands, the rivers and the banks where it will mix with other individuals to sleep in the evening.The Ibis feeds with locusts, and with its beak it captures **worms**, molluscs, fish, lizards and other crustaceans. Doing so, it reveals hidden things by catching earthworms. Its name in the Ekang languages therefore clearly means the hidden things revealer.

There is a game word that only a someone practicing the Ekang languages can see here.
Indeed, the word /ŋgema-/ can also be written /ŋgyema/. This word means: the tail and makes the connection with the baboon. But this tail is also the criterion of differentiation between men and monkeys. However the variant /* yem/- will *me*an in Ewondo: **To know**.
/ moghè / "hide, hide"

Ewondo : / ŋgem / variante Ewondo : /ŋgyem/ "tail".

Proverb: *Kale à á-sobo a-sù ŋkâ, ŋgyem e yənə.* "Such is hidding as the varan, but we see his tail."

Indeed the varan is hidden, but we still see his long tail.

Meaning: Expecting to hide something that everyone can see.

By the way, **Ewondo** : /ŋgemamemoghe/ "ibis" Could mean the seeying part of the hidden.
Quid? Who is the hidden? And What is hidden? Please think about it.

The reading and interpretation of  *km.t* "monkey" is interesting because it illuminates the collective consciousness on several levels.
The intellectual swindle that led Darwin to postulate that Man comes from the monkey and especially the raciological theory of Gobineau that led to associate the negro with the monkey.

Hollywood productions with blockburster like *Planet of the Apes* allow an easier understanding of the production that underlies it.

In addition to these epiphenomena, which result from a minimalist reading of the transliterated script *km.t*, we see that the monkey is associated with the resemblance and the vocables to designate it in the African languages make it the incarnation of the intelligence. So we are talking not about the monkey but about the maternal figure.

Historically, this reading can be confirmed by the Indian culture that makes *Hanuman* the god-ape, leader of wrestlers, god of wisdom. He is often represented with a club (gadā).

Curiously as *DHwty "Thot, Hermès Trismegistus"*, he among many other talents, *Hanuman* was grammarian; and the Ramayana says, "The sacred monkey is perfect; no one equals it in shastra, nor in learning, nor to understand the meaning of the scriptures, (nor to move at leisure). It is well known that Hanuman is the ninth author of grammar.

According to me, Buddah can be also assimilated to *Hanuman* because he is the enlightened one.

Ewondo: ndâba "Fontanelle" ☞**Indian**: Buddah
French: *daba* "an African tool" ☞**English**: dab
Ewondo: i-dimi "ignorance" ☞**English**: dimness

Indeed, the tonsure was normally done at the top of the head around the welding of the old fontanelle. One might like to think of the use of tonsure by some fraternities of the Middle Ages.

A characteristic of *Hanuman* is his complete and permanent fidelity to *Rama*, his spiritual Master. His whole life, his actions, his days and his nights are dedicated to serving his Master. In a word, *Hanuman* is the image of the perfect disciple. After the victory, *Rama* wanted to reward *Hanuman.* He refused, too happy to have been the champion of love and justice. This nobility is still illustrated today by the Hindu proverb "monkeys cry over others, never about themselves".

Hanuman. means in Ewondo whoever has members. Could it be possible that *Hanuman* and *human* are sharing the same root?

Ewondo: /i-nam/ "member"; **Ewondo**: /i-namà/ "whoever has members"

Originally *Hanuman* was the guardian of the properties and any founder of a new village had to erect his statue.
Ganesh a "God" with an elephant head,

Ewondo: I-kama "male elephant."
Ewondo: ngan "crocodile".

Then we better understand the ban that strikes the Indians to consume or to attack the simians. Note by the way that the word *simians* shares the the root of the items *same*, *similar*.

Hanumān meets Sītā (bas-relief of Prambanan Temple in Indonesia)

Note:

May you please note this:

English: ape
☞**Middle-Egyptian**: *imy-bAH* "ancestor"
☞**Middle-Egyptian**: *Hâpy, Nil* "from ancestor"
☞**Middle-Egyptian**: *api*"fly; jump as a monkey"

English: monkey
☞**Middle-Egyptian**: *mk*"protector"
☞**Ewondo**: kama "protector"

Ewondo: /*ùn*-ŋkôm/ "crocodile" **English**: come "reach"

Note: We often use the feet to walk. Here the link between the two languages is metonymic.

We can observe that the consonants skeleton *sm* involved in the reading of this script is used in western languages.

Ewondo: /à-sam/ "to gather" **English**: same "similar;like"

Ewondo: /à-lod/ "to pass" **English**: old "aged;ancient"

Note: we use the feet to pass and the foot is our ancestor he is the oldest. *Old* and *lod* are anagrams.

Ewondo: /*ko/ "foot" **English**: go "pass; retire, travel etc."

All the signs involved on the script were able to be read in western languages if they were using their african etymology. By the way we should have avoided what professor Bilolo often call Esperanto Egyptology. Words are invented by supposed specialist because they don't really understand what they are doing.

Among many meanings, the script ***km.t*** authorizes the level of understanding: black land.

The semites are an invention, unfortunately, modeled on the men of **sin** in African languages. In no case they do represent an operational category in physical anthropology or the least people who have lived in ***km.t***.

3.5.2.2. WHO WAS NAR-MR

- **Nr-mr**

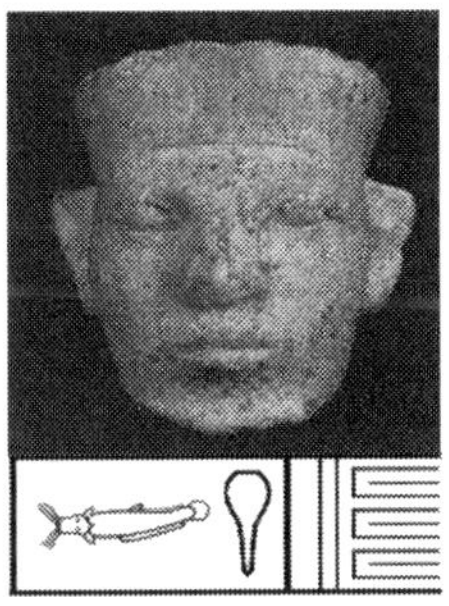

Often regarded as the first ruler of unified Egypt, historiographyively yet his identity still arouses debate. Many historians and Egyptologists in particular identify it with "***Menes"***. Just that name is reminiscent of a Greek, Portuguese or Spanish more than an African. It is therefore necessary to look a little on this subject.

The information on this pharaoh is essentially drawn from the famous "Narmer" palet (*see illustration n ° 15*). Palet of schist on which there is a victorious representation of a sovereign against a people that we have not yet apparently identified.

The historical interest of this pallet is not to be denied because there appear the first clearly defined African sacred characters, which are supposed to form the name of this king "Narmer" ("fish - chisel"). We will look at reading the script on the cartridge above.

Among the information concerning him, he is represented alternatively wearing the crown of the South (white crown of the kings of Upper Egypt) on one of the faces of the pallet and that of the North (red crown of the kings of Lower Egypt or the delta) on the other side.

According to some official sources, Meni is the name used on the documents from the XVIIIth dynasty (scarab of Hatshepsut and Thoutmosis III). Menes is the Greek variant used by the famous historian *mnjw-Htr* (Manethon Sebennythos).

> *"[...] The Royal List of Abydos, on the wall of a corridor of the Temple of Sethos I. at Abydos, gives in chronological order a series of seventy-six kings from* ***Mênês*** *to Sethos I"*[265]

In his book [266] published in 2001, the Egyptologist and CNRS researcher Bernadette Menu recorded the idea that ***Menes*** would be a kind of title that would mean, "the one who establishes", carried by the first two rulers of the first dynasty, "Narmer" and " Aha. Surprisingly, however, Menès is not an African name but an African name *greecised*.

The words ***mani, man, mwén, mwan***, etc. surprisingly are used in Africa. The famous Egyptologist, historian and African linguist who assisted Diop in Cairo in 1974 is called T. Obenga ***Mwené*** Ndzalé. In Ewondo it would be translated by / mwan dyal / or / mwan dzaŋ / "the child of the country". For an African, it is therefore a common name where the specialists are in the supposition and quite surprising assertions.

According to the official diktat, the name of the first ruler could mean "someone" or "person". It would be supposed here that it is possible that the priests, no longer knowing the name of the first African ruler, but having to quote him, used this expression (mênês / mani) instead.
Beyond all the suppositions, it is indeed a sovereign with negro characters certainly an African ruler.

Amazingly Africans were not first inquired about the semantics of the name attributed to this historical figure. Curiously we have not bothered to verify the accuracy of the statement by the search for historiographic facts some teeming in Africa across the historical continuum. Information that could easily

265. Manethon, Traduction de Aegyptiaca, Harvard University press, 1940-1964, p.XXI.
266. Bernadette Menu, Limites et exclusions, l'Harmattan, Paris 2001, p12

be found by questioning the political history of Africa through the writings of African ethnologists, anthropologists, sociologists and historians or simply the African royal institutions that are still established everywhere on the continent.

According to official speculation, it is possible that "***Narmer***" has changed its name after the unification and establishment of the pharaonic institution taking the title of ***Menes***, "the founder" was it Greek to take such a title? The pharaoh Narmer is -3185 to -3125 BOE.
According to historians, the Greeks were present in Egypt long before the conquest of Alexander the Great. It was from the Saite era in the seventh century that they settled many in the country. They were first mercenaries, commander of the army, then merchants, the Pharaoh Psammetic I (-664 to -610) and Amasis (-570 to -526) will give them a city, Naucratis in the Delta. A Greek, Chabrias, will even be counselor of the Pharaoh. After the fall of the second Persian domination, they settled permanently in the cities of Alexandria, Ptolemais etc ... In the Fayum, they owned large farms. The three kings of the "XXVI Dynasty", known as the Macedonian dynasty, never really lived in Egypt (Alexander spent less than six months there).

They managed from Babylon the Macedonian Empire. Egypt was administered by a local Satrap, Ptolemy I, who was a friend of Alexander the Great and later was proclaimed Pharaoh. Over the years the relationship between Greeks and Egyptians brought to a new society. The Egyptianized Greeks even worshiped pharaonic deities who mingled with their pantheon. A Greco-Egyptian art was born, the Greek chapels neighboring with Egyptian temples. Philippe III Arrhidée, who was mentally handicapped (or epileptic?), And Alexandre IV Aigos, a little boy, were surprisingly portrayed on the walls of Egyptian temples or monuments as virile and strong Pharaohs, making offerings to the Gods.

One may wonder wether between the Thinite Era (or the Archaic Era) and the XXVIth dynasty the misperception that leads to confuse an African name with a Greek name is normal. One may wonder if it Is not an intellectual swindle considering the time difference between the two domination periods ? This is stated because it seems that the two names were found on a seal where they are associated, but in a way that does not ensure that it is the same person, the name of ***Narmer*** being as an HR and that of ***Menes*** in a normal way.

According to Naomi Porat, the name "***Narmer***" would be attested throughout Egypt and southern Canaan.
In Egypt, his name was found on thirteen sites: three in Upper Egypt (Hierakonpolis, Nagada, and Abydos); eight in Lower Egypt (Tarkhan, Helwan, Zawiet al-Aryan, Tell Ibrahim Awad, Ezbet el-Tell, Minshat Abu Omar, Tell el-Farkha, and Kafr Hassan Dawood) and two in the desert (Wadi el-Qaash and Kharaga Oasis).

During the reign of "***Narmer***," she says, Egypt has an active economic presence in southern Canaan. Shards of pottery were found on several sites, some from jars made in Egypt and imported into Canaan, others made locally in Egyptian style. The latest findings suggest that Egypt's presence is not merely the result of trade but attests to colonization.
After 200 years of activity in Canaan, Egyptian presence peaks under Narmer's rule before declining after him

Analysis of the script nar-mr[267] 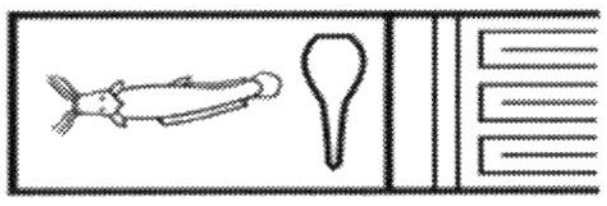

The "Narmer" Hr's name is written with two characters: nʿr, the fish (nar) and mr, the chisel (mr). Its name could mean, according to some literatures, the catfish that strikes.

The vocalization of the consonants skeleton nar-mr by "Narmer" can lead to a problem for the African and for any curious person because this name does not refer to anything known in the current African semantic and sociological field.

What does Narmer mean in Ewondo, Toupouri, Baoulé or Walof? I will have trouble to tell what it is in my mother tongue (Ewondo). Is it not legitimate to be a little surprised if this word serves to designate an African ruler as important as the "first" African ruler?

Since the works of Cheick Anta DIOP and Théophile Obenga, it is accepted that Pharaonic Egyptian is an African language. Thus, the search for the appropriate vocalization of this script according to the current dialects can easily allow Africans to reconnect with their past.

In order to overcome this historical handicap, I will use my language which is the Ewondo for illustrative purposes but this should not strictly prevent other Africans to reach the same conclusions as me, following the methodology that I have suggested until now.
Ancient Egypt was a monarchy.

Under the ideologically qualified theocratic regime of a priest-led theocracy, the fara was at once the chief administrator, the chief of the armies, the chief magistrate and the guarantor of the supreme worship of the country. Indeed, the fara had a mission to fulfill: to implement the Maat[268] on earth, that is to say to ensure harmony between men and heaven, to be a guarantor of the morality of his people, thus contributing to ensuring his eternity.
To exercise control over lands and resources, the fara relied on an administration made up of clerical civil servants who run their affairs on a daily basis. This administration is headed by his trusted man, the /*ùn*-nkukuma/[269] "***TA.ty*** (vizier)", who acts as the king's representative and coordinates land surveying, treasure, construction projects, legal system and archives.

1- In-depth study of the vocalized script "Narmer"

According to the method set up by Egyptologists, the reading of a script usually associates two milestones, the transliteration is to read the identification of the character with the "Latin" alphabet and the translation which gives the vocalization and the meaning. In the specific case of vocalization "Narmer" try to rigorously apply the Egyptian method.

Transliteration

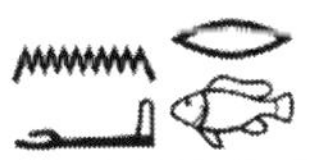					
N° of the Gardiner List	**N35**	**D36**	**D21**	**K5**	**U72**
Description / Signification	Representing a fillet of water (var.S3,S4)	Arm out-stretched, palm	Mouth	Fish, aquatic animal, smell bad	

267. Beckerath H0-H4
268. The maât is the unsurpassable law that governs the functioning of the universe in all its visible and invisible dimensions. It is often materialized by the golden spiral or in mathematics by the Fibonacci suites.
269.**ASUNẐOGHE** (hh) n.l, pl. basun-zoghe (elephant face). First, the first, avant-garde, in front. Asunẑoghe, vakh me bizi, first of all, give me something to eat. Asunẑoghe has zam bîa yia ye bo, the first thing to do. bôr asun-zoghe, those of the avant-garde. Bedu akokh, those of the rearguard, those who close the march. Syn.: Ousua, Osu, Bila, (Elephant's trunk). P.48 Fang-French dictionary.

	(hir.N16)	hand up (var.D38,D39) (curs.D37 à D44)		Fish petrocephalus bane	
Phonetic Value	*N*	*a*	*r*	*bs, rm, bw, aD, ad*	*mr*

According to classical Egyptology this script would be transliterated because the associated characters carry these equivalences in Gardiner's list and would vocalize "Narmer" by pure convention.
Thus, with the exception of the fish and the sign U72, the other signs may be considered as silent letters used to spell the name and therefore, put together these signs would serve to give a vocalization that refers to a signifier.
If the transliteration of this script does not seem to be an undue problem, the resultant vocalisation and translation of the term "Narmer" lead to a major difficulty because of the conventions of modern Egyptology such as the random distribution of vowels to form words.

Words that syntactically are correct but unfortunately do not always refer to a specific reality. In the present case of "Narmer" with the random distribution of vowels we create a term that does not refer to any known reality in the anthropology and sociology of the African.
As we have seen among specialists no one knows really what does that item means.

In addition, there does not seem to be any universal transcription of hieroglyphs. This is simply explained by the existence of difficult obstacles to accurately transcribe the hieroglyphs, obstacles that force extrapolations that make illegitimate any supremacy of one system over another.
By this simple fact, it is therefore imperative to deviate a little bit from the general method of hieroglyphs reading.

EGYPTOLOGY MEANINGS

Modern Egyptology offers these possible translations related to the vocalization of Narmer:

- nar-mer ="catfish - chisel".
- nar-mer "the striking catfish, the catfish striker"
- "The angry catfish". (I would like to know against whom …)
- Mer-Nar and interpreted by the "Beloved of Nar
- Nar-mer "electric catfish"
- etc.

Some translations are self-explanatory because how is it possible to talk about electricity when here the scribe does not use the sign K15?
The sign K15  of the Gardiner's list translates the idea of a fish that would deliver electric charges. Rigorously one should therefore discard some of these translations because of their crumbling to a questioning a tad tighter.

Endogenous Method

Many works have been done on ***mDw nTr*** but very often, those who do not speak an African language have arrogated authority on the subject. The few Africans who have been interested in the subject have rigorously relied on the diktat established by those who are authoritative. One of these errors is the transliteration that imposes a skeleton consonants as a norm and leads to intellectual acrobatics to read what is obvious in African languages.

Some dimensions are evacuated to the benefit of the only linguistic techniques leading to dead ends and sadness.
According to African sociological codes, a sovereign name is not a simple juxtaposition of devitalized letters or phonemes. (*See illustration N ° 14*)
A name in Africa is an energy, a personality, a multifaceted complex reality that is part of the dozen material and immaterial components that constitute the identity of an individual (M.E: *rn*). A name is an energetic signature. It must refer to a signified and furthermore is the living ontological connection between the named individual and his eponymous ancestor.

It can not therefore be understood that the methodology of vowels distributing to a devitalized consonants skeleton to constitute names of African rulers is adapted in such circumstances. On this account, why did the oracles dig their heads to choose series of names constituting for their ancient African rulers their official title?

If we dismiss the sociological argument in considering a name of sovereign in Africa, is not it wrong? As a general rule for an individual there are different names in Africa. In the particular case of the fang / Beti society here are the variants of names that a earthborn can bear before any contact with colonization and which still persist among traditionalists:

- *Name: dy-oé (h), pl. mi (h). glorious name,*
- *Nickname: dy-oé abara(hh).*
- ***The name the is passed on in a family**, i-yôlé (b).*
- *Name given to a girl born after the death of her father, **Meligha** (h). If it's a boy **Mevwène**.*
- *Name given to a boy or a girl born after the death of the mother, **Nsaghmebie** (mh).*
- *Name given to a homonym,Sakename Mwé. Two men are both called Wagha, each calls the other: a Mwé a mong.'.*
- *Taboo name: dy-oé **ikî** (hb).*
- ***The name vocalized by a series of onomatopoeia and allows to designate a person:** Ndan **is a personal code attributed to every man, and which is musicalized with the Nkùl. It is according to the minimalist vision of Laburthe Tolra "A drum motto"**[270]*

Example1 of ndan:

> "Amboa Ambo'o mi-ŋguiah, bwan bə tàn zokh-zokh. Bə-tàn be bələ, bə-nyî bəzâ à yi minkô"

Literally: Amboa daughter of Ambo'o and mi-ŋguiah, 5 big, strong children (elephant). To be stopped, 9 persons: 5 to hold her and 4 to bring the ropes to tie.

Other example of ndan: "fəgh bità, bità bi-nghənə ù-yap".

Literally: war strategy, only when the war is still distant.

No need to repeat that all this is drummed or vocalized by a series of very precise onomatopoeia and always the same all the time.

In the matriarchal regime, each child is designated by his eponymous name which is often a link to a deceased or living person followed by his mother's name

270. Laburthe Tolra, To the light? or, Ariel's desire: about the Beti of Cameroon, Karthala, Paris, 1999, p.208

Example:

Ndône Ŋgyeme "Ndôŋe *son of* **Ŋgyeme".**
Ndône be/ba Ŋgyeme "Ndôŋe *son of* **Ŋgyeme".**
Ndôŋe *belongs to* **Ŋgyeme, Ndôŋe ya be Ŋgyeme.**

The proper names of people as well as the names of places, clans or lineages contain clues that make it possible to particularise the lineages to which they belong.

> *"[...]traditional names are so specific to each ethnic group that it is easy to identify a person's language [or region of origin], according to the name it bears ..."*[271]
>
> *(Mayer 1992: 143)*

Generally speaking in the Ekang country, from the endogenous model, the attribution of the name means that:

> *"[...] in a given family, birth names circulate in a closed circuit, because each name is that of a living ascendant or deceased"*
>
> *(Mayer, 1992: 143)*

The attribution of the name in Ekang country and in Africa in general obeys to a certain number of specific criteria.
So, there are names that are attributed:

> *"Depending on the sex of the individual. This is why when a child is born, it is assigned (in the existing range of names) a name reserved specifically for men or women. Among the Fang, male anthroponyms are for example; Adzo, Eya, Ekomi, etc. By the way, anthroponyms; Manomba, Massounga, Oulabou, etc., are examples of names attributed to female members of the Punu"*
>
> *(Mayer, 1992).*

Marriage also brings its nominalization to this account, it should be noted that:

> *"The mothers-in-law never called their brutes by name. They always renamed them to a name that expressed the perception they had or the function they expected of them."*
>
> *(Elo Mintsa and Ngbwa Mintsa, 2003).*

There are also names reserved essentially for twins; the examples of Wora and Yeno among the Myenes, Nkal and Ndumè, Doue and Nkwendi among the Shiwe, Nzoghe and Ndzime among the Fang, etc., are perfect illustrations.

One can add to this expanded and deepened range, the name that a person takes after an initiation; Bwiti initiates, for example, acquire specific names (Koumbou). Could one manage ***km.t*** without being initiated of high rank? We answer each other by an absolute negation. Without knowing the hermetic codes of African initiatory societies how pretend to read a name of a great insider and appreciate all its semantic content? Seen from Africa we were really surprised because it is an approach that can escape even Africans traditionalists but not initiated.

Not to mention the lineage-specific names (Isaac Nguema, 1969) attributed to some individuals in addition to their usual name (the one received at birth) in the village of their maternal uncles (lineage of the mother) with regard to ethnic minorities. patrilineal descent or in the father's village, if the person concerned has received an endonym of the maternal lineage, etc.[272].

(Agyune Ndone -2006)

271.R Mayer. History of the Gabonese family.Libreville, ditions du LUTO, 2002 (1992), P.143
272.F. Angyene Ndone, Comparative Anthroponymy makina, fang and Sake: some specificities on the names of people in Makina of Gabon, 2006

By the way, claiming that one has read the name of a person without knowing the subtleties that accompany the nominalization in Africa, is to show a lot of pretensions and it would be rather advisable to be surrounded by modesty if not of curiosity.

If one refers to the cartouche which is supposed to belong to "Narmer", and if it is understood that this cartouche belongs to the sovereign who succeeded the unification of the two lands pompously called Upper and Lower Egypt, if it is understood that this sovereign was an African, one should undoubtedly find in the constitution of his name as well as the translation of his name sociological clues peculiar to the matriarchy and secondarily indices which base his role of sovereign, the exploits he has made. In ancient Africa up to the present day, apart from the sociological codes related to the name I briefly presented, the rulers answer at least three functions[273] by which they will be named in addition:

- Cultual Officiant: because of the burden that lies to the cult of ancestors.
- Uncle of the country because of the link with the land as the depository of the land right and responsible for the arbitration of land boundaries.
- Security Guarantor: link with lineage protection in wars involving the use of force.

So any serious research on the sovereign wrongly named Narmer should lead to exhuming those indissociable elements that are related to his function.

General Analysis of the Script

Let's analyze the script from another angle that of the signs that accompany the determinative.
On the far right, the scribe tells us that it is a fish, it is the determinative of the script (sign K1 of the Gardiner's list).
A fish whose skeleton consonant would be ***nar*** in the African languages if one sticks to the Egyptological transliteration.

So if you are rigorous with yourself, you must find out which fish you are talking about. Indeed, it is the fish that appears in the cartouche and visually indicates the name of the first ruler of the land of Egypt. As always when there is a doubt we will go to the sources.

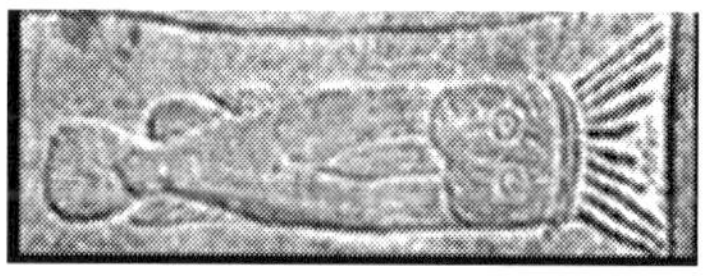

It is indeed, a fish with barbels on the mouth:
The catfish (Siluridae family) Ictalurus Melas (lat.) Royalfish (Eng.)
Whose mustaches give the impression of a crown.

So we just need to name this fish in African languages to find out which fish it is rigorously. I just want to point out that an African speaking language did not really need to do a transliteration to synchronize with the norm.

In fang Beti this fish is called ***ŋgol, ŋwoŋ, ŋgoŋ, ŋgwal*** etc. Just because it looks like a rope / ù-ŋkol /. Depending on the species it will have more specific names. We have here just the generic name that we could approach the term silurid.

Nothing very complicated for the nominalization of a common name to a proper noun in Ewondo, we have this equation to realize the antonomasia:

proper name = *common noun + copy of the last vowel*

273. T.Obenga, The Bantu: Languages, peoples, civilizations,Présence Africaine, Paris, 1985, P. 81 etc.

So at least, this pharaoh is called: ngwal + a = **Ngola** "incarnation of the catfish"

According to African thought, he embodied one of the qualificative of the eldest ancestor

Ewondo: /ŋgôl/ "elder; ancient; senior; head"
Ewondo: /ŋgôl/ "fence; barrier; the chief"
Ewondo: /ŋgôl-ŋgôl / "toothless; the chief"
Ewondo: /ŋgwal / "elephant trunk; the chief" ☞Repport to illustration N°19

READING VERIFICATION

To designate a young man in Basa'a land they often use the expression *àlogà*, which can also mean the one we fish.

Ewondo: /à.log/ "fishing" ☞is an anagram of / ngol / "catfish"
And the girl is **Basa'a**:/ ngo / "catfish; panther" ☞uterine succession requires.

☞ African toponyms formed from the root *Ngol*

So from **Ngol** the fish we can mention:

-Dongola (ancient African kingdom situated south actual Egypt), ↔Nubia called **Makurie**
-Angola (name of the country),
-Yaoundé Indeed, the original name of the capital of Cameroon is Ngola (Ewondo).
- etc.

"The Mfoundi department was originally established on what was then called "Ngola Ewondo". That is the "country" of the Ewondo, a tribe that is part of the great Beti-Fang ensemble. Formerly, this department went north of Mballa II, at the entrance of what is today the Etoudi district. He stopped south of the capital, at a place called Coron. But over time, the Mfoundi has expanded. Some knowledgeable observers speak of "an enlargement for electoral purposes. "Since then, the surrounding rural areas (Nkol Bisson, Nkolodom, Nkolmesseng and others Ekounou, Odza, Nkol Nkoumou) have become an integral part of the Mfoundi Department."

In the many names that make up the villages around Ngola we will have the term nkol which refers to the hill or a bute. Ngol / nkol (g <k) phoneticment we find here indisputable genetic links.

Angola

One can also notice the striking similarity with the name of the country Angola which returns indisputably to this same reality because this country is mainly populated of Ekang. Ovimbundu (37% of the population), Ambundu (25%) and Bakongo (13%)

Six Bantu languages have national language status10: umbundu (26%), kikongo (14%), kimbundu (8%), chokwe (7%), nganguela etkuanyama10,2. A total of 38 Bantu languages are spoken as mother tongue or second language by Angolans including Lingala.

*Pon**gola**, (north of Natal); A**ngola**[274], Nga**nguela**, Ga**ngela**;Do**ngola**, Do**ngola**wi (Soudan)*

274. TASTEVIN , C . A grafia etimológica dos nomes tribais africanos, Portugal em Afr.,n° 4, 1947, p. 332-338.

.Anthroponyms

In Beti fang, especially in central Cameroon we will find the same name Ngòla, **Ngóla** worn by men or women.**Ngola** Kiluanji (king of Ndongo in 1500); **Ngola** (Angola)[275]

Ethnonyms:

Bale**ngola**; Baso**ngola**; Le**ngola** (RDC)

Homophones and associated Meanings [276] –

NGOL: Catfish fish without scales, fin cutting (mesom). Ŋgol is the generic name.Here are ŋgol: sôma, anduma, mvar, deghe.

☞NGOL : Very long link prepared to make baskets.

☞NGOL : Deceived.

☞NGOLENGOLE : toothless, disarmed.

☞NGÔL (h) n.2, pl. miŋgôl.

1. High barrier to encircle the elephants. By ext.: entrenchment, terrace, redoubt (terms of siege). See akaŋ. Ŋgôl ô nẑokh, barrier for elephants. Nẑokh sa tu ŋgôl, the elephant crosses the barrier.

2. Senior. Nsime ye ŋgôl, youngest and youngest. Syn.: nyamôre, ntôl .: Elder, adult, notable.

☞NGOL METEKH: Beast without legs like a fish.

MORPHOLOGICAL VERIFICATION

In order to carry out the verification I will rely on the methodology of comparative linguistics. And for the purpose, let us compare the skeletons of consonants and determine the morphological links that exist between the different skeletons of consonants and the morphemes that result from them. [ŋgwol] / [ŋgwal] or [ŋgwoŋ] / [ŋgwaŋ] (l <ŋ)
The two terms being indifferently used one can postulate the presence of an isoglosse.

	mDw nTr	*Transliteration*	*vocalisation*
Middle-egyptian		*N'r*	*Nar*
Fang/Béti	-	*ŋwn*	*Ŋwoŋ*
	-	*ŋgl*	*Ŋgol*
	-	*ŋgŋ*	*Ŋgoŋ*

275. MONTENÉGRO , J. P., Esboço de um vocabulario toponímico angolano, Bol. Inst. Angola, n° 9,1957; galement n°" 10 à 12,14 à 17, 1958, etc.
276. Samuel Galley, Fang-French dictionary. and French Fang followed by a fang grammar, ditions Henri Messeiller , Neuchâtel, p.351

The term to designate the catfish in Ewondo is / Ngwal-kwas / that is to say the royalfish fish. The expression / Ngwal-kwas / can therefore be approximated to silurid. But as a general rule we say / Ngwal / to designate the catfish fish. The Frenchman also preserved the said semantax.
This linguistic precision is given to distinguish it from / kwas-bibas / "fish with scales".

This precision is useful here because we have two skeletons of consonants: ***Nar*** and ***mr*** and we name the fish / Ngwal-kwas /. So on the morphological level, whatever the morpheme uses, / Ngwal / or / kwas / we should justify it with the skeleton of consonant ***nar*** the term fish being a generic and the term catfish also referring to the idea of fish.

According to the comparative linguistic method, the following equations can be used for fang / beti languages:

☞ General equation: nar = ŋwoŋ or nar = ŋgoŋ or nar = ŋgol
☞ Comparison of morphemes

1. n - = * w- or n - = * g- (here we notice that the labio-velar and uvular ridges of the fang / beti) will split in time with Middle-egyptian to give a simple "n" , a "w" and a simple "g". This hypothesis is confirmed by words in Middle-egyptian that will refer to the agrarian world, telluric, or fishing, cultual or force and that are directly related to this script.

1. -r=-**ŋ** ou -r=-**l** (on a ici deux alvolaires) pas de l en pharaonique.

Schma de principe

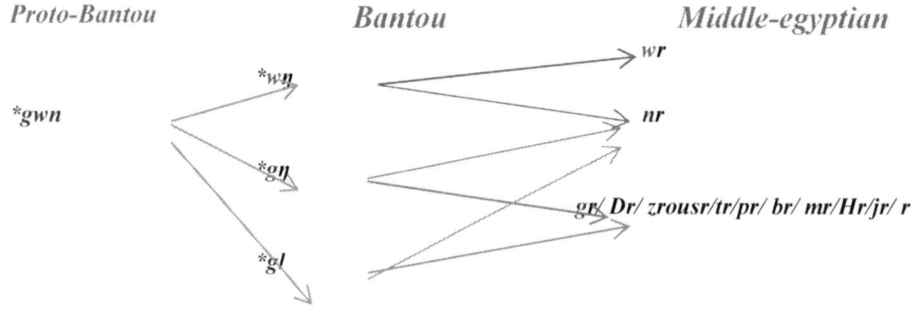

So we find here the words nr and rm which all refer to the word "fish". Just at this level you can read nar-mr as [Ngwal kwas] "catfish". Specify if it was necessary  ***rm*** "fish"

All other forms that are skeletons of consonants of words that refer to royalty and its attributes.
We note here two things, the root common to all these words is * r and we get all the forms from the form containing the velar consonant g / k to the simplest form with a simple r (word, mouth , hole, a)

Middle-Egyptian: ***pr*** "palace"; ***zr/sr*** "noble"; ***wr*** "great", ***Hr*** "chief, king" etc.

Listing of words with the consonants skeleton *wr*

1. *M.E.: Wr "swallow; martin"*[277]
2 *M.E.: wr "great; numerous, abundant; elder; great, important, eminent, superior"*[278]
3 *M.E.: wr "greateur, sufficiency, excess; a lot; how much?"*[279]

277.Faulkner p. 63, Wb. I p. 326, 11 symbol improperly associated with a certain horus the big one the bird is called in Ewondo: [kol-woro]. (Pyenonotus gabonensis) which therefore responds to the consonant skeleton hr-wr.
278. Faulkner p. 63, Wb. I p. 326, 11+ sq.
279. wr Faulkner p. 63, Wb. I p. 330, 7-11

4 M.E.: wr "great person, magnat; chief; leader"[280]
5 M.E.: rwt "judge, juge"
6 M.E.: rw "lion"
7 M.E.: rwyt "courthouse"
8 M.E.: rwd/rwD "force, firmness"

List of words with the skeleton consonant *nr*

1 M.E.: nr "chasing, harassing"
2 M.E.: nr "guardian, shepherd"
3 M.E.: rn "cartouche, king's name"
4 M.E.: nri "to be afraid, scary, scared"

5 Middle-Egyptian*: nri "protect"*
6. Middle-Egyptian*:nr.t "vulture to head bald" African royal symbol.*
7. Middle-Egyptian*:nar "catfish"*
8. Middle-Egyptian*:nwr "shaking"*
9 Middle-Egyptian*:inr "plate, rock, block, stone"*
10 Middle-Egyptian*:nrw "fear, dread, confusion, terror"*
11 Middle-Egyptian*: nrw "the scary, the terrible"*

Listing of words with the consonants skeleton *gr*

1. Middle-Egyptian: Hr "Horus, the king"
2. Middle-Egyptian:His face, face, face; view and look
3. Middle-Egyptian: Hr "on, above; in etc ".
4. Middle-Egyptian: Hr "distant, distant, far away"
5. Middle-Egyptian: Hr "terrible, terrifying"
6. Middle-Egyptian: Hr "terror, fright"
7. Middle-Egyptian: grw "silent"
8. Middle-Egyptian: Hrw "upper part, top, height"
9. Middle-Egyptian: "sun, day, high"
10. Middle-Egyptian: Hr.t "heavens, sky"
11. Middle-Egyptian: rx "to know; be aware of, be aware of, take notice of; learn; request!; knowledge, know; knowledge; opinion, opinion "
12. Middle-Egyptian: rx "wise man"
13. Middle-Egyptian:rx "to know"
14. Middle-Egyptian: gw / ngAw "bull"

I did not notice all the words but we observe that the words are related to the semantic field of African royalty.

Among many meanings that have come to light, I have noted some interesting:

Middle-Egyptian: rw.t "dance, palpitation"
Ewondo: NYAR (b) n.3, pl. menyar. Dance of men for the mourning of a chief. One man dances among the others who sing. It happens in broad daylight, women and children watch.Syn.: ŋgu, mefa.

Middle-Egyptian: gw/ngAw "bull , ox"

280. wr Faulkner p. 64, Wb. I p. 328, 14-16 sq

Ewondo: NYAR (h) n.3, pl. benyar. Buffalo, beef (Brachyceros, Bos pumilus). sa nyar, male buffalo (or nnôm ô nyar View lom nyar, ônsôm ô nyar.

We can see that we can read the consonant skeleton ***nar*** of the Pharaonic [nyar] en Ewondo

Passons au sign U72

Let's observe this little comparison

The land

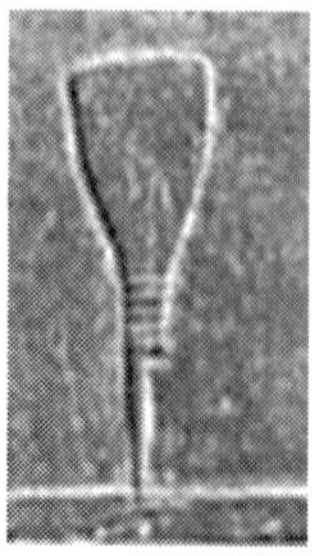

The chisel
mr

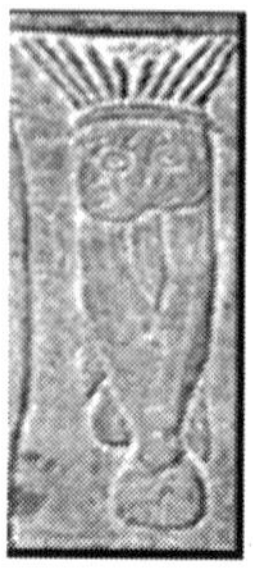

Cat-fish
nr

The sign U72 transliterate ***mr***: "chisel, carve".

I posit that the second element of the U72 transliterated U7 cartouche delivers a lot of information and that we will analyze at different levels

☞**ideogram**

it recalls the shape of the catfish we approached it by [ngwal kwas]
it recalls the shape of the country (km.t)
it recalls the shape of a chisel / chisel etc. Ewondo: [Mbokh]
he will remember the link. Ewondo: [ngol]; Ewondo: [nkol]; etc.
it recalls the shape of the head of an elephant with a trunk, Ewondo: [ikama]

☞**phonogram**

chisel ☞[mbokh] which will be vocalized in Ewondo like (priest, unit, feather, savior, hindrance, spiral etc.)

- Induce phonetic proximity mvogh (country)
- Induce phonetic proximity mbokh (lengthening, forge bellows perseverance, hole.)
  Induce phonetic proximity mfəkh (bag, smart etc.)
- Ebokh (footprint of an animal) Nanga-Ebogo, Ebogo Emerand Ebogha (hallucinogenic plant, banana bark for its yarn, food etc.)
- Bokh (flatten, agreatir, braid, twist, screw etc.)
- Boghe (fix nomads, be glued etc.)
- Boghda (getting up each other etc.)
- Etc

In reality one can not separate the ideoscriptic character from its phonoscriptic side it is necessary to use both aspects to have a good comprehension.

Example:

NGOL (m) n.3, pl. ***meŋgol.*** Very long link prepared to make baskets.

NGOL (b) n.3, pl. beŋgol (peu us).
Catfish fish without scales, with sharp fins (mesom). Ŋgol is the generic name. Here are ŋgol: sôma, anduma, mvar, deghe.

NGOL (h) n.2, pl. miŋgol. Elephant trunk (ŋgol ô nẑokh). Syn.: bîa.

As a phonogram, it comes to clarify the idea of link or unite that is guessed on the representation of the catfish. This idea is to be taken into account because of the rope knots that appear below and which has been previously described.

Hieroglyphic

Fang/Béti

Mr	***mer***		
Vl	*vul*	*Gather, sew, unite*	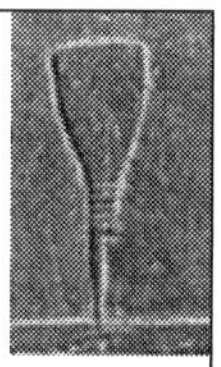
Vr/wr	*Vure, vura, wure*	*Bander, bandage, tie*	
Ngl	*Ŋgol*[281]	*Link for baskets*	
Yl	*yale*	*Bander, tie*	
Lr	*lare*		
**bkh*	*m.bokh*	*Chisel*	

So we have clear evidence that NARMER vocalization is not the right one. And it would be surprising to find an African with this name or to give an explanation of this term in a language spoken today. NARMER is a conventional construction linked to an erroneous redundant juxtaposition of phonemes.

Although this analysis brings a certain amount of information, it is however limiting and at this point almost useless.

This demonstration reinforces the idea mentioned above that the vocalization of the term catfish in African languages alone is enough to validate a good reading of this script.

The name of this sovereign is therefore the name of the catfish in our languages.
In fang / bet, the catfish or catfish is called ŋgwol or ŋgwoŋ.

Remember that to pass from the common name to the proper name we simply add the final vowel.
For many people in Africa it is therefore logically forbidden to consume catfish because consume catfish is symbolically consume the country and the sovereign. For example, the N'Dénian AGNI of Ivory Coast do not consume catfish, the fang beti do not consume the catfish of the well. In Benin idem

On the mythological level the catfish would have swallowed the virile member of ***wsir*** reason why many people do not eat this fish.

So let's remember that the sovereign is named **Ngola**. As the pharaoh his name is **Fara Ngola**.

Land depositary

We said earlier that ngwal was a generic name (silurid). To know the exact name of the catfish concerned, we need only remind ourselves that the sovereign is a Panther Man. So the catfish in question is a catfish with a spotted coat.

His name must feet the morphosyntactic requirements ***Nar mr***. Thus an excellent reading of the script reveals that the sovereign would be called precisely in Fang / beti ***Angwan*** or ***Mvas***, ***Mvar*** because that is how his name would vocalize in fang / beti.

Has the land bailee he is:

Ewondo : /*ùn*-Nya myar/ "the shrunk's mother" see illustration N°19

281.**ÑGOL** (m) n.3, pl. **meñgol.** Very long wire prepared to make baskets. (Dixit dictionary fang / french page 281)

The worship leader function

> MBÔKH (b) n.2, pl. mimbôkh. Prison, Prisoner, Slavery. N d a mbôkh, nda mimbôkh. Mbôkh ô wo, mbôkh abo, a Gabonese cangue species, a piece of very heavy wood attached to the arms or jam-be of the prisoners. Syn: ebony.
>
> B Ô K H (b) vb. 1. Agreatir the hollow of a tree (abôŋ) to take what is there. Azo mbôgha, see azo. Bôkh yṽi, take honey in a tree hollow. Bôkh fo, kô, befô, ŋgôm, take the bigvers of palms, a parrot, rats, a porcupine. - 2. Bôkh Bokon, separate banana offspring from their mother for replanting.
>
> B Ô K H (h) vb. Flatten, arrange flat. Bôkh nẑen, smooth the way. Bôkh nda, smooth the floor of the box. Bôkh enoŋ, make a bed. Bôkh sil, flatten the hair. Mbôgha si, groomed ground, flattened.

The presence of signs D21 and D36 may seem anecdotal in a kind of alphabetical approach of this script but in reality it is not so.

Indeed, the sign D21 represents the mouth and thus a hole returns the sound / -okh / in the African languages and indisputably indicating the numeral one: / fokh /, / ùm-bokh /.

The sign D36 representing the arm also returns the idea of the numeral one will find it in the script which indicates the numeral one in Egyptology. At this point it is vocalized /w'a/ "arme; one".

Clarias sp

synodontis schall fish

The catfish named in Ekang land /à-ŋgwàn/ this item shares many other meanings like simpers etc. But the most important meaning according to me is :

Nanga Eboko : *Korongon* « synodontis schall fish »
Ewondo : *Angomoro* « synodontis schall fish » litt. The catfish man
Eton : *Watwat* « synodontis schall fish » litt. The ancestor
Bamoun : *rohroh* « synodontis schall fish » litt. The ancestor
Bùlù : *Wo'o ambwang* « synodontis schall fish » litt. The ancestor
Manguissa: *ikohn*« synodontis schall fish »
Ewondo : /à-ŋgwàn/ « grouping ; ingathering ; rally ; muster etc. »

The Fara Angwan was indubitably the resposable of the alliance between the north and the south of Egypt.

☞Ethnonyms

Fangwe, Mangwe, Pangwe, Langwane(Tchad, Cmr, Gabon); Kwangwa(Zambia); Nganguela, Gangela, Gangwela, Bangwa, Bakwange(RDC, Angola); Bakangwane, Ngwane (Swaziland), Bamangwato, Bangwaketse, Mangwato, Ngwato (Botswana); Amangwane, Ngwane, (Afsud); Kipangwa(Tanzania) etc.

☞**Toponyms**

Nsangwini (Transval)

☞**Anthroponyms**

Angwan (Cmr); Ngwane (Swaziland)

The force can also mobilize vocables around the root * ngo. Indeed, this expression also makes it possible to point out the real name of the country as that of the leopard-men. This term designates the sovereign and the country at the same time. "Catfish" is a homonym and homophone of Proto-Bantu ****ngo***

Symbolic of the force: N'r mr: leopard-man

Why here the expression / ngón / is primordial?
Complementary and interchangeable

The word / ŋgoŋ / is a homophone of / ngon / = "panther / leopard" in the languages of the Congolese basin. Indeed at the Bafia people of central Cameroon, / ngon / is the totemic name of the panther / leopard.

The words catfish in our African languages thus indisputably designate the true name of the sovereign. He then posits the evidence that he is part of the African Brotherhood of Leopards.
Let's just remember that the "leopard, panther" is an animal consecrated to royalty in Ethiopian societies, and particularly in the region of Greats Lakes, the place where the Nile originates. Proto-bantu * ngo.

Words in the / ŋgoŋ family / for example:

Ewondo: aŋgône "TERROR fear, danger, (m), wôn (m)"

When it is symbol of Force and in its attributes related to the defense then it is not only catfish but the buffalo which is a word coming from the same Negro-African root as ***nar*** and which is associated with the royalty. On this account it is necessary to vocalize no in Ewondo, ***Nyara*** and the consonants skeleton *nr-mr* is vocalized ***nyar-moro*** "man-buffalo". This information is recorded in the so-called "***Narmer".***

Middle-Egyptian: rw.t "dance, palpitation"
Ewondo: N Y A R (b) n.3, pl. menyar. Dance of men for the mourning of a chief. One man dances among the others who sing. It happens in broad daylight, women and children watch. Syn.: ŋgu, mefa.

Middle-Egyptian: gw/ngAw "taureau , bœuf"
Ewondo: NYAR (h) n.3, pl. benyar. Buffalo, beef (Brachyceros, Bos pumilus). Esa nyar, male buffalo (or nnom ô nyarj-i ^ See elom nyar, ônsôm ô nyar.

We therefore see that we must read the consonants skeleton ***nar*** of the pharaonic / nyar / in Ewondo language.
The function related to the security of the political entity falls to the conquering ruler. This is one of the conditions that indigenous peoples must face in the event of an amnesty following a war.
Unquestionably the sovereign whose name is here must fulfill this condition.

His attribute of war is / Nyar-moro / that is to say the buffalo man. (See the so-called NARMER pallette)

Just as Louis XIV was called roi soleil "the Sun King". Without that being his name. François Mitterrand was "the quiet force"etc.

I want to clarify that this attribute that we call NDAN in Ewondo which can correspond to ***rn*** in pharaonic. It is not a name it refers to a qualification and in the southern regions of Cameroon this [ndan] can be vocalized by a sequence of onomatopoeia or drummed with the nkul[282] it's not just a slogan.

> **Ndaŋ** (b) n.l, pl. bedaŋ (vb daŋ b) . Winner, the one who triumphs, who is the strongest. N'daŋ byal, the one who enters the canoe. Ndaŋ ayo, the one who is the most great. Ndaŋ miŋkal, the one who faces the test fetish. The contr. of winner: ntu, betu the vanquished.

For example, in the lineage of atangana mbala their lineage ndan is:
/**nyar** mə-kom tə fulan à yi mod/ "the domestic buffalo does not look like the man"

Symbol of Multitude
The pharaoh being the head of several lineages, he is also the father of a multitude.
We have a catfish with the name :

Ewondo : ébul "small catfish" ☞ **Ewondo** : ébul "multitude, 9; nine"

So the pharaoh could have been called by the following names : Bulu, Bulà, Ibulu; Baal; Ebèlè; Ebalé; Bela; Bala; Belibi; Lobé; Luba; Imbula; mvulà; Mvele; Mvilà etc.

This is confirm par the historical continuum because, we can find all these names as anthroponyms or ethnonyms elsewhere in Africa.

Note : The equivalent of this name is the Ram "mutton; lamb" among the shepherds.
In a multicultural society like ancient Egypt, this symbol is represented by a hoe ***mr*** among the farmers.
Among the fishermen obviously the symbol is a fish ***rm*** .

The force symbol:

Because the cartouche only shows a catfish we assume then the fact that this catfish could be one of the dangerous species namely in Ewondo :

Ewondo : *anyia* "torpedo fish or electric catfish" or
Ewondo : *mbam* "torpedo fish or electric catfish"

These two catfish are able to deliver electric discharge. As we have seen; the mother's figure is associated to the ancestor. /*nya/ "mother and /mvam/ "ancestor"

As a symbol of strength the ruler can be identified by his ndan who is [Nyar-Moro] "buffalo man""[283]

But as son of God his name is absolutely in Fulfuldé :

282. The nkul is a horizontal drum
283. D.T. Niane, Soundiata or the Mandingo epic,Karthala, paris, 1960, p. 24

Fulfuldé : *njawdiri mbaala* " the ram" we assert that this is only one meaning.

In addition to this we can find this meaning in the different slips of meaning related to the consonant skeleton **nr** and ***mr.***

The Ewondo expression / Nyar-moro / meaning man-buffalo shares a phonetic proximity with the expression / nyamoro / with the elision of the "r" of / nyar /.
The / Nyamoro / being in african traditions the accomplished Man, the eldest. Expression that is naturally found elsewhere in Africa, especially in the country Bamana under the term nyamogo or in Fulfulde with nyaamugo or nyiri "food, to eat"

Here we have again one of the multiple senses of /ŋgôl/ "the elder; the oldest"

In addition to this detail, we find the anthropomorphic notion of the buffalo in the vast African cultural field; especially with the birth of **Sundiata Keita** whose mother was supposed to be a **buffalo** woman. Sogolon Kondé, the mother of Sundiata Keita was said to be an ugly, hunchbacked woman (hence her nickname "Sogolon Kedjou", "Sogolon the Ugly"), but endowed with powerful magical powers. The origin of his surprising union with the king **Nar Maghann** Konat, king of the Manding (or Manden) is related by the episode of the buffalo of Dô in Mandingo country[284].

From this history, one can extinguish from the collective memory in Mandé land the vocalization associated with the name of the pharaoh ***nar-mr***.

Nevertheless, I wanna emphasized that, we must pay attention to the details of this story because this tale in my opinion is a scrambled message for insiders.
The fact that Soundjata's mother is told to be hunchback is a clear reminder of the zebu or the **Ewondo** : Zamba ↔(**Middle-Egyptian**: *bs* "Bès").

Holy cow, female goddess of ancient Egyptians

According to Hathor, cow, female goddess and Sekhmet with lioness head, quartzite, c. 663 - 332 BC, Late Period, Ancient Egypt, (Marsailly / Blogostelle)

Holy cow, female Of Narmer's pallet.

The people on the pallet cannot claim to be the owners of the Egypt land. This will be ostentatious in African thought. The Egypt land belongs to **Hr** or Zamba. The occupants of the land are mere tenants who can not claim any ownership. For the purpose the is consider as a simple manserver.

So the union of the tale that she is preached with King Nar Maghann gives us her insider name, namely the **buffalo woman**.
This union simply recalls the **alliance** between the true owner **Zamba** and the manserver

Ewondo: nyar-minga "buffalo woman" ☞it we can see on the pallet above

284. Youssouf Tata Ciss et Wa Kamissoko, Soundiata, From origins to the founding of the empire, Karthala, Paris, 1988, p. 47-51

What is certain a woman was the owner of Egypt because her name tells that also.

Ewondo: /à-myar/ “to shrink an object in the middle, to make a narrowing”
For a better understanding may you report on Map N°5 or illustration N°19.

Ewondo: /*nya *u*-myar/ "the owner of Egypt here called the shrunk"

Note: As we have highlighted even though on the pallet the king is a man, but he embodies a name of a woman because he is just a deputy not the owner of the power. In a matriarchal society, power belongs to women, but it can be exercised by a man.
The power being of feminine incarnation, thus the word / *ùn*-nya / “mummy; barrier, protection; high etc” finds here all its justification.

THE SOVEREIGN IS THE COUNTRY

The sovereign being the country, I assert also that what has been proposed as being MANI, MENES is neither than its name is equivalent of the script NIWT.

		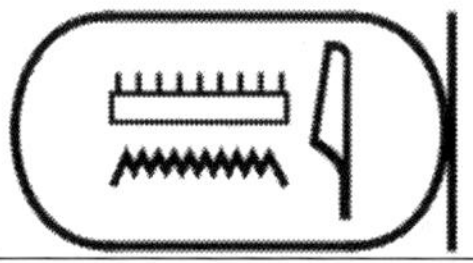	
Script read Niw.t meaning city	*Script transliterated* ***imnti*** *Meaning west*	*Mani/Mnès*	*Mani/Mnès*
Ùn-Nnàm = city Un-nám (meal)	*i-moŋ =left/ left-handed i-nam: member*	*Nnama*	*mi-nnam (bi bè =2)*

The country is a catfish, so it is a dish, a food, a meal, etc. and it is therefore natural and suitable for the word / *ùn*-nam / “catfish; meal” to be here homophone of / *ùn*-nam / “country” and Zàmbà because it is the name of the earth too. This is why many people in Africa forbid the consumption of catfish. Let us note by the way that until now the Ekang call the Head of State / *mor *ùn*-nàm / "the Man who embodies the country" ↔ **Middle-Egyptian** : Nr-mr.

The Mogho Naba (also Moro Naba, Morho Naba, Mogh-Naba or Moogo Naaba), literally head (naba) of the world (moro), is the monarch of Wogodogo (Ouagadougou), one of the Mossi Kingdoms located in present-day Burkina Faso. The kingdom takes its name from its historic capital, now the Burkinabé national capital of Ouagadougou. Although the most politically powerful of the Mossi Kingdoms, there was no Mossi "empire", and the Wogodogo king did not have authority over the other kingdom.

Note: The consubstantiality between the country and the pharaoh leads us to conceive in Africa that the Pharaoh embodies the country and the country is the Pharaoh.

We note here that the expression Moogo Naaba highlights several things including the morpheme *moogo* whose ananym is *(n)gomo.*
The chief and the country are the big tam-tam:
Ewondo: / nyà-ngom / "big tam-tam" which is confirmed by the illustration N ° 19. Egypt indeed has the shape of the African tam-tam.

We also discover the porcupine meaning associated with Egypt. We have already noted that during the study of the ***km.t*** script including the country of Akang by Kuma-Si.

Moreover, I had highlighted a singularity in the word associated with the religious confirmation namely:

Ewondo : / i-fyer Amougou / "confirmation"

Kékwe branch☞ ***gom-** ☚ Kweké branch

Tam-tam, porcupine; curare ← → Amougou Maghà ↔Mecca

Ewondo: / à-ngom / "reconciliation"

Amougou appears just as an ananym of "ngom" porcupine, tam-tam, reconciliation etc. "

Reminder

Since we postulated that there is no accidental homonymy but an homophony related to the convergence phenomenon to a same signified, the term / ifyer / giving the skeleton ***fr*** and ***nfr*** sharing the same root * **fr-** to the agent prefix **n** – /nyo/ "whose". So let's use an endogenous analysis on the script: 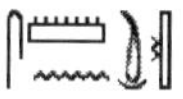 ***smn*** "to establish, perpetuate, etc.".

The sign U32  of the Gardiner's list is given as a pestle and a mortar. The pestle / mortar set is used in principle to crush.

Ewondo: /á-fir/ "to crush" thus offering phonetic link with / i-fyér /

The sign Y5  of the Gardiner's list is transliterated ***znt*** and often associated with the item "zenet" name given to an African game by Egyptologists. The two other associated signs transliterated ***sn***, I will not dwell on them too much.

I do say that the sign Y5  is read here /*ùn*-sɔŋ/ "piquant; sharp; biting". This reading allows a synonymy with:

Ewondo: /*ù*-myar/ "porcupine spines"

The African sacred animal that represents the best illustration of the idea of quills is indeed the porcupine. It can be found nowadays as the Ashanti royalty totem in Kumasi. There are many anthroponyms in Africa, toponyms associated with the porcupine. The porcupine is called:

Middle-Egyptian:  ***Hnty*** "porcupine"
Ewondo: /ŋgɔm/ "porcupine"

All the subtlety of the scribe is therefore based on the mindset consisting in using a word from the kekwé branch instead of a word from the kweke branch to express an idea.

The terms **ŋgɔm** and **mog** are indeed consonantal metathesis as seen above. So we say that, the consonantals skeleton **mog** is the Beti anthroponym equivalent of: Amougou in the expression /ifyer Amougou/ "confirmation".

Middle-Egyptian: 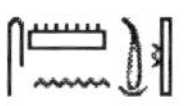 ***smn*** " to establish, found, perpetuate, etc..".
Ewondo:/Ifyer-Amougou /"religious confirmation".

The toponyms: *Ngoumou*; *Ngomo; Mongomo* do exist in Ekang land and elsewhere in Africa..

There is in the catfish family found in Africa a variety of small catfish. This catfish is called in Ekang country:
Ewondo: / i-fyer / "small catfish"

This term maintains a metonymic relationship with the word "buffalo" present in the so-called Narmer's pallet
Ewondo: / i-fir / or /i-fyar/" crusher; shrinker; buffalo"

Thus, the item of **Middle-Egyptian**: ***nar-mr*** can give in Ewondo:

Ewondo: "*ù*-myar *i*-fyer /" the small catfish spine "
Ewondo: / i-fyer Amougou / "alliance, the country of the alliance; the catfish of the alliance"

Egyptian bread cake (illustration n°35)

The resemblance between the sign O49⊗ and the bread cake above is pretty amazing. This further strengthens the reading / *ùn*-nàm / for country because / *ùn*-nám / is also the term to designate a food. The sign O49 is neither more nor less than this cake of bread.

ROYALFISH AND PROHIBITION

Mali, dogon land, village of Bamba.

> *"The pond of Antogo is a sacred place where it is forbidden to fish during the year, even in case of scarcity, except a day announced by the hogon, patriarch of the village. We mainly fish for **catfish**. However, this fishing is prohibited for women and villagers in Yanda, a neighboring village of Bamba, except for children whose mums are from Bamba. For years, Bamba and Yanda fought over the Antogo pond.*
> *Around the lake, 1000 villagers are standing or squatting, a trap in hand, a bag of the other listening to the prayers of the hogon, patriarch of the village. At his signal, they rush into the molasses molasses clearing a huge cloud of dust.*
> *The end of the fishing is announced by a gunshot on the public square. The fish are redistributed fairly by the hogon.*
> *Antogo fishing is a big celebration in Dogon country, a festival of communion, peace and festivities. The festivities continue in the evening with a wake that continues until the early morning".*

Finally, it should be noted that, including among Ojibwa Indians, catfish is a totemic animal. It seems indeed that the man was born in Africa.

Summary of achievements

We must rigorously rename this pallet : the pallet of Ngola, mvar or Nyara-moro because it is his real name. I also postulate that the characters used to designate the sovereign's name indicate the registers in which that name could be applied: 

Sign N35: Fillet of water-> denotes the area of water (catfish)
Sign D36: The Arm -> Area of power-justice (the ruler symbolized by the crown)
Sign D21: The Mouth / Hole -> Ritual Domain (Priesthood)

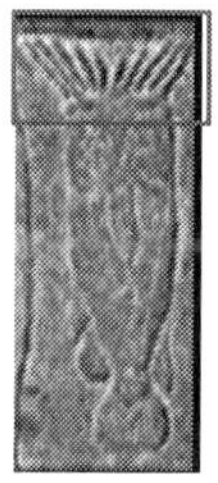

The analysis of this fish alone can read the whole script:

Let's carefully observe the representation on the right.

1. The fish barbels have been organized like a crown. This crown or diadem refers to the idea of sovereign. Thus one can even think that the name of the fish is the one of the sovereign.

2. The choice of a barbed fish here allows the sculptor to arrange them in a flared shape which indisputably recalls the lower Egypt (the delta) and the body of the fish Upper Egypt. So this fish shaped sculpture also denotes the name of the country.

3. The intersection of the mouth of the fish and the barbels represents a hidden shape of the link between the two territories. This precision is brought by the nodes visible on the script ***mr*** vocalised **mer** by the modern Egyptological current.

This precision on the country is given by this comparison.

Country
rmT

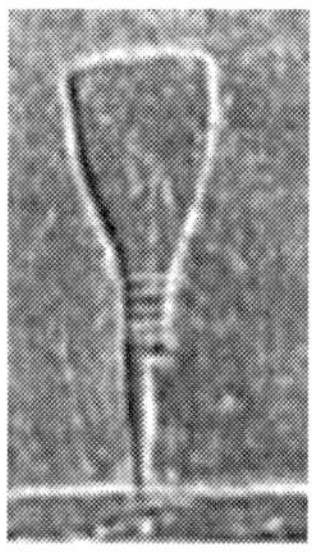

chisel
mr

Royal-fish
nr

Djembé
Zamba

Nar-mr incarnate the  ***smA-tAwy***

The signs k13  (cat-fish) and the sign U72 (ciseau) are actually intelligent representations of the country at the time of the sovereign of which he is the incarnation.

As we previously described on the sign  U72, the rope that serves as a link between the two elements of the chisel is called Ewondo: / nkol / "rope" but precisely /ngol/[285] homophone of / ngol / "catfish". So the sovereign being the unifier of the two lands and the land depositary he is the ultimate incarnation of the ***smA-tAwy***.

Ewondo: [à sama] "add; pool" ☞English: same "like"
Ewondo: [i.samba] or [n.sama] or [nsamba]: "together, file"
Ewondo: [zama] "earth where we bury their dead reciprocally][286]
Ewondo: [Zàmbà] "le creator"
M.E.: ***tAwy*** □/me si me-bè/; /yob y si/ "sky and earth"

285.Samuel Galley, Fang-French dictionary. and French Fang followed by a fang grammar, ditions Henri Messeiller, Neuchâtel, p.351
286.Samuel Galley, Fang-French dictionary. and French Fang followed by a fang grammar, ditions Henri Messeiller , Neuchâtel, p.382

It is said according to a kind of semantax that Dzô is the master of heaven and earth. The Fara being the representation of Dzô (heaven) on earth, he is therefore the master of heaven and earth.

Middle-Egyptian: ***smA-tAwy***
Ewondo: [si Zàmbà] "The land of the creator", paradise?

ŋgoŋ Associated expressions in fang / beti Languages

Many expressions are used in Ekang country. I recorded two:

- Ewondo: [Ma á-kə à ŋgon] "I go to ŋgoŋ", I go to my plantations; farms.
- Ewondo: [A-lih man i.səp ŋgoŋ] "clear a field in the land"

Let's go back to this definition

Ewondo: ŋgoŋ (h) n.2, pl. miŋgoŋ. Fish, lamprey. He looks like a snake. It is an ŋgol (Mastacembelus, Ophiocephalus) .Syn.: ŋwoŋ. He has the pointed mouth.

Fish caught in Lake Tanganyika, Zambia:

<u>Egyptology's Position</u>

Middle-Egyptian: xA.t "oxyrhynque[287]"
Ewondo: [ŋgóŋ] or [ŋgwaŋ]
Fish often associated with ***wsjr*** in ancient Egypt it was discovered on the west bank of the Nile.

It has been identified by specialists as being the Oxyrhynch because characterized by a "Tapered nose", "Sharp nose", because of the morphology of the Oxyrhynch, its head is finished with a slender snout! *Pemdj* for the Copts...

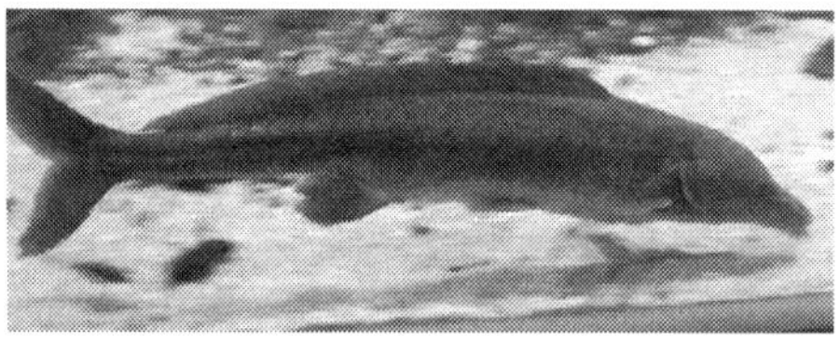

The oxyrhynchfish

In Africa we often find tapered nose or beaked fishes in Ekang land, here are the most common:

Ewondo : /ŋkore/ "beaked fish"
Ewondo : /i-koko/"beaked fish"

☞ **Ewondo** : / *ùn*-ŋkoré/ "the savior"
☞ **Ewondo** : /*ùn*-ŋkùghù/ "very old; ancestor"

287. Oxyrhynch: Mormyrus kannume (cf illustration n°18)

The sovereign and his subjects

mr.yt Crocodiles	*mr* sick	*mr*= at-tached/tied	*rm.tyw in*ha-bitants	*rm* fish	*nar* fish
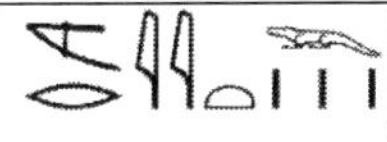					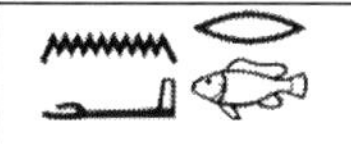
					Hole = a.long Ngol = catfish Ib, at bo, at ko
mi-*mbombokh* [mi.ŋ*kôm*]	[m.*boghbe*] *m.bokh* (Bedridden) the one who nests	*m.bokh* *n.kôm* prisoner	[mi.ŋ*kôm* *mi-m.bogh* *mi-n.tobo*	[à.*kom*] ***Ngol***	***KôaS*** ***ngoal***

3.5.2.3. PR AA

The great intelligence of classical Egyptology has been to invent an African language with its own words. Curiously this language is not spoken anywhere in Africa.

The tragedy of this invention is to lead Africans to abandon their own languages to give themselves a scholarly air by vocalizing some of these barbarisms at the expense of their own languages.

Among these many barbarisms, there is one which has returned to common jargon and which appears in dictionaries. The term pharaoh transliterates ***pr aA*** (Faulkner p 89, Wb., I p 516, 2-12)
To whom we have given the official semantic content of "greate house". Why would the Africans qualify their ruler as a Greate House? This question can challenge the most basic common sense.

Let us return below the logic of the censor decision makers:

Middle-egyptian: pr "house"
Middle-egyptian: aA "great"

An analysis outside the official normative framework might have already pointed out that the symbol O29 transliterated aA, this symbol can take the following meanings:

- Great; abondant, multiple, rich
- Notable; elder son
- Magnitude
- Column, pilar
- Here; there, over-there
- Door
- Moke, donckey

Etc.

Why then, the censors only retain the idea of magnitude? And not at random the idea of pillar? In order to say that the ruler, is the pillar of the family. In which way that have been disconcerting? Surprisingly, we did not ask Africans or African royalties, who still have a reputation for being the most accurate, before endorsing it as such. That's why this script should be revised by Africans.

What is challenging by the meanings imposed by Egyptologists is the fact that the words are often African but they are imposed in everyday language. No African ruler in the 21st bears the title of pharaoh except the braggarts.

Yet some specalists claim that Pharaoh is a masculine noun that comes from a loan made to the Latin Christian Pharao + onis, title of the kings of Egypt, taken by antonomasia[288] as the proper name of every king of Egypt, himself borrowed from the ancient Greek Φαραώ / Pharaṓ and this one from the biblical Hebrew פַּרְעֹה / par'ōh, in turn borrowed from the ancient Egyptian perāa designating first the "Greate house", that is to say the palace, then, by metonymy, the occupant of the palace, that is to say the sovereign.

The question seems important here because it is central to know how Africans were calling or called their rulers. Amazingly here, everyone is concerned, the Greeks, the Latins, the Hebrews but do we seek the advice of the African component. So reading this script is relatively easy for any African who knows the African royal lexicon. It is enough to give the name of the chief or the sovereign to read this script. The reading, if on the morpho-syntactic plane will be able to pose a problem but on the semantic plane that will be absolutely right. If we know the name of the ruler by metonymy we will know what his place is called or simply give the word used to designate the house of the ruler in this way we will avoid the shifts of meaning related to metonymy.

It is therefore important to bring back the true title and to get rid of the diktat imposing to Africans a knowingly biased reading of their own history.

If I do a poll in Africa to ask what means [per aa] or pharaoh in an endogenous language, I think we will have a hard time understanding what we are talking about.

I therefore offer to analyze the term separately from any knowledge that is constituted and erected as a norm.
I recommend to redo a little feedback in order to help a better understanding of the reading content.

If in pharaonic ***pr*** is the house, what house are we talking about and why?
Do Africans still designate nowadays the places where their rulers live as great house?
Here are some ideas that will interest us in this reading.

Firts part of the Script 𓉐𓏤

This part is composed of signs 𓉐O1 et 𓏤Z1 of the Gardiner's list.
This script is supposed to return the idea of house and is vocalized **per** with a consonantal skeleton ***pr.***

Middle-egyptian: 𓉐𓏤 *pr* "house"
Ewondo: /à-fér/ "to close" **Koonzimé:** epér "to close"
Ewondo: /i-fyér/ "çatfish" ↔ **Ewondo:** /*ko-/ "fish; place; house"
Ewondo:/*ùn*-mbèh/ "door" **Ewondo:**/*ùn*-mboâ/ "one"

Note : we say that in Ekang languages, a place is sharing an homophonic relationship with the word fish. So drawing a place can also be drawing a catfish. We have emphasized that the souvereign is the country and perforce the country is the souvereign. By the way, the country having the shape of a catfish then the souvereing is the catfish.

In the light of this small comparison, the consonant skeleton ***pr*** decided by the Egyptology appears compared to the Ewondo like the place closable by a door. But in a wider sense as a place by traversing

288. An antonomasia (from the Greek antonomazein "call of a different name", from anti- "instead of", and onomazein "to name", from onoma "name") is a figure of speech or a trope, in which a name or a periphrase stating its essential quality, is used as a common noun, or conversely, when a common name is used to signify a proper name. Some common antonomases end up being lexicalized and appear in the usual dictionaries ("a trash can," "a silhouette", "a Don Juan", "a harpagon", "a Bordeaux", "a Roquefort", "the macadam", "A gavroche", "a tartuffe", etc.)

all the meanings associated with the symbol 𓉐O1. But the scribe makes it clear that by all these meanings, we must retain the one of the field of the numeral one.

Fulfuldé: suudu - cuɗi "house"
Hausa: ɗākī "house"
IsiZulu: indlu "house"
Ɛʋɛgbɛ: xɔ̀ "house"
Yemba: ŋgyā "house"
Tk: ndʒo "house"
Kimbu: inyumba "house"
Oluganda: nyumba
kinyarwanda: zu "house"
Bamana: so "house"
Baoulé: swà "house"
wolof: kër "house"
Kwere: ukae "house"
Chuka: cii "house"

If we understand that the prefix ***ko-*** designates the foot in our languages and that Kai, ci, zu, di, da etc. it is food, so we understand that these words above mean neither more nor less than the place of food, that is, of the father or the mother. Indeed, in African thought, the father and the mother are the nourishing roots of the family tree.

We better understand the meaning of the word house in the languages spoken in Cameroon. In Ewondo we call /nda/ "house".

Among the siblings of Beti's languages: the Duala, Basa'a, Kongo there is a small change of word to say house. **Duala**: / ndabo /; **Basa'a**: / ndab /; **Lingàla**: / ndako /

According to the comparative linguistics skills, on the diachronic plane we can suggest the following evolution: Nda <ndab <ndabo <ndako.

But this comparative linguistics point of view is not totally true. From the Ewondo side, /ndako/ is the house of the fish, the house of the ancestor; ***km.t***.
/ndabo/ is the house of the bee also ***km.t.***
/ndab/ is the house of the bee version beer language (without the final vowel) meaning also ***km.t.***
Morphologically I have emphasized that we have HOUSE + Qualifier here is the semantax used by Ekang people.
Naming the same signified and sayint: it's the home of the bee, the home of the fish, or simply home, does that meant any evolution of the word? I think we should pay attention with linguistic certitudes and their rules savant and their savant's rules.

According to comparative linguistics they could assume this precision because the other variants ndab <ndabo <ndako are not more used in the language Ewondo but morpho-syntactically, we find the proto-variants that led to the term Ewondo current. Between the other variants and the Ewondo one notices that the Ewondo use only the lexeme. On the other hand, in the lexeme nda- the other languages have suffixed a particle serving to designate the place or the leg [* bo], [* ko]. "Foot".

One could have found in the suffixal variant a semantax inherited from Negro-egyptian often serving to designate a place.

Middle-Egyptian: 𓉐 pr "house, palace"
Ewondo: /nda /"house"

Have I have emphasized above, we can note in Ewondo that the mental structure that consists in naming a place always uses the structure: house + qualifier:

Example:

Ewondo: /nda à-wu/ " mourning house "
Ewondo: /nda *bi*-yem/ "house of the animals" ☞ cowshed.
Ewondo: /*i*-toun nda/ "portion of house" ☞ room
Ewondo: /nda mi-mbokh/"the house of prisoners" ☞prison; jail
Ewondo: /nda bàngà/ "house of marijuana" ☞ coffee shop.
Ewondo: /nda Zàmbà/ "house of God" ☞ church
Etc.

The fact that we have for ***pr aA*** the structure: **house + qualifier** is interesting

On the historical level

The word ***pr aA*** is to be associated with the function of **fara Ngola** (***nr-mr***).
The part of the country wrongfully called Lower Egypt had also a sovereign's place of life named:

Middle-Egyptian: ***pr bity*** "The bee's palace; the Bwiti or Bwètè temple"
Ewondo: /nda fôŋ/ "the house of the bee" ☞ **Ewondo**:/boe/⇔*bee*

The term **fon** is used in the west of the country to designate some local rulers until nowadays.
The same term is also used everywhere in Africa to designate some African lineages such as Fong or Fong Gbé of Togo / Benin or Fong of Cameroon etc.

Since the Ewondo do not have leaders but only spiritual guides, the term has become almost obsolete.

[289]BÔE (b) n.l, pl. bebôe. big species of bees and their honey. Voir ***abè***, ***fôfôñ***, mvem, ***ko***.

Let's remember what we noted above: Duala: / ndabo /; Basa'a: / ndab /; Lingala / ndako /. We find the words /abo/, /*ko/ which designate the bee.
If by habit one says ndabo, ndako to designate the house in fact it is a vestige of the name of the ruler's house.

On its delta side, before the alliance there was an instance called ***pr nsw*** because having at its head , the ***nsw***.

Middle-egyptian: ***pr nsw*** "palace of….."

Ewondo: /nda 'ndzoué/; /nda *i*-dzoué/, "authority";
Ewondo:/nda-tín/ "the house of the source"
Ewondo:/nda à-tíŋ/ "the Ark of the alliance";
Ewondo:/nda-'nti/ "house od the lord";
Ewondo:/nda-nswóaŋ/ "house of the summit " etc.

Following Nyar-moro (***nar-mr***) achievement, the country operates a union of lineages to constitute a political entity. Certainly the **fara Ngola** (*nar-mr*) concluded a ritual marriage with the panther (the daughter of the uncle of the country or his sister) in order to share with the inhabitants of the delta, the same common ancestor. That's how he became their leader *bity*.

---

289. Samuel Galley, Fang-French dictionary. and French Fang followed by a fang grammar, ditions Henri Messeiller , Neuchâtel, p.74

This ritual union with the panther is not necessarily carnal it is the condition to become bailee of the delta lands. Better than that, this union ensures its impartiality as the head of the political entity because he was sharing a common ancestor by the marriage with his woman lineage.

By this union in an exogamous matriarchal society, he committed an abominable act which directly puts him out of the social cell and reduces him to a mere role of man of the service: 𓊹𓍛 *Hm-ntr* in Ewondo [nkom Tara].

Being the power of ***nsw***, it becomes by merger of the two lineal authorities automatically *nsw-bity*. Ewondo : /ndzué Bwity/

It is important to inquiry these postulates because they will be illuminating for the purpose in any case. I hope…
Let's go back to the word ***bity*** which designates as we said the bee. In reality, the society of bees is a society built around the queen.
This little digression reminds us the notion of the matriarchy that governs the southern countries, especially Nubia "neighbours", ran by the **kndke** "kindake" (unfortunately called candaces). Kindake is a Basa'a term used to designate a sovereign woman.

Recall what are the sovereign's missions in an African political entity:

- War responsible
- Depositary of the land boundary
- Religious leader

We posited that the Ewondo simply qualified the house according to the purpose.

If a war context, the purpose used will be house + war chief
If the subject concerns the cult aspect then we will have: house + head of the cult
If the subject concerns the function land boundary we will have: house + head of the land boundary.

So how to read 𓉐𓉻***pr aA***?

Cultual Chief

> **Middle-Egyptian**: pr aA
> **Ewondo**: /nda mbogh/;
> **Ewondo**:/nda fara/ or more exactly
> **Ewondo**: [Nda nkóm]

The object represented by the sign O29 𓉻 is called [mbokh] in Ewondo it is a pole, a pillar. It is the famous mysterious object of the sign U72 that one finds on the script of ***nr-mr***. This object leads on the phonological level to the vocalization of the term related to the priesthood or worship. This item is a sacred item in the dogon's Kommo.

I emphasize here that the vocal [ivogh] means in Ewondo big, fat. The translation [ivogh nda] or [nda ivogh] is therefore not adapted here for ***pr aA***. Semantically it lacks thickness and height of mind.

The terms [mbokh] and [ivogh] being substantially identical morphologically because being labials.
> The letter ***b*** is a bilabial occlusive consonant voiced.
> The letter ***v*** is a fricative consonant labio-dental voiced.

The two words thus admit a single consonant skeleton without having the same semantics. Did some specialists knowingly take each other for granted? That's the question.

Note: The sign O29 is to be linked with the idea of **big** like in the script of the big snake : aApp « anaconda ». Although the door is an opening, it is also used to close. The idea of closing is curiously put forward with the transliteration ***pr.***

Solar boats are a symbolic representation of what our scientists today call "Worm Holes" - like a worm in an apple running through the apple faster than it go around the surface of the apple) But "snake hole" in reference to the Egyptian context since these "solar boats would be the hyeroglyphic illustration of the travelers of" trimensionnal doors". Nothing less !

The drawing of the "gods on the solar boat" represents their journey through " Worm Holes ". The drawings above speak by themselves.

Middle-Egyptian : ***pr aA*** then this script has to be read the big dor or the dor of the middle.

Supralignment authority

As representatives of the two lineages on this account *pr aA* can be vocalized in Ewondo [nda mvogh] but there is still not here an idea or size. But an idea that conveys the idea of a lineage assembly.

Pr aA can the be read [nda bot] "family", house of people; family's house.

Taking into account the constraint of the ***smA tAwy*** (the union) ***pr-aA*** appears as the house of unity; the unity palace.

Chief of the land boundary

Middle-Egyptian: : pr aA

If we remember that the land belongs to the panther. So here ***pr aA*** is the house of the panther.
Ewondo: / nda ze / "house of the panther"

Head of the War Authority

Middle-Egyptian: : pr aA
Ewondo: [nda ngoul] "house of the force"

On this account, if the sovereign has a name of war, the local can be identified at his /ndan/. Example: nar-mr say: [nda-nyar moro] "the house of the buffalo man"

The house of justice: [Nda *à*-dzó] or [nda *mə*-dzó]

On the semantic and morphosyntactic level we note that the skeleton of "consonants" ***aA*** refers to number of meanings[290]:

1 column, pillar
2 great; abundant, multiple; rich in; groin
3 noticeable; eldest son
4 magnitude
5 her; there, over there
6 door
7 Donckey

The ruler is the pillar of the family / ndî /
The ruler is the eldest of the family [nto]
He is the bourricot / mbokh / the man of the service, the one who carries the collective charge.
He is great by the authority he embodies, by his birthright, the respect he inspires.
He is the door because he is the incarnation of the country.
Etc.

The sign *A* Often noted *3* in some literatures [291] comes from the Negro-egyptian * kh of the blow on the diachronic plane this morpheme can thus take all the possible forms in our languages, forms related to its segmental or suprasegmental environment until its form the simpler that is the vowel when he flaunts the face of the weak consonant h in Middle-Egyptian.

When it is the unity palace or the palace of the unique [nda w'a] with a weling phenomenon because of the contact of two vowels.

Surprisingly the scribe does not write pr aA as well .
Nevertheless to name the numeral "1" it uses the script vocalized script / w'a / like the arm in Ewondo.
And for the supposed idea of magnitude we have: we notice that the two scripts have a common sign the sign D36 by the way the sign o29 and the sign T21 are either homophones or synonymous because they are associated with the sign of the arm .

When it is the sovereign house then we have / nda *ìn*-nti / or the great assembly / nda bot / (the b here just indicates a number prefix) the velar consonant k has amuït to become a simple dull alveolar occlusive.
Let's remember that the tenant of the house bears, the title of ***nsw-bity*** so ***pr AA*** logically is the house of the ***nsw-bity***.

Middle-egyptian: ***Pr aA*** will be able to read / nda i-dzoué / "house of the authority" or / nda bot / "the assembly"

Depositary of the land

In the panther form aA = [ze / ngu] we have nothing more than the velar form of the K, weaken in different variants: * z <* g <* k
The most archaic form is still used in Koonzime, a "Bantu" language spoken in eastern Cameroon by the Lomié side.

290.Gardiner p. 155, §205
291. Sign 3 found in the codification of transliterations proposed by Erman & Grapow (1926-1953), Gardiner (1957), Schenkel (1991), Hannig (1995), Allen (2000)

Basa'a: ngué/ndjé "panther"
Ewondo: zəə "panther"
kɔ́ɔnzim: nkœœ "panther"

In the variant aA = /mbokh/ we have nothing more than one of the most archaic variants from Negro-egyptian. Which is revealed here through semantics and philology.
AA * <* pA.

Conclusion

At the end, how to understand the entire script ***pr aA***?
To read the greate house is to reason in a foreign paradigm because it brings no information on the semantics level. If we want to imply that it is house lineage then as I demonstrated previously it does not make it a great house but a simple house. We can agree that it is the house of the houses.
Pr aA means not great house because the good translation will be [i.vôkh nda], which would give the inhabitant would be the big person. Indeed we could predict that ***Pr aA*** is the important person. In this case the word would not be [i.vôkh] but [mfaŋ mor]. But since the sovereign is ***Hm ntr***, he is not likely to have named himself that way.
I therefore retain the most sensible idea in my opinion which is: Humanity or the United Assembly, the only assembly and the accepted meanings related to the function of the sovereign. There are many ways to understand this script and the important one is fara nén or nyâ *ùn*-mbèh.

3.6.3. Reading Of Some Texts

Some illustrative texts are discussed in this section to become familiar with the Endogenous method. It is therefore possible that everyone can arrive at this stage.

3.6.3.1 Atwàna olugu nn

"The Great Anthem of Aton in Ewondo"

According to[292] this anthem was composed by Akəŋ-Atwana .
According to specialists, this hymn was composed by Akəŋ-Atwana.
It is composed of 120 verses, in 10 stanzas, written between the year V and IX of Akəŋ-Atwana (ca 1365 BC), probably recited or chanted during liturgies at the temple. It evokes the daily cycle of the sun (stanzas 1-5) and its revelation at Akhenaton (str 6-10). - Engraved at the entrance of the tomb of Awa ( ***ntr-itf Aÿ***) in Amarna on 13 columns, one with the title. - Incipit: "To worship Re-Horakhty who rejoices in the horizon in his name of light to-Sa (Shou) who is in the disc Atwana, that he lives forever and forever! The great living disc (Atwana) which is celebrating Jubilee ŋkomda (***sd***)

The Hymn to Atwana was written by ***Imn-Htp*** IV / Akəŋ-Atwan. Even if the latter did not necessarily write this text himself, he uses the same terms before Zamba Elo (***ws-jr*** or *Imn*) Some like CS Lewis believed to see a resemblance between the Psalm 104 of David taken from the Bible with the Hymn of Atwana. Miriam Lichtheim refuted this by commenting that similarities were more the result of a vaguely generic similarity (lexical field for example, religious nature of the text, poetic description of nature) between the Egyptian hymns and the biblical psalms. She added that specific interdependence between texts is highly unlikely.

292. Pierre Nillon, The true Bible of Moses,Anibw, 2009

The Great Hymn to Atwana, as a single copy, is engraved in the tomb of Awa (*Aÿ*), in the entrance hall, while there are several copies of the Small Hymn to Atwana apart from the intervention of third parties.

GREAT HYMN TO THE ATEN

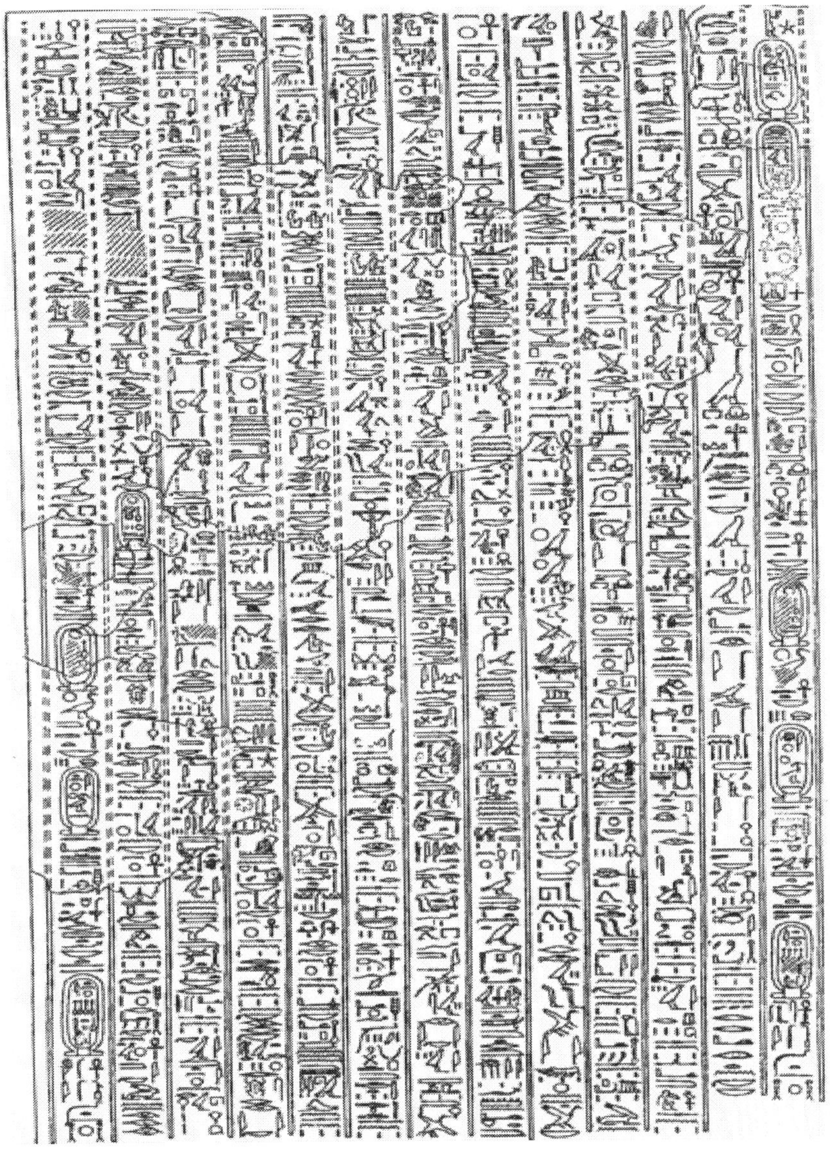

great anthem of Aten on a stele at Amarna

O-lúgû nənə Ntî

Great hymn of the Aten

I-káŋ Zàmbà-Elo ââ fón à un-nàm ùn-mfiè à dỳ–òè à-sè à-nə à-twàn dà ɲiŋ kwam ya kwam i-sə. I-ɲiŋ mfáŋ à-twàn ù nə à-bogh mə-kê, ùn-myè dy-óm i-sə à-twàn dà bômà. Saŋ, yop, ùn-myè si, saŋ,ndà atwànàà yà Àkəŋ-Àtwàn; í-kàŋ ùn-ŋkukumà ndà àtíŋ ya ndà bot yà ɲiŋ mââ ùn-myè mə si mə-bè, Anyəbə Eyo-'nkomboro, Zàmbà itàm à-ɲiŋ à yi mââ, ùn-myè bi-kàri,à-yàp i-nyiŋ ùn-ntombà i-diŋ ngà, 'minga yà mə si mə bè. Ànyəbe-Àtwànà à-nyiŋ à-kəl kouï à mə máná m' àtóŋ.

Nkukumà, mbəghə i-fəbgha à nómó Fàrà, Àwà à kobo nâ:

Wà koui ndulà à-bəŋ dỳòè. À mə wôàè àyop, Àtwàn dà nyiŋ, nkomo íɲiŋ; wà təbə à ùn-ŋkol mbo ngàlà ù dzàlgà ùn-nàm ù-sə ndulà dỳòè. Wà nyəp, à-nn, ùn-mfiè, à-yop si-ndón i-sə. Mi-miñkükü my myà fiè vom ù-sə bi-kom bi bi nə. ù nə ùn-ntəblə ùn-nlô dzôbò, wà dỳòè mi-nàm mi-sə à kələ koui à mi-nyiè màbà. Wà tindi bo à-su nyo-mwàn i-díŋ.

Wà yo, miñkükü mi mi nàmbghà fouà sí-ndÒn, ù nə à-sù mys mân. Zén dỳòè inə i-dímí. I yoŋ wà bombo à-wàè mbo nòmò, sí-ndon í-sə i nə à dibi hà-nə n'wouèn. Bor bə nə oyo'o à bi-toun bi ndà biābā. Nlo'o ntiŋdèn, mor təgə fə yəm í ùn-mànyàñ ùn-mbwà. To bà dyp à-kumà dàbà à mys màbà à-si, bə təgə fwà fə bàr yén. Bi mbgwəm bi-sə byà koui à-mə-sou màbà, nyo í-sə yà lôbèn.

Si ndōn i-sə í-nə à-míōŋ. Àŋgəŋ díbí. Un-komo wou à-bwo à mə-wàè. Wà vəbə kiri i-sə à ùn-ŋkol. Wà fiè, ù-kən-à-twàn à mós, wà dim díbí, wà kub ùn-mfiè wou. Mi-nàm mi-bèh mí nə à mbgwè. Bor bá vəbə, bə təbghə à mə kol bàmà. Wo wà bo nâ bə təbə. Nyol dzàbà n'wàrn, bà bwàr bi-y, moh màbà mà-kàŋ mə kouli mw. Si ndòn i-sə i-ngā komzèn à bi-si. Biyəm bi-sə bi diŋ'i bilouàg byàbà. Bil biloàgh bi-sə i-són.

Ànwàn à korəghə à mə doumbà, mə fàb minyân bə kân'à ngoul i-nyiŋ dỳòè. Bi-yəm bià yələ à mi-mbiè myàbà, i-bà bə-sə bà yələ, i bà bəsə bà bərbə, bà nyiŋ à mə-kùli mwé. Myàl mià dzàbən ivoundi, bə bərbən bə soussou à ndén. À mos à-sə à nə n'yàbən, wà kouli. Ù-sù, kwàs yà yələ ùn-mfà y'à-su dwé. Mi-'ñkükü my myà nyièn à nəm ngul i-són. Wà bo nâ ù-mwàn à-byàli à bingà, wà kom ndəm bə-fàm. Wà və ù-mwàn i-nyiŋ à-tw nyà, wà fur nyiè à-miôŋ dy-om yà tələ mi-ŋyÓn.

U nə un'nyà í-nyo à-nguənə à nyàŋ'à. Wà və i-komghà i-sə ùn-həbə. I-yòŋ i-komghà dyà dzəŋ ùn-həbə i-yoŋ dyà koui à-boum wà yà dyo à-nyu. Wà və dyo à-bim á-yàng'èn dwo. Màn ù-nwàn à nə à-ki dy. À highdən à i-bor à ki dy wà və niè ùn-həbə à nəm i-téré, wà və nyè i-nyiŋ. Ú ngà tigh à sù dyé ngut à-bogh i-komghà à sù yi nà à təbə ndulà, à bù -bor à-ki dy à *ùn*-nəm i-téré, à koui à-ki dyé, à ngà yon íyóŋ à-bogh à so yà, àngà koui à wulgu mə-kol mwé.

Akià à-boui à nə bi-komghà byé *ùn*-nsuàlén à mys mân, Zàmbà *ùn*-mbwà təgə *un*-mfə wà kom si ndōn i-sə à-kià *ùn*-nəm woué wà yi íyoŋ ù ngà bə i-tàm bor, mi-nzàñ bə-tit, í-dyóm i-sə yà nyiŋ à-si i wulghu mə-koul, í-dyóm i-sə i-nə à minkól i-yəlghə mə-fàb mi-'nyàèn mi-'nàm minkùs i-kàŋ, *un*-nàm Fang, wà tələ mor ù-sə à folà dy. Wà və nyè à-bim dà yènə nyè.

Mi-nkóbō mi nə mə-kyà mə-kyà mə fulu mà-bà, mə tèghè tŏbo dyàm dədà, i-kob nyol i-nə ùn-ŋkàn, ù ngà màn fyàl mə-yóŋ bə-dzín. U ngà kom í-bubué à si i-téré. Wà kulèn nyè à *ùn*-nsisim wu à soui í-nyiŋ bor yà *ùn*-nàm i-Kàmà. U ngà bo nàlà wà-mén. U nə myè wàbà. Wà simzən bwo, myè *mi*-'nàm *mi*-sə. Wà təbə *à*-sù dàbà *à*-twàn *à*-mos à-nén *ùn*-ntombà. Wà və *un*-nàm bə-dzin *u*-sə *í*-nyiŋ, two ù-yàp. U ngà təl i-bubué à-yop. Yà sus à su dàbà.

Yà və nyol ndén à-su i-bàrən mə fub mà-bà y mi-'nàm myàbà. Bi-kàngà by bi nə àfè. À mbè-mbè nti, i-bubué yà yop i-nə à-tom dwé à-su bə-dzín bə-bor. À tir í-sə yà *ùn*-nlóŋ yà wulu mə-kol mwé, à sù *ùn*-nding'èn si, i-bubué yà so à *ùn*-nsikiè. Mi-nkuküi myé myà nyê mə fub mə-sə. Wà təbə, bà nyiŋ, bà fê à-su dẁé. Wà tələ mə-bokh mbəŋ. Wà nê i-komghà dỳòè i-sə ùn-mbəŋ.

À-bogh à-vəp dà və à vəp, i-səp də bo nâ bor mə diŋ wà. Wà ngà kom si yé yop. Wà təbə é-nyè. Wà digi i-komghà dỳòè i-sə dys dẁé. U two ù nə borà-borà wà mén. Wà təbə à-nyol *ùn*-nyiŋ'èn à-twàn yà kuï ù fièghè à nə oyàp, à nə bàbi. Wà vàè bə towinà mi mbè-mbè mi nyol à tyènə é wà mén. U nə à borà-borà é wà mén, mi-'nààm, bi-fàs mi-nàm, mə-fub, mə-zén, mə-swé. Dys à-sə dà yénən à-su dwé. U-nə à-twàn y'à-mos à si-ndón. Wà yoè. Təghə i-komghèn dỳ–òè yà nyiŋ təghə wà səkh wà-mén i-tàm.

Təghə i-komgà dỳòè wà və i-nyiŋ yà yén wà. U two à *ùn*-nəm wàmà. I tələ kigh *ùn*-mfə à yəm wà à-tiènə é nyo *ùn*-mwàn Àkəŋ à-twàn. Wà lərə *ùn*-nyè mi-ntsoghèn myé, ngoul dỳòè. Si ndon yà so i-nyiŋ à wa w-òé. À kyà wà kom nyè u-ngà təbə, à ngà nyiŋ. U bombó à ngà wué. U-nə i-yəm i-nyiŋ. Byà nyiŋ à bə wà. Mys mà bəbə à-bəŋ dw–òè é à-tôŋ à kəl koui à-wâ dw-òé. Wà wâè à *un*-ómó. I-sié i-sə i ngà təbə. À mə-kuï mw-òé wà màn nâ bo dỳ-òm i-sə i fê à ù-nkukumà.

À-koghè à-ngà bi à-kol à-sə. Wà màn kom si ndón i-sə. Wà kuli nyè à-su nyo *ùm*-wàn à ngà so à bə fàr yà ndà à-tiŋ zàmbà. À nyiŋ'í à-mbgwè é si-ndon i-sə. Myè mi-'nàm mi-bè. Mwàn Zàmbà Elo, à-nyiŋ'í à-mbgwè é si-ndon i-sə. Myè bi-kàri Àkəŋ-Àtwán. Nâ i-bóŋ í-nyíŋ dyé i-tóbó à-yàp. Nâ *ùn*-mfàŋ *ùm*-ingà w-òè à díŋ, mingà yà mi-'nàm mi-bè *ùn*-nso'o à-bəŋ'ə à nyiŋ à-tóŋ à-kəl kuï à-mə máná mbè-mbè.

3.6.3.2. Tale Of Two Brothers In Ewondo, Papyrus Said Of Orbiney

Papyrus d'Orbiney dates from the late nineteenth dynasty. It is also called *the tale of the two brothers*. It can be approached at a logical content level to the Papyrus said Jumilhac. It is considered as one of the oldest known texts of Egyptology. It has been acquired in 1857 by the British Museum (BM 10183). The text was written at the end of the nineteenth dynasty, specifically under the reign of ***StXy*** II. It has given rise to numerous analyzes, according to comparative, psychoanalytical or allegorical approaches, which have not really made it possible to understand their nature. Egyptology, on the other hand, considers it most often as a literary text imbued with mythological data. The goal here is not to make an endogenous analysis but simply to translate it. We are not gonna deal with all these works, summarized in the works of W. Wettengel and S. Tower Hollis. It is interesting to approach such a text because it connects the reader to his ancestral past. It is brought here as an illustrative purpose.

Bə mbə lêghè nâ bo'benyaŋ bə-bèh.
Ùn-nya *ùm*-bokh é *i*-sia *ùm*-bokh,
Zambo i-yolé *ùn*-ntol, Bala â mbə i-yol ù-tokh.
A tələgə é Zambo, à mbə bələ ndah à yi *ùm*-ingà.

Ùm-à-'nyaŋ *ùm*-ongo, à mbə noŋbèn v'á-nə *ùm*-wan woé
A mbə i nyô bá dzo nyi nâ à á-sob mi-'mbor.
A mbə i nyô à doughèn à bi-sièghè é mi-'mbokh myé.
A mbə i nyô à mbe silbèn fə nâ, à-saŋ, à-fakh, à-bèh, i-nyo à á-waŋ nyè.
A mbə i nyô ba mbə boh nâ à á-boh bi-syé bi-sə ya mə-fub.

Kə i *ùm*-à-'nyaŋ mongo, à mbə i mot à nyeb bi-dzô.
Tèghè mot ùm-bokh à á-mbe v'á-nə nyè à ùn-nàm.
A mbə bələ ngoul Zamba.
A mvous, à-bui mə-louh à nga lor,
I-yoŋ *ùm*-à-'nyaŋ à mbə nyè à mvous *mi*-mbokh,

A-bô mi-yiène myé yà á-mos ù-sə.
A mbə dûghèn à ndah v'à mə-sui m'-àlou.
A two *ùn*-ŋkâèn é bidi bi-si mə-kyà mə-kyà mə-sə à-sù *ùm*-à-'nyaŋ
A yi ndzâghe, à yi à-bəŋ byôm bi-sə y'à mə fub.
I-Yòŋ té fə à-zu byo sui à si à mə-kol m'i *ùm*-à-'nyaŋ *ùm*-bokh.

Ùm-à-'nyaŋ à-twa fwà à si é *ùm*-inga woué.
A sui i-koré, i-yoŋ té à á-koui à ndâ à sou yi nâ
A kə zakhbə i-tam à zaŋ nda bi-yəm.
I yòŋ si ya fiébèn, à-mos à ákouî,
Bi-yánghà mbəlén, yòŋ té à á-suî à mə-kol me i *ùm*-à-'nyaŋ *ùm*-bokh
I yoŋ té ny'à á-və nyè bi-kay à sù bi-sié.
Ny'à á-kə é yi nyakh à mə-dîghà, à mə-sî.
A wulghu à mvous. Nyakh dyé i nga lərə nyè zén vom à-kyé da nyəb.
I yoŋ té ny'à á-vóghlo boh bəsə

ùn-mbol té à nga kə é boh à-fola à-bəŋ à-kyé à fê.

I nyakh i nga bə nyè u-su i nga man kui à-bəŋ à lor ndula bi-yoŋ bi-bè
Nyakh i bagh mə-sui i lor ndula bi-yoŋ bi-bè.
Ndə à-bokh u-sùm, manyaŋ ùm-bokh à kar nyè nâ
A-kom mə-ŋgoŋ mi-'mbokh à-sú ù-sum.
Si i-nə ùn-ŋkor. Ya nyeb à-su ù-sum.

"Wa yi so à fub fì, à su yi nâ bi kê ù-sum kiri" à nga dzô nyè.
Nda fə ùm-à-'nyaŋ mongo à nga man boh mi-'ntəb'lə mi-sə i ùm-à-'nyaŋ ùm-bokh
à nga dzô nyè nâ à-boh.
Veda i-yoŋ si i nga fibèn ndə à-mos bèh à nga koui, ndo bə nga kə à-fub
A yi fer bə mbə bələ.

Bə bô ù-sum ùn-nəm mi-ntag. A su wo i-si ya mə-kê u-loro ndula bi-yoŋ bi-bèh.
Veda à-mvous à-bui à-mos à nga lor, bə nga dzəmn fer à-fub.
Ndâ à nga dzo i 'm-à-'nyaŋ mongo na, à-kə mbil à kə noŋ fr à dyal.
I 'm-à-'nyaŋ mongo à nga kob ù-minga ùm-à-'nyaŋ mbokh à two à-si, à-áfənbèn
Nda à nga dzò nyi nâ: "A nâ: -korgho tə à si, ù və me fer .

A-su yi nâ mə yomdo mbi i-bêghè à-mou yi nâ madzaŋ à-yanga ma à-fub".
Ndâ ù-minga à nga yalèn nyè nâ: "Yâghə ùn-mbè i-kwène fi, noŋ'o à-bim wa yaŋ'ène dwo. Te wa tələ ma *ùn*-mfén à zén".
Ndâ fwa fə *ùm*-à-'nyaŋ *ùm*-òngò à ngà kə *ùn*-mfakh nda bi-yém byé.
A noŋ'o more *mi*-'ŋkyeŋ'e *ùm*-bokh.

A nòŋ mi-'ŋkyeŋ mi-fər.
A bəghə ù-vugh à yi bodo.
A ndaŋ'a à-'nsəŋ à yi byô.
Um-inga *ùm*-à-nyaŋ à nga sili à nyè nâ: "yi wa yəm i-taŋ ù-bəghə i-tul?"
A yalen nyè nâ: "mi-'mfəgh *mi*-bodo *mi*-lèh à yi *mi*-'mfəgh *mi*-ù.vugh *mi*-bèh.

Mi-'mfəgh *mi*-tán *mi* bərə ma à bi-tul".
Ndâ à á-ngà dzô nyè nâ: "- *U bələ à-bui ngul, ma yén ngul té à-mos ù-sə*".
A nga yi nâ à yəmèn é nyè à bi-dzô.
Ndâ à á-wogho à si à á-bəmə nyè.
A dzo nyè nâ: " - *za'a bi bo bi-sokh à-wolà da. Wa à yi ma, à su ùn-mfi woué.* »

Ma yi fə wo kom í *mi*-mbor mi á-nyeb"
Ndâ mongo à á-nga wu ù-lun va á-nə zə. I tièn é be-bəh bi-bukh à nga kar nyè.
Um-inga à nga ku fwa waŋ. Ndâ à á-dzo nyè nâ.
" - Tam yén nâ mə-ləbghə mə 'nnyâ, ù və mə mo,
nyo *ùn*-óm à-lebga i-sia, à və ma. *Un*-ntəbə ù-sù à yi mə man kouli mor.
Akya mor *ùn*-mbəh dyam wa dzô ma di?
Tə wa bar me dzô à-kya dyam té i-yoŋ i-fə".
"Ma bén do kar i mor *ùn*-mbokh. Da yi kigh kui à-nyu dàmà à su i bor *be*-vokh.
Nd' à ánga sóŋ *ùn*-mbəghə woué à á-kə à fub. Nd'à á-nga kə kob i *ùm*-à-'nyaŋ *ùn*-mbokh
Be dughèn kê bi-syé à su yi nâ bə man à-bokh i-syé dàbà.

Ndâ à-bokh à-lu à nga kui. *Ùm*-à-'nyaŋ *ùm*-bokh à tələ, à dughàn à-dyal é byal.
I *ùm*-à-'nyaŋ *ùm*-ɔngɔ à-mvùs à yi mi-'mbokh, à kâ bi-kay mə-fub.
A tələ *mi*-'mbokh bi-yəm ù-sù nâ à á-və boh i-bóŋ à dyal à nda biyem dy-àbà.
Ùm-inga *ùm*-à-'nyaŋ à á-nga bə sangh'èn à-sù i-mam à nga kar *ùm*-ɔngɔ.
Nda à nga noŋ ndəkh *ùn*-ŋkwàn bi-tè à su yi nâ à yéné ha nə mi-'n-wuwub mi nga tsakh nyè.

A mbə à á-bələ *ùn*-ntsoghèn nâ à á-kar *ùn*-nom nâ i *ùm*-à-'nyaŋ *ùm*-ɔngɔ ny' à á-tsagh nyè.
Un-nóm à ngâ kə soh é byl à luh i-téré v'á-nə à-mos ù-sə.
Nyin à nga nyî à nda dyé, à so fwa kob *ùn*-gal à bwo i-nɔŋ à tiène é ù-kwàn mi-'n-wuwub.
Tèghè à-foua fə nyi so və mə-ndim à su môh ha nə bi-yóŋ bi-vuakh.

Tèghè nyi so é *ùn*-mfyè à-su dyé.
Ndah i two foua v'à dibi
Nyè à-dzogho à-si à yòô.
Ndâ *ùn*-om à dzô nyè nâ: "-Za à-kobo yé wa à-béh?"
Ndâ à yalèn nyi nâ mor à dzikigh mə kɔbɔ à béh à tièn y *ùm*-o-'nyaŋ *ùm*-ɔngɔ.

I yòŋ à soh nòŋ fer à sù dẃé
A kob ma mə two fwa i-tam mə tindghə *ùn*-mfén. Ndâ à-sili mə nâ bi boh bi-sokh man à-wola. A kobgho é mə nala, mə tèghè voghlo mam mwé.
Ndâ mə nga sili nyè nâ: "-nga me nə *ùn*-nyoà? I *ùm*-ɔ-'nyoŋ à tələ wa ù-sû, ngâ à sié wa ùn-mbəŋ?". A á-kû à waŋ, à nga tsagh ma à su yi nâ te ma toŋ wa dyam té.
Nghə wa dzoghè nyè n'à á-nyiŋ, ma yi woué ma-mén.
Ndə, i-yoŋ à yi so à-lu i-téré mùna ! tə wa vɔghlo nyè. Mà á-wogh mintyé à su à-béh â á-boh ma *ùn*-ŋgoghè.

Ndo *ùm*-à-'nyaŋ ùn-nto(l) à nga man lingbèn à ntwo à-yokh v'à á-nə zə.
A man dzeb à-kóŋ dyé, à á-noŋ do nâ à á-zu woué.
Nda fwà *ùm*-à-'nyaŋ ùn-ntô à nga sɔbɔ fə à-mvous *ùn*-mbèh i-yakh nâ à woué *ùm*-à-'nyaŋ *ùm*-ɔngɔ.

I yoŋ *ùn*-nlô dzôbo à kə yi waè.
ùm-à-'nyaŋ *ùm*-ɔngo à-kà'a bi-kay mə-fùb.
A kya y' à-mos ù-sə.
A nga dûghèn à ndâ.

Nyakh i-wulughu nyè ù-sù. I-yɔŋ i nyakh i nga bə wulghu ù-sù ya nyiène mfâ ya ndə-bi-yəm, Bə-'ti
Ndâ i-nga kar mbâla woué nâ "kalgha é wa mén à mu yi nâ,
Ùm-à-'nyaŋ woué à-tələ ù-sù à yi à-kɔŋ nâ à woué wà. Kyelbe ù-sù vom à nə.
Ndâ à nga wogh dy-om nyakh ù-sù i-nga kar nyè.
I nyakh fwa i nyî, i kar nyè dyam dədah.

Ndâ à ngâ ləmbə à kol mbèh i-yakh
A yənə me-kôl mə *ùm*-à-'nyân *ùn*-ntôl à-tələ à mvous *ùn*-mbèh.
A bələ à-koŋ à wà.
Nyin à nga seghde ùn-mbəghə woué à sî,
nda à nga kui *ùn*-mbil à *ùn*-nsəŋ

Ùm-à-'nyân à nga tsirèn à nyè à mvous à yi à kóŋ dyé.
Ndâ ùm-à-'nyân ùm-ɔngɔ à-nga loué Zamba-Elo, nti yob yé si.
A dzo nyi nâ : "wa wa á-tigh à zaŋ *ùn*-wué à yi suswo mor"
Ndâ, Zamba à nga voghlo ngoglèn dyé í-sə.
Ndo Zamba à nga kuli maŋ à zang *ùm*-à-'nyân à yi nyè.

A dyali do à yi ngàn.
Bə-vogh bə koulghi *ùn*-mfakh wi bə vogh bə kulghi ùn-mfakh *ùm*-bokh.
I *ùm*-à-'nyân à nga bə ntô à á-nga bimi bi-yoŋ bi-bèh
A moh moué à mou, tèghè nyi woué
Nda *ùm*-à-'nyân mongo à kar nyè à tələ i kindigh vùagh nâ à

Tobo à kindigh dyé à kel kui i-yoŋ à-mos wà yi təbə à sù 'nkar Zamba.
A yi kar i-nyo à-kùh ùn-ŋkar
Ma yi kigh wou bar və i-nyiŋ dyàmà à kələ kui à mə-manà.
Ma yi kigh bar kui à folà wâ ù-nə
Ma yi kə ùn-mfâ yâ à-yar ùn-mbàŋ à-səŋ.

I-yoŋ *ù*-vièn wà yi fiè si, à-mos bêh à á-kui.
Zamba, nti yob yé si, à á-yâ təbə à ya yén bî bə-sə, mor *ùm*-bokh à yi i-mor *ùm*-bokh.
Ndâ i *ùm*-à-'nyân 'mɔngɔ à nga kɔbɔ yé i 'm-à-'nyân *ùn*-ntôl. A á-dzô nyi nâ:
"A mou dyé wa zu mə tóŋ à yi à-kóŋ à mvous, é ù-lakh nâ wa woué ma teghè kê voghlo i dyom
à-nyù dama d' á-dzó?

Ngâ mə nə wa i *ùm*-à-'nyân 'mɔngɔ?
U nga boh nâ mə tələ wa ha nə i-sia
U nga bo nâ mə tələ ùm-inga woué ha nə *ùn*-nya. Ngâ sə nala?
I yoŋ ù ve lom ma nâ mə kə bî noŋ fər
Nyo *ùn*-ngal ny'à á-kar mə nâ bi kə boh bi-sokh man à-wola

Bəbəghə à kya à á-kilèn mam à mvous, à bə wa
Nd' à á-man nyi tyé à-kià mam mə nga lorbèn à yi 'minga woé.
Ndâ fə à-man kan à dyoé Zamba Elo nâ ù zu mə wué à-sù
Minal m' i-wolwos ù-swan ùm-inga.
Ndâ manyaŋ à ne mɔngɔ à nga kə noŋ ù-koŋ i-sóŋ ù-nə à-vol. ndâ à á-nga tigh
Byəm bi fam byé. I yoŋ à-tigh byo nàlà nd'à áwuo byo ù-swé.

ŋgwal kwàs i-mín byo.
I *ùm*-à-'nyân *ùn*-ntô à kou ù-kwan à á-nga wu *ùn*-ntùgh.
A á-nga dzobèn aboui à *ùn*-nəm i-téré.
A á-nga bə à á-tələ, à á-yôno à yi ngul i-sə. Veda tèghe zam daŋ ù-su

Un-mfâ i *ùm*-à-'nyân *ùm*-ɔngɔ à-nga bə à su ngan yà zàŋ.
Ndo ùm-à-'nyân ùm-ɔngɔ à-nga loué nyè nd'à á-sili nyi nâ: "ù he tsogh i-dyom i-nə à-béh i-tiène
yi ma à mou dyé ù ndze tsogh i-di i-nə mbəŋ nghe kigh i-byom m'a boh à sù dwé?"
Tam à dûghèn à ndâ dyùé ù-tam kə sié bi-yem byé. A-mù mə sikigh fə dyam

Tobo a fola dah à yi wa.
Ma yi kə 'mfâ yâ à-bare ùn-mbàŋ à-səŋ.
A á-tələghə yi dyom ù nə zam bo à-su dàmà wa yi mə so tunghlu,
ngə ù-wogh nâ dyom i-boh ma. Ma yi toué ùn-nəm wama
I yoŋ té ma yi wo béré à ùn-mbaŋ à-seŋ

Nguə be tighe à-səŋ té, wa yi ku à-si. Zâ wo dzəŋ two ndo wa noŋ mi-mbuh zembgwa.
Te wa teghe.
Nyin wa kob wo ù four' wa à ndəgh mə-ndim voué
I yoŋ té ma yâ dughèn nyiŋ mə tam à á-kun i-bor bə nga kê ma.
Wa yâ wogh i-minlaŋ ba yi wo lê nâ dyom i bo ma.

Ba yi wu və i wangha mə-yogh ya fighèn.
Te wa tobo i-yóŋ nala wa yi bôbèn
Ndâ à nga kə à ùn-mbaŋ à-səŋ ùm-à-'nyân nyâ à dughèn à ndâ dyé moh mə bərə à ùn-nloh.
A nga bə nyôl ùŋ-wabèn ùn-mbulugh.
I-yóŋ à zâ kui à nda dyé

Nd'a á-nga wué ùm-inga woué
Nda nga wouo nyè à mvu
A tobo à-si à nga yon ùm-à-'nyân nyâ.
Abui mə-lou à nga lor à mvous dyam té.
Mor *ùn*-mfə tèghè fwa fə à yi nyè

A loro à bogh dyé à sə nâ à some tsit yâ *ùn*-ŋlóŋ.
A á-dughèn fwa son â à zou waè à *ùn*-mbaŋ à-səŋ vom
Un-nəm wé ù-nga bə bərə à-sam hé ye à-lùh.
A mvous à-bui mə lu, nd'à á-nga lóŋ nda mə-kogh à yi móh mwé à ùn-nsesekh ùn-mbaŋ à-səŋ.
Nda té i-nga bə *ùn*-ndzalèn y' à-bəŋ byóm.

I-yoŋ à á-nga kui à ndà dyé, ndâ à á-bomèn à yi mi-'ntomba mi-ntour i-bùlù.
Be two à dùlù nâ ba ke ba kom mam à ùn-nam ù-sə.
Ndâ foua ùn-ntomba nti ùm-bokh wa dzô nyè nâ:
"-Ha Bala *ùn*-nkàga ya ùn-nam i-bùlù yi ù-nə i-tam à-mù ù-nga man tub dyal dwé
A tələ à yi ùm-inga ùm-o-'nyôŋ?

Bəbəghə ù nga man nyi kilèn fəgh, à á-kələ kui ha nə à nga wué ùn-gal
Yi wa yalèn i bor bə-sə b'à á-kê wa ngâ?
Ndâ mi-nəm myaba mi nga yén fwa nyè ngoual ùn-mbəh y'à-boui.
Ndâ Zamba à á-nga dzô ùn-ŋkomboro nâ
Komo Bala à-bəŋ màn ùm-inga à su yi nâ te à təbə i-tàm à ù-ngoùm í-nyíŋ.
Ndâ *ùn*-nkomboro à á-nga man kom Bala *ùm*-inga à su dyé.

A á-nga bəh fwa à-bəŋ à-bui tèghè yi à kya té à ùn nam ù-sə.
A á-mbə bələ mvom mi-ntour mi-sə.
Nde bi-kari zembgwâ bi nga zu nyi yén bə nga dzô i yoŋ dyà nâ:
A yi wouében à yi ù-kəŋ.

Bala à nga diŋ nyè à lore dyom i-sə.
Ùm-inga à *ùm*-bə twò à ndâ dyé
Ny'à á-kələ à *ùn*-nsom à *ùn*-nlóŋ. A woué tsir à zâ dyo dzoghè à-su dyé.
Bala à nga təb'lə nyè nâ: -Tèghè kui à ùn-səŋ à-mui Zamba maŋ à yi wu vaè.
Wa yəm kigh à kya ba tùb.

U-nə i-vôm man *ùm*-inga ha nə i-boh bə-vogh.
Un-nəm wàmà ù bərə à-yob à-sàŋ yà ùn-mbaŋ à-səŋ
Ngue mor à á-kob wo byé nyè bya yâ foua dyo'èn.
Ndâ à nga yâè nyè ùn-nəm woué.
Bala à-kə à ùn-nsom va nə à-bui mə-mos mə-sə mə-vokh.

Ndô 'm-inga à kə à-wulu *ùn*-mfâ ya ùn-mbaŋ à-səŋ à mbgwâ nda dyàbà.

Ndâ à nga yen Zamba maŋ à mbə à bighden mi-ŋkomi myè. Ndâ à tùb ùn-mbil à kə sobo à ndah i-téré.
Ndâ Zamba maŋ à á-dzô à yi ùn-mbaŋ à-səŋ nâ à bi nyè à-sù dyé.
A-səŋ à nga nóŋ nyè ngop ùn-mfénèn i-sil.
Ndo Zamba màŋ à nga kə à yi ùn-'mfén té à ùn-nam Kama à kə y wô vom bore bi-sié

bə Fara nén bə nga bəh.
I-nyum ngwap i-sil té i-ngâ man kə à bi-yé bi fara.
Fara à á-nga womèn à yi bə bo-bisyé bẃ à su mə-nyum mə nga bə à bi-yé byé.
Fara à ùn-mbəh womghèn à-mos à-sə à yi bə-boh bi-syé boué à-su mə-nyum m'à-mbə təghə à-kya bə nə mo zam man vâè.

Ndô ùn-dzoué be-sob biyé bi fara ya kə foua à nsakh vôm b'à á-sob bi-yé bi fara.

A nga bə ùn-nəm mintié à su yi nâ bə- sob bi-yé bə ùn-mbəh womèn à yi fara à mos ù-se.
Ndâ à á-təbə tətələ à ùn-nsələghə à á-bəbəghə ngouap i-sil i nga bə à maŋ i-téré.
Ndâ à á-kar mor ùm-bokh nâ à-kə nyi dyo noŋ à-və dyo nyè.
A á-wogh mə-nyum mwé à-bəŋ ndâ à-ke é nguap i-sil té à-və dyo fara.

Ndo fara à nga loué bə yəmə à yi mi-ŋkəkəŋ myé.

Bə man fombo nguap i-sil ndâ ba yalen é fara nâ: "i-guap i-sil nyi i nə i-nyo man i-vôm ya ùn-nam Zamba -Eyo. Yebghe nâ bi-lom mor wé à sû dwé.

Etc.

IMPORTANT NOTE

This tale is typically a type of story told to children in Africa. We find the same mindset etc. However without entering into a tighter analysis of the concepts contained in the tale, one simply notices an important detail, the opening of the waters were already carried out in Africa curiously before Moses saved the chosen people.
A quoted below the extract of the story regarding the sea opening:

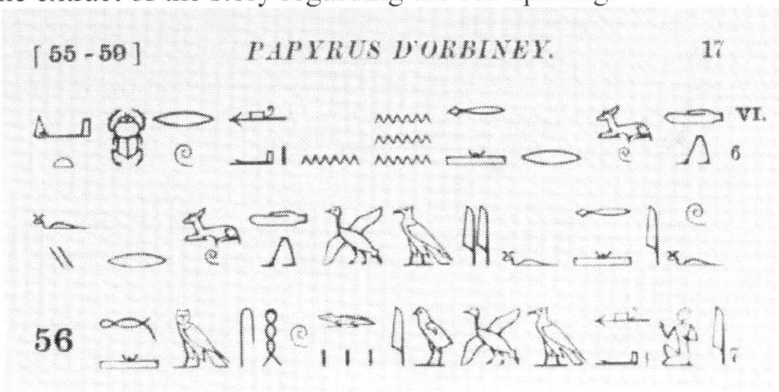

He gives birth to a large expanse of water to separate him from his brother. He fills it with crocodiles

Idealology

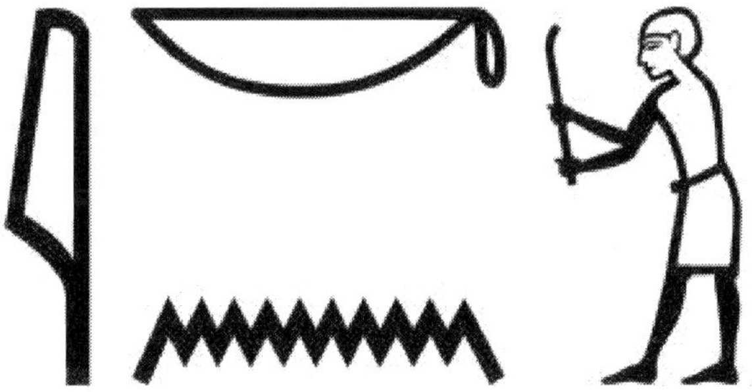

The real script

-The word  feather and the word divide are homophones in african languages.

Ewondo:/à-sè/ "feather"
Ewondo:/á-sè/ "to divide"

-We have also that the word to collect  and to divide  who are homophones.

Middle-Egyptian:  ***ikn*** "to collect water; to draw water"
Ewondo:/á-kwèn/ "to draw water"

Only an evil person who knows african languages can use this script above in a very close sense to confuse the credulous people. Indeed the verb *to separate* is pronounced almost like the verb *to collect*. This is where the trap was hidden here.

Middle-Egyptian:  ***ikn*** "to collect water; to draw water"
Ewondo:/á-kwèn/ "to draw water"

Ewondo: /á-kèèn/ "to divide, to separate, to divert"

Only a malevolent and fundamentally machiavellian-minded person may have used such an approach to fool. Because, collecting water with a container, it is necessary in certain cases, to separate it first of all. That's why the two ideas are very similar phonetically according to scribes. Only a person who knew it will taint his discourse of ideology by diverting things from their basic context. But unfortunatelly for him, he could not wipe of all the scripts.

Since this word exists before the arrival of the hypothetical Moses in Egypt, we must conclude here for a clear plagiarism that the iconography has ended up actually in books supposedly written by people who speak within of their own death.

Incredibly, one study states that the phenomenon recounted in the Bible has of natural origin.

> *"And Moses stretched forth his hand over the sea, and the LORD set back the sea all the night by an impetuous wind east, and he put the sea to dry, and the waters were divided. The children of Israel entered the midst of the sea, in its dry bed, the waters rising up on their right and left sides"*
> *(Exodus, verses 21 and 22).*

The event would have occurred not in the Red Sea on the occasion of a tsunami, as claimed by some researchers, but in the Nile Delta, on a shallow lagoon, after a whole night of strong continuous wind. "Our simulations fit well with the Exodus story," say Carl Drews and Weiqing Han of the University of Colorado.

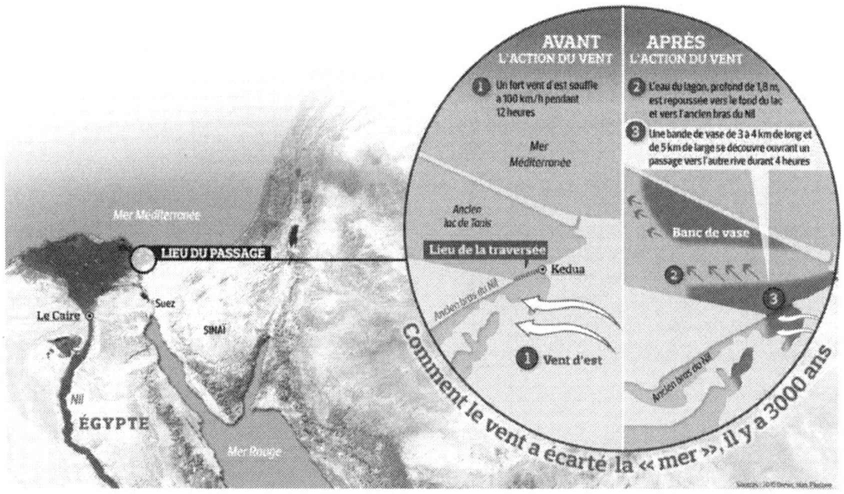

The imaginary crossing of the red sea by Carl Drews and Weiqing Han, of Colorado University

Treacherous Semites clamor, but unduly, the paternity of the pharaonic cultural heritage. But they understand nothing of ancient Egyptian civilization, its essence, its mysteries, its ultimate meaning. They keep in an irregular, crapulous way, the cultural riches whose singularity and deep understanding will escape them totally for eternity, that is to say without possibility of seizure of the essential: this is the reason why these these deceitful people can not exegete them and have for centuries deprived themselves of the cultural heritage of the Negroes, without a state of mind and in a barbarous and criminal way. It is the most monstrous example of heritage capture in history, the most outrageous intellectual scandal, the most spectacular and villainous extort of archaeological wealth and invaluable historical material goods of others. The sandal of scandals, a real crime against Civilization and against Humanity.

3.6.3.3 TRANSLATION IN EWONDO OF THE SEMNA'S HEADSTONE

The Semna's Stele is a red granite discovered on the west bank of the Nile and published by Lepsius ((*Denkmäller II, 136 i*). It is currently in the Berlin Museum under number 14753.

In the year 8 of his reign, Sn wsrt III crossed the second cataract and progressed to the end of Batn el-Hagar where he established his border, an event that he immortalized in the year 8 and in the year 16 on two stelae (Berlin 14753 and Berlin 1157). It is in the fortress that bears his name (sxm-xa-kA.w-Ra - Mighty is Khakaure) and of which he is probably the builder, that they were found.

This stele is interesting because it was used by ideologists in order to prouve that the ancient Egyptians were not blacks. Those who used some extracts of the below stele to support their argument consisting in arguwing that the Black Man has never be able of science and ancient Egypt was populated of peoples other than Blacks. The real populations of the ancient Egyptian civilization were semites or westerns.

For this stele we are going to read from left to right:

Year 8 stele

tAS rsw jry m HA.t- sp 8 xr Hm n(y) nsw.t bjty #a-kA.w-Ra d(w) anx D.t

Un-nyè yà si'ŋkyè; ùn-tələ̀n à mbù mwôm (8) ùn-ntum i-dzoé Fara mbokh Bwiti Ngoul Ikàŋ Elo à-nə i-nyíŋ kwàm.

Southern border, established in the year 8, under the Majesty of the King of Upper and Lower Egypt Khâkaourê, endowed with life, eternally.

r nHH r tm rd(w) sn sw NHs(y) nb m xd

yà kwàm i-sə. Two Man Ekang ùm-bokh təghè dàŋ zut ù-swé byal

and forever. So that no Nubian cross it, going down the stream

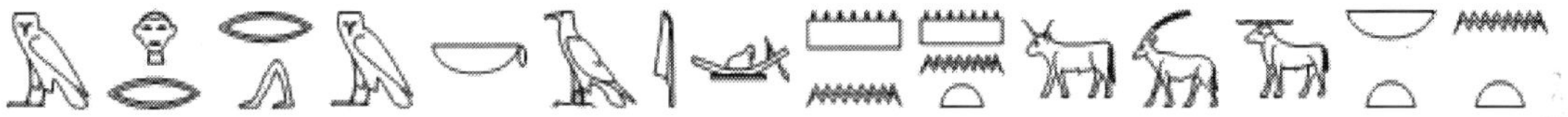

m Hrty, m kAj, mnmn.t nb.t n(y).t

two à si two byal two i-yəm i-tyèn à yî

by land or by boat, and no livestock belonging to

NHs(y).w, wpw- Hr NHs(y) jwt(y)=f(y) r jr.t swn.t m iqn,

Ekaŋ à sàlà y mor Kindigh ùm-bokh à yi so bo makir à Akono.

Nubians, except a Nubian who would come for trading in Iqen (Mirgissa),

m wp (w).t r(A)-pw. jr=t(w) [x.t] nb.t nfr Hna=sn, nn sw.t rd.t

ngə və i-nyo à nə ùn-lomèn. I dyóm i-sə i nə ùn-mbəŋ ya yi vəbèn à yi boh, tèghè yəbə or on mission. Anything (Breasted, Ancient records Vol I, p 294, note a) will be good with them, but without allowing

swA kAj n(y) NHs(y).w m xd Hr HH r nHH.

i-kàï yà b'I-kàŋ y'à-swa à ùn-ŋkyè Bikokh à kwan yà kwam i-sə.

that no boat belonging to Nubians descend downstream from Heh (Semna), eternally.

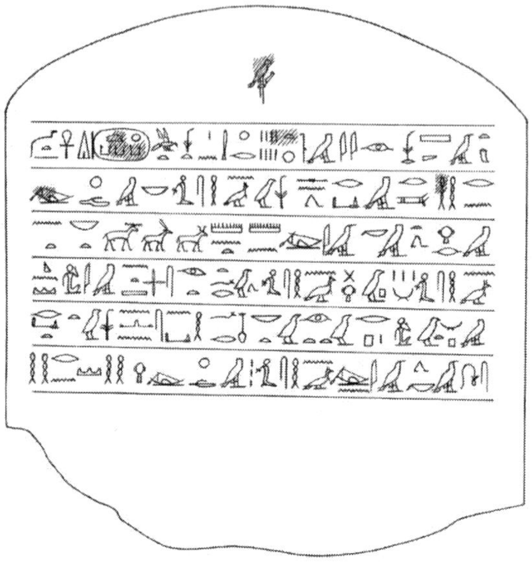

Drawing by Claude Obsomer, Catholic University of Louvain (UCL),

Year 16 stele

For this stele we are going to read from left to right:

anx Hr: nTry xpr.w, Nbty: nTry msw.t, n(y)-sw.t bjty #a-kA.w-Ra d(w) anx,
I-nyiŋ kwàm i-boh y Eyo, y mintour mə-kuli, bə-bondé mi-ntour Mə-si, Fara Bwiti Ngul i-kàŋ Elo à á-fon i-nyiŋ kwàm.

May Horus live: divine of forms, divine of birth, the king of Upper and Lower Egypt Khâkaourê, gifted with life,

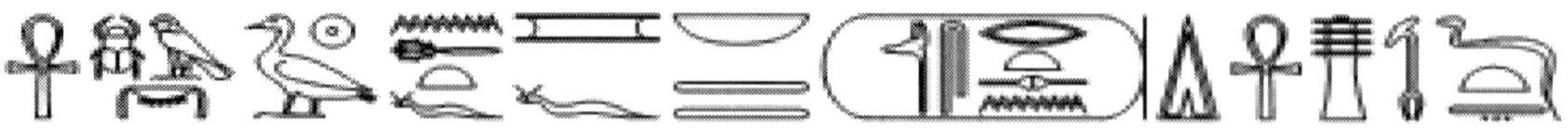

anx Hr-nbw: xpr(w), sA Ra, n(y) X.t=f, mr(y)=f, nb tA.wy ¤-n(y)-Wsr.t, d(=w) anx, Dd, wAs, D.t.

I-nyiŋ kwàm i-boh yé idzoué Kongo, à ngà byali ùm-wan Eyo, ùn-son wué, i-diŋ dyé, ùn-ŋkùkùmà yop yé si, Unsuru à á-fon é i-nyiŋ kwàm, ùn-nom ndî, i-vés kwam yà kwam i-sə.

Let the golden Horus live: the one who comes into existence, the son of Re, of his flesh, his beloved, the master of the Double-Country, Senousret, gifted with life, stability, power, eternally.

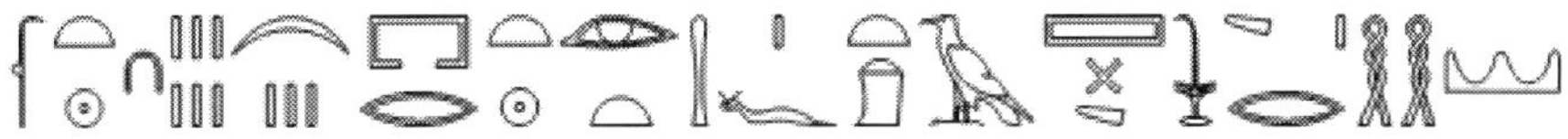

HA.t-sp 16, Abd 3 (-nw n(y)) pr.t (sw 1), jr.t Hm=f tAS rsy r HH

Ùn-təblə ùn-ŋkùkùmà i-tîghà ùn-nyè ya nkyè Bikok, i-səp ùn-mbù à-wôm-é-sàmen (16) ngwan làlà (3) y'àbokh mə-fê.

Year 16, 3rd month of peret (the first day), establish by His Majesty the south border of Heh

Iw Ir~n=i tAS=i xnt=i nt=i itf.w=i. iw rd~n=i

A-zén é byal, ma tigh ùn-nyê si é bə-tàrà. Mà və nâ,

I established my border as I traveled south of my fathers land. I did,

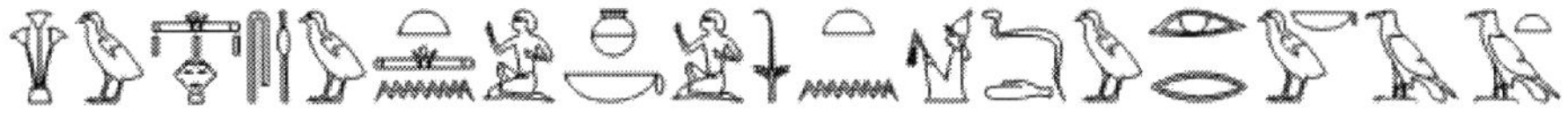

HAw Hr swD(w).t n=i. ink n(y)-sw.t Dd(w), irr(w): kAA(w).t

A dàŋ i-dyom i-sə i ngà vêbèn mà. Mə nə ùn-ntùm i-bùkh é màm: nàlà ngâ fwà

more than what was assigned to me. I am a king who speaks and acts: this is what conceives

ib=i pw xpr(w).t m-a=i, Adw r i T.t; sxm(w) r

ùn-nəm wàmà, wà bo ùn-yièn wàmà; ùn-ŋkâbgà ùn-yièn, ngoul à yi

my heart, which happens by my arm; eager to conquer, impetuous for

mar, tm(w) sDr(w) md.t m jb=f, xmt(w) twAw.w, aHa(w) Hr

i-bəkh, à ùn-nəm yà dyam dà bombo kigh;yà kə ù-sù i-bà bà yi, i-bà bà yi bà saghbà.

success, in the heart of which a business does not sleep, which goes to the front of those who claim, who leans on

sf, tm(w) sfn n xrwy pH sw, pHw pH=t(w)=f, grw gr=t(w),

i-yəghlə, i-sikigh bi-dzûghà à-su ùn-ziziŋ i-yɔŋ à á-kê; à-kêghè i-yɔŋ bà kê nyè, à kyghè i-yɔŋ bà kyê.

indulgence, which is not misguided to the enemy when he attacks, who attacks when attacked, who is silent when one is silent

wSbw md.t mj xpr(w).t jm=s Dr-nt(y).t jr gr m-xt pH, ssxm

à yalghèn à-dzô à-kyà à nə zam so à mou à tɔbɔ i-vol à mvus bi-tè nala wà wombèn

which responds to a situation according to what can result from it since to be silent after an attack is to affirm

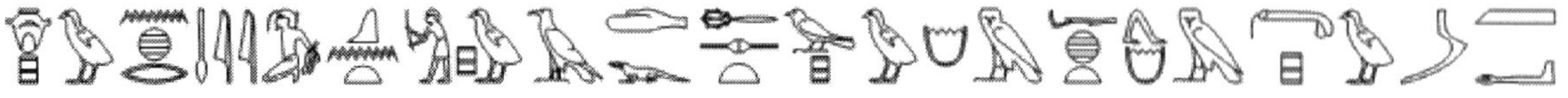

jb pw n(y) xrwy. on.t pw Ad, Xs.t pw Hm-xt, Hm pw mAa
ùn-nəm ùn-ziziŋ. A lərə à-yɔkh ù-nə yé 'mfi, à-tùb ù-ne waŋ; nàlà ù-nə afep-bibîe.
the heart of the enemy. Being aggressive is brave, withdrawing is cowardice, it's a real coward

Arw Hr tAS=f. Dr-nt(y).t [sD]m NHs(y) r xr r(A), jn
I-nyo à á-tub mvuar si dyé. A mu Ekang ba yəbə i-dzuari i-bugh ù-sù,
than one who is removed from its border. Since the Nubian obeys to submit to the (first) word,

wSb=f dd(w) Hm=f. Ad=t(w) r=f dd=f sA=f, Hm-xt(=tw) wA=f r Ad.
I-yalèn dzié yà boh nâ à-tub! I-yoŋ bà lərə nyè à-yokh à ngà lərə mvùs dyé, I-yoŋ bà kilèn nyè mvùs à ngà lərə à-yokh dyé,
it's him to answer that makes him back! That one shows oneself aggressive against him and he shows his back, that one recoils and he comes to be aggressive.

n rmT js nty Sf=t(w) s.t, Hrw.w pw sD-jb.w. jw
Bə sə i-bod bà və ù-lùghù; bə nə bi-sughlù mi-'nem ùm-bolèn.
These are not people that we respect, they are miserable with a broken heart.

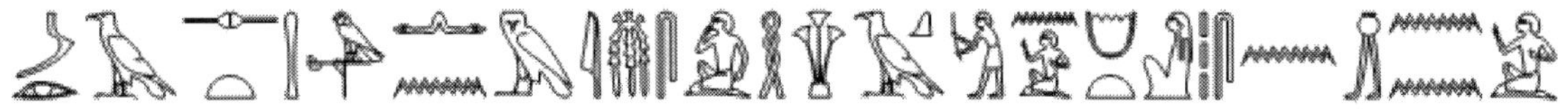

mAA~n s.t Hm=j, nn m j(w)ms, HAo~n=j Hm.wt=sn, jn~n=j
Mà ùn-ntum mə ngà yén bɔh, i sikigh minal, mə bî bingà bâbà, mə kəh
My Majesty saw them, it's not a lie, I captured their wives, I took them

Xr.w=sn, pr r Xnm.wt=sn, Hw kA.w=sn, wHA jt=sn
à-ləh dàbà, mə bərə à mə nyù màbà, mə woué nyakh dyàbà; mə tîghi fər dyàbà
their subjects, climbing to their wells, slaughtering their bulls, cutting their barley

rd xt jm. anx n=j jt=j, Dd=j m mAa.t nn xn n jm
mə dîghi fə. A bəbəla nâ i-sià wàmà à nyiŋ à-sù dàmà; i nə à bəbəla nguə ma kɔbɔ i sikigh və i- bômda.

and setting it on fire. As true as my father lives for me, it is in truth that I speak and this is not a speech.

n(y) aba pr(w) m r(A)=j. jr gr.t sA=j nb srwDt(y)=fy tAS
Ya koui mà à-nyù! E i ùn-mwan wàmà ù-sə à áyi mà bâlà ùn-nyiè nyi.
exaggeration out of my mouth! And as for any son of me who will strengthen this border.

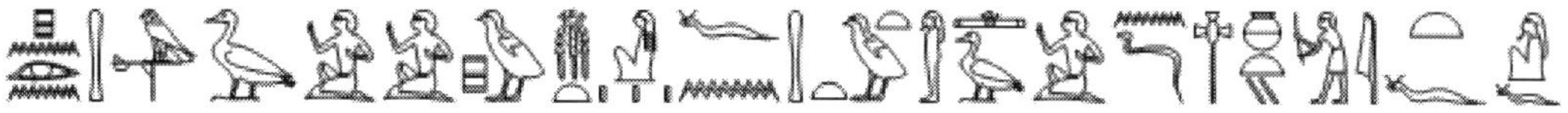

pn jr(w)~n Hm=j, sA=j pw ms=t(w)=f n Hm=j, twt sA-nDty-jt=f,
Yà ùn-tùm wàmà wà tələ; à yi bo ùm-wan wàmà, à byàli à-sù ù-lùghù wàmà; hà nə ùn-man ùm-bâlà i-sià.
which my Majesty has established, it will be my son, brought into the world for my Majesty, in the image of the Son-Protector-of-his-Father,

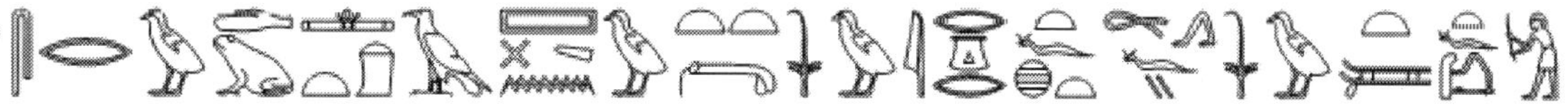

srwD(w) tAS n(y) wtt(w) sw. jr gr.t fxt(y)=fy sw, tmt(y)=f(y) aHA(w)
I-nyo à wôm ùn-mbâgh i-nyo à ngà və nyiè i-nyiŋ. E i-nyo à yi ligh tèghe fə kàmèn.
He who strengthens the border of the one who has begotten him. And whoever abandons him, who will not fight.

Hr=f, n sA=j js, n ms=t(w)=f js n=j. jsT gr.t rd~n Hm=j jr.t twt
Ùn-làn té, nyol té teghe fe bo ùn-mwan wàmà, à sikigh mə-ki màmà. A mù té ùn-ntùm wàmà wà télé i-fulèn
for her, it will not be my son, it will not have been born for me. So here is My Majesty had a picture made

n(y) Hm=j Hr tAS pn jr(w)~n Hm=j n-mrw.t rwd=Tn Hr=f, n-mrw.t aHA=Tn Hr=f.
dyàmà à ùn-nyiè si mə ngà télé, à sù yi nâ mi bughbèn i tələn à yi dyo, à sù yi nâ mi kàmèn à yi dyo.
My Majesty on this frontier established by My Majesty, so that you may be prosperous thanks to her, so that you may defend her.

3.7. THE BASIS OF A MATHEMATICAL APPROACH

The study of the Pharaonic led to highlight a multi-faceted African approach. A character or a script gives a great deepness at the level of the semantic fields.

The comparative study of the synchronies of current African languages reveals a constant. This constant means that starting from the same semantics, the different occurrences used to name a reality come together in a single script. These occurrences can be a character of the script or associations of characters of the script.

Languages	Family	House/cell	Numeral *one*	Lineage	Country
Ewondo	*Ndâ-bot*	*nda*	*mbokh*	*mvokh*	*mvokh*
Wolof	*mbokh*	*Pir/fer*	*mbokh*	*bu:m*	*pɛgg*
Basa'a	*Lì-hàà*	*ndab*	*mbokh*	*lòk*	*bɔ́k*

Duala	*n'dabo*	*ndabo*	*-ɔ̃*	*mwasa*	*mboa*
Lingala	*li-bótá*	*ndako*	*mɔ̌kɔ́*	*nkómbó*	*mboka*
Yemba	*atseŋɛ ŋ̄gyá*	*letshiɛ́ ḿmōk.*	*ùɛ̀-mɔ̀ʔ*	*mo7*	*ŋkɔ'ɔ́*
Nzimé	*mbɛ bûr*	*mbɛ̌r*	*mpwár*	*mbɔŋ*	*mbɛ̌r*

Chart 106: House in few african languages

The script 𓉐 ***pr*** for the house; fireplace, palace; temple is linked to the numeral one, the housene, the door and therefore to the family. By semantic shift it is linked to the country too. Because the country is shared by those who have the same ancestor.
For a better understanding of this script, the house is / nda / in Ewondo simply because this expression derives its etymology of the unit, the numeral one (da). Nda means therefore, what is one. / Inyo à nə da / "to be one". Thus giving the term / nda /.

Compared to other terms, we have ndako> ndabo> ndab> ndah> nda.
Ndako or ndabo means what is ONE accompanied by the noun class suffix of places / places which is -bo / -ko meaning foot. Rigorously house therefore appears according to this approach as the place of what is ONE or the place of Unity. Note that this nominal class suffix has disappeared in Ewondo. In the current synchrony, it can be predestined to mean not the place of house but the foundations of the house: Ewondo: / ako (l) nda / "foundations of the house".

After this aside let us translate the terms used in the other variants of languages taken in example above to say family in Ewondo although one could have taken the other senses.

Wolof : mbokh "family" equals the Ewondo: mbokh "unity, the cave, the canal" 𓉐
Basa'a: Lì-hàà "family" equals the Ewondo: iyàà "concession; grant" 𓉐
Duala : n'dabo "family" equals the Ewondo: ndabot "family" 𓉐
Lingala: li-bótá "family" equals the Ewondo: le-bot>bot "people" 𓉐
Yemba: *atseŋɛ ŋ̄gyá*"Family" equals the Ewondo: à.kol-nda "foundations" 𓉐
Kɔ́ɔnzime: mbɛ bûr "family" equals the Ewondo: ùmbè-bor "the men's door" 𓉐

It is clear that the terms used in the other sibling variants to designate the family appear in the same semantic field as the script. These terms are part of the lexicon Ewondo also. House can be identified by its fondations, its content, its limits, its door etc.
It appears then that the different speakers seem to have retained simply particularisms of the script from the point of view Ewondo and vice versa.
So the only way to allow easy understanding between the different speakers is indeed to use the ***mDw nTr*** as a common reference but also to list all occurrences in which appears a given character and the associated senses.

De facto for our script ***pr*** its most realistic approach will mobilize all its occurrences.

- Axyt closed room; mouthrie, slaughter-house
- Axt horizon
- Axty inhabiting the horizon; residing on the horizon (idiom.)
- scenes of carnage; places of execution
- iart remains (of a god)
- iwyt house; sanctuary; district; public places
- iwnyt hypostyle hall
- iWnn sanctuary
- ib refuge
- ibw refuge, shelter
- ipA private office
- ipAt private office
- ipip Epiphi
- ipt nsw king's harem
- ipt Harem; secret room
- imAw tent; hut
- imy-is advisor
- imy-wrt starboard; western part; West; phyl of priests
- imy-r aXnwty chamberlain
- imy-r pr intendant
- imy-st-a kicker; assistants
- imnt secret place, hidden place
- imHt Lower World, Beyond, World of the Dead
- imt-pr testament, donation
- iri create, do, act, etc.
- iry-at warden, room attendant
- iry-sA keeper of livestock pens
- ihw camp, encampment
- IHw barn
- is tomb, syringe; Chamber, Council Palace (idiom.); workplace, workshop, office
- Isbt throne; stool, folding seat; pedestal
- ist palace
- itrt row; chapel, niche, box
- itH fortress
- iDrt room, hallway
- aany camp, encampment
- afAy camp
- arrwt door; beating; lintel; judgment room; dwelling, dwelling
- aH palate; temple
- aHAt battlefield
- Ahat falls
- aXnwty courtroom
- at room, room; house
- wAHt processional station
- wAxy forecourt with columns, forecourt
-
- wAD an offering bread
- wy like ...!
- waaw privacy, privacy, loneliness; private apartments
- wabt place of embalming, grave, kitchen, refectory; table of offerings
- leg wart; lower leg; foot
- wbA open courtyard
- wmt thick door, gate
- wmtt thick wall (idiom.); thickness
- wryt a holy place
- wxA column room
- ws crack, split; small window
- wsxt hall, courtyard
- wDA warehouse, store
- wDAyt remains mythical of Amon-Re
- bAbAw hole
- bw up
- bbt hole, cavity; niche
- bnrw outside (idiom.)
- bnrt date cakes
- bxn castle, fortress
- Bxnt pylon
- bkyt enclosure
- pr house; fireplace, housene; palace; temple
- ready to be active
- Pra-aA Greate House, palace; Pharaoh
- pr-anx House of Life
- pr-wr National Shrine of Upper Egypt to Hieraconpolis
- pr-bity palace of the king of Lower Egypt
- pr-mDAt library, archives
- pr-nw national shrine from Lower Egypt to Dep
- pr-nbw House of Gold
- pr-nfr funeral workshop, embalming workshop
- King's Palace, Court; temple
- Pr-nsr National Shrine of Lower Egypt at Pe
- pr-hry upper part of a building
- pr-HD White House, treasure
- pr-xnty harem
- pr-Xry ground floor
- pr-Sna work establishment, workshop
- pr-dwAt cloakroom
- pr-dsr House red, treasure
- pr-Dt funeral area
- take out, ride
- pry hero, champion
- pry wild bull
- pryt houses, mansions

- PRYT crisis
- prw movement; procession; final result
- prw surplus, surplus
-
- prw land emerging from the flood
- prsn a loaf
- ready procession
- ready Peret
- fruit, seed
- prt-a activity
- prt-xrw invocative offering
- psn a loaf
- pqr Peqer, pregnant with Osiris in Abydos
- m-Xnw inside
- Marw gazebo
- miaHat falls, cenotaph
- Annex, Annex
- mabAyt House of Trento
- maHat falls, cenotaph
- mnw fortress
- mnnw fortress
- mnqb cool place; Chapel
- LCSA Street; driveway
- mrt street
- mxAw stake
- mxtmt barn
- mXr warehouse, store, barn, attic
- msxn remains
- msqt Milky Way
- mkAt support, pedestal
- mkt protection; correct position; appropriate location
- stable MDT
- mDt cattle put in the barn
- nAt weaving workshop
- nbt-Hwt Nephthys
- nfrw end, back part
- nmt slaughterhouse
- nht shelter, refuge
- nxn Nekhen, Hieraconpolis
- n.times.n chapel
- ns language
- nst seat, throne
- r-awy door, gate
- r-pr temple, chapel
- rwyt court of law, hall
- rwt door, gate
- rwty lion's den
- rwty outside, outside (n.)
- rwdw staircase, grave well
- rd foot
- rdw staircase; markets; grave well
- h yard
-
- Hai down; fall; reach; drip, drop by drop; get to; to grasp the meaning of; load, melt on, fall on; attack
- hAyt portal, gantryhrmw poulailler, enclos à volaille
- HA Outside (n.)
- HAi outer parts
- HAp secret place
- HAt falls
-
- ha palace
- Hwt temple; funerary chapel; mansion, "castle"; administrative district; field
- Hwt-AAT palace; administrative center; temple, funeral chapel
- Hwt-anx: Castle of Life
- Hwt-wryt: the Greate Residence
- Hwt-pat: "Castle of the Patricians"
- Hwt-nTr: temple
- Hwt-Hry-ib: Athribis
- Hwt-kA: chapel
- Hnkyt: bed; beer, funeral bed
- Htt: mine, career
- HD: chapel (idiom.)
- xA: office, room, administration
- xAwt altar room
- xAty office
- shelter xyt
- XWT sanctuary
- xbt execution spot
- xpr come to exist, to happen; to become; to change to; greatir; to happen, to happen, to happen; take place; to be effective; to pass, to have passed; come; Carry on; to exist, to be; do (passive); to provoke, to make arrive
- xm chapel, sanctuary
- xnw place of rest, residence
- thirteenth harem
- xnrt prison fortress; council chamber
- xnty the one, what is ahead; whoever is at the head of; etc.
- xnty anteroom
- xr grave, necropolis
- xtm chest; shop; strong
- xtmyt sealed room, closing
- xtt-servant, servant
- inside XNW; house, dwelling; Residence
- Xnt burial chamber
- sA barn
- swnw tower
- swtwt walk around; to travel, to travel

- SBA door
- to teach, to teach; education
- sbxt portal
- spri miss the target; expel
- sfxy fortress
- snwt shrine, chapel
- snhAt shelter, refuge
- srx a government department
- SH tent; greate room
- sxwy slaughterhouse
- sxm temple, sacred place
- sS threshold
- SSY nest
- SSD window
- ssdt reliquary, shrine
- sqbbwy shower room
- st seat, throne; place, place, place; foundations, land; service, office, administration; shop; rank, function
- sti successor
- stp-sA palace
- sDr room, apartment
- Saa granary, barn with grain
- SAat empty, empty space (n.)
- Sps tomb chapel, gravestone
- Sna turn around; maintain order; hold; to rush
- sna worker
- Snaw store, warehouse; kitchen workshop
- Snat store
- SNWT attic
- Sr mouth, obstruct, block
- Sspt chapel; flag; bedroom
- SS nest
- StAyt sanctuary of the god
-
- gs-pr administrative district, temple
- tA oven
- tp-Hwt roof
- tpHt cavern, hole
- tnnt shrine in Memphis
- Tart cabin; entrenched camp, fortress
- TPhT cave; hole
- Tnnt Tjenenet, sanctuary in Memphis

-Dwarf, Lower World, Beyond, World of the Dead; low room, crypt

-dwAtyw inhabitants of the Lower World (Douat), dead

-DAdw courtroom

-DbAt cloakroom, dressing room; sarcophagus

-Drwt room,

-Dsrw sacred place

Let's analyze a simple mathematical script to understand this phenomenon:

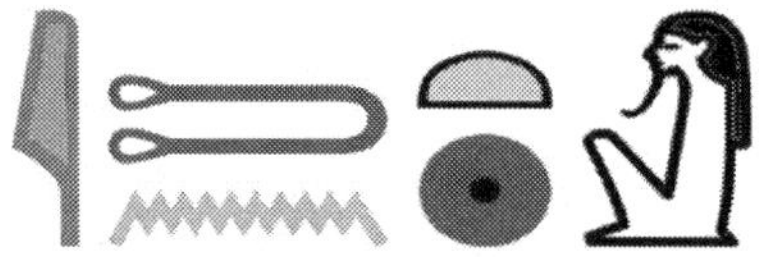

Observe and note that this script is composed of 6 characters.

The sign A40 : the semanteme will be left aside for this exercise
The sign N5 : the dterminative will be left aside for this exercise
The sign X1 : The gender tag will be left aside for this exercise.

We will focus on the signs that we call phonetic complements in the authorized reading. And who are used to read this script.

The sign M17 : reed in flower often transliterated by: I
The sign V13 : Rope for attaching animals, often transliterated: T
The sign N35 : Water ripples, transliterated: n

Egyptology will therefore logically propose the ***Itn*** transliteration. As we have just seen, this approach is reductive. Because it is based solely on an alphabetical dimension and not on the concepts underlying these characters.

Let's build a logical table to list all the possible cases related to the script above. Indeed to vocalize family in the previous example, the different varieties of languages have not respected a precise sense of reading. We had the impression of a random approach.
A truth table is a mathematical table used in logic, especially classical propositional calculus and Boolean algebra to semantically represent logical expressions and calculate the value of their function relative to each of their functional arguments (each value combination assumed by their logical variables).

Truth tables can be used in particular to say if a proposition is true for all legitimately imputed values, that is: if a proposition is "logically valid"
With three characters one can have 2^3 that is to say 8 possible logical cases. Because by combining the characters they can take the value or the state "true" or false.
Since African languages operate by consonant metathesis, it will be necessary not only to combine but also to introduce a column for permutations.

I	T	N
0	0	0
1	0	0
0	1	0
1	1	0
0	0	1
1	0	1
0	1	1
1	1	1

conv. Sense	Permutations
i	
t	
it	ti
n	
in	ni
tn	tn
itn	Int/tni, tin/nti, nit

Chart 107: logical table

Why do we have to ask for this analysis? Simply because if we remember the example on the family each retains a simple aspect of the script. So the analysis of this script from this angle gives all possible cases. The combinations that can be made from this example and the permutations because the Kweke and Kekwe languages operate by metathesis of consonants skeletons.

Which conclusions can be drawn from this?

Each unique character is a concept in itself.
The association of two characters gives a higher level of information
The association of all characters gives rise to philosophical concepts.

In this example, 7 combinations are generated in one reading direction and then 8 other combinations by permutations. 4 words per permutation of 2 characters (2!) And 6 words per permutation of 3, that is 3! (3x2x1)

Or 15 possible words to read this script. But in reality we are below reality because each case can produce words by variation of affixes (duel, semanteme, masculine, feminine, etc.) vowel variation (e, a, i, u, o, y, w), by variation of tones (5 max), clicks etc. and in addition synonyms (variation of the consonants while keeping the same signified). So by combining them again we have a very high number of possibilities offered by the reading of this script.
It is therefore in this that we can multiply to infinity on reading by association of ideas.
That is why to read this script a single symbol is enough for the one who knows his language and who understood the way the system is designed and what it implies.

	value	Words
	**i**	*Tín, *ŋtoŋ, n̲yô, *tetom, *ndolom, *kol, fès, fi, zir, etc.*
	**t**	*Toŋ, *bokh, í-bom, *tiŋ, ngarakh, nyakh, baal, loba etc.*
	**it**	*Tín- Toŋ; *toa; tara; *yo,*
	n	**tóŋ , *tn, *so,máŋ, *komi, *kumde etc.*
	**in**	*Tín-u.Toŋ etc.*
	**tn**	*u-ntol, nyamoro, nkor*
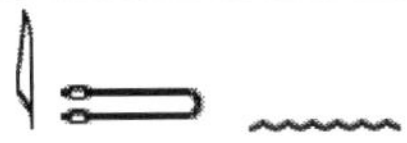	**itn**	*Tin, *nti, atone, *kn, *bôkh, *mos, kiri, zamba, Râ, Dzô, *tour etc.*

Chart108: demultiplexing table

On associations of two characters one can combine the meanings or perceive the new idea that results. Both steps are right.

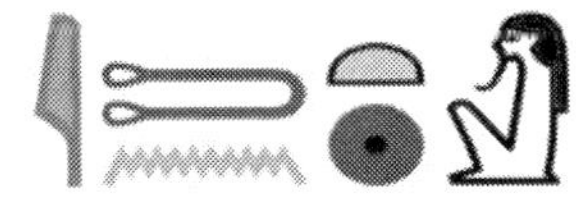

The reading ***itn*** offered to this script is therefore minimalist.
If I gladly dismissed the three characters on the right:

The sign A40 : The semanteme
The sign N5 : The determinative
The sign X1 : The gender tag

We can take them here to read this script from the African point of view.

All I need to do is look for the sign A40 𓀭: the semantics and read Tara "the father" so that my reading is right without having to stuck the left signs.

The sign A40 𓀭: the semanteme could have been read also in Ewondo: Tin "the strain"
All the words that I can use to describe the semanteme appear in my combinatorial analysis.

This is equally true for the sign N5𓇳: the determinative. It is enough for me to recognize the sun here to name it in my language without however being false since by synonymy, no matter the reading mobilized, my intention will be necessarily right.

So according to whether we are talking about cosmogony, cosmology, cosmoscript, cosmocracy, botany, farming, the agrarian field or water, we are going philosophically to be able to mobilize the indicated terms and use a single and same script. One can also move from one register to another by using the link that unites the elements of the script since they refer to the same idea taken individually or by combinatorial analysis.

We will thus distinguish levels of comprehension and readings

A level for the simple vulgum pecus
A level for initiates who are able to recognize signs and their deep meaning.
A level for the scribes those who understand the concepts that are behind the signs.

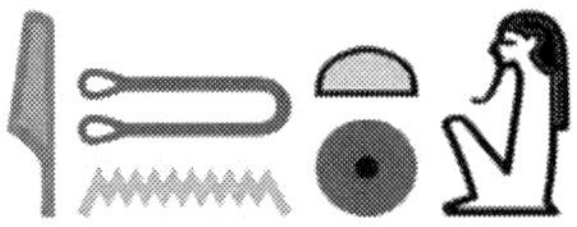

This script records water, air, earth, and fire.
This script records the plant, the animal, the aquatic world
This script records life etc.

I will stop here on the introduction to this mathematical approach which seems to me rich by some aspects. I will try to deepen it in my next productions.

Computering Help: Zipf 's Law

The question of ancient texts is not limited to reading graphs, it goes beyond and uses mathematical and statistical concepts. The challenge here is to read whole texts faster but also tomorrow preceded by the computerization of these texts. It is therefore useful to think of a technical and scientific means that allows the passage to the path of computing.
It is therefore useful to be interested in the mathematical[1] approaches that have been carried out among human languages.

According to this point of view, long old texts of the ancient Egyptian can be analysed by ranking words according to frequencies, to verify whether the current texts follow the mathematical relation known as Zipf 's Law, where the word frequencies (f) vary with the rank (z) according to the relation:

1. Harris, Z., Mathematical Structures of Language. New York: Wiley, 1968.

$$f_z = \frac{C}{z^{\alpha}} \;, z = 1, 2, 3 \ldots n$$

where $\alpha \approx 1$.

The approach of Lin and. al. in finding the mutual information amongst words in the corpus and then computing a similarity between the words based off of this. The mutual information between two words in the text, x and y, is termed
I(x,y) and is defined as

$$I(x, y) = \sum_x \sum_y p_{xy} \log \frac{p_{xy}}{p_x p_y}$$

where $p_{xy}(x, y)$ is the frequency of word pair (x, y) and p_x and p_y are the frequencies of words x and y in the texts. Two different arrays of mutual information were calculated for the word distance one and two pair frequencies. Finally, a blended mutual information is calculated using different weightings of the one and two distance mutual information.
The blended mutual information, I_B, is:

$$I_B = \sqrt{I_1^2 + (WI_2)^2}$$

where I_1 and I_2 are the mutual information for distance one and two word pairs, respectively and the weight (W) takes a value between 0 and 1. It is difficult to find an objective value for W.
For the blended mutual information the cosine similarity measure, S, is calculated where
S is defined as:

$$S_{xy} = \sum_z \frac{2 I_B(x,z) I_B(y,z)}{I_B(x,z)^2 + I_B(y,z)^2}$$

where z is all words in the corpus where $z \neq x, y$

In order to more clearly see how the words relate to each other, it is possible to graphically visualize the similarity relationships using the distance metric derived by Gower[2]. This distance metric is used to convert comparison metrics such as correlation or similarity among variables to metric distances:

$$d_{ij} = \sqrt{2(1-s_{ij})}$$

where d_{ij} is the similarity between words i and j. These distances can then be plotted onto a minimum spanning tree

It is evident that no language has ever been fully deciphered using purely statistical or mathematical techniques but African people have to awake and computer their past. So computering skill can just be helpful.
Africans must wake up and in this way in order to take back their historical heritage. On this account, it is important for them to do this computing work also.

---

2. Gower, J.C. 'Some distance properties of latent root and vector methods used in multivariate analysis', Biometrika 53 (1966), 325–41

3.8. SOME SCIENTIFIC NOTIONS

Zoba liboso, mayele na sima

(African wisdom)

The one and only way that will lead to the end of the regression system confining Africans to repetition and servile imitation is to regain control of their own educational system.
This inevitably involves the creation of endogenous documents containing current knowledge in an African language. I therefore proposed the translation into the language chosen for this work of some scientific notions taken according to my mood. Science is advancing every day, and technical jargon is becoming denser and more diversified to deceive and maintain a tight barrier with people embroiled in phantasmagorical beliefs. The notions are diverse and varied from a technical point of view, but I remain convinced that standardization work with experts makes it easy to pursue what is only a small beginning here.

Ewondo	**Lexicon**
ùn-nsoŋ; *ùn*-ntóŋ	*Angle*
Ùn-mbokh	*Spiral*
ùn-ntén	*Straight line*
i-yalà	*Length*
ùn-ntié; à-bim; à-nén	*size*
ndàm	*Width*
à-yab; *ùn*-ntòŋ	*Height*
à-twan	*Point*
i-yani	*Plan*
ùn-yarba	*Concave*
i-ŋgoŋbo	*Convexity*
Zaŋ; *ùn*-nəm	*Center*
ùn-ŋgoul	*Strength*
fìm	*Energy*
Ngap ngoul ; à-bim-ngoul	*Intensity*
kî	*Power*
ùn-mfakh	*Side*
ùn-mbîrga; *ùn*-nsora	*Axis*
i-ndəndam	*Area*
à-bum	*Volume*
nguilghedigha	*Circle*
ŋguə; yə	*disk*
i-yamla	*Median*
ùn-mbem	*Height*
i-lara ùn-nkilna	*Hypotenuse*
Ngap *ùn*-nsoŋ	*bisectrix*
kuŋe	*orthocenter*
ùn-nəm	*Center*
ùn-nkol, i-salgha	*Rope*
I-toun i-sal nguilghediga; *ùn*-ntəm; ŋgàngà	*Ray*
i-sal nguilghediga	*Diameter*
ùn-mfan, *ùn*-mbeŋy	*Bow*
Ngwara *ùn*-ntén	*Tangent*
Ngwara *ùn*-ntén ŋguilghedigha	*Tangent of the circle*
ùn-nka	*Circumference*
mə-kê; mə-tari	*Origin*

ùn-ŋkyeŋbe i-vəghə *ùn*-ntén	*Abscissa*
tətələ i-vəghə *ùn*-ntén	*orderly*
i-lumbà; Ùn-ndeŋbə	*Slope*
mi-ʻmbias	*Symmetry*
Mekan m'vora mə-nê,	*Right angle*
i-vora	*square*
mi-nəŋbe mi-nsoŋ	*Opposing angles*
ùn-nsoŋ ya ùn-nem ŋguilghediga	*Inscribed angle*
i-tsəndghà	*Dilation*
à-bim i-tséndgha	*Homothety report*
ŋgôe	*Fish bone*
ù-titəŋ	*Diamond*
i-dəŋ	*Oscillation; swaying; rocking*
ndén	*flood or rush of water in a river*
i-béré	*Elevation, rise*
i-kakh	*Projection*
i-sokh	*Cone*
à-kan	*Right angle*
ùn-ntsn nguilguedigà	*Ellipse*
i-Vîghda	*Cylinder*
i-ŋgoŋbe; i-bouma	*Sphere*
ndəm; i-yəmə	*Number*
ŋgura i-yəmə ne-vyokh	*Whole number*
ndəm bi-yəyəmə	*Imaginary number*
i-bulgha	*Coefficient*
ndəm i-kyeŋbe ù-ntn	*Abscissa*
i-vêgha i-lûmba ùn-ntn	*Angular coefficient*
i-vêgha i-lûmba ùn-nsɔŋ	*Tangent*
ùn-ntén	*Right*
à-bôbôn	*Pole*
à-bôbôn i-kilèn	*Inversion pole*
i-vêgha i-lûmba à-bôbôn	*Tangent at the top*
i- kurgà *ùn*-ntén	*Asymptote*
ù-ntîghèn i-sokh	*Conical section*
ù-nsoŋ mi-ʻ *ùn*-ntén mi bèh	*Angle of two straight*
à-twan *ùn*-ndaŋba ùn-ntén	*Distance from a point to a line*
ùn-mbwam mi-ʻ*n*tén	*Beam of rights*
Tèghè mə-mâna à-twànà	*Point to infinity*
Tèghè mə-mäna *ùn*-ntén	*Right of infinity*
A-twan ùn-nsaŋ	*Multiple point*
ùn-mbias à-twàn	*Double point*
À-kəŋ bi Vughle	*Theory of envelopes*
i-yarbèn	*Curvature*
I-vwé i-yarbèn	*Rectification of a curve*
i-sal bi-sókh	*Diameter in conics*
Təm bi-sókh	*Center in conics*
Ùn-ndaŋba (A,B)	*Interval (A, B)*
i-vyəlèn	*Derivative*
Ùn-mfer i-vyəlèn ùn-mfob'lâ	*Derived zero of a constant*
I-twé bəbəla Amougou	*Theorem of Amougou*
tàŋ à-tiŋ mi-ʻnsamba	*A mathematical function*

tətələ à-tiŋ mi-'nsamba	*A continuous function*
Fyar tətələ à-tiŋ mi-'nsamba	*A discontinuous function*
A-tiŋ mi-'nsamba à mə-tari	*A primitive function*
ùn-mvenbe à-tiŋ mi-'nsamba	*A constant function*
ùn-mvyəlèn à-tiŋ mi-'nsamba	*A differentiable function*
i-yè à-tiŋ mi-'nsamba	*Increasing function*
Ùn-ntogbèn à-tiŋ mi-'nsamba	*Decreasing function*
Ùn-mfassèn à-tiŋ mi-'nsamba à ngwar à-twan	*Study of a function near a point*
À-fola i-yarbèn i-tîèn à yi à-twan i-lûmba	*Position of the curve with respect to the tangent at a point*
À-twan i-fwalèn	*Inflection point*
Ngap i-bugbèn	*Coefficient of development*
i-nyè à-tiŋ mi-'nsamba	*Bounded function*
i-vyəlèn ù-sù	*Derivative first*
i-vyəlèn bê	*Derivative second*
Mvouara à-tiŋ mi-'nsamba	*Pair function*
i-fàrà à-tiŋ mi-'nsamba ùn-mbias	*Odd function*
i-fàs ùn-nlàn i-kurgà	*Find an asymptotic direction*
Ùn-ŋkwar, ùn-ntsén, zeus	*Oblique*
Ùn-ŋkwar i-kurgà ùn-ntén	*Asymptote oblique*
i-kurgà ù-ntén à-bùm	*Asymptote cubic*
Ù-tokh; ù-bə	*Infinitely small*
Un-ntol ù-tokh	*Infinitely small principal*
Ngap i-fas ùn-ntol	*Report to the main part*
À-kàlà ù-tokh	*Infinitely small comparable*
A-tokh mi-ntié	*Infinitely small order*
i-ŋgura	*Infinitely great*
À-yala à-tokh	*Order of infinitely small*
À-yala i-ŋgura	*Order of infinitely great*
Zén; i-kə	*Direction*
Ùn-mfakh zén; Ùn-mfakh i-kə	*way*
Ùn-ntouar à-koŋ	*Free vector*
Ùn-nlarèn mə-kɔŋ	*Related Vectors*
Bi-fialèn mə-kɔŋ	*Sliding vectors*
Mə tari; mə-kê; ùn-zwandam	*Thin; fine*
i-twa, i-bêgha	*Support*
À-boui bi-tséndé; à-tiŋ mi-'nsamba	*Multiple variable function*
Ùn-nyê à-tiŋ mi-'nsamba	*Limit of a function*
Ndùghrù; à-yɔŋ	*Heat*
i-ndàmbèn; ndàm; i-vɔghbo	*dilation*
Ndùghrù i-nyàmghà	*Heat of fusion*
À-bim ndùghrù	*Amount of heat*
Ùn-ntsütsüe ndùghrù	*Heat of vaporization*
i-ndəndam ùn-mfob'lâ ndùghrù	*Isothermal surface*
Ùn-mfogh nənə	*Absolute zero*
Ngarga ndùghrù	*Thermal resistance*
i-bubuar ngarga ndùghrù	*insulation*
À sal à-yɔŋ	*Thermally insulate*
Ngura nyol	*Homogeneous body*
Ùn-mbamèn nyol	*Heterogeneous body*

Ùn-nsin ndùghrù-mvura	*Thermodynamic equilibrium*
Tèghè i-dzê ndùghrù	*adiabatic*
i-si	*Job; work*
i-kalikh	*wheelbarrow*
i-sob'lo	*shovel*
i-vuas; i-digh; i-vuegh mə-sɔŋ	*rake*
Ngap à-vol i-tyène é à-bokh; ùn-mfəp; ù-sélé	*Acceleration*
À-vol, ùn-mfol; ndén	*Speed*
tom ; i-toum ; i-wôme ; i-wône	*atom*
i-yinbà	*Pressure*
i-kwènə tàŋ; i-koulèn taŋ; I-tié taŋ; à-tiŋ	*Equation*
i-kwènə tàŋ ù-nyiè ù-sù	*First degree equation*
i-kwènə tàŋ ù-nyiè ù-sù i tiènə é i-lərə dyà	*First degree equation to an unknown*
Ùn-ŋkwènə tàŋ ùn-nyiè bêh	*Equation of the second degree*
Ùn-ŋkul'nə tàŋ ùn-nyiè bêh é bi-lərə bi-bèh	*Second degree equation with two unknowns*

Biblioscript

MOUTHR, Jules. *La symbolique maçonnique, 2ème dition.* Paris , 1953.

BOUTAMMINA, Nas E. *Comprendre la Renaissance - Falsification et fabrication de l'Histoire de l'Occident,.* Books on Demand, 2015.

BRAUDEL, Fernand. *Civilisation matrielle, conomie et capitalisme : XVe-XVIIIe siècle.* Paris: Armand Colin, 1979.

BREHIER, EMILE. *Histoire de la philosophie, 1er fascicule supplmentaire : "La philosophie en Orient.* Paris: FELIX ALCAN, 1938.

Bruly, Bouabr. *Solution Bruly aux poids-d'or ou Livre africain,* . 1972.

BUDGE, Sir Ernest Alfred WALLIS. *Egyptian Hieroglyphic dictionary.* London: Harrisson and sons , 1920.

BWEMBA-BONG. *Quand l'africain tait l'or noir de l'Europe: L'Afrique, actrice ou victime de la « traite des Noirs » ?* . Paris: Menaibuc, 2005.

HAGÈGE, C., *Les pronoms logophoriques* . Paris: Bulletin de la Socit de Linguistique de Paris, 1974.

ROCHE Carole, Pierre BORDREUIL. *D'Ougarit à Jrusalem: recueil d'tudes piscriptiques et archologiques offerts à Pierre Bordreuil.* Paris: Editions De Boccard, , 2008.

CATALAN, Claude. *Au-delà du matrialisme les ralits immatrielles.* ditions Baudelaire, 2016.

CAUVILLE, Sylvie. *Dendara V-VI.* Peeters publishers, 2004.

CAVALLI-SFORZA, L. Luca. *Gènes, peuples, et langues Vol. 94:.* "Proc. Natl. Acad. Sci_ .-U_: , 1997.

CSAR, Jules. *Csar : La guerre des Gaules.* Paris: ditions Palo, 2004.

Ch., BONNET. *La ville nubienne de Kerma.* Lausanne: Favre, 2014.

CHAMPOLLION-Figeac, Jacques-Joseph. *gypte ancienne.* Paris: Firmin Didot Frères, Editeurs, 1839.

CHAROLLES, Michel. *L'anaphore : problèmes de dfinition et de classification.* Verbum, , 1991.

CHASSANG, Marie-Nicolas Bouillet et Alexis. *« Kutsami » dans Dictionnaire universel d'histoire et de goscript,.* Paris: Hachette, 1878.

CHRETIEN Jean-Pierre, PERROT Claude Hlène. *Afrocentrismes. L'histoire des Africains entre Egypte et Amrique.* Paris: Khartala, 2000.

CLARK, John Henrik. *Can African People Save Themselves?*, 1990.

COMPIEGNE, Louis DUPONT, marquis de. *Afrique Equatoriale Gabonais Pahouins Gallois II* . Paris : Okenda Bangouens Osyeba , 1875-1876.

CRISTO, Albert Di. *La prosodie de la parole.* Solal Editeurs, 2013.

D., BAUMANN h et WESTERMANN. *Les peuples et les civilisdations de l' Afrique.* 1957.

D., BONNET Ch. & Valbelle. *Les temples gyptiens de Doukki Gel-Pnoubs, « le jujubier » (Soudan).* Lausanne: Favre, 2018.

D., CREISSELS. *Aperçu sur les structures phonologiques des langues ngroafricaines.* Grenoble: Ellug, 1994.

—. *Description des langues ngro-africaines et thorie syntaxique.* Grenoble: Ellug, 1991.

OKASHA, Daly, El. *Egyptology: The Missing Millennium : Ancient Egypt in Medieval Arabic Writings.* London: UCL press, 2005.

Dan, XU. *Introduction to Chinese Syntax.* Kindle Edition, 1995.

NANJIRA, Daniel Don. *African Foreign Policy and Diplomacy : From Antiquity to the 21st Century,.* ABC-CLIO, 2010.

DANILOU, Alain. *Mythes et dieux de l'Inde.* Paris: Champs Flammarion, 1994.

De GANAY, S. *« Toponymie et anthroponymie de l'Afrique noire », Onomástica, n° 2, vol. 2.* . Paris, 1948.

de, SAUSSURE F. *Cours de linguistique gnrale.* Paris: Payot & Rivages, 1972.

DGARDIN, F. *Les homonymes et les homoscriptes de la langue française.* Paris: L. Lefort, 1857.

DEMEKE, Girma Awgichew. *The Origin of Amharic,* . Addis Abeba: Centre français des tudes thiopiennes, tudes thiopiennes n° 5, 2009.

DESCES, J-P. *Quelques concepts relatifs au temps et l'aspect pour l'analyse des textes in Etudes Cognitives : smantique des catgories de l'aspect et du temps. Warszawa, Acadmie des Sciences de Pologne- Institut de slavistique.* 1994.

DIOP, Cheick Anta. *Antriorit des civilisations nègres – mythe ou vrit historique ?* Paris. Prsence africaine, 1967.

—. *Parent gntique de l'Egyptien pharaonique et des langues ngro-africaines.* Dakar: IFAN, 1977.

DIOP, Cheikh ANTA. *Alerte sous les Tropiques : articles 1946-1960 culture et dveloppement en Afrique Noire.* paris: prsence africaine, 1990.

DIOP, Cheikh Anta. *L'Afrique noire prcoloniale : Etude compare des systèmes politiques et sociaux de l'Europe et de l'Afrique Noire, de l'Antiquit à la formation des Etats modernes.* Paris: Prsence Africaine, 1960.

—. *Les fondements conomiques et culturels d'un Etat fdral d'Afrique noire.* paris: prsence africaine, 1974.

—. *Nations nègres et culture.* Paris: Prsence Africaine, 1954.

DO-NASCIMENTO, Jos. *La renaissance africaine comme alternative au dveloppement : Les termes du choix politique en Afrique.* Paris: L'Harmattan, 2008.

DUBOIS-CHARLIER, F. *Structures verbales.* Aix, 2002.

EBOUSSI, Fabien. *Le Bantou problmatique.* Paris: Prsence africaine, 1986.

EKOU, Sophie. *Sagesse d'Afrique,* . Paris: Hachette , 2016.

ELONO, Raphaël. *La cosmogonie des beti be nanga: Croyances ancestrales – Traditions* . Paris: Editions Edilivre, 2014.

Emilienne-Nadège, MEKINA. *DESCRIPTION DU FANG-NZAMAN, LANGUE BANTOUE DU GABON : Phonologie et classes nominales ,.* Strasbourg: Universit, 2012.

Engelbert, MVENG. *L'art africain d'hier et l'Afrique d'aujourd'hui.* Paris: prsence africaine, 1963.

ERIKSON, Erick. *Childhood and society.* New York: Norton, 1950.

ERIKSON, Erik. *Adolescence et crise. La quête de l'identit.* Paris: Flammarion, 1972.

ERMAN A., Grapow H. *Wörterbuch der aegyptischen Sprachen.* Leipzig, 1926-1950.

ESSONO, J-M. *L'Ewondo : Langue bantu du Cameroun, Phonologe, Morphologie, Syntaxe,* . Yaound: Presses de l'Universite Catholique d' Afrique , 2000.

ESSONO, Ongun. *Grammaire de l'Ewondo. Essai de grammaire mthodique.* Yaound: Cerdotola, 2012.

F., GRIFFITH. *The Meroitic Inscriptions of Shablûl and Karanóg.* Pennsylvanie: University of Pennsylvania, , 1911.

FABRE, Baylon C. et P. *Initiation à la linguistique* . Paris: Nathan universit, 1975.

FAGG W, LIST H. *Merveilles de l'Art Nigrien.* . Paris :: Chêne, , 1963.

FAÏDARO, Dominique. *Parlons Fon langue et culture du Bnin.* Paris,: l'Harmattan, 2001.

FANON, Frantz. *Les damns de la terre.* Paris: ditions Maspero, 1970.

Farina, Jules. *Grammaire de l'ancien gyptien(hiroglyphes),* . Paris: Payot, 1927.

FAULKNER, Raymond Oliver. Rainer Hannig, s.d.

FERROl, Gilles. *Dictionnaire de sociologie, dition 2010.* Paris: Armand Colin diteur, , 2010.

François-Xavier FAUVELLE, Jean-Pierre CHRTIEN, Claude Hlène PERROT. *Afrocentrismes. L'histoire des Africains entre Egypte et Amrique.* Paris: Khartala, 2000.
Frantz, FANON. *crits sur l'alination et la libert.* paris: La dcouverte, 2015.
Fritz MORGENTHALER, Florence WEISS, Marco MORGENTHALER,. *Conversations au bord du fleuve mourant: ethnopsychanalyse chez les Iatmouls de Papouasie/Nouvelle-Guine,* . Paris: Editions Zo, , 1987.
G., LECLANT J. & TRIGGER B. «Meroitic Newsletter (Bulletin d'informations Meroitics) n° 13, .» *(Bulletin d'informations Meroitics)* , 1973.
GALLEY, Samuel. *Dictionnaire Fang-Français. et Français Fang suivi d'une grammaire fang.* Neuchâtel: ditions Henri Messeiller , 1964.
GENY, Michel Fartzoff - Evelyne. *Pouvoir des hommes, signs des dieux dans le monde antique,* . paris: pufc, , 2002.
Georges, NIANGORAN BOUAH. *L'univers akan des poids à peser l'or.* Abidjan: Les nouvelles ditions africaines, 1983.
Gervais, MENDO ZE. *Ecce Homo Ferdinand Lopold Oyono. Hommage à un classique africain.* Paris: Khartala, 2006.
Gilombe, MUDIJI MALAMBA. *Le langage des masques africains : tude des formes et fonctions symboliques des "Mbuya" des Phende,* . Kinshasa: Facults catholiques de Kinshasa, , 1989.
Gorges BALANDIER et Jacques Maquet. *Dictionnaire des civilisations africaines.* Paris: F.Hazan, 1968.
Goujard), CATON (trad. Raoul. *De l'agriculture.* Paris: Socit d'dition Les Belles Lettres, coll. « Universits de France », 1975.
GRAFFIN, R. et Pichon, J. *Grammaire ewondo,* . Paris: rue Lhomond, 1966.
GRELL, Chantal. *L'gypte imaginaire de la Renaissance à Champollion.* paris: Presses de l'Universit de Paris-Sorbonne, , 2001.
GRIAULE, Marcel. *Le renard pâle.* Ant-Matière, 1991.
GRIAULE, Marcel. *Dieu d'eau: entretiens avec Ogotemmêli.* Paris: Fayard, 1948.
Günter DREYER. *Recent Discoveries at Abydos Cemetery U, dans « The Nile Delta in Transition : 4th-3th Millenium BC »,* . Tel Aviv: dit. M. Van Den Brink, 1992.
Gustave LEFEBVRE, Serge Sauneron,. *Grammaire de l'gyptien classique,.* Le Caire: Ifao/inst.Franc.Archeol., 1955.
H., GREENBERG J. *Languages of Africa.* La Hague: The Netherlands, Mouton & Co, 1966.
Hampat BA et Germaine DIETERLEN. 1961.
Hancock, GRAHAM. *Magiciens des dieux.* Pygmalion,, 2015.
HARDY, Georges. *Une conquête morale: l'enseignement en A.O.F.* Paris: Armand colin, 1917.
Heine, Michael TELLINGER and Johan. *Temples of the African Gods.* Zulu Planet Publishers, mars- 2010.
http://fr.chabad.org/library/article_cdo/aid/995025/jewish/La-crmonie-des-Kapparot.htm. 28 novembre 2017. http://fr.chabad.org/library/article_cdo/aid/995025/jewish/La-crmonie-des-Kapparot.htm (accès le 11 28, 2017).
https://news.yale.edu/2017/06/20/yale-archaeologists-discover-earliest-monumental-egyptian-hieroglyphs. s.d. https://news.yale.edu/2017/06/20/yale-archaeologists-discover-earliest-monumental-egyptian-hieroglyphs (accès le dcembre 15 , 2016).
Huerta, Blas Castellón. «Cúmulo de símbolos, la serpiente emplumada.» *Arqueología Mexicana núm. 53,* , 2002 JANVIER - FVRIER Castellón Huerta, Blas, "Cúmulo de símbolos. La serpiente emplumada", : pp. 28-35.
Innes, Gordon. *A Mende-English dictionary.* London: Cambridge University Press, 1969.
IPGRI. « Les fruitiers forestiers comestibles du Cameroun, Bioversity International.» s.d.
IZARD, Michel. *Moogo, L'mergence d'un espace tatique ouest-africain au XVIe siècle.* paris: Karthala, 2003.
J., PERROT. *Les langues dans le monde ancien et moderne : Afrique subsaharienne, pidgins et croles, tome 1,* . Paris: CNRS, 1981.
Jacques EVOUNA. *La morphologie nominale en Ewondo : La double prfixation.* Connaissances et Savoirs, 2017.
Jean BOTTERO. *Msopotamie. L'criture, la raison et les dieux.* Paris: Gallimard, Première parution en 1987.
Jeune, CHAMPOLLION Le. *Grammaire Egyptienne, Principes gnraux de l'criture sacre gyptienne applique a la representation de la langue parlee.* Paris: FIRMIN DIDOT FRERES, 1836.
JOSEPHE, Flavius. *Contre Apion.* Paris: Ernest Leroux - diteur, 1911.
K, BALIHUTA. *Langue et culture des Bantu.* Paris : Prsence africaine, 1975, .
KADIMA, M. *Le système des classes en bantu,* . Leuven,: Vander, 1969.
KANE, ABDOULAYE. *Les Systèmes de numration parle des groupes ouest-atlantique et Mand: contribution à la recherche sur les fondements et l'histoire de la pense logique et mathmatique en Afrique de l'Ouest.* Lille 3: ANRT, 1988.
KAUFMANN, Jean-Claude. *L'invention de soi. Une thorie de l'identit ?* Paris: A. Colin, coll. Individu et socit, 2004.
Kelman, Gaston. *Je suis noir et je n'aime pas le manioc.* Paris: Max Milo, 2004.
KHALDUN, Ibn. *Discours sur l'Histoire universelle Al Muqqadima ou Prolgomènes.* Paris: Sindbab, 1377.
Ki-Zerbo, Joseph. *Eduquer ou Prire.* Paris: l'Harmattan, 1990.
—. *Histoire de l'Afrique Noire.* Paris: Hatier, 1980.
KI-ZERBO, Joseph. *Histoire et conscience nègre Prsence africaine.* Paris: Prsence Africaine, 1957.
KŒMPFER, Engelbert. *Histoire naturelle, civile et ecclsiastique de l'Empire du Japon.* La Haye: P. Gosse, & J. Neaulme, 1729.
L., HOMBURGER. *Noms des parties du corps dans les langues ngroafricaines,* . Paris: H. Champion, s.d.
La CPÈDE (Bernard Germain Etienne de La Ville sur Illon, M. le comte de). *Histoire gnrale, physique et civile de l'Europe : depuis lesdernières annes du cinquième siècle jusque vers le milieu du dix-huitième.* Bruxelles: P.J de MAT, 1826.
LACOUTURE, Jean. *Champollion, Une vie de lumières.* Paris: Grasset, 1988.
LANGE, Marie-France. *Le choix des langues enseignes à l'cole au Togo: quels enjeux politiques?, dans Politique africaine,* . Paris, 1987.
LARR, Christian. *L'Hritage spirituel de l'ancienne Egypte* . Paris: Diffusion rosicrucienne, 1998.
LEPROHON, Ronald J. *The Great Name : Ancient Egyptian Royal Titulary.* Atlanta: Society of Biblical Lit, 2013.
LONIS, Raoul. «Les thiopiens du pseudo-Scylax ; mythe ou ralit goscriptique ?» *Revue Française d'Histoire d'Outre-mer, tome 66, n°242-243,* 1er et 2e trimestres 1979: 101-110.
LOUMPET-GALITZINE, Alexandra. *Njoya et le royaume bamoun : les archives de la Socit des missions vangliques de Paris.* paris: Karthala, 1917-1937.
LUBICZ, R. A. Schwaller de. *The Egyptian Miracle : An Introduction to the Wisdom of the Temple.* Inner Traditions International, 1963.
LUBICZ, SCHWALLER de. *Propos sur l'Esotrisme et The symbol* . Paris : Dervy-Livres, 1977.
Luc BOUQUIAUX. «L'Homme,.» *Revue française d'anthropologie (N° 179),* Juillet/Septembre 2006: 227-235.
M. le comte de, Bernard Germain Etienne de La Ville sur Illon- LA CEPEDE. *Histoire naturelle des quadrupèdes-ovipares, Tome premier.* Paris,: Edition du Temple de la Gloire, 1819,.
M. NGESSIMO. *GLIMPSES OF AFRICAN CULTURES, Echos des cultures africaines,.* Paris: L'Harmattan, 2011.
M., VOLTZ. *Le langage des masques chez les Bwaba et les Gurunsi de Haute-Volta.Ouagadougou : Universit de Ouagadougou,* . Ouagadougou: Universit de Ouagadougou, 1981.
MACGREGOR, J.K. *L'Afrique dans l'Antiquit* . s.d.
Marcelle, WEISSEN-SZUMLANSKA. *Origines atlantiques des anciens Egyptiens,* . Paris: Les Editions des Champs-Elyses, 1965.
Mark COLLIER et Stephen Quirke. *The UCL Lahun Papyri, I-III,* . Oxford: Oxford, Archeopress, , 2002, 2004 et 2006.
MARLÈS, Jean Lacroix de. *Histoire Gnrale De l' Inde Ancienne et Moderne Depuis l' an 2000 Avant J.C. jusqu' à Nos Jours - Prcde d' Une Notice Goscriptique et De Traits Spciaux Sur La Chronologie, La Religion, La Philosophie, La Lgislation, La Littrature, Tome 1.* Paris : Emler frères, Librairies-diteurs, 1828.
MARTIN, Louis-Claude de Saint. *Chart naturel des rapports qui existent entre Dieu.* Rochefort sur Mer: l'homme et l'Univers, 1946.

MASPRO, Gaston. *Des formes de la conjugaison en gyptien antique, en dmotique et en copte.* Paris, : Hachette Livre BNF, , 1871.
MAYER., R. *Histoire de la famille gabonaise. Libreville.* ditions du LUTO, , 2002 (1992).
MBOCK, Dibombari. *Kongo.* Lulu, 2013.
MBOLI, Jean Claude. *ORIGINE DES LANGUES AFRICAINES - Essai d'application de la mthode comparative aux langues africaines anciennes et modernes,.* Paris: L'Harmattan, 2010.
MBOLO. « Collecte et analyse de donnes pour l'amnagement durable des forêts : joindre les efforts.» s.d.
MESSINA Jean-Paul et Van Slageren Jaap. *Histoire du christianisme au Cameroun -des origines à nos jours.* Paris: Karthala, 2005.
Michèle, FRUYT. «D'Africus uentus à Africa terra Revue de Philologie, no 50,.» 1976,: p. 221-238.
MINTSA, Jean Joseph. *Le Complet de La Langue Fang.* Morrisville: Lulu Press, 2008.
MONTENGRO, J. P. *Esboço de um vocabulario toponímico angolano.* Bol. Inst. Angola, 1958.
MORROW, Susan Brind. *The Silver Eye: Unlocking the Pyramid Texts.* Londres: Head of Zeus , 2016.
MOTAZE, Louis Paul. *L'Afrique et le dfi de l'extension de la scurit sociale : L'example du Cameroun.* Pyramide-Papyrus Presse, , 2009.
Mouhamadou NISSIRE SARR. *Funrailles et reprsentations dans les tombes de l' Ancien et du Moyen Empires Egyptiens,* . Berlin: Lit Verlag , 2001.
N., DELBECQUE. *Linguistique cognitive* . Bruxelles: De boeck.Duculot, 2002.
NDIAYE, Tidiane. *Le gnocide voil enquête historique.* Paris: Gallimard, 2008.
NDLOVU-GATSHENI, Sabelo J. *Joshua Mqabuko Nkomo of Zimbabwe: Politics, Power, and Memory.* Prtoria: Springer International Publishing AG- dition , 2017.
NDONE, F. Angyene. *Anthroponymie compare makina, fang et Sak : quelques spcificits sur les noms de personnes chez les Makina du Gabon,* . 2006.
NDONGO, Jacques FAME. *Un regard africain sur la communication: à la dcouverte de la gomtre circulaire.* Yaound: Editions St Paul, 1996.
NGOM, Gilbert. *L'gyptien et les langues bantu : le cas du duala* . Paris : Prsence africaine, 1989.
NIANE, D.T. *Soundiata ou l'pope mandingue.* paris: Karthala, 1960.
NJOGU, Kimani. *Cultural Production and Change in Kenya : Building Bridges.* Twaweza Communications Kimani Njogu -G. Oluoch-Olunya, 2007.
OBENGA, Thophile. *L'Afrique dans l'antiquit – Egypte pharaonique-.* Paris: Afrique noire, 1973.
—. *L'Etat fdral d'Afrique noire : la seule issue.* paris: l'Harmattan, , 2012.
—. *L'Afrique dans l'Antiquit - Egypte ancienne-Afrique noire.* Paris: Prsence Africaine, 1973.
—. *Les Bantu: Langues, peuples, civilisations.* Paris: Prsence Africaine, 1985.
—. *Origine commune de l'gyptien ancien, du copte et des langues ngro-africaines modernes.* Paris,: L'Harmattan,, 1993.
OMOTUNDE, Jean Philippe. *Les racines africaines de la civilisation europenne.* Paris: Menaibuc, 2004.
OUM, NDIGI. *Les Basa du Cameroun et l'antiquit pharaonique gypto-nubienne : recherche historique et linguistique comparative sur leurs rapports culturels à la lumière de l'gyptologie,.* Paris: ANRT, 1997.
OVIDE FONTAINE, Jean Cottraux- Robert Ladouceur. *Cliniques de thrapie comportementale.* Mardaga, 1995.
P.F, CALAME-GRIAULE G et LACROIX. *smiotica, scripts et signs africains T 1 No 3.* 1969.
PATHE DIAGNE, Ki-Zerbo. *Histoire Gnrale de l'Afrique, volume 1.* Paris: Unesco, 1980.
Pauline, Marie, Cline FREALLE,. *Les Animaux dans les masques d'Afrique de l'Ouest, thèse de doctorat vtrinaire,* . Paris: ANRT, 2002.
PRIN, Pierre Rich et Patrick. *Dictionnaire des Francs - Les temps Mrovingiens.* Bartillat, 1996.
Peter S. BELLWOOD, Colin RENFREW. *Examining the Farming/language Dispersal Hypothesis, McDonald Institute for Archaeological Research,* . Cambridge: University of Cambridge, 2002.
PFOUMA, Oscar. *L'Harmonie du monde: anthropologie culturelles des couleurs et des sons en Afrique depuis l'Egypte ancienne.* Paris: Editions Mnaibuc, 2000.
Philippe LABURTHE-Tolra, P. *Initiations et socits secrètes au Cameroun: les mystères de la nuit.* Paris: Karthala ditions, , 1992.
Plumelle-URIBE, Rosa Amelia. *La Frocit blanche: Des non-blancs aux non-aryens, gnocides occults de 1492 à nos jours.* Paris: Albin Michel, 2001.
PORAT. *Local Industry of Egyptian Pottery in Southern Palestine during the Early Bronze Period,* . s.d.
PURY, Albert DE. «« Les "Smites" n'existent pas »,.» *Le Temps stratgique, no 89,*, septembre/octobre 1999: p. 6-16.
R. P. G. HULSTAERT, M. S. C. *La ngation dans les langues congolaises.* Bruxelles: Institut Royal, 1950.
RĀZĪ, FAKHR al-Dīn Muḥammad ibn 'Umar. *Trait sur les noms Divins, Traduction depuis l'arabe : Maurice Glotton. Hritage spirituel,* . Paris: dition al-Burak , 1986.
RGNIER-DESMARAIS. *Trait de la grammaire française.* Paris: Jean-Baptiste Coignard, s.d.
Rmi CEILLIER, Laurent Etienne RONDET. *Histoire gnrale des auteurs sacrs et ecclsiastiques, Volume 1, livre troisième,* . Paris: Chez Louis Vivès, 1865.
RENAUD H.P.J. *La première mention de la noix de kola dans la matière mdicale des Arabes.* Hespris, vol. 8,, 1928.
RIALLAND, Annie. *Tonologie africaine et modlisation prosodique.* paris 8, 2000.
RICHARD CHABY, Karen Gulden. *Mots et Noms de l'Egypte Ancienne: Volume 2 : Français – Egyptien.* Paris: Reli, 2014.
—. *Mots et Noms de l'Egypte Ancienne: Volume 2 : Français - Egyptien.* BOD books on demand, 2014.
RILLY, Claude. *L'criture Meroitic.* Villejuif: CNRS-LLACAN, s.d.
Roland BAUCHOT, Cassian BON, Patrick DAVID, Jean-Pierre GASC. *Les serpents.* Paris: Artmis, 2005.
RÖMER, Thomas. *L'Invention de Dieu.* Paris: Seuil, 2014.
ROMMELAERE, Catherine. «Les chevaux du Nouvel Empire gyptien.» *Connaissance de l'Egypte ancienne* , 1991.
ROSE, Martin. *Une hermneutique de l'ancien testament : Comprendre, se comprendre, faire comprendre.* Genève: Elabor et Fides, 2003.
S, BATTESTINI. *Ecriture et texte , Contribution africaine* . Paris : Prsence africaine., 1997.
S. BAR, D. KAHN, J.J. SHIRLEY. *Egypt, Canaan and Israel . History, Imperialism, Ideology and Literature.* Danvers: BRILL, 2011.
SAUSSURE, Lopold de. *Psychologie de la Colonisation Française dans ses Rapports avec les Socits Indigènes.* Paris: Flix Alcan, 1899.
SCHEPS, Ruth. *Introduction in ,La Science sauvage.* Paris: Seuil, 1993.
SNÈQUE. *Œuvres complètes de Snèque le philosophe, Tome 2.* paris: Hachette, 1861
SERTIMA, Ivan Van. *Early America Revisited.* New Brunswick (U.S.A.) and London (U.K.) : Transaction Publishers, 1998.
SERTIMA, Ivan Van. *They Came Before Columbus, The African Presence in Ancient America.* New York: Random House, 1976.
SHAVIT, Yaacov. *History in Black: African-Americans in Search of an Ancient Past,* . New-york,: Routledge, s.d.
STAPLETON, w. *Comparative Handbook of congo languages.* London: Hannah Wade" Printing Press, Baptist Missionary Society, 1903.
STUART, Munro-Hay. *Aksum : An African Civilization of Late Antiquity.* Edinburgh: Edinburgh - University Press, 1991 .
SURGET, Alain. *L'œil d'Horus.* Paris: Flammarion, 1998.
T, BYNON. *Historical linguistics.* Cambridge: Cambridge University Press, , 1977.
TABI-MANGA, J. *De la grammaire èEwondo à une thorie du mot.* Paris: Didier rudition, coll. Linguistique, 1992.
TACITUS, Publius Hornlius. *Mœurs des Germains.* Paris: CreateSpace Independent Publishing Platform, 1871.
TASTEVIN, C . *A grafía etimológica dos nomes tribais africanos,* . Portugal em Afr., 1947.
TELLINGER, Michael. *African Temples of the Anunnaki.* Zulu Planet Publishers, mai 2013.
TOLRA, Laburthe. *Vers la lumière ? ou, Le dsir d'Ariel: a propos des Beti du Cameroun.* Paris: Karthala, 1999.
UNESCO. *Histoire gnrale de l'Afrique, Etudes et documents, le peuplement de l'Egypte* . Paris: Unesco, 1978.
UNESCO. *Le peuplement de l'Egypte ancienne et le dchiffrement de l'criture Meroitic, acte du colloque du caire, du 28 janvier au 3 fevrier 1974,.* Paris: Unesco, 1974.

VALBELLE D., Charles Bonnet. *De la vigne au jujubier.* Lausanne: Favre, 2014.
VANGU, Emmanuel VANGU. *Sexualit, initiations et tapes du mariage en Afrique -Au heart des rites et des symboles.* Paris: Editions Publibook, 2012.
VAUCELLE, Louis. *Chronologie des monumens antiques de la Nubie : d'après l'interprtation des Des Legendes Royales .* Paris: Hachette Livre, 2013.
VERGOTE, Jozef. *phonetic historique de l'gyptien. Les consonnes.* Louvain: Bureaux du Muson, 1945.
VIVIANE Koenig, Marie-Aude Waymel. *Dieux et Gnies de l'gypte ancienne.* Paris: Livres de Poche, 2004.
VIVIEN, J. et Faure, J. J. *Fruitiers sauvages d'Afrique - Espèces du Cameroun.* . Editions NGUILA-KEROU, France, , 1995.
VOLOKHINE, Youri. *Le porc en gypte ancienne : mythes et histoire à l'origine des interdits alimentaires.* Liège: Presses Universitaires de Liège, 2014.
VOLTAIRE. *Œuvres complètes de Voltaire: Dictionnaire philosophique Tome septième.* Paris: Hachette Livre, 1817.
W, BLEEK. *De nominum generibus linguarum Africae australis, Copticae, Semiti-Arum Aliarumque sexualium.* Bonn : Thèse, 1851.
W., DIXON R. M. *The rise and fall of languages.* Cambridge: Cambridge University Press, 1997.
WEBER, Max. *conomie et Socit,.* Paris: Plon, 1922 .
Y.,MOÑINO. *Le proto-gbaya : Essai de linguistique comparative sur vingtet-une langues d'Afrique centrale.* Paris: Peeters, 1995.
Y.PERROT. *Linguistique gnrale.* Paris: PUF, 1954.
YOUSSOUF TATA CISS et Wa Kamissoko, Soundiata. *Des origines à la fondation de l'empire, .* Paris: Karthala, 1988.
YOYOTTE, P. Vernus et J. *Les pharaons.* Paris: MA ditions, 1988.
ZRIBI-HETZ, Annie Rialland- P. Sauzet et A. *Thories linguistiques et langues sub-sahariennes.* Paris: L'Harmattan, 2004.

Achevé d'imprimer en mai 2020 par LEN S.A.S. – 93400 St Ouen
Dépôt légal : mai 2020
Imprimé en France